RADIATION BIOLOGY IN CANCER RESEARCH

The University of Texas System Cancer Center
M. D. Anderson Hospital and Tumor Institute
32nd Annual Symposium on Fundamental Cancer Research

Published for
The University of Texas System Cancer Center
M. D. Anderson Hospital and Tumor Institute
Houston, Texas, by Raven Press, New York

The University of Texas System Cancer Center
M. D. Anderson Hospital and Tumor Institute
32nd Annual Symposium on Fundamental Cancer Research

Radiation Biology in Cancer Research

Edited by

Raymond E. Meyn, Ph.D.
Department of Physics
The University of Texas System Cancer Center
M. D. Anderson Hospital and Tumor Institute
Houston, Texas

H. Rodney Withers, M.D., Ph.D.
Department of Radiotherapy
The University of Texas System Cancer Center
M. D. Anderson Hospital and Tumor Institute
Houston, Texas

Raven Press ■ New York

Raven Press, 1140 Avenue of the Americas, New York, New York 10036

Library of Congress Cataloging in Publication Data

Symposium on Fundamental Cancer Research, 32d, Anderson
Hospital and Tumor Institute, 1979.
Radiation biology in cancer research.

Published for the University of Texas System Cancer
Center M D Anderson Hospital and Tumor Institute.

Includes bibliographical references and indexes.
1. Cancer–Radiotherapy–Congresses. 2. Radiobiology–Congresses. I. Meyn, Raymond E.
II. Withers, Hubert Rodney, 1932- III. Anderson
Hospital and Tumor Institute, Houston, Tex.
IV. Title (DNLM: 1. Neoplasms--Radiotherapy--Congresses. 2. Radiobiology–Congresses.
3. Radiation injuries–Congresses. W3 SY5177 32d
1979r /WZ269 S994 1979r)
RC271.R3S95 1979 616.9'94'06424 79-54643
ISBN 0-89004-402-3

This volume is a compilation of the proceedings of The University of Texas System Cancer Center M. D. Anderson Hospital and Tumor Institute 32nd Annual Symposium on Fundamental Cancer Research, held February 27 and 28 and March 1 and 2, 1979, in Houston, Texas.

The material contained in this volume was submitted as previously unpublished material, except in the instances in which credit has been given to the source from which some of the illustrative material was derived.

Great care has been taken to maintain the accuracy of the information contained in the volume. However, the Editorial Staff and The University of Texas System Cancer Center cannot be held responsible for errors or for any consequences arising from the use of the information contained herein.

Editors' Foreword

The 32nd Annual Symposium on Fundamental Cancer Research, the subject of this volume, was very successful both in terms of attendance and quality of the presentations. This year's topic, "Radiation Biology in Cancer Research," was indeed timely as radiation biology had not been a topic of the Annual Symposium since 1964. It is not surprising, therefore, that so many new developments in this field were presented that are important to both our understanding of the mechanisms of radiation carcinogenesis and the treatment for cancer by radiation. The various sessions gave special emphasis to the basic mechanisms of radiation injury, biological responses to low doses and low dose rates, variations in radiation response, kinetic changes after irradiation, and tissue response to radiation. The last two sessions were devoted to potential therapeutic application of radiobiological findings to human cancer radiotherapy. We hope that this symposium resulted in new ideas and sharpened focus on the radiobiological aspects of cancer cause and treatment.

As cochairmen we acknowledge our appreciation to those many individuals who provided guidance and advice in all matters pertaining to this symposium and this volume. The symposium organizing committee members from The University of Texas System Cancer Center included: Arthur Cole, Physics; Peter M. Corry, Physics; David J. Grdina, Radiotherapy; Walter N. Hittelman, Developmental Therapeutics; Ronald M. Humphrey, Science Park; Lester J. Peters, Radiotherapy; Robert J. Shalek; Physics; and Howard D. Thames, Jr., Biomathematics. The planning committee was given excellent guidance and direction by the advisory committee: J. D. Chapman, William C. Dewey, Herman D. Suit, M. M. Elkind, John F. Fowler, Robert Painter, E. L. Powers, and G. F. Whitmore.

Special thanks and appreciation are given to Frances Goff and her staff for the many functions that they expertly and carefully planned and conducted. Also our thanks to the following companies for providing funds for the hospitality rooms wherein the speakers and symposium attendees could meet and discuss mutual interests: Beckman Instruments, Inc.; Atomic Energy of Canada Limited; Varian; and Siemens Corporation. We are especially grateful to the National Cancer Institute and the Texas Division of the American Cancer Society for their continued support. We also wish to thank The University of Texas Health Science Center at Houston, Graduate School of Biomedical Sciences, for its assistance. The Publications Office staff aided invaluably in all the matters pertaining to the information, announcements, and publications of the symposium.

Special acknowledgement is accorded Walter Pagel, who played a most important role in the completion of this volume, and Leslie Wildrick.

Raymond E. Meyn, Ph.D.
H. Rodney Withers, M.D., Ph.D.
Co-editors

Contents

Responses to Low Doses and Low Dose Rates

Variations in Radiation Response

Kinetic Changes after Irradiation

Contributors

Tikvah Alper
Birkholt, Crableck Lane
Sarisbury Green
Hampshire SO3 6AL, England

Elwood Armour
Department of Physics
The University of Texas System Cancer Center
M. D. Anderson Hospital and Tumor Institute
Houston, Texas 77030

G. W. Barendsen
Radiobiological Institute
Organization for Health Research TNO
Rijswijk, The Netherlands

Howard Barkley
Department of Radiotherapy
The University of Texas Svstem Cancer Center
M. D. Anderson Hospital and Tumor Institute
Houston, Texas 77030

Barthel Barlogie
Department of Developmental Therapeutics
The University of Texas System Cancer Center
M. D. Anderson Hospital and Tumor Institute
Houston, Texas 77030

Joel S. Bedford
Department of Radiology and Radiation Biology
Colorado State University
Fort Collins, Colorado 80523

Eleanor A. Blakely
Department of Biophysics and Medical Physics and Biology and Medicine Division
Lawrence Berkeley Laboratory
University of California
Berkeley, California 94720

Antone L. Brooks
Inhalation Toxicology Research Institute
Lovelace Biomedical and Environmental Research Institute
Albuquerque, New Mexico 87115

P. M. Busse
Washington University School of Medicine
St. Louis, Missouri 63110

J. Donald Chapman
Department of Radiation Oncology
Cross Cancer Institute
Department of Radiology
University of Alberta
Edmonton, Alberta, Canada T6G 1Z2

Ruth Chen
Department of Physics
The University of Texas System Cancer Center
M. D. Anderson Hospital and Tumor Institute
Houston, Texas 77030

Edward P. Clark
Department of Radiology and Radiation Biology
Colorado State University
Fort Collins, Colorado 80523

Kelly H. Clifton
Department of Human Oncology and Radiology
Wisconsin Clinical Cancer Center
University of Wisconsin Medical School
Madison, Wisconsin 53792

Arthur Cole
Department of Physics
The University of Texas System Cancer Center
M. D. Anderson Hospital and Tumor Institute
Houston, Texas 77030

C. A. Combs
Department of Radiology
Stanford University School of Medicine
Stanford, California 94305

Peter Corry
Department of Physics
The University of Texas System Cancer Center
M. D. Anderson Hospital and Tumor Institute
Houston, Texas 77030

Ronald A. Coss
Department of Radiology and Radiation Biology
Colorado State University
Fort Collins, Colorado 80523

David L. Denman
Department of Radiology and Radiation Biology
Colorado State University
Fort Collins, Colorado 80523

Lyle A. Dethlefsen
Section of Radiation and Tumor Biology
Department of Radiology
University of Utah Medical Center
Salt Lake City, Utah 84132

William C. Dewey
Department of Radiology and Radiation Biology
Colorado State University
Fort Collins, Colorado 80523

M. M. Elkind
Division of Biological and Medical Research
Argonne National Laboratory
Argonne, Illinois 60439

David Ewing
Department of Radiation Therapy and Nuclear Medicine
Hahnemann Medical College and Hospital
Philadelphia, Pennsylvania 19102

Stanley B. Field
Cyclotron Unit
Medical Research Council
Hammersmith Hospital
London W12 0HS, England

Susan E. Fletcher
Department of Physics
The University of Texas System Cancer Center
M. D. Anderson Hospital and Tumor Institute
Houston, Texas 77030

Sterling H. Fly
Board of Regents
The University of Texas System
Houston, Texas 77030

John F. Fowler
Gray Laboratory of the Cancer Research Campaign
Mount Vernon Hospital
Northwood, Middlesex HA6 2RN, England

Michael H. Fox
Department of Radiology and Radiation Biology
Colorado State University
Fort Collins, Colorado 80523

Allan J. Franko
Department of Radiology
Stanford University School of Medicine
Stanford, California 94305

Michael L. Freeman
Department of Radiology and Radiation Biology
Colorado State University
Fort Collins, Colorado 80523

Bryan M. Furlong
Department of Radiology
Stanford University School of Medicine
Stanford, California 94305

Mary Esther Gaulden
Department of Radiology
The University of Texas Health Science Center at Dallas
Dallas, Texas 75235

Chester A. Gleiser
Section of Experimental Animals
The University of Texas System Cancer Center
M. D. Anderson Hospital and Tumor Institute
Houston, Texas 77030

Michael Goitein
Department of Radiation Medicine
Massachusetts General Hospital
Department of Radiation Therapy
Harvard Medical School
Boston, Massachusetts 02114

Mario Gonzales
Department of Radiotherapy
The University of Texas System Cancer Center
M. D. Anderson Hospital and Tumor Institute
Houston, Texas 77030

D. T. Goodhead
MRC Radiobiology Unit
Harwell, Didcot OX11 0RD, England

David J. Grdina
Section of Experimental Radiotherapy
Department of Radiotherapy
The University of Texas System Cancer Center
M. D. Anderson Hospital and Tumor Institute
Houston, Texas 77030

George M. Hahn
Department of Radiology
Stanford University School of Medicine
Stanford, California 94305

R. B. Hawkins
Washington University School of Medicine
St. Louis, Missouri 63110

Robert C. Hickey
The University of Texas System Cancer Center
M. D. Anderson Hospital and Tumor Institute
Houston, Texas 77030

Donald P. Highfield
Department of Radiology and Radiation Biology
Colorado State University
Fort Collins, Colorado 80523

Walter N. Hittelman
Department of Developmental Therapeutics
The University of Texas System Cancer Center
M. D. Anderson Hospital and Tumor Institute
Houston, Texas 77030

J. W. Hopewell
Churchill Hospital Research Institute
University of Oxford
Headington, Oxford 0X3 7LJ, England

Shirley Hornsey
Cyclotron Unit
Medical Research Council
Hammersmith Hospital
London W12 0HS, England

David H. Hussey
Department of Radiotherapy
The University of Texas System Cancer Center
M. D. Anderson Hospital and Tumor Institute
Houston, Texas 77030

John H. Jardine
Section of Experimental Animals
The University of Texas System Cancer Center
M. D. Anderson Hospital and Tumor Institute
Houston, Texas 77030

Robert F. Kallman
Department of Radiology
Stanford University School of Medicine
Stanford, California 94305

Scott D. Kelley
Department of Radiology
Stanford University School of Medicine
Stanford, California 94305

Hannah L. Kemper
Department of Radiology
Stanford University School of Medicine
Stanford, California 94305

Ann R. Kennedy
Department of Physiology
Harvard University School of Public Health
Boston, Massachusetts 02115

John F. R. Kerr
Departments of Pathology
University of Queensland Medical School and Royal Brisbane Hospital
Herston, Brisbane
Queensland, Australia

A. J. van der Kogel
Radiobiological Institute
Organization for Health Research TNO
Rijswijk, The Netherlands

H. Dieter Kogelnik
Strahlentherapeutische Klinik
Allgemeines Krankenhaus der Stadt Wien
Vienna, Austria

Henry I. Kohn
Department of Radiation Therapy
Harvard Medical School
Shields Warren Laboratory
New England Deaconess Hospital
Boston, Massachusetts 02115

Thomas Michael Koval
Division of Radiation Oncology
Allegheny General Hospital
Pittsburgh, Pennsylvania 15212

Gloria C. Li
Department of Radiology
Stanford University School of Medicine
Stanford, California 94305

John B. Little
Department of Physiology
Harvard University School of Public Health
Boston, Massachusetts 02115

Kathryn A. Mason
Section of Experimental Radiotherapy
Department of Radiotherapy
The University of Texas System Cancer Center
M. D. Anderson Hospital and Tumor Institute
Houston, Texas 77030

Jane B. Marmor
Department of Radiology
Stanford University School of Medicine
Stanford, California 94305

Raymond E. Meyn
Department of Physics
The University of Texas System Cancer Center
M. D. Anderson Hospital and Tumor Institute
Houston, Texas 77030

Rupert G. Miller
Department of Radiology
Stanford University School of Medicine
Stanford, California 94305

James B. Mitchell
Department of Radiology and Radiation Biology
Colorado State University
Fort Collins, Colorado 80523

Robert C. Murry
Department of Radiology
The University of Texas Health Science Center at Dallas
Dallas, Texas 75235

Frank Q. H. Ngo
Department of Biophysics and Medical Physics and Biology and Medicine Division
Lawrence Berkeley Laboratory
University of California
Berkeley, California 94720

John Nove
Laboratory of Radiobiology
Harvard University School of Public Health
Department of Radiation Therapy
Harvard Medical School
Boston, Massachusetts 02115

Robert B. Painter
Laboratory of Radiobiology
University of California
San Francisco, California 94143

Lester J. Peters
Section of Experimental Radiotherapy
Department of Radiotherapy
The University of Texas System Cancer Center
M. D. Anderson Hospital and Tumor Institute
Houston, Texas 77030

Theodore L. Phillips
Department of Radiation Oncology
University of California
San Francisco, California 94143

Douglas W. Pounds
Department of Radiology
Stanford University School of Medicine
Stanford, California 94305

E. L. Powers
Laboratory of Radiation Biology
Department of Zoology
The University of Texas
Austin, Texas 78712

G. Peter Raaphorst
Department of Radiology and Radiation Biology
Colorado State University
Fort Collins, Colorado 80523

Diane Rapacchietta
Department of Radiology
Stanford University School of Medicine
Stanford, California 94305

Gilbert L. Raulston
Section of Experimental Animals
The University of Texas System Cancer Center
M. D. Anderson Hospital and Tumor Institute
Houston, Texas 77030

Harald H. Rossi
Radiological Research Laboratory
Department of Radiology
College of Physicians and Surgeons of Columbia University
New York, New York 10032

W. L. Russell
Biology Division
Oak Ridge National Laboratory
Oak Ridge, Tennessee 37830

David Schoenfeld
Department of Radiology
Stanford University School of Medicine
Stanford, California 94305

Jeffrey Searle
Departments of Pathology
University of Queensland Medical School and Royal Brisbane Hospital
Herston, Brisbane
Queensland, Australia

Marguerite A. Sognier
Department of Developmental Therapeutics
The University of Texas System Cancer Center
M. D. Anderson Hospital and Tumor Institute
Houston, Texas 77030

William Spanos
Department of Radiotherapy
The University of Texas System Cancer Center
M. D. Anderson Hospital and Tumor Institute
Houston, Texas 77030

Ira J. Spiro
Department of Radiology and Radiation Biology
Colorado State University
Fort Collins, Colorado 80523

G. G. Steel
Radiotherapy Research Unit
Divisions of Biophysics and Radiotherapy
Institute of Cancer Research
Sutton, Surrey, England

T. C. Stephens
Radiotherapy Research Unit
Divisions of Biophysics and Radiotherapy
Institute of Cancer Research
Sutton, Surrey, England

Herman D. Suit
Department of Radiation Medicine
Massachusetts General Hospital
Department of Radiation Therapy
Harvard Medical School
Boston, Massachusetts 02114

Masaji Takahashi
Department of Radiology
Stanford University School of Medicine
Stanford, California 94305

Cornelius A. Tobias
Department of Biophysics and Medical Physics and Biology and Medicine Division
Lawrence Berkeley Laboratory
University of California
Berkeley, California 94720

L. J. Tolmach
Washington University School of Medicine
St. Louis, Missouri 63110

Stephen P. Tomasovic
Department of Radiology and Radiation Biology
Colorado State University
Fort Collins, Colorado 80523

R. L. Ullrich
Biology Division
Oak Ridge National Laboratory
Oak Ridge, Tennessee 37830

Ralph R. Weichselbaum
Laboratory of Radiobiology
Harvard University School of Public Health
Department of Radiation Therapy
Harvard Medical School
Boston, Massachusetts 02115

H. Rodney Withers
Section of Experimental Radiotherapy
Department of Radiotherapy
The University of Texas System Cancer Center
M. D. Anderson Hospital and Tumor Institute
Houston, Texas 77030

Rosemary S. L. Wong
Department of Radiology and Radiation Biology
Colorado State University
Fort Collins, Colorado 80523

Tracy C. H. Yang
Department of Biophysics and Medical Physics and Biology and Medicine Division
Lawrence Berkeley Laboratory
University of California
Berkeley, California 94720

Radiation Biology in Cancer Research, edited by
Raymond E. Meyn and H. Rodney Withers.
Raven Press, New York © 1980.

Welcome Address

Sterling H. Fly, Jr.

Board of Regents, The University of Texas System, Houston, Texas 77030

On behalf of the Board of Regents of The University of Texas System, I am pleased to welcome each of you to the 32nd Annual Symposium on Fundamental Cancer Research. This year scientists have gathered here in Houston to exchange, discuss, and debate new ideas, and to share information that will ultimately enhance our understanding of the malignant disease process.

The University of Texas System has long supported the need for rapid exchange of information in all areas of science and in cancer specifically. During the past 31 years, this series of symposia has come to represent a major facet of this commitment, for it has become increasingly apparent that the control of cancer is dependent upon the collaboration of scientists of many interrelated disciplines.

This reasoning was the impetus behind the creation of the symposia series. Beginning in 1946, the symposium has grown from a small local meeting to a conference of international stature. The first meetings attracted investigators primarily from Houston, Austin, and Galveston, and the papers presented were on a variety of topics of general interest. In 1952, the format was changed to focus on a single theme that reflected prevalent areas of interest and progress to cancer specialists. Successive symposia have covered a wide range of research approaches to the cancer problem. The particular importance of radiation biology—the topic of this symposium—is underscored by the fact that it was also the subject of symposia in 1958 and 1964.

At this time, I would like to take the opportunity to extend my warmest congratulations to Dr. Mortimer M. Elkind of the Argonne National Laboratory. Dr. Elkind was named the 28th recipient of the annual Bertner Memorial Award for his distinguished contributions to cancer research in the field of radiation biology. The award honors the late E. W. Bertner, who was M. D. Anderson Hospital's first acting director and first president of the Texas Medical Center.

On behalf of the Board of Regents, I also congratulate Dr. Craig W. Spellman of the University of New Mexico, the recipient of the eighth annual Stone Memorial Award. The award, which is given to a student for outstanding research in the field of biomedical sciences, was named in honor of the late Dr. Wilson S. Stone, former vice chancellor of The University of Texas System.

A geneticist of international note, Dr. Stone was instrumental in the development of the sciences within The University of Texas System.

To the National Cancer Institute and the Texas Division of the American Cancer Society, I express a special note of thanks and appreciation for their continued cooperation and support of this important scientific meeting.

Once again, I extend a hearty welcome to you all. I trust the information shared here in the following days will aid you with your timely work in the laboratory and bring us one step closer to meeting the all important challenge of controlling cancer.

Radiation Biology in Cancer Research, edited by
Raymond E. Meyn and H. Rodney Withers.
Raven Press, New York © 1980.

Keynote Address: Survival Curve Models

Tikvah Alper

Birkholt, Crableck Lane, Sarisbury Green, Hampshire S03 6AL England

In experiments involving the killing of cells, it is customary to analyze results in terms of the parameters of some chosen model of cell survival. For many years, extensive use was made of parameters proper only to a survival curve that approximates to exponential killing at high dose, viz inactivation dose, D_o, and extrapolation number, n. But according to the currently popular linear-quadratic model, or, indeed, to any multihit model, those parameters are meaningless because killing never proceeds exponentially. If it does, single-hit inactivation must eventually be involved; so it is a matter of importance to establish whether the older assumption is justified.

EXPONENTIAL SURVIVAL CURVES

The several equations in use to describe shouldered survival curves were formulated on the basis of widely differing assumptions; but no comparable disagreement exists in the interpretation of an exponential curve, represented by $f = e^{-\lambda D}$, where f is surviving fraction and D the dose. In terms of the target concept, on which all currently used survival curve models are based, such a relationship implies single-hit action. However, this does not mean that only a single vital target is involved. There may be several, all subject to inactivation by single energy deposits, and all requiring to retain function if the cell is to proliferate successfully. In that case, exponential survival is given by

$$\ln f = -(\lambda_1 + \lambda_2 + \lambda_3 \ . \ . \ . \ )D \text{ or } \ln f = -\Sigma\lambda D \qquad (1)$$

An important aspect of equation 1 is that repair processes and modifying agents may act differently on different components, a concept that we were constrained to introduce many years ago (Alper and Gillies 1958) to account for the dependence of the oxygen enhancement ratio (OER) for *Escherichia coli* B on *post*-irradiation culture conditions (Table 1). This was the origin of my model attributing cell death to events not only in DNA but also in a component of a chemically different nature (Alper 1963). Suppose equation 1 refers to anoxic cells. If the overall OER were m, equation 1 would be modified to give

$$\ln f = -m\lambda D = -(m_1\lambda_1 + m_2\lambda_2 + \ . \ . \ . \ ) \qquad (2)$$

with m_1, m_2, etc., referring to intrinsic OERs proper to the individual components.

TABLE 1. *Dependence of OER on post-irradiation culture conditions,* Escherichia coli B *

Medium	Incubation Temperature Immediately After Irradiation (°C)	D_0, grays Anoxic	Oxygenated	OER
A†	45	120 ± 3.0	32.5 ± 1.0	3.67 ± 0.15
B†	45	90 ± 3.4	27.1 ± 0.6	3.32 ± 0.13
A	37	87.2 ± 1.7	30.1 ± 1.3	2.90 ± 0.14
A + NaCl, 4 g/l	37	63.8 ± 1.3	23.4 ± 0.4	2.73 ± 0.08
B	37	24.7 ± 0.8	12.2 ± 0.5	2.02 ± 0.21
A	19	21.1 ± 1.1	11.8 ± 0.5	1.79 ± 0.11
B	19	9.8 ± 0.4	5.7 ± 0.3	1.63 ± 0.11

* (Alper 1961)
† Medium B contained NaCl, 4 g/liter and had a higher peptone content than A.

Dittrich (1960) showed how a spuriously exponential survival curve could be arrived at by compounding several curves with "variable hit numbers and a given target" or "variable targets with a given hit number." But it would require great contrivance to put together a group of shouldered curves to give spurious exponential survival to levels much below 10^{-2}, and it would be implausible to interpret in that way all the numerous examples of exponential survival, particularly convincing for some strains of bacteria whose survival has been reduced to very low levels (e.g., Figure 1). Whether or not single-hit inactivation

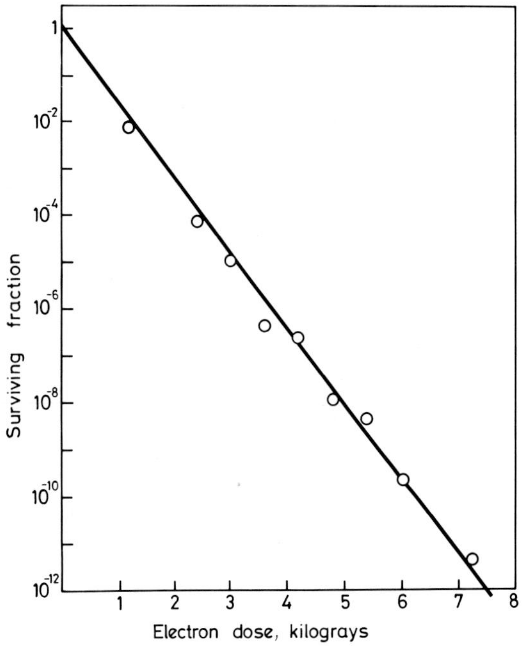

FIG. 1. Survival curve for freeze-dried *Brucella abortus,* about 10^{12} organisms per container. Unpublished observations of M. Sterne, T. Alper, and G. Trim.

plays a role in the killing of mammalian cells is clearly important for our understanding of the mechanisms of radiobiological action. It is therefore worth attempting to decide whether shouldered survival curves, particularly for mammalian cells, really do terminate in exponential regions.

MULTITARGET Models

The equation chosen by Puck and Marcus (1956) to describe their shouldered survival curve for HeLa cells was governed by the then general acceptance of exponential killing at high dose. They stressed this feature of their results, in accordance with which they adopted the model

$$f = [1 - (1 - e^{-D/D_0})^n]^m \qquad (3)$$

This applies to a population such that each member contains m loci or sites, *all* of which are essential for survival. Each site is postulated to consist of a target divided into n parts, the integrity of *any one* of which is sufficient for the functioning of the site; and each part of the target requires only one hit for inactivation. Puck and Marcus described this incorrectly as an "n-hit model"; but, since they calculated m to be 1, their expression became

$$f = 1 - (1 - e^{-D/D_0})^n \qquad (4)$$

i.e., the single-hit multitarget equation, with $n = 2$ in the case of HeLa cells. Equation 4 was used for many years as a description of mammalian cell survival curves, even after the abandonment of reference to "*the* mammalian cell survival curve" with extrapolation number two.

The high dose approximations to equations 3 and 4 are of the form

$$f = \text{constant} \cdot e^{-\lambda D} \qquad (5)$$

and it is important to note that any model giving that approximation must incorporate the principle of single-hit inactivation. It has never been correct to refer to numbers derived from back-extrapolation of exponential regions of survival curves as "hit numbers" since, strictly speaking, such regions do not exist if multiple hits are required for inactivation of a target.

Initial Nonzero Slopes

Equations 3 and 4 predict zero slope to the survival curve at zero dose, a result hardly ever realized in practice, even with cell populations as nearly homogeneous as careful synchronization can make them (See Alper 1975). However, nonzero slope at zero dose is easily accommodated by the introduction of a one-hit term, e.g.,

$$f = e^{-\lambda_1 D}\{1 - (1 - e^{-\lambda_2 D})^n\} \qquad (6)$$

The introduction of that term does not affect the important issue, mentioned above, of whether or not the equation approximates to the form of equation 5 at high dose. If it does, inactivation is attributable eventually only to single energy deposits, though two modes of single-hit inactivation must then be involved.

At very low dose, equations such as 6 approximate to

$$\ln f = -\lambda_1 D \tag{7}$$

DO MAMMALIAN CELL SURVIVAL CURVES TERMINATE EXPONENTIALLY?

The Linear-Quadratic Model

The main challenge to the previous rather general assumption of an exponential terminal region has come from the proposal that survival (at least of mammalian cells) is described by the equation

$$\ln f = -(\alpha D + \beta D^2) \tag{8}$$

Theoretical derivations of that formulation were propounded on the basis of two quite different sets of assumptions. Kellerer and Rossi (1972) based their rationale on the conclusion that, for a variety of effects on animal and plant tissues, the relative biological effectiveness (RBE) for fast neutrons varied inversely with (neutron dose)$^{1/2}$. On the assumption that, at high linear energy transfer (LET), lethal events accumulated linearly with dose, they deduced that at low LET, i.e., for effects caused by electrons, lethal events accumulated as the square of the dose. This suggested to them that inactivation of a critical site required the passage of two separate electrons. Chadwick and Leenhouts (1973) derived the same equation on the assumption that a cell is killed by the induction of a double-strand break in the DNA. This, they suggested, might happen as the result of a single event occurring simultaneously in both strands, or separate events in the two strands, provided those events occurred close enough together.

Use of the Linear-Quadratic Model to Determine Initial Slope

The weight of the evidence supports the concept of an initial nonzero slope, even for synchronized cells, and that parameter, rarely determined in earlier work on the survival of mammalian cells, is of great importance in the two major applications of cellular radiobiology: radiotherapy and the evaluation of radiological hazards. Equation 8 has been shown to be an admirable fit to many cell survival curves, at least in the low-dose region, and the evaluation of the parameter α is an elegant method for determining initial slope. It has been shown by Chapman (1979) that this can be very neatly done by a graphical method: equation 8 can be rewritten

$$\ln f/D = -\alpha - \beta D \tag{9}$$

so a plot of ln *f/D* against dose gives α as an intercept on the ordinate. This method of visual presentation of the parameters of equation 8 makes for much easier communication and comprehension than if they are presented simply as the best fitting parameters as calculated, usually, of course, by computer.

Inadequacy of Curve-Fitting for Validating Models

Despite the many demonstrations that experimental data on mammalian cell survival are well described by equation 8, it is incautious to claim support for either of the theoretical derivations of that model on those grounds. When no law or principle governing the relationship between two variables is known, a standard method is to find the constants in the polynomial

$$y = A + Bx + Cx^2 + Dx^3 + \dots$$

y and x being the dependent and independent variables, respectively. If that equation is fitted to a survival curve characterized by negative slope at zero dose, with no pronounced curvature near the origin, A is zero, and the logarithm of surviving fraction is negative, so the data points may very well be fitted by equation 8, a fairly large radius of curvature making the two terms of the polynomial sufficient within the limits imposed by experimental error—provided survival is not reduced to levels much below 10^{-2}.

Figure 2 has been drawn to illustrate the difficulty in deciding whether mammalian cell survival data of even the highest standard are best fitted by a linear-quadratic equation even to quite low survival levels, or one predicting a terminal exponential region. In Figure 2 are reproduced data of Hall (1975) from a carefully constructed experiment on tightly synchronized V 79 hamster cells in late S stage. Hall took particular pains to avoid the artefacts known to give "continuously bending" survival curves. The last two measured survivals were after doses of 12 and 14 Gy. If those two points were actually in an exponential

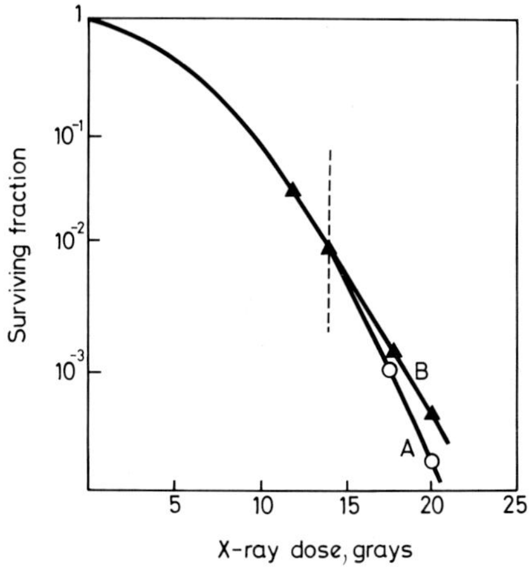

FIG. 2. From data of Hall (1975) for V 79 Chinese hamster cells irradiated in late S stage. Vertical dashed line marks highest dose for which fractional survival was shown. ▲, upper two points experimental; lower two hypothetical, on assumption of exponential survival after all doses exceeding 12 grays. ○, survival at 17.5 and 20 grays calculated from parameters fitted by Hall to equation 8.

region, higher doses would have given curve B, whereas if the parameters calculated to fit the linear-quadratic equation held accurately for all doses the survival curve would continue as A. Even the best experimental techniques could hardly be expected to show such a difference as significant; even if points at higher doses had been precisely as marked by the triangles on curve B, a slight change in the parameters would doubtless have given a perfectly satisfactory fit by equation 8. It says much for the accuracy of Hall's work that he found the continued curvature required by that model to "strain the fit" at higher doses.

Cell survival in vivo can be measured to levels much lower than is usually possible in vitro, and curves (on semilogarithmic plots) for some clonogenic cells in vivo certainly appear to terminate in straight lines covering many powers of ten (e.g., Hewitt and Wilson 1961, Withers 1967). But such survival curves cannot be used to test the validity of the linear-quadratic model, first because the necessary degree of experimental accuracy is not attainable; second, even if it were, survival at low dose is not measurable, at least for survival curves on clonogenic cells in organized tissues; third, those cells cannot be synchronized, so it is possible to argue that the terminal region appears straight because of heterogeneity in the population (cf. Scott 1975).

Evidence for Terminal Exponential Regions

Although the question cannot be settled by curve-fitting, there are some lines of evidence that seem to me to support the conclusion that the linear-quadratic model can be regarded only as a low-dose approximation to shouldered survival curves. The early investigations of Elkind and Sutton (1959, 1960) with Chinese hamster ovary cells, and subsequent work by Elkind and his collaborators, demonstrated features relevant to this problem: (1) When cells were exposed to a conditioning dose followed by an interval for full recovery, the secondary survival curves had terminal regions that were accurately parallel with those of the single-shot curves. (2) The value of $D_2 - D_1$ was independent of the magnitude of the conditioning dose once the latter had exceeded a certain minimum. Neither of these features could pertain to continuously bending survival curves. (The symbols D_2 and D_1 are used to represent, respectively, the sum of the two split doses and the single dose required to give the same effect.)

It is interesting to recall the first intimation of the high extrapolation number for the survival curve for stem cells of intestinal mucosa: this was deduced by Hornsey and Vatistas (1963) from their observation that $D_2 - D_1$ for intestinal death in mice was constant, about 4 Gy, for values of conditioning dose ranging from about 5 to 10 Gy. From the clone-counting techniques later developed by Withers and Elkind (1969,1970), it is now known that the corresponding range of fractional survival is more than 100; and the estimate of the value of 4 Gy has been confirmed by the construction of full secondary survival curves (e.g., Withers and Elkind 1969).

The substantial evidence of wholly exponential survival curves for some mammalian cell lines (Table 2) demonstrates that cell killing at low LET cannot

TABLE 2. *Examples of exponential survival curves for asynchronous populations of mammalian cells*

Origin	Cell Line or Tissue	D_0, Gy	Lowest Survival Measured	Reference
Human, freshly cultured	Skin, Spleen, Ovary	1.0*	10^{-4}	Puck et al. 1957
Human, freshly cultured	Skin	1.04†	10^{-1}	
	Lung	1.08†	10^{-1}	Norris and Hood 1962
Human, freshly cultured	Skin fibroblasts	1.32	10^{-3}	Weichselbaum et al. 1976
Human, freshly cultured	Skin fibroblasts	1.26	10^{-4}	Cox et al. 1977
Human	Burkitt's lymphoma	0.62	5×10^{-3}	C. Sato, cited by Okada 1975
Mouse	Leukemia (established in vitro)	0.62	5×10^{-3}	Caldwell et al. 1965
Mouse	Lymphoma L5178Y/S (established in vitro)	0.40	10^{-4}	Ehmann et al. 1974
Mouse	Normal hemopoietic (irradiation in vitro)	1.04	4×10^{-3}	Silini and Maruyama 1965

* Estimate from revision of dosimetry.
† Values of D_0 decreased with age of culture.

occur by a two-hit process of target inactivation at least in those cases. Mammalian cells that survive exponentially at all stages of the cell cycle might be claimed as exceptions to a rule that already excludes not only whole classes of cells—bacteria and haploid yeast—but almost all mammalian cells when they are in mitosis; but a rule with so many exceptions, and one that cannot account for some important phenomena, may forgivably be regarded as more honored in the breach than the observance. The more plausible hypothesis is that single-hit inactivation is the basic mechanism of cell killing by radiation. I propose to analyze the significance of survival curve shoulders on that basis.

REPAIR MODELS FOR SHOULDERED SURVIVAL CURVES

For many years survival curve shoulders have been widely assumed to show that damaging events (sublethal damage) must accumulate before proliferative capacity is lost. Little attention has been given to an alternative type of model attributing the shoulder of a cell survival curve to a repair process operating at the outset of irradiation but becoming less and less effective as the dose increases, until inactivation continues without concomitant repair. This idea was suggested by Powers (1962), and it has been formalized by Orr et al. (1966) and Laurie et al. (1972). The same concept has been applied to derivations of theoretical survival curves for bacteria irradiated by ultraviolet light (Ginsberg and Jagger 1965, Haynes 1966), and many photobiologists take for granted an interpretation that has made little impact in the field of ionizing radiation biology. The general principles of repair models are illustrated in Figure 3.

Provided a shouldered survival curve has nonzero initial slope, it is characterized by that slope, the final slope, and the extrapolation number: exactly the same parameters as those of equation 6. Thus the curve should have low- and high-dosc approximations of

$$\ln f = -\lambda_1 D \text{ and } \ln f = -\lambda D + \ln n$$

respectively, and the general expression for the survival curve might be written

$$\ln f = \lambda D + (\text{a function of } D) \tag{10}$$

where the function has the property of approximating to $(\lambda - \lambda_1)D$, when D is very small, and $\ln n$ when D is large. The function used by Haynes (1966),

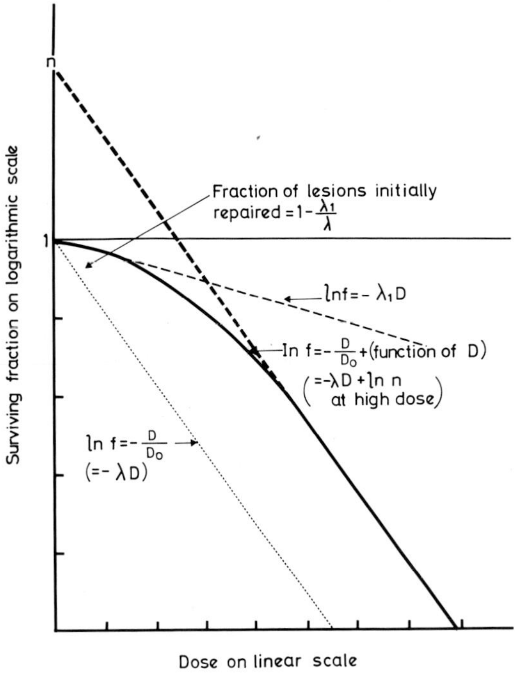

FIG. 3. The principles of repair models.

for example, was $\alpha(1 - e^{-\beta D})$. His terms α and β would correspond respectively to ln n and $\lambda_1/\ln n$, in the notation used here.

It is convenient to have a name for the postulated agent(s) of repair, and for the lesions subject to repair of that kind, in order to distinguish the latter from lesions that are subject to the process now generally known as repair of potentially lethal damage. There is some similarity between the basic assumptions of repair models and the hypothesis of Sinclair (1973) that there is a factor, which he called Q, regulating the magnitude of survival curve shoulders as cells proceed through the cycle. The postulated agent, type of repair, and specific lesions affected thereby will therefore be referred to as Q factor, Q repair, and Q lesions, respectively (cf. Alper 1977).

Implications for Mechanisms of Cell Killing

If the reasoning behind repair models proves to be valid, there are some important implications with respect to mechanisms:

1. Single-hit inactivation is the basic mode of killing at all dose levels. Various lines of evidence suggest that several modes of single-hit inactivation may occur concurrently (cf. equation 1).

2. Extrapolation number n has nothing to do with hit or target number. It is associated only with the relationship between the operation of Q factor and the accumulation of Q lesions. In the simplest form of association, n might be related to the concentration of Q factor.

Evidence Supporting a Repair Model

Since there is a strong case for retaining the belief in terminal exponential regions to shouldered survival curves, the choice must lie between multitarget and repair models. The most important difference between these is in the significance assigned to n, the extrapolation number. The following are some of the grounds for regarding repair models as the more plausible:

1. The variability of extrapolation numbers (a) between cell lines with a common pedigree, (b) with progress through the phases of growth, in some instances, (c) with progress from one mitosis to the next, and (d) with minute differences in methods of culture. Values of n for mammalian cells in vitro were reported by Elkind et al. (1961), for example, to be variable from experiment to experiment, despite careful standardization of techniques.

2. Extrapolation numbers to survival curves may be changed by a variety of post-irradiation treatments. For example, exposure of Chinese hamster cells to actinomycin D (Elkind et al. 1967b) or to acriflavine (Arlett 1970) after irradiation markedly reduced n; the addition of highly polymerized DNA to irradiated mouse L cells was found by Miletić et al. (1964) to effect a slight

but significant increase in n. It seems implausible that a target number should be capable of retrospective alteration.

3. With some cell lines, the value of n is greater if the cells are irradiated while in contact than if irradiated as single cells (Durand and Sutherland 1972). It is relevant to the argument that when cells were grown as spheroids, and separated before irradiation, the property conferring the larger value of n was lost gradually, over a period of hours.

Exponential Survival Curves for Mammalian Cells According to Repair Models

According to these models, the manifestation of a shoulder to the curve is evidence of repair of a special kind, Q repair, so cells lacking Q factor should be killed exponentially. As shown by Table 2, freshly cultured cells originating in organized tissues are particularly liable to exponential killing, i.e., on this interpretation, to be defective in Q factor—a plausible consequence of the drastic change in their growth conditions. A consequent implication is that considerable doubt must attach to any assumption that survival curves for cells freshly cultured from solid tumors are a reflection of their survival in situ.

Exponential survival curves should also be observed for cells in which the Q repair system remains uniformly effective throughout the dose range used. In such cases, of course, the repair would be indistinguishable from what is commonly known as repair of potentially lethal damage.

SOME DISQUIETING RESULTS OF DOSE FRACTIONATION

The choice of model for mammalian cell survival curves will determine the interpretation of the Elkind recovery phenomenon. According to multi–sublethal lesion models, the process consists of the sequential repair of all $n - 1$ out of n lesions postulated as necessary for cell killing; whereas, according to repair models, the recovery interval permits resynthesis or reconstitution of the factor(s) involved in Q repair. But, irrespective of the model, the accepted basis for predicting or interpreting results of dose fractionation rests on the observation, due in the first instance to Elkind and his colleagues, that the process of recovery after a conditioning dose restores to still-viable cells precisely the characteristics that determine the shape of the single-shot survival curve.

If that happens after each of many dose fractions, an exponential curve will emerge from a plot of fractional survival against dose when survival is always measured after delivery of equal dose fractions preceded by intervals for full recovery. This prediction was found by Elkind and Sutton (1960) to be fulfilled when V 79 cells in lag phase (two hours after plating) were exposed to five fractions each of 5.05 Gy separated by 22 hours, or four fractions each of 6.32 Gy. That such results will hold for many more fractions has been the assumption underlying the interpretation of results of multifraction experiments on animal tissues, including those designed to give an estimate of initial slopes of survival curves for the relevant clonogenic cells. But when McNally and de Ronde (1976) measured survival of V 79 cells in stationary stage after each of a larger number of smaller dose fractions (1.5 or 2 Gy, separated by six-hour intervals), survival was exponential only for the total dose delivered in the first five fractions (Figure 4). Fractional survivals after further dose fractions were considerably less than would be expected on the basis of an extrapolation of the results of the first few fractions. McNally and de Ronde confirmed what Figure 4 suggests, namely that cells that had survived five dose fractions gave single-shot survival curves with extrapolation numbers considerably reduced. If this happened also in some cells in vivo, survival curves deduced from multi-fraction experiments could well appear to be continuously bending, in contrast with the single-shot survival curves when those can be established. For example, Douglas and Fowler (1976) deduced this property for mouse skin, whereas single-shot survival curves by the clone-counting method for mice of the same strain had shown exponential inactivation over many powers of ten (Withers 1967). But, apart from its relevance to the interpretation of experimental results, the phenomenon observed by McNally and de Ronde has clear implications also for the results of conventional fractionated regimens in radiotherapy. Are some tumors easily controlled because the cell's capacity for recovery is gradually reduced? Or, are some unexpectedly severe tissue reactions attributable to a

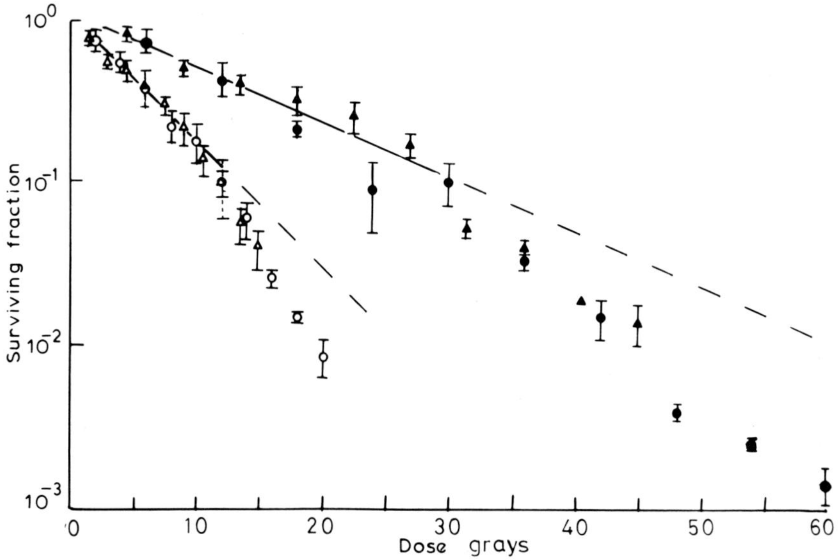

FIG. 4. Data of McNally and de Ronde (1976) on V 79 Chinese hamster cells irradiated in plateau phase. Counts were made after each fraction of 4.5 Gy (▲) or 6 Gy (●) to anoxic cells, and after each fraction of 1.5 Gy (△) or 2 Gy (○) to oxygenated cells, each dose fraction being preceded by a recovery interval of six hours. Solid lines, extrapolated by dashed lines, fitted by method of least squares to data for fractional survival above 10^{-1}.

subtle biochemical peculiarity that results in such a loss in recovery capacity? Clearly, much more information is needed.

"Over-repair" Stimulated by Radiation

Repair or recovery after a conditioning dose may also be substantially more than would be predicted from the classical experiments on Chinese hamster cells. This phenomenon has been studied more in algae than in other classes of cells, although its occurrence in sublines of HeLa cells was reported as early as 1961 by Lockhart et al. (Figures 5, 6). Exposure of *Chlamydomonas reinhardii* to a conditioning dose results in an increased value of D_0 for the surviving cells (Hillová and Drášil 1967, Bryant 1974, 1975, 1976; Figure 7), while Horsley and Laszlo (1971) found that in *Oedogonium cardiacum* a conditioning dose increased the extent of Q repair above that seen in unirradiated cells: the value of D_Q for the secondary survival curve was increased by a factor of 1.5 to 2. Somewhat similar results were reported by Howard and Cowie (1975, 1978) for the desmid *Clostiferum monoliferum.* Increased resistance was shown to occur also in bacteria after preliminary exposure to ultraviolet (UV) light or nalidixic acid as well as to ionizing radiation (Pollard and Achey 1975). All these treatments temporarily inhibit DNA synthesis. Howard and Cowie (1978) likewise found over-repair to be induced by preliminary doses of UV as well as of ionizing radiations, so it is a resonable tentative hypothesis that temporary

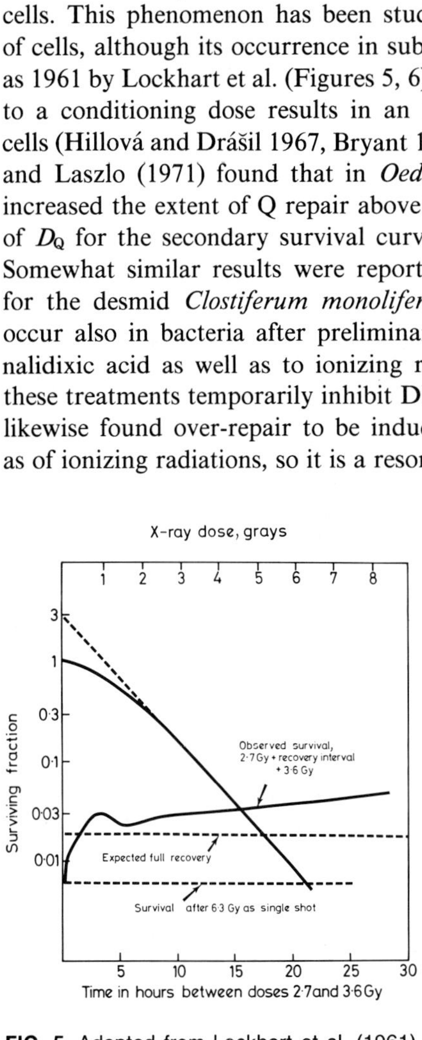

FIG. 5. Adapted from Lockhart et al. (1961). Single-shot survival curve for HeLa cells, subline S3-1, and fractional survival after delivery of a conditioning dose of 2.7 Gy followed by recovery intervals of various lengths, then delivery of 3.6 Gy.

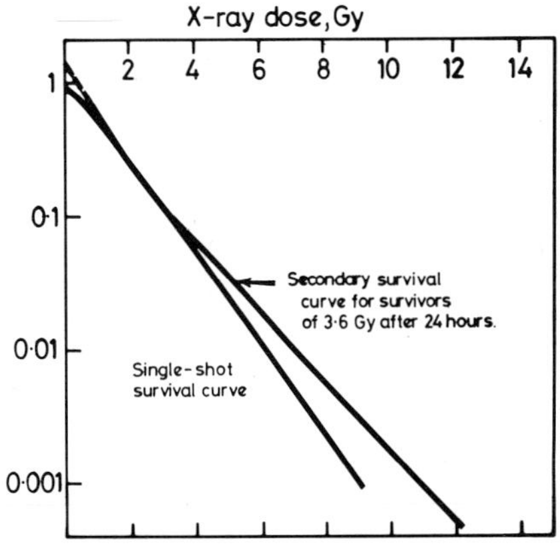

FIG. 6. Adapted from Lockhart et al. (1961). Single-shot survival curve for HeLa cells, subline S3-3, and secondary survival curve taken 24 hours after a conditioning dose of 3.6 Gy. The original diagram showed also that survivals of 3.6 Gy + 40 hr + 3.6 Gy and 3.6 Gy + 24 hr + 3.6 Gy were equal.

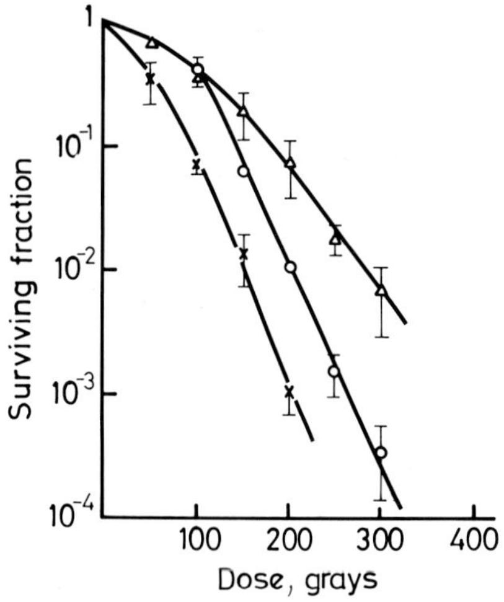

FIG. 7. After Bryant (1975): results with *Chlamydomonas reinhardii.* ○, Survival after single shots; △, Survival of cells exposed to 150 Gy, allowed three hours for recovery then exposed to further doses. Fractional survival related to numbers of cells surviving the conditioning dose. X, same as △, but chloramphenicol (100 μg/ml) and cycloheximide (10 μg/ml) present during the recovery interval.

inhibition of DNA synthesis may have this result in some cell lines. Inhibition of protein synthesis prevents the manifestation of induced radioresistance (Horsley and Laszlo 1973, Bryant 1975, Howard and Cowie 1978), although it was found not to affect normal Q repair in Chinese hamster cells (Elkind et al. 1967a), or in HeLa (Berry 1966) or in *Oedogonium* (Horsley and Laszlo 1973), both lines having given evidence of radiation-stimulated repair. In contrast, Bryant (1975) found that, with *Chlamydomonas,* the value of *n* was reduced for survivors of a conditioning dose if they were treated with chloramphenicol and cycloheximide during the recovery interval (Figure 7).

There may be an analogy between the induction of increased radioresistance and the production of "protein X," a membrane protein synthesized de novo in bacteria when DNA synthesis has been inhibited (Inouye and Pardee 1970, Gudas and Pardee 1976). The analogy is strengthened by the fact that the protein was not synthesized in one of the repair-deficient strains that Pollard and Achey had found noninducible for increased radioresistance.

As far as I know, the results of Lockhart et al. (1961) have not been discussed in the context of possible implications for radiotherapy. Do surviving tumor cells in vivo sometimes respond to early dose fractions by developing increased resistance? The results summarized in this section demonstrate a possibly very important biochemical response to radiation, even if its effects are manifested only in some cell lines. This is also further evidence, if indeed such evidence is needed, that progress in mammalian cell radiobiology requires the continuing support of radiation research on naturally free-living cells.

SUMMARY AND CONCLUSIONS

There are grounds for believing that shouldered survival curves for mammalian cells approximate to the exponential at high dose, and strong indications that the commonly observed relative insensitivity at low dose is attributable to a special mode of biochemical repair, rather than to target multiplicity. The correct interpretation of survival curve shape is of theoretical importance, but it has a bearing also on clinical aspects of cellular radiobiology. If, for example, the local control of some tumors is rendered difficult by virtue of very effective Q repair, biochemical means might be sought for coping with that problem.

It is hardly necessary to emphasize the importance of further research into aspects of dose fractionation that do not conform with the classical pattern. But there is also a moral to be pointed. I was struck by the statement made a few years ago by two eminent radiobiologists that cellular radiobiology had begun only in 1956, the year of the first survival curve for mammalian cells. One of the dictionary meanings of "cellular," however, is "divided into compartments"; so, although *cell* radiobiology dates back almost to the discovery of ionizing radiation, it may be true to say that, regrettably, it has become more and more *cellular* since 1956, with less and less attention paid by some "mission-oriented" research workers and research institutes to anything other than mammalian cell radiobiology. Naturally free-living cells like bacteria, fungi, and algae were used to investigate almost every radiobiological phenomenon of importance, before techniques for counting clones of mammalian cells were established; and they remain the most suitable test systems for investigating a good many of the biological effects of radiation. The goals of research aimed at improving the results of radiotherapy will be achieved the more slowly, the greater the tendency, already well marked in some quarters, to regard work on cells other than mammalian as irrelevant.

REFERENCES

Alper, T. 1961. Variability in the oxygen effect observed with micro-organisms. Part II. *Escherichia coli* B. Int. J. Radiat. Biol. 3:369–377.

Alper, T. 1963. Lethal mutations and cell death. Phys. Med. Biol. 8:365–385.

Alper, T., ed. 1975. Cell Survival after Low Doses of Radiation. Institute of Physics and John Wiley & Sons, London.

Alper, T. 1977. Elkind recovery and "sub-lethal damage": a misleading association? Br. J. Radiol. 50:459–467.

Alper, T., and N. E. Gillies. 1958. Dependence of the observed oxygen effect on the post-irradiation treatment of micro-organisms. Nature 181:961–963.

Arlett, C. F. 1970. Influence of post-radiation conditions on the survival of Chinese hamster cells after γ-irradiation. Int. J. Radiat. Biol. 17:515–526.

Berry, R. J. 1966. Effects of some metabolic inhibitors on X-ray dose-response curves for the survival of mammalian cells in vitro, and on early recovery between fractionated X-ray doses. Br. J. Radiol. 39:458–463.

Bryant, P. E. 1974. Change in sensitivity of cells after split dose recovery. A further test of the repair hypothesis. Int. J. Radiat. Biol. 26:499–504.

Bryant, P. E. 1975. Decrease in sensitivity of cells after split-dose recovery: evidence for the involvement of protein synthesis. Int. J. Radiat. Biol. 27:95–102.

Bryant, P. E. 1976. Absence of oxygen effect for induction of resistance to ionising radiation. Nature 261:588–590.

Caldwell, W. L., L. F. Lamerton, and D. K. Bewley. 1965. Increased sensitivity in vitro of murine leukaemia cells to fractionated X-rays and fast neutrons. Nature 208:168–170.

Chadwick, K. H., and H. P. Leenhouts. 1973. A molecular theory of cell survival. Phys. Med. Biol. 18:78–87.
Chapman, J. D. 1980. Biophysical models of mammalian cell inactivation by radiation, *in* Radiation Biology in Cancer Research (The University of Texas System Cancer Center 32nd Annual Symposium on Fundamental Cancer Research), R. E. Meyn and H. R. Withers, eds. Raven Press, New York, pp. 21–32.
Cox, R., J. Thacker, and D. T. Goodhead. 1977. Inactivation and mutation of cultured mammalian cells by aluminum characteristic ultrasoft X-rays. Int. J. Radiat. Biol. 31:561–576.
Dittrich, W. 1960. Treffermischkurven. Zeitschrift fuer Naturforschung 15b:261–266.
Douglas, B. G., and J. F. Fowler. 1976. The effect of multiple small doses of X rays on skin reactions in the mouse and a basic interpretation. Radiat. Res. 66:401–426.
Durand, R. E., and R. M. Sutherland. 1972. Effects of intracellular contact on repair of radiation damage. Exptl. Cell Res. 71:75–80.
Ehmann, U. K., H. Nagasawa, D. F. Petersen, and J. T. Lett. 1974. Symptoms of X-ray damage to radiosensitive mouse leukaemic cells: asynchronous populations. Radiat. Res. 60:453–472.
Elkind, M. M., W. B. Moses, and H. Sutton-Gilbert. 1967a. Radiation response of mammalian cells grown in culture. VI. Protein, DNA, and RNA inhibition during the repair of X-ray damage. Radiat. Res. 31:156–173.
Elkind, M. M., and H. Sutton. 1959. X-ray damage and recovery in mammalian cells in culture. Nature 184:1293–1295.
Elkind, M. M., and H. Sutton. 1960. Radiation response of mammalian cells grown in culture. I. Repair of X-ray damage in surviving Chinese hamster cells. Radiat. Res. 13:556–593.
Elkind, M. M., H. Sutton, and W. B. Moses. 1961. Post-irradiation survival kinetics of mammalian cells grown in culture. J. Cell. Comp. Physiol. 58 (Suppl. 1):113–134.
Elkind, M. M., H. Sutton-Gilbert, W. B. Moses, and C. Kamper. 1967b. Sub-lethal and lethal radiation damage. Nature 214:1088–1092.
Ginsberg, D. M., and J. Jagger. 1965. Evidence that initial ultraviolet lethal damage in *Escherichia coli* strain $15T^{-}A^{-}U^{-}$ is independent of growth phase. J. Gen. Microbiol. 40:171–184.
Gudas, L. J., and A. B. Pardee. 1976. DNA synthesis inhibition and the induction of protein X in *Escherichia coli.* J. Molec. Biol. 101:459–477.
Hall, E. J. 1975. Biological problems in the measurement of survival at low doses, *in* Cell Survival After Low Doses of Radiation, T. Alper, ed. Institute of Physics and John Wiley & Sons, London, pp. 13–14.
Haynes, R. H. 1966. The interpretation of microbial inactivation and recovery phenomena. Radiat. Res. Suppl. 6:1–29.
Hewitt, H. B., and C. W. Wilson. 1961. Survival curves for tumor cells irradiated *in vivo.* Ann. N.Y. Acad. Sci. 95:818–827.
Hillová, J., and V. Drášil. 1967. The inhibitory effect of iodoacetamide on recovery from sublethal damage in *Chlamydomonas rheinhardii.* Int. J. Radiat. Biol. 12:201–208.
Hornsey, S., and S. Vatistas. 1963. Some characteristics of the survival curve of crypt cells of the small intestine of the mouse deduced after whole body X-irradiation. Br. J. Radiol. 36:795–800.
Horsley, R. J., and A. Laszlo. 1971. Unexpected additional recovery following a first X-ray dose to a synchronized cell culture. Int. J. Radiat. Biol. 20:593–596.
Horsley, R. J., and A. Laszlo. 1973. Additional recovery in X-irradiated *Oedogonium cardiacum* can be suppressed by cycloheximide. Int. J. Radiat. Biol. 23:201–204.
Howard, A., and F. G. Cowie. 1975. Survival-curve characteristics in a desmid, *in* Cell Survival after Low Doses of Radiation, T. Alper, ed. Institute of Physics and John Wiley & Sons, London, pp. 3–10.
Howard, A., and F. G. Cowie. 1978. Induced resistance in *Closterium:* Indirect evidence for the induction of repair enzyme. Radiat. Res. 75:607–616.
Inouye, M., and A. B. Pardee. 1970. Changes of membrane proteins and their relation to deoxyribonucleic acid synthesis and cell division of *Escherichia coli.* J. Biol. Chem. 245:5813–5819.
Kellerer, A. M., and H. H. Rossi. 1972. The theory of dual radiation action. Curr. Topics Radiat. Res. Qtly 8:85–158.
Laurie, J., J. S. Orr, and C. J. Foster. 1972. Repair processes and cell survival. Br. J. Radiol. 45:362–368.
Lockhart, R. Z., M. M. Elkind, and W. R. Moses. 1961. Radiation response of mammalian cells

grown in culture. II. Survival and recovery characteristics of several sub-cultures of HeLa S3 cells after X-irradiation. J. Natl. Cancer Inst. 27:1393–1404.

McNally, N. J., and J. de Ronde. 1976. The effect of repeated small doses of radiation on recovery from sub-lethal damage by Chinese hamster cells irradiated in the plateau phase of growth. Int. J. Radiat. Biol. 29:221–234.

Miletić, B., D. Petrović, A. Han, and L. Šašel. 1964. Restoration of viability of X-irradiated L-strain cells by isologous and heterologous highly polymerized deoxyribonucleic acid. Radiat. Res. 23:94–103.

Norris, G., and S. L. Hood. 1962. Some problems in the culturing and radiation sensitivity of normal human cells. Exptl. Cell Res. 27:48–62.

Orr, J. S., S. E. Wakerley, and J. M. Stark. 1966. A metabolic theory of cell survival curves. Phys. Med. Biol. 11:103–108.

Pollard, E. C., and P. M. Achey. 1975. Induction of radioresistance in *Escherichia coli.* Biophys. J. 15:1141–1153.

Powers, E. L. 1962. Considerations of survival curves and target theory. Phys. Med. Biol. 7:3–28.

Puck, T. T., and P. I. Marcus. 1956. Action of X-rays on mammalian cells. J. Exptl. Med. 103:653–666.

Puck, T. T., D. Morkovin, P. I. Marcus, and S. J. Cieciura. 1957. Action of X-rays on mammalian cells. II. Survival curves of cells from normal human tissues. J. Exptl. Med. 106:483–500.

Sato, C., cited by S. Okada, 1975. Repair studies at the molecular, chromosomal and cellular levels: A review of current work in Japan, *in* Radiation Research: Biomedical, Chemical and Physical Perspectives, O. F. Nygaard, H. I. Adler, and W. K. Sinclair, eds. Academic Press, New York, pp. 694–702.

Scott, O. C. A. 1975. The interpretation of survival curves. Studia Biophysica 53:47–50.

Silini, G., and Y. Maruyama. 1965. X-ray and fast neutron survival response of 5-bromo-deoxycytidine-treated bone-marrow cells. Int. J. Radiat. Biol. 9:605–610.

Sinclair, W. K. 1973. N-Ethylmaleimide and the cyclic response to X-rays of synchronous Chinese hamster cells. Radiat. Res. 55:41–57.

Weichselbaum, R. R., J. Epstein, J. B. Little, and P. L. Kornblith. 1976. In vitro cellular radiosensitivity of human malignant tumors. Eur. J. Cancer 12:47–51.

Withers, H. R. 1967. The dose-survival relationship for irradiation of epithelial cells of mouse skin. Br. J. Radiol. 40:187–194.

Withers, H. R., and M. M. Elkind. 1969. Radiosensitivity and fractionation response of crypt cells of mouse jejunum. Radiat. Res. 38:598–613.

Withers, H. R., and M. M. Elkind. 1970. Microcolony survival assay for cells of mouse intestinal mucosa exposed to radiation. Int. J. Radiat. Biol. 17:261–267.

Mechanisms of Radiation Injury

Radiation Biology in Cancer Research, edited by
Raymond E. Meyn and H. Rodney Withers.
Raven Press, New York © 1980.

Biophysical Models of Mammalian Cell Inactivation by Radiation

J. Donald Chapman

Department of Radiation Oncology, Cross Cancer Institute, Department of Radiology, University of Alberta, Edmonton, Alberta, Canada T6G 1Z2

Proliferative or mitotic cell death by ionizing radiation results from chemical alterations in vital cellular molecules that often require one or more cell divisions for expression. This process has been described by various mathematical equations, each based upon a specific physical or molecular model. The requirements for an acceptable biophysical model to describe radiation-induced mammalian cell killing are threefold. First, the mathematical expression resulting from the biophysical model must accurately describe the observed cellular response. Second, the physical and molecular postulates of the model should have the capacity to integrate current knowledge about radiation-induced events in cells that occur at widely varying times and are observed by widely diverse techniques and specialties (see Figure 1). This knowledge includes:

1. the spatial distribution of absorbed energy by various radiations at the nanometer level,
2. the rapid free-radical processes that can occur within charged-particle tracks,
3. the diffusion-controlled free-radical reactions that occur in cells,
4. the processes that result in the decay of radicals in target molecules, particularly those that can result in altered chemical structures,
5. the enzymatic processes that can remove toxic radiation products and can restore chemically altered vital molecules to their normal structure, and
6. the physiological mechanisms whereby altered cell molecules result in cell death.

Although no comprehensive model that encompasses all these phenomena has been proposed, specialists have developed an impressive level of understanding of the specific events that occur at these different times. Third, a requirement for any acceptable model is that it be amenable to experimental verification: it has been said that the most useful models proposed by scientists are those that can be disproved by their experiments.

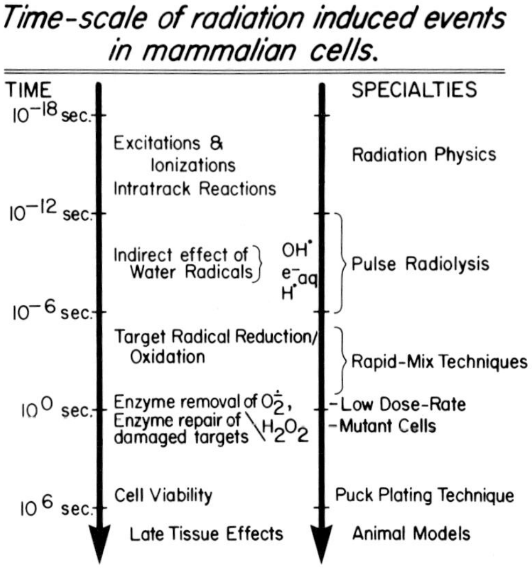

FIG. 1. A schematic of the time-scale of radiation-induced events in mammalian cells and the specialties or techniques that can provide information about the processes peculiar to each time.

MATHEMATICAL EXPRESSIONS FOR RADIATION INACTIVATION

Most of today's radiation biologists were nurtured on the classical target theory. An elegant presentation of this theory, including its single-hit single-target, multihit single-target, single-hit multitarget, and multihit multitarget ramifications for both homogeneous and heterogeneous cell populations, can be found in Elkind and Whitmore (1967). The equation from this theory most widely used for mammalian cell survival data is the single-hit multitarget equation,

$$S/S_0 = 1 - (1 - e^{-k_1 D})^n \tag{1}$$

where S/S_0 = fraction of cells that survive a radiation dose, D;
k_1 = ultimate exponential rate of cell kill at high radiation dose; and
n = number of targets per cell.

This model defines the radiation target(s) as a specific sensitive volume whose location within the cell is unspecified. This equation has been found to grossly underestimate radiation-induced cell killing by X rays at doses between 0 and 400 rad for both asynchronous (Chapman et al. 1975b) and synchronized (Gillespie et al. 1975) cell populations. In fact, the tenor of the sixth L. H. Gray Memorial Conference on "Cell Survival after Low Radiation Doses: Theoretical and Clinical Implications" (Alper 1975) was that a single-hit component was required to accurately describe mammalian cell killing at low radiation doses.

Bender and Gooch (1962) proposed that mammalian cells were inactivated by two independent mechanisms, one following single-hit kinetics and a second following single-hit multitarget kinetics as shown in equation 2.

$$S/S_0 = e^{-k_1 D}[1 - (1 - e^{-k_2 D})^n] \quad (2)$$

where S/S_0 = fraction of cells that survive a radiation dose, D;
k_1 = rate of cell kill by single-hit mechanism;
k_2 = ultimate exponential rate of cell kill at high radiation dose by multitarget mechanism; and
n = number of targets per cell associated with multitarget mechanism.

This equation has found wide acceptance, particularly among those who have investigated mammalian cell killing with radiations of high linear energy transfer (LET) (Barendsen et al. 1963, Todd 1966, Wideroe 1966). Even when experiments are technically designed so that one can assume little or no error in the determinations of S_0, there is unfortunately not enough information in most given experiments of 8–12 survival determinations over two to three decades of cell kill to specify all three parameters of equation 2 (k_1, k_2, and n) with significant precision. Consequently, approximations (such as fixing the value of n based upon intuition or long experience) are used to reduce the independent parameters to two, a manageable number. It is safe to say that all values of k_1, k_2, and n used to describe cell inactivation and derived by the famous "eye-ball" procedure should be treated with suspicion.

Another expression to describe cell inactivation that is gaining acceptance among radiation biologists is known as the linear-quadratic equation.

$$S/S_0 = e^{-\alpha D - \beta D^2} \quad (3)$$

where S/S_0 = fraction of cells that survive a radiation dose, D;
α = rate of cell kill by single-hit mechanism; and
β = rate of cell kill by double-hit mechanism.

This mathematical expression has been proposed on empirical grounds (Sinclair 1966), as a consequence of the dual-action theory based on microdosimetry (Kellerer and Rossi 1972) and as a consequence of a molecular model that defines the lethal lesion in cells as unrepaired double-strand breaks in DNA (Chadwick and Leenhouts 1973). This expression can adequately describe cell inactivation of synchronized cell populations (Hall 1975, Gillespie et al. 1975) and asynchronous cell populations (Chapman et al. 1975b), at least over the first two to three decades of cell kill. In fact, it gives a better fit to most sets of survival data tested than any of the two-parameter models.

A fourth equation to describe mammalian cell inactivation has been suggested by Green and Burki (1972) and is a consequence of a "repair-saturation" model discussed by Haynes (1966).

$$S/S_0 = (a + 1) / (a + e^{k_1 D}) \tag{4}$$

where S/S_0 = fraction of cells that survive a radiation dose, D;
k_1 = ultimate exponential rate of cell kill at high radiation dose; and
a = a parameter related to the probability of repair.

The repair capacity of this model has usually been associated with a finite quantity of repair enzyme but could just as easily be envisaged as a finite pool of radical reducing species in close proximity to the targets. Either entity would be depleted by the first few hundred rads of radiation and would follow typical enzyme kinetics for restoration. In her keynote address to this conference, Alper (1980, see pp. 3–18, this volume) has made a strong case for a "repair-saturation" model. The most important biophysical consequence of this model is that all basic lesions generated by all radiations that result in cell killing are induced by single-hit kinetics only.

As stated previously, only two independent parameters can be obtained with a precision of ±10% or less from typical 8–12 point mammalian cell survival curves generated over two to three decades of cell kill. And these only when precautions are taken to minimize cell heterogeneity and enzymatic repair during the radiation exposure. The former can be accomplished by using synchronized or stationary phase cultures and the latter by using dose rates in excess of 150 rad/min or temperatures of ≤4°C during irradiation. Analyses of several sets of our synchronized mammalian cell survival data generated over the past five years indicate that a single-hit parameter is required to accurately describe cell killing at low dose by low-LET as well as by high-LET radiations. Furthermore, equation 3 with only two independent parameters (assuming no error in the determination of S_0) can describe most of these data as well as a correct equation could be expected to (Gillespie et al. 1975).

BIOPHYSICAL MECHANISMS OF RADIATION INACTIVATION

Let us assume that mammalian cells can be inactivated by at least two independent mechanisms, characterized by single-hit and multihit kinetics. It has been shown that several factors that influence cellular radiation response have different and independent effects on the single- and multiple-hit mechanisms (Chapman et al. 1976). Additional evidence comes from cellular studies on the radiation chemical mechanism of bromodeoxyuridine (BUdR) radiosensitization. Figure 2(a) shows survival data for G_1-phase Chinese hamster cells, with and without incorporated BUdR, irradiated with 250 kV X rays. These experiments were technically quite difficult in that the cellular monolayers from which mitotic cells were selected were incubated for 24 hours with 10 μM BUdR. Mitotic cell populations were incubated for two hours for cells to divide and progress into G_1 phase. Data are also shown for both BUdR-labeled and unlabeled cells irradiated in the presence of 1 M dimethyl sulfoxide (DMSO), a

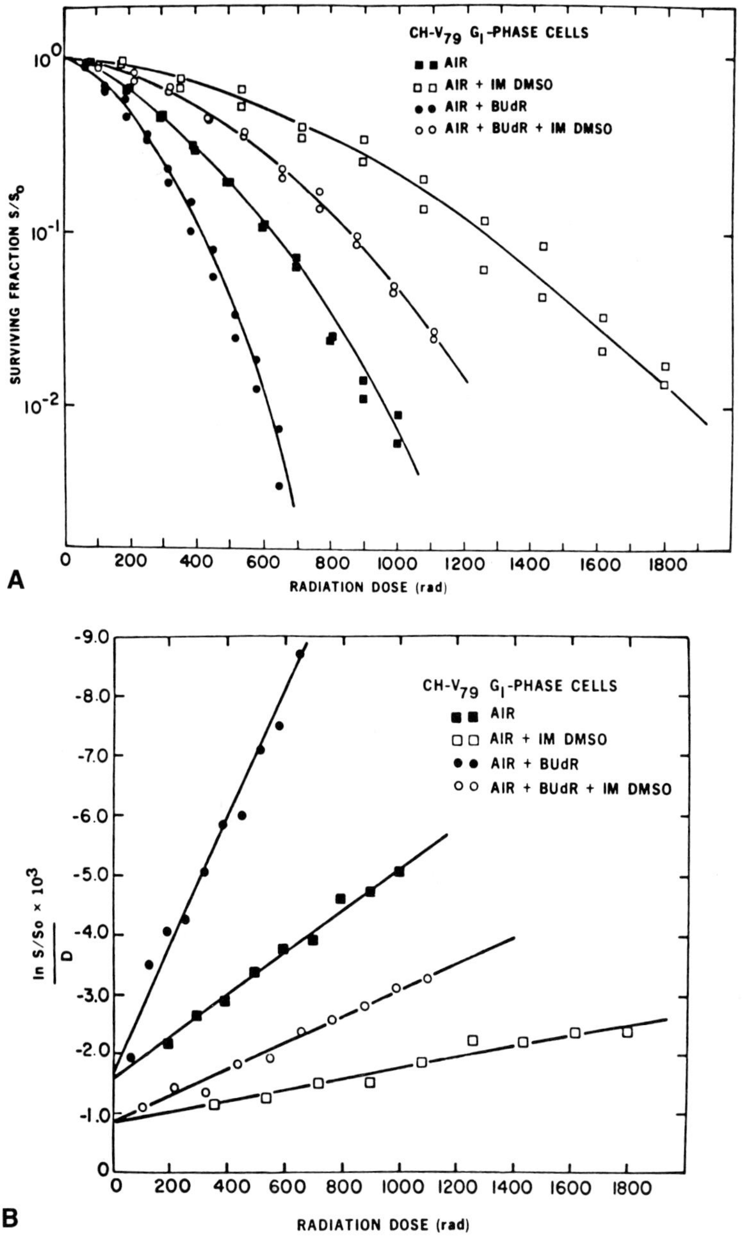

FIG. 2. a, Radiation inactivation data for BUdR-labeled and unlabeled G_1-phase Chinese hamster cells irradiated with 250 kV X rays in the presence and absence of 1 M DMSO. **b,** Average values of the surviving fractions from a, plotted as the $(\ln S/S_o)/D$ versus radiation dose.

radioprotective agent that acts by scavenging the indirect action of OH in cells (Chapman et al. 1973). For each case, data are plotted from two independent experiments performed on different days, and remarkably good agreement is observed in spite of the complicated experimental procedures. BUdR is shown to radiosensitize hamster cells in air and DMSO is shown to radioprotect both BUdR-labeled and unlabeled cells. Both results confirm previous reports. In Figure 2(b), averages of the data shown in Figure 2(a) are plotted as ($\ln S/S_0$)/D versus radiation dose, a procedure that results in a graphical linearization of equation 3:

$$(\ln S/S_0)/D = -\alpha - \beta D \tag{5}$$

The slope of each plot is the quadratic inactivation constant (β) and the intercept with the zero dose axis is the linear inactivation constant (α). This type of cell-inactivation plot assists in making the independent parameters of cell inactivation readily apparent. It is of interest to note from Figure 2(b) that the radiosensitization by BUdR incorporated into G_1-phase hamster cells is exclusive to the quadratic mechanism. On the other hand, the radioprotection effected by DMSO on both labeled and unlabeled cells operates on both of the mechanisms. Here then is a clear example in which a chemical modification of cellular DNA (BUdR incorporation) results in sensitization of one mechanism and not the other.

Radiosensitization of cells by incorporated BUdR has been attributed to increased damage in the DNA by solvated electron attack (Zimbrick et al. 1969). In the presence of 1 M DMSO, where most indirect effects of OH have been removed, BUdR was found to radiosensitize cells just as well as when OH radicals were present. This result indicates that OH radicals are not involved in BUdR radiosensitization and is consistent with enhancement by the indirect action of solvated electrons.

Additional evidence for the independent and distinctive nature of the linear and quadratic mechanisms comes from studies on the radioprotection of mammalian cells by DMSO (Chapman et al. 1979). Figure 3 shows the linear and quadratic mechanisms (as ratios of unprotected control values) for G_1-phase hamster cells irradiated with both low-LET and high-LET radiations in the presence of various concentrations of DMSO. Evidence has been published that indicates that DMSO effects cellular radioprotection by scavenging the indirect action of OH (Chapman et al. 1975a). The data in Figure 3 indicate that a large component of the linear mechanism ($\sim$ 80%) of cell inactivation can be protected by very high concentrations of DMSO, suggesting that the indirect action of OH is dominant for this mechanism of cell kill by both X rays and charged-particle beams, some with median LETs in excess of 100 keV/μm. As much as 75% of the quadratic mechanism of cell inactivation produced by X rays (or 50% of $\sqrt{\beta}$) can be eliminated by DMSO as well, but lower concentrations are required for this protection, suggesting that lower OH concentrations and consequently lower energy densities are involved on the average

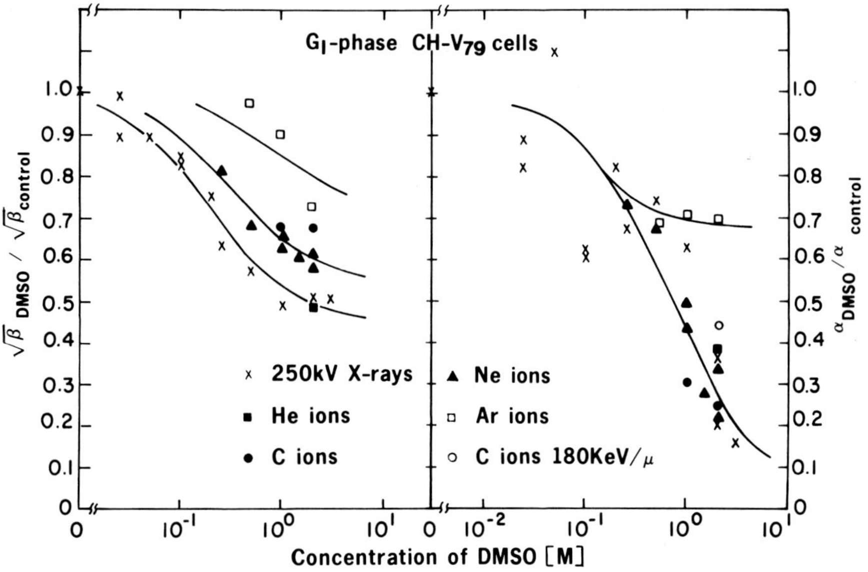

FIG. 3. Values of the double-hit ($\sqrt{\beta}$) *(left)* and single-hit (α) *(right)* inactivation coefficients for G_1-phase Chinese hamster cells, plotted as ratios of the control values, for various concentrations of DMSO, a potent OH scavenger. (Reprinted from Chapman et al. 1979, with permission from Springer-Verlag.)

in quadratic events. This radiation-chemical result indicates that some energy-dense event associated with the dissipation of absorbed energy from X rays is responsible for the observed single-hit killing. Tobleman and Cole (1974) have shown that when a 10-keV electron beam is stopped in the nuclei of mammalian cells, the resultant cell killing is characterized by single-hit kinetics, little or no recovery from sublethal damage, and a low oxygen enhancement ratio (OER). The importance of slow electrons or electron track-ends in cell killing by the linear mechanism was consequently recognized (Paretzke 1976). The idea that several ionizations within a distance of 10 nm are necessary for effective cell killing was a conclusion of Barendsen's study (1964) with alpha particles of various energies.

Figure 4 shows two-dimensional representations of several independent tracks of 1-keV electrons generated with a Monte-Carlo simulation program similar to that described by Paretzke (1976). The average track length of such electrons in water is 25.5 ± 4.0 nm with an average LET of 40.2 ± 6.0 keV/μm. If an initial energy of 700 eV is used, the average track length becomes 11.3 ± 2.3 nm with an average LET of 64.3 ± 13.5 keV/μm. These values of LET are approaching those found to give maximum RBE from other charged-particle beams (Barendsen et al. 1963, Todd 1966). Figure 5 shows values for the single-hit mechanism of cell killing of mitotic, G_1-phase, and stationary-phase hamster cells as a function of median LET for various energetic particle beams produced

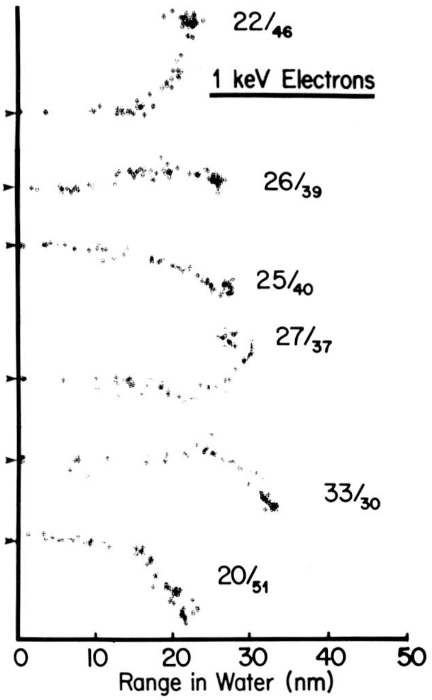

FIG. 4. Two-dimensional representations of 1-keV electron tracks generated with a Monte-Carlo simulation program. The first number (large type) opposite each track-end is the range at which the electron stops and the second number (smaller type) is the LET of that particular tract in keV/μm.

at the 184-inch cyclotron and the BEVALAC at the Lawrence Berkeley Laboratory. Cells were irradiated in stirred suspensions by the unattenuated portion and by the spread Bragg peaks of the various charged particles investigated (see Chapman et al. 1977). The median values of LET for the spread peaks were computed by Curtis (1977). The data show that the radiosensitivities of these homogeneous cell populations, which are dramatically different for low-LET radiations like 250 kV X rays, become quite similar at LET in excess of 60 keV/μm.

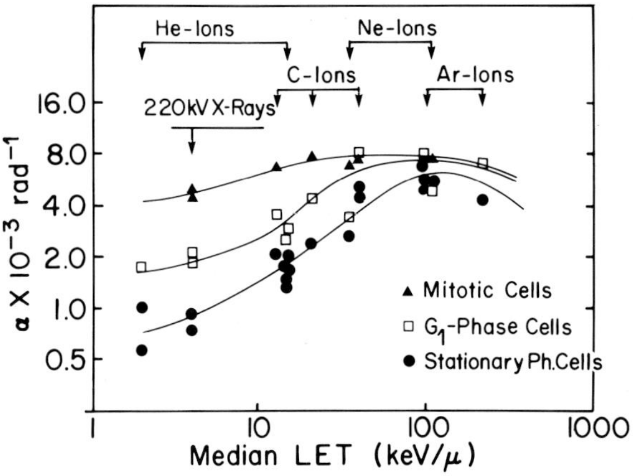

FIG. 5. Single-hit inactivation coefficients for homogeneous populations of mitotic, G_1-phase, and stationary-phase Chinese hamster cells irradiated with 220-kV X rays and various charged-particle beams as a function of median LET in keV/μm.

The simulated electron track studies indicate that slow electrons of 700 eV (or track-ends) would produce densely ionizing events over a dimension of ~ 10 nm. In matter consisting mostly of water, this absorbed energy would be expected to generate at least 20 electrons (strongly reducing species), which may or may not become solvated before reaction with cellular molecules, and at least 20 hydroxyl radicals and hydrogen atoms (strongly oxidizing species). Such events could be responsible for mammalian cell killing by single-hit mechanisms with X rays and other charged particles and explain why such a large proportion of single-hit kills can be protected by an OH scavenger at high concentrations (see Figure 3). The dimension associated with these physical and chemical events is significantly greater than the dimensions of the DNA double helix but coincides with the dimension of the basic nucleosome structure of cellular DNA. These nucleoprotein subunits are composed of 160–200 base pairs of supercoiled DNA, along with a somewhat larger amount of histone, and are packaged into a structure of approximately 10 nm in diameter (Finch and Klug 1976, Worcel and Benyajati 1977). Some 200 base pairs was found to be the distance over which sublesions of the quadratic mechanism could interact to produce a lethal event and DNA double-strand breaks (Gillespie et al. 1976, Dugle et al. 1976). This structural dimension of DNA is consequently implicated in cell inactivation by both the linear and quadratic mechanisms.

A growing volume of experimental evidence, which includes the data presented in this paper, indicates that mammalian cells are killed by ionizing radiation through at least two independent mechanisms. These mechanisms can be distinguished not only on the basis of dose kinetics but also on the basis of different radiation-chemical characteristics that are a consequence of the energy densities associated with the respective physical events. Single-hit cell kill results from energy-dense events, involves high local concentrations of water-free radicals over a dimension of ~ 10 nm, exhibits an OER of 1.6–1.7, and shows evidence of little or no repair. Multiple-hit cell kill results from at least two independent energy deposition events, involves lower concentrations of water-free radicals (or larger diffusion distances), exhibits an OER of 3.0–3.5, and shows complete repair of sublethal lesions under appropriate conditions. Such an understanding of cell inactivation has interesting implications for radiation oncology.

IMPLICATIONS FOR RADIATION ONCOLOGY

Radiation oncologists treat most of their patients with daily radiation fractions of ~ 200 rad, five days a week, up to a limiting total dose determined by normal tissue tolerance in and/or around the treatment volume. Human kidney cells in aerated culture have linear and quadratic inactivation constants of 1.73×10^{-3} rad^{-1} and 3.47×10^{-6} rad^{-2}, respectively. For these cells, 500 rad must be given before the contributions to cell killing (from the two mechanisms) are near equal. As the dose decreases, the proportion of total cell kill by the linear mechanism increases, and at 200 rad over 70% results from this mechanism. In order to provide a better scientific basis for radiation oncology, radiation biologists must become more concerned with characterizing the single-hit mechanism of cell inactivation, which is dominant in most therapy protocols. In light of the different characteristics already described between the linear and quadratic mechanisms of cell kill, it is unlikely that our (radiation biologists') persistent use of D_o, which is an approximation of $1/\sqrt{\beta}$, to characterize cell radiosensitivity has been of great value to our medical colleagues. Experiments are required to determine the single-hit radiation sensitivities of the clonogenic cells in normal and tumor tissues. If these radiosensitivities for cells in vivo are similar to those determined with human cell lines in vitro, some consequences of interest in radiation oncology arise, of which only two are described.

The single-hit mechanism of tumor cell kill could be as much as 25% greater than that of normal cell kill. Such a difference in radiation sensitivity would not be considered significant from survival curves generated over two to three decades of cell kill with relatively high radiation doses, especially if the multihit sensitivities were similar. Nevertheless, this difference, amplified by successive daily fractionation with 200 rad, could result in a 10- to 100-fold difference in cell kill after a treatment with 4,000 to 6,000 rad. Much of the present success of current radiotherapy protocols might then be attributed to differences in single-hit radiosensitivities between tumor and normal tissue cells that become significant with daily fractionation schemes.

Low dose-rate radiotherapy is considered to have some radiobiological advantages over conventional fractionated treatment (Hall 1972). At low dose rate ($\leq$ 0.5 rad/min) all cell killing will result from single-hit lesions and any difference between tumor and normal tissue cells will be maximized. In current radiotherapy practice, low dose rates are usually associated with interstitial and intracavitary radiation techniques. Modern megavoltage therapy units are designed to provide dose rates of about 200 rad/min over large fields and consequently would prove very cost-ineffective if routinely utilized for external beam low dose-rate therapy. Using the inactivation constants for human kidney cell killing (given above), one can show that three fractions of 70–80 rad per day with four-hour intervals between fractions for repair could result in most of the proposed benefits of low dose-rate radiation. This technique would not be as taxing on radiotherapy equipment, and some early studies indicate that the clinical results obtained might warrant wider use of such hyperfractionation procedures (Littbrand and Edsmyr 1976, Douglas 1977).

ACKNOWLEDGMENTS

I thank Dr. C. J. Gillespie for many hours of fruitful discussions on mechanisms of radiation-induced cell killing and for writing the computer program to generate the electron tracks shown in Figure 4. The assistance of A. P. Reuvers with the BUdR studies, E. A. Blakely and K. C. Smith with the heavy charged-particle experiments performed at the Lawrence Berkeley Laboratory, and Shirley Dawson in preparing the manuscript is appreciated.

REFERENCES

Alper, T, ed. 1975. Cell Survival after Low Doses of Radiation: Theoretical and Clinical Implications. The Institute of Physics, and John Wiley & Sons, London.
Alper, T. 1980. Survival curve models, *in* Radiation Biology in Cancer Research (The University of Texas System Cancer Center 32nd Annual Symposium on Fundamental Cancer Research), R. E. Meyn and H. R. Withers, eds. Raven Press, New York. pp. 3–18.
Barendsen, G. W. 1964. Impairment of proliferative capacity of human cells in culture by alpha-particles with differing linear-energy-transfer. Int. J. Radiat. Biol. 8:453–466.
Barendsen, G. W., H. M. D. Walter, J. F. Fowler, and D. K. Bewley. 1963. Effects of differential ionizing radiations on human cells in tissue culture. III. Experiments with cyclotron-accelerated alpha-particles and deuterons. Radiat. Res. 18:106–119.
Bender, M. A., and P. C. Gooch. 1962. The kinetics of x-ray survival of mammalian cells in vitro. Int. J. Radiat. Biol. 5:133–145.
Chadwick, K. H., and H. P. Leenhouts. 1973. A molecular theory of cell survival. Phys. Med. Biol. 18:78–87.
Chapman, J. D., E. A. Blakely, K. C. Smith, and R. C. Urtasun. 1977. Radiobiological characterization of the inactivating events produced in mammalian cells by helium and heavy ions. Int. J. Radiat. Oncol. Biol. Phys. 3:97–102.
Chapman, J. D., S. D. Doern, A. P. Reuvers, C. J. Gillespie, A. Chatterjee, E. A. Blakely, K. C. Smith, and C. A. Tobias. 1979. Radioprotection by DMSO of mammalian cells exposed to x-rays and to heavy charged-particle beams. Envir. Radiat. Biophys. 16:29–41.
Chapman, J. D., D. L. Dugle, A. P. Reuvers, C. J. Gillespie, and J. Borsa. 1975a. Chemical radiosensitization studies with mammalian cells growing in vitro, *in* Radiation Research, Biomedical, Chemical, and Physical Perspectives. Academic Press, New York, pp. 752–760.
Chapman, J. D., C. J. Gillespie, A. P. Reuvers, and D. L. Dugle. 1975b. The inactivation of Chinese hamster cells by x-rays: The effects of chemical modifiers on single- and double-events. Radiat. Res. 64:365–375.
Chapman, J. D., A. P. Reuvers, J. Borsa, and C. L. Greenstock. 1973. Chemical radioprotection and radiosensitization of mammalian cells growing *in vitro*. Radiat. Res. 56:291–306.
Chapman, J. D., A. P. Reuvers, S. D. Doern, C. J. Gillespie, and D. L. Dugle. 1976. Radiation chemical probes in the study of mammalian cell inactivation and their influence on radiobiological effectiveness, *in* Proceedings of the Fifth Symposium on Microdosimetry. Commission of European Communities, Luxembourg, pp. 775–798.
Curtis, S. 1977. Calculated LET distributions of heavy ion beams. Int. J. Radiat. Oncol. Biol. Phys. 3:87–91.
Douglas, B. G. 1977. Preliminary results using superfractionation in the treatment of glioblastoma multiforme. J. Can. Assoc. Radiol. 28:106–110.
Dugle, D. L., C. J. Gillespie, and J. D. Chapman. 1976. DNA strand breaks, repair, and survival in x-irradiated mammalian cells. Proc. Natl. Acad. Sci. USA 73:809–812.
Elkind, M. M., and G. F. Whitmore. 1967. The radiobiology of cultured mammalian cells. Gordon and Breach, New York, pp. 7–51.
Finch, J. T., and A. Klug. 1976. Solenoidal model for superstructure in chromatin. Proc. Natl. Acad. Sci. USA 73:1897–1901.
Gillespie, C. J., J. D. Chapman, A. P. Reuvers, and D. L. Dugle. 1975. The inactivation of Chinese hamster cells by x-rays: Synchronized and exponential cell populations. Radiat. Res. 64:353–364.
Gillespie, C. J., D. L. Dugle, J. D. Chapman, A. P. Reuvers, and S. D. Doern. 1976. DNA damage and repair in relation to mammalian cell survival: Implications for microdosimetry, *in* Proceedings of the Fifth Symposium on Microdosimetry. Commission of European Communities, Luxembourg, pp. 799–813.
Green, A. E. S., and J. Burki. 1972. A note on survival curves with shoulders. Radiat. Res. 60:536–540.
Hall, E. J. 1972. Radiation dose-rate: A factor of importance in radiobiology and radiotherapy. Br. J. Radiol. 45:81–97.
Hall, E. J. 1975. Biological problems in the measurement of survival at low doses, *in* Cell Survival after Low Doses of Radiation: Theoretical and Clinical Implications. The Institute of Physics, and John Wiley & Sons, London, pp. 13–24.

Haynes, R. H. 1966. The interpretation of microbial inactivation and recovery phenomena. Radiat. Res. Suppl. 6:1–29.
Kellerer, A. M., and H. H. Rossi. 1972. The theory of dual radiation action. Curr. Top. Radiat. Res. Q. 8:85–158.
Littbrand, B., and F. Edsmyr. 1976. Preliminary results of bladder carcinoma irradiated with low individual doses and a high total dose. Int. J. Radiat. Oncol. Biol. Phys. 1:1059–1062.
Paretzke, H. G. 1976. An appraisal of the relative importance for radiobiology of effects of slow electrons, *in* Proceedings of the Fifth Symposium on Microdosimetry. Commission of the European Communities, Luxembourg, pp. 41–60.
Sinclair, W. K. 1966. The shape of radiation survival curves of mammalian cells cultured in vitro, *in* Biophysical Aspects of Radiation Quality. I.A.E.A., Vienna, pp. 21–48.
Tobleman, W. T., and A. Cole. 1974. Repair of sublethal damage and oxygen enhancement ratio for low-voltage electron beam irradiation. Radiat. Res. 60:355–360.
Todd, P. 1966. Reversible and irreversible effects of densely ionizing radiations upon the reproductive capacity of cultured mammalian cells. Med. College Virginia Q. 1:2–14.
Wideroe, R. 1966. High-energy electron therapy and the two-component theory of radiation. Acta Radiol. 4:257–278.
Worcel, A., and C. Benyajati. 1977. Higher order coiling of DNA in chromatin. Cell 12:83–100.
Zimbrick, J. D., J. F. Ward, and L. S. Myers. 1969. Studies on the chemical basis of cellular radiosensitization by 5-bromouracil substitution in DNA. I. Pulse- and steady-state radiolysis of 5-bromouracil and thymine. Int. J. Radiat. Biol. 16:501–523.

Radiation Biology in Cancer Research, edited by
Raymond E. Meyn and H. Rodney Withers.
Raven Press, New York © 1980.

Mechanisms of Cell Injury

Arthur Cole, Raymond E. Meyn, Ruth Chen, Peter M. Corry, and Walter Hittelman*

*Departments of Physics and *Developmental Therapeutics, The University of Texas System Cancer Center M. D. Anderson Hospital and Tumor Institute, Houston, Texas 77030*

Ionizing radiation induces many forms of cellular damage, such as reproductive death, interphase death, division delay, chromosome aberrations, mutation, and transformation. It is generally accepted that the molecular basis for some of these cellular responses involves damage to the nuclear DNA. For example, among experiments that strongly implicate DNA as a critical target in cell reproductive death are the "suicide" experiments of Ragni and Szybalski (1962), Burki and Okada (1968), Burki (1976), and Kirsch and Zelle (1969), which utilize lethal radioactive label in the DNA. On the other hand, Alper (1971) has suggested that the nuclear membrane is the site of oxygen-dependent radiation damage, while Dewey and Highfield (1976) and Schneiderman et al. (1977) suggest that damage to a division-specific protein contributes to division delay.

One way to investigate whether damage to DNA constitutes a critical or causal step in the expression of a cell response is to administer irradiation or chemical treatments using different experimental conditions. If DNA damage and cell response vary with the experimental conditions in the same manner, a causal relationship is suggested, although not proved. Some modifiers of radiation response that have been extensively studied by many investigators include: oxygen concentration, chemical enhancing and protecting agents, cell temperature before, during, and after irradiation, cell cycle stage during irradiation, cell repair incubation, and radiation dose, dose rate, dose fractionation, dose distribution, and linear energy transfer (LET).

Since ionizing radiation or chemicals may induce many types of DNA damage including single-strand breaks (SSB), double-strand breaks (DSB), base damage, DNA-DNA and DNA-protein cross-links (see Cerutti 1974, for a brief review), correlations between cell responses and various forms of DNA damage are necessary to specify which types of DNA damage are most likely to be involved in cellular responses. Cole et al. (1975) have presented evidence that "simple" DNA lesions such as repairable SSB probably do not result in cell death or other cell responses. On the other hand, complex lesions such as DNA DSB probably do contribute to cellular responses. These conclusions were based on the observation that the relative biological effectiveness (RBE) of both cell responses and the induction of DSB was found to increase with the LET of the radiation, whereas the induction of DNA SSB was found to decrease with LET. Similar results were reported earlier for breaks induced in bacteriophage DNA (Christensen and Tobias 1972).

Another type of radiation study utilizes radiation microbeams (Cremer et al. 1976, Ohnuki et al. 1972) or ionizing particle beams of limited penetration to irradiate selected regions of a cell. The latter technique, which was probably first used by Haskins (1938) and later applied by Davis (1954), Hutchinson (1955), Preiss and Pollard (1961), and others to irradiate dry biological material was developed by Cole (1959) to irradiate fully hydrated organisms (Cole 1961, 1965, Cole and Langley 1963, Cole et al. 1963, 1974, 1977, Zermeno and Cole 1969, Tobleman and Cole 1974, Datta et al. 1976, Cole and Datta 1976b, Humphrey et al. 1978). These studies provide another correlation between DNA damage and biological responses: the location within the cell where the susceptible sites for the different effects are found.

METHODS AND RESULTS

This paper briefly summarizes many studies carried out in our laboratories. Details either have been or will be reported in other publications. Our objective here is to examine possible correlations between radiation-induced cell responses and DNA damage for a variety of experimental conditions.

Dose Response

The analysis of the dependence of biological response on radiation dose forms a cornerstone in the study of radiation biology. Early discussions may be found in Lea's prophetic book published over 30 years ago (Lea 1946). Later discussions may be found in the book by Elkind and Whitmore (1967), while contemporary discussions are presented in this volume by Alper (1980, pages 3–18), Chapman (1980, pages 21–32), Rossi (1980, pages 185–194), Tobias (1980, pages 195–230), and Goodhead (1980, pages 231–247). To simplify, two types of responses usually are observed, one in which an increment in response is linearly proportional to a dose increment and another in which the response is a nonlinear function of dose. The linear dose response is interpreted to result from damage induced by a single particle traversal of the radiation-susceptible region, while the nonlinear response results from multiple particle traversals of the susceptible region. Neary (1965) and Kellerer and Rossi (1971) have proposed that the nonlinear component involves the independent induction of similar molecular lesions that must interact to yield the observed response. Alternative interpretations invoke the intervention of cell repair systems (Alper 1979, Tobias 1979). Some models, such as the multitarget model, predict a negligibly small response at doses below a threshold value. During the last several years, many response curves have been fitted to the simple quadratic relationship

$$\text{Response} = \alpha D + \beta D^2$$

where α and β are rate constants for the linear and nonlinear (quadratic) dose (D) dependencies. This relationship yields, for an unaffected or surviving population, ln survival $= -\alpha D - \beta D^2$.

The presence of a linear component means that any amount of radiation, no matter how small, will produce a finite effect, and this suggests that some critical molecular lesions can never be properly repaired. Based on studies using low energy electron beams (Tobleman and Cole 1974), Cole has proposed that this may be true for any form of ionizing radiation and even during the most radioresistant phases of the cell cycle. Figure 1 plots some results from a recent low dose resolution study of cell survival carried out in our laboratory (Chen et al. 1978, Cole et al., manuscript in preparation), which illustrates an initial linear component with $\alpha = 1.05 \pm 0.13$ KR^{-1} detectable down to 10 R and

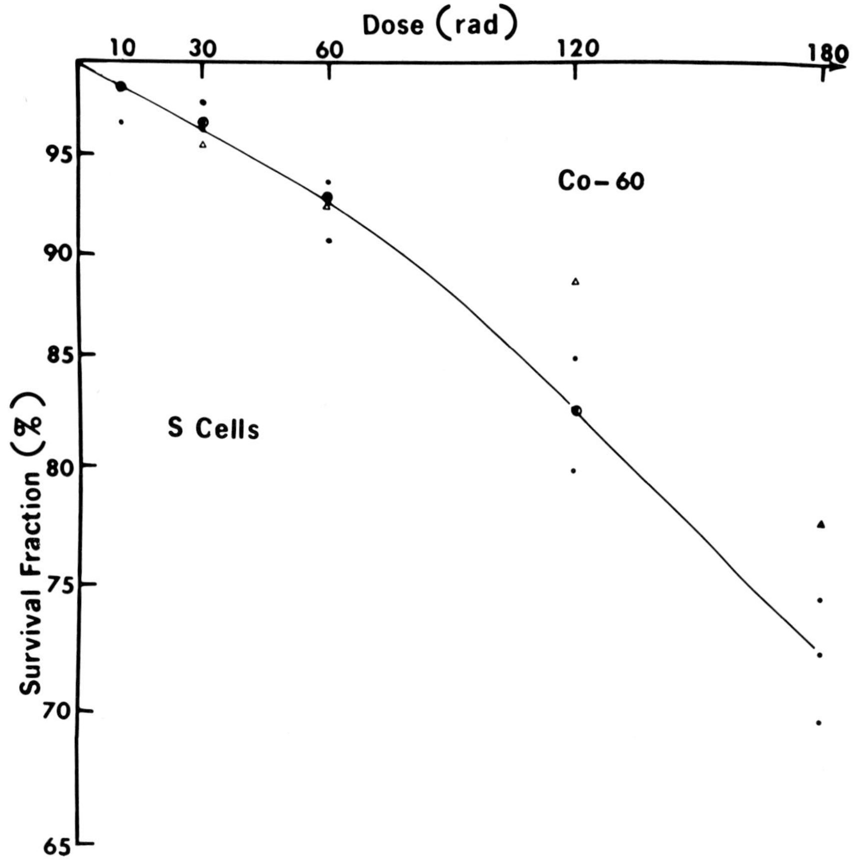

FIG. 1. Portion of high-resolution low-dose low-LET lethality study of S phase CHO cells. Data summarize about 10^6 colony counts. Circles are data for single-dose experiments; triangles are data for split-dose experiments with three hour repair incubation between equal doses. The single-dose data were fitted to the equation, In Survival $= -(1.05 \pm 0.13)D - (4.27 \pm 0.32)D^2$ and split-dose data to the equation, In Survival $= -(0.94 \pm 0.20)D - (2.06 \pm 0.36)D^2$.

1% lethality. The experiment summarized in Figure 1 involved the assay of nearly 10^6 individual clones and utilized radiation of minimal LET to irradiate Chinese hamster ovary (CHO) cells synchronized by cell elutriation to select for the most radioresistant phase of the growth cycle.

Repair of Radiation Damage

That a large fraction of radiation damage in mammalian cells can be repaired was shown in the classic paper of Elkind and Sutton (1959). Figure 2 from Tobleman and Cole (1974) illustrates a consequence of this repair: greater cell survivals are observed when the dose is split into two equal fractions separated by a two-hour incubation. This repair of sublethal damage, which exhibits half-repair times of about 30 minutes, is assumed to demonstrate the repair or reversal of molecular (DNA) lesions prior to the damage being confirmed by interactions with subsequent radiation products. Other results from the study of Tobleman and Cole (1974) showed that when low-energy electron track-ends were used to irradiate the cell nucleus, the log survival versus dose curves became much straighter (i.e., the value of α increased), split-dose recovery was much reduced, and the oxygen enhancement illustrated in Figure 2 was much reduced. In

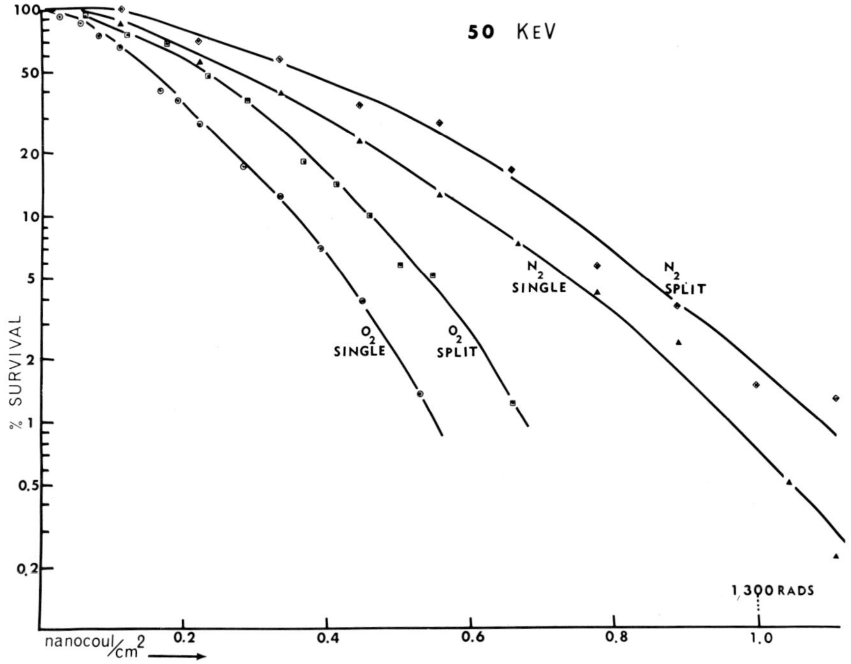

FIG. 2. Representative data for lethality induced by penetrating 50-KeV electron beams. Single- and split-dose responses with and without oxygen are shown. Reproduced from Tobleman and Cole (1974), with permission of Academic Press.

effect, electron track-ends behaved like relatively high LET particles, which induce a greater proportion of irreparable damage. Thus, low-LET radiation would be expected to induce many irreparable lesions, since about 10% of the absorbed dose is contributed by secondary electron track-ends (see Discussion).

Extensive studies of the induction and repair of DNA SSB and DSB have been made in our laboratories (Corry and Cole 1968, Cole et al. 1970, 1974, 1975, 1978, 1979, Corry and Cole 1973, Corry et al. 1977, detailed reports of recent work in preparation). These studies used alkaline and neutral sucrose gradient sedimentation techniques (McGrath and Williams 1966, Kaplan 1966) and alkaline elution techniques (Kohn and Grimek-Ewig 1973, Kohn et al. 1974, 1976).

Figure 3 illustrates the general type of result obtained for native DNA from CHO cells lysed in 1% sarkosyl, 0.1 M Na_2SO_4, 0.05 M EDTA, and 5% sucrose at pH 9.6 on top of a 5% to 22% sucrose gradient containing 5 M NaCl, 0.1 M Na_2SO_4, 0.05 M EDTA, and saturated with chloroform and dieth-

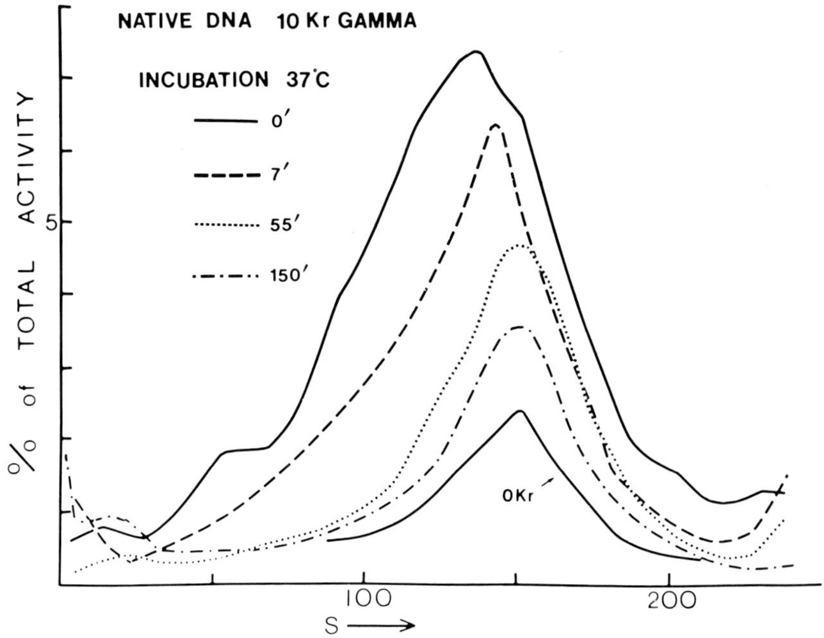

FIG. 3. Representative sedimentation profiles plotted as percent of total activity per fraction for native DNA from cells lysed on top of high salt, pH 9.6, sucrose density gradients. At zero dose, about 30% of the DNA label appeared in the profile representing freely sedimenting DNA; the rest appeared in a non-freely-sedimenting fraction (NFSF) found on top of the gradient. Moderate cell radiation doses caused activity from the NFSF to appear in sedimentation profiles at slightly reduced S values compared with zero-dose profiles. Repair incubations first led to shifts of the profiles to higher S values followed by a shift of activity back to the NFSF.

ylether at pH 9.6. Centrifuge rotation speeds of 10^4 Krpm were used to minimize anomalous sedimentation (Chen and Cole 1979). At zero dose, only 30%–40% of the label appeared as freely sedimenting DNA, the remainder appeared as a non-freely sedimenting fraction (NFSF), which for the most part floated at the top of the gradient. Doses of 0 to 20 KR before cell lysis caused increasing amounts of DNA to be released from the NFSF to sediment freely to a slightly smaller molecular weight region of the profile than zero-dose controls. Repair incubation after irradiation and before lysis first caused a shift of the profiles back to a higher molecular weight region and then a shift of label back to the NFSF. Thus, two types of repair were observed: the first a repair of nascent DSB induced in DNA subunits, and the second probably a reconstruction of large (membrane-associated) DNA arrays that do not sediment freely in our high-density high-salt gradients.

Figure 4 summarizes some of the high-resolution studies of the first repair process. The profiles constitute an average for five independent experiments, each reproducible to about ± 1% per fraction, and show activity per fraction expressed as a percent of the activity that sediments freely rather than as a percent of the total activity including the NFSF. Number average molecular

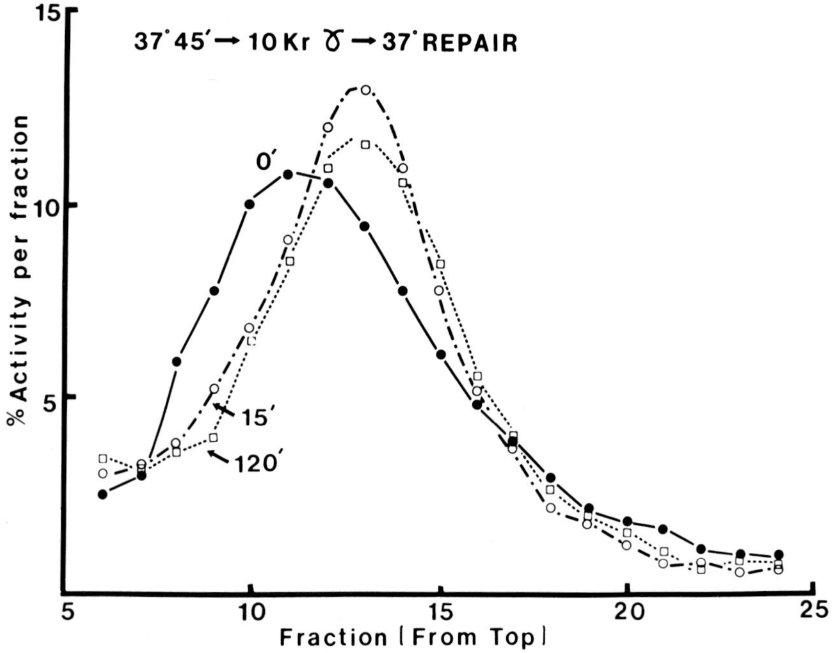

FIG. 4. High-resolution sedimentation profiles (average for five experiments) plotted as percent of freely sedimenting activity per fraction. Profiles for 10 KR gamma doses followed by 0-, 15-, and 120-minute repair incubations are shown.

weights, Mn, were calculated using the relationship of Lange et al. (1977) and the number of DSB per molecule, DSBm, calculated as:

$$\text{DSBm} = (\text{Mn for zero-dose profile/Mn for test profile}) - 1.$$

The value of Mn for zero-dose profiles was close to 10^9 daltons. Cells were irradiated with low-LET gamma rays and high-LET 3-MeV alpha particles following cell incubation pretreatments in media at 37° and 43°C. Figures 5 and 6 plot the fraction of the initially induced DSB which remained unjoined after various postirradiation repair incubation periods. Following moderate doses ($\leq$ 20 KR) of low-LET radiation, nascent DSB were rejoined such that less than 10% of the initial DSB remained, while after moderate doses ($\leq$ 5 KR) of high-LET radiation, some 40% of the initial DSB remained unjoined. Initial half-repair times obtained from these and other data were about 8 minutes and 10 minutes for low- and high-LET irradiations, respectively. Preirradiation hyperthermic incubation significantly inhibited strand rejoining and also enhanced radiation lethality (Cole and Corry, unpublished data).

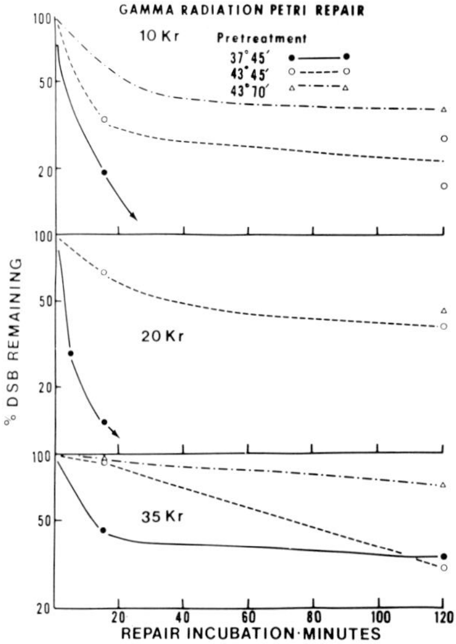

FIG. 5. Repair of DNA double-strand breaks induced by gamma rays. Cells were irradiated in suspension, plated on plastic Petri dishes for repair incubation and removed with trypsin to place on gradients. For doses of less than 20 KR, at least 90% of the DSB were rejoinable. Pre-irradiation treatments of 37°C, 45 minutes; 43°C, 45 minutes; and 43°C, 70 minutes were utilized.

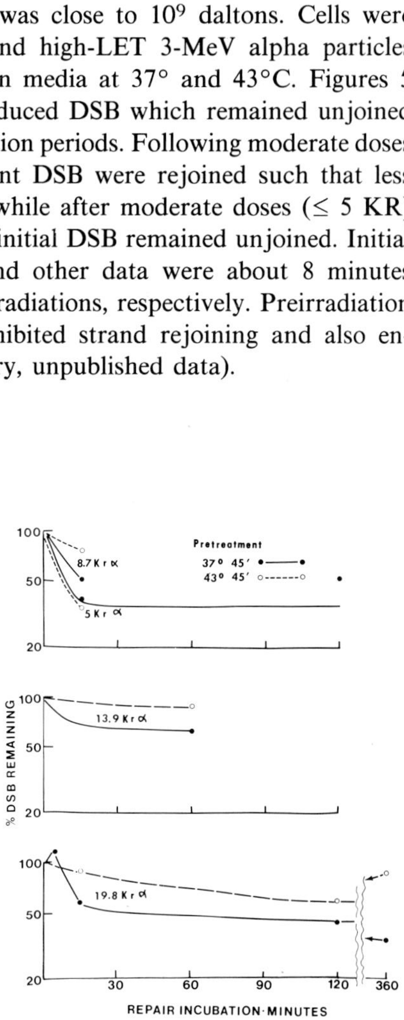

FIG. 6. Repair of DNA double-strand breaks induced by alpha particles. Cells were irradiated either on wetted membrane filters or submerged thin Mylar foils. Repair incubations were made in plastic Petri dishes (for cells removed from membrane filters) or on the Mylar foils. About 40% of the DSB appear to be nonrejoinable.

Some data for the second repair process, i.e., restoration of the NFSF, are summarized in Figure 7, which shows that restoration, which proceeded with a half-repair time of about 30 minutes (supportive detailed data not shown here), was inhibited by high doses and by preirradiation hyperthermia. Alpha radiation was more effective both in causing the initial breakdown of the NFSF and in inhibiting its subsequent restoration.

The induction and repair of total DNA breaks (SSB plus DSB) was studied with the alkaline elution technique, which uses tetrapropylammonium hydroxide at pH 12.1 to slowly elute denatured DNA from cells lysed on PVC membrane filters (Cole et al. 1978, 1979a, manuscript in preparation). Complete elution profiles for incubated samples were compared with profiles obtained for unincubated samples exposed to various radiation doses (≥ 10 R) to estimate damage remaining following the tested repair incubation periods. Figures 8 and 9, which summarize some of the data, show that following low- and high-LET radiation, about 1% and 20%, respectively, of the initial damage remained after long incubation periods of up to six hours. These results were similar to those published by Ritter et al. (1977) and Roots et al. (1979). Initial half-repair times were about 6 minutes and 13 minutes for low- and high-LET irradiation, respectively.

Analysis of both the sucrose gradient and alkaline elution studies indicated that, of the total breaks assayed by alkaline techniques, for low-LET irradiation about 1 in 40 originate from DSB, and for high-LET irradiations about 1 in 4 originate from DSB. With this information, the data for unrejoinable total breaks can be corrected for the contribution from DSB to derive the unrejoinable

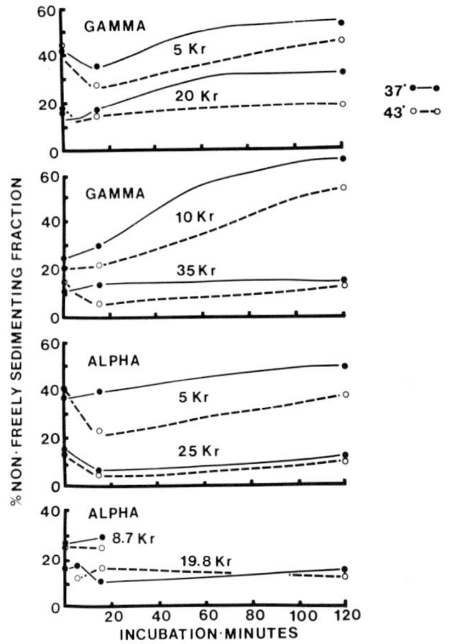

FIG. 7. Restoration of the NFSF after gamma and alpha irradiations. Initial half-repair times of about 30 minutes (based on more detailed data not shown) following the lower doses of gamma rays were observed.

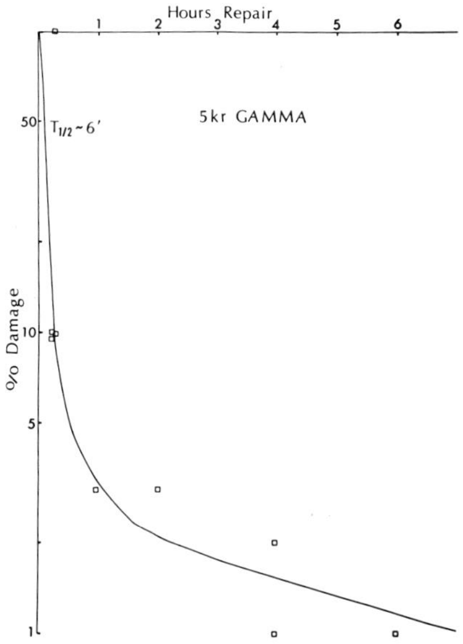

FIG. 8. Repair of DNA total strand breaks induced by gamma rays and assayed by alkaline elution. About 1% of the breaks were not rejoined.

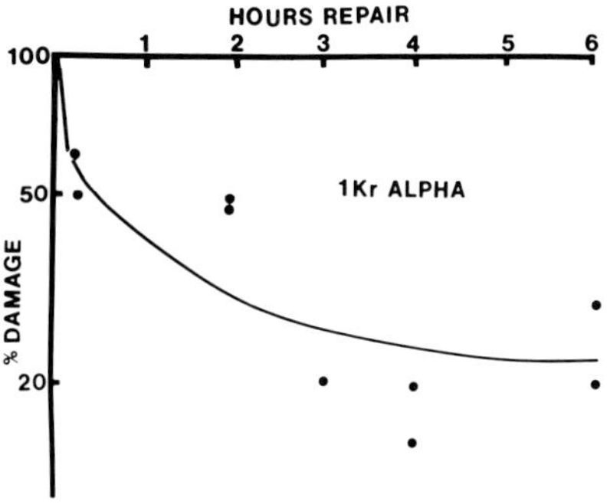

FIG. 9. Repair of DNA total strand breaks induced by alpha particles and assayed by alkaline elution. About 20% of the breaks were not rejoined.

SSB component. The results are shown in Table 1, which demonstrates that a portion of all breaks, whether SSB or DSB or whether induced by low- or high-LET radiation, are not repairable. As described later in the discussion section, we propose that the irreparable breaks represent complex DNA damage that contains multiple proximate lesions induced by the high-LET component (heavy particles or electron track-ends) of any form of ionizing radiation.

TABLE 1. *Percent unrejoinable breaks*

	Low LET	High LET
SSB	$\gtrsim 0.75$	$\simeq 10$
DSB	< 10	$\simeq 40$

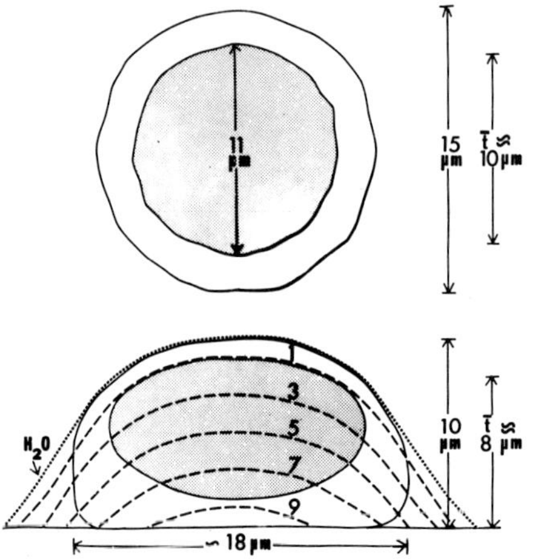

FIG. 10. Anticipated dose configuration for beams that penetrate from 1 to 10 μm into CHO cells freshly plated on membrane filters.

Partial Cell Irradiation

Electron and alpha-particle beams of limited penetration have been used to irradiate fully hydrated cells to selected depths. Beams were directed either from above into cells plated on wetted membrane filters, or from below to irradiate submerged cells plated on 4-μm thick Mylar foils. Figure 10 illustrates the extent of cell exposure anticipated for beams that penetrate 1 to 10 μm into a CHO cell freshly plated on a membrane filter. From measured depth-dose distributions for the beams (Cole 1969, Datta et al. 1976), the energy deposited in various cell compartments can be calculated and models of anticipated beam effectiveness generated. Figure 11 shows results for beam penetration

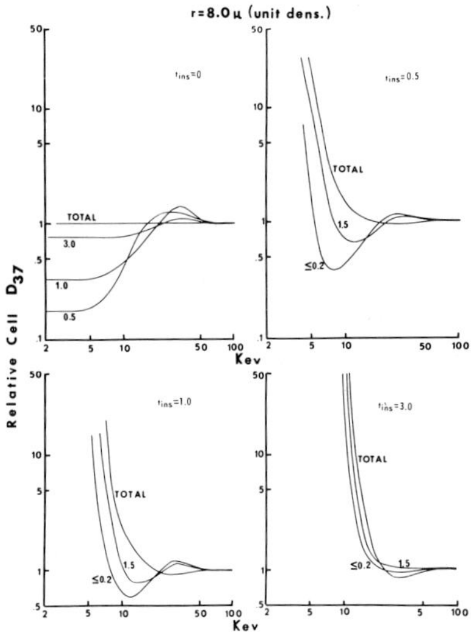

FIG. 11. Anticipated electron beam penetration responses for cells of 8 μm radius containing an insensitive coat thickness, t_{ins}, of from 0 to 3 μm and, immediately within the coat, a sensitive region of from less than 0.2 μm thick up to a thickness encompassing the total inner core. The mean lethal cell dose, D_{37} (total energy deposited per cell divided by the mass of the cell) is plotted versus the electron penetration (energy). Distinct minima are anticipated if a thin, sensitive region is located immediately within an insensitive coat.

responses anticipated for cells that contain various insensitive coat thicknesses, t_{ins}, which immediately surround sensitive regions of various thicknesses (indicated by the dimension in μm placed beside each curve). These plots of average cell dose (total energy deposited in the cell/cell mass) versus beam penetration illustrate a general result found for various cell configurations; that is, that distinct minima in the plots of D_{37} dose (i.e., a dose affecting all but 37% of the population) are found when the sensitive region is constrained to a thin region ($\lesssim$0.2 μm) located immediately within an insensitive coat region. In this case, the minimal D_{37} values occur for beams that penetrate just beyond the sensitive region, since a maximum fraction of total energy deposited in the cell will be deposited within the sensitive region. Shorter or longer beam penetrations waste a larger fraction of the energy within insensitive cell compartments.

Lethality

Figure 12, which summarizes some studies on lethality induced by electron and alpha-particle beams (Cole and Datta 1976a, Zermeno and Cole 1969, Cole et al. 1979b), shows distinct D_{37} minima for interphase cells that implicate radiation-sensitive regions less than about 0.2 μm thick located about 1 μm below the cell surface. A broad sensitive region is implicated by the flatter beam penetration responses observed for metaphase cells. That similar responses were seen for the electron and alpha-particle beams refutes the supposition that the minima simply reflect a greatly enhanced biological efficiency associated with the relatively high LET of electron track-ends that terminate in the nucleus. Although electrons exhibit an increased LET toward the track-end, alpha parti-

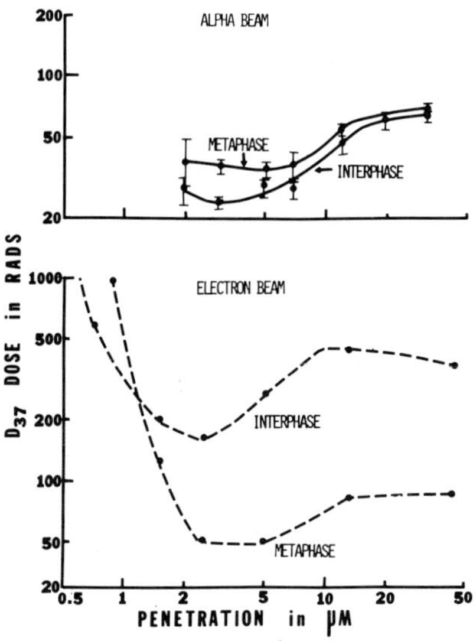

FIG. 12. Plots of mean lethal cell doses, D_{37}, versus penetration of electron and alpha particle beams. The deep minima in the plots for interphase cells implicated a thin sensitive region near the nuclear periphery. The flatter responses for metaphase cells implicated a more diffuse sensitive region.

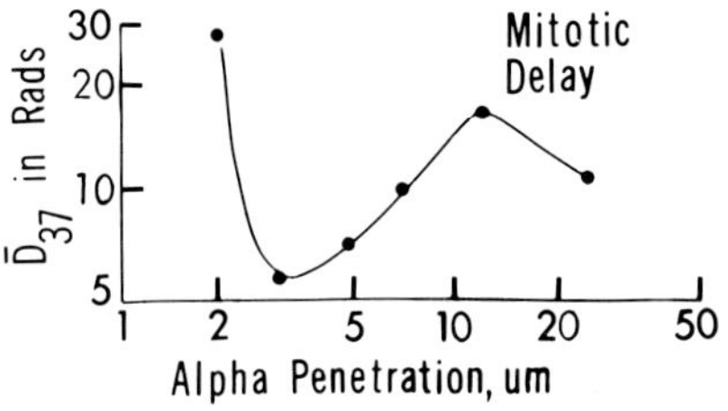

FIG. 13. Plot of mean cell dose, D_{37}, which causes a G_2 block in all but 37% of the cells, versus penetration of alpha particle beams. The deep minimum implicates a thin sensitive region near the nuclear periphery.

cles exhibit a *reduced* LET toward the track-end; thus, it appears that the minima result from the spatial distribution of the sensitive region rather than from an enhanced effectiveness of track-ends.

Division Delay

Studies on division delay (G_2 block) induced by alpha-particle beams (Cole et al. 1977, and Cole, in preparation) are summarized in Figure 13. The experiments utilized the mitotic shake-off technique developed by Dewey and Highfield (1976) and Schneiderman et al. (1977) and exhibit beam penetration responses very similar to that presented for lethality. Thus our results suggest that the sites for inducing both G_2 block and lethality are located in a thin region at the nuclear periphery. Similar conclusions were reached by Munro, who used his meticulous technique of alpha-particle irradiation of single cells (Munro 1970a, 1970b) and by Warters and Hofer (1977) using ^{125}I as a localized label. On the other hand, many other investigators have suggested other structures or regions (for example, Gaulden and Perry 1958, Rustad 1970, Dewey and Highfield 1976, Schneiderman et al. 1977).

Mutation

Figure 14 summarizes studies on mutation of CHO cells to 6-thioguanine resistance induced by alpha particle beams of 4-μm and 24-μm penetrations

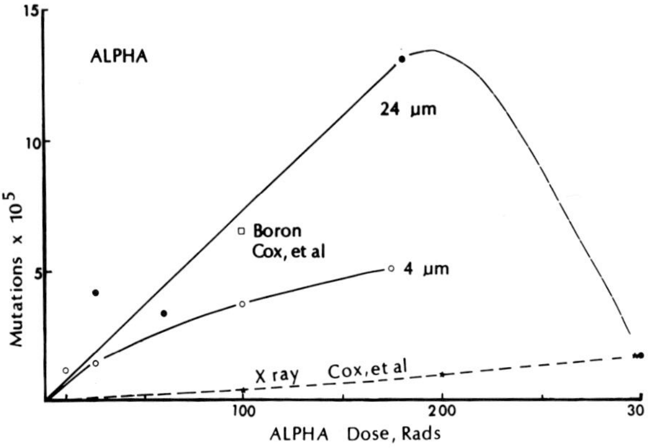

FIG. 14. Plots of 6-thioguanine mutation induction versus dose for alpha particle beams of 4-μm and 24-μm penetration. X-ray and penetrating boron ion beam data from Cox et al. (1976) are shown for comparison. The comparable responses for the 4- and 24-μm beams implicate sensitive sites in the nuclear interior.

(Humphrey et al. 1978, Cole et al., in preparation). Assay methods reported by Chu (1971) and Hsie et al. (1975) were utilized. Our data show that, unlike data for lethality or division delay, the short penetration beam was not more efficient than the fully penetrating beam and imply that the sites for this mutation are not located at the nuclear periphery but in the nuclear interior. The result was somewhat pleasing since there was no reason to assume *a priori* that specialized sites associated with this mutation would be constrained to genetic material at the nuclear periphery. As will be discussed later, the radiation action cross section (effective target area) for the mutation sites was very small compared to the cross sections for less specific responses such as lethality or division delay; i.e., many potential (peripheral) sites are probably involved in the latter responses. Figure 14 illustrates the gratifying agreement of our data for penetrating alpha irradiation with those of Cox et al. (1977) for a penetrating (boron) particle beam of equivalent LET.

Chromosome Damage

We are currently studying alpha particle–induced chromosome gaps, breaks, and exchanges in interphase cells using the method of premature chromosome condensation (Hittelman and Rao 1974a, 1974b, Hittelman et al. 1979). Current data for penetrating beams are included in Table 2 and discussed in a later section. Since results for short penetration beams are not yet available, the location of the related sensitive sites can not be deduced, although one of the authors (A. C.) speculates that the results will implicate sites at the nuclear periphery.

DNA Damage

Distinct dose minima of the penetration responses were observed for electron beam induction of DNA SSB and DSB as shown in Figure 15. Thus, responses for cell lethality, division delay, and DNA scission were all qualitatively similar, implicating the same thin sensitive region at the periphery of the nucleus. This result also implies, but does not prove, that DNA damage is a precursor lesion for the cellular responses.

In the DNA damage studies (Cole et al. 1974), it was observed that electron beams that penetrated a short distance into the nucleus yielded double-peaked sedimentation profiles that contained both degraded and unaffected DNA fractions. This was not surprising since, as shown in Figure 10, only a small volume of the nuclear DNA was irradiated. What was remarkable was that approximately one half of the total DNA was degraded for beams that penetrated as little as 0.1 μm below the nuclear surface. This suggested that every nuclear DNA molecule must have had a significant fraction of its length closely associated with the nuclear membrane. These and previous conclusions are summarized in the schema presented in Figure 16, which proposes that a significant fraction (about 10%) of each chromosomal DNA subunit molecule of 10^9 daltons is

TABLE 2. *Optimum alpha particle RBE values and action cross sections and projected areas of cell components*

Effect	RBE α/λ	σ μm^2	Area μm^2	Component
DNA single-strand break (SSB)	0.13	4,400	3,200	Total nuclear DNA
DNA double-strand break (DSB)	1.5	600		
Release of DNA molecule from NFSF	1.6	500	330	Cell
SSB_{unrej}	1.7	330		
DSB_{unrej}	6.0	200		
Division delay (G_2 block)	>2.2	200		
Interphase (G_2) chromosome breaks	6.7	170	160	Nucleus
Interphase (G_2) chromosome gaps	2.4	73	145	Dispersed DNA from one chromosome
Lethality—low dose	12	65		
—high dose	2.0			
Interphase (G_2) chromosome exchange	2.5	11	10	Condensed chromosome
Cell transformation	5–10	0.16	1	Centriole
Mutation—low dose	15	1.6×10^{-3}	1×10^{-3}	Gene (10^6d DNA)
—high dose	9			

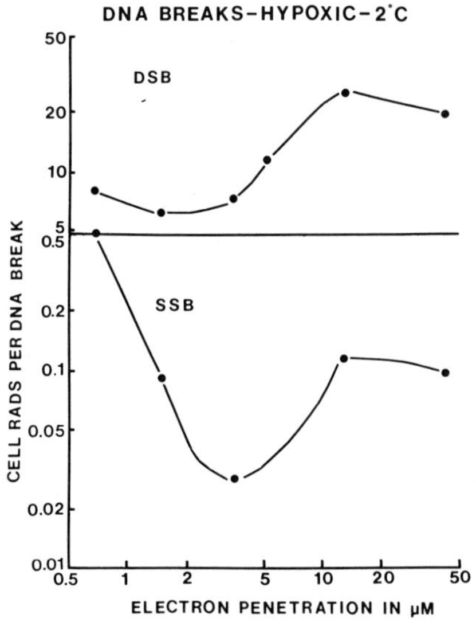

FIG. 15. Plots of dose per break for DNA single-strand and double-strand breaks versus electron beam penetration. The distinct minima, similar to that for survival, implicate sensitive regions near the nuclear periphery.

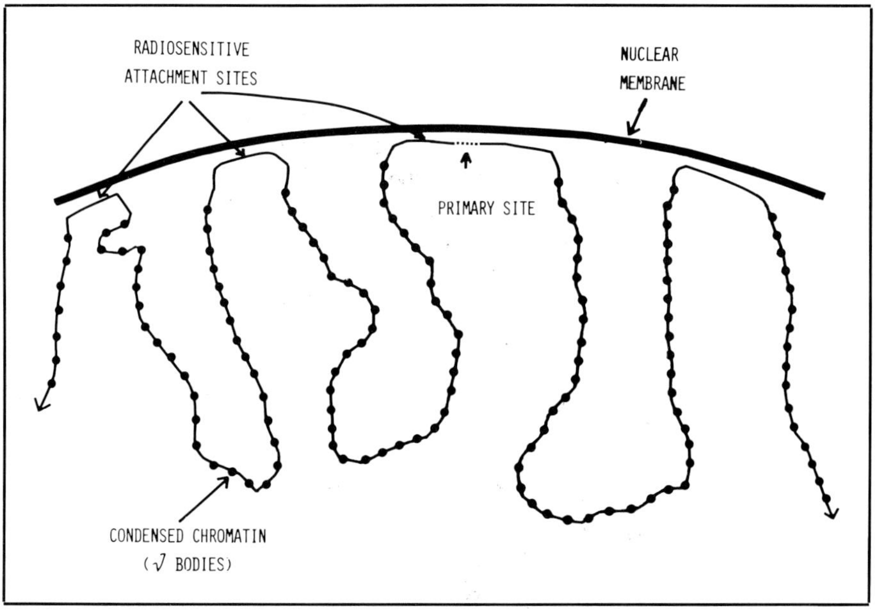

FIG. 16. Scheme to rationalize experimental results. Each DNA molecule of about 10^9 daltons has 10 to 100 association sites at the nuclear membrane. About 10% of the total DNA at or near the association sites is more radiosensitive than the remainder, which is mostly condensed on nucleosomes.

associated in varying degrees with the nuclear membrane and that regions at or near the association sites are more radiosensitive than the bulk DNA, which is condensed on nucleosomes (ν bodies) first discerned by Olins and Olins (1974). The increased sensitivity could be due to: (1) susceptibility to indirect radiation damage because of a more open and extended DNA structure (Ansevin 1977), (2) susceptibility to indirect damage because of a lack of protective associated nucleoprotein, (3) transfer of energy or active radicals from the nuclear membrane to the DNA, (4) reduced accessibility of repair enzymes, or (5) susceptibility to the formation of complex DNA damage (such as DNA-lipoprotein cross-links) that is irreparable.

A speculative possibility is that a "scaffold" protein recently described by Paulson and Laemmli (1977), to which DNA is attached in metaphase cells, becomes associated with the nuclear membrane of interphase cells, and that regions of DNA–scaffold protein association are radiosensitive.

It should be noted that Blackburn et al. (1978) have shown that a small fraction (~0.1%) of the total DNA is tightly bound to nuclear membrane components and that this fraction does not appear to be radiosensitive; thus, another (larger) fraction of less tightly bound DNA would have to be postulated to account for the effects described above.

Action Cross Section and RBE

In particle beam studies, one can specify the number of particles incident on the cell per unit area, N_0, which corresponds to the induction of one event per cell (strand break, chromosome aberration, mutation, transformation) or which corresponds to the induction of a response in all but 37% of the population (lethality, division delay). The reciprocal of N_0 is the action cross section, σ, which becomes a measure of the projected area, A, of the susceptible target sites in the ideal case in which a site traversal by an incident particle produces the effect with unity probability. If a site traversal produces the effect with a probability, p, less than unity, then the measured cross section will be $\sigma = pA$. With first principles, σ also may be calculated from the cell average dose, D_0, that corresponds either to the induction of one event per cell or to a response in all but 37% of the cells.

$$\sigma\ (\mu m^2) = 16\ \Delta E\ (keV)/\bar{t}\ (\mu m) \cdot D_0\ (rads)$$

where ΔE is the average energy deposited per particle in cells with mean thickness t.

One advantage of using 0.5- to 3-MeV alpha-particle beams to induce molecular and cellular responses is that such a large amount of energy is deposited in a target (~500 eV in 3 nm) that the target will probably contain multiple lesions and be irreversibly damaged. Thus, p for many events can be expected to approach unity and the measured cross section taken as an estimate of target size. Furthermore, of the total energy deposited by a moderate energy (1 MeV)

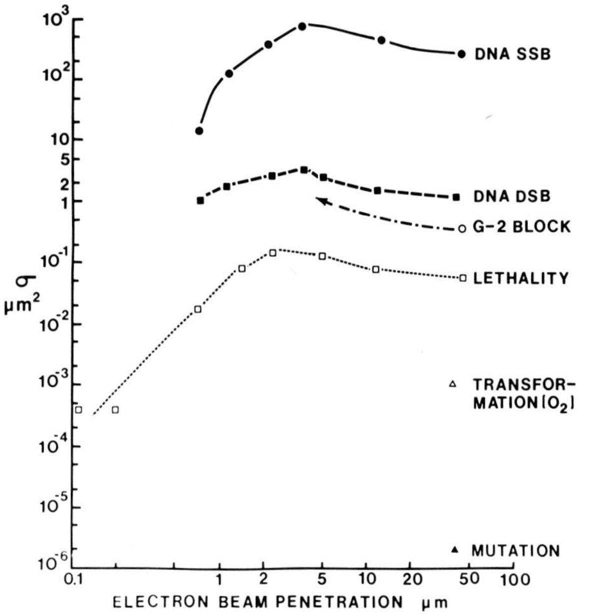

FIG. 17. Plots of electron action cross sections versus beam penetration for various responses. Maximum cross sections for DNA breaks and lethality (plot for G_2 block is for an anticipated, not an observed, response) occur for electron beams of 4-μm penetration, which exhibit maximum LET values near the nuclear periphery.

alpha particle, 75% is constrained within 1 nm, and 90% within 3 nm of the particle track (Cole, unpublished results). Thus, small cross sections may be determined with accuracy.

Cross section versus beam penetration plots for electron and alpha particle beams are shown in Figures 17 and 18. Maximum cross sections for electrons occurred for 2- to 4-μm beams, which penetrated just below the nuclear surface,

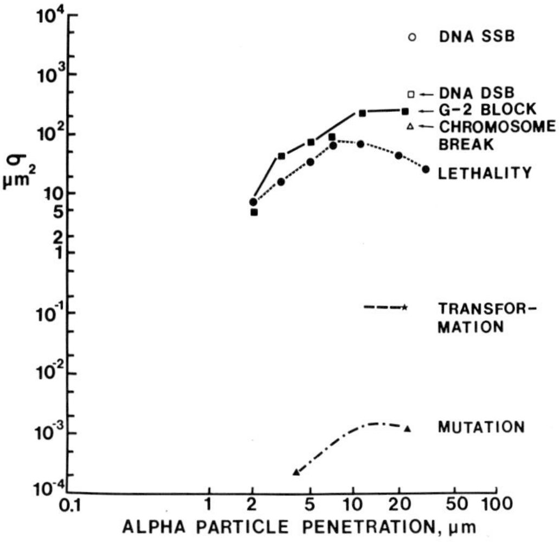

FIG. 18. Plots of alpha particle action cross sections versus beam penetration for various responses. Maximum cross sections for lethality, G_2 block, and mutation occur for alpha particle beams of 10- to 15-μm penetration, which exhibit maximum LET values of 230 KeV/μm in the nuclear region.

whereas maximum cross sections for alpha particles occurred for fully penetrating 10- to 15-μm beams. These results support the assumption that the sensitive sites were at the nuclear periphery, since maximum particle LET values of about 30 KeV/μm for electrons and 230 KeV/μm for alpha particles occurred in that region (i.e., at the track-ends of electrons) and from 6 to 9 μm from the track-ends of alpha particles.

In general, maximum cross sections for alpha-particle beams were about 300 times larger than those for electron beams; however, the partly penetrating electron beams only irradiated half of the nuclear surface and only about half of the electrons incident to the cell penetrated into the nucleus. Thus, if a particle beam with LET properties equivalent to those of electron track-ends could be passed through the total nucleus, we would anticipate cross section values four times larger than those for the short-range electrons and the maximum alpha particle cross sections would be about 75 times greater. This ratio of 75 is roughly equal to the square of the ratio of LET values for the two beams, i.e. $(230/30)^2 = 59$, and supports an assumption that the interaction probability, p, increases with the square of the LET of the traversing particle (Cole and Datta 1976a).

The only exception to the general analysis presented above was observed for the induction of DNA SSB, where the ratio $\sigma_\alpha/\sigma_{e^-}$ was about 6, corresponding to an anticipated ratio of only 1.5 for the simulated fully penetrating beams. Thus, for SSB only a small dependence of p on LET was observed over the range 30 to 230 KeV/μm. This relatively low efficiency of high-LET beams for inducing SSB is also expressed in the low RBE (alpha particles compared with gamma rays) of 0.13 listed in Table 2. The reason for this result is that sufficient energy (>50 eV) is deposited by each low-energy electron traversal of a DNA molecule to induce a simple lesion, such as an SSB, with high probability. Traversals of DNA by particles of higher LET may be expected to induce complex damage containing among other lesions a number of closely spaced SSB, but the effect will be assayed as one or at most two SSB (i.e., a DSB) by sedimentation or elution methods, which do not detect very small DNA fragments.

DISCUSSION

Figures 5 through 9, which summarize the kinetics for the repair of DNA breaks and the NFSF, present curving plots that exhibit initial fast repair rates followed by progressively slower rates. The plots ultimately flatten to reveal an unrepairable component. We propose that the high initial rates correspond to the repair of relatively simple lesions (utilizing DNA ligases in some cases) whereas increasingly complex lesions require increasingly longer repair times (utilizing excision repair and other enzyme systems), finally leaving unrepaired those complex lesions that the cell cannot master.

As illustrated in Figure 19, complex damage involving multiple proximate DNA lesions would be expected with high probability when a high-LET particle

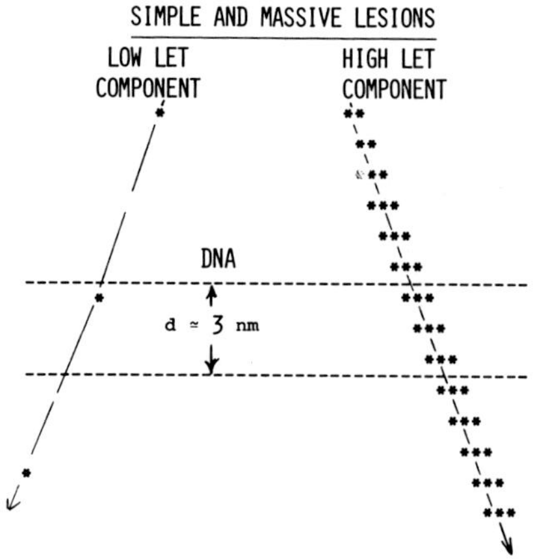

FIG. 19. Representative distributions of energy exchange events for traversals of a DNA molecule by particles with low and high LET. A 0.5-nm hydration sheath, which contributes active radicals, is added to the 2-nm DNA diameter to form a 3-nm diameter sensitive cylinder. Multiple damage sites can occur when a high LET particle traverses the DNA.

traverses a DNA molecule. The asterisks in Figure 19 symbolize average energy exchanges of about 60 eV, each of which is capable of inducing simple lesions such as a backbone scission or sugar or base damage. Table 3 presents estimates based on published data (Cole 1969), which show that a significant fraction (5% to 15%) of the dose of a minimal LET radiation is deposited by the relatively high LET component at the track-ends of secondary electrons. Thus, low-LET radiations would be expected to induce a significant number of irreparable complex lesions, some of which would be assayed as unrejoinable SSB and DSB.

Table 2 lists calculated RBE values and maximum alpha particle cross sections for a variety of end points. Projected areas of various cell components are also listed so that comparison with target cross sections may be made.

The cross section for DNA SSB of 4,400 μm^2 was larger than the projected area of the total nuclear DNA. This means that, on the average, more than one SSB was induced for each traversal of a DNA duplex molecule by an

TABLE 3. *Estimated energy deposits in DNA plus 0.5 nm water sheath for random traversal by primary and secondary particles*

	% Dose	Energy Deposits (eV)
Low LET (e^-, γ, X)	85	≤100
	10	120
	2.5	175
	2.5	300
200 KeV/μm Alpha particles	0.6	≤100
	13	350
	54	600
	24	1600
	9	6000

alpha particle. This would be expected since each molecular traversal deposited about 500 eV, which is much greater than the 60 eV average energy spent in DNA by low-LET radiation to induce an SSB. The excess of SSB induced per alpha particle traversal must be due to the production of occasional DNA DSB. The optimum cross section for DSB was about 600 μm^2, which was about 1/5 of the projected area of the cell DNA. Thus, about one in five DNA traversals generated a DSB, which, therefore, contributed an excess 20% to the SSB induced per traversal. The projected area of the nuclear DNA was 20 times the projected area of the nucleus; hence an average of 20 DNA molecules were traversed by a particle that traversed the nucleus. This means that about 24 SSB and 4 DSB were generated for each traversal of the nucleus by a maximal LET (200 KeV/μm) alpha particle.

Table 2 also lists RBE and cross-section values for induction of nonrejoinable DNA SSB and DSB. Although high-LET radiation induced total SSB with relatively low efficiency, it induced both unrejoinable SSB and unrejoinable DSB with relatively high efficiency. We have proposed that unrejoinable breaks represent a class of complex damage containing multiple proximate lesions. These sites that harbor various types of lesions in close proximity may contain one or more strand scissions on one strand (SSB_{unrej}), two or more strand scissions on opposite strands (DSB_{unrej}), or other damage not expressed as strand scissions. All three forms may contribute to biological damage. The significance of this speculation is buttressed by the computations presented in Table 4, which lists the number of initial and unrejoinable lesions induced by alpha-particle and gamma-ray doses that produce the same effect, 50% survival. It is seen that vastly different numbers of SSB_{total} were produced by the D_{50} doses and similar but unequal numbers of SSB_{unrej}, DSB_{total}, and DSB_{unrej} were obtained, whereas the same value, 9, was calculated for the combined number of unrejoinable SSB and DSB. It is conceivable that a portion of these unrejoinable lesions are lethal by themselves and a portion must interact with other similar lesions to produce lethality, as proposed in the theory of dual radiation action of Kellerer and Rossi (1971).

The optimum α particle cross section for cell lethality was 65μm^2, which is about 0.4 times the projected area of the nucleus. Thus, an average of 1/0.4, or 2.5, nuclear traversals and 60 SSB and 10 DSB (plus other DNA lesions)

TABLE 4. *Number of lesions induced by high and low LET radiations by D_{50}* (50% survival) doses

	D_{50} (α), 42 R	D_{50} (γ), 250 R
SSB_{total}	56	1000
SSB_{unrej}	5.6	7.5
DSB_{total}	9	20
DSB_{unrej}	3.6	2
$SSB_{unrej} + DSB_{unrej}$	9.2	9.5

were accumulated for each lethal event. This may be compared with the 1000 SSB and 16 DSB and other lesions accumulated per lethal event for low-LET radiation. It is obvious that a large number of DNA lesions can be tolerated by the cell.

Division delay (G_2 block) exhibited an optimum alpha particle cross section of about 200 μm^2, which is not significantly different from the projected area of the nucleus, but is substantially smaller than the projected area of the cell. This result indicated that a single alpha particle traversal anywhere through the nucleus induced a (150-minute) division delay. The optimum cross section for G_2 block is obviously larger than the projected area of a gene or a centriole and requires a target that is dispersed throughout the whole nuclear area. Such targets could involve (1) a large fraction of the chromatin associated with the nuclear periphery, (2) dispersed peripheral chromatin from one or a limited number of chromosomes, (3) peripheral chromatin associated with the NFSF, (4) the nuclear membrane, or (5) dispersed cytoskeletal structures associated with the nuclear periphery. Since 24 SSB and 4 DSB plus other lesions were induced for each alpha particle traversal of the nucleus, a portion of these total lesions might be responsible for division delay. Table 2 shows that induction of nonrejoinable DNA SSB and DSB exhibited cross sections of 330 μm^2 and 200 μm^2, which were comparable to that for division delay. This similarity implicates these types of complex lesions in division delay. Furthermore, Figures 8 and 9 show that DNA repair kinetics decelerated abruptly at times comparable to the division delay periods.

Another possible precursor lesion responsible for division delay was that involving the release of a DNA subunit molecule from the NFSF. The cross section for this effect was about 500 μm^2, which means that 500/160, or three, DNA molecules were released per nuclear transversal. The repair times of the lesions involved in this effect were comparable to the division delay periods of 90 minutes induced by low-LET radiation, although this comparison must be made with caution since the tenfold higher doses required for the studies on the NFSF were supralethal. The RBE values were also similar for the two effects.

Similarly, it could be argued that lesions in the NFSF contribute to cell lethality since about seven lesions were generated per 50% lethal dose of alpha particles and about 40 NFSF lesions were generated by a 50% lethal dose of low-LET radiation. However, the fraction of these lesions that were nonrestorable has not yet been determined. For low-LET radiation, the kinetics for restoration of the NFSF were similar to those for repair of sublethal damage. Inhibited restoration of the NFSF occurred following high-LET radiation, a result consistent with the observation that much reduced repair of sublethal damage occurs following high-LET radiation.

The optimum alpha particle cross sections for induction of interphase chromosome (G_2) aberrations as assayed by premature chromosome condensation techniques were about 170 μm^2 for breaks, 73 μm^2 for gaps, and 11 μm^2 for exchanges. The technique allows for the assay of aberrations produced within 15 minutes after irradiation, which thus limits modification of precursor lesions by cell (repair) processes. On the other hand, most of the DNA SSB, and 60% of the DNA DSB, could be rejoined within the 15-minute preparation period. Both the cross section and RBE values for the induction of DSB_{unrej} were comparable to those for the induction of chromosome breaks, strongly implicating DSB_{unrej} as a precursor lesion for this effect. The data for induction of gaps and exchanges does not provide much background for the prediction of precursor lesions; however, the similar RBE values obtained from these aberrations suggested that gaps, rather than breaks, contained the precursor lesions involved in chromosome exchanges.

Our studies on cell mutation indicate a target(s) within the nucleus that presents a cross section of about $1.6 \cdot 10^{-3}$ μm^2, which corresponds to the size expected for one or several genes. This relatively large cross section implies that (1) an alpha particle traversal anywhere through a rather large gene ($\sim 10^6$ daltons DNA) inactivated it, (2) a number of genes made up the target and a mutation occurred when any one of these was damaged, or (3) damage migrated from a relatively large area to a specific locus. At present, additional data are required to distinguish among these possibilities.

The data currently available from other laboratories on cell transformation (Borek 1976, Borek et al. 1978) indicated that a cross section for high-LET particles of about 0.16 μm^2 is associated with this effect. This value corresponds to the area presented by many genes (~200) and implicates multiple or diffuse sites (perhaps many control regions) of the genome. We have not yet obtained results with short-penetration beams to determine the location of sites involved in cell transformation.

To date the studies with short-penetration particle beams have implicated sensitive sites for DNA damage, lethality, and division delay that were located in the nuclear periphery. We have presented a scheme in Figure 16 that rationalizes these interpretations. A further speculation that the DNA nuclear membrane association results from a condensation of chromosomal "scaffold" protein (Paulson and Laemmli 1977) onto the nuclear membrane was presented. If the specialized regions where chromosomal DNA associates with scaffolding protein are radiosensitive, then these regions will be present as a diffuse chromosome core in metaphase cells and as a collapsed peripheral shell in interphase cells.

In summary, the effects of many parameters (dose, dose fractionation, particle LET, beam fluence, beam penetration, cell repair incubation, cell hyperthermia, cell growth cycle stage) were similar for both the induction of complex irreparable DNA damage and the induction of biological responses. Studies of effects of oxygen concentration (Cole 1965, Cole et al. 1974, Tobleman and Cole 1974) and chemical modifiers (Cole et al. 1963, 1970, Corry and Cole 1968) have been limited and were not discussed in detail in this paper, but results have been consistent with the conclusions stated above.

ACKNOWLEDGMENTS

Major support for the research reported in the paper was provided by contract DOE-76-S-05-2832 from the Office of Health and Environmental Research of the Department of Energy. Other support was through grants CA 04484 and GM 23252 from the National Institutes of Health/Department of Health, Education and Welfare.

A large number of investigators have contributed to the studies. Those who have made significant contributions beyond that indicated by the reference listing include: Allen Ansevin, Douglas Vizard, Ruthan Langley, Sandra Robinson, Frank Shonka, Ira Kristal, Veronica Willingham, Susan Fletcher, W. Grant Cooper, Bobbye Morrow, Darla Holland, and Chris Grell.

REFERENCES

Alper, T. 1971. Cell death and its modifications: the roles of primary lesions in membranes and DNA, *in* Biophysical Aspects of Radiation Quality. International Atomic Energy Agency, Vienna.

Alper, T. 1980. Survival curve models, *in* Radiation Biology in Cancer Research (The University of Texas System Cancer Center 32nd Annual Symposium on Fundamental Cancer Research), R. E. Meyn and H. R. Withers, eds. Raven Press, New York, pp. 3–18.

Ansevin, A. T. 1977. Limited capacity of chromosomal proteins to protect eukaryotic DNA from damage by ionizing radiation. (Abstract) Radiat. Res. 70:616.

Blackburn, G. R., D. P. Highfield, W. C. Dewey. 1978. Studies on the nuclear envelope as a radiation sensitive site in mammalian cells: The absence of an effect on the binding of nuclear membrane-associated DNA. Radiat. Res. 75:76–90.

Borek, C. 1976. In vitro cell transformation by low doses of X-irradiation and neutrons, *in* Biology of Radiation Carcinogenesis, J. M. Yuhas, R. W. Tennant, and J. D. Reyan, eds. Raven Press, New York, pp. 309–326.

Borek, C., E. Hall, and H. Rossi. 1978. In vitro transformation of hamster cells by monoenergetic neutrons and heavy ions. (Abstract) Radiat. Res. 74:536.

Burki, H. J. 1976. Critical DNA damage and mammalian cell reproduction. J. Mol. Biol. 103:559–610.

Burki, H. J., and S. Okada. 1968. A comparison of the killing of mammalian cells induced by decay of incorporated tritiated molecules at −196°C. Biophys. J. 8:445.

Cerutti, P. A. 1974. Effects of ionizing radiation on mammalian cells. Naturwissenschaften 61:51–59.

Chapman, J. D. 1980. Biophysical models of mammalian cell inactivation by radiation, *in* Radiation Biology in Cancer Research (The University of Texas System Cancer Center 32nd Annual Symposium on Fundamental Cancer Research), R. E. Meyn and H. R. Withers, eds. Raven Press, New York, pp. 21–32.

Chen, R., and A. Cole. 1979. Speed dependence and wall effect for sedimentation of mammalian cell duplex DNA in neutral sucrose gradient. (Abstract) Biophys. J. 25:256a.

Chen, R., A. Cole, and S. Robinson. 1978. Initial slope of the survival curve for low-dose low-LET radiation. (Abstract) Radiat. Res. 74:495.

Christensen, R. C., and C. A. Tobias. 1972. Heavy ion induced single and double strand breaks in ΦX-174 replicative form DNA. Int. J. Radiat. Biol. 22:457–477.

Chu, E. 1971. Induction and analysis of gene mutations in mammalian cells in culture, *in* Chemical Mutagens: Principles and Methods for Their Detection, A. Hollaender, ed. Plenum Press, New York, pp. 411–444.

Cole, A. 1959. Electron-gun irradiator for microorganisms. (Abstract) Radiat. Res. 11:428.

Cole, A. 1961. Study of radiation susceptible structures in microorganisms with monoenergetic electron beams of 0.5 to 150 kev energies. Abstracts of the International Biophysics Congress, pp. 95–96.

Cole, A. 1965. The study of radiosensitive structure with low voltage electron beams, *in* Cellular

Radiation Biology (The University of Texas System Cancer Center 18th Annual Symposium on Fundamental Cancer Research). Williams and Wilkins Co., Baltimore, pp. 267–271.

Cole, A. 1969. Absorption of 20-eV to 50,000-eV electron beams in air and plastic. Radiat. Res. 38:7–33.

Cole, A., R. Chen, and I. Kristal. 1977. Division delay in CHO cells induced by partly penetrating alpha particles. (Abstract) Radiat. Res. 70:640–641.

Cole, A., W. G. Cooper, F. Shonka, P. M. Corry, R. M. Humphrey, and A. T. Ansevin. 1974. DNA scission in hamster cells and isolated nuclei studied by low-voltage electron beam irradiation. Radiat. Res. 60:1–33.

Cole, A., P. M. Corry, and R. Langley. 1970. Effects of radiation and other agents on the molecular structure and organization of the chromosome, *in* Genetic Concepts and Neoplasia (The University of Texas System Cancer Center 23rd Annual Symposium on Fundamental Cancer Research). Williams and Wilkins Co., Baltimore, pp. 346–379.

Cole, A., and R. Datta. 1976a. Model for irreparable cell damage induced by ionizing radiation of different qualities. (Abstract) Biophys. J. 16:184a.

Cole, A., and R. Datta. 1976b. Comparison of radiosensitive sites in mitotic and interphase CHO cells using partly penetrating alpha particle irradiation (Abstract) Radiat. Res. 67:632.

Cole, A., R. M. Humphrey, and W. C. Dewey. 1963. Low voltage electron beam irradiation of normal and BUdR-treated L-P_{59} mouse fibroblast cells in vitro. Nature 199:780–782.

Cole, A., and R. Langley. 1963. Study of the radiosensitive structure of T2 bacteriophage using low energy electron beams. Biophys. J. 3:189–197.

Cole, A., R. E. Meyn, and R. Chen. 1979a. Relative contribution of unrepaired single strand and double strand DNA breaks to biological damage. (Abstract) Biophys. J. 25:155a.

Cole, A., R. E. Meyn, R. Chen, and S. Fletcher. 1978. Repair of DNA strand breaks induced by low doses of low and high LET radiations. (Abstract) Radiat. Res. 74:553.

Cole, A., S. Robinson, and R. Datta. 1979b. Distribution of radiosensitive sites in mitotic and interphase CHO cells using track-end alpha particle irradiation. Radiat. Res. (in press).

Cole, A., F. Shonka, P. M. Corry, and W. G. Cooper. 1975. CHO cell repair of single strand and double strand breaks induced by gamma and alpha radiations, *in* Molecular Mechanisms for the Repair of DNA, II., P. Hanawalt and R. Setlow, eds. Plenum Press, New York, pp. 665–676.

Corry, P. M., and A. Cole. 1968. Radiation induced double strand scission of the DNA of mammalian metaphase chromosomes. Radiat. Res. 36:528–543.

Corry, P. M., and A. Cole. 1973. Double strand rejoining in mammalian DNA. Nature New Biol. 245:100–101.

Corry, P. M., S. Robinson, and S. Getz. 1977. Hyperthermic effects on DNA repair mechanisms. Radiology 123:475–482.

Cox, R., J. Thacker, D. T. Goodhead, and R. J. Munson. 1977. Mutation and inactivation of mammalian cells by various ionizing radiations. Nature 267:425–427.

Cremer, C., T. Cremer, C. Zorn, and L. Schoeller. 1976. Effects of laser UV-microirradiation ($\lambda = 2573$ Å) on proliferation of Chinese hamster cells. Radiat. Res. 66:106–121.

Datta, R., A. Cole, and S. Robinson. 1976. Use of track-end alpha particles from ^{241}Am to study radiosensitive sites in CHO cells. Radiat. Res. 65:139–151.

Davis, M. 1954. Irradiation of T-1 bacteriophage with low-voltage electrons. Arch. Biochem. Biophys. 49:417–423.

Dewey, W. C., and D. P. Highfield. 1976. G-2 Block in Chinese hamster cells induced by X-irradiation, hyperthermia, cycloheximide, or actinomycin-D. Radiat. Res. 65:511–528.

Elkind, M. M., and H. Sutton. 1959. X-ray damage and recovery in mammalian cells in culture. Nature 184:1293–1295.

Elkind, M. M., and G. F. Whitmore. 1967. The Radiobiology of Cultured Mammalian Cells. Gordon and Breach Science Publishers, New York.

Gaulden, M. E., and R. P. Perry. 1958. Influence of the nucleolus on mitosis as revealed by ultraviolet microbeam irradiation. Proc. Nat'l. Acad. Sci. USA 44:553–559.

Goodhead, D. T. 1980. Models of radiation inactivation and mutagenesis, *in* Radiation Biology in Cancer Research (The University of Texas System Cancer Center 32nd Annual Symposium on Fundamental Cancer Research), R. E. Meyn and H. R. Withers, eds. Raven Press, New York, pp. 231–247.

Haskins, C. P. 1938. Apparatus for studying the biological effects of cathode rays. J. Appl. Phys. 9:553–561.
Hittelman, W. N., and P. N. Rao. 1974a. Premature chromosome condensation. I. Visualization of X-ray induced chromosome damage in interphase cells. Mutat. Res. 23:251–258.
Hittelman, W. N., and P. N. Rao. 1974b. Premature chromosome condensation. II. The nature of chromosome gaps produced by alkylating agents and ultraviolet light. Mutat. Res. 23:259–266.
Hittelman, W. N., M. A. Sognier, and A. Cole. 1980. Direct measurement of chromosome damage and its repair by premature chromosome condensation, *in* Radiation Biology in Cancer Research (The University of Texas System Cancer Center 32nd Annual Symposium on Fundamental Cancer Research), R. E. Meyn and H. R. Withers, eds. Raven Press, New York, pp. 103–123.
Hsie, A. W., P. A. Brimer, T. J. Mitchell, and D. G. Gosslee. 1975. The dose-response relationship for ethyl methanesulfonate-induced mutations at the hypoxanthine guanine phosphoribosyl transferase locus in Chinese hamster ovary cells. Somatic Cell Genetics 1:247–261.
Humphrey, R., A. Cole, V. Willingham, and R. Chen. 1978. Mutation of CHO cells induced by partly and fully penetrating alpha particles. (Abstract) Radiat. Res. 74:538.
Hutchinson, F. 1955. Use of charged particles to measure skin thickness and other surface properties. Ann. N.Y. Acad. Sci. 59:494–502.
Kaplan, H. S. 1966. DNA strand scission and loss of viability after X-irradiation of normal and sensitized bacterial cells. Proc. Nat'l. Acad. Sci. USA 55:1442–1446.
Kellerer, A. M., and H. H. Rossi. 1971. RBE and the primary mechanism of radiation action. Radiat. Res. 47:15–34.
Kirsch, R. E., and M. R. Zelle. 1969. Biological effects of radioactive decay: the role of the transmutation effect, *in* Advances in Radiation Biology, Vol. 3, L. G. Augenstein, R. Mason, and M. Zelle, eds. Academic Press, New York, p. 177.
Kohn, K. W., L. C. Erickson, R. A. Ewig, and C. A. Friedman. 1976. Fractionation of DNA from mammalian cells by alkaline elution. Biochemistry 15:4629.
Kohn, K. W., C. A. Friedman, R. A. Ewig, and Z. M. Iqbal. 1974. DNA chain growth during replication of asynchronous L1210 cells. Alkaline elution of large DNA segments from cells lysed on filters. Biochemistry 13:4134.
Kohn, K. W., and R. A. Grimek-Ewig. 1973. Alkaline elution analysis, a new approach to the study of DNA single-strand interruptions in cells. Cancer Res. 33:1849.
Lange, C. S., D. F. Liberman, R. W. Clark, and P. Ferguson. 1977. The organization and repair of DNA in the mammalian chromosome. I. Calibration procedures and errors in the determination of the molecular weight of a native DNA. Biopolymers 16:1063–1081.
Lea, D. E. 1946. Actions of Radiations on Living Cells. University Press, Cambridge.
McGrath, R. A., and R. W. Williams. 1966. Reconstruction in vivo of irradiated *E. coli* deoxyribonucleic acid: The rejoining of broken pieces. Nature 10:534–535.
Munro, T. R. 1970a. The relative radiosensitivity of the nucleus and cytoplasm of Chinese hamster fibroblasts. Radiat. Res. 42:451–470.
Munro, T. R. 1970b. The site of the target region for radiation-induced mitotic delay in cultured mammalian cells. Radiat. Res. 44:748–757.
Neary, G. J. 1965. Chromosome aberrations and the theory of RBE. I. General considerations. Int. J. Radiat. Biol. 9:477–502.
Ohnuki, Y., D. E. Rounds, R. S. Olson, and M. W. Berns. 1972. Laser microbeam irradiation of the juxtanuclear region of prophase nuclear chromosomes. Exp. Cell Res. 71:132–144.
Olins, A., and D. Olins. 1974. Spheroid chromatin units (ν bodies). Science 183:330–332.
Paulson, J. R., and U. K. Laemmli. 1977. The structure of histone-depleted metaphase chromosomes. Cell 12:817–828.
Preiss, J. W., and E. Pollard. 1961. Localization of β-galactosidase in cells of *E. coli* by low-voltage electron bombardment. Biophys. J. 1:429–435.
Ragni, G., and W. Szybalski. 1962. Molecular radiobiology of human cell lines. II. Effects of thymidine replacement by halogenated analogues on cell inactivation by decay of incorporated radiophosphorus. J. Mol. Biol. 4:338.
Ritter, M. A., J. E. Cleaver, and C. A. Tobias. 1977. High LET radiations induce a large proportion of non-joining DNA breaks. Nature 266:653–655.
Roots, R., T. C. Yang, L. Craise, E. A. Blakely, and C. A. Tobias. 1979. Impaired repair capacity

of DNA breaks induced in mammalian cellular DNA by accelerated heavy ions. Radiat. Res 78:38–49.

Rossi, H. H. 1980. Dual radiation action in carcinogenesis, *in* Radiation Biology in Cancer Research (The University of Texas System Cancer Center 32nd Annual Symposium on Fundamental Cancer Research), R. E. Meyn and H. R. Withers, eds. Raven Press, New York, pp. 185–194.

Rustad, R. C. 1970. Variations of the sensitivity to X-ray induced mitotic delay during the cell division cycle of the sea urchin egg. Radiat. Res. 42:498–512.

Schneiderman, M. H., L. A. Braby, and W. C. Roesch. 1977. Division delay after low X-ray doses and treatment with cycloheximide. Radiat. Res. 70:130–140.

Tobias, C. A., E. A. Blakely, F. Q. H. Ngo, and T. C. H. Yang. 1980. The repair-misrepair model in cellular radiobiology. *in* Radiation Biology in Cancer Research (The University of Texas System Cancer Center 32nd Annual Symposium on Fundamental Cancer Research), R. E. Meyn and H. R. Withers, eds. Raven Press, New York, pp. 195–230.

Tobleman, W. T., and A. Cole. 1974. Repair of sub-lethal damage and oxygen enhancement ratio for low voltage electron beam irradiation. Radiat. Res. 60:355–360.

Warters, R. L., and K. G. Hofer. 1977. Radionuclide toxicity in cultured mammalian cells. Elucidation of the primary site for radiation-induced division delay. Radiat. Res. 69:348–358.

Zermeno, A., and A. Cole. 1969. Radiosensitive structure of mitotic and interphase Don-C hamster cells as studied by low voltage electron beam irradiation. Radiat. Res. 39:669–684.

Radiation Biology in Cancer Research, edited by
Raymond E. Meyn and H. Rodney Withers.
Raven Press, New York © 1980.

The Role of DNA Damage and Repair in Cell Killing Induced by Ionizing Radiation

Robert B. Painter

Laboratory of Radiobiology, University of California, San Francisco, California 94143

One of the main areas of discussion in the last M. D. Anderson Hospital symposium devoted to radiobiology was the critical target for cell killing by ionizing radiation. This 1979 symposium has begun with two chapters whose content assumes that DNA is the important target; in this paper I hope to substantiate this assumption further.

At the earlier symposium, Szybalski and Opara-Kupinska (1964) presented a wealth of data on *Bacillus subtilis,* comparing the sensitization afforded by bromodeoxyuridine to various damaging agents when measured by loss of reproductive integrity or by inactivation of transforming principle (i.e., the DNA) from the same cells. However, there were some in the audience who were unconvinced. Since that time, elaboration of the mechanism of bacterial radiosensitive mutants has made it obvious that DNA is the principal radiosensitive target in prokaryotes. In the same symposium, Howard-Flanders and Boyce (1964) reviewed the then brand-new information on thymine-dimer excision, which was originally reported by Setlow and Carrier (1964) and by Boyce and Howard-Flanders (1964) in the same issue of the *Proceedings of the National Academy of Sciences.* The era of DNA repair had begun.

Although highly detailed mechanisms of DNA repair have been elaborated in prokaryotes since 1964, the data are not so clear for mammalian cells. The first evidence for mammalian DNA repair was also reported in 1964 (Rasmussen and Painter 1964), but progress in this area has been much slower than in the bacterial systems. Excision repair occurs in mammalian cells, but there are many peculiarities, e.g., excision of ultraviolet (UV) light-induced thymine dimers plays an obligatory role in the prevention of cell killing in human cells (Cleaver 1968, Maher et al. 1979), but not in mouse cells. Essentially no dimers are excised from the DNA of mouse cells (Klímek 1966), which survive equal exposures to UV as well as or better than do human cells (Painter 1970, Rauth 1970). Moreover, the first step in repair of dimers is a complete mystery; cells completely defective in excision repair of dimers (from patients with xeroderma pigmentosum [XP] group A) contain enzymes that can perform the first several steps in repair (Mortelmans et al. 1976), including the incision step. Because at least seven complementation groups (Bootsma 1978) perform the steps preceding or accompanying the incision step, we know that a very complicated biochemical system is involved.

Even though our knowledge of DNA repair systems in mammalian cells does not yet allow us to make the correlations between DNA repair and survival that are possible in prokaryotes, there is extremely good evidence that DNA is the principal target for killing of mammalian cells by ionizing radiation.

Mutant human cells sensitive to ionizing radiation are not as abundant as those sensitive to UV light or chemicals. However, ataxia telangiectasia, a genetic disease, is unequivocally associated with deficient DNA repair after X irradiation. Cells from some patients with this disease have a reduced extent of X-ray-induced repair replication, especially irradiated hypoxic cells (Paterson 1978). There is another class of patients whose cells are radiosensitive but do not exhibit deficient repair replication. Because they are otherwise phenotypically similar to those whose cells are deficient in repair replication, it is logical to assume that another DNA repair process is defective in these patients, as is true of XP. Patients with XP can be classified into two major groups: those who are deficient in dimer excision and those who are not (the so-called variants). After much early confusion, Lehmann et al. (1975) showed that the variants are defective in a DNA repair process that occurs during or after replication on parental DNA that still contains dimers ("postreplication repair").

A radiosensitive cell, L5178Y S/S, has been shown to exhibit an exaggerated mitotic delay (G_2 block), which has been interpreted as evidence that the target for enhanced sensitivity lies in some structure other than DNA (Ehmann et al. 1974). Although this interpretation is still possible, the recent work of Warters and Hofer (1977) shows that ^{125}I incorporated into DNA causes a prolonged G_2 delay, whereas ^{125}I on the cell membrane does not, suggesting that mitotic delay is due to DNA damage. Therefore, the sensitivity of L5178Y S/S cells may very well be due to a defect in repair of DNA or chromatin.

The strongest evidence that DNA is the main target for radiation-induced killing of mammalian cells also derives from work with ^{125}I (Warters et al. 1977). Warters and co-workers labeled the DNA of Chinese hamster ovary cells with ^{125}I-deoxyuridine and obtained, as had others (Burki et al. 1973, Chan et al. 1977), data showing that ^{125}I decays in DNA kill cells very efficiently (D_o = 100 decays/cell; D_q = 0). In contrast, when the cells' membranes were covered with trypsin-inactivated ^{125}I-concanavalin A, the D_o for cell killing was 12,000 decays/cell and the inactivation curve had a very large shoulder (D_q = 10,000 decays/cell) (Figure 1). When these data were replotted in terms of dose to the nucleus and compared with cell killing induced by ^3H-labeled DNA (Figure 2), it became evident that the killing by ^{125}I on the cell membrane can be accounted for completely by gamma radiation reaching the cell nucleus, whereas the more efficient killing by ^{125}I in DNA is due to local effects of Auger electrons. Hofer et al. (1977) have also presented convincing data to support the concept that the effects of ^{125}I in DNA are due to the highly localized radiation and not to transmutation or charge-induced molecular fragmentation. Thus, the interpretation of these experiments is that local irradiation

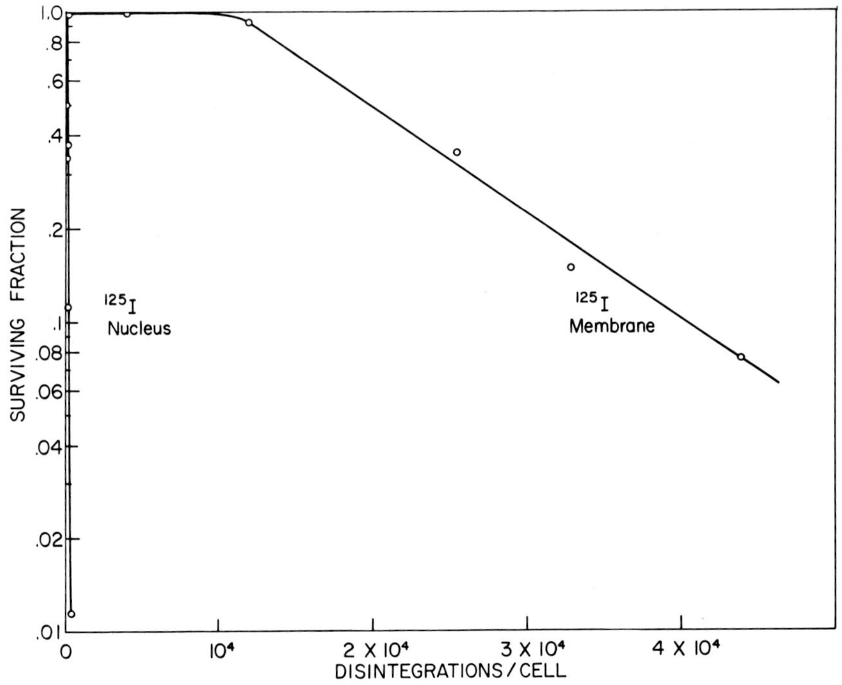

FIG. 1. Effect of ^{125}I decays in DNA or on the membrane on reproductive integrity of Chinese hamster ovary cells. (Reproduced from Warters et al. 1977, with permission of North Holland Publishing Co.)

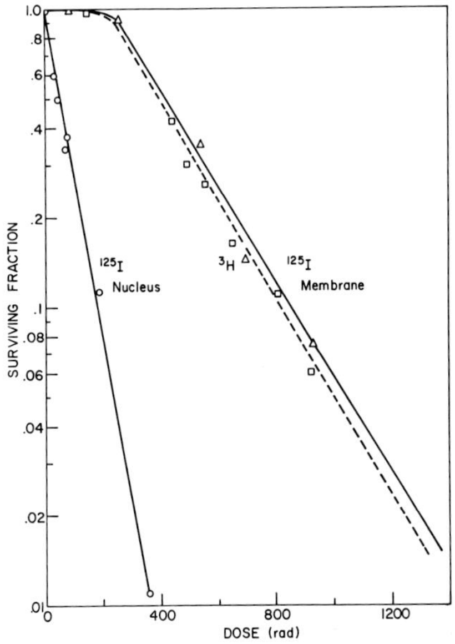

FIG. 2. Effects of ^{125}I decays in DNA or on the membrane and of ^{3}H in DNA on reproductive integrity of Chinese hamster ovary cells, plotted as function of dose to the whole nucleus. (Reproduced from Warters et al. 1977, with permission of North Holland Publishing Co.)

of DNA by ^{125}I is extremely effective in killing mammalian cells, whereas large doses of intense ^{125}I radiation to the plasma membrane are completely ineffective. Hofer et al. (1975) effectively eliminated the cytoplasm as a radiosensitive site as well by showing that ^{67}Ga-citrate, which concentrates in the cytoplasm, is also extremely ineffective in cell killing.

Given that DNA damage is an important cause of cell killing, the next question is what is the principal lesion responsible for this effect? The discussion of this point will be restricted to those effects observable in dividing cells. (The cause of the extreme radiosensitivity of cells such as oocytes and some lymphocytes is an even greater mystery.) It is generally accepted that chromosome aberrations are the principal cause of cell death in irradiated populations of dividing cells (Sasaki and Norman 1967, Carrano 1973a,b), although the exact time for the expression of killing is controversial (Bedford et al. 1978). One of the most convincing arguments for this concept is, unfortunately, in an unpublished thesis (Grote 1972). By following the fate of irradiated individual diploid baby hamster kidney cells, Grote found an absolute correlation between chromosome aberrations and failure to form a colony. When she later attempted to repeat these results with an aneuploid line, the 100% correlation did not hold (S. J. Grote and S. Wolff, personal communication). This suggests that cell lines whose chromosomes are already abnormal are capable of existing with additional aneuploidy to a greater extent than diploids, and it may explain the lack of 100% correlation between nonreciprocal chromosome aberrations and cell death that has been reported for other in vitro systems.

A chromosome aberration requires at least one double-strand break in DNA, assuming, as do most geneticists and molecular biologists, that the DNA in a chromosome is one large molecule. Therefore, one can state with some degree of assurance that the ultimate DNA lesion that precedes a chromosome aberration is an unrepaired double-strand break. The mechanisms of formation of double-strand breaks may be several, but there is considerable evidence that many of them are the result of failure to repair double-strand breaks originally formed at the time of irradiation. The best evidence derives from the work of Ritter et al. (1977), who showed that the relative efficiency of increasing linear energy transfer (LET) radiation for forming "nonrejoining" strand breaks (breaks that fail to rejoin within 12 hours as measured in alkaline sucrose gradients) correlates extremely well with the efficiency of these radiations for single-hit cell killing (Figure 3). Because it is known from work with bacteriophage DNA (Christensen et al. 1972) that double-strand breaks increase with increasing LET, the work of Ritter et al. (1977) strongly suggests that the nonrepairable breaks are double-strand breaks formed by traverse of a single particle through the DNA molecule.

A similar example is found in studies with ^{125}I in DNA. In bacteriophage and bacterial DNA, ^{125}I produces about one double-strand break per decay (Krisch et al. 1977). In repair-proficient bacterial cells, about two of the first three ^{125}I-induced double-strand breaks per genome can be repaired, but no

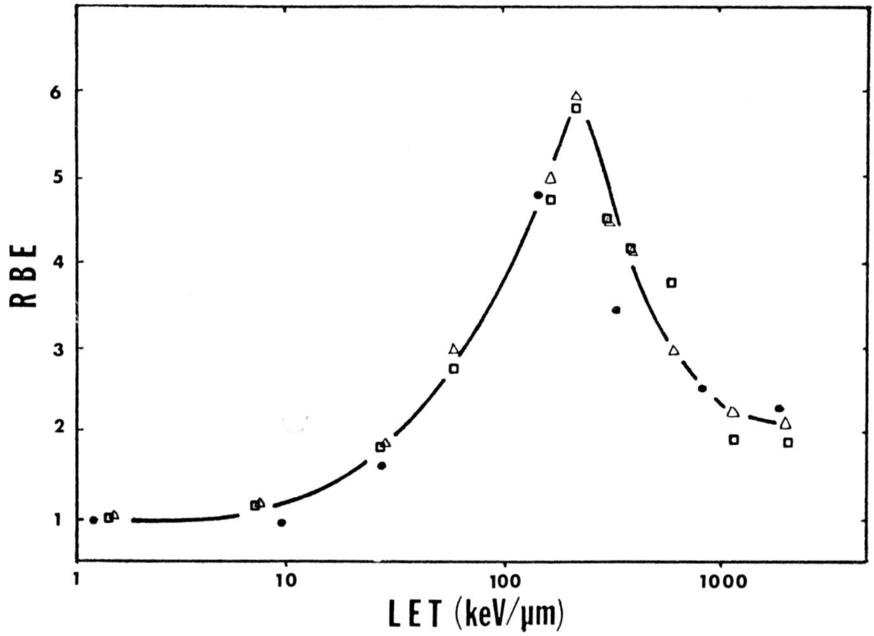

FIG. 3. Relative efficiencies for nonrejoining break induction at various LETs. The number of breaks per rad times the fraction of nonrejoining breaks was calculated for each LET, normalized to X rays, and superimposed onto data for single-hit cell killing. ●, Nonrejoining breaks; □, initial slopes of survival curves (Todd 1975); △, 80% survival levels (Todd 1975). (Redrawn from Ritter et al. 1977.)

more (Krisch et al. 1976). In mammalian cell DNA, in which ^{125}I is a potent killer (Burki et al. 1973, Warters et al. 1977, Chan et al. 1977), about one half of all strand breaks induced by ^{125}I ($\geq 6 \times 10^2$ decays/cell) fail to rejoin during the six-hour repair period when 80–100% of X-irradiated strand breaks have done so (Table 1 and Figure 4; Painter et al. 1974). These experiments measured single-strand breaks because methods for measurement of double-strand breaks in mammalian cells are not yet adequate; nevertheless, in light of the work of Krisch et al. (1977) we are confident that at least one half of all the breaks measured after ^{125}I decays in DNA are double-strand breaks. This shows another excellent correlation between lethality and failure to repair strand breaks.

Nonrepaired (or nonrepairable) double-strand breaks are probably of the kind termed "complex lesions" that are discussed in the preceding chapter (Cole et al. 1980, see pages 33 to 58, this volume). An example of this kind of break (Figure 5) is that a single photon or particle cleaves off a thymine base in one strand and, five nucleotide pairs away, causes a break at the 3′-4′ sugar bond in the other strand. The loss of thymine is followed by a break at the adjoining 3′OH-5′ phosphoryl (either by chemical hydrolysis or by action of an apyri-

TABLE 1. *Induction and repair of single-strand breaks by ^{125}I in DNA or by X-irradiation of frozen Chinese hamster cells*

Treatment	Breaks/cell × 10⁻⁴ Frozen	Incubated	% Repaired
0.6 × 10⁴ decays/cell	9.0	5.4*	40
10 krad	5.1	1.3*	75
1.0 × 10⁴ decays/cell	7.6	4.2*	45
10 krad	5.1	0.0*	100
2.0 × 10⁴ decays/cell	13.5	7.0*	48
20 krad	10.2	3.0*	71
3.8 × 10⁴ decays/cell	16.4	7.8†	52
30 krad	15.3	0.0†	100
5.8 × 10⁴ decays/cell	21.9	15.4*	30
25 krad	12.8	3.4*	73
15.2 × 10⁴ decays/cell	71.7	36.2‡	50
70 krad	63.8	12.7‡	80

* 3-hour incubation.
† 5-hour incubation.
‡ 6-hour incubation.

midinic endonuclease). The repair system is suddenly confronted with a double-strand break with a base missing at one end and a damaged deoxyribose at the other. Although an exonuclease would be capable of initiating repair in the upper strand if only a single-strand break had occurred, the local denaturation

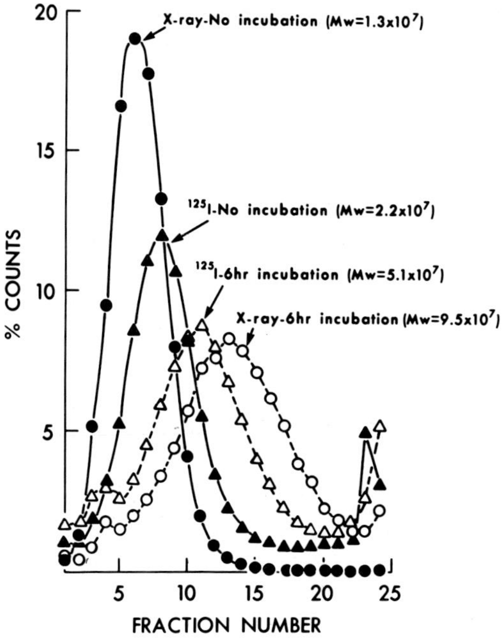

FIG. 4. Alkaline sucrose gradient profiles of DNA from frozen ^{125}I-containing cells and from frozen X-irradiated cells, with or without 6-hour incubation after thawing. (●) Cells irradiated with 60 krad, no incubation; (▲) ^{125}I-containing cells, no incubation; (△) ^{125}I-containing cells, 6-hour incubation; and (○) cells irradiated with 60 krad, 6-hour incubation. (Reproduced from Painter et al. 1974.)

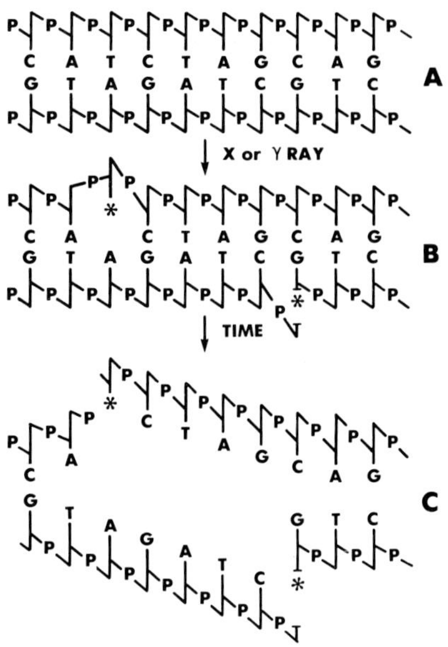

FIG. 5. Possible complex double-strand break in DNA. A, undamaged polynucleotide; B, immediately after irradiation DNA has lost a pyrimidine (thymine) and, five nucleotides away, a single-strand break has formed at the 3′-4′ bond in deoxyribose; C, with time the apyrimidinic site is converted to a break (enzymatically or chemically), forming a double-strand break.

caused by the double-strand break has removed the template for repolymerization. Moreover, the lower strand has a sugar group, probably an aldehyde, as the terminus of the broken strand, and the exonuclease may not be capable of acting at this site. Even if the break is somewhat stabilized by chromatin structure (nucleosomes or higher order), the repair system will have a difficult substrate to contend with. Of course, even more concentrated damage can be imagined and probably happens, especially at the very high LET generated in the Bragg peak of heavy nuclei.

It is possible that DNA synthesis plays a role in the formation of unrepaired double-strand breaks. In many cell types a maximum radiosensitivity to X radiation occurs at the G_1/S border (Terasima and Tolmach 1963, Sinclair and Morton 1966). Initiation of DNA replication represents a time when maximum lability of parental strands is expected, both because of the local denaturations necessary for the multiple initiations going on at replicon origins and because DNA replication probably requires nicking of parental strands to facilitate unwinding during DNA replication (Grivell et al. 1975, Strayer and Boyer 1978). X irradiation at the G_1/S border exposes the whole genome to the possibility of DNA synthesis-associated induction of double-strand breaks, because once committed to the S phase, mammalian cells always complete it. If, on the other hand, irradiation occurs later during S, a large part of the DNA has already completed replication and this "cooperative" induction of putative double-strand breaks will occur only in the unreplicated part of the DNA. Thus, if DNA replication does enhance double-strand break formation, and replication fixes damage, one would expect the probability of killing to decrease directly as a function of the fraction of S period completed, as observed in several cell types

(Sinclair 1968a). The fact that treatment with agents that delay the onset of the S phase causes even higher radiosensitivity (Weiss and Tolmach 1967, Sinclair 1968b) is consistent with this idea, in that excess origins become primed for subsequent initiation under these conditions (Painter and Schaefer 1971, Taylor et al. 1973).

Although this discussion has centered around unrepaired double-strand breaks, it is probable that other lesions in DNA are also involved in cell killing. The recent work of Maher et al. (1979) on the mechanism of potentially lethal damage repair after UV irradiation of human cells shows that repair of base damage (pyrimidine dimers) is solely responsible for this phenomenon. These data indicate that unrepaired base damage induced by X radiation may play a role in cell killing. Mattern et al. (1975) have shown that X-ray-induced thymine damage is rapidly repaired, but this repair is never complete. These data were collected after very high doses of ionizing radiation, and it is probable that at biological doses ($\leq$ 1,000 rads) a high fraction of base damage is repaired. These observations do suggest, however, that saturation of the base excision repair system may play an important role in so-called interphase death, when dividing cells exposed to very high doses of ionizing radiation die before reaching the next mitosis, i.e., before chromosome aberrations can manifest their effects.

ACKNOWLEDGMENTS

This work was supported by the U.S. Department of Energy. I thank Nomi Sinai and Mary McKenney for their aid and patience in preparation of the manuscript.

REFERENCES

Bedford, J. S., J. B. Mitchell, H. G. Griggs, and M. A. Bender. 1978. Radiation-induced cellular reproductive death and chromosome aberrations. Radiat. Res. 76:573–586.

Bootsma, D. 1978. Xeroderma pigmentosum, *in* DNA Repair Mechanisms, P. C. Hanawalt, E. C. Friedberg, and C. F. Fox, eds. Academic Press, New York, pp. 589–601.

Boyce, R. P., and P. Howard-Flanders. 1964. Release of ultraviolet light-induced thymine dimers from DNA in *E. coli* K-12. Proc. Natl. Acad. Sci. USA 51:293–300.

Burki, H. J., R. Roots, L. E. Feinendegen, and V. P. Bond. 1973. Inactivation of mammalian cells after disintegrations of ^{3}H or ^{125}I in cell DNA at −196°C. Int. J. Radiat. Biol. 24:363–375.

Carrano, A. V. 1973a. Chromosome aberrations and radiation-induced cell death. I. Transmission and survival parameters of aberrations. Mutat. Res. 17:341–353.

Carrano, A. V. 1973b. Chromosome aberrations and radiation-induced cell death. II. Predicted and observed cell survival. Mutat. Res. 17:355–366.

Chan, P. C., E. Lisco, H. Lisco, and S. J. Adelstein. 1977. Cell survival and cytogenetic responses to ^{125}I-UdR in cultured mammalian cells. Curr. Top. Radiat. Res. Q. 12:426–435.

Christensen, R. C., C. A. Tobias, and W. D. Taylor. 1972. Heavy-ion-induced single- and double-strand breaks in ϕX-174 replicative form DNA. Int. J. Radiat. Biol. 22:457–477.

Cleaver, J. E. 1968. Defective repair replication of DNA in xeroderma pigmentosum. Nature 218:652–656.

Cole, A., R. E. Meyn, R. Chen, P. M. Corry, and W. Hittelman. 1980. Mechanisms of cell injury, *in* Radiation Biology in Cancer Research (The University of Texas System Cancer Center 32nd Annual Symposium on Fundamental Cancer Research), R. E. Meyn and H. R. Withers, eds. Raven Press, New York, pp. 33–58.

Ehmann, U. K., H. Nagasawa, D. F. Petersen, and J. T. Lett. 1974. Symptoms of X-ray damage to radiosensitive mouse leukemic cells: Asynchronous populations. Radiat. Res. 60:453–472.

Grivell, A. R., M. B. Grivell, and P. C. Hanawalt. 1975. Turnover in bacterial DNA containing thymine or 5-bromouracil. J. Mol. Biol. 98:219–233.

Grote, S. J. 1972. Radiation-induced genetic damage and lethality in mammalian cells. Ph.D. dissertation, University of London.

Hofer, K. G., C. R. Harris, and J. M. Smith. 1975. Radiotoxicity of intracellular ^{67}Ga, ^{125}I and ^{3}H. Nuclear versus cytoplasmic radiation effects in murine L1210 leukaemia. Int. J. Radiat. Biol. 28:225–241.

Hofer, K. G., G. Keough, and J. M. Smith. 1977. Biological toxicity of Auger emitters: Molecular fragmentation versus electron irradiation. Curr. Top. Radiat. Res. Q. 12:335–354.

Howard-Flanders, P., and R. P. Boyce. 1964. The repair of ultraviolet photoproducts in DNA of bacteria, *in* Cellular Radiation Biology (The University of Texas System Cancer Center 18th Annual Symposium on Fundamental Cancer Research, 1964). Williams and Wilkins Co., Baltimore, pp. 52–63.

Klímek, M. 1966. Thymine dimerization in *L*-strain mammalian cells after irradiation with ultraviolet light and the search for repair mechanisms. Photochem. Photobiol. 5:603–607.

Krisch, R. E., F. Krasin, and C. J. Sauri. 1976. DNA breakage, repair and lethality after ^{125}I decay in *rec*$^{+}$ and *rec*A strains of *Escherichia coli.* Int. J. Radiat. Biol. 29:37–50.

Krisch, R. E., F. Krasin, and C. J. Sauri. 1977. DNA breakage, repair, and lethality accompanying ^{125}I decay in microorganisms. Curr. Top. Radiat. Res. Q. 12:355–368.

Lehmann, A. R., S. Kirk-Bell, C. F. Arlett, M. C. Paterson, P. H. M. Lohman, E. A. de Weerd-Kastelein, and D. Bootsma. 1975. Xeroderma pigmentosum cells with normal levels of excision repair have a defect in DNA synthesis after UV-irradiation. Proc. Natl. Acad. Sci. USA 72:219–223.

Maher, V. M., D. J. Dorney, B. Konze-Thomas, A. L. Mendrala, and J. J. McCormick. 1979. DNA excision repair processes in human cells can eliminate the cytotoxic and mutagenic consequences of ultraviolet irradiation. Mutat. Res. (In press).

Mattern, M. R., P. V. Hariharan, and P. A. Cerutti. 1975. Selective excision of gamma ray damaged thymine from the DNA of cultured mammalian cells. Biochim. Biophys. Acta 395:48–55.

Mortelmans, K., E. C. Friedberg, H. Slor, G. Thomas, and J. E. Cleaver. 1976. Defective thymine dimer excision by cell-free extracts of xeroderma pigmentosum cells. Proc. Natl. Acad. Sci. USA 73:2757–2761.

Painter, R. B. 1970. The action of ultraviolet light on mammalian cells, *in* Photophysiology, vol. 5, A. C. Giese, ed. Academic Press, New York, pp. 169–189.

Painter, R. B., and A. W. Schaefer. 1971. Variation in the rate of DNA chain growth through the S phase in HeLa cells. J. Mol. Biol. 58:289–295.

Painter, R. B., B. R. Young, and H. J. Burki. 1974. Non-repairable strand breaks induced by ^{125}I incorporated into mammalian DNA. Proc. Natl. Acad. Sci. USA 71:4836–4838.

Paterson, M. C. 1978. Ataxia telangiectasia: A model inherited disease linking deficient DNA repair with radiosensitivity and cancer proneness, *in* DNA Repair Mechanisms, P. C. Hanawalt, E. C. Friedberg, and C. F. Fox, eds. Academic Press, New York, pp. 637–650.

Rasmussen, R. E., and R. B. Painter. 1964. Evidence for repair of ultraviolet damaged deoxyribonucleic acid in cultured mammalian cells. Nature 203:1360–1362.

Rauth, A. M. 1970. Effects of ultraviolet light on mammalian cells in culture. Curr. Top. Radiat. Res. 6:195–248.

Ritter, M. A., J. E. Cleaver, and C. A. Tobias. 1977. High-LET radiations induce a large proportion of non-rejoining DNA breaks. Nature 266:653–655.

Sasaki, M. S., and H. Norman. 1967. Selection against chromosome aberrations in human lymphocytes. Nature 214:502–503.

Setlow, R. B., and W. L. Carrier. 1964. The disappearance of thymine dimers from DNA: An error-correcting mechanism. Proc. Natl. Acad. Sci. USA 51:226–231.

Sinclair, W. K. 1968a. Cyclic X-ray responses in mammalian cells in vitro. Radiat. Res. 33:620–643.

Sinclair, W. K. 1968b. The combined effect of hydroxyurea and X-rays on Chinese hamster cells in vitro. Cancer Res. 28:198–206.

Sinclair, W. K., and R. A. Morton. 1966. X-ray sensitivity during the cell generation cycle of cultured Chinese hamster cells. Radiat. Res. 29:450–474.

Strayer, D. R., and P. D. Boyer. 1978. Integrity of parental DNA during replication. J. Mol. Biol. 120:281–295.

Szybalski, W., and Z. Opara-Kubinska. 1964. Radiobiological and physicochemical properties of 5-bromodeoxyuridine-labeled transforming DNA as related to the nature of the critical radiosensitive structures, *in* Cellular Radiation Biology (The University of Texas System Cancer Center 18th Annual Symposium on Fundamental Cancer Research, 1964). Williams and Wilkins Co., Baltimore, pp. 223–240.

Taylor, J. H., A. G. Adams, and M. P. Kurek. 1973. Replication of DNA in mammalian chromosomes. II. Kinetics of ^{3}H-thymidine incorporation and the isolation and partial characterization of labeled subunits at the growing point. Chromosoma 41:361–384.

Terasima, T., and L. J. Tolmach. 1963. Variations in several responses of HeLa cells to x-irradiation during the division cycle. Biophys. J. 3:11–33.

Todd, P. W. 1975. Heavy-ion irradiation of human and Chinese hamster cells in vitro. Radiat. Res. 61:288–297.

Warters, R. L., and K. G. Hofer. 1977. Radionuclide toxicity in cultured mammalian cells. Elucidation of the primary site for radiation-induced division delay. Radiat. Res. 69:348–358.

Warters, R. L., K. G. Hofer, and C. R. Harris. 1977. Radionuclide toxicity in cultured mammalian cells: Elucidation of the primary site of radiation damage. Curr. Top. Radiat. Res. Q. 12:389–407.

Weiss, B. G., and L. J. Tolmach. 1967. Modification of x-ray-induced killing of HeLa S3 cells by inhibitors of DNA synthesis. Biophys. J. 7:779–795.

Radiation Biology in Cancer Research, edited by
Raymond E. Meyn and H. Rodney Withers.
Raven Press, New York © 1980.

Introduction of the Ernst W. Bertner Memorial Award Recipient

Robert C. Hickey

The University of Texas System Cancer Center M. D. Anderson Hospital and Tumor Institute, Houston, Texas 77030

The Bertner Award is conferred annually on one who has made distinguished contributions to cancer research.

This award was established in 1950 and honors Doctor E. W. Bertner, the first acting director of M. D. Anderson Hospital and Tumor Institute and the first president of the Texas Medical Center. It is funded by the former Bertner Foundation and The University Cancer Center.

The award medallion symbolizes the goals of cancer research: prevention and cure. The hands of Hygeia emerge from a star in which a serpent, symbol of medical wisdom, is fed. Hygeia, daughter of Aesculapius, represents hygiene and disease prevention. The star denotes our state and the Texan for whom the award is named.

The recipient of the 1979 Ernst W. Bertner Memorial Award is Doctor Mortimer M. Elkind.

Dr. Elkind's home base is the Argonne National Laboratory in Chicago, where he is a Senior Biophysicist in the Division of Biological and Medical Research and the Assistant Director for Research Training. Prior to this he had had a series of appointments at the Hammersmith Hospital, London; Brookhaven National Laboratory, Upton, Long Island; National Cancer Institute; Donner Laboratory in Berkeley; and others.

His training initially was in mechanical and electrical engineering, with degrees from Cooper University School of Engineering, Polytechnic Institute of Brooklyn, and the Massachusetts Institute of Technology; his doctorate degree is from the Massachusetts Institute of Technology.

Thus, Doctor Elkind entered the field of radiation biology from a background of engineering and physics. Since then, he has made several milestone contributions to the field, the most important being the first characterization of the biophysics and repair kinetics of radiation-induced sublethal injury in mammalian cells. Although this is his most widely acclaimed achievement, it is supported by major contributions in other areas including radiation- and drug-induced DNA damage and repair, an early description of the effect of hyperthermia on repair of sublethal damage, a detailed analysis of the relationship between the repair of potentially lethal and sublethal damage, repair of potentially lethal injury after "single-hit" inactivation, and in radiation transformation. The quality of his research has been acknowledged by major awards from the Atomic Energy Commission (E. O. Lawrence Award) and the International Commission on Radiation Units (L. H. Gray Medal), among others.

The work for which he is best known is his classic investigation of the nature of sublethal injury in irradiated mammalian cells, first published in 1959. The biophysical principles of the killing of cells through the accumulation of sublethal injury and of the quantitation of the extent and time course of repair of the sublethal injury (Elkind repair) are basic to our understanding of almost every aspect of conventional and experimental radiotherapy. Without the characterization of this phenomenon, subsequent understanding in such important areas as dose fractionation, low dose-rate irradiation, differentials in sensitivity of the cells of various tissues and tumors, and variations in radiation response as cells progress through the division cycle would not have been possible.

Early in the investigation of drug radiation interactions, Elkind emphasized and illustrated in his experiments with actinomycin D and nitrogen mustards that understanding the cellular basis for the interaction could be the only way to quantify the total effect and ultimate clinical application of such interactions. The importance of this approach is underlined by the ineffectiveness of the empirical approach decided upon by the National Cancer Institute several years ago, and which is now being reviewed. Similar considerations of the importance of an understanding of the cellular basis of interactions with radiation have led him to investigate the cellular mechanism for the interaction of hyperthermia and radiation. More recently, with the emerging interest in repair of potentially lethal damage as a factor determining the therapeutic ratio in radiotherapy, he has pioneered the study of repair of potentially lethal damage at low doses and with high linear energy transfer beams, in which cell survival is a simple exponential function of dose and has been presumed to reflect irreparable single-hit injury. It is repair of potentially lethal damage at these doses that is of primary interest to radiotherapy. Also, his recent studies have focused on the cellular mechanisms for radiation transformation, once again with a view to understanding the cellular mechanisms rather than merely demonstrating consequences of the phenomenon.

In 25 years of outstanding work in the field of radiation biology, Dr. Elkind has elucidated cellular mechanisms of radiation effect that have formed the basis for much of our understanding in clinical radiotherapy and public health aspects of radiation, as well as contributed to our understanding of basic cellular biology.

Radiation Biology in Cancer Research, edited by
Raymond E. Meyn and H. Rodney Withers.
Raven Press, New York © 1980.

The Ernst W. Bertner Memorial Award Lecture

Cells, Targets, and Molecules in Radiation Biology

M. M. Elkind

Division of Biological and Medical Research, Argonne National Laboratory, Argonne, Illinois 60439

Radiation biology encompasses a series of dualities.

To begin with, the absorption of ionizing radiation is both *random* and *discrete.* All molecules in a cell are susceptible to damage, but only a few are hit. Further, energy absorption events are large enough to break any molecular bond.

Second, randomness and discreteness imply *hits* and *targets.* Because radiant energy is deposited in volumes of atomic dimensions, targets consisting of large molecules, or aggregates of molecules, are in principle resolvable from measurements of radiation reponse.

Third, the stochastics of radiant energy deposition require the distinction between *microdosimetry* and *macrodosimetry.* Whereas every unit volume of a cell may be essentially equally vulnerable, large regions between absorptions initially remain undamaged. After "small" doses, it is conceivable, therefore, that no biological effect is produced simply because no damage is registered in a susceptible target.

Fourth, the physical-chemical effects that follow the initial energy absorption events may result in *indirect damage* in addition to the *direct damage* registered in a target. Indirect action is mediated largely by the radiolytic products of water and effectively increases target sizes.

Fifth, the net effect of a dose of radiation reflects both *damage* and *repair.* Damage is proportional to dose. At the level of the elementary biological structure, the cell, repair is operational and depends upon our ability to measure biological change as well as the properties of the radiation, the time course of its delivery, and the inherent repair capabilities of the cell. A return to the undamaged biological state, if it occurs at all, may only be specified relative to what is measurable and, hence, is an operational notion.

And last, radiation biology is fundamental to *public health* as well as to *radiation therapy.* It has been clearly established that radiation is mutagenic and carcinogenic. It also has been clearly demonstrated that cancer may be cured by radiation. This last duality illustrates the broad-ranging importance of radiation biology in the current day.

As the world's resources shrink and its population grows, potential hazards to man increase due to various types of environmental pollutants. The need for a sustained and vigorous pursuit of the fundamentals of radiation action should be evident. To this end, I offer an overview of damage/repair processes in mammalian cells and some speculations about mechanisms, rather than a series of reminiscences and anecdotes, which one might assume are called for by the occasion. I will deal principally with radiation lethality (i.e., suppression of division). The choice of this end point, which is easily measured, is only in part a matter of convenience. Cell killing, critical in the treatment of cancer, also has a central role in the alteration of other cell properties. To transform a cell, to mutate it, to change its differentiation, in each case viability must be conserved. Hence, to understand the radiation induction of altered cell properties in general, one must also understand how radiation affects viability.

CELLULAR DAMAGE AND REPAIR

Inactivation Models

The probability of survival of a mammalian cell in a population of cells may be expressed as the product of two terms:

$$S = e^{-\alpha D}\, M(D), \qquad (1)$$

where S is the surviving fraction, α is a constant, and D is the dose; M(D) is discussed below. The factor $exp(-\alpha D)$ is referred to as the "single-hit" mode of killing and describes the fact that most survival curves (Elkind 1975, 1977) have an initial, nonzero slope (i.e., $-\alpha < 0$). The factor M(D) represents a damage accumulation mode of inactivation, and, in general, M(D) is represented by expression (2a) or (2b), β, γ, and n being constants,

$$M(D) = \begin{cases} e^{-\beta D^2} & \text{(2a)} \\ 1 - (1 - e^{-\gamma D})^n. & \text{(2b)} \end{cases}$$

Where (2b) is found to apply, γ ($\equiv 1/{}_nD_o$) is interpreted to be the sensitivity of each of n targets in a cell, all n of which must be inactivated to kill the cell. Setting $\alpha = (1/{}_1D_o)$, the reciprocal D_o of the final slope of a survival curve that involves the term (2b) is

$$D_o = ({}_1D_o)({}_nD_o)/\{({}_1D_o) + ({}_nD_o)\}. \qquad (3)$$

The fraction of cells killed, K, is given by

$$K = 1 - S = aD + bD^2 + cD^3 \ . \ . \ . \qquad (4)$$

where $a = \alpha$ and the coefficients of the remaining terms depend upon which expression for M(D) is used.

Whichever expression for M(D) fits a given set of survival data, each requires that damage must be accumulated to produce an effect. If the damage is repairable, less killing might result if the dose is delivered in a way that favors repair (e.g., in fractions over a period of time, or at low dose rate). In contrast, from Poisson statistics $exp(-\alpha D)$ implies that one "hit" in a cell's sensitive "target" suffices to inactivate. Consequently, in this mode of cell killing damage need not be accumulated, and it follows that repair processes would be without effect on the initial slope of a survival curve.

Dose-Effect Curves and Cancer Research

In the context of damage and repair processes, the equations for survival (1) and killing (4) are schematized in Figure 1. In the left panel, where the ordinate is logarithmic and the abscissa is linear, it is assumed that M(D) is given by (2b). A survival curve with zero initial slope (i.e., $\alpha = 0$) is sketched as curve A. The corresponding killing curve is A′ in the right panel, which also starts with zero slope, although here both axes are linear. Curves B and B′ are the corresponding pair of curves for $-\alpha < 0$, or a > 0, equations 1 and 4.

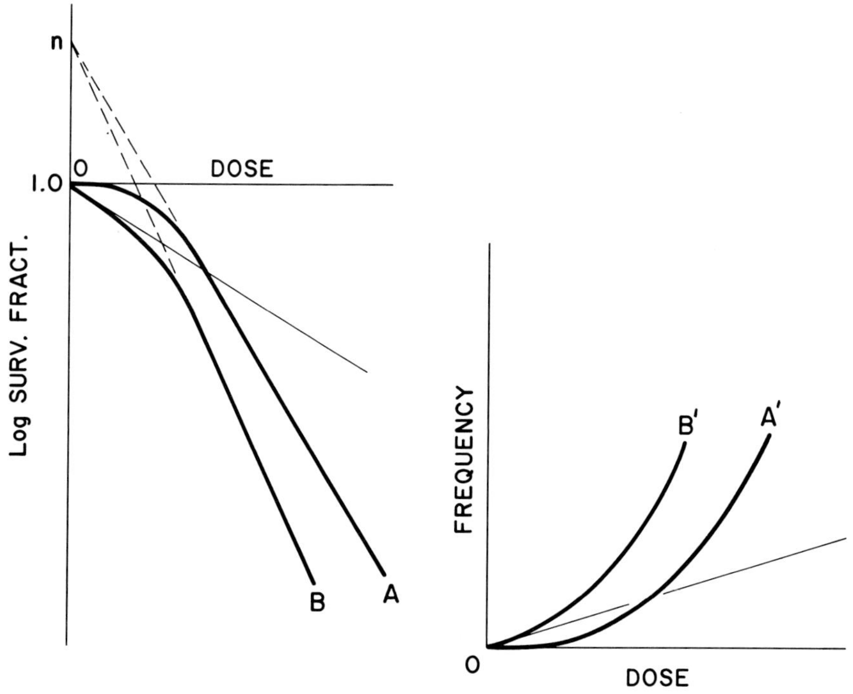

FIG. 1. Schematic representations of zero (A, A′) and nonzero (B, B′) initial slopes for cell survival (left panel) and the induction of effects like cell killing, mutagenesis, and oncogenesis (right panel). The left panel represents a logarithm-linear plot, and the right panel is a linear-linear plot. (Reproduced from Elkind 1975, with permission of John Wiley & Sons.)

Taken together, these two pairs of curves symbolically characterize the relationship of radiation biology to cancer research. In the clinic, the objective is to reduce the probability of cell survival to a level low enough to sterilize a tumor without inflicting an unacceptable level of damage in the incidentally irradiated normal tissue. Toward this end, a series of moderately large dose fractions are employed that frequently sum up to a large total dose. The repair of sublethal damage (Elkind and Sutton 1959, 1960), to be illustrated presently, requires that the total dose to obtain a given surviving fraction be increased as the number of fractions is increased (Elkind 1960). If the single-dose survival curve has the shape A, the total dose versus number of fractions would be expected to increase without limit (Elkind et al. 1968b, Elkind 1976). In contrast, if curve B applies, it may be inferred (Dutreix et al. 1973) that when each dose fraction is not large enough to surpass the initial linear portion of curve B (i.e., $M(D) \simeq 1$), the total dose will approach a maximum as the number of fractions is increased (see Elkind 1976 for the additional details).

Hazards to public health due to radiation result in general from small total doses. Hence, induced effects—e.g., cell killing, Figure 1, or mutation, neoplastic transformation, etc.—are usually plotted on linear scales. A zero initial slope, curve A′, implies no effect from a small enough dose. Further, in this case the theoretical possibility is frequently noted that at a low dose rate the curvature of A′ would be reduced so that induction frequencies even for large doses would approach zero. In contrast, when the induction curve has shape B′, it is inferred that even if repair at low dose rate reduces the incidence at high doses, the incidence would never be less than that traced by the initial slope. This follows from the assumption that the limiting slope of curve B′ reflects single-hit damage and is therefore not dose-rate dependent. Since effects directly proportional to dose (exponential for small to large doses or linear for small doses) are interpreted not to involve damage accumulation, a role for repair is precluded because in a surviving cell (or in a nonmutated cell, a nontransformed cell, etc.) no subeffective damage has been registered.

From the foregoing it is evident that repair processes expressible at a cellular level—proliferative ability, mutation, transformation, etc.—can have a critical influence on both qualitative and quantitative responses. The discussion to follow will be confined to cell survival. The importance of this cellular end point to cancer therapy is evident since the ultimate objective is tumor sterilization. But cell survival is also of critical importance in other connections. Cell viability must be maintained for the expression of a mutated or a transformed phenotype. Consequently, even if repair influences only cell survival, the expression of changes in other cell properties may be affected and in particular those of importance to public health such as mutation and transformation.

Repair Relative to Damage Accumulation

Mammalian cells exposed to X rays or γ rays frequently have single-dose survival curves of shape B in Figure 1. In reference to equation 1, cells surviving doses for which a significant decrement in survival results from damage accumulation [M(D) less than ~ 0.2] are very likely sublethally damaged. Dose fractionation (Elkind and Sutton 1959, 1960) or irradiation at reduced dose rates (Ben-Hur et al. 1972, 1974) may be used to demonstrate the repair of sublethal damage. Because the survival response of cells depends upon their "age" in their growth cycle (see Sinclair and Morton 1964 for the age-response pattern of Chinese hamster cells), it is necessary to account for partial synchronization when dose fractionation is used. Hence, an experiment using synchronized cells is discussed as an illustration of repair of sublethal damage.

In Figure 2, the age-response patterns of V79 Chinese hamster cells for two acute doses of X rays are shown (see legend for details). A two-dose fractionation sequence was started at 3½ hours, at about the middle of the S phase. Hence, fractionation data with asynchronous cells may be expected to approximate those in Figure 2 because of the partial synchronization that results from a reasonably large first dose (Elkind and Redpath 1977). The prompt increase in two-dose survival has been shown to result from a redevelopment of the shoulder on the survival curve of those cells surviving the first dose (Elkind

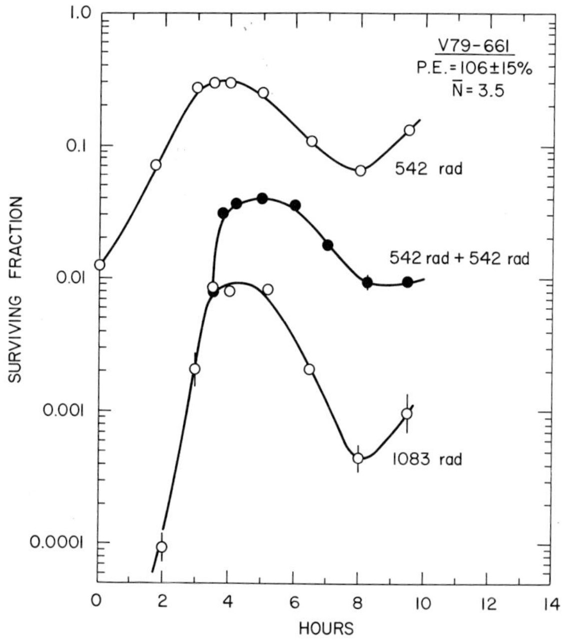

FIG. 2. Single-dose and two-dose survival patterns for V79 Chinese hamster cells irradiated (50 kV X rays, 1,800 rads/min) at different ages (or times) in their growth cycle. Cells were synchronized by hydroxyurea (Sinclair 1967) and hence zero hour on the abscissa corresponds to the G_1-S border. The dose fractionation was started at 3⅓ hours, which is close to the time of greatest resistance in the cell cycle. P.E. = plating efficiency and $\bar{N}$ = average cell multiplicity of colony forming units. Standard errors in survival are shown where they are larger than the size of symbols. (Modified from Elkind and Redpath 1977; similar results have been reported by Sinclair and Morton 1964, Elkind and Sinclair 1965, and Elkind et al. 1967.)

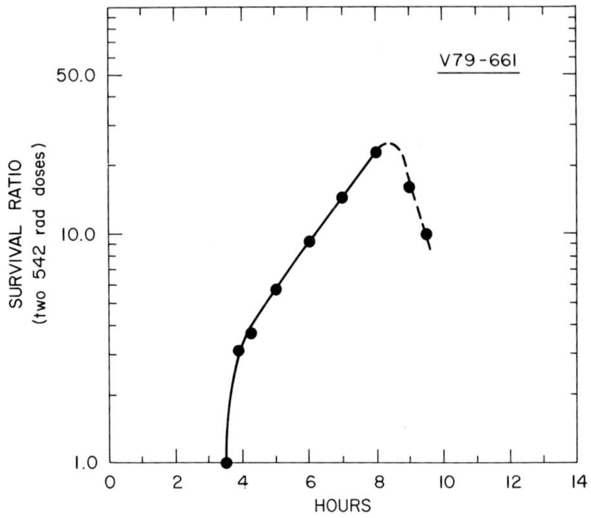

FIG. 3. A plot of the survival ratio of the two-dose to single-dose data in Figure 2. (Modified from Elkind and Redpath 1977.)

et al. 1967), and the displacement of all of the two-dose data upward from the single-dose curve for the same total dose traces the progressive effect of the repair of sublethal damage. From the plot of the ratio of two-dose to single-dose survival in Figure 3, one may infer that the repair of sublethal damage takes longer than 4½ hours to complete. This is in accord with an estimate of the time course obtained from the kinetics of the loss of damage interaction between X rays and actinomycin D (Elkind et al. 1967). The downturn in the survival ratio curve in Figure 3 results from a shift in the age-phase relationship of the two-dose to single-dose curves in Figure 2. Because cells surviving the first 542-rad dose are delayed in their aging, the minimum in the two-dose curve is at a later time than for the single-dose curve. The maximum in Figure 3 results from this displacement and not from a "relapse" in the first-dose survivors (see Elkind and Redpath 1977 for further discussion of this point).

Potentially Lethal Damage

When a change in cell cultivation conditions *after* irradiation results in a change in survival, one may infer that potentially lethal damage plays a role. Many of the studies of this form of damage have demonstrated increases in survival after single doses when postirradiation conditions suboptimal for growth and, apparently, more favorable for repair were used (e.g., see Phillips and Tolmach 1966, Belli and Shelton 1969, Little 1969, and Hahn and Little 1972). Also, the assay procedures used to demonstrate enhanced survival after irradiation frequently employ confluent or stationary-phase cultures (Hahn and Little 1972) and hence may not apply to cells throughout the cycle as in actively growing cultures. A brief postirradiation incubation in anisotonic phosphate-buffered saline (PBS) results in a reduction in the survival of actively growing cells in the absence of a concomitant effect on the viability of unirradiated

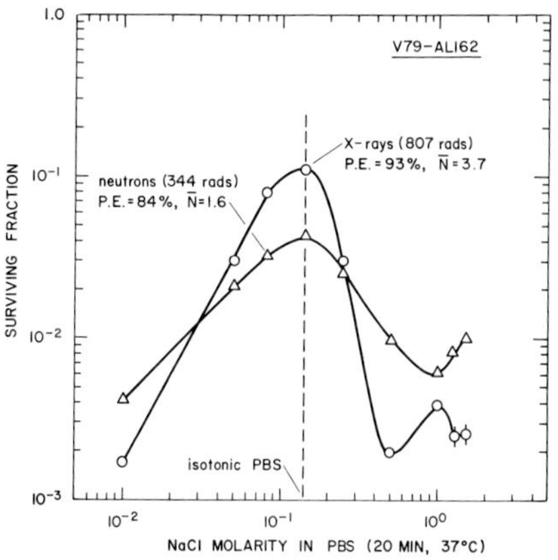

FIG. 4. The dependence of the survival of V79 Chinese hamster cells on the NaCl molarity of PBS added for 20 minutes immediately following a fixed dose of 50-kV X rays or fission-spectrum neutrons obtained from the JANUS reactor at the Argonne National Laboratory. The doses chosen resulted in about the same single-cell surviving fractions for isotonic PBS posttreatment (or medium). The range of NaCl concentrations shown did not reduce the survival of unirradiated cells. Other details as in Figure 2. (Reproduced from Utsumi and Elkind 1979b, with permission of International Journal of Radiation Biology.)

cells (Utsumi and Elkind 1979a, 1979b). The indications are that anisotonic PBS incubation posttreatment results in the enhanced expression of damage, which, therefore, is ordinarily only potentially lethal, as the next several figures illustrate.

Following an acute X-ray dose, the survival of Chinese hamster cells is further reduced with time if they are incubated in anisotonic PBS (Utsumi and Elkind 1979a) (Figure 4). The rate of reduction depends upon the NaCl molarity and the incubation temperature. For cells incubated for 20 minutes at 37°C after X-ray and fission-neutron exposure, Figure 4 shows the dependence of survival on the NaCl concentration of PBS. Maximum survival (the same achieved in medium) results if isotonic (0.14 M NaCl) PBS is used. Althought the range of survival change is less for neutrons than for X rays, it is clear that the qualitative result is the same after both radiations.

The survival curves of aerobic and hypoxic cells treated after irradiation with anisotonic PBS are shown in Figure 5. It is clear that for mid-to-large doses, doses for which M(D) in equation 1 contributes significantly to survival decrements, appreciably more potentially lethal damage is made evident by postirradiation anisotonicity than for small doses. Of particular importance is the fact that the oxygen enhancement ratios (OERs) are approximately equal for isotonic and anisotonic treatment. This suggests that essentially the same spectra of radiochemical lesions are involved for the two levels of damage expressed. Hence, it appears likely that the enhanced expression of potentially lethal damage does not result from nonspecific stress of anisotonic PBS exposure.

Repair of Potentially Lethal versus Sublethal Damage

A further indication that anisotonicity reveals radiation damage that is ordinarily repaired is demonstrated in Figure 6. Following a fixed X-ray dose, cells were incubated in isotonic medium at the temperatures and for the times shown. They were then treated with hypertonic PBS, after which medium was returned for colony formation. A rapid, strongly temperature-dependent loss of the susceptibility of cells to enhanced damage expression is evident. Consistent with the

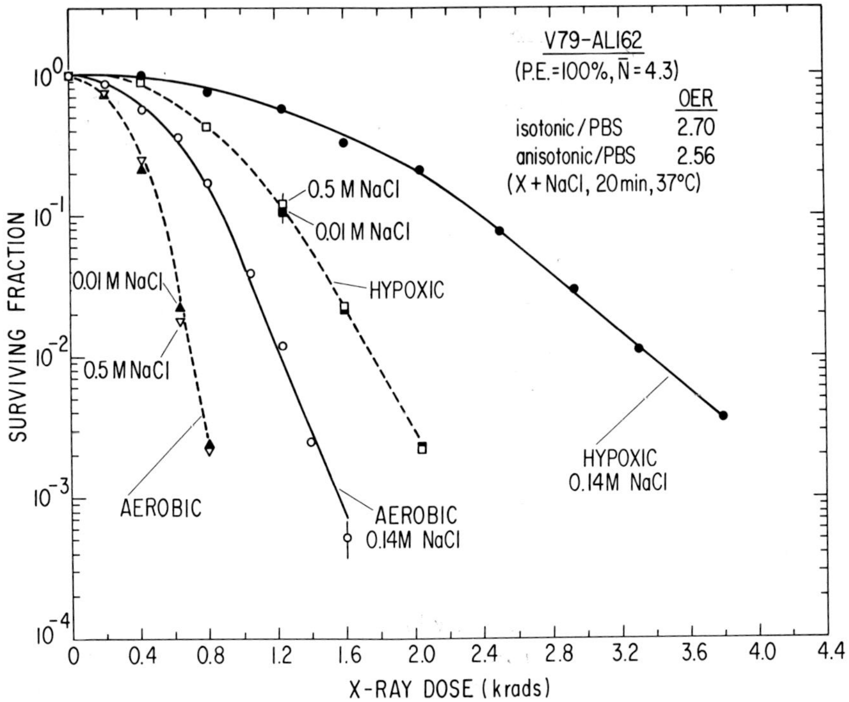

FIG. 5. The influence of hypoxia on the survival curves of X-irradiated asynchronous Chinese hamster cells treated after irradiation with (aerated) PBS solutions containing the NaCl concentrations shown. The survival curve parameters are as follows: aerobic irradiation, isotonic posttreatment, $D_o = 138$ rads, $D_q = 393$ rad, *ln* n = 2.85; hypoxic irradiation, isotonic posttreatment, $D_o = 375$ rad, $D_q = 1{,}050$ rad, *ln* n = 2.80; aerobic irradiation, anisotonic posttreatment, $D_o = 70.0$ rad, $D_q = 248$ rad, *ln* n = 3.54; and hypoxic irradiation, anisotonic posttreatment, $D_o = 182$ rad, $D_q = 619$ rad, *ln* n = 3.40. (D_o, the 1/e dose along the exponential, terminal part of the curve; D_q, the shoulder width corrected for single cells; and *ln* n, the natural log of the extrapolation number.) Other details as in Figure 2. (Reproduced from Utsumi and Elkind 1979a, with permission of Academic Press.)

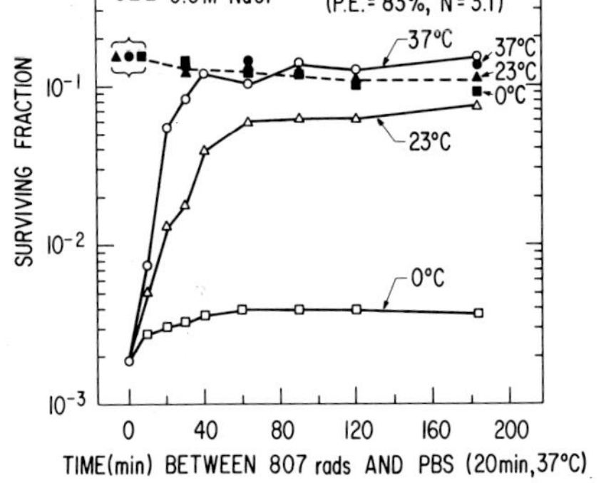

FIG. 6. Repair of the potentially lethal damage in X-irradiated (807 rads) Chinese hamster cells that is made evident by hypertonic PBS posttreatment. The open symbols show the time dependence of the loss of the expression of additional lethal damage; each point denotes the time at which a 20-minute, 37°C hypertonic treatment was started following postirradiation incubation in complete medium at the temperatures shown. The closed symbols refer to treatments with 37°C isotonic PBS for 20 minutes. Other details as in Figure 2. (Reproduced from Utsumi and Elkind 1979a, with permission of Academic Press.)

results in Figure 6, it was found that by 60 minutes after irradiation, the entire survival curve is no longer modifiable by hypertonic PBS.

The data in Figure 6 suggest that the repair of potentially lethal damage, expressible by anisotonicity, is appreciably more rapid than the repair of sublethal damage, shown in Figure 3. This implies that these two forms of damage differ. Further support for this likelihood is contained in Figure 7, where survival curves after single- or two-dose X irradiation are shown for isotonic and hypertonic PBS treatments after each dose. It is clear that the reduced survival resulting from hypertonic PBS treatment is not accompanied by a loss of the ability of cells to repair sublethal damage.

Hence, in the mid-dose range [M(D) less than ~ 0.2], where damage accumulation can be expected to make an appreciable contribution to cell killing, it is clear that the repair of sublethal damage is independent of the amount of potentially lethal damage expressed. Further, independence of repair implies that

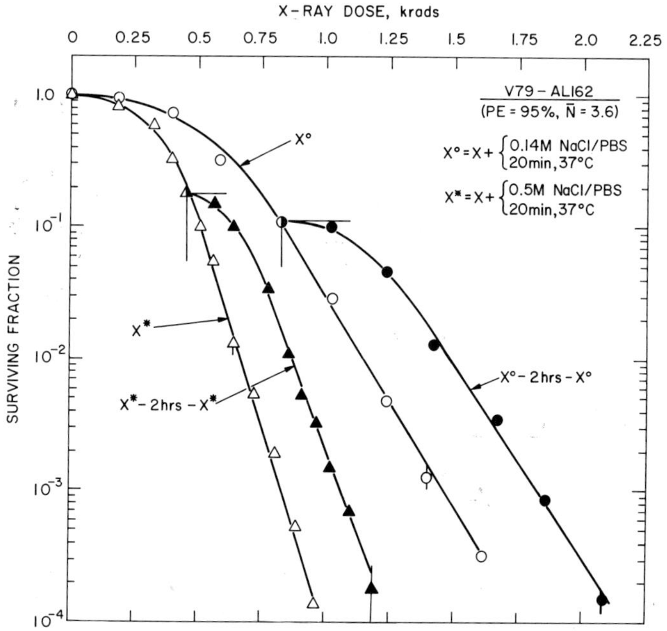

FIG. 7. Single-dose and two-dose survival curves of X-irradiated Chinese hamster cells treated immediately after each dose with either isotonic or hypertonic PBS at 37°C. The first doses for the fractionation survival curves were: isotonic PBS, 834 rads; and hypertonic PBS, 457 rads. During the 2-hour fractionation interval, cells were incubated in growth medium at 37°C. The single-cell survival curve parameters for isotonic PBS posttreatment are: single-dose D_o = 135 rads, D_q = 347 rads, and *ln* n = 2.57; and two-dose D_o = 137 rads, D_q = 170 rads, and *ln* n = 1.24. For hypertonic posttreatment, the parameters are: single-dose D_o = 71 rads, D_q = 259 rads, and *ln* n = 3.65; and two-dose D_o = 75 rads, D_q = 104 rads, and *ln* n = 1.39. Other details as in Figure 2. (Reproduced from Utsumi and Elkind 1979a, with permission of Academic Press.)

these two forms of damage are molecularly distinct as well as operationally distinct at a cellular level.

Is "Single-Hit" Damage Repairable?

The detection of damage modification after mid-to-high doses, i.e., for surviving fractions lower than ~0.1, is frequently facilitated by the relatively large changes in survival which may be effected. This is clearly the case in connection with both sublethal damage and potentially lethal damage. For example, because the amount of sublethal damage that may be repaired decreases as the size of the first dose is decreased, the repair of sublethal damage becomes increasingly more difficult to demonstrate as the first dose approaches zero. This is because, in principle, the sublethal damage accumulated approaches zero: $M(D) \rightarrow 0$. In the instance of potentially lethal damage, the enhanced expression of which may result mainly in a D_o change, the survival difference becomes progressively less as the small dose region is approached (see Figure 5). Consequently, it is generally difficult to examine whether or not the conventional interpretation of no repair associated with an exponential (or linear) dose dependence—$exp(-\alpha D)$—is fulfilled experimentally. This difficulty is even more apparent for end points like mutation and transformation induction because low-frequency events are usually associated with large uncertainties.

In spite of these difficulties, one may test for repair in the region of small doses with the aid of an extrapolative procedure. Figure 8 shows pairs of survival curves for isotonic *versus* hypertonic PBS postirradiation treatment. (As noted earlier, an isotonic posttreatment does not influence survival.) Although the relative change in survival for fission-spectrum neutrons is less than for 50-kV X rays, it is clear that in both cases the postirradiation treatment enhances

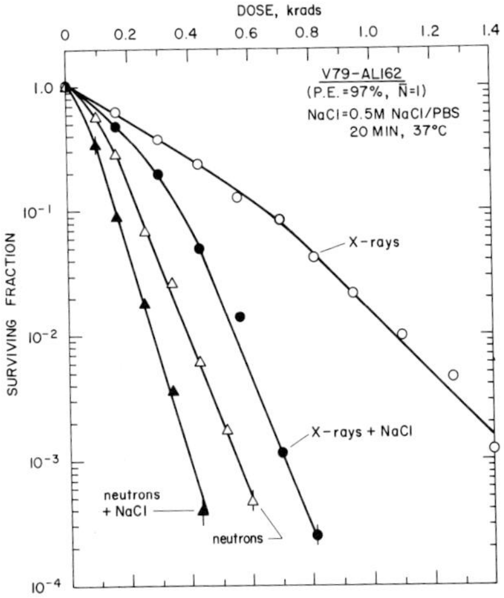

FIG. 8. Survival curves of V79-AL162 Chinese hamster cells following single doses of 50-kV X rays or fission-spectrum neutrons. Immediately after irradiation, cells were incubated in either isotonic PBS (○,△) or hypertonic (0.5 M NaCl) PBS (●,▲). The survival curve parameters are: X rays + isotonic PBS, $D_o = 148$ rads and $D_q = 280$ rads; X rays + hypertonic PBS, $D_o = 70$ rads and $D_q = 233$ rads; neutrons + isotonic PBS, $D_o = 68$ rads and $D_q = 83$ rads; and neutrons + hypertonic PBS, $D_o = 48$ rads and $D_q = 53$ rads. Other details as in Figure 2. (Reproduced from Utsumi and Elkind 1979b, with permission of *International Journal of Radiation Biology*.)

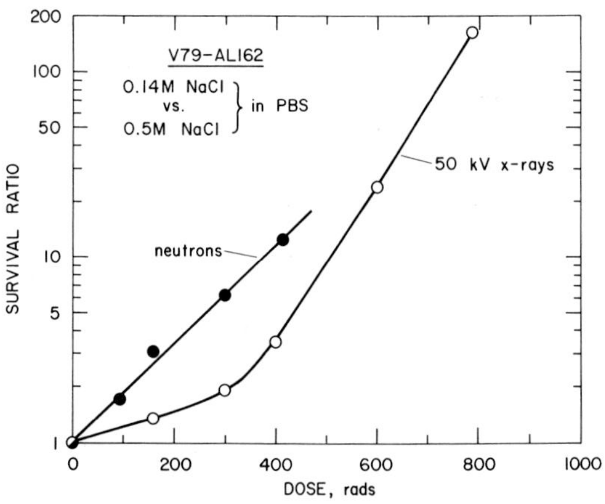

FIG. 9. Survival ratios as a function of dose for Chinese hamster cells treated immediately after irradiation with isotonic versus hypertonic PBS. The survival ratios are obtained from the data in Figure 8. (Reproduced from Utsumi and Elkind 1979b, with permission of *International Journal of Radiation Biology.*)

the expression in the mid-to-high dose ranges of what is otherwise only potentially lethal damage.

To decide whether or not the initial part of each curve in Figure 8 has remained unchanged, one may reason as follows. For each radiation, the survival ratio may be plotted as a function of dose. If the initial part of either survival curve is unaltered by the postirradiation treatment, the plot of the survival ratio would have a zero initial slope, indicating that in the small-dose region, survival is not reduced by hypertonicity. From Figure 9, which shows plots of this type, it seems clear that neither curve intersects the abscissa at a dose greater than zero. Thus, the inference follows that hypertonic postirradiation treatment results in enhanced lethality even in the dose region where survival is controlled by the first factor in equation 1. Results similar to those shown in Figures 8 and 9 have also been obtained with hypotonic posttreatment.

Two separate kinds of experiments have also been performed (Utsumi and Elkind, results to be reported) that support the foregoing inferences. In the first, synchronized cells irradiated in late S phase with X rays or with fission-spectrum neutrons were examined. In both instances, clear modifications of the initial parts of the survival curves are evident when the repair of potentially lethal damage is inhibited by anisotonicity or by other means (e.g., caffeine). In the second, a radiation-sensitive variant of V79 Chinese hamster cells was examined. The neutron and X-ray responses of these cells are modified neither in the high nor mid survival regions by the tonicity changes that are effective against parental cells, suggesting that the sensitivity of the variant results from an inability to repair the same sector(s) of damage whose expression is caused in parental cells by anisotonicity.

Hence, data obtained with repair-proficient and repair-deficient cells are internally consistent and lead to the conclusion that the "single-hit" survival factor $exp(-\alpha D)$, as well as the damage accumulation survival factor M(D) in equation 1, ordinarily reflect the net damage remaining after the repair of the portion that is only potentially lethal. As discussed elsewhere (Elkind 1977), this conclusion is not in conflict with hit-target theory. The conventional inference that $exp(-\alpha D)$, or, equivalently, a linear dose dependence, represents single-hit killing and, therefore, is not subject to biological modification corresponds to an interpretation of a limiting case of a more general mathematical theorem in which a specific accumulation of events in a target also yields an exponential survival curve.

Lastly, as has been noted, by 60 minutes after an acute X-ray dose, the *entire* survival curve is no longer modifiable by anisotonicity. Hence, one may infer that low dose-rate irradiation involving exposures of the order of, or longer than, 60 minutes will result in survivals only minimally modifiable by anisotonicity. From this inference, the more general implication may be drawn that the initial slope of a survival curve could become shallower with decreasing dose rate. In a given case, if the expression of survival involves postirradiation conditions that limit the amount of potentially lethal damage that is repaired, or that limit the rate of repair, then at a reduced dose rate more repair could occur and, as a result, α in equation 1 could decrease.

Cell Killing and DNA Damage Due to Nonionizing Radiation

The characteristic discrete and random nature of energy deposition by ionizing radiation is associated with absorption events large enough to break any molecular bond. Thus, the structure evident in an age-response pattern (see Figure 2) reflects the age-dependent ability of cells to cope with radiation damage in sensitive targets rather than an age-dependent variation in the amount of energy deposited in such targets. The correlation of X-ray resistance with the DNA synthetic phase, the potentiation of X-ray killing by DNA interactive chemicals like actinomycin D (Elkind et al. 1964, 1967, 1968a), and the sensitization that results from the incorporation of halogenated pyrimidines into DNA (Djordjevic and Szybalski 1960, Erickson and Szybalski 1961a, 1961b, Mohler and Elkind 1963, Shipley et al. 1971) all point to an involvement of DNA in cell killing by ionizing radiation.

However, DNA damage is also implicated in lethality due to other radiations. For example, DNA damage is very likely involved in lethality due to far-ultraviolet (UV) (254-nm) light (Rauth and Whitmore 1966, Setlow and Setlow 1972, Cleaver 1974, Han et al. 1975), in part, at least, because it is readily absorbed by nucleic acids and produces copious quantities of DNA photo products (Table 1). And the photosensitivity that results from replacement of thymidine by 5-bromodeoxyuridine (BUdR) appears clearly related to DNA damage because BUdR is incorporated exclusively into DNA and because light exposure results in extensive DNA damage (Table 1; also see Figure 12). For these reasons, it is instructive to consider the interactive effects of nonionizing and ionizing radiations in the context of damage-repair processes.

With cells synchronized by the mitotic selection technique (i.e., zero hours

TABLE 1. *Summary of cell and DNA damage-repair data*

Treatment	Surv. Curve	SLD Repair	PLD Repair	SSB/D_o	SSB Repair	SSB in Survivors	DSB/D_o	DSB Repair	DSB in Survivors	$\overline{TT}$ per SSB
Far-UV (254 nm)	thresh-old	slow (next cycle)	none	~100*	slow to none	?	<<SSB	probably slow	?	~5,000*
BUdR/FL	expo-nential	none	none	~50,000†	rapid	?	~250‡	probably rapid	?	?
X rays	thresh-old	$T_{1/2} \simeq$ 90 min	$T_{1/2} \simeq$ 15 min	~1,000§	rapid	<1:300‖ unre-paired	~40§	probably rapid	<1:12‖ unre-paired	~5 #

SLD = sublethal damage; PLD = potentially lethal damage expressible by postirradiation treatment with anisotonic PBS (Utsumi and Elkind 1979a,b).

SSB = single-strand break (i.e., alkali-labile lesion); DSB = double-strand break; $\overline{TT}$ = thymine-containing dimer; BUdR/FL, the fluorescent light exposure of cells containing template DNA labeled with 5-bromodeoxyuridine.

* See Elkind and Han (1978).

† See Ben-Hur and Elkind (1972b).

‡ Estimated from the number of single-strand DNA breaks in the strand complementary to the one labeled with BUdR (Hagan and Elkind 1979).

§ See Elkind and Redpath (1977).

‖ See Elkind (1978).

\# Based on data obtained with bacteria (Wang and Smith 1979) and translated to mammalian cells with the assumption that photoreactivatable lesions are approximately equal to thymine-containing dimers.

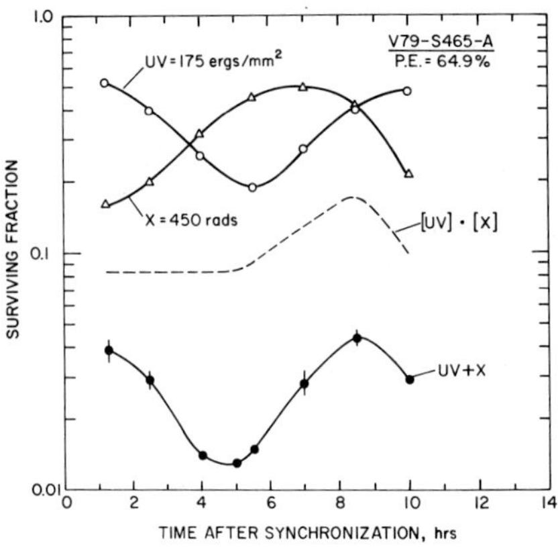

FIG. 10. Survival of Chinese hamster cells after single or combined exposures to UV light and X rays as a function of position in the growth cycle. From the top down: single-dose UV survival, O; single dose X-ray survival, △; predicted survival through the growth cycle after a combination of UV plus X rays based upon independent action, [UV]·[X]; and observed survival versus cell age from a combined treatment UV + X, ●. Other details as in Figure 2. (Reproduced from Han and Elkind 1977, with permission of *International Journal of Radiation Biology.*)

corresponding to the beginning of the cycle in this case), Figure 10 shows the relationship between the X-ray and the far-UV light age-response patterns for survival. Both irradiations suggest an involvement of DNA because of the large changes during the S phase (~ 2–9 hours), but because the patterns are almost mirror images of each other, the action of the two radiations on DNA must differ.

To examine the possibility of interactive damage, the age responses of these radiations must be taken into account. The dashed curve in Figure 10 is obtained by forming the product of the individual survivals as a function of time and predicts the combined age-response if the two radiations act independently. The lowest curve shows, in fact, what is observed.

For synchronized cells at their mid-S phase, Figure 11 shows how the

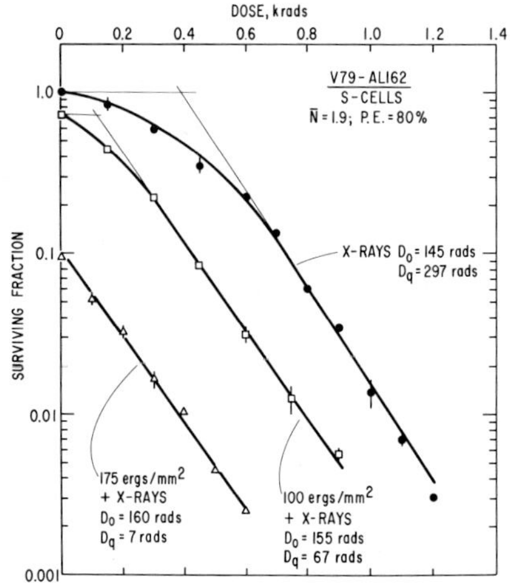

FIG. 11. X-ray survival curves for synchronized Chinese hamster cells irradiated in the middle of the S period (e.g., at 5 hours in Figure 10). Graded X-ray doses were administered immediately after the UV doses shown. Other details as in Figure 2. (Reproduced from Han and Elkind 1977, with permission of *International Journal of Radiation Biology.*)

X-ray survival curve is altered by a first exposure to far-UV light. These data indicate that the shoulder of the X-ray survival curve is progressively removed by increasing doses of this nonionizing radiation. Experiments in which the effect of X rays on the far-UV light survival curve was examined (Han and Elkind 1977) and on the rate of repair of interactive damage (Han and Elkind 1978) support the interpretation that nonlethal far-UV damage is operationally equivalent to sublethal X-ray damage.

Although at a molecular level effects of far-UV light differ from those of X rays, most notably in the production of pyrimidine dimers in DNA, far-UV light also differs in its lesser ability to produce DNA alkali-labile lesions (or single-strand breaks; see Elkind and Han 1978, and Table 1). In contrast, the fluorescent light (FL) exposure of cells containing BUdR results in much higher yields of DNA breaks, as shown in Figure 12. The rates of induction of single-strand breaks per 2×10^8 daltons are compared in Figure 12 by normalizing dose in each case by the corresponding D_o dose. Although BUdR has a small effect on the X-ray breakage efficiency (Ben-Hur and Elkind 1972b), a major difference in the induction of breaks by BUdR/FL compared to X rays is nevertheless apparent. A second point of importance is the fact that one line suffices to fit all of the BUdR/FL data when dose is expressed in units of D_o. Growing cells in 1×10^{-5} M BUdR results in an appreciably steeper FL survival curve than growing cells in 1×10^{-6} M BUdR (Ben-Hur and Elkind 1972a). But when the consequent change in D_o is taken into account, the DNA breakage data are fitted by a single line (Figure 12). Thus, a connection between DNA breakage and cell killing is implicated.

In relation to the effects of nonionizing radiations, the foregoing may be summarized as follows.

1. Operationally, far-UV light nonlethal damage is equivalent to sublethal X-ray damage.
2. BUdR/FL cell killing very likely results from DNA lesions and may be

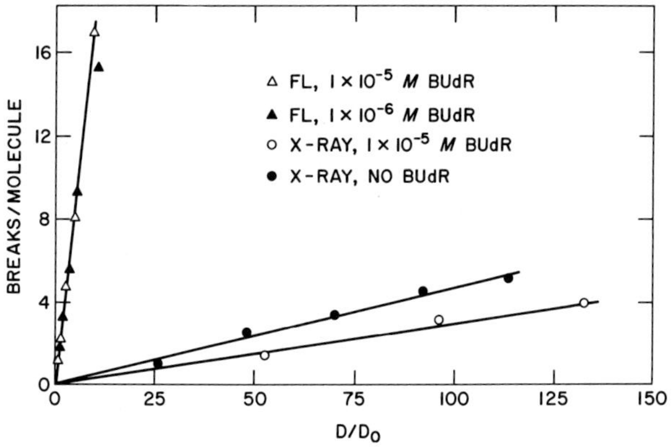

FIG. 12. DNA single-strand breakage (per 2×10^8 daltons) as a function of dose normalized by the D_o dose for flourescent-light (FL) exposure of cells grown for ~ 20 hours in BUdR, and for cells X irradiated with and without prelabeling with BUdR. (Reproduced from Ben-Hur and Elkind 1972b, with permission of the Rockefeller University Press.)

due to single-strand breaks or to a lesion in DNA induced in proportion to the induction of single-strand breaks.

3. BUdR/FL is appreciably more efficient than X rays per unit amount of cell killing in producing single-strand breaks in DNA.

Anisotonicity versus Lethality due to Nonionizing Radiation

In view of the distinction already made between potentially lethal and sublethal damage due to ionizing radiation, it is relevant to consider whether nonionizing radiation cell killing is modifiable by anisotonic PBS postirradiation treatment. Figure 13 shows that after a far-UV light dose, or after a near-UV exposure of cells containing BUdR, essentially no alteration in survival is effected by treatment with PBS solutions containing NaCl concentrations of from 0.01 M to 0.5 M. (The dashed curve represents survival following 807 rad X rays, Figure 4, which produces about the same survival for isotonic PBS as do the nonionizing radiations.) Further, it has been shown that 0.5 M NaCl PBS has essentially no effect on the far-UV and BUdR/near-UV survival curves (Utsumi and Elkind 1979b).

Thus, the following points are suggested:

1. Anisotonicity-related, potentially lethal damage is not involved in far-UV light lethality. This point supports the differences noted between X-ray-induced potentially lethal and sublethal damage and the operational equivalence between nonlethal far-UV light damage and sublethal X-ray damage.

2. Anisotonicity-related, potentially lethal damage is not involved in BUdR/near-UV light (or, equivalently, BUdR/FL) lethality. Since BUdR/FL lethality very likely results from DNA lesions, and in view of the efficiency with which these lesions are produced (see Figure 12), one may infer that either BUdR/FL-induced DNA lesions differ in some important respect from X-ray-induced

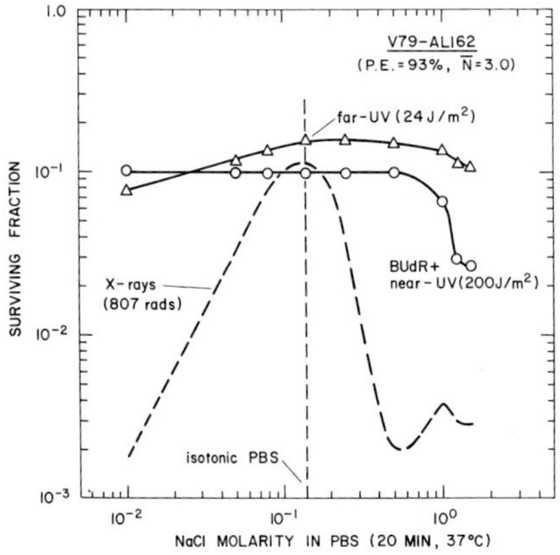

FIG. 13. The influence of postirradiation treatment with anisotonic PBS on the survival of Chinese hamster cells following 24 J/m² of far-UV light (primarily 254 nm) or 200 J/m² of near-UV light (filtered, Westinghouse sun lamps). In the latter instance the cells had been made light sensitive by prior growth in BUdR (see text). Other details as in Figure 2. (Reproduced from Utsumi and Elkind 1979b, with permission of *International Journal of Radiation Biology.*)

lesions or the modes of killing of these two treatments differ (Hagan and Elkind 1979).

DNA DAMAGE AND REPAIR

Several generalizations may be stated about cell killing by ionizing radiation.

First, repair-competent cells are able to repair potentially lethal damage starting essentially from levels of damage corresponding to zero dose. This is the case not only for sparsely ionizing radiation such as X rays and γ rays, but also for a more densely ionizing radiation, fission-spectrum neutrons. Therefore, biophysical theory intended to connect biological action with microdosimetry (Kellerer and Rossi 1972) will underestimate target sizes (Elkind 1977) unless the effects of repair are taken into account.

Second, cells exposed to ionizing radiation suffer both potentially lethal and sublethal damage. The damage accumulation mode of cell killing [M(D) in equation 1] may result from the saturation of a repair system by damage otherwise only potentially lethal or from the radiation inactivation of a repair system of potentially lethal damage (Alper 1977). Since the potentially lethal damage expressible by anisotonicity operationally appears to be unrelated to sublethal damage, this suggests that there are classes or types of potentially lethal damage that are not the same as the potentially lethal damage operationally equivalent to sublethal damage.

Third, if it is accepted that lesions in DNA give rise to both potentially lethal and sublethal damage, it is likely that one or both of the following hold: the same types of lesions act in different targets, or different lesions act in the same targets.

To place the preceding in perspective, Table 1 summarizes the results of damage-repair studies relative to cells and DNA. Although not all the same kinds of information are available for each radiation, certain relationships that are consistent and some that are not are suggested. (Cell cycle fluctuations in response and variations of similar magnitude are considered second-order effects for purposes of the discussion to follow.)

Consistencies

1. The ability of cells to repair sublethal damage correlates with the presence of a shoulder on the survival curve.
2. For each radiation, the number of DNA single-strand breaks (alkali-labile lesions) induced per D_o is too large for one such lesion to be *the* hit produced by a "mean lethal dose." (See Elkind and Whitmore 1967 for a discussion in the context of hit-target theory.)
3. With the exception of far-UV light (because the data are not available), DNA double-strand breaks are also produced in considerable excess per D_o.

Inconsistencies

4. The ability of cells to repair potentially lethal damage does not correlate with the shape of the survival curve (compare far-UV light and BUdR/FL with X rays).
5. The number of single- or double-strand breaks in DNA per D_0 does not correlate with the ability, or inability, of cells to repair sublethal or potentially lethal damage (compare far-UV light and BUdR/FL).
6. If cells have an ability to repair DNA breaks, they are not necessarily able to repair cellular damage (compare far-UV light and BUdR/FL).

It is evident from the foregoing that systematic relationships between damage and repair at a cell level are not correlatable with damage-repair relationships at a molecular level.

An important set of entries on X-ray data in Table 1 refers to DNA breaks in surviving cells (Elkind 1978). Briefly, the estimates that fewer than 1 in 300 single-strand breaks and 1 in 12 double-strand breaks remain unrepaired were arrived at as follows: Because technical difficulties preclude the measurement of single-strand breaks by sedimentation methods unless supralethal X-ray doses that reduce the surviving fraction to appreciably less than 0.001 are used and because it is not possible to separate survivors from cells destined to die, DNA damage-repair measurements essentially trace the course of events in killed cells. Remarkably enough, as indicated in the table (see also Ben-Hur and Elkind 1972b, Han and Elkind 1977), the repair of single-strand breaks, and probably of double-strand breaks as well, is rapid.

To circumvent this difficulty in part, V79 cells were irradiated with repeated dose fractions of ~ 1000 rad until a dose of 50.8 krad was accumulated. Between each fraction, the cells were incubated for several days, permitting the nonsurvivors to lyse and the survivors to repopulate. From an analysis of the sedimentation patterns obtained with the further irradiation of the progeny of the 50.8-krad survivors compared to those from control cells, it was possible to arrive at an upper-limit estimate of the proportion of unrepaired single-strand breaks that were propagated. Also, because one double-strand break is produced for every 25 single-strand breaks, an upper limit estimate could be made for the persistence of double-strand breaks. Thus, for X rays a further point in Table 1 is that most, if not all, of the breaks in the DNA of the progeny of surviving cells are repaired. Since the repair in nonsurvivors is completed before the first division after irradiation (Elkind and Han 1978), repair is probably equally as rapid in surviving cells. Consequently, probably few, if any, breaks are propagated.

DNA DAMAGE AND LETHALITY

In spite of the differences evident among the damage-repair processes just reviewed, some inferences about X-ray killing may be drawn for biophysical modeling. These are only qualitative or semiquantitative inferences, since it is of questionable usefulness to offer detailed speculations before additional data are available.

To begin with, the properties of potentially lethal damage and its repair are sufficiently different from sublethal damage and its repair to suggest, as already noted, that either different targets or different lesions in the same target are involved. As the following indicates, at present it is not possible to choose between the two.

Hypotonic and hypertonic buffer solutions (see Figure 5) have the same qualitative effect on cell survival. Thus, a mechanism of expansion and contraction of some vital structure(s) related to the genome is implicated in the inhibition of the repair of potentially lethal damage. Such a structure could be the nuclear envelope and/or the protein matrix of the nucleus (Berezny and Coffey 1975) to which the genome may be attached at different points. I refer to these proposed points as "anchor points."

The kinetics of repair of single-strand DNA breaks (Ben-Hur and Elkind 1972b, Han and Elkind 1977) is close to that for the repair of potentially lethal damage (see Figure 6). Hence, it is a possiblity that the repair of strand continuity returns the DNA to a state in which anisotonic treatment—stretching or contraction of DNA-related structures—is without effect. Alternatively, the anchor point itself, or the DNA in it, could be compromised by the anisotonic treatment while the DNA is being repaired, with the result that survival is decreased.

Actinomycin D affects the X-ray survival curve of Chinese hamster cells (Elkind et al. 1967, 1968a) in a manner qualitatively similar to far-UV light (see Figure 11). As a result of its intercalation into DNA, actinomycin D introduces a strain in the duplex because each intercalation results in a small amount of unwinding (Waring 1968). Although the nature of the distortion differs, the pyrimidine dimers produced by far-UV light (Table 1) also distort the duplex by causing intrastrand straightening and associated local denaturation (Hagen et al. 1965, Salgnik et al. 1967). The influences on the X-ray response of actinomycin D and far-UV light persist for appreciably longer periods than does the susceptibility of cells to the enhanced expression of potentially lethal damage (several to many hours compared to less than one hour). In contrast, the kinetics of the repair of single-strand X-ray breaks is only slightly influenced, if at all, by either actinomycin D (Elkind and Chang-Liu 1972b) or far-UV light (Han and Elkind 1977). These various factors all point to an effect on DNA whose repair is appreciably slower than the repair of strand discontinuities. Furthermore, this effect is not influenced by anisotonicity. Hence, a lesion different from a strand break that is sublethal and that leads to a distortion of the genome is suggested, presumably at the level of the secondary structure of DNA.

MISREPAIR AND LETHALITY

A last major point in Table 1 involves the excessive amount of DNA damage per D_0 and the indications that single- and double-strand breaks may be completely repaired in cells surviving X radiation. This means that, to survive, a cell must repair a very large number of breaks (not to mention other lesions). To illustrate, a D_0 dose reduces survival by at most a factor 1/e [expression 2b for equation 1 assumed]. However, statistically it is not possible that the genomes of essentially all of the survivors of a D_0 dose increment escape suffering large numbers of lesions. Hence, repair of DNA damage appears to be required for survival and must be effected with enough fidelity to be compatible with proliferative ability.

From the foregoing, the notion arises that lethality may be the result of *misrepair.* That is, the presumed complete restitution of DNA breaks following an X-ray dose occasionally results in a configuration that is incompatible with sustained proliferation. Since ~1,000 single- and ~40 double-strand breaks (plus other lesions) are registered per D_0, the frequency of misrepair would not have to be large. An alternate hypothesis compatible with the apparent large excess of DNA damage per D_0 is that only a small fraction of the genome, ~1/1000th for a single-strand break or ~1/40th for a double-strand break, is sensitive to ionizing radiation. The detection of repair or its absence in a small fraction of the genome is probably beyond the capabilities of currently available methods.

Of the two possibilities just described, misrepair is favored because of the following additional observations. After high-survival doses of X rays (100–800 rads), the sedimentation of DNA released from mammalian cells is altered discontinuously in manner analogous to the insertion of a single-strand break in a closed circular superhelical duplex (Elkind 1971, Elkind and Chang-Liu 1972a, Cook and Brazell 1975, 1976a). Cells can reverse the changes involved in this process as rapidly (Elkind and Chang-Liu 1972a, Cook and Brazell 1975) as they can repair potentially lethal damage. From target considerations, the molecular size of this structure is estimated to be around 3×10^9 daltons. A unit of about this size has been identified in the maturation of newly synthesized DNA (Hagan and Elkind, unpublished results). Cook and Brazell (1975, 1976a, 1976b) have shown that the sedimentation of this material in the presence of DNA intercalators is reminiscent of the sedimentation of a circular supercoiled DNA structure (Waring 1970), and they have proposed that this results from a "topological constraint" that prevents the free rotation of this DNA unit (Cook and Brazell 1975). The number of such units per cell would be about the same as the number of single-strand breaks per D_0. The points of topological constraint could also be the genome anchor points to the nuclear protein matrix. Hence, in the range of biologically relevant doses, a large fraction of these constrained DNA units would suffer a single-strand break, and since these units appear to be completely repairable (Cook and Brazell 1975), an infrequent misrepair could be a likely cause of reproductive death. Anisotonic treatment immediately after irradiation could enhance lethality by increasing the frequency of misrepair. Alternatively, the anchor points, or points of topological constraint, themselves could represent the target equivalent to the small fraction of the genome in which a hit is potentially lethal. Misrepair, due to an accumulation of sublethal damage—damage which distorts the duplex and, hence, the tertiary of higher-order structure of these ~3×10^9 dalton units—would also seem to be a reasonable cause of cell death.

PERSPECTIVE AND PROSPECTS

Radiation biology has reached a critical stage both in its importance in human affairs and in its position at the leading edge of fundamental science. On the one hand, society needs answers—answers to questions that limit the effectiveness of cancer therapy and answers to questions of the degree of the hazard to people from widespread radiation exposures (from nuclear power, diagnostic X rays, etc.). On the other hand, the fundamental insights from which such answers will come are not likely until the molecular and cell biology of the genome—its structure, replication, segregation, and control of phenotypic expression including cell proliferation—is more clearly understood. Thus, at this juncture, radiation biology, as well as other areas of science, characterizes an ultimate duality; that is, the strong coupling between *fundamental science* and improvements in *human welfare.*

ACKNOWLEDGMENTS

Over the years that I have worked in radiobiology, my research benefited from collaboration with a number of scientists. In the order of my first contact with them, my principal collaborators were C. A. Beam, R. Z. Lockart, Jr., T. Alescio, A. Han, G. F. Whitmore. W. K. Sinclair, F. Mauro, H. R. Withers, K. Sakamoto, W. U. Shipley, E. Kano, E. Ben-Hur, B. Bronk, R. D. Ley, B. F. Kimler, F. Q. H. Ngo, M. Hagan, and H. Utsumi. The specific illustrative material discussed comes from studies with E. Ben-Hur, A. Han, and H. Utsumi. Some of the research described, as well as the writing of this review, was supported by the United States Department of Energy.

REFERENCES

Alper, T. 1977. Elkind recovery and "sub-lethal damage": A misleading association? Br. J. Radiol. 50:459–467.

Belli, J. A., and M. Shelton. 1969. Repair by mammalian cells in culture. Science 165:490–492.

Ben-Hur, E., B. Bronk, and M. M. Elkind. 1972. Thermally enhanced radiosensitivity of cultured Chinese hamster cells. Nature New Biol. 238:209–210.

Ben-Hur, E., and M. M. Elkind. 1972a. Survival response of asynchronous Chinese hamster cells exposed to fluorescent light following 5-bromodeoxyuridine incorporation. Mutat. Res. 14:237–245.

Ben-Hur, E., and M. M. Elkind. 1972b. Damage and repair of DNA in 5-bromodeoxyuridine labelled Chinese hamster cells exposed to fluorescent light. Biophys. J. 12:636–647.

Ben-Hur, E., M. M. Elkind, and B. V. Bronk. 1974. Thermally enhanced radioresponse of cultured Chinese hamster cells: Inhibition of repair of sublethal damage and enhancement of lethal damage. Radiat. Res. 58:38–51.

Berezney, R., and D. S. Coffey. 1975. Nuclear protein matrix: Association with newly synthesized DNA. Science 189:291–293.

Cleaver, J. E. 1974. Repair processes for photochemical damage in mammalian cells. Adv. Radiat. Biol. 4:1–75.
Cook, P. R., and I. A. Brazell. 1975. Supercoils in human DNA. J. Cell Sci. 19:261–279.
Cook, P. R., and I. A. Brazell. 1976a. Conformational constraints in nuclear DNA. J. Cell Sci. 22:287–302.
Cook, P. R., and I. A. Brazell. 1976b. Characterization of nuclear structures containing superhelical DNA. J. Cell Sci. 22:303–325.
Djordjevic, B., and W. Szybalski. 1960. Genetics of human cell lines. III. Incorporation of 5-bromo- and 5-iododeoxyuridine into the deoxyribonucleic acid of human cells and its effect on radiation sensitivity. J. Exp. Med. 12:509–531.
Dutreix, J., A. Wambersie, and E. Bounik. 1973. Cellular recovery in human skin reactions: Application to dose fraction number—overall time relationship in radiotherapy. Eur. J. Cancer 9:159–167.
Elkind, M. M. 1960. Cellular aspects of tumor therapy. Radiology 74:529–541.
Elkind, M. M. 1971. Sedimentation of DNA released from Chinese hamster cells. Biophys. J. 11:502–520.
Elkind, M. M. 1975. The initial shape of cell survival curves. A summary and review of the conference, *in* Proceedings of the 6th L. H. Gray Conference, 1974. The Institute of Physics and John Wiley and Sons, Bristol, pp. 376–388.
Elkind, M. M. 1976. Fractionated dose radiotherapy and its relationship to survival curve shape. Cancer Treatment Reviews 3:1–15.
Elkind, M. M. 1977. The initial part of the survival curve: Implications for low-dose, low-dose-rate radiation responses. Radiat. Res. 71:9–23.
Elkind, M. M. 1978. DNA damage and mammalian cell killing, *in* DNA Repair Mechanisms (ICN-UCLA Symposia on Molecular and Cellular Biology, 1978), P. C. Hanawalt, E. C. Friedberg, and C. F. Fox, eds. Academic Press, New York, pp. 477–480.
Elkind, M. M., and C.-M. Chang-Liu. 1972a. Repair of a DNA complex from X-irradiated Chinese hamster cells. Int. J. Radiat. Biol. 22:75–90.
Elkind, M. M., and C.-M. Chang-Liu. 1972b. Actinomycin D inhibition of repair of a DNA complex from Chinese hamster cells. Int. J. Radiat. Biol. 22:313–324.
Elkind, M. M., and A. Han. 1978. DNA single-strand lesions due to "sunlight" and ultraviolet light: A comparison of their induction in Chinese hamster and human cells, and their fate in Chinese hamster cells. Photochem. Photobiol. 27:717–724.
Elkind, M. M., and J. L. Redpath. 1977. Molecular and cellular biology of radiation lethality, *in* Cancer: A Comprehensive Treatise, Vol. 6, F. F. Becker, ed. Plenum Publishing Corp., New York, pp. 51–99.
Elkind, M. M., K. Sakamoto, and C. Kamper. 1968a. Age-dependent toxic properties of actinomycin D and X-rays in cultured Chinese hamster cells. Kinetics 1:209–224.
Elkind, M. M., and W. K. Sinclair. 1965. Recovery in X-irradiated mammalian cells, *in* Current Topics in Radiation Research, Vol. 1, M. Ebert and A. Howard, eds. North-Holland Publishing Co., Amsterdam, pp. 165–220.
Elkind, M. M., and H. A. Sutton. 1959. X-ray damage and recovery in mammalian cells in culture. Nature 184:1293–1295.
Elkind, M. M., and H. A. Sutton. 1960. Radiation response of mammalian cells grown in culture. I. Repair of X-ray damage in surviving Chinese hamster cells. Radiat. Res. 13:556–593.
Elkind, M. M., H. Sutton-Gilbert, W. B. Moses, and C. Kamper, 1967. Sub-lethal and lethal radiation damage. Nature 214:1088–1092.
Elkind, M. M., and G. F. Whitmore. 1967. The Radiobiology of Cultured Mammalian Cells. Gordon and Breach Science Publishers, New York, 600 pp.
Elkind, M. M., G. F. Whitmore, and T. Alescio. 1964. Actinomycin D: Suppression of recovery in X-irradiated mammalian cells. Science 143:1454–1457.
Elkind, M. M., H. R. Withers, and J. A. Belli. 1968b. Intracellular repair and the oxygen effect in radiobiology and radiotherapy. Front. Radiat. Ther. Oncol. 3:55–87.
Erickson, R. L., and W. Szybalski. 1961a. Molecular biology of human cell lines. I. Comparative sensitivity to X-rays and ultraviolet light of cells containing halogen-substituted DNA. Biochem. Biophys. Res. Commun. 4:258–261.
Erickson, R. L., and W. Szybalski. 1961b. Molecular biology of human cell lines. III. Radiosensitizing properties of 5-iododeoxyuridine. Cancer Res. 23:122–130.

Hagan, M., and M. M. Elkind. 1979. Changes in repair competency after 5-bromodeoxyuridine pulse-labeling and near-ultraviolet light. Biophys. J. 27:75–85.

Hagen, U., K. Keck, H. Körger, F. Zimmerman, and T. Lücking. 1965. Ultraviolet light inactivation of the priming ability of DNA in the RNA polymerase system. Biochim. Biophys. Acta 95:418–425.

Hahn, G. M., and J. B. Little. 1972. Plateau-phase cultures of mammalian cells: An in vitro model for human cancer. Curr. Top. Radiat. Res. 8:39–83.

Han, A., and M. M. Elkind. 1977. Additive action of ionizing and nonionizing radiations throughout the Chinese hamster cell cycle. Int. J. Radiat. Biol. 31:275–282.

Han, A., and M. M. Elkind. 1978. Ultraviolet light and X-ray damage interaction in Chinese hamster cells. Radiat. Res. 74:88–100.

Han, A., M. Korbelik, and J. Ban. 1975. DNA-to-protein crosslinking in synchronized HeLa cells exposed to ultraviolet light. Int. J. Radiat. Biol. 27:63–74.

Kellerer, A. M., and H. H. Rossi. 1972. The theory of dual radiation action. Curr. Top. Radiat. Res. 8:85–158.

Little, J. B. 1969. Repair of sub-lethal and potentially lethal radiation damage in plateau phase cultures of human cells. Nature (London) 224:804–806.

Mohler, W., and M. M. Elkind. 1963. Radiation response of mammalian cells grown in culture. III. Modification of X-ray survival of Chinese hamster cells by 5-bromodeoxyuridine. Exp. Cell Res. 30:481–491.

Phillips, R. A., and L. J. Tolmach. 1966. Repair of potentially lethal damage in X-irradiated HeLa cells. Radiat. Res. 29:413–432.

Rauth, A. M., and G. F. Whitmore. 1966. The survival of synchronized L cells after ultraviolet irradiation. Radiat. Res. 28:84–95.

Salgnik, R. I., V. F. Drevick, and E. A. Vasyuninia. 1967. Isolation of ultraviolet-denatured regions of DNA and their base composition. J. Mol. Biol. 30:219–222.

Setlow, R. B., and J. K. Setlow. 1972. Effects of radiation on polynucleotides. Annu. Rev. Biophys. Bioeng. 1:293–346.

Shipley, W. U., M. M. Elkind, and W. B. Prather. 1971. Potentiation of X-ray killing by 5-bromodeoxyuridine in Chinese hamster cells: A reduction in capacity for incurring sublethal damage. Radiat. Res. 47:437–449.

Sinclair, W. K. 1967. Hydroxyurea: Effect on Chinese hamster cells grown in culture. Cancer Res. 27:297–308.

Sinclair, W. K., and R. A. Morton. 1964. Recovery following X-irradiation of synchronized Chinese hamster cells. Nature (London) 203:247–250.

Utsumi, H., and M. M. Elkind. 1979a. Potentially lethal damage versus sublethal damage: Independent repair processes in actively growing Chinese hamster cells. Radiat. Res. 77:346–360.

Utsumi, H. and M. M. Elkind. 1979b. Potentially lethal damage: Qualitative differences between ionizing and nonionizing radiation, and implications for "single-hit" killing. Int. J. Radiat. Biol. 35:373–380.

Wang, T.-C. V., and K. C. Smith. 1979. Enzymatic photoreactivation of *Escherichia coli* after ionizing radiation: Chemical evidence for the production of pyrimidine dimers. Radiat. Res. 76:540–548.

Waring, M. J. 1968. Drugs which affect the structure and function of DNA. Nature (London) 219:1320–1325.

Waring, M. J. 1970. Variation of the supercoils in closed circular DNA by binding of antibiotics and drugs: Evidence for molecular models involving intercalation. J. Mol. Biol. 54:247–279.

Radiation Biology in Cancer Research, edited by
Raymond E. Meyn and H. Rodney Withers.
Raven Press, New York © 1980.

Repair of Radiation Damage In Vivo

Raymond E. Meyn, David J. Grdina,* and Susan E. Fletcher

*Department of Physics and *Section of Experimental Radiotherapy, Department of Radiotherapy, The University of Texas System Cancer Center M. D. Anderson Hospital and Tumor Institute, Houston, Texas 77030*

Repair of the DNA strand breaks produced by ionizing radiation has been studied in great detail in cells in culture, and models describing the relationships between DNA damage, DNA repair, and cell survival have been proposed. Many of these models have been discussed by other speakers at this symposium (see chapters in this volume by Chapman, Cole, Painter, Rossi, Tobias, and Goodhead). However, since radiation response in the whole animal is known to differ significantly from the in vitro response and only a few comparable DNA repair studies have been attempted in in vivo systems, progress towards development of appropriate models for radiation response in vivo has been slow. This situation has been exacerbated by the lack of adequate methodology. Since most DNA damage assays require radioactively labeled DNA, investigations in vivo have been hampered by the difficulty of labeling animal tissues.

This problem has been circumvented by combining alkaline elution, a technique for the sensitive detection of DNA strand breaks, and a nonradioactive assay for DNA. Using this combined approach, we have begun an investigation of the degree of DNA damage produced by gamma rays in normal and tumor tissue irradiated in vivo.

MATERIALS AND METHODS

Female C_3Hf/Bu mice, 8 to 10 weeks old, from a specific pathogen-free breeding colony were used in these experiments. Fibrosarcoma (FSA) tumors were originally induced in C_3H mice by injection of methylcholanthrene (Suit and Suchato 1967). Tumors were stored in liquid nitrogen, and only tumors of the fifth generation were used. Tumor cells were injected subcutaneously into the hind legs of mice and experiments were performed with tumors about 12 mm in diameter.

Chinese hamster ovary (CHO) cells were used in some experiments. These cells were grown in culture in McCoy's 5A medium supplemented with 20% fetal calf serum. Cultures were incubated at 37°C in a 5% CO_2, 95% air, water-saturated environment. Cells were harvested for experiments by trypsinization (0.025% trypsin) for five minutes at 37°C. Mice were whole-body irradiated with various doses of ^{137}Cs-γ-rays at a dose rate of 295 rad/min. CHO cells and normal cells and tumor cells from mice were irradiated in suspension using a high dose-rate (5,500 rad/minute) ^{137}Cs source.

Immediately following irradiation, mice were sacrificed and placed in an ice bath. The tumor was removed and finely minced with ophthalmic scissors. The mince was added to a beaker of 0.025% trypsin in solution A (8.0 g NaCl, 0.4 g KCl, 1.0 g glucose, and 0.35 g $NaHCO_3$ per liter) containing 5 mM EDTA and stirred for 20 minutes at room temperature. Control experiments using cultured cells showed that no DNA repair took place under these conditions. After stirring, the undigested tissue was allowed to settle to the bottom of the beaker and the upper two thirds of the suspension was removed and mixed with an equal volume of McCoy's medium containing 20% fetal calf serum. The suspension was then passed through a stainless steel mesh (200 wires/inch) and centrifuged three times at 225 *g* for five minutes. The remaining cell pellet was resuspended in medium. The cells were counted with a hemacytometer. Viability was determined by trypan blue exclusion to be routinely greater than 95%. The yield of viable cells was about 10^8 per gram of excised tumor tissue.

Bone marrow was obtained from the same tumor-bearing mice by expelling the cells with medium forced through a 26-gauge needle from the tibias and femurs into a beaker (Milas and Mujagic 1973). The cell viability was greater than 90%.

The alkaline elution technique used was that of Kohn et al. (1976). Three to four million cells were filtered onto 47 mm diameter 2-μm pore polyvinylchloride filters. After several washes with cold phosphate-buffered saline, the cells were lysed with 10 ml of 2 M NaCl, 0.04 M EDTA, and 0.2% Sarkosyl (pH 10.0). The lysis solution was allowed to flow through the filter by gravity. The filter was then washed with 10 ml of 0.02 M EDTA (pH 10.3) and eluted in the dark with 0.1 M tetrapropylammonium hydroxide and 0.02 M EDTA (pH 12.1). A pump speed of 0.05 ml/minute was used. Fractions were collected every 90 minutes for 15 hours. DNA retained on the filter was removed by extensive vortexing in 5 ml of the eluting solution. The nonradioactive assay used was essentially that published by Parodi et al. (1978). Briefly, 1-ml samples were removed from each fraction, and 100 μl each of 100% TCA and 0.2% bovine serum albumin was added. The resulting precipitate was centrifuged and reprecipitated twice with cold 95% ethanol containing 0.1 M potassium acetate. The final precipitate was dried and 30 μl of a 40% aqueous solution of 3,5-diaminobenzoic acid dihydrochloride was added. The samples were then incubated for 30 minutes at 70°C. We added 1.5 ml of 0.6 N perchloric acid to each sample and read the fluorescence at 520 nm with an excitation wavelength of 436 nm.

RESULTS

In order to use the alkaline elution technique as described by Kohn et al. (1976) for studies on cells from tissues irradiated in vivo, it was necessary to modify the method so that DNA from nonradioactive cells could be assayed. These modifications included using a larger filter (47 mm vs. 25 mm) so that more cells could be lysed to provide more DNA and thereby permit use of the microfluorometric DNA determination as described by Parodi et al. (1978). As a test of these modifications, CHO cells were labeled with ^{14}C-thymidine, irradiated with 1,000 rad, and subjected to the modified alkaline elution technique. The DNA in each fraction from the elution was assayed both by the microfluorometric determination and by liquid scintillation counting of the ^{14}C label. The results of this test are shown in Figure 1. The close agreement in the elution profiles using the two different DNA assay procedures validates this modified technique.

The degree of DNA strand breakage produced in FSA tumor cells and bone marrow cells irradiated in vivo were compared by irradiating a tumor-bearing mouse whole-body, preparing cell suspensions of the tumor cells and bone marrow cells, and subjecting the cells to alkaline elution analysis. The results of this analysis are shown in Figure 2 along with control profiles for the DNA from cells prepared from comparable tissues from an unirradiated mouse. Some of the cells from these latter suspensions were irradiated with 1,000 rad in vitro and alkaline elution profiles for them are also shown in Figure 2.

Several important points are evident from the results presented in Figure 2. First, the profiles for the unirradiated controls, while reasonably good, are not quite as high as those for CHO cells in Figure 1. Thus, the cell suspension preparation methods may introduce some DNA damage. Second, cells irradiated in vivo have, in general, less DNA damage than do cells irradiated in vitro.

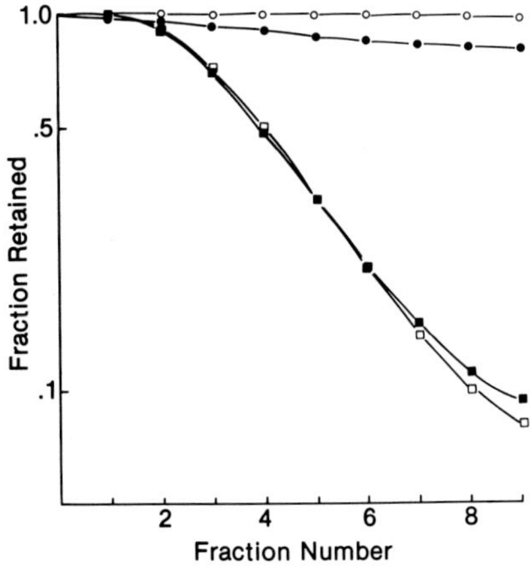

FIG. 1. A comparison of the radioactive (open symbols) and fluorometric (closed symbols) assays for DNA when applied to alkaline elution of DNA from unirradiated (circles) and γ-irradiated (squares) CHO cells.

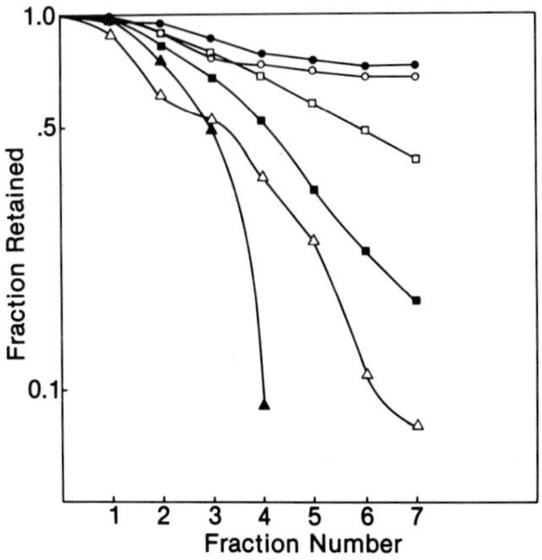

FIG. 2. Alkaline elution profiles for DNA from mouse fibrosarcoma tumor cells (open symbols) and bone marrow cells (closed symbols) irradiated with 1,000 rads either in vivo (squares) or in vitro (triangles). Profiles for unirradiated cells are also shown (circles).

Third, tumor cells have less DNA damage than do bone marrow cells irrespective of whether they were irradiated in vitro or in vivo.

The results from several experiments such as those shown in Figure 2, including some at 500 rad, are summarized in Figure 3 as plots of strand scission factor (SSF) versus dose. SSF is calculated by taking the ratio of the natural logarithms of the values of the fraction of DNA retained for the irradiated and control samples at the fifth fraction from the alkaline elution profiles. The plots shown in Figure 3 demonstrate a linear relationship between the degree of DNA strand breakage and radiation dose. Furthermore, calculation of the slopes of the lines for the various cells and irradiation conditions shown in Figure 3 allows a quantitative (although relative) indication of radiation sensitivity. That is, bone marrow cells are two to four times more sensitive than are tumor cells irrespective of whether they are irradiated in vitro or in vivo, and cells irradiated in vitro are two to four times more sensitive than cells irradiated

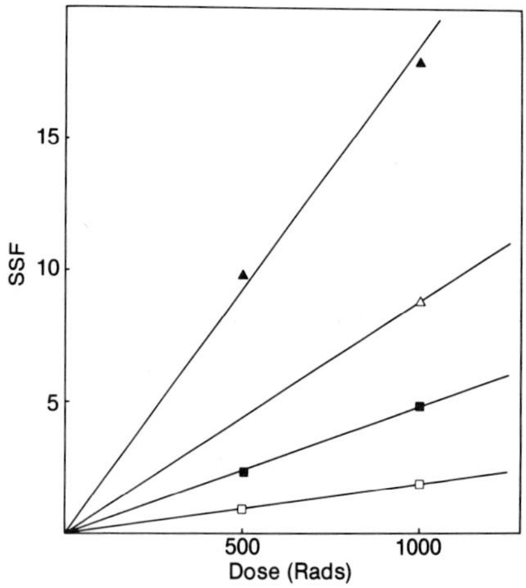

FIG. 3. Strand scission factor (SSF) as a function of dose in rads calculated from alkaline elution profiles as described in the text. The plots shown in this figure are for DNA from mouse fibrosarcoma tumor cells (open symbols) and bone marrow cells (closed symbols) irradiated either in vivo (squares) or in vitro (triangles).

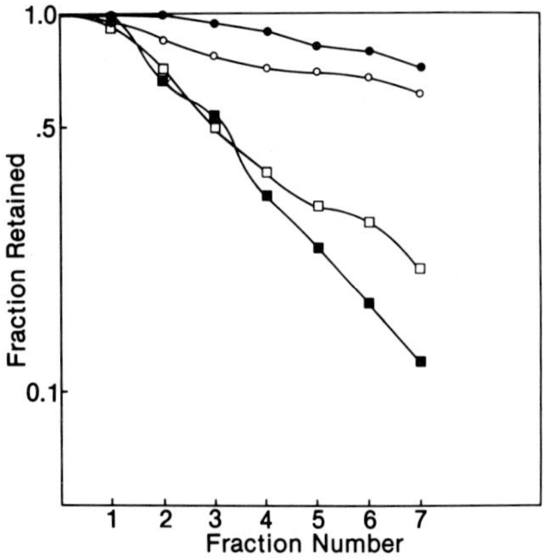

FIG. 4. Alkaline elution profiles for DNA from mouse FSA tumor cells (open symbols) and bone marrow cells (closed symbols) isolated from mice which were irradiated with 1,000 rads and either sacrificed immediately (squares) or one hour after irradiation (circles).

in vivo irrespective of whether the cells are of tumor or normal tissue origin.

The ability of FSA tumor and bone marrow cells to repair DNA strand breaks in vivo was tested by irradiating a tumor-bearing mouse whole-body with 1,000 rad, allowing the animal to live for one hour following irradiation, and then sacrificing the animal and subjecting the DNA from the cells to alkaline elution analysis. The results of this experiment (Figure 4) demonstrate that virtually all of the measurable DNA strand break damage is repaired in vivo under these conditions by both cell populations.

DISCUSSION

A review of the previous reports concerning repair of DNA strand breaks in vivo shows that, while cells from a variety of animal tissues have, in general, similar sensitivities in terms of efficiency of strand break production and capacity for rejoining, some important differences exist. Wheeler and Lett (1972) reported that single-strand breaks were produced with equal efficiency in the DNA of differentiated neurons, a permanently nondividing tissue, and of cultured dividing cell lines such as CHO. They further noted, however, that the rejoining process in the neurons was quite different from that in cultured cell lines. Karran and Ormerod (1973) investigated the production and rejoining of radiation-induced strand breaks in the DNA of cells from the thymus, spleen, bone marrow, and muscle of rats and the lymphocytes and erythrocytes of chickens. They found that, while certain cells from most of these tissues had some capacity for rejoining, the chicken erythrocytes did not. Ono and Okada (1974) compared the degree of DNA strand breakage and rejoining in mouse thymocytes and hepatocytes and described several important differences. The number of breaks was greater when the cells were irradiated in vitro compared to in vivo; and rejoining was detected in thymocytes in vivo and in vitro, but in hepatocytes only in vivo.

The DNA repair kinetics of rat cerebellar neurons and intracerebral 9L tumor cells were simultaneously compared in the same rat in the work of Wang and Wheeler (1978). The nondividing neurons repaired their radiation-induced DNA damage to the same extent as did the dividing tumor cells; however, they did so at a slower rate. This study by Wang and Wheeler represents the most extensive one to date of the possible differences in DNA strand break production and repair in normal and tumor tissue irradiated in vivo. The method of analysis that they used consisted of alkaline sucrose gradient sedimentation of the cellular DNA in zonal rotors followed by fluorometric determination of the amount of DNA per gradient fraction using a modification of the technique published by Kissane and Robins (1958). This use of zonal rotors provides a very sensitive and quantitative method for the determination of single-strand breaks in the DNA of either dividing or nondividing mammalian cells. An inherent limitation of this method is that only one DNA sample can be sedimented per rotor per run.

It was with this limitation in mind that we modified the alkaline elution technique as originally described by Kohn et al. (1976) for use with unlabeled DNA from cells of animal tissue origin. The modifications are similar to those described by Brambilla et al. (1977) and have been used by these authors to detect the DNA damage produced in mouse tissues in vivo by chemical carcinogens (Parodi et al. 1978). The basic alkaline elution technique itself has been shown in the extensive work of Kohn and co-workers (see review by Kohn 1978) to be a reliable, simple, and fast method for the detection of several types of DNA damage. A limitation of the method is that it does not yield molecular weight values as does alkaline sucrose gradient sedimentation; however, it is very sensitive and the DNA strand breaks produced by doses of ionizing radiation of as low as 30 rad can be easily detected.

Our preliminary results, presented here, demonstrate the utility of this approach when applied to questions concerning the sensitivity and DNA repair capacity of cells from tissues irradiated in vivo. The sensitivity of the overall technique is such that in other experiments (not shown) the effects of 200 rad in vivo were detected. One conclusion from the results presented in Figures 2 and 3 is that cells irradiated in vitro are more sensitive to ionizing radiation in terms of the amount of DNA damage produced than are the same cells when irradiated in vivo. This finding confirms the earlier work of Ono and Okada (1974) and may be due to differences in the cellular environment in these two irradiation conditions, which may influence the yield of DNA strand breaks. One of these environmental factors could be the oxygen tension, a factor known to affect the yield of DNA damage (Koch and Painter 1975). This effect may explain the difference in the yield of breaks shown in the FSA tumor cells, since this tumor has been shown to contain a large hypoxic fraction (Grdina et al. 1976). This idea would not explain the difference in the in vitro and in vivo results for the bone marrow cells, however, and we must consider that cell-cell interactions may also play a role. Such interactions have been shown to cause important modifications in radiation sensitivity (Durand and Sutherland 1972).

The difference in the yield of DNA strand breaks between the bone marrow cells and tumor cells when irradiated in vitro (Figures 2 and 3) was unexpected. It may be due to: (1) differences in fast-repair capabilities; (2) induction of DNA degradation in the bone marrow cells but not in the tumor cells, an effect reported for chicken erythrocytes (Karran and Ormerod 1973); or (3) a continued protective effect of the previous hypoxic state of the tumor cells. The first two possibilities seem to be ruled out by the finding that the alkaline elution profiles have returned to control levels within one hour after irradiation (Figure 4), but a more detailed analysis of the repair kinetics of these two cell populations is warranted. The last possibility is suggested by the work of Bedford and Mitchell (1974) and could perhaps be tested in model systems. In any event, it will be important to verify this difference by examining other mouse tissues.

In conclusion, it appears to us, on the basis of these preliminary results, that the application of the alkaline elution techniques to cells derived from animal tissues is valid and thus represents a new, sensitive, relatively simple and rapid method for investigating DNA damage and repair in vivo. Ultimately, we hope to extend these studies to other questions concerning the influence of cellular environment on radiation sensitivity, inherent differences in sensitivity from tissue to tissue, and relationships between DNA damage, repair, and cell survival.

ACKNOWLEDGMENTS

This investigation was supported by Grant Number CA 04484, awarded by the National Cancer Institute, Department of Health, Education and Welfare.

We gratefully acknowledge the technical assistance of Ms. Sandra Jones, Ms. Velma Holiday, and Ms. Mollie McCune. We also wish to thank Ms. Audrey Johnson for her help in preparation of this manuscript.

REFERENCES

Bedford, J. S., and J. B. Mitchell. 1974. The effect of hypoxia on the growth and radiation response of mammalian cells in culture. Br. J. Radiol. 47:687–696.

Brambilla, G., M. Cavanna, L. Sciaba, P. Carlo, S. Parodi, and M. Taningher. 1977. A procedure for the assay of DNA damage in mammalian cells by alkaline elution and microfluorometric DNA determination. Ital. J. Biochem. 26:419–427.

Durand, R. E., and R. M. Sutherland. 1972. Effects of intercellular contact on repair of radiation damage. Exp. Cell Res. 71:75–80.

Grdina, D. J., I. Basic, S. Guzzino, and K. Mason. 1976. Radiation response of all populations irradiated in situ and separated from a fibrosarcoma. Radiat. Res. 66:634–643.

Karran, P., and M. G. Ormerod. 1973. Is the ability to repair damage to DNA related to the proliferative capacity of a cell? The rejoining of X-ray-produced strand breaks. Biochim. Biophys. Acta. 299:54–64.

Kissane, J. M., and E. Robins. 1958. The fluorometric measurement of deoxyribonucleic acid in animal tissues with special reference to the central nervous system. J. Biol. Chem. 233:184–188.

Koch, C. J., and R. B. Painter. 1975. The effect of extreme hypoxia on the repair of DNA single-strand breaks in mammalian cells. Radiat. Res. 64:256–269.

Kohn, K. W. 1978. DNA as a target in cancer chemotherapy: Measurement of macromolecular DNA damage produced in mammalian cells by anticancer agents and carcinogens, *in* Methods in Cancer Research, vol. 16, pt. A, V. T. De Vita and H. Busch, eds. Academic Press, New York, pp. 291–345.

Kohn, K. W., L. C. Erickson, R. A. G. Ewig, and C. A. Friedman. 1976. Fractionation of DNA from mammalian cells by alkaline elution. Biochemistry 15:4629–4637.

Milas, L., and H. Mujagic. 1973. The effect of splenectomy on fibrosarcoma "metastases" in lungs of mice. Int. J. Cancer 11:186–190.

Ono, T., and S. Okada. 1974. Estimation in vivo of DNA strand breaks and their rejoining in thymus and liver of mouse. Int. J. Radiat. Biol. 25:291–301.

Parodi, S., M. Taningher, L. Santi, M. Cavanna, L. Sciaba, A. Maura, and G. Brambilla. 1978. A practical procedure for testing DNA damage in vivo, proposed for a pre-screening of chemical carcinogens. Mutat. Res. 54:39–46.

Suit, H. D., and D. Suchato. 1967. Hyperbaric oxygen and radiotherapy of fibrosarcoma and of squamous cell carcinoma. Radiotherapy 89:713–719.

Wang, T. S., and K. T. Wheeler. 1978. Repair of X-ray-induced DNA damage in rat cerebellar neurons and brain tumor cells. Radiat. Res. 73:464–475.

Wheeler, K. T., and J. T. Lett. 1972. Formation and rejoining of DNA strand breaks in irradiated neurons in vivo. Radiat. Res. 52:59–67.

Radiation Biology in Cancer Research, edited by
Raymond E. Meyn and H. Rodney Withers.
Raven Press, New York © 1980.

Direct Measurement of Chromosome Damage and Its Repair by Premature Chromosome Condensation

Walter N. Hittelman, Marguerite A. Sognier, and Arthur Cole*

*Departments of Developmental Therapeutics and *Physics, The University of Texas System Cancer Center M. D. Anderson Hospital and Tumor Institute, Houston, Texas 77030*

One of the main topics of this symposium is the effect of radiation on biological tissue. Radiation damage manifests itself in cells in a variety of ways, including the induction of mutations, cell transformation, DNA damage and repair, chromosome damage, cell cycle progression delay, loss of reproductive survival, and long-term effects. While we have learned a lot about these radiation effects in recent years, the common molecular events that may link these various manifestations of radiation damage are not well understood. The purpose of the following communication is to begin to provide an understanding of some of the molecular and cellular events that may form a common basis for the varied expression of radiation damage.

There is an extensive body of literature that suggests correlations between many of the reported cellular manifestations of radiation damage. For example, after treatment of cells with radiation or drugs, one can detect DNA damage and its repair, chromosome damage, and repair of sublethal and potentially lethal damage, as measured by survival studies. While these various processes probably have a common linkage, direct correlations have been technically difficult to provide between the various types of damage, their repair, and cell survival. Nevertheless, cells that have a particular DNA repair deficiency are often observed to exhibit increased levels of chromosome damage and decreased cell survival after treatment with the appropriate agent (e.g., Scott et al. 1974). In addition, Dewey and co-workers (1970) and Carrano (1973) have shown that there exists a close correlation between the level of chromosome aberrations observed at mitosis and cell survival.

On the other hand, DNA repair–deficient cell lines do not show increased chromosome damage and lowered cell survival in response to all damaging agents. The effect seems to be more specific. For example, cells derived from xeroderma pigmentosum patients are quite sensitive to ultraviolet radiation with respect to chromosome aberrations and cell survival; however, they are normal in their sensitivity to ionizing radiation (Cleaver 1977). This finding points out that if correlations are to be made between DNA repair capacity, chromosome

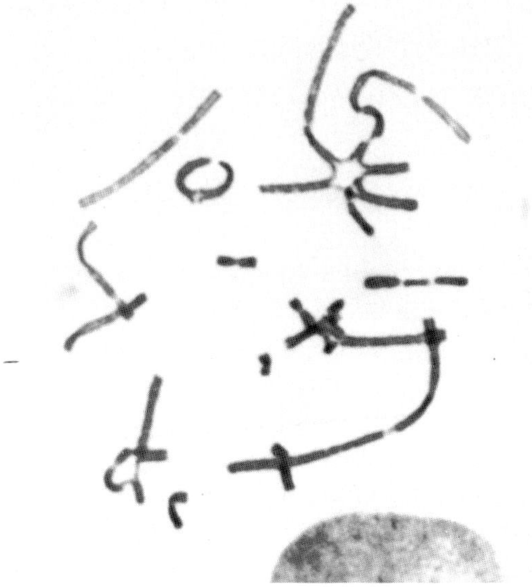

FIG. 1. CHO mitotic cell exhibiting numerous exchanges, breaks, and gaps.

damage, and cell survival, one particular type of lesion at a time must be focused on.

Our laboratory has been interested in elucidating the molecular mechanisms of chromosome aberration formation. Figure 1 illustrates the types of chromosome lesions we are interested in. They include chromosome intrachanges and interchanges, breaks, and gaps. An understanding of the underlying molecular changes that lead to chromosome lesions has been thwarted because basic chromosome structure has not yet been elucidated. However, recent studies have made great advances in this area (for review, see Olins and Olins 1978). The current model for chromosome structure is that the unit chromatin thread represents an association of histones and DNA into a pattern of beads on a string (nucleosomes). With further histone modification and acquisition of non-histone chromosomal proteins, this thread can be coiled into solenoids and further coiled into thick chromatin fibers and clumps. At mitosis, the chromatin masses are further coiled and are packaged into a structure that resembles chromatin threads tacked on to a protein scaffold (Laemmli et al. 1978). We have been interested, therefore, in determining how a lesion at the DNA or chromatin level can be translated into a chromosome aberration.

TECHNIQUES FOR MEASURING DNA AND CHROMOSOME DAMAGE

Our approach to this problem has been to try to identify the particular DNA lesions important in the formation of chromosome aberrations. One difficulty

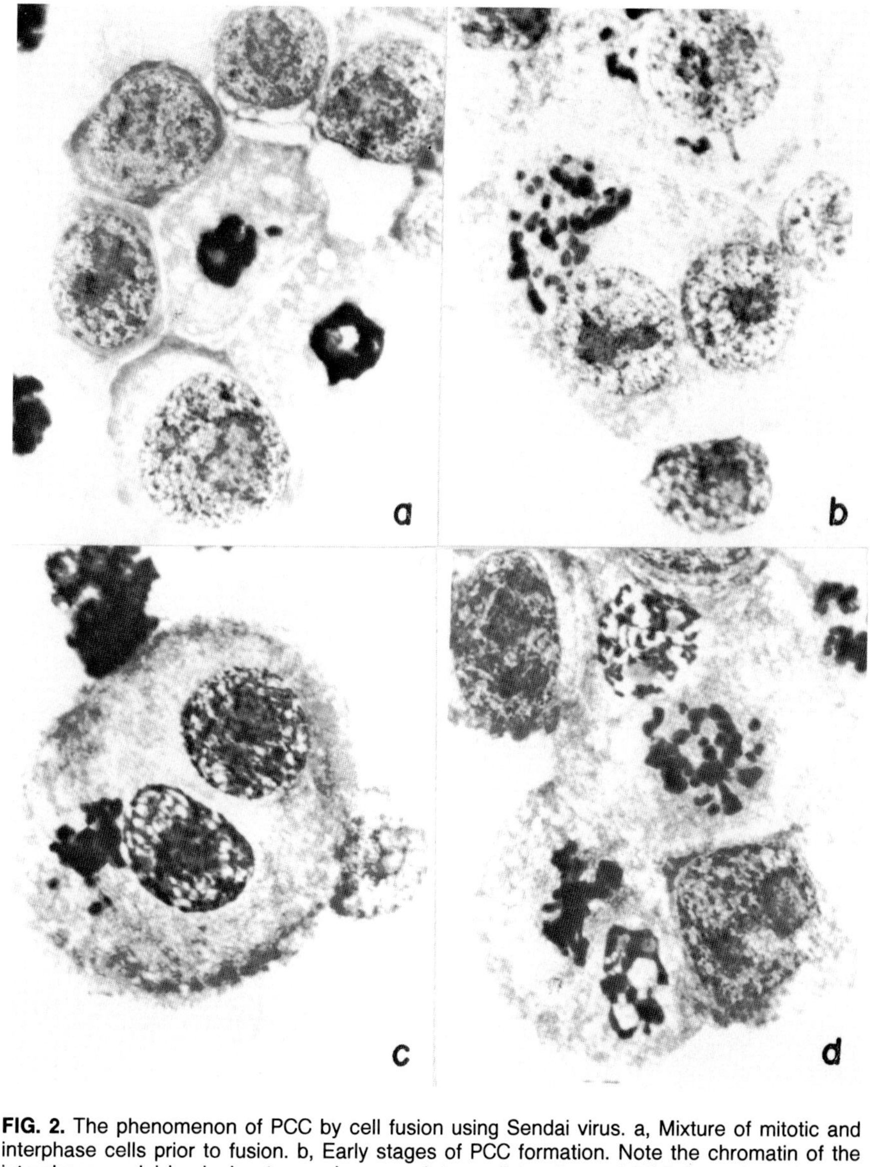

FIG. 2. The phenomenon of PCC by cell fusion using Sendai virus. a, Mixture of mitotic and interphase cells prior to fusion. b, Early stages of PCC formation. Note the chromatin of the interphase nuclei beginning to condense. c, Intermediate stage of PCC formation. Note increased chromatin condensation and breaking down of the nuclear membrane. d, Late stage of PCC formation. The interphase chromatin is condensed and the nuclear membrane is broken down. Note the continued association of the nucleoli with the PCC. Also note adjacent cells with uncondensed nuclei that were not involved in fusion with mototic cells.

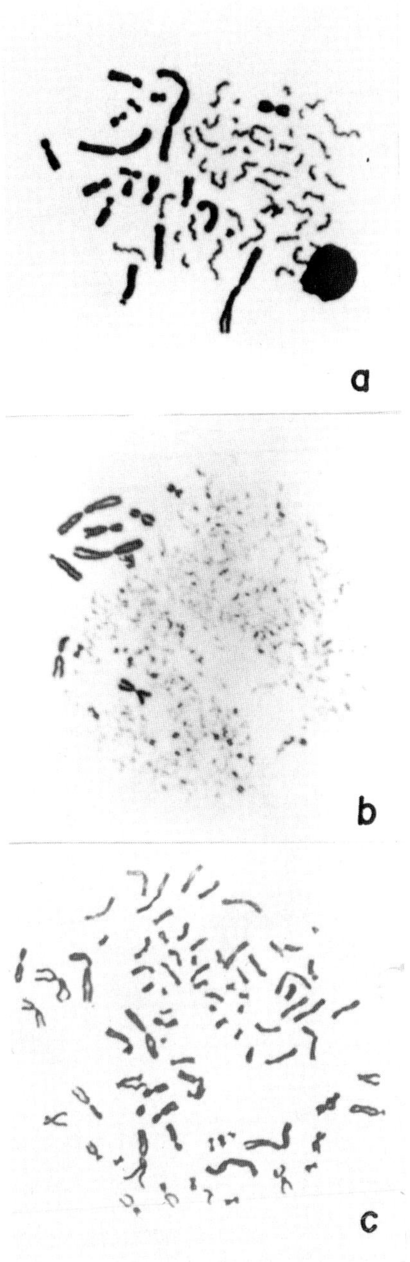

FIG. 3. PCC of human bone marrow cells from various stages of the cell cycle. a, G_1 PCC exhibiting a single chromatid per chromosome. b, S PCC exhibiting a pulverized appearance with both single (prereplicative) and double (postreplicative) elements. c, G_2 PCC exhibiting two chromatids per chromosome.

with this type of approach in the past has been that DNA damage and chromosome damage could not be viewed under the same experimental conditions at the same time. For example, the commonly used techniques for the measurement of DNA damage (alkaline and neutral sucrose gradients) are relatively insensitive, and massive radiation or drug doses are required. These massive doses obliterate chromosomes and make chromosome analysis impossible. At the same time, traditional cytogenetic techniques require cells to enter mitosis in order to be able to observe chromosome aberrations. Thus there is a time lag between the induction of damage in interphase cells and the visualization of damage when these cells reach mitosis. As a result, aberration frequencies observed in mitosis might not reflect the initial aberration frequencies due to repair of damage and cell progression delay of damaged cells. For these reasons, it has been difficult to make direct correlations between DNA damage and chromosome damage.

In an attempt to overcome these problems, we have turned to two recently developed techniques that now allow us to concurrently measure damage at both the DNA and chromosome levels. In order to measure damage at the DNA level, we have adopted the alkaline elution technique. The alkaline elution technique, as first described by Kohn and Ewig (1973) and lucidly described by Raymond Meyn in the preceding chapter, is very useful in these studies because it not only detects DNA damage at cytologically relevant doses, but it is capable of distinguishing various types of DNA lesions (breaks, alkali-labile sites, and DNA-DNA and DNA-protein cross-links) induced by a variety of agents. The procedures used in these experiments are essentially the same as those described by Meyn et al. (1980, see pages 95–102, this volume).

In order to look at chromosome damage directly in interphase cells, we have used the technique of premature chromosome condensation. In 1970, Johnson and Rao discovered that if a mitotic cell is fused with an interphase cell using Sendai virus, the interphase chromatin is induced to condense into discrete chromosome units, the prematurely condensed chromosomes or PCC (Figure 2). As shown in Figure 3, cells from G_1 phase give rise to PCC with one chromatid per chromosome, whereas G_2 PCC exhibit two chromatids per chromosome. S phase PCC appear pulverized, yet both single (prereplicative) and double segments (postreplicative) can be recognized (Johnson and Rao 1970).

RESULTS

Detection of Chromosome Aberrations by the PCC Technique

The first step in these studies was to show that aberrations could be visualized directly in the PCC. If we irradiated interphase cells with ionizing radiation just prior to fusion, we found we could immediately visualize chromosome damage in the G_1 and G_2 PCC (Hittelman and Rao 1974a, Waldren and Johnson 1974). As shown in Figure 4, the same sorts of aberrations are visualized in

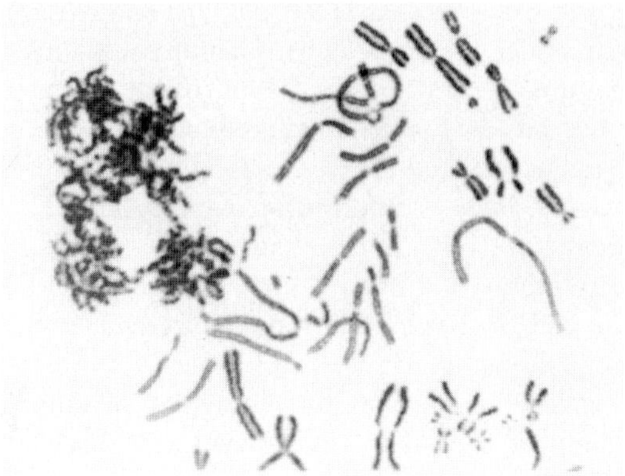

FIG. 4. G_2 PCC from γ-irradiated CHO cell. Note the visualization of an exchange, breaks, and gaps.

the PCC as are observed in mitosis (i.e., exchanges, breaks, and gaps). On the other hand, when we treated interphase CHO cells with ultraviolet radiation (100 ergs/mm^2) and then immediately fused these cells with mitotic cells to induce PCC, we did not observe an increase in the frequency of chromatid exchanges or breaks in the G_2 PCC (Table 1). However, we did observe an increase in the frequency of gaps (Hittelman and Rao 1974b). These results supported the notion that ultraviolet light does not induce chromosome aberrations directly, but rather the cells must pass through S phase to translate the DNA lesions into chromosome aberrations (Bender et al. 1974, Chu 1965). The increase in the frequency of gaps probably reflected an effect of ultraviolet light on the ability of the chromosomes to condense properly.

The next experiment was designed to probe the relationship between the structural integrity of the chromosome and the ability to form chromosome aberrations. Experiments by St. Amand on grasshopper neuroblasts (1965) and by Dewey and Miller (1969) on CHO cells had shown that aberrations could not be visualized immediately if cells already in mitosis were irradiated. However, when these irradiated mitotic cells were allowed to divide and proceed through the next division cycle, aberrations were apparent at the next mitosis. These

TABLE 1. *Comparative estimates of ultraviolet radiation (100 erg/mm^2) damage to CHO cells by scoring PCC and mitotic chromosomes*

Treatment	Type of Chromosomes Scored	No. Cells Scored	Aberrations per Cell		
			Gaps	Breaks	Exchanges
+UV	PCC	50	0.38	0.10	0
−UV	PCC	50	0.16	0.10	0.02
+UV	Mitotic	100	0.07	0.01	0.01
−UV	Mitotic	100	0.02	0.01	0

(Reproduced from Hittelman and Rao 1974b, with permission from Elsevier.)

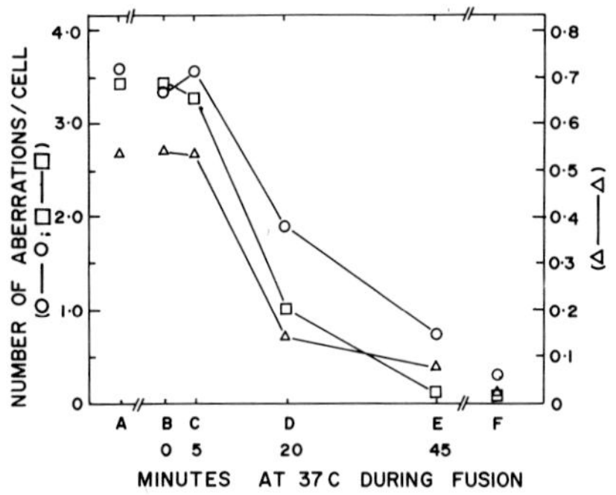

FIG. 5. The frequency of chromatid aberrations in CHO G_2 PCC as a function of the time of γ irradiation during cell fusion. Cells were treated with 217.5 rad γ rays: A, before addition of virus; B, after exposure to 4°C for 15 min; after incubation at 37°C for: C, 5 min; D, 20 min; and E, 45 min. F, no irradiation. O-O, gaps; □-□, breaks; △-△, exchanges. (Reproduced from Hittelman and Rao 1974a, with permission of Elsevier.)

observations suggested that even though chromatin damage probably existed in the mitotic chromosomes, this damage was hidden in the intact structure and could not be expressed until after the chromatin had unwound during interphase. To test this notion we irradiated CHO cells at various times during the fusion procedure while the interphase chromatin was being condensed into PCC (Hittelman and Rao 1974a). As shown in Figure 5, the frequency of exchanges and breaks remained constant for the first five minutes at 37°C during fusion but dropped off significantly by 20 minutes when the chromatin had become fairly condensed. The break and exchange frequencies continued to drop until 45 minutes after fusion, when they approached the frequencies of the unirradiated controls. It was interesting to note that the mitotic inducer cells never showed aberrations even though they were irradiated along with the interphase cells. These observations confirmed the earlier observations that aberrations can be hidden in the condensing chromosomes and that the chromatin must be somewhat decondensed for aberrations from DNA damage to be expressed.

The Sensitivity of the PCC Technique—Measurement of Chromosome Repair

The next experiment was designed to test the sensitivity of the PCC technique for detecting chromosome aberrations in interphase cells by comparing aberration frequencies observed directly in PCC of cells irradiated in G_2 with those observed when these G_2 cells reached mitosis. Asychronous CHO cells were irradiated with 217.5 rad of γ rays. A portion of the irradiated cultures were immediately fused to induce PCC while another portion was allowed to proceed through the cell cycle and collected in mitosis with Colcemid 1½ hours later. We scored both G_2 PCC and accumulated mitoses for chromatid aberrations; thus, in both G_2 PCC and mitotic populations we were measuring the damage of cells treated in G_2. When the frequencies of the various classes of aberrations were compared (Figure 6), nearly twice as many chromatid breaks were observed in PCC compared to that measured in mitosis, while only 1.25 times as many exchanges and gaps were observed (Hittelman and Rao 1974a).

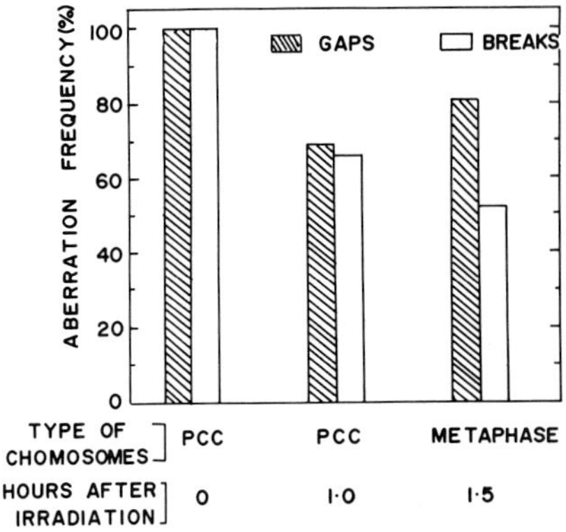

FIG. 6. The frequency of gaps and breaks in PCC as compared to those in metaphase chromosomes after γ irradiation. The number of aberrations (gaps and breaks) scored in PCC at one hour after irradiation and in metaphase chromosomes is expressed as a percentage of that observed in PCC at 0 hours after irradiation. (Reproduced from Hittelman and Rao 1974a, with permission of Elsevier.)

The fact that the G_2 PCC exhibited higher aberration levels than those observed in mitosis might be explained by one or both of the following: (a) The G_2 cells could repair some of the chromosome damage prior to entry into mitosis, or (b) the more damaged cells were preferentially delayed in G_2 and were not included in the mitotic population. In order to directly measure the rate of chromosome repair in G_2 cells, we γ-irradiated CHO cells and then incubated the cells at 37°C for various periods of time prior to fusion. As shown in Figure 7, the chromosome aberration frequencies remained constant for 30 minutes and then the break and gap frequencies began declining by 45 minutes postirradiation. By 60 minutes, the break and gap frequencies declined by nearly one half. On the other hand, the exchange frequency remained constant. These results suggested that chromatid breaks and gaps can be repaired with a half-time on the order of an hour, but exchanges, once formed, cannot be repaired. It is interesting to note that the half-repair time for chromosome

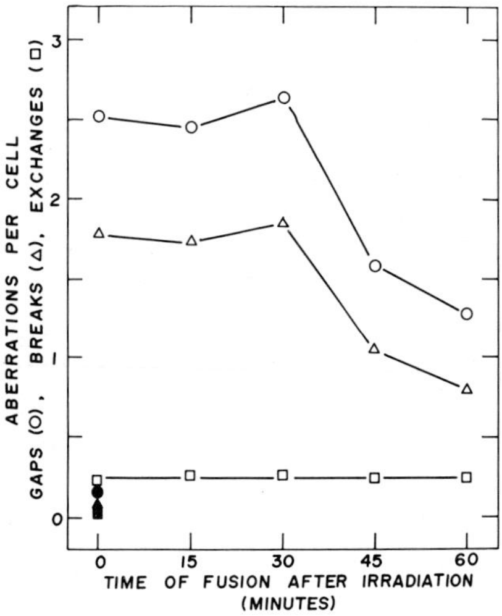

FIG. 7. Determination of the rate of chromosome repair after γ irradiation. The frequency of chromatid aberration in G_2-PCC is a function of the time elapsed between γ irradiation and cell fusion.

breaks and gaps is on the same order of time reported for repair of sublethal damage by Elkind in an earlier chapter (1980, see pages 71–93, this volume). Thus, there is a hint that repair of sublethal damage might reflect the repair of chromosome aberrations.

These experiments also reveal something about the mechanism of chromosome exchange formation. It has often been suggested that chromosome exchanges are simply the result of the misrepair of chromatid breaks (Evans 1962, Sax 1940). If this were the case, one might expect the exchange frequency to rise with the repair of breaks. However, such a finding was not observed. These results therefore support the notion (Neary and Savage 1966) that only lesions appearing close in time and space can interact to form a chromosome exchange. The above experiment also suggested that much of the difference in aberration frequencies in G_2, determined directly by the PCC technique, and in mitosis is due to repair of some of the chromosome damage as the cell goes to mitosis. However, in light of our findings that (1) the exchange frequency was lower in the mitotic cells than that observed in G_2 PCC and (2) exchanges are not repaired, the results suggest that some of the more damaged cells were blocked or delayed in G_2. This phenomenon has been subsequently observed in a variety of drug-treated cell populations (Hittelman and Rao 1974c, 1975, Rao and Rao 1976).

In an earlier chapter, Cole et al. (1980, see pages 33–58, this volume) suggested that after treatment with higher linear energy transfer (LET) radiation, a higher proportion of lesions might be nonrepairable. Using the PCC technique we tested this notion measuring the repair of chromosome damage after treatment of CHO cells with α particles. We observed that only about one third of the chromatid breaks could be repaired in an hour's time compared to nearly one half repaired after γ irradiation. Thus, these preliminary findings are consistent with Cole's hypothesis.

Comparison of DNA and Chromosome Repair after γ-irradiation

In the above experiments we found that chromosome repair did not begin until 30 min after γ irradiation. Since we were interested in the correlation between DNA damage and repair and chromosome damage and repair, we decided to measure, under the same conditions, the rate of repair of DNA damage utilizing the alkaline elution technique. As shown in Figure 8, significant DNA repair is observed within two minutes after irradiation. In fact, considerable DNA repair takes place during the irradiation period (eight minutes) since the presence of 0.2 M EDTA and pyrophosphate during irradiation results in more DNA damage than if they are added immediately after irradiation (Figure 8, bottom two curves). The presence of a fast-repair component after γ irradiation has been previously reported (Volculetz et al. 1977, Leontjeva et al. 1976). This fast-repair component was hypothesized to result from polynucleotide ligase activity since it is ATP-dependent, can be inhibited by the addition of EDTA and pyrophosphate, and repairs DNA to the same extent and with the same

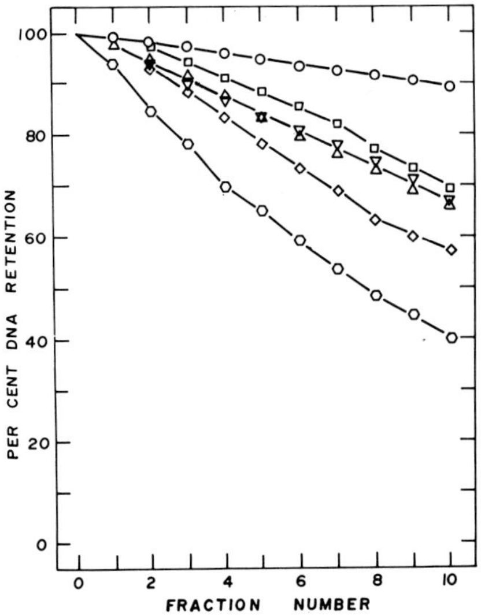

FIG. 8. Determination by the alkaline elution technique of the rate of DNA repair after 300 rad γ irradiation. ⬡-⬡, irradiation in the presence of EDTA and pyrophosphate, no repair; ◇-◇, irradiation, no repair; ▽-▽, irradiation, 2-minute repair; △-△, irradiation, 5-minute repair; □-□, irradiation, 15-minute repair; O-O, EDTA and pyrophosphate alone, no irradiation.

kinetics as polynucleotide ligase acting on irradiated DNA in vitro (Jacobs et al. 1972, Mathelet et al. 1978). Whatever the biochemical nature of this fast-repair component, these experiments shed some light on the nature of the DNA lesions important for chromosome aberration formation. Since the chromosome aberration frequency remains constant for at least 30 minutes after irradiation while more than half of the DNA lesions are repaired within five minutes, these results suggest that the fast-repairing DNA lesions (probably single-strand breaks and alkali-labile regions) are not important in the formation or repair of chromosome breaks. These results support the notion that the more slowly repairing DNA lesions (perhaps more complex in nature) are the key lesions in chromosome damage.

Bleomycin Damage and Repair

While this symposium is devoted to radiation biology, I would like to switch gears and report some of our results using the radiomimetic drug bleomycin. We have taken the same approach in these studies as with those reported above, i.e., measure at the same doses and treatment conditions the initiation and repair of both DNA and chromosome damage.

In the initial experiments we measured the induction of chromosome aberrations in CHO cells as a function of the dose of bleomycin. All cultures were treated for 30 minutes with the appropriate dose of bleomycin. Half the treated cultures were immediately fused with untreated mitotic cells to induce PCC while the other half of the cultures were washed free of bleomycin and incubated in drug-free medium, and the cells treated in G_2 were collected in mitosis with Colcemid. As shown in Figure 9, significantly higher frequencies of breaks and

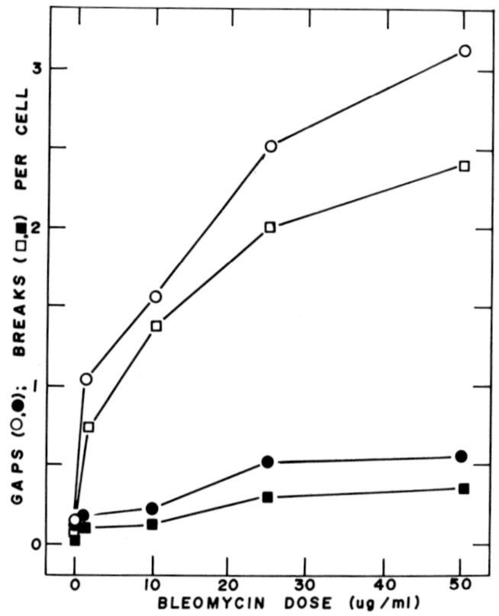

FIG. 9. Bleomycin dose response for chromatid gaps and breaks in cells treated in G_2 and measured either in G_2 PCC or mitotic preparations. Open symbols, G_2 PCC; closed symbols, mitotic cells.

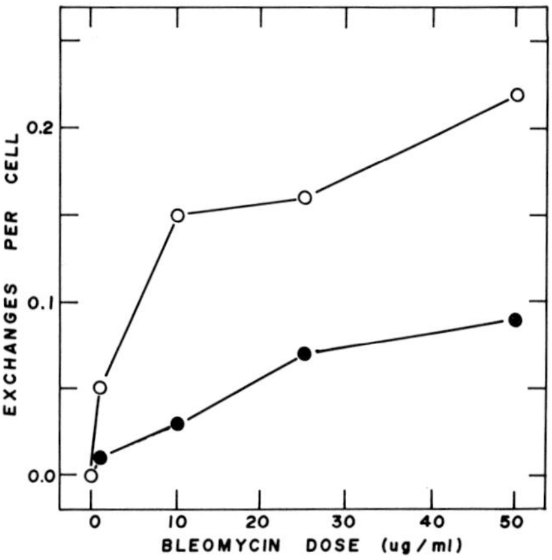

FIG. 10. Bleomycin dose response for chromatid exchanges induced in G_2 CHO cells and measured in G_2 PCC (O-O) or mitotic preparations (●-●).

gaps were observed in the G_2 PCC immediately after bleomycin treatment than when the damage to the G_2 cells was assayed in mitotic preparations. Similarly, the dose-response curve of the G_2 PCC was much steeper than that determined from mitotic cell preparations (which began to plateau past 25 μg/ml). Figure 10 illustrates the relative chromatid exchange rates determined in the same experiment. While the G_2 PCC still exhibited significantly more damage than that observed in mitosis, the exchange frequency did not plateau quite as fast as with breaks and gaps. These results further support the finding that the PCC technique is much more sensitive for measuring the initial levels of chromosome damage than conventional techniques.

As in the case of our studies with ionizing radiation reported above, the differences in the aberration frequencies determined directly in the G_2 cells by the PCC technique and those measured when the G_2 cells reach mitosis can result from either (1) repair of some damage prior to entry into mitosis or (2) G_2 progression delay of the more heavily damaged cells. Repair measurements with a protocol similar to that reported above for γ rays showed one third to one half of the breaks and gaps after a 30-minute bleomycin exposure (25 μg/ml) can be repaired within an hour, whereas exchanges were not repaired (Hittelman and Rao 1974c).

Under the same experimental conditions and with the alkaline elution technique as a probe, we found that greater than half of the bleomycin-induced DNA lesions were repaired within five minutes of drug removal; after this,

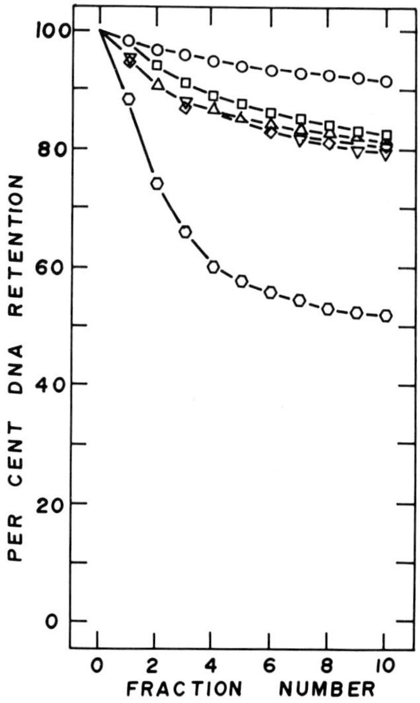

FIG. 11. The kinetics of DNA repair after bleomycin treatment. CHO cells were treated with bleomycin (25 μg/ml for 30 minutes) and then incubated in drug-free medium for 0 (⬡-⬡), 5 (▽-▽), 10 (△-△), 15 (◇-◇), 30 (□-□), or 60 minutes (O-O).

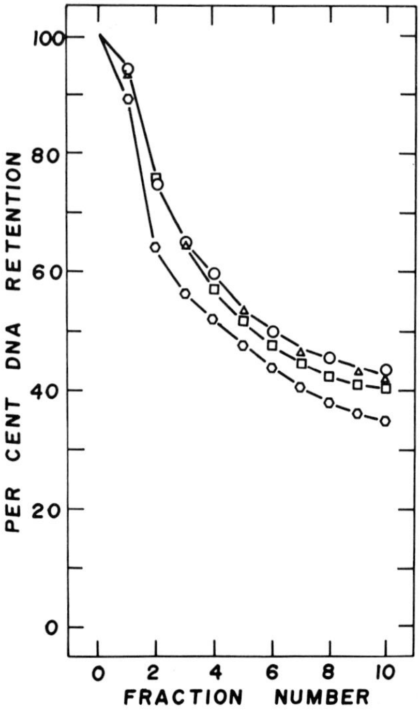

FIG. 12. The effect of treatment duration on the elutability of DNA from bleomycin-treated cells. Cultures were treated with 25 μg/ml bleomycin for 5 (△-△), 10 (O-O), 15 (□-□), or 30 minutes (⬡-⬡).

DNA repair continued at a slower rate until most of the damage was repaired by 60 minutes (Figure 11). We further found that, as with γ rays, the fast-repair component could be inhibited by EDTA.

We next treated cells with 25 μg/ml bleomycin for various periods of time (5–30 minutes) and found (Figure 12) that the apparent amount of DNA damage only slightly increased with duration of treatment. One possible explanation for this result was that only a limited number of sites in the cell were accessible for damage by bleomycin. However, if we added fresh bleomycin during the harvesting procedure, we found that more damage could be induced. We therefore favor the alternative explanation that during bleomycin exposure at this dose there is a balance between the induction of new damage and the fast repair of existing bleomycin-induced damage, and the slow increase in apparent damage is due to the accumulation of the more slowly repairing and irreparable lesions. It will now be interesting to determine the relationship between bleomycin treatment duration and chromosome aberration levels in order to better understand the importance of the slowly repairing DNA lesions in chromosome aberration formation.

The bleomycin experiments reported above generate information similar to that obtained with ionizing radiation, i.e., the DNA lesions that are rapidly repaired are not important in the formation and repair of chromosome breaks. This is not to say, however, that interference with this fast DNA-repair process does not allow these lesions to be transformed into lesions important for chromosome aberrations. For example, Meyn and co-workers (unpublished results) and Corry et al. (1977) have shown that hyperthermia might inhibit a portion of the fast-repair process after bleomycin treatment and ionizing radiation. Along these lines, Braun and Hahn (1975) have shown enhanced cell killing when bleomycin treatment is combined with hyperthermia. Hyperthermic treatment might therefore allow the cell to transform unrepaired single-strand breaks or alkali-labile sites into double-strand breaks that could then be expressed as chromosome aberrations. This might reflect the molecular basis for potentially lethal damage.

Another way to correlate DNA and chromosome lesions is to look at the effect of purported repair inhibitors on both DNA and chromosome repair. The protocol for such an experiment is shown in Figure 13. In order to determine the effect of agents on induction of damage, cells are treated with bleomycin in the presence or absence of the particular inhibitor and then the cells are either fused to induce PCC or the DNA damage is assayed by the alkaline elution technique. To determine the effect of an inhibitor on the repair of bleomy-

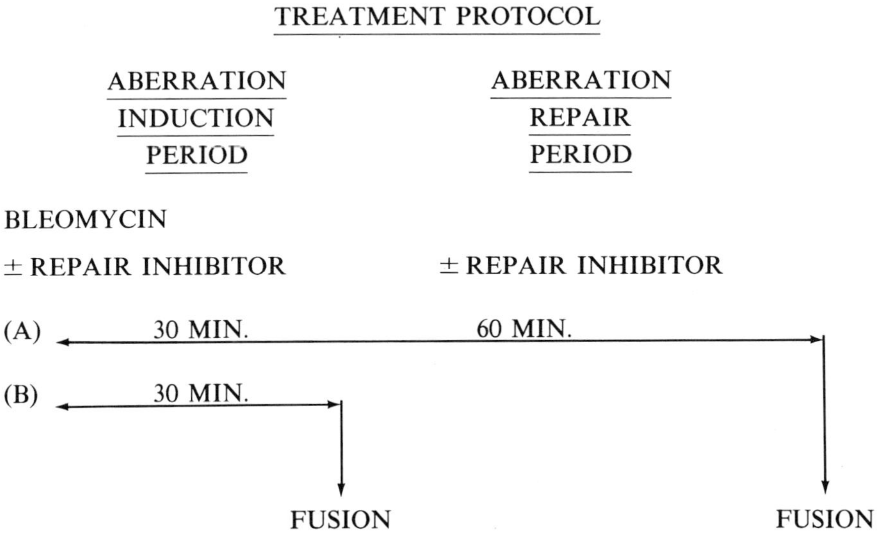

FIG. 13. Experimental protocol for determining the effect of repair inhibitors on the induction and repair of bleomycin-induced damage. (Reproduced from Sognier et al. 1979, with permission of Elsevier.)

TABLE 2. *Effect of hydroxyurea on repair of bleomycin-induced chromosome damage*

Treatment			Aberrations per Cell			
Induction Period	Repair Period*	No. Cells	Gaps	Breaks	Exchanges	Percent Breaks Repaired†
Control	—	70	0.5	0.2	0.00	
Hydroxyurea	+, In medium+ hydroxyurea	61	0.4	0.2	0.00	
Bleomycin + hydroxyurea	—	51	1.2	1.3	0.11	0
Bleomycin + hydroxyurea	+, In medium	53	0.8	0.4	0.04	81.8
Bleomycin + hydroxyurea	+, In medium+ hydroxyurea	55	0.7	0.5	0.11	72.7

* One hour at 37°C.

† Percentage of breaks repaired = [(No. breaks initially induced − No. breaks repaired in 1 hr) ÷ (No. breaks initially induced − No. breaks in control)] × 100.

(Reproduced from Sognier et al. 1979, with permission of Elsevier.)

cin damage, the cells are washed free of bleomycin after treatment and incubated for various periods of time in the presence or absence of inhibitor.

Table 2 illustrates the results observed at the chromosome level when hydroxyurea (10 mM) was employed as the potential repair inhibitor. The results shown here indicate that breaks and gaps are repaired to nearly the same extent in an hour's time whether or not hydroxyurea was present. At the DNA level under the same treatment conditions (Figure 14), hydroxyurea only slightly slowed the repair rate after bleomycin treatment, but the differences in repair rates could nearly be accounted for by induction of new lesions by hydroxyurea alone (open circles). Thus, in these experiments hydroxyurea did not significantly inhibit the repair of bleomycin damage at either the DNA or chromosome level. Similar results were obtained when hycanthone was used as a repair inhibitor (Sognier et al. 1979, and paper submitted for publication).

A completely different result was observed when cycloheximide and streptovitacin A, both potent inhibitors of protein synthesis, were tested as repair inhibitors. As shown in Table 3, cycloheximide effectively blocked the repair of chromatid breaks and gaps in G_2 cells as determined by the PCC technique. However, at the DNA level, as measured by alkaline elution, cycloheximide only slightly retarded the repair rate after bleomycin treatment (Figure 15). These results are consistent with the notion that the DNA lesions responsible for chromosome aberrations represent only a small percentage of the total number of DNA lesions produced by bleomycin. Thus, even if cycloheximide blocked the repair of these specific lesions, the effect would not be apparent in the alkaline elution technique if cycloheximide did not block the repair of the other DNA lesions. For example, analysis of the types of lesions induced in PM2 DNA (Lloyd et. al. 1978) have indicated that single-strand breaks, alkaline-labile sites, and double-strand breaks occur in the proportion of 5:5:1. It is further expected that single-strand breaks are rapidly repaired by a ligase reaction that is not

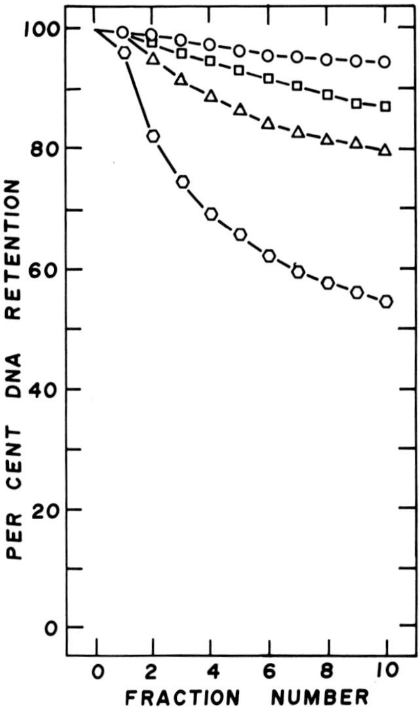

FIG. 14. The effect of hydroxyurea on the repair of bleomycin-induced DNA damage. Cells were treated with bleomycin (25 μg/ml for 30 minutes) in the presence of 10 mM hydroxyurea and allowed to repair for 0 minutes (⬡-⬡) or 60 minutes with (△-△) or without (□-□) hydroxyurea. Hydroxyurea alone for 1½ hours (O-O).

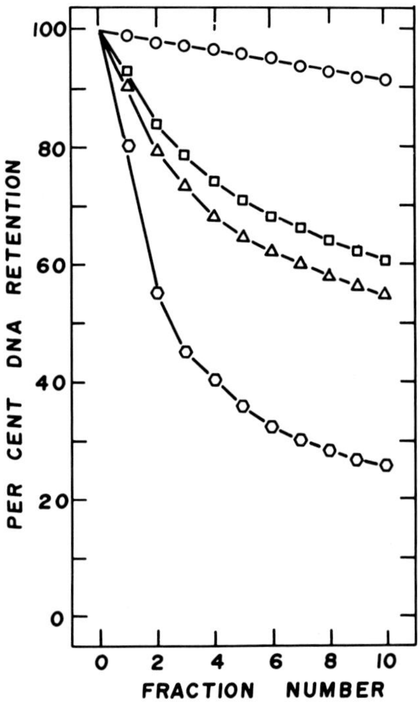

FIG. 15. Effect of cycloheximide on the repair of bleomycin-induced DNA damage. Cells were treated with bleomycin (25 μg/ml cycloheximide) and allowed to repair for 0 minutes (⬡-⬡) or for 60 minutes with (△-△) or without (□-□) cycloheximide. Cycloheximide (25 μg/ml) alone for 2 hours (O-O).

TABLE 3. *Effect of cycloheximide on repair of bleomycin-induced chromosome damage*

Treatment		Cell	Aberrations per Cell			Percent Breaks
Induction Period	Repair Period*	No.	Gaps	Breaks	Exchanges	Repaired
Control	—	59	0.1	0.1	0.00	
Bleomycin	—	64	1.3	2.1	0.24	
Bleomycin + cycloheximide	—	48	1.3	2.9	0.24	0
Bleomycin + cycloheximide	+, in medium	53	0.8	1.4	0.20	53.6
Bleomycin + cycloheximide	+, in medium+ cycloheximide	51	0.9	2.6	0.30	10.8

* One hour at 37°C.
(Reproduced from Sognier et al. 1979, with permission from Elsevier.)

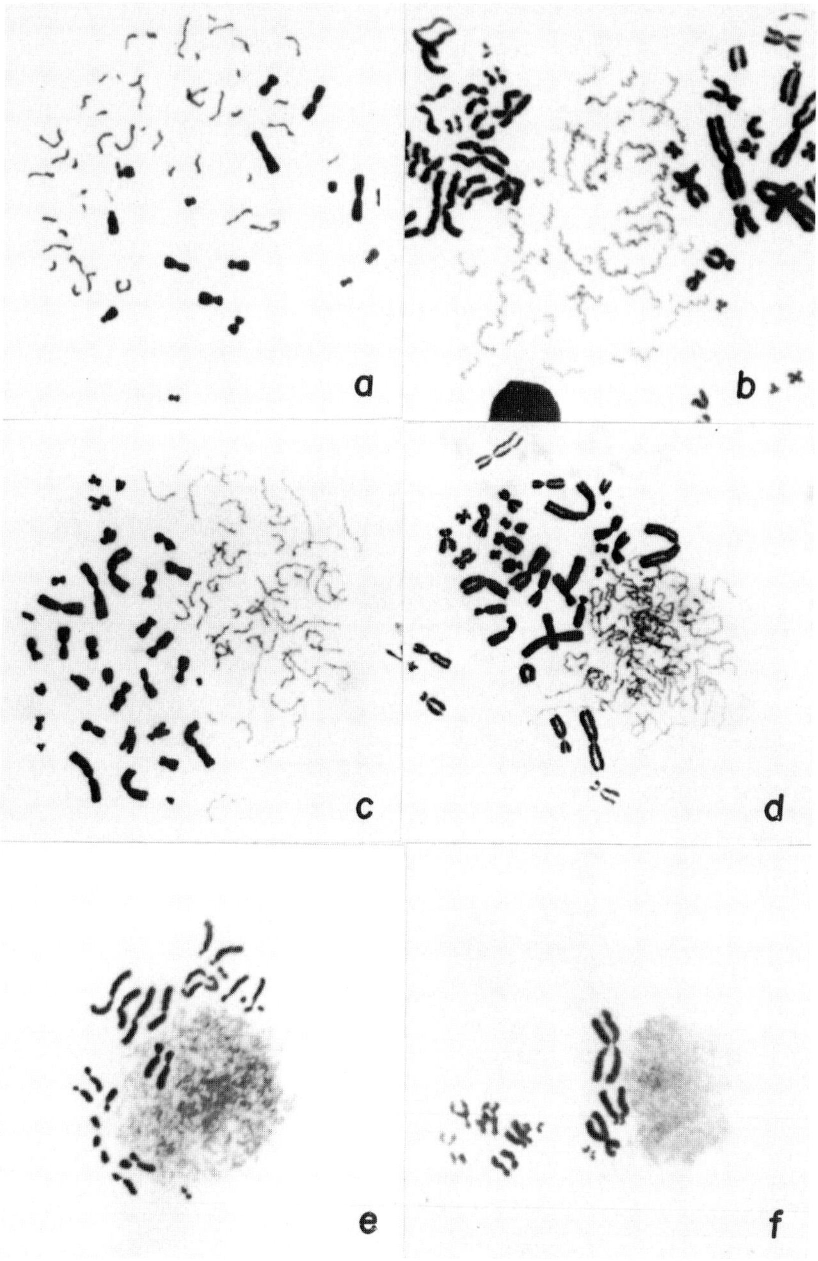

FIG. 16. G_1 PCC of human bone marrow cells showing varying degrees of chromosome condensation. The darkly stained chromosomes are from mitotic cells.

cycloheximide sensitive (Evans et al. 1976). Thus, if cycloheximide only blocked double-strand break repair, this effect would be difficult to detect by alkaline elution. However, until we can accurately detect double-strand breaks at the low doses required for chromosome aberration studies, a direct correlation between chromatid break and gap repair and double-strand break repair cannot be proved. An alternate explanation for these results is that since both histone and non-histone proteins are important in the integrity of chromosomes, protein synthesis might be important in the repair of the chromosome structure.

Future Directions

The studies reported here are only the beginning of attempts to define the underlying molecular basis for DNA damage and repair, chromosome damage and repair, and repair of sublethal and potentially lethal damage. One of the problems with the use of exponentially growing populations for these experiments is that if long repair periods are employed, cell cycle changes may occur during the time of the experiment and differential cell cycle specificities may confuse the analysis. For this reason, we have chosen to look at non-growing or quiescent cell populations in which cells can be held at a uniform point in the cell cycle for any desired time period.

Using the PCC technique, we have found that the G_1 phase of the cell cycle can be mapped by the morphology of the G_1 PCC (Figure 16). Early G_1 cells exhibit very condensed PCC, while late G_1 cells give rise to extended G_1 PCC (Hittelman and Rao 1976, 1978a). Further, we have found that quiescent normal cells come to rest in early G_1 phase, whereas transformed populations tend to accumulate in late G_1 phase (Hittelman and Rao 1978a) (Figure 17). Quiescent normal cell populations are therefore ideal for these types of studies because they are stable synchronous cell populations that yield short and easily countable G_1 PCC after cell fusion. Figure 18 illustrates the neocarzinostatin-induced chromosome damage visualized in the G_1 PCC of quiescent, normal human PA2 cells.

Finally, since this monograph's title is *Radiation Biology in Cancer Research,* I thought it might be appropriate to illustrate another situation in which the PCC technique is useful in the visualization of chromosome damage in interphase cells. It is well established that most cancer therapeutic agents induce chromosome aberrations, probably as a prelude to reproductive death in cells. In recent years we have been monitoring bone marrow cell populations of leukemia patients undergoing therapy for their disease, and we have been able to determine the levels of chromosome aberrations in these cells. Figure 19 illustrates the type of chromosome damage that can be visualized by the PCC technique in the bone marrow cells of a patient undergoing chemotherapy. In this particular case, the damaged cells were observed one day after the initiation of remission

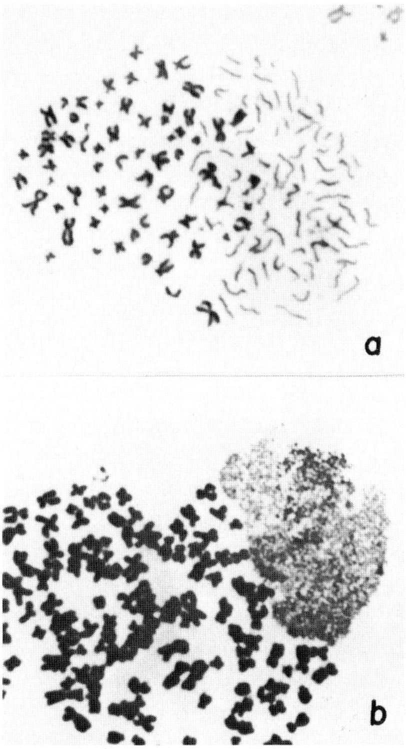

FIG. 17. Typical G_1 PCC from plateau phase cultures of (a) normal 3T3 cells and (b) transformed SV-3T3 cells.

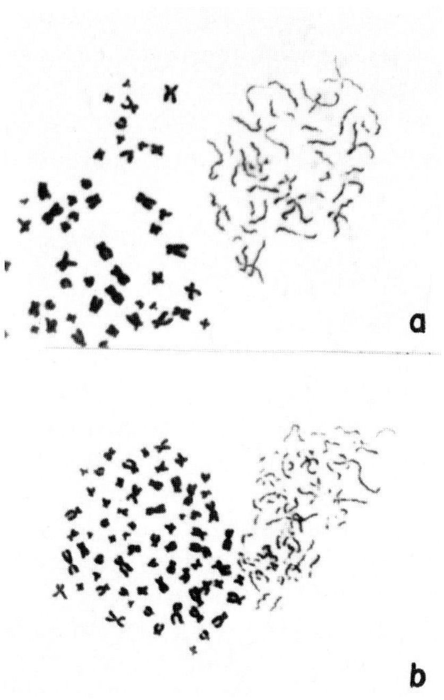

FIG. 18. G_1 PCC of quiescent normal human PA2 cells. a, Untreated cell; b, after treatment with neocarzinostatin. Note the extent of chromosome breakage.

induction therapy, and the G_2 PCC illustrate the high frequency of exchanges typical after Adriamycin treatment (Hittelman and Rao 1975). While these studies are still preliminary, we have found that effective therapy is accompanied by significant chromosome damage in the patient's bone marrow cells, whereas resistant disease is correlated with little chromosomal changes during therapy. Thus, by monitoring bone marrow cells by the PCC technique for chromosome damage as well as for proliferative changes, we have found that we can detect resistant and relapsing leukemic disease, often months before any clinical evidence of leukemic relapse appears (Hittelman and Rao 1978b).

I have tried to show how a combination of techniques—alkaline elution for detection of DNA damage and repair and PCC analysis for chromosome damage—can be used to elucidate the common molecular lesions that link damage at the DNA and chromosome levels to eventual loss of reproductive capability of cells. Further, the PCC technique can be used effectively to monitor chromosome changes associated with effective cancer therapy.

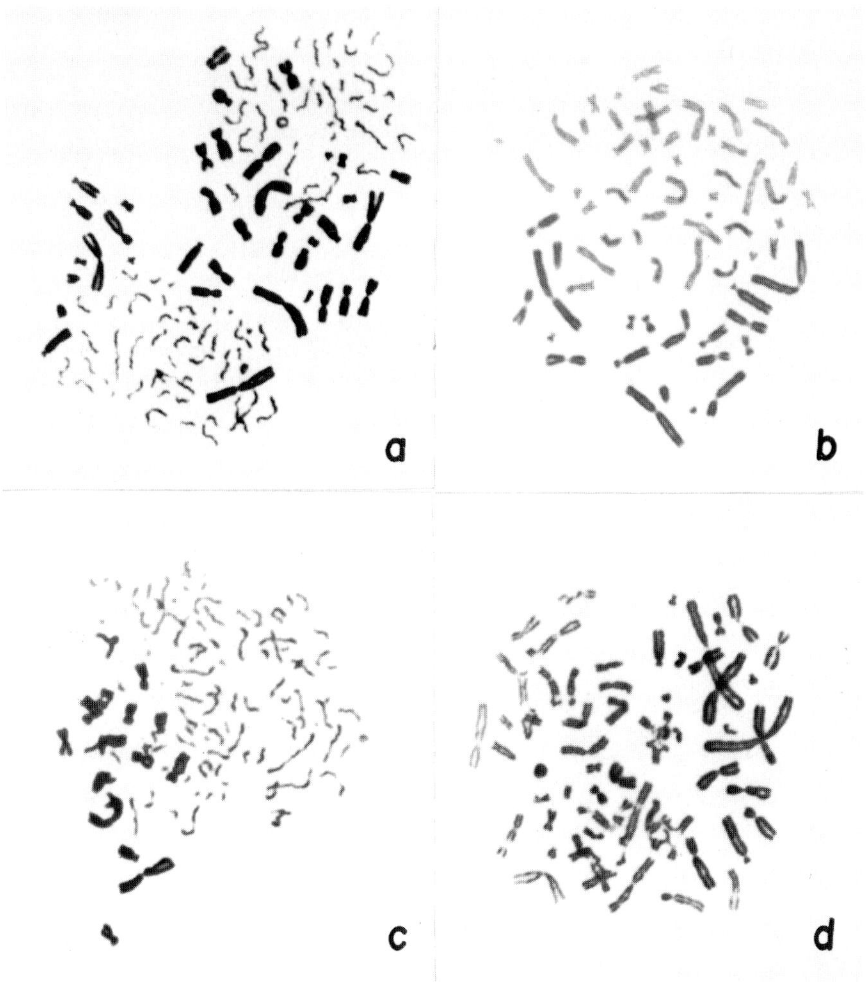

FIG. 19. PCC of bone marrow cells obtained from a leukemia patient undergoing remission induction therapy. a, G_1 PCC; b, G_2 PCC prior to initiation of therapy. Note ring chromosome. c, G_1 PCC; d, G_2 PCC obtained one day after initiation of therapy with Adriamycin, cytosine arabinoside, vincristine, and prednisone. Note extensive chromosome damage in the PCC.

ACKNOWLEDGMENTS

We thank Dr. Raymond Meyn for help in setting up the alkaline elution technique and Dr. Potu N. Rao for helpful discussions. Supported in part by Grant number CA-14528, awarded by the National Cancer Institute and Grant number GM-23252 awarded by the National Institute of General Medical Sciences, National Institutes of Health, Department of Health, Education and Welfare.

REFERENCES

Bender, M. A., H. G. Griggs, and P. L. Walker. 1974. Mechanisms of chromosomal aberration production. I. Aberration induction by ultraviolet light. Mutat. Res. 20:387–402.

Braun, J., and G. M. Hahn. 1975. Enhanced cell killing by bleoymcin and 43° hyperthermia and the inhibition of recovery from potentially lethal damage. Cancer Res. 35:2921–2927.

Carrano, A. V. 1973. Chromosome aberrations and radiation-induced cell death. II. Predicted and observed cell survival. Mutat. Res. 17:355–366.

Chu, E. H. Y. 1965. Effects of ultraviolet radiation on mammalian cells. I. Induction of chromosome aberrations. Mutat. Res. 2:75–94.

Cleaver, J. 1977. Human disease with *in vitro* manifestation of altered repair and replication of DNA, *in* Genetics of Human Cancer, J. J. Mulvihill, R. W. Miller, and J. F. Fraumeni, Jr., eds. Raven Press, New York, pp. 355–363.

Cole, A., R. E. Meyn, R. Chen, P. M. Corry, and W. Hittelman. 1980. Mechanisms of cell injury, *in* Radiation Biology in Cancer Research (The University of Texas System Cancer Center 32nd Annual Symposium on Fundamental Cancer Research), R. E. Meyn and H. R. Withers, eds. Raven Press, New York, pp. 33–58.

Corry, P. M., S. Robinson, and S. Getz. 1977. Hyperthermic effects on DNA repair mechanisms. Radiology 123:475–482.

Dewey, W. C., S. C. Furman, and H. H. Miller. 1970. Comparison of lethality and chromosomal damage induced by γ-ray in synchronized Chinese hamster cell in vitro. Radiat. Res. 43:561–581.

Dewey, W. C., and H. H. Miller. 1969. γ-Ray induction of chromatid exchanges in mitotic and Gl Chinese hamster cells pretreated with Colcemid. Exp. Cell Res. 57:63–70.

Elkind, M. M. 1980. Cells, targets, and molecules in radiation biology, *in* Radiation Biology in Cancer Research (The University of Texas System Cancer Center 32nd Annual Symposium on Fundamental Cancer Research), R. E. Meyn and H. R. Withers, eds. Raven Press, New York, pp. 71–93.

Evans, H. H., S. R. Littman, T. E. Evans, and W. N. Brewer. 1976. Effects of cycloheximide on thymidine metabolism and on DNA strand elongation in *Physarum polycephalum.* J. Mol. Biol. 104:169–194.

Evans, H. J. 1962. Chromosome aberrations induced by ionizing radiations. Int. Rev. Cytol. 13:221–321.

Hittelman, W. N., and P. N. Rao. 1974a. Premature chromosome condensation. I. Visualization of γ-ray induced chromosome damage in interphase cells. Mutat. Res. 23:251–258.

Hittelman, W. N., and P. N. Rao. 1974b. Premature chromosome condensation. II. The nature of chromosome gaps produced by alkylating agents and ultraviolet light. Mutat. Res. 23:259–266.

Hittelman, W. N., and P. N. Rao. 1974c. Bleomycin induced damage in prematurely condensed chromosomes and its relationship to cell cycle progression in CHO cells. Cancer Res. 35:3027–3035.

Hittelman, W. N., and P. N. Rao. 1975. The nature of adriamycin-induced cytotoxicity in Chinese hamster cells as revealed by premature chromosome condensation. Cancer Res. 35:3027–3035.

Hittelman, W. N., and P. N. Rao. 1976. Premature chromosome condensation. Conformational changes of chromatin associated with phytohemagglutinin stimulation of peripheral lymphocytes. Exp. Cell Res. 100:219–222.

Hittelman, W. N., and P. N. Rao. 1978a. Mapping G1 phase by the structural morphology of the prematurely condensed chromosomes. J. Cell. Physiol. 95:333–342.

Hittelman, W. N., and P. N. Rao. 1978b. Predicting response or progression of human leukemia by premature chromosome condensation of bone marrow cells. Cancer Res. 38:416–423.

Jacobs, A., A. Bopp, and V. Hagen. 1972. In vitro repair of single strand breaks in gamma irradiated DNA by polynucleotide ligase. Int. J. Radiat. Biol. 22:431–435.

Johnson, R. T., and P. N. Rao. 1970. Mammalian cell fusion: Induction of premature chromosome condensation in interphase nuclei. Nature 226:717–722.

Kohn, K. W., and R. A. G. Ewig. 1973. Alkaline elution analysis, a new approach to the study of DNA single-strand interruptions in cells. Cancer Res. 33:1849–1853.

Laemmli, V. K., S. M. Cheng, K. W. Adolph, J. R. Paulson, J. A. Brown, and W. R. Baumbach. 1978. Metaphase chromosome structure: The role of nonhistone proteins. Cold Spring Harbor Symp. Quant. Biol. 42:351–360.

Leontjeva, G. A., Y. A. Manzighin, and A. L. Gazier. 1976. The ultrafast repair of single strand breaks in DNA of gamma-irradiated Chinese hamster cells. Int. J. Radiat. Biol. 30:577–580.

Lloyd, R. S., C. W. Haidle, and R. R. Hewitt. 1978. Bleomycin-induced alkaline-abile damage and direct strand breakage of PM2 DNA. Cancer Res. 38:3191–3196.

Mathelet, M., L. Clerici, F. Campagnari, and M. Talpaert-Borle. 1978. The activity of mammalian polynucleotide ligase on γ-irradiated DNAs. Biochim. Biophys. Acta 518:138–149.

Meyn, R. E., D. J. Grdina, and S. E. Fletcher. 1980. Repair of radiation damage in vivo, *in* Radiation Biology in Cancer Research (The University of Texas System Cancer Center 32nd Annual Symposium on Fundamental Cancer Research), R. E. Meyn and H. R. Withers, eds. Raven Press, New York, pp. 95–102.

Neary, G. J., and J. R. K. Savage. 1966. Chromosome aberrations and the theory of RBE: II. Evidence from track-segment experiments with protons and alpha particles. Int. J. Radiat. Biol. 2:209.

Olins, D. E., and A. L. Olins. 1978. Nucleosomes: The structural quantum in chromosomes. Am. Sci. 66:704–711.

Rao, A. P., and P. N. Rao. 1976. The cause of G2-arrest in Chinese hamster ovary cells treated with anticancer drugs. J. Natl. Cancer. Inst. 57:1139–1143.

Sax, K. 1940. An analysis of γ-ray-induced chromosomal aberrations in tradescantia. Genetics 25:41–68.

St. Amand, W. 1956. Mitotic inhibition and chromosome breakage induced in grasshopper neuroblasts by γ-irradiation at known mitotic stages. Radiat. Res. 5:65–78.

Scott, D., M. Fox, and B. W. Fox. 1974. The relationship between chromosomal aberrations, survival, and DNA repair in tumor cell lines of differential sensitivity to γ-rays and sulphur mustard. Mutat. Res. 22:207–221.

Sognier, M. A., W. N. Hittelman, and P. N. Rao. 1979. Effect of DNA repair inhibitors on the induction and repair of bleomycin-induced chromosome damage. Mutat. Res. 60:61–72.

Volculetz, N., M. Stoian, M. Tudor, M. T. Sandulescu, and M. Stan. 1977. DNA single strand breaks induced by ionizing radiation in mammalian cells and the kinetics of repair processes. Studia. Biophys. 61:9–16.

Waldren, C. A., and R. T. Johnson. 1974. Analysis of interphase chromosome damage by means of premature chromosome condensation after γ- and ultraviolet-irradiation. Proc. Natl. Acad. Sci. USA 71:1141.

Radiation Biology in Cancer Research, edited by
Raymond E. Meyn and H. Rodney Withers.
Raven Press, New York © 1980.

The Relation between Depressed Synthesis of DNA and Killing in X-Irradiated HeLa Cells

L. J. Tolmach, R. B. Hawkins, and P. M. Busse*

Washington University School of Medicine, St. Louis, Missouri 63110

Several operationally distinct phenomena might be treated in a discussion of radiation-induced perturbations of cellular growth and proliferation kinetics. These include inhibition of macromolecular syntheses, particularly DNA synthesis, delay in reaching mitosis, aberrant behavior during mitosis or at cytokinesis, cell fusion, cell death, and cell disintegration. Our primary concern is with cell killing, and hence the relevance of the other kinetic perturbations to death is of particular interest. This discussion deals mainly with the relation of one manifestation of damage induced in HeLa cells by 220 kV X rays—the inhibition of DNA synthesis in the irradiated generation—to cell killing.

X-RAY-INDUCED INHIBITION OF DNA SYNTHESIS

Inasmuch as damage residing in DNA is strongly implicated in cell lethality (Painter 1980, see pages 59 to 68, this volume), and perturbations in the rate of DNA replication could reflect lethal lesions in the template, it was felt that study of radiation-induced depression of the rate of synthesis might provide information relevant to the lesions. Initial studies (Weiss 1971) on both synchronous and randomly dividing cultures irradiated with doses generally limited to 1 krad showed that the rate of labeled thymidine incorporation into acid-insoluble form declines immediately upon irradiation if the cells are in S phase (Figure 1); that if cells are in G_1 when irradiated, synthesis is depressed when they reach S; that the depression is of only limited extent, at least after small doses; that maximal depression apparently is maintained only transiently; and that the shape of the dose-response curve depends on the time after irradiation at which it is measured.

Kinetic Changes after Large Doses and Long Times

To provide more complete characterization of the kinetic changes induced, and to maximize the effects, we measured the rate of synthesis in randomly

* Present address: St. Louis University School of Medicine St. Louis, Missouri 63104

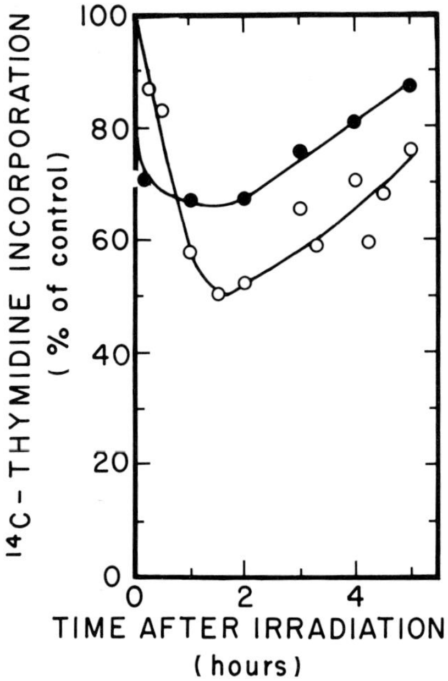

FIG. 1. Incorporation of labeled thymidine (30-minute pulses) into irradiated cultures at close intervals following exposure to either 400 rad (●) or 1,000 rad (○). At 1,000 rad, the plotted points represent average values for five independent experiments for both random and synchronous cultures in S; at 400 rad, average values for three experiments are shown. (Reproduced from Weiss 1971, with permission of Academic Press.)

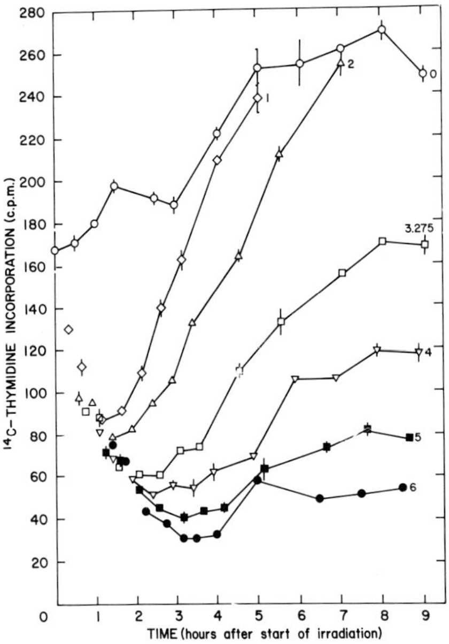

FIG. 2. Time-course of the rate of DNA synthesis as measured by sequential pulse-labeling for 20 minutes with [^{14}C]thymidine following irradiation with different doses of X rays. Radioactivity was measured in a low-background (2 counts per minute) Geiger counter. The vertical bars show the range of counts per minute for the individual dishes of the pairs used in the determination of each point; when absent, the two values fell within the point. The numbers to the right of each curve identify the X-ray dose in krad. (Reproduced from Tolmach and Jones 1977, with permission of Academic Press.)

dividing cultures following X-ray doses of up to 8 krad; measurements were continued for as long as 12 hours after exposure (Tolmach and Jones 1977). Within one to three hours after irradiating nonsynchronized cultures, the rate reaches a minimum whose depth increases with dose (Figure 2), and then begins to climb approximately linearly with time, at a dose-dependent rate (Figures 2 and 3). The deceleration of synthesis, in contrast, is independent of dose (Figure 2) and is roughly exponential with time, the half-time being of the order of 1.2 hours (Figure 4). It has been shown that X rays temporarily inhibit the initiation of replicons (Watanabe 1974, Makino and Okada 1975, Walters and Hildebrand 1975, Painter and Young 1975). The curves of Figures 2 and 4 are consistent with prompt inhibition of the initiation of replicons by X-ray doses smaller than 1 krad, the dose-independent decline in rate representing the completion of clusters already initiated before exposure. In contrast, inhibition of DNA synthesis induced by treatment with inhibitors of protein synthesis, which effectively prevents chain elongation (Gautschi and Kern 1973), occurs more rapidly and with different kinetics (Figure 4).

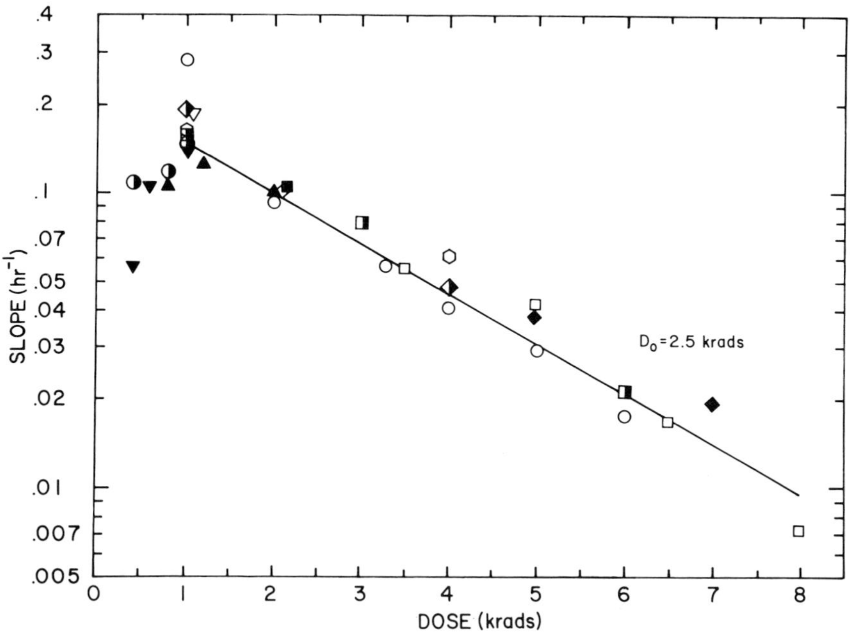

FIG. 3. Slopes of the ascending limbs of several rate-time curves (like Figure 2) as a function of dose. Each different symbol refers to a different experiment. The line has been fitted by eye. D_o is the dose required to reduce the slope by the factor 0.632. (Reproduced from Tolmach and Jones 1977, with permission of Academic Press.)

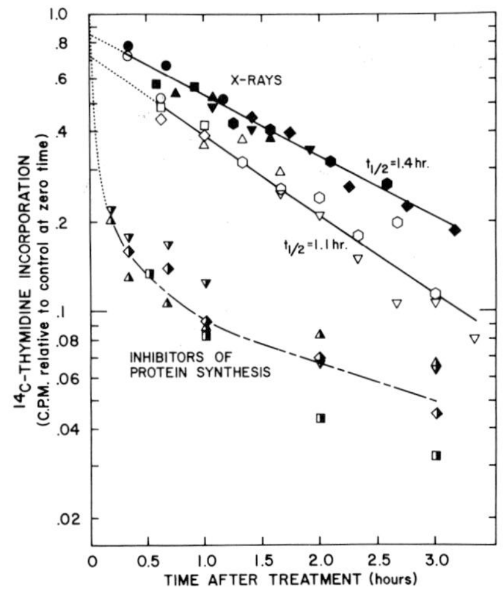

FIG. 4. The descending portions of the curves of Figure 2, plotted on semilogarithmic coordinates relative to the value for the unirradiated control at zero time (the start of treatment), are given by the solid symbols. Each different symbol refers to a different dose (key: ●, 1; ■, 2; ▲, 3.28; ▼, 4; ⬢, 5; ◆, 6 krad). The open symbols show similar data from a separate experiment in which a pulse-labeling time of 10 minutes was used (key: ○, 1; □, 1.5; ◇, 2; △, 3; ⬡, 5; ▽, 7 krad). The straight lines have been fitted by eye. The lower curve (half-filled symbols) refers to measurements of DNA synthesis in cultures treated with inhibitors of protein synthesis, viz., puromycin (20 μg/ml) (squares, triangles, diamonds), or cycloheximide (8 μg/ml) (inverted triangles). (Reproduced from Tolmach and Jones 1977, with permission of Academic Press.)

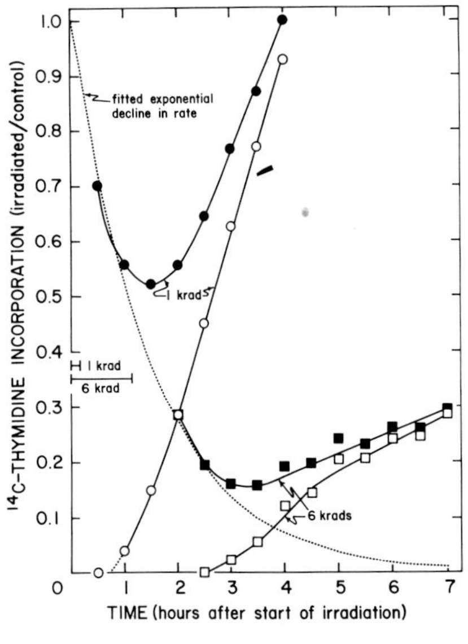

FIG. 5. Subtraction of the descending limb of the rate curve from the measured values. The solid symbols show the results of a typical experiment in which doses of 1 and 6 krad were delivered. The durations of the exposures are shown by the bars. The dotted line is an exponential function fitted to the measured points lying on the descending limbs. The open symbols show the difference between measured values and the exponential curve. (Reproduced from Tolmach and Jones 1977, with permission of Academic Press.)

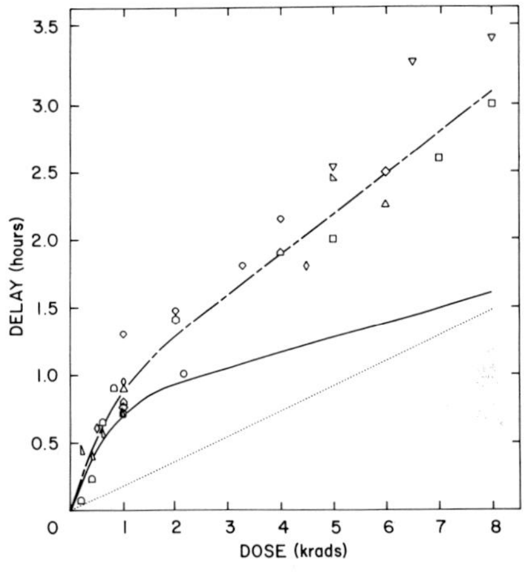

FIG. 6. Delay in the inception of recovery in the rate of DNA synthesis, as a function of dose. Subtractions like those shown in Figure 5 were carried out in 13 different experiments (different symbols) to yield the values shown. The dashed curve was fitted by eye. The solid curve shows this corrected for the time required to deliver the dose (dotted line). (Reprinted from Tolmach and Jones 1977, with permission of Academic Press.)

The Hiatus in Initiation

If the increase in rate that begins one to three hours after the start of irradiation (depending on the dose) reflects the reinitiation of replicons, i.e., if the descending and ascending limbs of the rate curve represent essentially independent processes, it should be possible to subtract the descending limb from the overall curve to determine the duration of the hiatus in replicon initiation. Two representative subtractions are shown in Figure 5; data for a large number of such determinations are shown in Figure 6 from which it appears that the delay in the inception of recovery is strongly dose dependent at low doses, but less so at doses above 1.5 krad. We have not been able to determine whether the delay ultimately reaches a maximal, dose-independent value at high dose. Independent determination of the duration of the hiatus after 1 krad was afforded by splitting a dose of 2 krad into two equal parts separated by increasing intervals and noting the interval that shows rate increases indicative of recovery. The data in Figure 7 suggest that the hiatus lasts for more than 60 but less than 90 minutes. However, a more sensitive test indicated that it may be shorter; that is, if the total dose is administered in several discrete fractions spaced closely enough to prevent reinitiation, the onset of recovery should be delayed so that the two limbs of the rate curve are separated, and the slope of the recovery curve should not be altered. Such behavior was observed (Figure 8) when we delivered a total dose of 5 krad in 1-krad increments at hourly intervals, but the data indicate that some recovery took place during the intervals; i.e., the minimum in the fractionated-dose curve lies above rather than below that for the single-

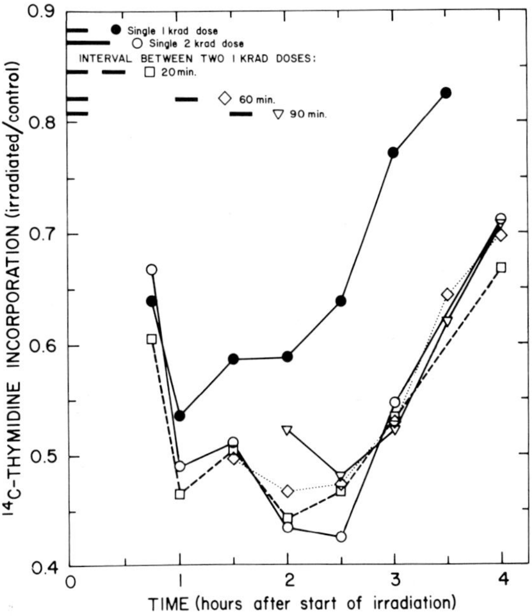

FIG. 7. Time-course of the rate of DNA synthesis, relative to unirradiated controls, after irradiation with one (solid circles) or two (open symbols) doses of 1 krad separated by increasingly longer intervals. The bars show the times of irradiation. (Reprinted from Tolmach and Jones 1977, with permission of Academic Press.)

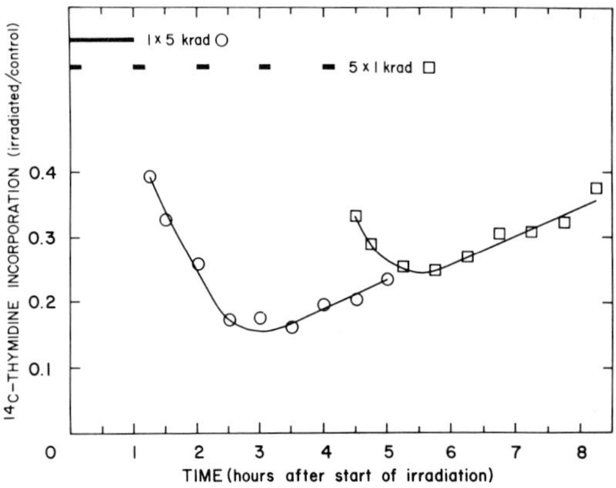

FIG. 8. Partial separation of the development phase from the recovery phase of the X-ray-induced depression in rate of DNA synthesis. The time-course of the rate, relative to unirradiated controls, is shown after a single dose of 5 krad (○) and after five doses of 1 krad (□) delivered at the times shown by the bars. (Reprinted from Tolmach and Jones 1977, with permission of Academic Press.)

dose curve. Even intervals as short as 30 minutes gave evidence of some recovery. This suggests that the duration of complete inhibition may have been overestimated in the experiment of Figure 7; populational heterogeneity might account for the discrepancy (Tolmach and Jones 1977).

Autoradiographic grain counts obtained from samples labeled during the period when the rates were rising (three and six hours after irradiation with 1 or 2 krad) indicated that the rate of synthesis rises in individual cells, i.e., that the increases measured on a per culture basis do not represent merely changes in the age composition of the population (Tolmach and Jones 1977).

Measurements at Long Times after Irradiation

When measurements of the rate of synthesis were continued for 12 hours after irradiation with 1 to 6 krad, it became evident (Tolmach and Jones 1977) that the curves pass through maxima 7 to 10 hours after the start of the irradiation and then begin a second decline (Figure 9). The peak rate after 1 krad is greater than the rate in unirradiated cultures; this can be attributed to the accumulation of cells in S, brought about by the slowing of replication, together with unimpeded entry of G_1 cells into S. The second decline in rate would then reflect passage into G_2. This sequence cannot occur after the larger doses, however, as replication proceeds too slowly for the cells to reach G_2 by the time the rate begins to fall. We previously suggested that the decline reflects the imminence of cell death, even though cells remain alive and metabolically active for at least 16 hours after irradiation with doses as high as 8 krad (Tolmach and Jones 1977). That this explanation is incorrect, however, is shown by the data presented in Figure 10 for synchronous cells that had been prelabeled with a very low level of [^{14}C]thymidine. Panel A shows that cell loss from the monolayer, which accompanies metabolic death, does not begin until 16 to 20 hours after irradiation at G_1/S. Furthermore, the accumulation of label, administered to the cultures at a higher level subsequent to exposure at G_1/S, continued at a constant rate

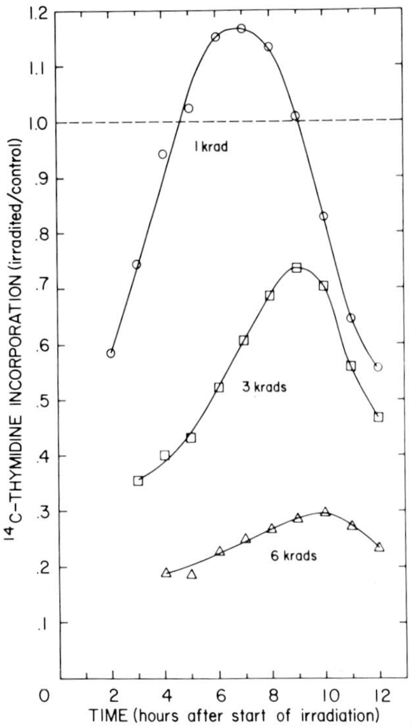

FIG. 9. Time-course of the rate of DNA synthesis relative to unirradiated controls after irradiation with three different doses. Measurements of [^{14}C]thymidine incorporation in 20-minute pulses were carried out over a long enough time to reveal a second depression. (Reproduced from Tolmach and Jones 1977, with permission of Academic Press.)

in cells that remained attached during the period when the curves of Figure 9 pass through their maxima and also when cells later begin to die (Figure 10, panel B). The decline in rate might arise from the arrest of cells in G_2. It may be noted, also, that the cells remaining attached synthesize essentially a full complement of DNA even after 8 krad, albeit at a greatly diminished rate.

Chemically Imposed Alterations of Radiation-Induced Kinetic Changes

The kinetics of the deceleration of replication shown in Figure 4 are consistent with a single type of radiation-induced lesion's being responsible for the temporary inhibition of replicon initiation, with consequent (transient) decline in the rate of synthesis. Two additional findings associated with the rate depression are pertinent to analysis of the relevance of this phenomenon to cell death. First, recovery of the rate of replication requires that DNA synthesis take place. As mentioned above, irradiation even in early G_1 is followed by a depression of synthesis when the cells reach S, as shown in Figure 11. Furthermore, if replication is temporarily blocked with fluorodeoxyuridine (FUdR) for as long as six hours (Figure 12), or with cytosine arabinoside, the typical transient depression in rate is observed after the block is bypassed (Saha and Tolmach 1976); larger depressions induced by higher doses are similarly postponed by

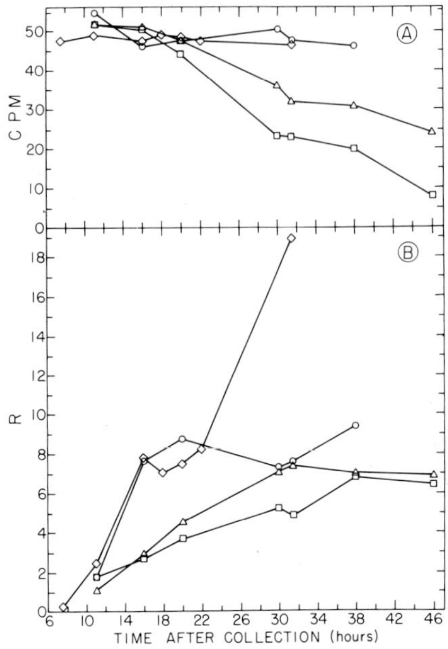

FIG. 10. Incorporation of [^{14}C]thymidine in HeLa cells following exposure to X rays. A monolayer culture of randomly dividing cells was incubated 13 hours in medium F-10 with 10% calf and 5% fetal calf serum to which was added 0.025 μCi (62 Ci/mole) of [^{14}C]thymidine per milliliter of culture. The culture flasks were then shaken to dislodge rounded cells four times at 15-minute intervals; at each time, the medium was decanted, discarded, and replaced with unlabeled medium. After 45 minutes incubation, the flasks were shaken and a suspension of mitotic cells bearing about 0.04 d.p.m. of ^{14}C per cell was collected. 0.25 μCi (62 Ci/mole) of [^{14}C]thymidine was added per milliliter of medium to one half of the suspension, and both halves were plated in replicate 35-mm plastic dishes. Dishes were irradiated with 0 (diamonds), 1 (circles), 4 (triangles), or 8 (squares) krad at a dose rate of about 90 rad per minute at times beginning 8 to 9.5 hours after plating. At the times shown on the abscissa, labeling was terminated by removing the medium and rinsing with cold buffer. The cells were then fixed for two hours or more in 0.2 M perchloric acid and dried with 70% ethanol. The bottom of each dish was assayed for affixed ^{14}C label in a low-background (about 1.8 counts per minute) gas-flow Geiger counter. Panel A shows the activity of ^{14}C on dishes with cells bearing only label that was incorporated prior to mitotic collection (CPM); this measures the loss of cells from the monolayer. If CPM* denotes the ^{14}C on a dish with cells that were also incubated in labeled medium after mitotic selection, then R, defined by R = (CPM* − CPM)/CPM, which is plotted in panel B, measures thymidine incorporation per unit of preexisting DNA in cells that remain attached to the dish at time of termination.

FUdR treatment (Tolmach and Jones 1977). This indicates that the depression does not result simply from inactivation of a component of the initiation system: there should have been sufficient time for its replacement. Postirradiation treatment with FUdR during S has little if any effect on survival.

Second, while DNA synthesis is necessary for the repair of the damage that causes the depression in the rate of replication, it is not sufficient. If cells irradiated with doses of the order of 1 krad are incubated in the presence of 1 mM

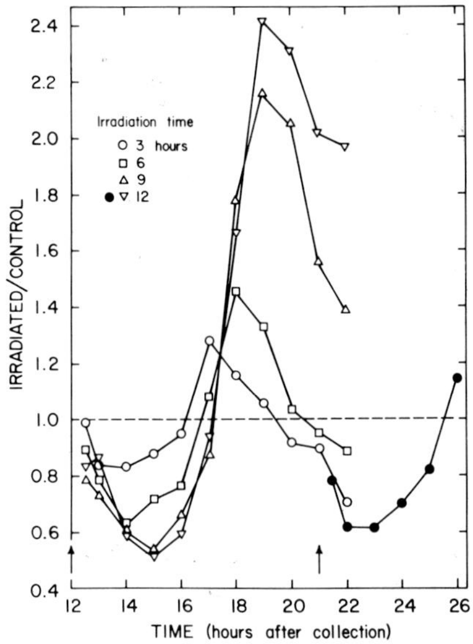

FIG. 11. Ratio of incorporation of [^{14}C]thymidine by cultures irradiated at different times in G_1 to incorporation by unirradiated cultures. FUdR (1 μM) was added to collected mitotic cells (zero hours), which were irradiated with 1 krad at 3, 6, 9, or 12 hours. The FUdR block was bypassed at 12 hours, and cultures were pulse labeled for 20 minutes at the times shown by the open symbols. The solid circles show incorporation by blocked cultures irradiated at 12 hours but not relieved from the block until 21 hours. The arrows indicate the times at which the FUdR blocks were bypassed. (Reproduced from Saha and Tolmach 1976, with permission of Academic Press.)

caffeine (which does not kill a significant number of cells under the conditions employed), DNA synthesis continues and development of the depression is largely suppressed (Walters et al. 1974, Boynton et al. 1974), as shown in Figure 13 for synchronous cells. When the caffeine is removed, however, an apparently identical transient depression occurs, as is evident in the data for randomly dividing cultures shown in Figure 14. Caffeine also suppresses the depression

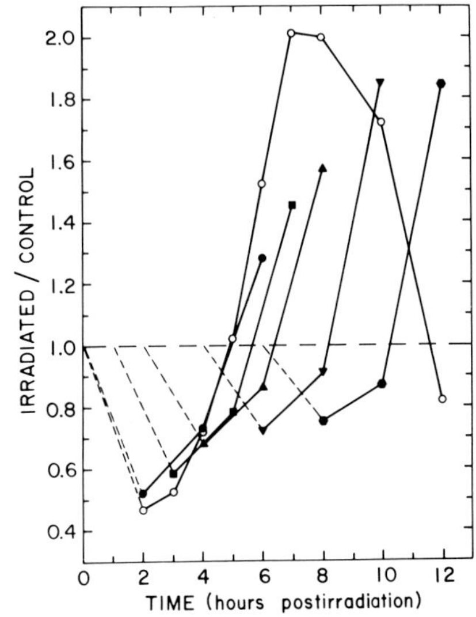

FIG. 12. Chemically induced delay of the radiation-mediated depression of DNA synthesis. Synchronous cultures were irradiated with 1 krad during S, 10.75 hours after mitotic collection. Ratios are given for the incorporation of [^{14}C]thymidine in 20-minute pulses by irradiated cultures compared to unirradiated cultures. The solid symbols show the effect of treatment with 1 μM FUdR for 0 (●), one (■), two (▲), four (▼), or six (⬢) hours after irradiation. The FUdR inhibition was bypassed with 10 μM thymidine. The open circles refer to untreated irradiated cells. The gradual shift of the ratios to higher values with longer exposure to FUdR is presumably artifactual, arising from accelerated passage through S as a result of the progressively longer inhibition of synthesis. Similar results have been obtained with randomly dividing cultures. (Reproduced from Saha and Tolmach 1976, with permission of Academic Press.)

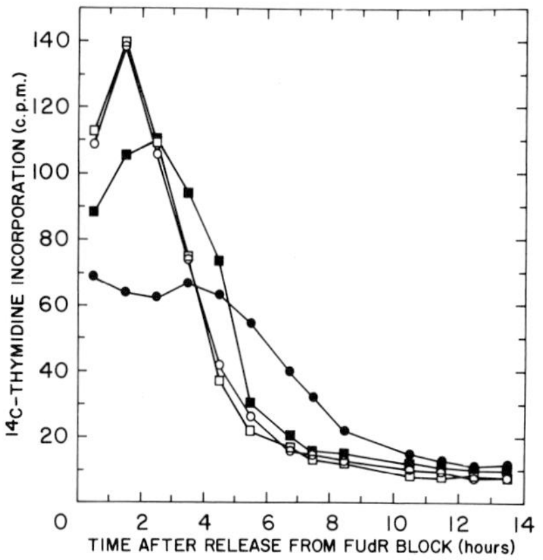

FIG. 13. Suppression of X-ray-induced depression of the rate of DNA synthesis in synchronously growing HeLa cells. Cultures resynchronized at G_1/S by treatment with FUdR were irradiated with 1 krad (●), treated with 1 mM caffeine only (□), or subjected to both treatments (■) and pulse labeled with [^{14}C]thymidine at hourly intervals after release from FUdR block (zero time). O: untreated controls. Vinblastine (0.03 μg/ml) was present to prevent cells from entering the next generation. The caffeine and vinblastine were added at the time of release from the block. Each point represents the mean of duplicate cultures. (Reproduced from Tolmach et al. 1977, with permission of Academic Press.)

after larger exposures, but higher concentrations must be used (Tolmach et al. 1977). This action of caffeine must be interpreted as preservation of damage, that is, the prevention of repair. Caffeine might simultaneously initiate spurious replication forks (Lehmann 1972, Buhl and Regan 1974) that permit synthesis to continue, though if so, it is not clear why the rate is often maintained by caffeine at or close to the control value (Figures 13 and 14, Lehmann 1972); perhaps some factor limits the overall rate of synthesis, irrespective of the number of replication forks. Alternatively, the replication system might simply ignore

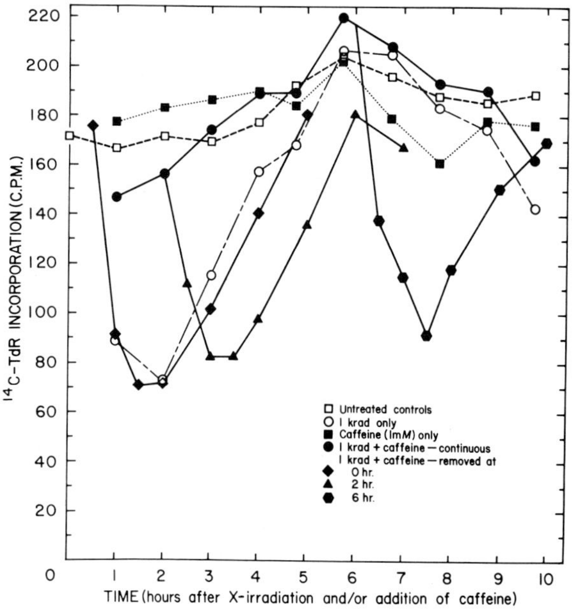

FIG. 14. Caffeine-mediated delay in the expression of X-ray-induced inhibition of DNA synthesis. Replicates of a randomly dividing culture of HeLa cells were irradiated with 1 krad and treated as indicated in the figure. Duplicate cultures were pulse labeled for 20 minutes with [^{14}C]thymidine at the time shown by each point. (Reproduced from Tolmach et al. 1977, with permission of Academic Press.)

the lesions if caffeine is present. In any case, caffeine does not simply hide a depression; it prevents one.

Taken together, these observations implicate the DNA template in the inhibition of replication. The location of the lesion in DNA has been more definitively indicated by the experiments of Povirk and Painter (1976) in which X-ray-induced inhibition of replicon initiation was mimicked by 313-nm irradiation of bromodeoxyuridine-containing DNA.

EFFECT OF CAFFEINE ON X-RAY-INDUCED CELL KILLING AND G_2 ARREST

Our initial findings with caffeine led us to conclude that the lesions responsible for depressing the rate of DNA synthesis have no relevance to cell killing: the depression is prevented by caffeine, which was presumed not to affect survival. (If there were a relation between the two, one might expect the suppression of the radiation-induced depression of synthesis to be accompanied by a decrease in cell killing.) Indeed, the simultaneous suppression and preservation of damage

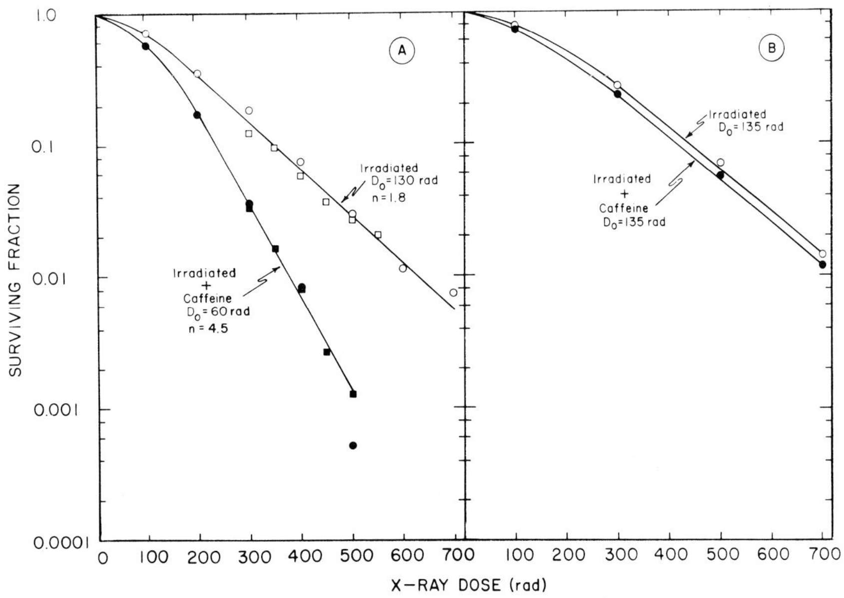

FIG. 15. Effect of 1 mM caffeine on the survival of X-irradiated HeLa cells. Panel A: A randomly dividing population was trypsinized, plated, and irradiated four hours later with the doses indicated. Cultures were then treated with caffeine for 24 hours. All data points are the means of duplicate dishes. Open and solid symbols represent control and caffeine-treated cultures, respectively. Circles and squares are data from two separate experiments. Panel B: Cultures were treated with caffeine for 21 hours immediately preceding irradiation. D_o is the mean lethal dose in the exponential portion of the curve; n is the extrapolation number. (Reproduced from Busse et al. 1977, with permission of Academic Press.)

suggested that the depression is only indirectly related to lesions in DNA (Tolmach et al. 1977). However, subsequent study of the action of caffeine on X-ray-induced lethality has prompted a reexamination of these conclusions.

Effects of Caffeine on Cell Killing

We found, first, that despite scattered reports to the contrary (Rauth 1967, Arlett 1970, Walters et al. 1974), postirradiation treatment with caffeine markedly enhances the lethal action of X rays (Figure 15) on certain cell lines (Busse et al. 1977). Using synchronous cultures of HeLa or Chinese hamster ovary (CHO) cells, and a constant 24-hour treatment time, the enhancement was observed to be more effective following irradiation in the latter part of the cycle than earlier (Figure 16). Further detailed kinetic examination (Busse et al. 1978) showed that this age response is the result of a rapid and extensive killing of irradiated cells during G_2, in contrast to a much smaller (and more variable) killing during G_1 and a still smaller response in S (Figure 17). We found also that although cells respond to caffeine during G_2 irrespective of their age at irradiation, the amount of enhanced killing is not the same for irradiations carried out at all cell ages; it is apparently greater in older cells. This is shown in Figure 18, which depicts the result of an experiment in which

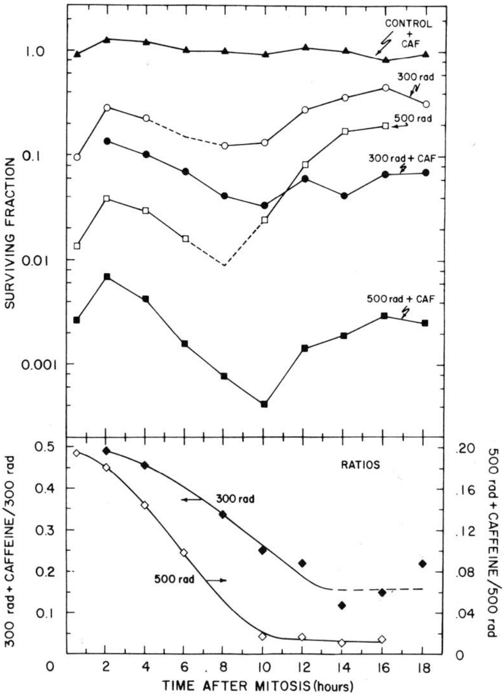

FIG. 16. Age response for caffeine-enhanced killing of HeLa cells by X rays. Collected mitotic cells were subjected to 300 (○) or 500 (□) rad at two-hour intervals throughout the cell cycle. The open symbols show survival after irradiation only. Solid circles and squares show survival after a 24-hour postirradiation treatment with 1 mM caffeine. Solid triangles show the lack of an effect of 1 mM caffeine on unirradiated cells. Dashed portions of the curves show survival values from repeat experiments. The ratios (lower panel) reflect the reduction in survival resulting from the caffeine treatment. The arrows point to the appropriate ordinate for each curve. (Reproduced from Busse et al. 1977, with permission of Academic Press.)

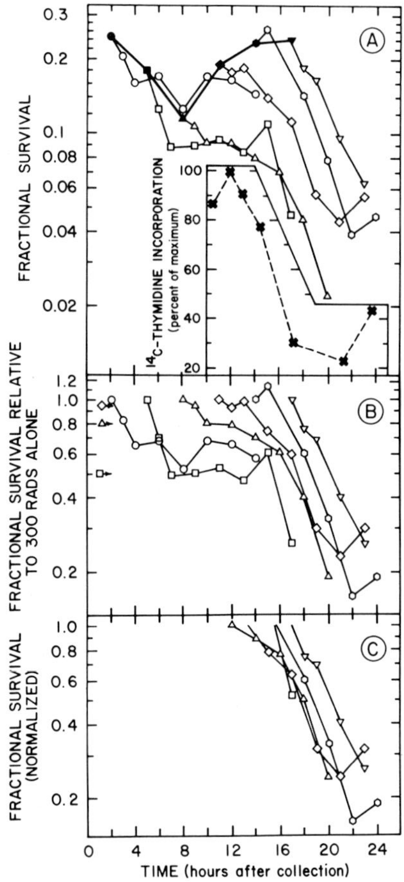

FIG. 17. Age-specific enhancement of X-ray-induced killing of HeLa cells by caffeine. Portions of a synchronous population were irradiated with 300 rad at three-hour intervals beginning two hours after mitotic collection. The solid symbols (heavy lines) in panel A represent survival after irradiation alone. Caffeine (1 mM) was added to each series immediately after irradiation and removed at the times indicated by the open symbols. The inset shows the rate of DNA synthesis as measured by the incorporation of [^{14}C]thymidine during 20-minute pulses, expressed as percentage of the maximum rate. In panel B, the survival after each irradiation has been set equal to 1.0, so that the values plotted show the enhanced cell killing from caffeine treatment. The enhanced cell killing in G_2 is more clearly discerned (panel C) after the G_1 component of killing has been removed by normalizing the survival level reached in G_1 (indicated by the arrows at left in panel B) to 1.0. (Reproduced from Busse et al. 1978, with permission of Academic Press.)

synchronous cells were irradiated at different times, caffeine was added, and survival was followed long enough to make evident the plateau levels achieved. We take this age response to reflect the progressively shorter times between irradiation and G_2, not an inherently greater sensitivity to caffeine in older cells. That is, repair of potentially lethal damage appears to occur during the cycle (Phillips and Tolmach 1966), and we presume that such repair is inhibited by caffeine. This interpretation has been strengthened by the finding that when caffeine was added at progressively later times after irradiation in G_1, sensitivity was lost in about nine hours (Figure 19), and even faster if cells were irradiated later in the cycle, in late S/early G_2 (Busse et al. 1978).

Effect of Caffeine on G_2 Arrest

The finding that the killing of irradiated HeLa and CHO cells by caffeine takes place most extensively in G_2 prompted examination of the effect of caffeine

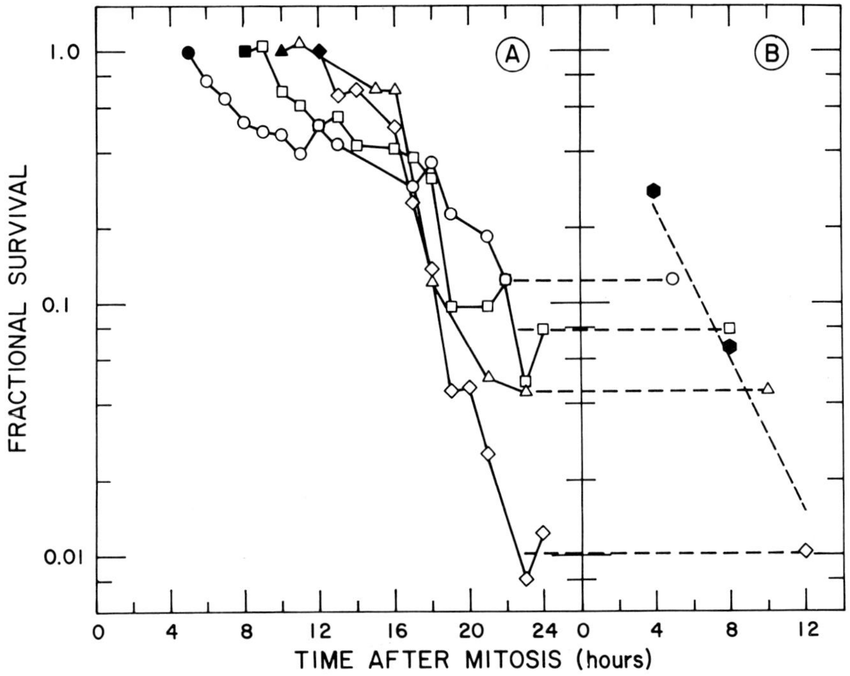

FIG. 18. Caffeine-enhanced cell killing as a function of cell age at the time of irradiation and duration of treatment. Portions of a population of synchronous cells were irradiated with 400 rad, 5 (●), 8 (■), 10 (▲), or 12 (◆) hours after mitotic collection. Survival after irradiation alone at each time has been set equal to 1.0 (filled symbols in panel A). Caffeine (1 mM) was added to each series immediately after irradiation and removed at the times indicated by the open symbols in panel A. The estimated plateau values have been plotted in panel B as a function of the time (age) at which cells were irradiated. The open symbols are from the data in panel A (dashed horizontal lines); each of the filled hexagons is from a separate experiment. (Reproduced from Busse et al. 1978, with permission of Academic Press.)

on X-ray-induced G_2 arrest in HeLa cells, since caffeine had been shown to shorten arrest in CHO cells (Walters et al. 1974). Using time-lapse cinemicrography (Tolmach et al. 1978), we found (Tolmach et al. 1977) that the duration of arrest is markedly shortened in these cells also (Figure 20). The enhancement of killing by caffeine during G_2 could reflect the premature truncation of a repair process that occurs during G_2 arrest. It is pertinent that in at least one cell line (EMT6) that does not show significant enhancement of radiation-induced killing by caffeine (Busse et al. 1977), the duration of G_2 arrest is only slightly reduced in the presence of the drug (Busse 1979). Alternatively, caffeine might prevent repair of damage while cells progress past a point in G_2 after which repair can no longer occur, and independently shorten G_2 arrest; i.e., the shortening of G_2 arrest mediated by caffeine could be unrelated to the enhancement of cell killing.

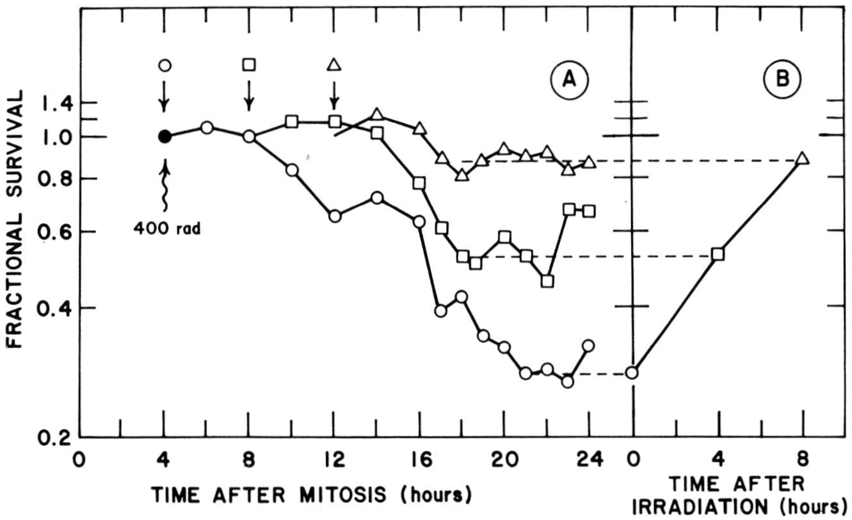

FIG. 19. Effect of delaying treatment of irradiated cells with 1 mM caffeine. Portions of a synchronous population were irradiated with a dose of 400 rad four hours after mitotic collection. The data in panel A trace the time-course of the enhancement of killing subsequent to the addition of caffeine at the times indicated by the upper arrows. Survival after irradiation alone (0.05) has been set equal to 1.0 (filled circle). Estimated plateau values in panel A (dashed lines) are plotted in panel B as a function of time after irradiation at which the caffeine was added. (Reproduced from Busse et al. 1978, with permission of Academic Press.)

DISCUSSION

We interpret the findings presented here in terms of a limited number of assumptions. Certain of these are widely accepted, and others have been previously discussed (sometimes negatively) in the papers cited above, and doubtlessly elsewhere. The assertions that follow are based largely on the experimental data summarized in the previous sections.

Damage of various kinds (strand breaks, cross-links, base alterations) is introduced in DNA by X rays. At least some of the lesions are potentially lethal: they ultimately kill the cell if not repaired. On recognition by a cellular surveillance system (or systems), some of the lesions are repaired. Cell progression through S stops, or is slowed, as the result of the inhibition of replicon initiation. The inhibition could be a direct consequence of the lesion, or it might reflect the activation of the repair process (e.g., the repair system could preempt part of the initiation system). The inhibition is relaxed after repair has occurred. Repair occurs only slowly in G_1, probably more rapidly in S, and possibly still faster in G_2, but it occurs scarcely at all during S in the presence of inhibitors of DNA synthesis, such as FUdR, possibly because thymidine is not available for repair, or because repair, which might involve the elimination of cross-

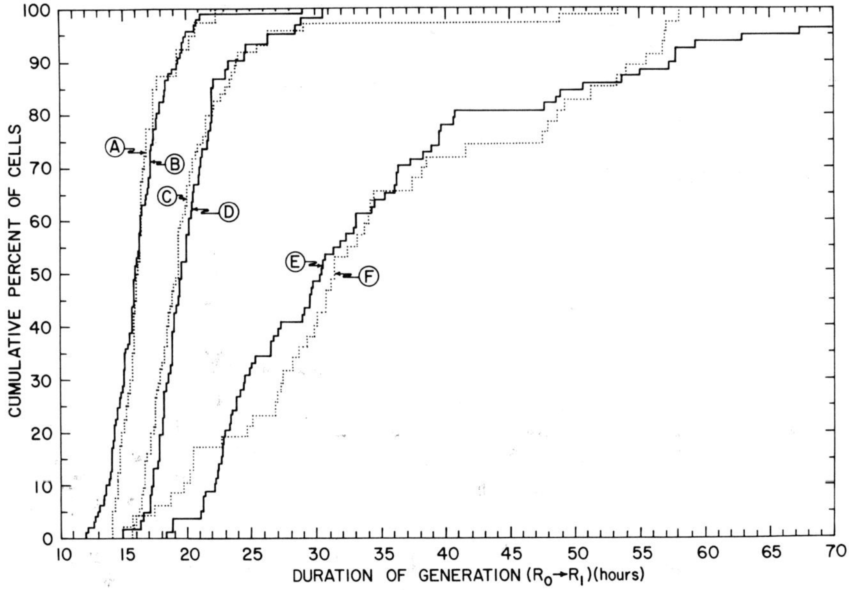

FIG. 20. Suppression of X-ray-induced division delay in HeLa cells by 1 mM caffeine. The cumulative number of cells reaching mitosis during the irradiated generation as a function of time ($R_0 \rightarrow R_1$) was determined from time-lapse film analysis. A, untreated control; B, cells treated with 1 mM caffeine; E and F, cells irradiated with 1 krad; C and D, cells irradiated with 1 krad and treated with 1 mM caffeine. D and E were obtained in one experiment and C and F in a second. As randomly dividing cultures were employed, the cells were exposed to caffeine for variable portions of the generation; this probably accounts for the absence of any apparent delay in cells treated with caffeine only (curve B compared with curve A), since caffeine at this concentration delays progression only through G_1. (Reproduced from Tolmach et al. 1977, with permission of Academic Press.)

links, is facilitated by the replication process. Nevertheless, postirradiation treatment with FUdR during S has no appreciable effect on survival, presumably because the lesions are repaired after the inhibitor is removed.

Because replicon initiation is inhibited only in those regions of DNA that bear lesions (Povirk 1977), the inhibition presumably stems directly from lesions in DNA; that is, damage appears not to lead to systemic inhibition of replicon initiation throughout the genome. Synthesis of the replicons already initiated at the time of irradiation is completed, giving rise to the descending limbs in the curves of Figure 2. The dose dependence of the duration of the hiatus in initiation (see Figure 6) is consistent with the saturation of the replicating regions with damage.

Resumption of initiation follows repair of some lesions. Repair may be autocatalytic, at least it does not occur in the absence of DNA synthesis, and as it proceeds its rate increases, thereby increasing the rate of synthesis. It is not certain that the rate returns to normal by the end of S, so it is not clear whether the types of lesions repaired in S and in G_2 are the same; they need not be.

In any case, repair evidently continues during G_2, and we attribute at least a portion of G_2 arrest to the time required for that repair. That is, the cell does not enter mitosis until certain lesions are eliminated. The large dispersion observed in the duration of arrest (Figure 20) can arise from the progressive occurrence of repair among the arrested cells.

We suggest that caffeine inhibits the repair process(es) in such a fashion that the cell behaves as if it contained no reparable damage. That is, caffeine seems to mask the lesions from the surveillance system, possibly by binding to them or to the surveillance apparatus. The former seems more likely because greater concentrations of caffeine are needed to suppress the effect after higher doses; i.e., the damage seems to be titrated. Although the radiation-induced perturbations in progression through S and G_2 are thereby suppressed, the cell continues to harbor damage; it is revealed as a depressed rate of replication when the caffeine is removed during S, or as increased killing if the caffeine is allowed to act past a point in G_2 at which the cell receives some signal having to do with entrance into mitosis. Progress of the cell through mitosis while it bears unrepaired lesions eventually leads to its reproductive death, perhaps because chromosome abnormalities resulting from the lesions cause either improper allocation of genetic material to daughter cells or fusion of sister cells after division. Because treatment of irradiated cells with caffeine during S alone does not kill them, the repair that is suppressed during S presumably takes place in G_2.

Certain aspects of this scheme are amenable to further study. For example, it will be of interest to determine whether cell lines in which caffeine fails to shorten X-ray-induced arrest in G_2, or to enhance killing, show caffeine-mediated suppression of the radiation-induced inhibition of DNA synthesis. Certain other features of the model can be examined only with considerable difficulty. Thus, it might be feasible to inhibit the putative repair during S with caffeine (prevent the slowing of synthesis) and then remove the caffeine when the cells reach G_2. Under such conditions G_2 arrest might be prolonged while the damage preserved during S is being repaired. G_2 arrest might also be accompanied by a burst of DNA synthesis. However, because of the dispersion in cell age that is inevitably introduced during growth, it may not be possible to carry out the required manipulations at the appropriate times.

ACKNOWLEDGMENT

These investigations were supported by Grant Number CA 04483 awarded by the National Cancer Institute, Department of Health, Education and Welfare.

REFERENCES

Arlett, C. F. 1970. The influence of post-irradiation conditions on the survival of Chinese hamster cells after gamma-irradiation. Int. J. Radiat. Biol. 17:515–526.

Boynton, A. L., T. C. Evans, and D. A. Crouse. 1974. Effects of caffeine on radiation-induced mitotic inhibition in S-180 ascites tumor cells. Radiat. Res. 60:89–97.

Buhl, S. N., and J. D. Regan. 1974. Effect of caffeine on post-replication repair in human cells. Biophys. J. 14:519–527.

Busse, P. M. 1979. The effects of caffeine on the survival of X-irradiated HeLa S3 cells. Ph.D. Dissertation, St. Louis University, St. Louis, Missouri.

Busse, P. M., S. K. Bose, R. W. Jones, and L. J. Tolmach. 1977. The action of caffeine on X-irradiated HeLa cells. II. Synergistic lethality. Radiat. Res. 71:666–677.

Busse, P. M., S. K. Bose, R. W. Jones, and L. J. Tolmach. 1978. The action of caffeine on X-irradiated HeLa cells. III. Enhancement of X-ray-induced killing during G_2 arrest. Radiat. Res. 76: 292–307.

Gautschi, J. R., and R. M. Kern. 1973. DNA replication in mammalian cells in the presence of cycloheximide. Exp. Cell Res. 80:15–26.

Lehmann, A. R. 1972. Effect of caffeine on DNA synthesis in mammalian cells. Biophys. J. 12:1316–1325.

Makino, F., and S. Okada. 1975. Effects of ionizing radiations on DNA replication in cultured mammalian cells. Radiat. Res. 62:37–51.

Painter, R. B., and B. R. Young. 1975. X-ray-induced inhibition of DNA synthesis in Chinese hamster ovary, human HeLa, and mouse L cells. Radiat. Res. 64:648–656.

Painter, R. B. 1980. The role of DNA damage and repair in cell killing, *in* Radiation Biology and Cancer Research (The University of Texas System Cancer Center 32nd Annual Symposium on Fundamental Cancer Research, 1979). Raven Press, New York, pp. 59–68.

Phillips, R. A., and L. J. Tolmach. 1966. Repair of potentially lethal damage in X-irradiated HeLa cells. Radiat. Res. 29:413–432.

Povirk, L. F. 1977. Localization of inhibition of replicon initiation to damaged regions of DNA. J. Mol. Biol. 114:141–151.

Povirk, L. F., and R. B. Painter. 1976. The effect of 313 nanometer light on initiation of replicons in mammalian cell DNA containing bromodeoxyuridine. Biochim. Biophys. Acta 432:267–272.

Rauth, A. M. 1967. Evidence for dark-reactivation of ultraviolet light damage in mouse L cells. Radiat. Res. 31:121–138.

Saha, B. K., and L. J. Tolmach. 1976. Delayed expression of the X-ray-induced depression of DNA synthetic rate in HeLa S3 cells. Radiat. Res. 66:76–89.

Tolmach, L. J., and R. W. Jones. 1977. Dependence of the rate of DNA synthesis in X-irradiated HeLa S3 cells on dose and time after exposure. Radiat. Res. 69:117–133.

Tolmach, L. J., R. W. Jones, and P. M. Busse. 1977. The action of caffeine on X-irradiated HeLa cells. I. Delayed inhibition of DNA synthesis. Radiat. Res. 71:653–665.

Tolmach, A. P., A. R. Mitz, S. L. VonRump, M. D. Pepper, and L. J. Tolmach. 1978. Computer-assisted analysis of time-lapse cinemicrographs of cultured cells. Comput. Biomed. Res. 11:363–379.

Walters, R. A., L. R. Gurley, and R. A. Tobey. 1974. Effects of caffeine on radiation-induced phenomena associated with cell-cycle traverse of mammalian cells. Biophys. J. 14:99–118.

Walters, R. A., and C. E. Hildebrand. 1975. Evidence that X-irradiation inhibits DNA replicon initiation in Chinese hamster cells. Biochem. Biophys. Res. Commun. 65:265–271.

Watanabe, I. 1974. Radiation effects on DNA chain growth in mammalian cells. Radiat. Res. 58:541–556.

Weiss, B. G. 1971. Perturbation of precursor incorporation into DNA of X-irradiated HeLa S3 cells. Radiat. Res. 48:128–145.

Radiation Biology in Cancer Research, edited by
Raymond E. Meyn and H. Rodney Withers.
Raven Press, New York © 1980.

Oxygen-Dependent Sensitization of Irradiated Cells

David Ewing and E. L. Powers*

*Department of Radiation Therapy and Nuclear Medicine, Hahnemann Medical College and Hospital, Philadelphia, Pennsylvania 19102; and * Laboratory of Radiation Biology, Department of Zoology, The University of Texas, Austin, Texas 78712*

Observations that oxygen increases the response of cells exposed to ionizing radiation can be traced back to the beginning of this century. The earliest of these seems to have been made by Schwartz (1909), although he did not interpret his results in terms of different oxygen concentrations. Investigations by Holthusen (1921) with *Ascaris* eggs, by Petry (1921) with seeds, by Crabtree and Cramer (1933) with a transplantable murine carcinoma, and by Mottram (1935) with rats are among those having special importance. However, the studies begun in the 1950s by L. H. Gray and his colleagues clearly represent a milestone in radiation biology. Gray's 1961 review is an excellent survey of information to that time.

When Gray surveyed the literature, clear evidence already showed that in very dry biological systems, for example, bacterial spores (Powers et al. 1960), more than one kind of oxygen-dependent sensitizing process existed. Recent work with cells irradiated in suspension (Alper 1963, Tallentire et al. 1972, Shennoy et al. 1975, Ewing and Powers 1976) has confirmed this 20-year-old observation: oxygen operates in more than one way in affecting the radiation sensitivity of the cell.

Oxygen sensitizes all cells to irradiation. Its effects have been studied with many experimental techniques and with many different organisms. Perhaps in part because of this diversity, definitive studies do not show whether oxygen sensitizes all organisms through identical chemical pathways. Some experimental information supports this assumption; other data contradict it. The survey below will examine the chemical mechanisms involved in oxygen-dependent sensitizations of cells irradiated in suspension.

SENSITIZATION BY DIFFERENT O_2 CONCENTRATIONS

The effects of varying O_2 concentrations have been studied in several cellular systems by a number of authors. Most of the early studies recognized a dependence of radiation sensitivity upon O_2 concentration that increases very sharply to a plateau with no microstructure in the response. The studies on *Bacillus megaterium* spores (Tallentire et al. 1972, Ewing and Powers 1976) demonstrated the important fact that there is an intermediate region of sensitivity in which there is a plateau at low [O_2] at approximately half the sensitivity of the plateau seen at high [O_2]. While previous authors have not demonstrated this intermediate effect in their studies, their results are indeed consistent with those two sets of experiments. In Figure 1, we have normalized the radiation sensitivity within several systems and have plotted the degree to which different [O_2] sensitize within the maximal O_2 effect. Note that the two spore studies are central, with the Ewing and Powers (1976) study demonstrating a smooth relationship between [O_2] and radiation sensitivity. As pointed out previously, the Manchester spore studies are consistent with the Austin studies, except that the plateau

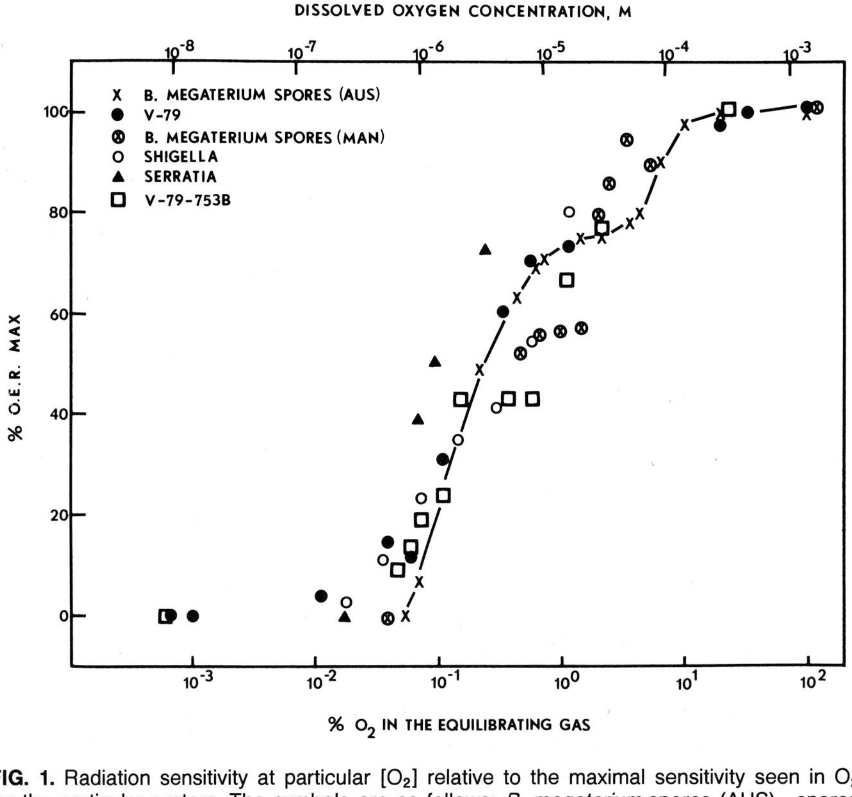

FIG. 1. Radiation sensitivity at particular [O_2] relative to the maximal sensitivity seen in O_2 for the particular system. The symbols are as follows: *B. megaterium* spores (AUS)—spores irradiated in H_2O with 50 kVp X rays (Ewing and Powers 1976); V-79—Chinese hamster cells irradiated in culture medium with 250 kVp X rays (Chapman et al. 1974); *B. megaterium* spores (MAN)—spores irradiated in phosphate buffer with ^{60}Co γ rays (Tallentire et al. 1972); *Shigella*—irradiated in phosphate buffer with 200 kVp X rays (Howard-Flanders and Alper 1957); *Serratia*—irradiated in buffer with 200 kVp X rays (Dewey 1963); V-79-753B—Chinese hamster cells irradiated in culture medium with ^{60}Co γ rays (Millar et al. 1979). The solid line is the response for spores as presented by Ewing and Powers (1976).

occurs at a different level, this being explained, perhaps, by the difference between ^{60}Co gamma rays and 50 kVp X rays. The remarkable aspect of Figure 1 is that two different vegetative bacteria and two sets of experiments on V79 mammalian cells are consistent with the response demonstrated by the bacterial spores. So, while there may be differences in absolute sensitivity among the variety of systems, the O_2 effect, when looked at as in Figure 1, is truly the same in the several systems, perhaps indicating a unity in the mechanisms involved in the O_2 effect, whether in mammalian cells or bacterial spores. Indeed, in the most recent observations on hamster cells (Millar et al. 1979), indicated as open squares on Figure 1, the presence of a plateau at low $[O_2]$ is acknowledged. The difference between the spore and the mammalian cells in this instance is that the plateau appears at O_2 concentrations of a factor of 10 below those at which the plateau appears in the spore system.

The relative amount of sensitization seen at a particular O_2 concentration is very nearly independent of cell type or the suspending medium at the time of irradiation. This indicates to us that the intensive investigations in bacterial systems, whether spores or vegetative cells, are proper models for construction of experiments in mammalian cells to search for similar effects.

SEPARATION OF O_2-DEPENDENT DAMAGE INTO COMPONENTS

Bacterial Spore Studies

The first recognition of oxygen's having more than one kind of action was made in a dried system. Using dried bacterial spores, Powers et al. (1960) showed that oxygen's sensitization could be experimentally resolved into at least two major classes of damage. In the experiments, the exchange of gases took several minutes, and, based on this reference time-scale, these authors established the following definitions: (a) *Class I* damage is independent of oxygen; its magnitude is the same whether or not O_2 is present; (b) *Class II* damage is oxygen-dependent and short-lived; it can be observed only when oxygen is present during irradiation; and (c) *Class III* damage, the "postirradiation" O_2 effect, is produced during either oxic or anoxic exposures; however, the development of this kind of damage is very slow in this dry biological system. There is good evidence that the kind of damage designated as *Class III* involves a reaction between O_2 and a radiation-induced cellular radical (Powers 1966).

Tallentire and Powers (1963) later showed that intracellular water protects against both kinds of oxygen-dependent damage. Spores irradiated in suspension or under saturated water vapor conditions show no *Class III* and only a reduced amount of *Class II* damage; under these conditions, O_2 introduced a few minutes after anoxic irradiation did not increase the sensitivity. Tallentire and Powers did propose, however, that two kinds of oxygen-dependent damage probably operate in the irradiated wet spore. The review article by Powers and Tallentire (1968) summarizes the work with dried biological systems.

Recent work has made it very clear that organisms irradiated in suspension do indeed show more than one kind of oxygen-dependent damage. Using ^{60}Co γ rays, Tallentire et al. (1972) found evidence for at least two components of oxygen-dependent damage in bacterial spores.

Later experiments with 50 kVp X rays showed that the sensitization of spores by oxygen could be separated into at least three components (Ewing and Powers 1976) that are O_2-concentration dependent (Figure 2). *t*-Butanol is an effective scavenger of OH radicals and the oxygen-dependent damage removed with this alcohol present is designated as the "•OH component." Two other components of damage are designated as the "low-O_2" and "high-O_2" components, to emphasize that different concentrations of oxygen can produce different kinds of damage (Ewing 1978a).

An important difference has not yet been resolved between the spore experiments with 50 kVp X rays (Ewing and Powers 1976) and those with ^{60}Co γ rays (Tallentire et al. 1972). In both sets of data, the changes in response occur over very similar oxygen concentrations, just as they do for most organisms (cf. Figure 1). Furthermore, the changes themselves are qualitatively the same: as the $[O_2]$ is raised, the responses increase, reach a plateau, then increase

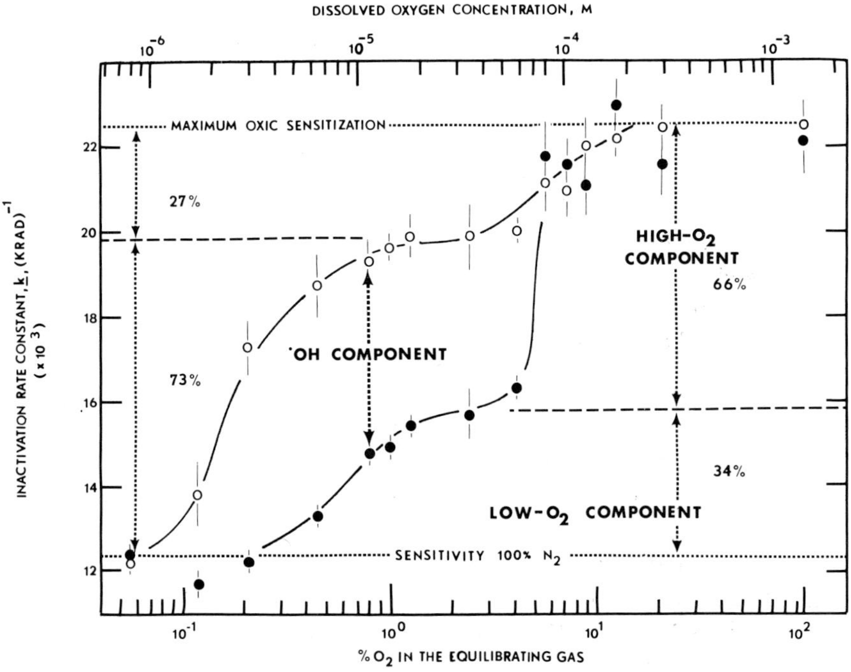

FIG. 2. The radiation sensitivity of *B. megaterium* spores as a function of O_2 concentration (Ewing and Powers 1976). Irradiation was with 50 kVp X rays. Different $[O_2]$ were prepared by adding measured amounts of N_2 to a cylinder containing O_2. The resulting $[O_2]$ was measured with a gas chromatograph.

again to a maximum value. However, with γ rays, the plateau occurs when about 30% of the maximum sensitization has been reached (cf. Figure 4, Tallentire et al. 1972). With 50 kVp X rays, the plateau comes at 70%. The basis for this discrepancy is not known, although differences other than photon energy exist between the two sets of experiments; e.g., the γ-ray work was in phosphate buffer whereas the X-ray experiments used water-suspended spores. This "water-versus-buffer" question would not be important when γ rays are used (Tallentire, personal communication, quoted in Ewing 1975), but it would be important with 50 kVp X rays (Ewing 1975).

Recent work with bacterial spores and 1,4-diazobicyclo [2.2.2]-octane (DABCO), a quencher of singlet oxygen ($O_2^*\ {}^1\Delta_g$), suggests that an additional fourth component of damage may now have been recognized (Barber and Centilli, unpublished results). Over a range of oxygen concentrations, DABCO has a protective effect that is additive to that seen through OH radical removal; this suggests a component of sensitization involving singlet O_2, although the results of further experiments are needed for confirmation.

The oxygen-dependent sensitization of bacterial spores can also be resolved into time components if the radiation dose is delivered in a very brief interval. With pulsed techniques that allowed examination of incidents in very short times after irradiation, Weiss and his colleagues (Weiss and McDonald 1976, Weiss and Santomasso 1977, Jones and Weiss 1977) noted that in wet spores, as in dry spores, O_2 delivered after anoxic irradiation can increase sensitivity. Complementary work by Tallentire and his colleagues (Tallentire et al. 1977, Stratford et al. 1977) showed that after anoxic irradiation of spores in suspension, the decay of the potentially lethal species can be resolved into two components; the radical half-lives associated with these two reactions are 9 seconds and 120 seconds. Spores irradiated in O_2 show only one first-order radical decay process ($T_{1/2} = 9$ seconds). Presumably, this is the faster of the two processes observed in anoxia.

Very recent results, also by Tallentire and his co-workers with this same rapid-mix method (Tallentire et al. 1979), provide additional information about these two long-lived components. When an OH radical scavenger is present during anoxic irradiation, the production and subsequent two-component decay is unaffected. Thus, $^{\bullet}OH$ appears not to be involved in either of the two O_2 actions. When these authors use N_2O (present during irradiation), the initial survival level is lowered for "zero time before O_2 introduction," and it remains at the same reduced level even when the introduction of O_2 is delayed. They suggested that $^{\bullet}OH$ is able to react with the two O_2-sensitive "species" to "fix" damage by excluding the possibility that the radical sites could decay to a harmless state. Limited data, for which they used an OH radical scavenger along with N_2O, show that the effect of N_2O can be partially blocked. We should note that according to these observations $^{\bullet}OH$-dependent O_2 sensitization operates only in the presence of N_2O—an agent that supposedly acts only by increasing $^{\bullet}OH$ yield. If that is the only action, we ask why $^{\bullet}OH$ scavengers do not affect O_2 sensitization processes in the absence of N_2O. The difference is the removal of e_{aq}^- in the N_2O case and its presence in N_2O absence. Tallentire and co-workers' experiment with acetone, supposedly removing e_{aq}^-, does not give the answer, for it is used at 1 M, a concentration that also effectively removes all $^{\bullet}OH$ ($k_{acetone} + {}^{\bullet}OH = 6.8 \times 10^7\ M^{-1}$ seconds $^{-1}$).

Although both laboratories whose work is cited above worked with *B. megaterium* spores, their results are not entirely compatible. Weiss and Santomasso (1977) used a single three-nanosecond pulse of electrons, giving a total dose of either 400 or 600 krad to spores mounted on wet-membrane filters. After anoxic irradiation, they observed the decay of an oxygen-sensitive radical having a half-life of either 10.5 seconds or 7.4 seconds, respectively, depending on the radiation dose they used. In contrast, Tallentire and his co-workers irradiated spores in suspension with two-microsecond pulses of electrons, giving a total dose of 600 krad at about 0.8 krad/pulse. The total exposure lasted up to two seconds. After anoxic irradiation, they observed the decay of two oxygen-sensitive "species." Apparently, Weiss and Santomasso saw one but not both the radicals observed by Tallentire et al. Neither set of authors has attempted to explain the difference in results.

Vegetative Bacterial Cell Studies

Although experiments with dried vegetative bacteria showed two classes of oxygen-dependent damage (Webb 1964), tests with bacteria irradiated in suspension have not. Epp and his colleagues, whose experiments have been recently summarized (Epp et al. 1976), used a double-pulse technique with both *Escherichia coli* and *Serratia marcescens.* They irradiated oxygenated cells; then, by varying the time before a second electron pulse, they allowed different amounts of O_2 to diffuse back into the cells. Their analysis showed an upper limit of $\sim 10^{-4}$ seconds for the lifetime of the radical that can react with O_2 to cause damage. However, they did find a discontinuity in the graph of "decade spacing" (i.e., relative decrease in logarithm of fractional survival) versus interpulse time. This "bump" might be taken as evidence for more than one kind of oxygen-dependent damaging process, although this is not certain.

Michael et al. (1973) used a gas explosion technique to deliver O_2 at controlled times to anoxically irradiated cells. They found a postirradiation O_2 effect and estimated the half-life as 500 microseconds for the radical that reacts with O_2 to cause damage. This value is different from that obtained by Epp and co-workers, described above, who also worked with vegetative bacteria. Michael and colleagues did not, however, find evidence for more than one kind of reaction involving O_2. Shenoy et al. (1975) have used the liquid rapid-mix method in somewhat similar studies. This procedure, which showed two oxic sensitization components with mammalian cells, also failed to demonstrate more than one kind of oxygen-dependent damage in wet bacteria.

This difference in the measures of radical lifetimes from the two experimental methods is significant and it merits further comment. Michael and co-workers, irradiating under anoxic conditions and introducing O_2 afterward, measured a radical half-life of ~500 microseconds; Epp and co-workers, who irradiated in O_2 to radiolytically bind the dissolved O_2 and thereby achieve anoxia before the second pulse of radiation, found an upper limit to the radical lifetime of about 10^{-4} seconds. The upper limit to the lifetime of the radical found by Michael's group is about ten times longer than that found by Epp's (1976). We note here that the conflict in their results could be based on the different experimental techniques. The different conditions of irradiation may not have produced the same radical populations, and the two laboratories may not, in fact, have been studying the same kind of O_2-dependent damage. This suggestion is based on published work with dried bacterial spores. In these spore experiments, Powers et al. (1960) irradiated in anoxia, as Michael and co-workers did, and produced Class III damage. When Powers and co-workers irradiated in O_2, as Epp and colleagues did, they produced both Class II and Class III kinds of O_2-dependent damage in spores. Powers and Held (1979) have, in fact, recalled an earlier suggestion of Ewing (unpublished results) that may emphasize the importance of the different irradiation procedures these two laboratories used with vegetative bacteria. As they pointed out, Stratford et al. (1977), using wet bacterial spores, observed the postirradiation decay of two O_2-sensitive radicals following anoxic exposure. However, when they irradiated in O_2, only one radical, the one with the shorter half-life, was seen. In terms of procedure, these conditions of irradiation duplicate those of Michael (anoxic) and Epp (oxic), and Epp did, in fact, observe a shorter radical half-life (based on his estimate of upper limit to the radical lifetime) than did Michael. While these experiments with vegetative bacteria have not resolved O_2-dependent damage into separate components, it may yet be possible to do so with different experimental techniques and, perhaps, different methods of analysis.

In Vitro Mammalian Cell Studies

In contrast to the results with vegetative bacteria, information collected with mammalian cells after pulsed exposures also shows that sensitization by oxygen can be resolved into components. However, the time scales for these effects are much shorter than those noted above for bacterial spores. This difference in radical lifetimes is a major point of conflict between those studies with bacterial spores and these with mammalian cells. Tallentire et al. (1977) proposed that the long lifetimes they observed might have resulted from a relative dryness of the spore core, the presumed site for radiation-induced damage; this dryness would exist even though the spores were suspended in water. Thus, we might infer that the chemical processes through which O_2 sensitizes are the same in both spores and mammalian cells, even though the reactions themselves are considerably slower in spores.

Using a liquid fast-flow, rapid-mix method with mammalian cells, Shenoy et al. (1975) found that O_2—at any concentration—delivered two milliseconds *before* irradiation gave a constant amount of sensitization. (With this mixing technique, two milliseconds is the shortest possible time between O_2 contact and the radiation pulse.) Greater sensitization was obtained by allowing longer O_2 contact times before irradiation. The authors discussed their results in terms of "fast" and "slow" O_2 effects and suggested that the most plausible interpretation was that damage was produced at two sites within the cells; the "fast" versus "slow" resolution represented the times needed by O_2 to diffuse to these different sites.

Watts et al. (1978) have recently applied the gas-explosion method to in vitro mammalian cell experiments. This procedure allows a resolution of about one millisecond between O_2 delivery time and the radiation dose, whereas in the liquid rapid-mix method, described just above, the best resolution is about three milliseconds. These two methods for delivering O_2 to the cells give different results for the shortest (pre-irradiation) O_2 contact time necessary to achieve the maximum amount of oxic sensitization. However, in these preliminary experiments, the gas explosion technique probably also shows two postirradiation O_2 effects, just as the liquid rapid-mix method did. Watts did not focus on this point, however; we infer this from our examination of their Figure 2.

Michaels et al. (1978) and Ling et al. (1978) have also recently studied O_2-dependent damage in mammalian cells irradiated in vitro. Using their double-pulse method, first applied to bacteria (Epp et al. 1973), they found an upper limit of about three milliseconds for the lifetime of the radical that can react with O_2 to cause damage. This is about thirty times longer than the upper limit found with vegetative bacteria with this same technique, although, as they pointed out, this limit is compatible with the greater size of the mammalian cells and, logically, therefore, with the greater distance O_2 must diffuse before it can reach the target radicals, presumably located near the cells' centers. The difference in upper limits they saw with bacterial and mammalian cells does not mean that the lifetimes of the O_2-sensitive radicals actually involved are different; as they discussed, they measured the maximum lifetime these radicals could have, not the radical's half-life (Ling et al. 1978).

Some recent work with chemical model systems, where biologically important molecules are irradiated in vitro, also suggests that O_2 may have more than one action. Held and Powers (in press) and Powers and Held (1979) have extracted wild-type DNA from a strain of *B. subtilis,* irradiated the DNA, and then measured the loss of transforming ability in a *trp*$^-$ recipient cell. As they described in an earlier paper (Held et al. 1978), they again found that O_2 at high concentrations ($\sim 10^{-3}$ M) protects DNA irradiated under these conditions (relative to the response after anoxic exposures). They also noted that the sensitivity changes little over a wide range of O_2 concentrations except around 10^{-3} M, where the sensitivity drops, and in the region around 10^{-6} M, where the sensitivity peaks sharply. An important property of this peak is that it is removed by addition of $^{\bullet}OH$ scavengers, indicating $^{\bullet}OH$ involvement in the O_2 effect at low concentrations of O_2, just as in the wet spore experiment of Ewing and Powers (1976).

Studies by Michaels and Hunt (1977a, 1977b) may be relevant to these results with transforming DNA. They irradiated single-stranded polynucleotides to study the reactions of O_2 at the radical site formed by $^{\bullet}OH$ attack. Using pyrimidines (polycytidylic acid and polyuridylic acid), they found evidence for two polynucleotide radicals that react with O_2 at different rates. They proposed that at low O_2 concentrations the absolute rate constant is about 5.8×10^8 M^{-1} seconds^{-1}, while at high O_2 concentrations, the rate constant for the poly C-$OH^{\bullet} + O_2$ reaction is about 1.8×10^8 M^{-1} seconds^{-1}. Their results with purine polynucleotides and with double-stranded polymers are considerably more complex, and, as they discuss, further experimental work is needed to identify the reactions that are occurring.

The studies cited in this section provide conclusive evidence that O_2 has more than a single chemical pathway through which it effects sensitization in several biological systems. The studies discussed in the following section will summarize what has been learned concerning the chemical nature of these components of damage.

EFFECTS ADDED CHEMICALS HAVE ON OXIC SENSITIZATION: PROTECTORS AND O_2

When water is irradiated, the three primary radiolytic products are the hydrated electron (e^-_{aq}), the hydrogen atom ($^{\bullet}H$), and the hydroxyl radical ($^{\bullet}OH$). Information concerning these radicals has accumulated rapidly since the 1960s, and radiation biologists have tried to associate reactions of these transients with specific biological end points. Only the hydroxyl radical has been clearly implicated in causing cell death after irradiation (Johansen and Howard-Flanders 1965, Sanner and Pihl 1969, Powers and Cross 1970, Chapman et al. 1975), although the information from different organisms suggests that different processes may be involved. Some years ago, Adams and Dewey (1963) noted that chemical additives that reacted well with hydrated electrons are generally radiation sensitizers. This observation, that e^-_{aq} removal increases the radiation sensitivity, suggests in itself that the e^-_{aq} may be playing a protective role in reducing the amount of radiation-induced damage. This concept is central to the electron sequestration model of Powers (1972), which deals with mechanisms of radiation damage and the chemical basis for the actions of some radiation sensitizers.

Results from studies with bacteria, bacterial spores, and mammalian cells have led to different conclusions about the involvement of $^{\bullet}OH$ in damage and sensitization. These differences may be due, in part, to the absence of comparative studies among the three types of cells. Careful studies are urgently needed before we can understand how to apply information from one of these biological systems to another.

Bacterial Spore Results

When bacterial spores are irradiated in anoxia or in high concentrations of O_2, the addition of a scavenger to remove OH radicals does not necessarily provide protection (Powers et al. 1972, Ewing 1975). *t*-Butanol, for example, will not protect spores irradiated under either of these two conditions (as long as no other sensitivity-modifying agent is present); in fact, high concentrations of *t*-butanol (>1 M) will increase the anoxic response (Ewing 1975), an observation that has not yet been explained.

Some OH-radical scavengers, however, have been found that do protect spores irradiated either in air (Ewing 1976a) or in anoxia (Ewing 1976b). Figure 3 illustrates the results in anoxia. This protection is not a simple function of •OH removal, but it seems to be correlated with the ability of the scavenger to form an α-carbon radical (i.e., a reducing radical) after reaction with •OH (Ewing 1976b). The results emphasize both the importance of this correlation and the fact that •OH removal itself does not necessarily provide protection. CO_2 is a protector (cf. Figure 3), but it does not react with OH radicals; in fact, CO_2 is one of the very few e^-_{aq} scavengers that is not a radiation sensitizer. Formate at 2×10^{-2} M reduces the response to the minimum level reached by any of the tested compounds, a decrease in k of ~20%. *t*-Amyl alcohol at 2.8×10^{-2} M and *t*-butanol at 9.6×10^{-2} M scavenge OH radicals as efficiently as that formate concentration, but neither of these two alcohols provides protection. This is clear evidence that •OH removal in itself will not invariably protect spores in the absence of sensitizers.

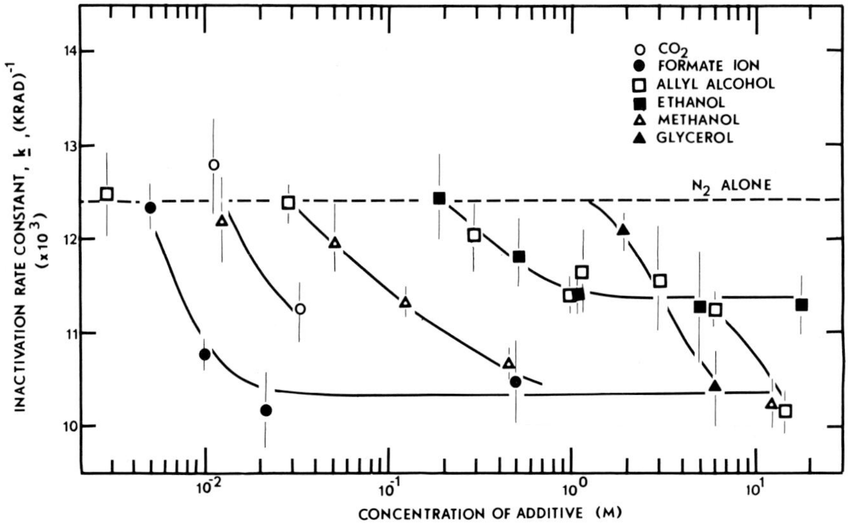

FIG. 3. Anoxic radiation sensitivity of *B. megaterium* spores, suspended in various concentrations of the additives noted. Irradiation was with 50 kVp X rays. (Reproduced from Ewing 1976b, with permission of Academic Press.)

In a set of experiments to test the importance of forming a reducing radical, methanol (a protector) and *t*-amyl alcohol (a nonprotector) were used simultaneously. At selected concentrations, where *t*-amyl alcohol scavenges •OH more efficiently than methanol, methanol's ability to protect spores was reduced. This supports the hypothesis that methanol is not itself the protector; instead, the protecting agent is formed after a reaction with a water-derived radical, in this case the •OH (Ewing 1976b).

Figure 2 shows that over a range of O_2 concentrations, *t*-butanol provides protection. In contrast to what was found after anoxic irradiation, tests have shown that this protection in low O_2 concentrations can be specifically attributed to a simple removal of OH radicals (Figure 4). Different additives used at the same •OH scavenging efficiency protect spores equally well (Ewing 1978a).

Additives have also been tested for effects on the low- and high-O_2 components of damage (cf. Figure 2). Under anoxic conditions, methanol reduces the response (Ewing 1976b). When spores are irradiated in 0.8% O_2 ($\sim 10^{-5}$ M dissolved), methanol removes the O_2-dependent •OH damage (cf. Figure 4). As shown in Figure 5, higher concentrations of methanol tested in 0.8% O_2 reduce the response further (Ewing, unpublished results). But the amount of the protection in 0.8% O_2 (i.e., the $-\Delta k$) and the methanol concentrations over which this

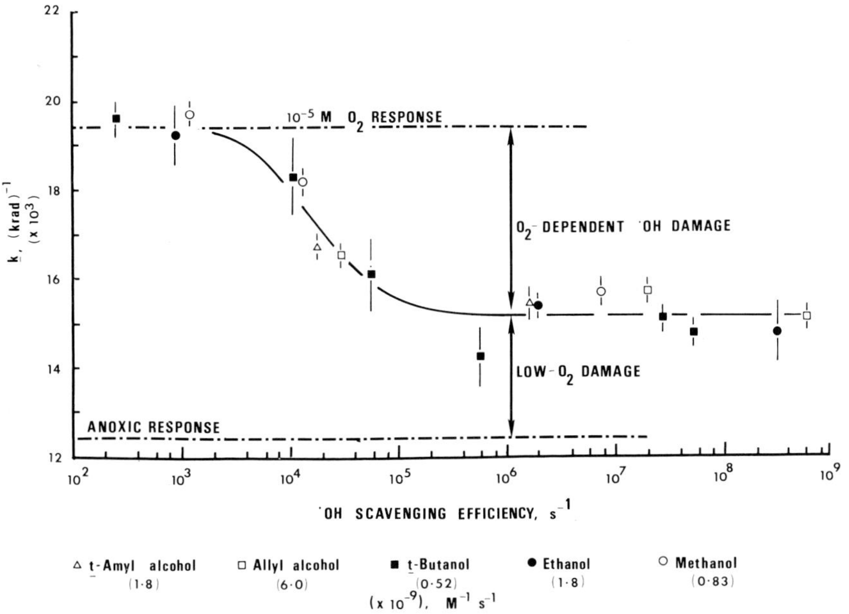

FIG. 4. Changes in the radiation sensitivity of *B. megaterium* spores irradiated in 10^{-5} M O_2 with various concentrations of several alcohols also present. The abscissa shows the •OH scavenging efficiency, the product of the specific alcohol concentration and the bimolecular rate constant for its reaction with •OH. (Reproduced from Ewing 1978a, with permission of Academic Press.)

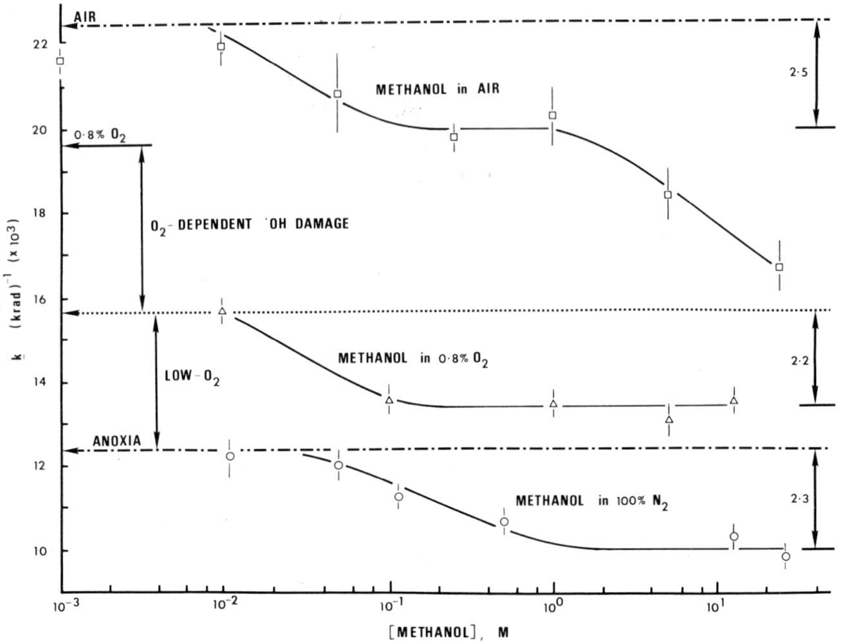

FIG. 5. Changes in the radiation sensitivity of *B. megaterium* spores, irradiated in water with 50 kVp X rays, under three reference conditions when different concentrations of methanol (MeOH) are present. The lower horizontal line shows the response in anoxia with no additive present; the symbols ○ shows the protective effects MeOH has in anoxia (Ewing 1976b). The middle horizontal line shows the response from the low-O_2 component of damage; in the [O_2] used for these tests, MeOH has already removed the ·OH component of damage (cf. Figure 4) at concentrations lower than those shown in this figure. The symbols △ show the reduction in the response with higher [MeOH] (Ewing, unpublished results). The upper horizontal line shows the response in air with no additive present; the symbols □ show the reduction in this response when MeOH is added at higher and higher concentrations (Ewing 1976a).

protection occurs suggest that this is simply the same protection seen with methanol in anoxia. From this, we conclude that methanol does not affect the low-O_2 kind of damage. Methanol tested in air (Ewing 1976a) first shows a reduction in the overall response, parallel to that seen in anoxia; higher methanol concentrations reduce the sensitivity even more. Thus, methanol is able to reduce, although not eliminate, damage attributable to the high-O_2 component.

The results with ethanol (Figure 6) are more complex. Like methanol, ethanol protects spores irradiated in anoxia, although the amount of protection (the $-\Delta k$) is not as great as with methanol. Again like methanol, ethanol does not appear to protect against low-O_2 damage, although as the ethanol concentration increases above 2 M, the response rises sharply (Ewing, unpublished results). In air, ethanol protects against the high-O_2 component, as methanol does, but high ethanol concentrations again increase the response. This increase seems

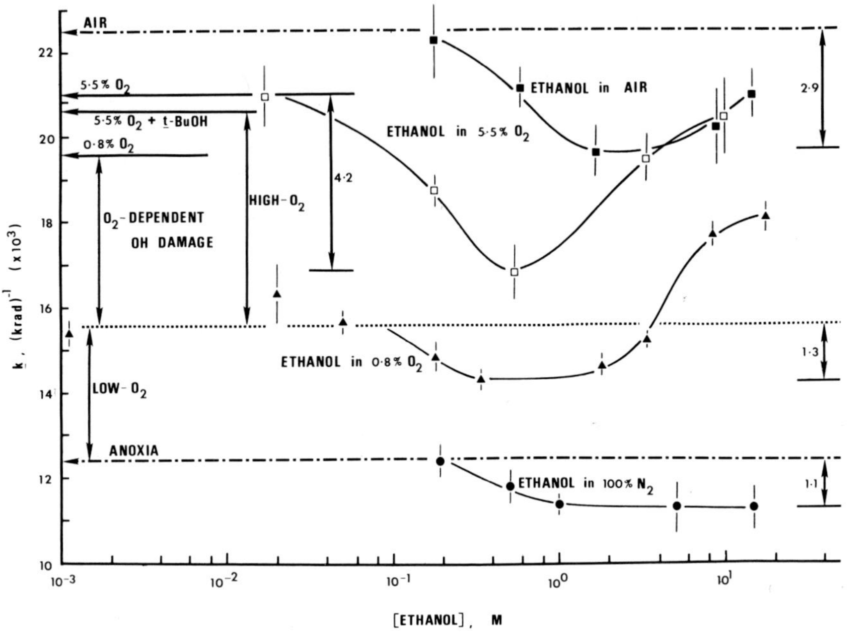

FIG. 6. Changes in the radiation sensitivity of *B. megaterium* spores, irradiated in water with 50 kVp X rays, under four reference conditions when different concentrations of ethanol (EtOH) are present. The lower horizontal line shows the response in anoxia with no additive present; the symbols ● show the protective effects EtOH has in anoxia (Ewing 1976b). The middle horizontal line shows the response from the low-O_2 component of damage; in the [O_2] used for these tests, EtOH has already removed the •OH component of damage (cf. Figure 4) at concentrations lower than those shown in this figure. The symbols ▲ show the effects on the low-O_2 component of oxygen-dependent damage when different [EtOH] are used (Ewing, unpublished results). The uppermost horizontal line shows the response in air with no additive present; the symbols ■ show the changes in radiation sensitivity when different [EtOH] are present. Arrows pointing to the ordinate show the response in 5.5% O_2 with no additive present and also the response in 5.5% O_2 when 10^{-1} *t*-butanol is added. The symbols □ show the changes in response when different [EtOH] are tested in 5.5% O_2.

to mirror the increase seen with ethanol in 0.8% O_2; that is, in air the turnabout in protection at high ethanol concentrations is probably due to ethanol's unexpected ability to increase damage through the low-O_2 component.

Tests with ethanol have also been run at an intermediate O_2 concentration. In 5.5% O_2 (~8×10^{-5} M dissolved), the •OH component is negligibly small. Presumably, the low- and high-O_2 components are both operating. In increasingly higher ethanol concentrations, the response in 5.5% O_2 is reduced; it passes through a minimum, then it increases again (cf. Figure 6). These results are qualitatively the same as those observed when ethanol was tested in air. However, two important quantitative differences are apparent. First, in 5.5% O_2, ethanol begins to provide protection at lower concentrations than it did when tested in air; second, the maximum amount of protection (the $-\Delta k$) is greater in 5.5% O_2 than it is in air. We find, in fact, that the ratio of ethanol concentrations for 50% protection (air: 5.5% O_2) is 0.6 M/0.15 M = 4.0. This is the same as the ratio of O_2 concentrations used in the two tests, 0.209/0.055 = 3.8. These results might suggest that O_2 and ethanol are competing for a single damaged cellular site. A lower O_2 concentration correspondingly reduces the ethanol concentration required for the same level of protection; it also increases the amount of protection that is possible.

Such a competition between ethanol and O_2 (specifically the high-O_2 component in these experiments) was not found in studies with *E. coli* by Johansen and Howard-Flanders (1965). From their experiments, they concluded that ethanol could interfere with the formation of $R^\bullet$ (the damaged cellular site), but ethanol and O_2 did not compete for reactions at that site. This result with bacterial cells does not complement that found by Weiss and Santomasso (1977), who pulse-irradiated spores suspended in water or pure ethanol. In ethanol, they found no decrease in the yield of the oxygen-sensitive radical (i.e., ethanol did not interfere with the formation of the damaged site), although the half-life of the radical in ethanol was considerably reduced. (No value was given for the reduced half-life.) Whether these results in pure ethanol (Weiss and Santomasso 1977) are compatible with those illustrated in Figure 6, which show an O_2-ethanol competition, is unknown. Pulse-irradiation studies at lower O_2 concentrations would clarify this point.

Glycerol is the additive that has the greatest protective ability in the spore system (Webb and Powers 1963, Ewing 1976a). At sufficiently high concentrations, glycerol protects spores irradiated in anoxia, and, even when irradiated in air, glycerol reduces the response to the same protected level seen in 100% N_2. With spores, no other additive has been found that can eliminate all oxygen-dependent damage.

From these spore results, in which radiation protectors have been tested in anoxia and in various O_2 concentrations, we may make these general observations:

1. Although OH radicals are clearly damaging under some experimental conditions, in anoxia—with no other sensitivity-modifying agent present—OH radical removal per se does not provide protection. The same generalization holds true for spores irradiated in high [O_2]: $^\bullet$OH removal in itself does not provide protection.

2. However, over a limited range of O_2 concentrations (roughly 10^{-6} to 10^{-4} M), simple $^\bullet$OH removal will reduce the response. Thus, although O_2 is not known to affect the initial yield of $^\bullet$OH and although O_2 and the $^\bullet$OH do not react, $^\bullet$OH damage becomes "temporarily" important. And its occurrence requires the presence of O_2 within a specific concentration range; at [O_2] > 10^{-4} M, $^\bullet$OH removal no longer provides protection.

3. Some additives have been found that protect spores irradiated under anoxic conditions. These agents probably function by forming an α-hydroxy radical,

which is the actual protector. These agents, which can all form reducing radicals, provide protection in anoxia: formate ion, methanol, CO_2, ethanol, allyl alcohol, glycerol, 1-propanol, and 2-propanol. If these agents protect spores in anoxia through the formation of a reducing radical, a reasonable inference is that the damage being repaired or prevented arises through an oxidation reaction.

4. Those agents that form α-hydroxy radicals and protect spores in anoxia also protect against one or both the low- and high-O_2 components of damage. (All OH radical scavengers that have been tested, including one sensitizer [Ewing 1978a], can remove the $^\bullet$OH component of oxygen-dependent damage and thereby provide protection.) Allyl alcohol (Ewing, unpublished results), methanol, ethanol, and glycerol all protect against high-O_2 damage; this implies, again, following the reasoning in item 3, that an oxidation reaction is involved in the damage from this specific component of oxygen's sensitization. This same conclusion, that the high-O_2 component probably involves an oxidation reaction, has been reached through other spore studies (Simic and Powers 1974, Ewing 1978a). The fact that methanol and ethanol do not protect against low-O_2 damage suggests that the process leading to sensitization here does not involve an oxidation step or process.

Bacterial Vegetative Cell Results

The vegetative cell studies most easily compared to the spore experiments described in the preceding section are those by Sanner and Pihl (1969) and by Johansen and Howard-Flanders (1965). Both investigations examined possible roles that radical scavenging agents have in reducing radiation sensitivity.

Sanner and Pihl (1969) used *E. coli* B, suspended in distilled water and irradiated in either liquid or frozen states, with or without selected additives. They found that both ethanol and glycerol protect bacteria irradiated anoxically in liquid suspension at 0°C; a 1 M concentration of either additive will reduce the response to about 65% of that seen without the additive. They also used acetone, an agent they found relatively poor at reducing the sensitivity of *E. coli* irradiated in anoxia.

In contrast, Tallentire and Jacobs (1972), working with bacterial spores, found that acetone was an effective radiation *sensitizer,* although very high concentrations were required for an effect. Figure 7 shows their results with these and several other compounds. They have plotted the relative sensitivity against either the $^\bullet$OH scavenging efficiency (left panel) or the e^-_{aq} scavenging efficiency (right panel). If the protection they observed must arise from either $^\bullet$OH or e^-_{aq} scavenging, clearly OH radical scavenging is responsible. However, the data points (left panel) do not themselves establish a well-defined relationship, although this is not strictly required since all the additives may not necessarily provide protection through OH radical scavenging. Perhaps more serious is the fact that the data points also show scatter around the theoretical line (shown in

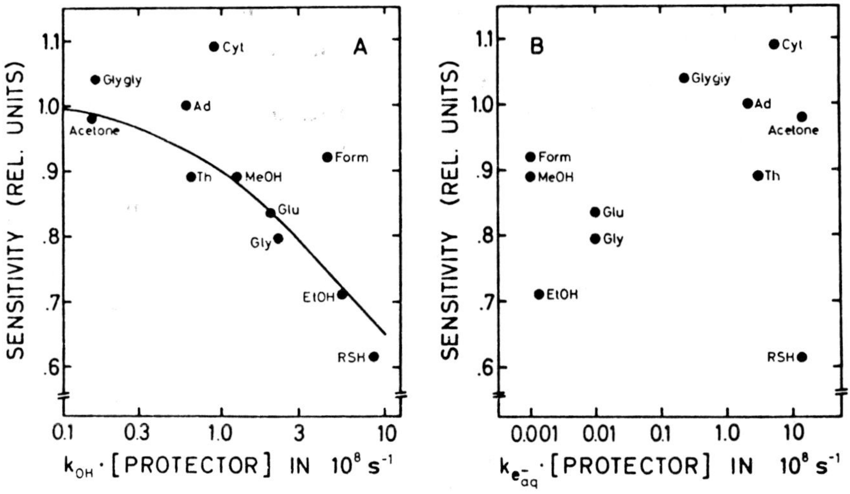

FIG. 7. Ability of different compounds to protect *E. coli* B at 0°C, as a function of their rate of interaction with •OH (A) and with e^-_{aq} (B), respectively (Sanner and Pihl 1969). The relative sensitivity observed in the presence of the compounds is plotted versus the product of the protector concentration and the respective second-order rate constants for the interaction of the different protectors with •OH and with e^-_{aq}. The fully drawn curve in A is a theoretical curve calculated as described in the text. The radiation sensitivity in the absence of added compounds was set equal to 1. All values are based on dose-effect curves. Abbreviations: Ad, adenine; Cyt, cytosine; EtOH, ethanol; Form, sodium formate; Glu, glucose; Gly, glycerol; Glygly, glycylglycine; MeOH, methanol; RSH, cysteamine; Th, thymine. (Reproduced from Sanner and Pihl 1969, with permission of Academic Press.)

Figure 7), which was drawn based on the assumption that •OH is responsible for 45% of the lethal damage in anoxia.

From these results, Sanner and Pihl concluded that •OH removal will protect bacteria irradiated under anoxic conditions; they estimated that under these conditions about 50% of the total damage arises through reactions of hydroxyl radicals. They noted that this estimate agrees well with conclusions from Webb's work (Webb 1964). He mounted *Staphylococcus aureus* cells on membrane filters and dried them; the anoxic radiation sensitivity dropped by about 50% when the equilibrium vapor pressure was reduced to about 1 Torr. Further drying had little effect on the response. However, this amount of protection, seen in either *E. coli* or *Staphylococcus,* is considerably greater than the protection observed when bacterial spores are irradiated under similar conditions. Tallentire and Powers (1963) found that drying produced only about a 25% reduction in anoxic response. (It is important to recall that in both the spore and *Staphylococcus* systems, drying *decreases* the radiation sensitivity only under anoxic conditions; in both organisms, the two classes of oxygen-dependent damage *increase* greatly as water is removed.)

Sanner and Pihl (1969) also concluded that no amount of protection, including that from the sulfhydryl cysteamine, was greater than that expected from •OH scavenging alone.

Johansen and Howard-Flanders (1965), in a slightly earlier study, used *E. coli* B/r, irradiated in buffered saline at 2°–5°C. (It is not known if the overall results would have changed if distilled water, rather than buffer, had been used; or, alternatively, if buffer, rather than water, had been used in the experiments described by Sanner and Pihl in 1969.) With bacterial spores, phosphate buffer is itself a slight radiation sensitizer (Ewing 1975), and the actions of *p*-nitroacetophenone in a low O_2 concentration are different in water compared with buffer (Ewing 1977).

Johansen and Howard-Flanders (1965) studied several radiation protectors in some detail. They concluded that the effects of the sulfhydryl mercaptoethanol are twofold: first, this agent could interfere with the formation of a damaged (oxygen-sensitive) site within the cell by scavenging water-derived radicals; second, mercaptoethanol could provide protection by successfully competing with O_2 for reaction at this damaged site. Figure 8 shows their results with five protective agents. Curiously, they included nitric oxide (NO) among the protectors. Although NO has complex effects on radiation sensitivity, their own results clearly show (Figure 4 of Johansen and Howard-Flanders 1965) that the sensitiv-

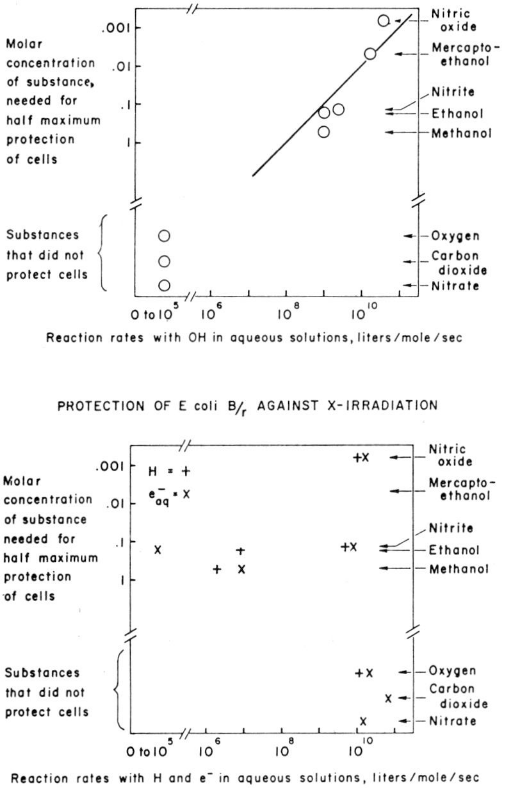

FIG. 8. Protection of *E. coli* B/r (Johansen and Howard-Flanders 1965). The effective concentrations of various substances in protecting bacteria against x-irradiation are plotted against data for the reaction rates of these substances with hydroxyl radicals (top panel) and with the reducing species (bottom panel). Oxygen at 2 $\times$ 10^{-3} M, carbon dioxide at 7 $\times$ 10^{-2} M, and sodium nitrate at 8 $\times$ 10^{-1} M did not protect bacteria; these substances are plotted below the intercept in both figures. (Reproduced from Johansen and Howard-Flanders 1965, with permission of Academic Press.)

ity at all NO concentrations is greater than that seen in its absence, although a peak in the response may indicate two actions of NO. This is parallel to the earlier observation of Powers et al. (1960) that NO has two actions in the dry spore.

The method of analysis used by Johansen and Howard-Flanders is slightly different from that of Sanner and Pihl; but, again, if the protection the former team observed in air must arise from scavenging either $^{\bullet}OH$ or a combination of e^{-}_{aq} and $^{\bullet}H$, the data clearly favor an involvement of OH radicals. As was the case with the anoxic study (cf. Figure 7), the fit of the data to the expected line is not extremely good. Johansen and Howard-Flanders, in fact, called the fit "reasonably good" and suggested that "uncertainties in the concentrations of the added substances within the cell" might account for some of the scatter (Johansen and Howard-Flanders 1965). From this analysis, they concluded that OH radicals contribute about half the lethal damage when these bacteria are irradiated under aerobic conditions.

These two investigations with vegetative bacteria provide a consistent view of the role played by OH radicals both in O_2 and in anoxia. They presented several chemical models that were reasonable reflections of the state of radiation chemical knowledge of that time.

The data from the two bacterial studies can be compared by plotting them on the same graph. Figure 9 shows such a plot after a recalculation of the aerobic data of Johansen and Howard-Flanders (1965), without the nitrous oxide point, to conform to the analysis method used by Sanner and Pihl (1969). More recent values of the reaction rate constants (Ross and Ross 1977) were used, and the points have shifted somewhat from their original positions. The line in this figure is the same as that from Figure 7 (Sanner and Pihl 1969). The fit of the aerobic data points to the theoretical line is no worse than the fit of the original anaerobic points.

In Vitro Mammalian Cell Results

Much of the groundwork on the oxic sensitization of mammalian cells and on the possible roles played by OH radicals comes from the work of Chapman and his colleagues. They tested several radiation protectors over a range of concentrations, usually up to the limit set by toxicity of the added agent (Chapman et al. 1975). Dimethylsulfoxide (DMSO) is one of the very few compounds that can be used at high concentrations in cultures of mammalian cells, and for this reason, unfortunately, DMSO must be used in the studies concerning OH radical involvement when cells are irradiated in vitro. Chapman and his colleagues found that DMSO can provide considerable protection for cells irradiated in air; DMSO also protects anoxically irradiated cells, although the magnitude of the effect is much smaller. DMSO has also been tested in bacterial spores (Ewing 1978b), where, in contrast, it was found to be a very potent radiation sensitizer. However, spores treated with DMSO and then washed before

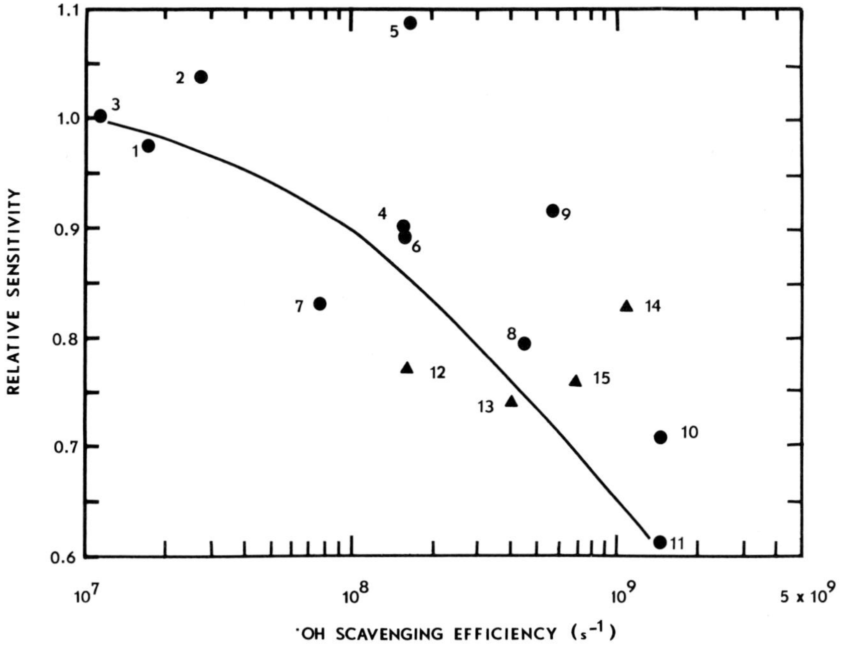

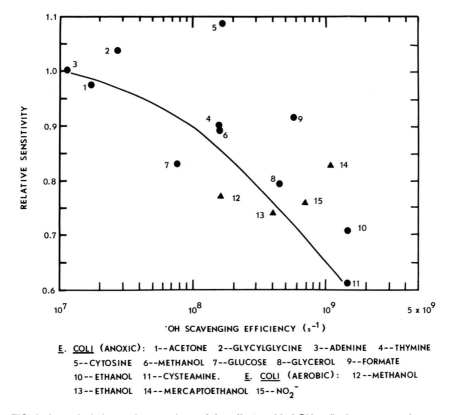

E. COLI (ANOXIC): 1--ACETONE 2--GLYCYLGLYCINE 3--ADENINE 4--THYMINE 5--CYTOSINE 6--METHANOL 7--GLUCOSE 8--GLYCEROL 9--FORMATE 10--ETHANOL 11--CYSTEAMINE. E. COLI (AEROBIC): 12--METHANOL 13--ETHANOL 14--MERCAPTOETHANOL 15--NO_2^-

FIG. 9. A recalculation and comparison of the effects added OH radical scavengers have on the anoxic (Sanner and Pihl 1969) and aerobic (Johansen and Howard-Flanders 1965) radiation sensitivity of *E. coli.* More recent values of the scavengers' reaction rate constants with OH radicals were used (Ross and Ross 1977) for this comparison, and some of the data points are shifted from their original sites (cf. Figures 7 and 8). More recent values for e_{aq}^- scavenging (Anbar et al. 1973) were not sufficiently different from those originally used to warrant our replotting those data. The aerobic test with NO was omitted from this comparison.

irradiation in water still showed virtually the same response noted if DMSO had not been removed before irradiation. Thus, this sensitization seems attributable to changes DMSO causes in spore "physiology." These unidentified changes, while clearly not toxic, seem responsible for the greatly increased responses to irradiation, both in O_2 and in anoxia, when DMSO was present. To our knowledge, this "washed out" experiment has not been done with mammalian cells.

Chapman and his co-workers (1973) have also observed that cysteamine was, in fact, a better protector of mammalian cells than DMSO. Lower concentrations of cysteamine, tested in O_2 and in anoxia, gave as much protection as higher concentrations of DMSO; in addition, the maximum amount of protection, in both gases, was greater from cysteamine. They concluded that, like DMSO, cysteamine protected by scavenging •OH; however, another radiation chemical process, perhaps repair through hydrogen donation, is also possible with cysteamine. This may conflict with the results Sanner and Pihl (1969) observed with bacteria. They found that all the protection from cysteamine could be accounted for solely by •OH removal.

Other additives that Chapman and his colleagues tested showed only small abilities to protect at concentrations lower than the limits set by toxicity. The results of their survey (Chapman et al. 1975) are illustrated in Figure 10. (The cells were irradiated in culture medium.) To analyze and to test for an involvement of OH radicals in lethal processes, the authors implicitly assumed that these compounds would all protect to the same minimum response level—the level seen with only DMSO—if toxicity were not a limiting factor. They plot the reciprocal of the additive concentration for 50% of the effect DMSO had against the rate constant for •OH scavenging by the specific additive. The result, shown in the right panel of Figure 10, is a straight line having the expected slope of +1.0. The relationship provided the basis for the conclusion that, in air, OH radical removal protects against O_2-dependent damage. Based on these and earlier data (Chapman et al. 1973), they estimated that for irradiation in air about 62% of the lethal damage in mammalian cells results from the actions of OH radicals. The small amount of protection they saw when DMSO-treated cells were irradiated anoxically (Figure 10) was attributed to OH radical removal, although anoxic experiments, like those for air shown in Figure 10, were not

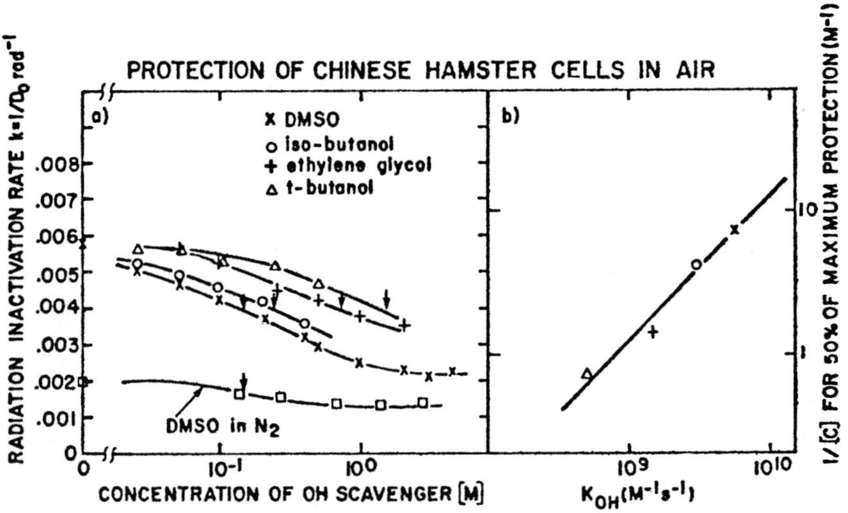

FIG. 10. a) The chemical radioprotection by various •OH scavengers of the multitarget inactivation rate of air-saturated Chinese hamster cells; b) The correlation between the reciprocal of protector concentration effecting 50% of the maximum radioprotection and the absolute rate constants of •OH with the specific chemical protector (Reproduced from Chapman et al. 1975, with permission of Academic Press).

reported. The authors estimated that in anoxia about 30% of the damage results from the effects of OH radicals, an estimate that seems somewhat high.

These data by Chapman and his colleagues (Figure 10) can be reanalyzed and compared with the results from the aerobic and anaerobic bacterial studies, which were collected in Figure 9. Table 1 shows the calculated values for "relative sensitivity" and "•OH scavenging efficiency" that were calculated from the original mammalian cell data (Chapman et al. 1975). Comparison with the bacterial cell data in Figure 9 shows a remarkably good agreement.

These results can also be compared with those in Figure 4, where, with bacterial spores, •OH removal is clearly responsible for the observed protection. According to Figure 4, a kC of about 10^4 $second^{-1}$ will achieve 50% of the maximum amount of protection, a scavenging efficiency almost 10^5 times lower than that needed for protection through •OH removal in the two organisms compared in Figure 9 and Table 1. (The calculated point for spores is "relative sensitivity" = 0.89; "$kC_{50\%}$" = 10^4 $second^{-1}$.)

This comparison of data from spores, vegetative bacteria, and mammalian cells raises perplexing questions and provides few satisfactory answers. •OH removal does not protect spores irradiated under either anoxic or well-oxygenated conditions; however, over an intermediate range of O_2 concentrations—in spores—simple •OH removal will protect spores. In contrast, results from several studies indicate that •OH removal will protect vegetative bacteria irradiated under either anoxic or well-oxygenated conditions. The amount of protection seen with vegetative bacteria under these two conditions seems to be a simple function of the efficiency for OH radical removal; equal •OH scavenging efficiencies will produce equal amounts of protection. In spores, •OH damage is clearly oxygen dependent (or, more generally, "sensitizer dependent"). In spite of this apparently unique origin for an •OH involvement, we do not understand why the •OH scavenging efficiency needed to protect vegetative bacteria is about 10^5 times greater than that needed in spores for the same relative amount of protection.

The same conclusion regarding the importance of OH radical removal has also been reached from studies with mammalian cells: it seems that simple OH radical removal provides protection. In this case, however, the maximum amount of protection seen in the anoxic studies is much less than that protection

TABLE. 1. *•OH Scavenging efficiency for half maximal protection of Chinese hamster cells irradiated in vitro*

Air/Anoxia	Additive	Relative Sensitivity	$k_{\cdot OH}C$ ($\times 10^{-8}$)
Air	DMSO	0.71	8.4
Air	Iso-butanol	0.71	7.2
Air	Ethylene glycol	0.71	11.0
Air	*t*-Butanol	0.71	6.6
Anoxia	DMSO	0.91	8.4

seen in experiments that used air-equilibrated cells. We do not understand why these two experimental conditions would give similar results with vegetative bacteria but different results with mammalian cells, especially since, in both kinds of cells, the role played by OH radicals is believed to be the same. On the other hand, under aerobic conditions, the same OH scavenging efficiency does give about the same relative amount of protection with either vegetative bacteria or mammalian cells.

With the information presently available from these three biological systems, it appears that O_2 need not sensitize these cells through the same chemical pathways. However, before accepting this conclusion, we should carefully reexamine the information on which it is based. We should remember that there have been no comparative studies in which the same additives were rigorously tested in all three systems. This is partly due, of course, to toxicity problems with vegetative bacteria and mammalian cells. We urgently need this kind of experimental data. With spores, we found that various OH radical scavengers did not always have the same effects, and consequently they could not be used interchangeably. In spite of the correlation drawn between protection and OH radical scavenging in both vegetative bacteria and mammalian cells, there is surely enough experimental uncertainty to emphasize the need for additional data.

SUMMARY

This survey has focused primarily on O_2 effects in three biological systems, all tested in suspension: bacterial spores, vegetative bacterial cells, and mammalian cells. We have examined information from these systems that shows that O_2 has more than one process through which it can act, and we have looked at the effects various protectors have on oxygen's ability to sensitize. While selecting from among the many studies within these guidelines, we have largely ignored the studies that test O_2 in combination with other radiation sensitizers. There is considerable information from these studies, but, within the intentionally limited scope of this survey, we cannot cover this information here.

Studies with bacterial spores provide clear evidence that multiple components to oxygen-dependent radiation sensitization exist. Studies with mammalian cells also show that at least two oxygen-dependent sensitization processes can be distinguished, although we have not yet learned how to relate the components of sensitization from these two very different organisms. Similar studies with vegetative bacteria in suspension have not resolved oxic sensitization into components, although different experimental techniques may yet do so. It is essential to emphasize that the observation noted almost 20 years ago with very dry bacterial spores now clearly applies to mammalian cells irradiated in vitro: there is more than one kind of oxygen-dependent damage.

We have examined the roles water-derived radicals might play in radiation sensitivity and, specifically, in sensitization by O_2. We find that, among the primary radiolytic products, OH radicals are clearly implicated in damage in all three biological test systems. However, we must exercise great care in drawing conclusions here, since the specific roles proposed for OH radicals are different in these organisms.

In bacterial spores, •OH removal in itself does not provide protection in anoxia or in high concentrations of O_2 if there is no other sensitivity-modifying agent present. (Many organic and inorganic sensitizers have effects that can be partially or completely removed by agents that scavenge OH radicals.) However, with spores, •OH removal over a limited intermediate range of O_2 concentrations will provide protection. Results of tests with those agents that protect in anoxia and in air show that although these agents scavenge OH radicals, that in itself is not the protecting step; instead, •OH scavenging probably results in the formation of the actual protector.

In bacteria, results of surveys to test the effects various radical scavengers have on radiation sensitivity and on the sensitization by O_2 have provided the basis for the supposition that •OH removal will provide protection both in anoxia and in the presence of O_2. Many authors have suggested that OH radicals react with a cellular target molecule and leave a radical site; this is the site that can then react with O_2 to cause damage. It is widely believed and often expressly stated that DNA is the likely cellular target for OH radical attack.

In mammalian cells, there are severe difficulties in using radical scavenging agents at the necessary high concentrations. Nevertheless, based on the information that can be obtained, a reaction scheme similar to that proposed for bacteria has been suggested for O_2-dependent sensitization; again, it is expressly stated that DNA is the likely target for cellular damage.

A reanalysis of the data from these biological systems suggests that these conclusions may not be as firm as we had thought. From the results with the different kinds of cells, we see that the proposed roles for oxygen-dependent sensitization, and especially for the involvement of OH radicals, are not complementary. Before we accept the conclusion that O_2 operates through different chemical pathways in these organisms, we should carefully reexamine the data on which our conclusions have been based.

We must remember when we use an OH radical scavenger and observe protection that we have not proved that •OH removal is the specific reaction responsible for the protection; neither have we proved that OH radicals are damaging to irradiated cells. These conclusions may be entirely true, but our simple observation of protection has not established them. We must also remember that for many years most analyses were based on the assumption that only one kind of O_2-dependent sensitization process exists. Clearly O_2 can sensitize bacterial spores, mammalian cells, and very likely bacterial cells as well, through more than one chemical or physical process. Our taking the simplest case of assuming only one effect of O_2, building models, and drawing conclusions may not have been as profitable as we had hoped. Perhaps it is time to discard this simplest case in our model building and take a more realistic, although necessarily more complex, view of how O_2 acts to sensitize cells.

ACKNOWLEDGMENTS

We are grateful to Peggy Centilli, Charles Dort, and Kathryn Held for their considerable assistance in the preparation of this paper. The spore research reported herein was supported by NIH GM-13557 and DOE EY-76-05-3408.

REFERENCES

Adams, G. E., and D. L. Dewey. 1963. Hydrated electrons and radiobiological sensitization. Biochem. Biophys. Res. Commun. 12:473–477.

Alper, T. 1963. Lethal mutations and cell death. Phys. Med. Biol. 8:365–385.

Anbar, M., M. Bambenek, and A. B. Ross. 1973. Selected Specific Rates of Reactions of Transients from Water in Aqueous Solution. I. The Hydrated Electron. U.S. Department of Commerce, NSRDS-NBS 43.

Chapman, J. D., A. P. Reuvers, J. Borsa, and C. L. Greenstock. 1973. Chemical radioprotection and radiosensitization of mammalian cells growing in vitro. Radiat. Res. 56:291–306.

Chapman, J. D., D. L. Dugle, A. P. Reuvers, B. E. Meeker, and J. Borsa. 1974. Studies on the radiosensitizing effect of oxygen in Chinese hamster cells. Int. J. Radiat. Biol. 26:383–389.

Chapman, J. D., D. L. Dugle, A. P. Reuvers, C. J. Gillespie, and J. Borsa. 1975. Chemical radiosensitization studies with mammalian cells growing in vitro, *in* Radiation Research: Biomedical, Chemical, and Physical Perspectives, O. F. Nygaard, H. I. Adler, and W. K. Sinclair, eds. Academic Press, New York, pp. 752–760.

Crabtree, H. G., and C. Cramer. 1933. The action of radium on cancer cells. II. Some factors determining the susceptibility of cancer cells to radium. Proc. Roy. Soc. B 113:238–250.

Dewey, D. L. 1963. The x-ray sensitivity of *Serratia marcescens*. Radiat. Res. 19:64–87.

Epp, E. R., H. Weiss, N. D. Kessaris, A. Santomasso, J. Heslin, and C. C. Ling. 1973. Oxygen diffusion times in bacterial cells irradiated with high-intensity pulsed electrons: New upper limit to the lifetime of the oxygen-sensitive species suspected to be induced at critical sites in bacterial cells. Radiat. Res. 54:171–180.

Epp, E. R., H. Weiss, and C. C. Ling. 1976. Irradiation of cells by single and double pulses of high intensity radiation: Oxygen sensitization and diffusion kinetics. Curr. Top. Radiat. Biol. Q. 11:201–250.

Ewing, D. 1975. Two components in the radiation sensitization of bacterial spores by *p*-nitroacetophenone: The •OH component. Int. J. Radiat. Biol. 28:165–176.

Ewing, D. 1976a. Effects of some •OH scavengers on the radiation sensitization of bacterial spores by *p*-nitroacetophenone and O_2 in suspension. Int. J. Radiat. Biol. 30:419–432.

Ewing, D. 1976b. Anoxic radiation protection of bacterial spores in suspension. Radiat. Res. 68:459–468.

Ewing, D. 1978a. Additive sensitization of bacterial spores by oxygen and *p*-nitroacetophenone. Radiat. Res. 73:121–136.

Ewing, D. 1978b. Effects of dimethylsulfoxide (DMSO) on the x-ray sensitivity of *B. megaterium* spores. (Abstract) Radiat. Res. 74:516.

Ewing, D., and E. L. Powers. 1976. Irradiation of bacterial spores in water: Three classes of oxygen-dependent damage. Science 194:1049–1051.

Gray, L. H. 1961. Mechanisms involved in the initiation of radiobiological damage in aerobic and anaerobic systems, *in* The Initial Effects of Ionizing Radiations on Cells, R. J. C. Harris, ed. Academic Press, New York, pp. 21–44.

Held, K. D., and E. L. Powers. 1979. Effects of varying O_2 concentration on the x-ray sensitivity of transforming DNA. Int. J. Radiat. Biol. (in press).

Held, K. D., R. W. Synek, and E. L. Powers. 1978. Radiation sensitivity of transforming DNA. Int. J. Radiat. Biol. 33:317–324.

Holthusen, H. 1921. Beiträge zur biologie der strahlenwirkung. Untersuchungen an askarideneiern. Pflüger Arch. 187:1–24.
Howard-Flanders, P., and T. Alper. 1957. The sensitivity of microorganisms to irradiation under controlled gas conditions. Radiat. Res. 7:518–540.
Johansen, I., and P. Howard-Flanders. 1965. Macromolecular repair and free radical scavenging in the protection of bacteria against x-rays. Radiat. Res. 24:184–200.
Jones, W. B. G., and H. Weiss. 1977. An ESR study of the transient signals produced in wet spores irradiated by intense 3 nsec pulses of electrons. (Abstract) Radiat. Res. 70:657.
Ling, C. C., H. B. Michaels, E. R. Epp, and E. C. Peterson. 1978. Oxygen diffusion into mammalian cells following ultrahigh dose rate irradiation and lifetime estimates of oxygen-sensitive species. Radiat. Res. 76:522–532.
Michael, B. D., G. E. Adams, H. B. Hewitt, W. B. G. Jones, and M. E. Watts. 1973. A posteffect of oxygen in irradiated bacteria: A submillisecond fast mixing study. Radiat. Res. 54:239–251.
Michaels, H. B., E. R. Epp, C. C. Ling, and E. C. Peterson. 1978. Oxygen sensitization of CHO cells at ultrahigh dose rate: Prelude to oxygen diffusion studies. Radiat. Res. 76:510–521.
Michaels, H. B., and J. W. Hunt. 1977a. Reactions of oxygen with radiation-induced free radicals on single-stranded polynucleotides. Radiat. Res. 72:18–31.
Michaels, H. B., and J. W. Hunt. 1977b. Radiolysis of the double-stranded polynucleotides poly (A+U) and DNA in the presence of oxygen. Radiat. Res. 72:32–47.
Millar, B. C., E. M. Fielden, and J. J. Steele. 1979. Radiosensitizers and the oxygen effect in mammalian cells. Presented at the Association of Radiation Research (Great Britain), North East Wales Institute in Wrexham, Clwyd, Wales. Elsevier, Amsterdam. (in press).
Mottram, J. C. 1935. On the alteration in sensitivity of cells towards radiation produced by cold and by anaerobiosis. Br. J. Radiol. 8:643–651.
Petry, E. 1921. The conditions for the biological activity of Röntgen rays. Biochem. Z. 119: 23–44.
Powers, E. L. 1966. Contributions of electron paramagnetic resonance techniques to the understanding of radiation biology, *in* Electron Spin Resonance and the Effects of Radiation on Biological Systems, NAS-NRC Nuclear Science Series, Washington, D.C., pp. 137–159.
Powers, E. L. 1972. The hydrated electron, the hydroxyl radical, and hydrogen peroxide in radiation damage in cells. Isr. J. Chem. 10:1199–1211.
Powers, E. L., and M. Cross. 1970. Nitrous oxide as a sensitizer of bacterial spores to x-rays. Int. J. Radiat. Biol. 17:501–514.
Powers, E. L., and K. D. Held. 1979. The oxygen effects in radiation biology and radiation chemistry. Presented at the Association of Radiation Research (Great Britain), North East Wales Institute in Wrexham, Clwyd, Wales. Elsevier, Amsterdam. (in press).
Powers, E. L., R. C. Richmond, and M. Simic. 1972. OH radicals in radiation sensitization. Nature New Biol. 238:260–261.
Powers, E. L., and A. Tallentire. 1968. The roles of water in the cellular effects of ionizing radiations. Actions Chimiques et Biologiques des Radiations 12:3–67.
Powers, E. L., R. B. Webb, and C. F. Ehret. 1960. Storage, transfer, and utilization of energy from x-rays in dry bacterial spores. Radiat. Res. (Suppl. 2):94–121.
Ross, A. B., and F. Ross. 1977. Selected Specific Rates of Reactions of Transients from Water in Aqueous Solution. III. Hydroxyl Radical and Perhydroxyl Radical and Their Radical Ions. U.S. Department of Commerce, NSRDS-NBS 59.
Sanner, T., and A. Pihl. 1969. Significance and mechanism of the indirect effect in bacterial cells. The relative protective effect of added compounds in *Escherichia coli* B, irradiated in liquid and in frozen suspension. Radiat. Res. 37:216–227.
Schwartz, G. 1909. Desensibilisierung gegen Röntgen- und radiumstrahlen. Muenchener Medizinische Wochenschrift 56:1217–1218.
Shenoy, M. A., J. C. Asquith, G. E. Adams, B. D. Michael, and M. E. Watts. 1975. Time-resolved oxygen effects in irradiated bacteria and mammalian cells: A rapid-mix study. Radiat. Res. 62:498–512.
Simic, M., and E. L. Powers. 1974. Correlation of the efficiencies of some radiation sensitizers and their redox potentials. Int. J. Radiat. Biol. 26:87–90.
Stratford, I. J., R. L. Maughan, B. D. Michael, and A. Tallentire. 1977. The decay of potentially lethal oxygen-dependent damage in fully hydrated *Bacillus megaterium* spores exposed to pulsed electron irradiation. Int. J. Radiat. Biol. 32:447–455.
Tallentire, A., D. J. W. Barber, R. L. Maughan, B. D. Michael, and I. J. Stratford. 1979. Manipulation

of long-lived free radical damage in wet bacterial spores. Presented at the Association of Radiation Research (Great Britain), North East Wales Institute in Wrexham, Clwyd, Wales. Elsevier, Amsterdam. (in press).

Tallentire, A., and G. P. Jacobs. 1972. Radiosensitization of bacterial spores by ketonic agents of differing electron-affinities. Int. J. Radiat. Biol. 21:205–213.

Tallentire, A., A. B. Jones, and G. P. Jacobs. 1972. The radiosensitizing actions of ketonic agents and oxygen in bacterial spores suspended in aqueous and non-aqueous milieux. Isr. J. Chem. 10:1185–1197.

Tallentire, A., R. L. Maughan, B. D. Michael, and I. J. Stratford. 1977. Radiobiological evidence for the existence of a dehydrated core in bacterial spores, *in* Spore Research 1976, A. N. Barker, J. Wolf, D. J. Ellar, G. J. Dring, and G. W. Gould, eds. Academic Press, New York, pp. 649–659.

Tallentire, A., and E. L. Powers. 1963. Modification of sensitivity to x-irradiation by water in *Bacillus megaterium.* Radiat. Res. 20:270–287.

Watts, M. E., R. L. Maughan, and B. D. Michael. 1978. Fast kinetics of the oxygen effect in irradiated mammalian cells. Int. J. Radiat. Biol. 33:195–199.

Webb, R. B. 1964. Physical components of radiation damage in cells, *in* Physical Processes in Radiation Biology, L. Angenstein, R. Mason, and B. Rosenberg, eds. Academic Press, New York, pp. 267–285.

Webb, R. B., and E. L. Powers. 1963. Protection against actions of x-rays by glycerol in the bacterial spore. Int. J. Radiat. Biol. 7:481–490.

Weiss, H., and J. C. McDonald. 1976. Response of anoxic and oxic wet spores to intense 3 nsec pulses of electrons. (Abstract) Radiat. Res. 67:535–536.

Weiss, H., and A. Santomasso. 1977. A post irradiation sensitizing effect of oxygen in anoxic wet spores exposed to intense 3 nsec duration pulses of electrons. (Abstract) Radiat. Res. 70:656–657.

Radiation Biology in Cancer Research, edited by
Raymond E. Meyn and H. Rodney Withers.
Raven Press, New York © 1980.

Relative Responses of Mammalian and Insect Cells

Thomas Michael Koval

Division of Radiation Oncology, Allegheny General Hospital, Pittsburgh, Pennsylvania 15212

Based upon numerous previous studies on various insect species (Ducoff 1972), it seems apparent that insects have considerably more inherent resistance to the lethal effects of ionizing radiation than do species of vertebrates. This apparent resistance is believed to be a consequence of the nondividing nature of most of the cells composing adult insects (O'Brien and Wolfe 1964). The range of lethal exposures to ionizing radiation lies between 200 and 900 R for most mammals (Bond et al. 1965) and 1,000 and 100,000 R for most adult insects (Ducoff 1972, Casarett 1968). More generally, O'Brien and Wolfe (1964) have stated that adult insects are at least 100 times less sensitive to the lethal effects of radiation than are vertebrates. This phenomenon has been explained by two facts: (1) the generalization of Bergonie and Tribondeau (1906), which establishes the sensitivity of cells to irradiation as being directly proportional to their reproductive activity and inversely proportional to their degree of differentiation, and (2) in insects very little cell division occurs in the larval and adult states (O'Brien and Wolfe 1964). Supporting this idea is the fact that dividing cells of insects, those of the gonads, can be sensitive to quite low doses of irradiation. In addition, Carlson (1969) has found that X-ray exposures as low as 8 R and 16 R administered to early prophase grasshopper neuroblast cells were sufficient to delay the cells from reaching metaphase. These findings all substantiate the hypothesis that the relative radioresistance to the lethal effects of ionizing radiation of adult insects is due to the limited amount of cell division they undergo. For this reason, and because there has been no good way to accurately determine the lethal effects of radiation on mitotically active somatic cells of insects, this explanation has remained relatively unchallenged.

CELL CULTURE STUDIES

Mammalian cells have been subjected to long-term passage in culture since Carrel's pioneering experiments in the early 1900s (Carrel 1912, White 1954). Quantitative radiation studies with somatic cells of mammals have been going on since the survival experiments of Puck and Marcus (1955). In the late 1950s and early 1960s, insect cell lines were established (Grace 1962, Hink 1972). However, very few quantitative studies have been published on the response of insect cell lines to radiation—especially ionizing radiation. Mosquito, *Aedes albopictus,* cells have been observed to be generally less sensitive than mammalian cells regarding radiation-induced mitotic delay and induction of chromosomal aberrations (Lee 1974, Blakely 1975). Trosko and Wilder (1973) found that cultured *Drosophila* cells possess repair mechanisms that are induced by ultraviolet (UV) light, but they have not investigated survival or repair after ionizing radiation exposure.

The TN-368, *Trichoplusia ni,* cabbage looper cell line was chosen for most of the studies described in this discussion because it had growth characteristics similar to those of several common mammalian cell lines. Characteristics of these cells are given in the next section. Growth experiments showed that the *T. ni* cells underwent a dose-dependent division delay following X-irradiation (Figure 1), just as do mammalian cells; however, the doses producing similar delays were much greater for the insect cells (Koval et al. 1975, Elkind and Whitmore 1967). An X-ray exposure of 10,000 R caused a 24-hour division delay in the *T. ni* cells, which was followed by a logarithmic recovery (Koval et al. 1975), while the growth of mouse leukemic cells (L5178Y) exposed to the same amount of X-radiation was permanently arrested (Watanabe and Okada 1967). This suggests that the insect cells do not sustain as much damage as the mammalian cells or that they have an increased capacity to repair the radiation damage.

At 24 hours after a 10,000 R X ray exposure, most of the *T. ni* cells are spherical and many others have numerous protoplasmic extensions around the entire cell membrane, giving them a spiny appearance (Koval et al. 1975). There

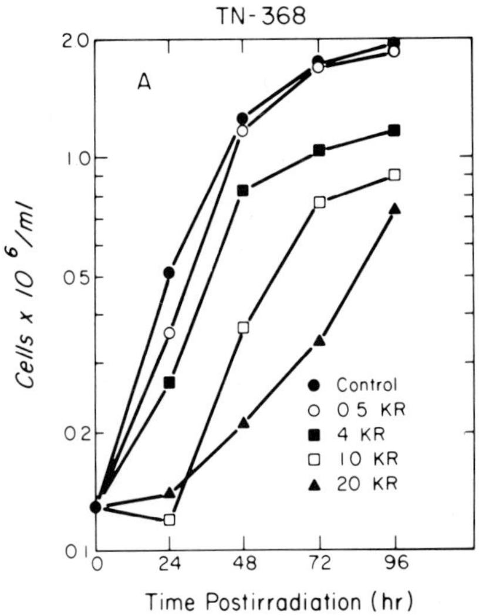

FIG. 1. Population growth curves for TN-368 cells following various X-ray exposures. (Reproduced from Koval 1978, with permission of Birkhäuser Verlag.)

is also an increase in the size of many of the cells during the postirradiation period, which is probably analogous to "giant cell" formation in irradiated mammalian cell cultures as noted by Tolmach and Marcus (1960) and Elkind et al. (1963), among others.

An interesting feature of the *T. ni* cells is that a population of cells that has recovered from a 10,000 R X irradiation attains a maximum population density of only about one-half the maximum density of an untreated population (Koval et al. 1975). It was thought that perhaps the X-rayed cells depleted the medium of some nutrient(s) essential for cell division, preventing the culture from attaining the higher density. However, media from both untreated and X-rayed cultures still retain the ability to support cell growth and reproduction. An alternate explanation for the cessation of cell division is that toxic substances in the growth medium may accumulate and inhibit cell reproduction. Analyses have indicated that irradiated cells produce more ammonia in the growth medium than control cells (Koval et al. 1976), which corresponds to an increased pH in the medium (Koval et al. 1975). A third possible reason for the reduced cell density in the X-irradiated cultures is that the cells may have suffered some irreparable damage. Results have shown that cell cultures X-rayed with 10,000 R do not reach a population density as great as control cultures for more than 25 passages after the irradiation, but there is evidence that recovery may be occurring since the maximum cell density does generally increase with succeeding passages (Figure 2). This suggests that the surviving population of cells has been genetically altered sufficient to decrease the maximum allowable cell density.

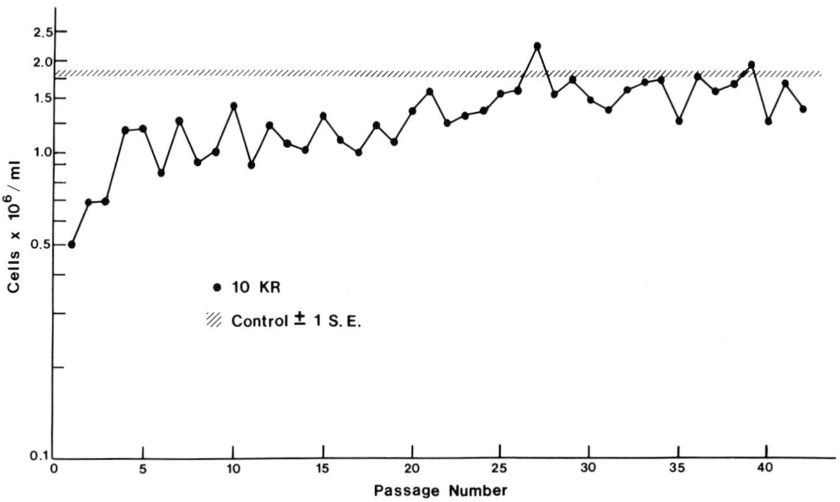

FIG. 2. Cell density of TN-368 cells upon subsequent subculturing following a 10,000 R X irradiation. Cultures were passaged at two- or three-day intervals.

Insect cells appear to utilize a transamination reaction in their intermediary metabolism whereby glutamic acid and pyruvic acid are converted to α-ketoglutaric acid and alanine (Koval et al. 1976, Landureau and Jolles 1969, Grace and Brzostowski 1966, Bursell 1963). It has been suggested that amino acid transamination for use in the tricarboxylic acid cycle and its resultant stimulation of respiration favor cell growth and viability in γ-irradiated *Escherichia coli* and DNA repair in UV-irradiated *E. coli* (Swenson et al. 1971, Swenson et al. 1975). The X-rayed TN-368 cells have a higher capacity for transamination (Figure 3) (Koval et al. 1976) and a greater rate of respiration, probably because of the uncoupling of oxidative phosphorylation (Koval et al. 1975), than untreated cells. These alterations may be partially responsible for the ability of TN-368 cells to survive high doses of X irradiation. It is noteworthy to mention that the hemolymph of insects contains very high concentrations of free amino acids (Wyatt 1961) in comparison to the blood of mammals (Gerke et al. 1968), which, following the above reasoning, may contribute to the apparent radiation resistance of insects.

Based on the cell culture studies mentioned above, it is clear that the radioresistance of the insect cells is not due to their mature nondividing nature as was previously suggested. Since the culture studies were performed on logarithmically growing cells, the radioresistance in this particular instance cannot be accounted

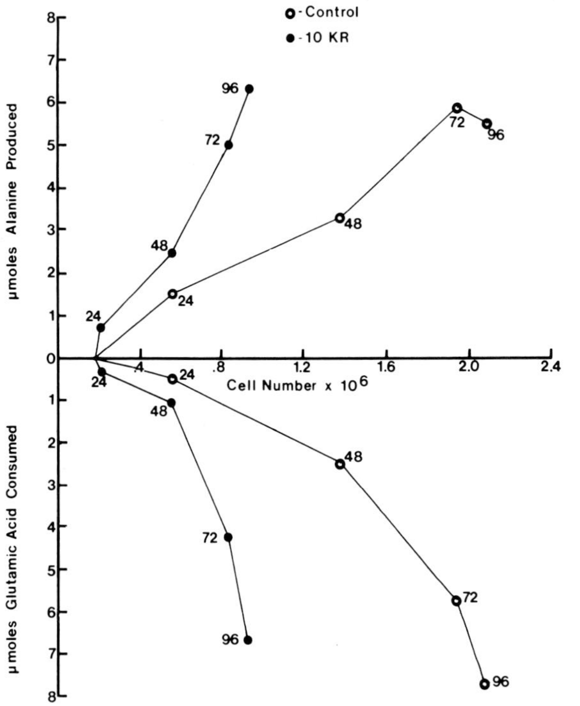

FIG. 3. Changes in glutamic acid and alanine in the growth medium of TN-368 cells. Numbers by the points represent the hours postirradiation. (Reproduced from Koval et al. 1975, with permission of Academic Press.)

for on the basis of a lack of mitosis; it must be due to other factors. While differences in intermediary metabolism have been suggested as conferring some resistance to the insect cells (Koval et al. 1976), these alone are probably not sufficient to account for their survival abilities.

Unrepaired or incorrectly repaired damage to cellular DNA should influence the ability of a cell to carry out its normal functions. Cells that can most efficiently repair DNA damage might be expected to have a greater probability of survival than cells with reduced DNA-repair capabilities. Therefore, it seems reasonable to investigate the assumption that insect cells in culture may be more resistant to the lethal effects of radiation because of very efficient DNA repair processes.

SURVIVAL AND REPAIR STUDIES

TN-368 Cell Line

The TN-368 cell line was derived from minced adult ovaries of the cabbage looper, *T. ni* (Hink 1970). The cells grow in a loosely attached monolayer generally requiring only moderate agitation to detach them from the bottom surface of polystyrene or glass flasks. The culture is characterized by round or oval cells with one to three protoplasmic extensions. Most of the cells have a diploid number of 82–95 chromosomes, but some are tetraploid and have a chromosome number of 160–180 (Hink 1972). As determined biochemically by the diphenylamine reaction (Burton 1956), the average DNA content of the TN-368 cells is about 6.8×10^{-12} g/cell (Koval et al. 1977). The population doubling time is about 16 hours, and the generation time is about 14.5 hours, with the following times allocable to each phase of the cell cycle: G_1, 1 hour; S, 4.5 hours; G_2, 8.5 hours; and M, .6 hour (Lynn and Hink 1978). The cells are usually maintained at 28°C in a humidified incubator in 25 cm^2 polystyrene tissue culture flasks containing 5 ml of TNM-FH medium (Koval et al. 1977).

V-79-4 Cell Line

The V-79-4 cell line was derived from lung tissue of a male Chinese hamster, *Cricetulus griseus,* and has been in culture for several years (Elkind and Whitmore 1967). The cells grow in a tightly attached monolayer and are generally treated with a 0.01% trypsin, 0.1% methyl cellulose solution to detach them from the surface of their growth vessel. Most of these fibroblast cells have 21–23 easily recognizable chromosomes (Kato 1968). The DNA content of V-79 cells was determined biochemically to be approximately 14.8×10^{-12} g/cell (Koval et al. 1977), though Kimball et al. (1971) indicated the amount to be about $11–12 \times 10^{-12}$ g/cell by cytofluorometry. The generation time is about 14 hours, with the following times allocable to the phases of the cell cycle: G_1, 5–6 hours; S, 6 hours; G_2, 2–3 hours; and M, .5–1 hour (Elkind and Whitmore 1967,

Kimball et al. 1971). The cells are usually maintained in minimum essential medium (MEM) (Ham 1972) supplemented with 10% fetal calf serum and grown in a 5% CO_2 environment in a humidified incubator at 37°C.

UV Studies

Survival

The development of the single-cell plating technique was essentially the beginning of quantitative radiobiology using cultured mammalian cells (Puck and Marcus 1955). The following experiments represent the initial application of this technique using insect cells (Koval 1978). Survival curves based on colony formation allow the determination of the surviving fraction of cells after a given radiation exposure. Roper and Drewinko (1976) have shown that colony formation is the only technique that gives a reliable estimate of cell survival.

Aliquots of 0.25 ml of TN-368 cell dilutions in Hanks' balanced salt solution (HBSS) (containing 15 mg/ml glucose) were placed in the center of Lux 60 mm diameter polystyrene dishes and exposed to 254-nm light. After the exposures, 2.5 ml of conditioned TNM-FH containing 0.1 M BES (N, N-bis (2-hydroxyethyl)-2-aminoethane sulfonic acid) was added to each dish (Koval 1978). Three-milliliter aliquots of V-79 cell dilutions in MEM with 5% fetal calf serum were placed in 60 mm diameter Corning polystyrene dishes for five hours, which allowed the cells to attach. The medium was then poured from each dish before exposing the cells to UV light. Prior to incubation of the cells, 2.5 ml of MEM containing 10% fetal calf serum was added to each dish. In order to prevent any photoreactivation, all irradiations were performed in the dark using only a dim red light when necessary. After incubating for seven days, the colonies were stained and counted, and plating efficiencies were calculated (Koval 1978).

The survival curves in Figure 4 indicate that the TN-368 cells are more than seven times more resistant than V-79 cells to the lethal effects of UV light. The D_o, extrapolation number, and D_q were 47.5 J/m^2, 6.8, and 32.5 J/m^2, respectively, for the TN-368 cells and 6.5 J/m^2, 2.0, and 2.5 J/m^2, respectively, for V-79 cells (Koval et al. 1977). Insect cells are frequently polyploid in vivo, and lepidopteran cells generally increase further in ploidy when cultured in vitro. The fact that the extrapolation number of the TN-368 cells is more than three times that of the V-79 cells may reflect an increased chromosome redundancy in the insect cells, meaning more "targets" or redundant genes must be inactivated in order to result in cell death.

Unscheduled DNA Synthesis (UDS)

UDS has been defined as the incorporation of radioactive precursors into the DNA of cells at a time when the cells are not normally involved in DNA

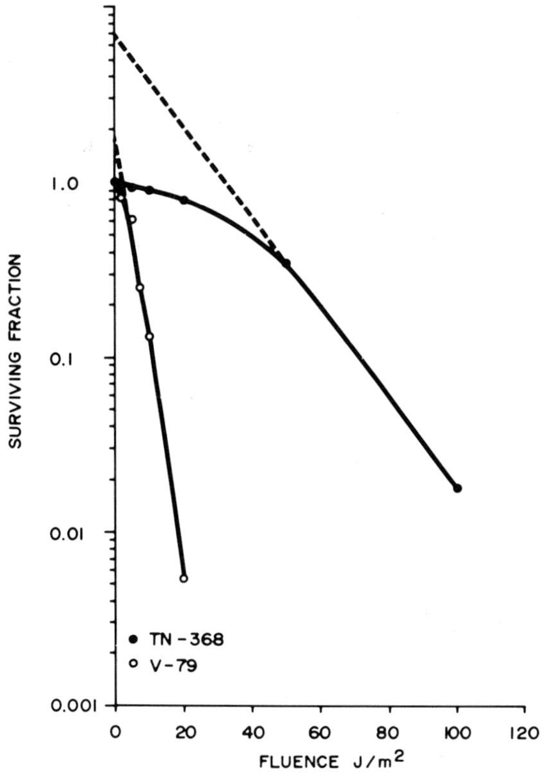

FIG. 4. UV survival curves for TN-368 and V-79 cells. (Reproduced from Koval et al. 1977, with permission of The University of Texas Press.)

synthesis (non–S-phase cells) (Elkind and Whitmore 1967). The process generally involves the replacement of bases in excised regions of parental DNA strands that have been damaged by ionizing or UV irradiation or by one of several chemical mutagens (Shaeffer and Merz 1971). UDS is usually measured autoradiographically by the incorporation of tritiated thymidine into the DNA of cells whose normal semiconservative DNA replication has been inhibited by hydroxyurea (Cleaver 1969). This method is a sensitive and quantitative assay and was used in this study to compare the amounts of DNA repair synthesis in the two cell lines.

Approximately 24 hours after cells were seeded on glass coverslips in Petri dishes, freshly prepared hydroxyurea solutions were added so that the final concentrations in the TN-368 and V-79 dishes were 0.05 M and 0.005 M, respectively. After six hours, the medium was suctioned from each dish and, in the absence of light except for a dim red one, the dishes were exposed to 0, 10.0, 20.0, or 50.0 J/m² of UV light. Fresh medium (2.5 ml) containing 3% calf serum, hydroxyurea as described above, and 4 μCi/ml tritiated thymidine (S.A. = 5 Ci/mM) was added to each dish. TN-368 cells were incubated at 28°C and V-79 cells at 37°C. Coverslips of each cell type at each dose were removed from incubation in the dark at 0, one, two, and four hours after irradiation, rinsed, fixed, dipped in Kodak NTB2 emulsion, developed, and stained.

There was a greater amount of tritiated thymidine incorporation into the V-79 cells than into the TN-368 cells at each dose of UV light (Figure 5), but since the V-79 cells have 1.5–2 times as much DNA per cell, the amount of incorporation was about the same for a given amount of DNA in both cell types. The average number of silver grains per cell increased proportionally with the dose of UV light and postirradiation time for both cell types. This suggests that the repair process was not saturated in the dose range studied and that after increasing amounts of damage there were correspondingly greater amounts of repair.

The study of Trosko and Wilder (1973), which demonstrated dimer removal in *Drosophila melanogaster* cells, along with the present UDS experiments, indicate that dimer removal and UV-induced excision repair are similar in both insect and mammalian cells. Since the survival of the TN-368 cells is about seven times greater than that of the V-79 cells following UV irradiation and the UDS is about the same in both cell types, some factor other than UDS or excision repair must be responsible for the greater survival of the TN-368 cells in this instance. Recently, Boyd and Setlow (1976) have reported the presence of two separate postreplication repair pathways in *D. melanogaster* cells. Therefore, chromosome redundancy or an error-free postreplication repair system in the insect cells may be responsible for increased survival.

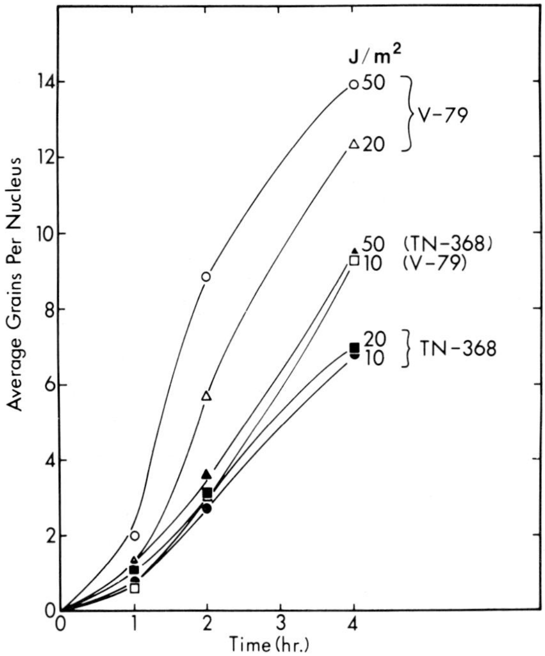

FIG. 5. The average amount of unscheduled DNA synthesis as a function of time following fluences of 10–50 J/m² for TN-368 and V-79 cells. (Reproduced from Koval et al. 1977, with permission of The University of Texas Press.)

X-ray Studies

Survival

For survival experiments, TN-368 cells were diluted in medium, plated, irradiated on ice with various doses of X rays, and incubated for seven days at 28°C. V-79 cells were irradiated on ice as a single-cell suspension, diluted in medium, plated, and incubated for seven days at 37°C (Koval et al. 1978). Colonies were stained and counted as described previously (Koval 1978).

Based upon the survival curves in Figure 6, the TN-368 cells appear to be over 30 times more resistant to X rays than the V-79 cells. The D_{37} value for TN-368 cells is about 10,000 R, while the D_{37} for the V-79 cells is approximately 325 R. The value of 325 R is very near the D_{37} observed by Hart et al. (1975) for the V-79 cells.

UDS

Cells were plated on glass coverslips in Petri dishes. Freshly prepared hydroxyurea solutions were added 24 hours later so that the TN-368 dishes were at 0.05 M and the V-79 dishes were at 0.005 M. (These concentrations reduced normal DNA synthesis to about the same level in each cell type.) About six hours later, the dishes were placed on a block of ice and X-rayed with 0, 500, 1,000, or 1,500 R, after which the medium was suctioned off and replaced with

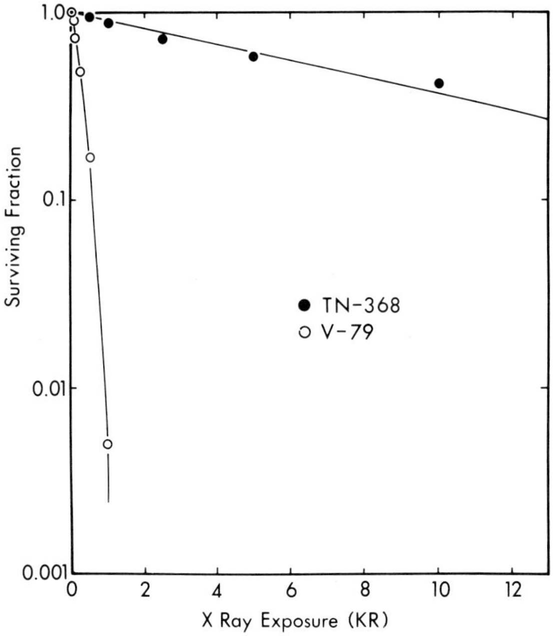

FIG. 6. X-ray survival response of cultured TN-368 and V-79 cells. (Reproduced from Koval et al. 1978, with permission of Elsevier/North-Holland Biomedical Press.)

10 ml of fresh medium containing 3% calf serum, hydroxyurea as described above, and 4 μCi/ml [^{3}H] thymidine (S.A. = 5 Ci/mM). Coverslips were removed from each dish at 0, 1, 2, and 4 hours postirradiation and prepared for autoradiography as described earlier (Koval et al. 1978).

The TN-368 cells incorporated over 10 times more tritiated thymidine via UDS than the V-79 cells after each of the three X-ray exposures (500, 1,000, and 1,500 R) investigated (Figure 7). Both the rate and extent of UDS were much greater in the TN-368 cells compared to the V-79 cells. Since the V-79 cells have 1.5–2 times as much DNA per cell, the TN-368 cells actually incorporated approximately 20 times more tritiated thymidine for a given amount of DNA.

It is possible that the large amount of UDS in TN-368 cells is a result of a greater amount of DNA damage in the TN-368 cells than in the V-79 cells, but that is not consistent with the knowledge that the TN-368 cells are many times more resistant to the lethal effects of X rays, as determined by survival studies. Therefore, it is believed that the resulting high yield of UDS in TN-368 cells after X irradiation is due to their having (1) a larger average size for X-ray–repaired regions or (2) a greater number of repaired regions per unit length of DNA. Further experimentation is necessary to distinguish between these two possibilities. In both bacterial and mammalian systems, the repair of γ-ray–induced lesions involves the insertion of one to three nucleotides (Hart and Trosko 1976). There is no reason to believe the number of inserted bases is any different in insect cells. If the size of the repaired regions turned out to be the same as for mammalian cells, this would indicate a greater number of

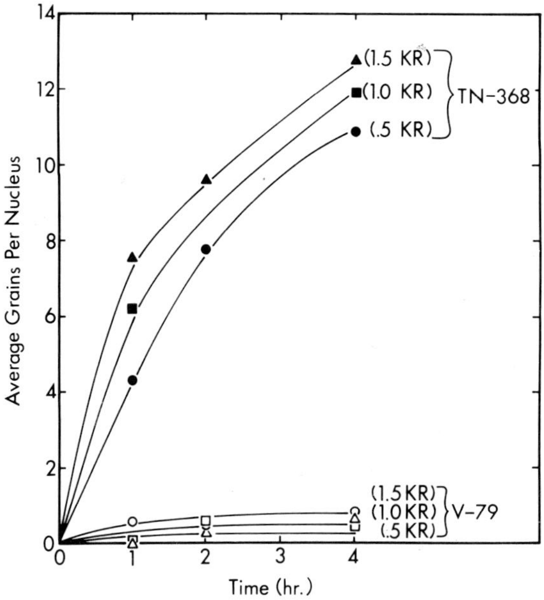

FIG. 7. X-ray–induced unscheduled DNA synthesis in TN-368 and V-79 cells following exposures to 500–1500 R. (Reproduced from Koval et al. 1978, with permission of Elsevier/North-Holland Biomedical Press.)

regions are repaired per unit length of DNA in the TN-368 cells and could partially account for the 30-fold increase in radiation resistance.

DNA Single-Strand Break Repair

Single-strand breaks (SSB) in DNA may arise from treatment of cells with either physical or chemical agents (Hart et al. 1978). The inability of cells to repair such damage has been associated with numerous biological dysfunctions including cell death, aging, and chromosomal aberrations (Hart and Trosko 1976). The most common method of studying the repair of these breaks is to isolate and sediment large single-stranded DNA from cells by gently lysing them directly on top of an alkaline sucrose gradient. This method, the technique of McGrath and Williams (1966), and its modifications allow the sedimentation of denatured, single-stranded DNA that has been subjected to minimal shearing degradation. The rejoining of DNA SSBs has been demonstrated in all systems examined including bacteria (Town et al. 1973), cultured mammalian cells (Lett et al. 1967), and isolated plant protoplasts (Howland et al. 1975). UDS is a generalized measurement of DNA repair capacity, and, in that radioisotopes incorporated during this procedure (especially during the first hour) may be due to SSB repair as well as excision repair of various types of DNA lesions, it was necessary to evaluate SSB repair separately in order to distinguish between these two repair processes.

Details of the SSB measurements have been described elsewhere (Koval et al. 1979). Briefly, cells were grown in medium containing [^{3}H]-thymidine or [^{32}P]-orthophosphate for 36–40 hours, X-rayed on ice with 10,000 R, and sampled after various incubation times (the TN-368 cells at 28°C and the V-79 cells at 37°C). ^{3}H-labeled X-rayed cells and ^{32}P-labeled control cells were layered together and lysed on top of alkaline sucrose gradients for each incubation time. Fractions were collected on filter paper and counted in a liquid scintillation counter. A minor modification of the above procedure permitted a direct, concurrent evaluation of the SSB repair capacity of the two cell systems. TN-368 cells were labeled with [^{3}H]-thymidine and V-79 cells with [^{32}P]-orthophosphate so that aliquots from each cell type were placed on the same gradient after duplicate treatments. SSB calculations were done according to those of Howland et al. (1975).

Values near 2×10^{8} daltons were obtained for the weight-average molecular weight of single-stranded DNA for both unirradiated TN-368 and V-79 cells. The rate of DNA SSB rejoining was found to be slightly higher in TN-368 cells than in V-79 cells (Figure 8), as most of the breaks in TN-368 cells were repaired within 10 minutes and in V-79 cells within 20 minutes. Under the experimental conditions, there were about 1.3 DNA SSBs per 10^{8} daltons of DNA from TN-368 cells and 2.2 per 10^{8} daltons of DNA from V-79 cells. The number of electron volts (eV) required to produce a SSB was estimated to be about 77 for TN-368 cells and 48 for V-79 cells. The value of 48 eV is

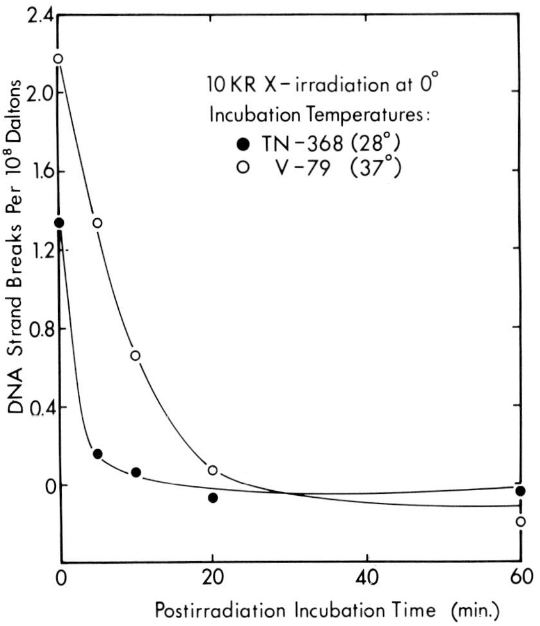

FIG. 8. DNA single-strand break rejoining in TN-368 and V-79 cells. (Reproduced from Koval et al. 1979, with permission of Taylor & Francis Ltd.)

very close to the accepted value of 44 eV per SSB in mammalian cells (Setlow and Setlow 1972, Lehman and Omerod 1970) (the value of 77 eV is near the 85 eV per break reported for plant cells (Howland et al. 1975)). Based on the previously described calculations and values for DNA content (Koval et al. 1978), there were approximately 5.5 breaks/rad/TN-368 cell and 15.7–19.4 breaks/rad/V-79 cell.

Experiments with irradiated DNA solutions have revealed that organic solutes, such as amino acids, sulphydryl compounds, etc., not only scavenge primary radicals, but may also completely alter the character of radiation damage to DNA (Blok and Lohman 1973). Since it is known that insects have very large free amino acid pools in their hemolymph in comparison to mammals (Koval et al. 1976), it is reasonable to expect that they may also have larger intracellular amino acid pools than mammals. If they do, large amounts of free radical scavenging by amino acids in the insect cells could account for the smaller number of SSBs produced in the TN-368 cells than in the V-79 cells in this study. Notwithstanding, it would still be difficult to correlate strand breaks with lethality since the breaks are rapidly repaired in both cell systems.

Reports have indicated that lepidopteran cells may have holocentric chromosomes (White 1973). If this is true, chromosome translocations produced by ionizing radiation need not necessarily result in cell death, whereas in cells with monocentric chromosomes, translocations would frequently result in cell death. Thus, this mechanism must also be considered in explaining the survival of some insect cells.

SUMMARY, CONCLUSIONS, IMPLICATIONS

The experiments described above quantitatively demonstrate that the insect cell line, TN-368, is about 30 and 7 times more resistant to the lethal effects of X irradiation and UV radiation, respectively, than the mammalian cell line, V-79. In general, the TN-368 cells have 0.5–0.67 the DNA content of the V-79 cells, and it is distributed over about four times the number of chromosomes. There is approximately 20 times the amount of UDS per unit amount of DNA in TN-368 cells four hours following 0.5–1.5 kR X irradiations than in the V-79 cells. The amount of UDS per unit amount of DNA is about the same in both cell types four hours after 10.0–50.0 J/m^2 UV irradiations. TN-368 cells suffer slightly fewer DNA SSBs and repair them a little quicker than V-79 cells, but repair is nearly complete by 20 minutes postirradiation with 10 kR of X rays in both cell types.

The fact that UDS is much greater in TN-368 cells than in V-79 cells after X irradiation while DNA SSB repair is only slightly greater indicates that an excision repair mechanism acting on endonuclease-sensitive sites in the DNA may be partially responsible for the greater radioresistance of the TN-368 cells. Conversely, since the UDS after UV irradiation for both cell types is about the same, some factor other than UDS or excision repair must be responsible for the greater survival of the TN-368 cells after UV exposure. It therefore appears that, although probably playing an important role, DNA repair processes alone cannot be used to account for the radioresistance of insect cells.

Ataxia telangiectasia (AT) is a human autosomal recessive neurovascular disease whose victims are prone to immunological defects and malignancy. Cells from patients with AT are defective in their ability to form colonies following exposure to γ radiation (Taylor et al. 1975) and to repair γ- or X-ray–damaged DNA by an excision repair process (Paterson et al. 1976, Weichselbaum et al. 1978). Therefore, the relationship between V-79 cells and TN-368 cells is analogous to that between AT cells and normal human cells. Recently, AT cells were found to have slightly reduced levels of endonucleolytic activity when compared to controls (Sheridan and Huang 1978). Perhaps TN-368 cells have excessive endonucleolytic activity. The study of cultured insect cells may provide a unique system for identifying molecular mechanisms of radiation resistance.

The above results may have a significant bearing on current theories of mammalian aging. Hart and Setlow (1974) have correlated life span with DNA repair capacity in eight mammalian species and have proposed that an organism's capacity to repair DNA damage is directly related to its life span. The present study demonstrates that insect cells have a tremendous capacity to repair DNA, and yet, in vivo, insects have very short relative life spans. Therefore, though I do not dispute that the correlation exists for mammals, the TN-368 studies caution that other contributing factors may perform a critical role in mediating an organism's life span.

A final related note is that some groups of insects have not changed appreciably during recent periods of geological history. In general, insects have relatively short life spans and a large number of generations. The number of offspring produced per generation in insects is also usually quite large. It seems reasonable, therefore, to expect that insect phenotypes would change quite rapidly in an evolutionary perspective. Efficient DNA repair mechanisms would be expected to minimize the DNA mutation rate by maintaining the genetic fidelity of cells. The fact that some insects have changed only minimally through the years supports the contention that insects possess very efficient DNA repair mechanisms, which allowed them to maintain their genetic integrity.

ACKNOWLEDGMENTS

I thank Drs. W. C. Myser, R. W. Hart, W. F. Hink, and H. S. Ducoff for helpful discussions and advice during the course of this research.

REFERENCES

Bergonie, J., and L. Tribondeau. 1906. De quelques résultats de la radiothérapie et essai de fixation d'une technique rationelle. C.R. Acad. Sci. 143:983–985. (English translation in Radiat. Res. 11 (1959) 587–588.)

Blakely, E. A. 1975. Growth and viability of *Aedes albopictus* cell line in vitro after cesium-137 gamma irradiation. Ph.D. thesis. University of Illinois. (Diss. Abstr. 36:2096–B.)

Blok, J., and H. Lohman. 1973. The effects of γ-radiation in DNA. Curr. Top. Radiat. Res. Q. 9:165–245.

Bond, V. P., T. M. Fliedner, and J. O. Archambeau. 1965. Mammalian Radiation Lethality. Academic Press, New York.

Boyd, J. B., and R. B. Setlow. 1976. Characterization of postreplication repair in mutagen-sensitive strains of *Drosophila melanogaster.* Genetics 84:507–526.

Bursell, E. 1963. Aspects of the metabolism of amino acids in the tsetse fly, *Glossina* (Diptera). J. Insect Physiol. 9:439–452.

Burton, K. 1956. A study of the conditions and mechanism of the diphenylamine reaction for the colorimetric estimation of deoxyribonucleic acid. Biochem. J. 62:315–323.

Carlson, J. G. 1969. A detailed analysis of x-ray induced prophase delay and reversion of grasshopper neuroblasts in culture. Radiat. Res. 37:1–14.

Carrel, A. 1912. On the permanent life of tissues outside of the organism. J. Exp. Med. 15:516–528.

Casarett, A. P. 1968. Radiation Biology. Prentice-Hall, Inc., Englewood Cliffs, New Jersey.

Cleaver, J. E. 1969. Repair replication of mammalian cell DNA: Effects of compounds that inhibit DNA synthesis or dark repair. Radiat. Res. 37:334–348.

Ducoff, H. S. 1972. Causes of death in irradiated adult insects. Biol. Rev. 47:211–240.

Elkind, M. M., A. Han, and K. W. Volz. 1963. Radiation responses of mammalian cells grown in culture. IV. Dose dependence of division delay and postirradiation growth of surviving and nonsurviving Chinese hamster cells. J. Natl. Cancer Inst. 30:705–721.

Elkind, M. M., and G. F. Whitmore. 1967. The Radiobiology of Cultured Mammalian Cells. Gordon and Breach, New York.

Gerke, C. W., R. W. Zumwalt, D. L. Stalling, and L. L. Wall. 1968. Quantitative Gas-Liquid Chromatography of Amino Acids in Proteins and Biological Substances. Analytical Biochemistry Laboratories, Inc., Columbia, Mo.

Grace, T. D. C. 1962. Establishment of four strains of cells from insect tissues grown in vitro. Nature 195:788–789.

Grace, T. D. C., and H. W. Brzostowski. 1966. Analysis of the amino acids and sugars in an insect cell culture medium during cell growth. J. Insect Physiol. 12:625–633.

Ham, R. G. 1972. Cloning of mammalian cells, *in* Methods in Cell Biology, Vol. V, D. M. Prescott, ed. Academic Press, New York, pp. 37–74.

Hart, R. W., R. E. Gibson, J. D. Chapman, A. P. Reuvers, B. K. Sinha, R. K. Griffith, and D. T. Witiak. 1975. A radioprotective stereostructure-activity study of cis- and trans-2-mercaptocyclobutylamine analogs and homologs of 2-mercaptoethylamine. J. Med. Chem. 18:323–331.

Hart, R. W., K. Y. Hall, and F. B. Daniel. 1978. DNA repair and mutagenesis in mammalian cells. Photochem. Photobiol. 28:131–155.

Hart, R. W., and R. B. Setlow. 1974. Correlation between deoxyribonucleic acid excision repair and life-span in a number of mammalian species. Proc. Natl. Acad. Sci. USA 71:2169–2173.

Hart, R. W., and J. E. Trosko. 1976. DNA repair processes in mammals, *in* Interdisciplinary Topics in Gerontology, Vol. 9, R. Cutler, ed. S. Karger AG, Basel, Switzerland, pp. 134–167.

Hink, W. F. 1970. Established insect cell line from the cabbage looper, *Trichoplusia ni.* Nature 226:466–467.

Hink, W. F. 1972. A catalog of invertebrate cell lines, *in* Invertebrate Tissue Culture, Vol. II, C. Vago, ed. Academic Press, New York, pp. 363–387.

Howland, G. P., R. W. Hart, and M. L. Yette. 1975. Repair of DNA strand breaks after gamma-irradiation of protoplasts isolated from cultured wild carrot cells. Mutat. Res. 27:81–87.

Kato, R. 1968. The chromosomes of forty-two primary Rous sarcomas of the Chinese hamster. Hereditas 59:63–119.

Kimball, R. F., S. W. Perdue, E. H. Y. Chu, and J. R. Ortiz. 1971. Micorphotometric and autoradiographic growth and decline of Chinese hamster cell cultures. Exp. Cell Res. 66:17–32.

Koval, T. M. 1978. A simple method for obtaining high plating efficiencies with cultured insect cells. Experientia 34:674–675.

Koval, T. M., R. W. Hart, W. C. Myser, and W. F. Hink. 1977. A comparison of survival and repair of UV-induced DNA damage in cultured insect versus mammalian cells. Genetics 87:513–518.

Koval, T. M., R. W. Hart, W. C. Myser, and W. F. Hink. 1979. DNA single strand break repair in cultured insect and mammalian cells after x-irradiation. Int. J. Radiat. Biol. 35:183–188.

Koval, T. M., W. C. Myser, R. W. Hart, and W. F. Hink. 1978. Comparison of survival and unscheduled DNA synthesis between an insect and a mammalian cell line following x-ray treatments. Mutat. Res. 49:431–435.

Koval, T. M., W. C. Myser, and W. F. Hink. 1975. Effects of x-irradiation on cell division, oxygen consumption, and growth medium pH of an insect cell line cultured in vitro. Radiat. Res. 64:524–532.

Koval, T. M., W. C. Myser, and W. F. Hink. 1976. The effect of x-irradiation on amino acid utilization in cultured insect cells. Radiat. Res. 67:305–313.

Landureau, J. C., and P. Jolles. 1969. Études des exigences d'une lingnée de celludes d'insectes (souche EPa). I. Acides amines. Exp. Cell Res. 54:391–398.

Lee, C. K. 1974. Some characteristics of replication and radiation response of *Aedes albopictus* cell line in vitro. Ph.D. Thesis. University of Illinois. (Diss. Abstr. 35:6056-B).

Lehmann, A. R., and M. G. Omerod. 1970. Double-strand breaks in the DNA of a mammalian cell after x-irradiation. Biochim. Biophys. Acta 217:268–277.

Lett, J. T., I. Caldwell, C. J. Dean, and P. Alexander. 1967. Rejoining of x-ray induced breaks in the DNA of leukaemic cells. Nature 214:790–792.

Lynn, D. E., and W. F. Hink. 1978. Cell cycle analysis and synchronization of the TN-368 insect cell line. In Vitro 14:236–238.

McGrath, R. A., and R. W. Williams. 1966. Reconstruction in vivo of irradiated *Escherichia coli* deoxyribonucleic acid: The rejoining of broken pieces. Nature 212:534–535.

O'Brien, R. D., and L. S. Wolfe. 1964. Radiation, Radioactivity and Insects. Academic Press, New York.

Paterson, M. C., B. P. Smith, P. H. M. Lohman, A. K. Anderson, and L. Fishman. 1976. Defective excision repair of x-ray-damaged DNA in human (ataxia telangiectasia) fibroblasts. Nature 260:444–447.

Puck, T. T., and P. I. Marcus. 1955. A rapid method for viable cell titration and clone production with HeLa cells in tissue culture: the use of x-irradiated cells to supply conditioning factors. Proc. Natl. Acad. Sci. USA 41:432–437.

Roper, P. R., and B. Drewinko. 1976. Comparison of in vitro methods to determine drug-induced cell lethality. Cancer Res. 36:2182–2188.

Setlow, R. B., and J. K. Setlow, 1972. Effects of radiation on polynucleotides. Ann. Rev. Biophys. Bioeng. 1:293–346.
Shaeffer, J., and T. Merz. 1971. A comparison of unscheduled DNA synthesis, D_o, cell recovery, and chromosome number in x-irradiated mammalian cell lines. Radiat. Res. 47:426–436.
Sheridan, R. B., and P. C. Huang. 1978. Apurinic and/or apyrimidinic endonuclease activity in ataxia telangiectasia cell extracts. Mutat. Res. 52:129–136.
Swenson, P. A., J. E. Ives, and R. L. Schenley. 1975. Photoprotection of *E. coli* B/r: Respiration, growth, macromolecular synthesis and repair of DNA. Photochem. Photobiol. 21:235–241.
Swenson, P. A., R. L. Schenley, and J. M. Boyle. 1971. Interference with respiratory control by ionizing radiations in *Escherichia coli* B/r. Int. J. Radiat. Biol. 20:213–223.
Taylor, A. M. R., D. G. Harnden, C. F. Arlett, S. A. Harcourt, A. R. Lehman, S. Stevens, and B. A. Bridges. 1975. Ataxia telangiectasia: A human mutation with abnormal radiation sensitivity. Nature 258:427–429.
Tolmach, L. J., and P. I . Marcus. 1960. Development of x-ray induced giant HeLa cells. Exp. Cell Res. 20:350–360.
Town, C. D., K. C. Smith, and H. S. Kaplan. 1973. Repair of x-ray damage to bacterial DNA. Curr. Top. Radiat. Res. Q. 8:351–399.
Trosko, J. E., and K. Wilder. 1973. Repair of UV-induced primidine dimers in *Drosophila melanogaster* cells in vitro. Genetics 73:297–302.
Watanabe, I., and S. Okada. 1967. Reproductive death of irradiated cultured mammalian cells and its relation to mitosis. Nature 216:380–381.
Weichselbaum, R. R., J. Nove, and J. B. Little. 1978. Deficient recovery from potentially lethal radiation damage in ataxia telangiectasia and xeroderma pigmentosum. Nature 271:261–262.
White, M. J. D. 1973. Animal Cytology and Evolution. Cambridge University Press, Cambridge, England.
White, P. R. 1954. The Cultivation of Animal and Plant Cells. Ronald Press, New York.
Wyatt, G. R. 1961. The biochemistry of insect hemolymph. Ann. Rev. Entomol. 6:75–102.

TMK present address:
Department of Radiation Therapy
Hahnemann Medical College
Philadelphia, Pennsylvania 19102

Radiation Biology in Cancer Research, edited by
Raymond E. Meyn and H. Rodney Withers.
Raven Press, New York © 1980.

Biophysical Mechanisms of Radiogenic Cancer

Harald H. Rossi

Radiological Research Laboratory, Department of Radiology, College of Physicians and Surgeons of Columbia University, New York, New York 10032

There can be no doubt that many, often very complex, processes intervene between the absorption of radiant energy and the manifestation of radiogenic cancer. It might, therefore, be believed that considerations based on the initial event have little relevance to the final outcome. Nevertheless, it seems that radiological physics and specifically microdosimetry not only can furnish insight into the relation between tumor frequency and the absorbed doses of various radiation but can also provide other information relating to the process of radiation carcinogenesis. The assurance with which such conclusions can be drawn vary from near certainty to reasonable speculation. I believe it will be in keeping with the spirit of this symposium if I present conclusions that range over a spectrum of assurance.

Direct applications of microdosimetry are limited to individual cells. It is essential, therefore, to distinguish between two types of cell response, which I will term "*autonomous*" and "*non-autonomous.*" The autonomous radiation response of a cell is one that is not influenced by the irradiation of additional cells in an organism or other ensemble of cells. It is thus determined exclusively by the radiation received by the cell under consideration. It would seem that autonomous response must occur in in vitro studies when cells are plated on an inert substrate at considerable separation. It should be remembered, however, that, apart from possible intercellular diffusion of molecular products, cell cultures—particularly in transformation experiments—often are irradiated some time after plating when they may already have formed microcolonies of two or more cells. Since in either case cellular interaction is conceivable, the notion of autonomous response could be an idealization. However, it seems likely that it is often at least approximated to an adequate degree.

In the following discussion, cell responses will be considered to be autonomous unless otherwise specified. The discussion is also limited to eukaryotic cells, usually mammalian.

THE AUTONOMOUS CELL RESPONSE

Whatever ionizing radiation cells may be exposed to, they receive energy as a result of traversals by charged particles. An energy deposition due to the appearance of a single charged particle or its secondaries (delta rays, associated spallation products, etc.) or both is termed an *event.* The *event frequency,* the mean number of events per unit of absorbed dose, is a quantity that can be established on the basis of physical (microdosimetric) measurements with considerable accuracy for all radiations and for a wide range of cellular or subcellular volumes.

When the cell is exposed to low linear energy transfer (LET) radiations (e.g., X and γ rays), many events may occur before biological effects are readily observable. However, in the case of high-LET radiations (e.g., 1-MeV neutrons) it is often relatively easy to quantitatively determine biological effects when the event frequency in the cell is near 1 and sometimes considerably less than 1. In the latter instance most cells receive no energy and the energy received by the others is independent of the absorbed dose that only determines the *number* of cells that receive any energy. It is obvious that under these conditions the number of cells affected in any way must be proportional to the absorbed dose regardless of the dependence of effect on absorbed energy. It may be assumed without loss of generality that under these conditions a single event produces one or more *lesions* that ultimately result in the biological effect observed.

When the absorbed dose is of the order of 10 mGy (1 rad), neutron secondaries will traverse only a very small percentage of the cells irradiated; but there are several traversals in each cell when this dose is due to X or γ rays. At such doses, the relative biological effectiveness (RBE) of neutrons is of the order of 50 with the result that for equal biological effect the event frequencies differ by a factor of the order of 10,000. There is, therefore, an entirely disparate dependence of energy deposition on absorbed dose: in the case of neutron irradiation, the energy absorbed in cells is independent of the absorbed dose that determines the number of cells receiving energy, while in the case of gamma radiation all cells receive energy essentially proportional to absorbed dose. It follows that an investigation of RBE as a function of absorbed dose should disclose the dependence of biological effectiveness of (at least the low-LET) radiation on energy concentration in the cell.

Such investigations routinely disclose a dependence of RBE on the inverse square root of the dose of high-LET radiation, which is indicated by a slope of −½ for the line relating the logarithm of RBE to the logarithm of neutron dose (Figure 1).

This observation has motivated the formulation of the theory of dual radiation action (Kellerer and Rossi 1972), which postulates that lesions are formed by the combination of pairs of *sublesions.* These are presumed to be molecular alterations that are produced at a rate proportional to the absorbed dose and substantially independent of radiation quality. A more detailed analysis also discloses that sublesions combine over distances on the average of the order of 1 μm.

It should be pointed out that some of the tenets of the theory have occasionally been misunderstood. The dependence of RBE on the inverse square root of

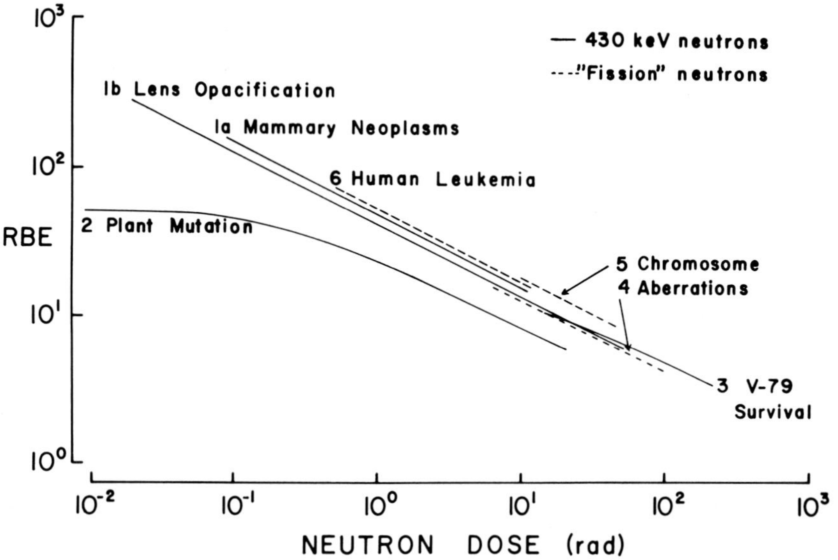

FIG. 1. Logarithmic plot of the RBE of low-energy neutrons vs. neutron dose. (Shellabarger et al. 1974, Bateman et al. 1972, Sparrow et al. 1972, Hall et al. 1973, Biola et al. 1971, Awa 1975, Rossi 1977).

the dose of high-LET radiation obtains only when the event frequency is much less than 1 for high-LET radiation and much more than 1 for low-LET radiation. When the first condition is not met, the RBE tends towards a fixed value near 1, whereas violation of the second condition results in very high RBE values that are ultimately also independent of dose. The full dependence is shown in Figure 2; a slope as steep as −½ is never reached when the quality of the radiations under comparison does not differ substantially. This may be the case, for instance, when neutrons having energies in the therapeutic range are compared with orthovoltage X rays. The only generally applicable statement is, therefore, that the RBE vs. high-LET dose curve is never steeper than −½.

Another source of confusion arises from the adoption of the "site" concept, in which it is postulated that sublesions combine with fixed probability in a subcellular volume having dimensions of the order of 1 μm. This concept leads to the relation:

$$\epsilon(D) = k(\zeta D + D^2) \qquad (1)$$

where $\epsilon(D)$ is the yield of radiation-induced lesions, D the absorbed dose, k a constant, and ζ a microdosimetric quantity (the mean of the square of the specific energy deposited in individual events).

Although this approximation is found to be sufficient in many cases, it may be expected to be inadequate when the hypothetical site is not traversed by single (nonassociated) particles. Experiments with low-energy X rays have dem-

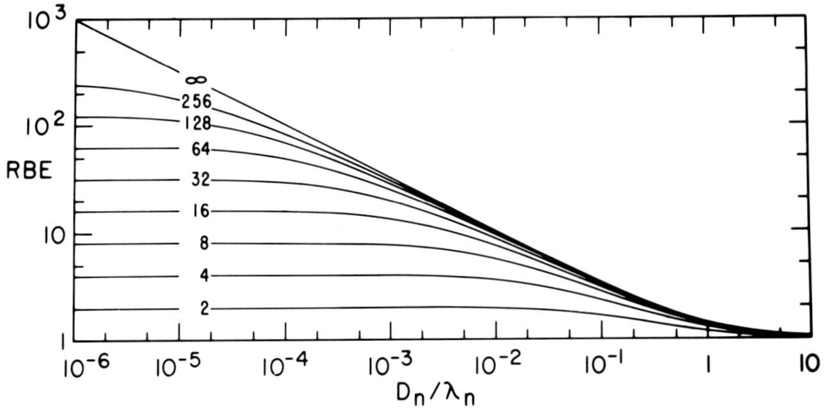

FIG. 2. Logarithmic representation of the dependence of RBE vs. dose of high-LET radiation dose (D_n). D_n is expressed in multiples of λ_n. λ_n is the ratio of the linear components of the dose-effect curves. (Reproduced from Kellerer and Rossi 1972, with permission of North-Holland Publishing Co.).

onstrated a marked increase of RBE when the photoelectron range decreases below values of the order of 1 μm (Cox et al. 1977). An experiment designed to investigate the distribution of initial separations of interacting sublesions has indicated that this distribution is highly skewed, with a rapid drop occurring at small separations (<0.1 μm) followed by an extension to larger distances (Rossi et al. 1978). This is in qualitative agreement with the X-ray observations. A recently published generalized formulation of dual radiation action does not employ the site concept, but it should be noted that it leads to a dependence of lesion yield on absorbed dose that is formally the same as that in equation 1, although the interpretation of ζ is more complex (Kellerer and Rossi, 1978).

In a recent review of the theory of dual radiation action additional modifying factors are considered (Rossi 1979). With regard to the subject of this monograph, the most important of these is competition between sublesions. It may be expected that this process can occur because the theory attributes a wide variety of effects to the interaction of pairs of sublesions. Therefore, an individual sublesion might be a partner in various lesions that cause different effects. In particular such a sublesion might enter a combination that causes cell lethality or it could be a partner in an interaction that results in a lesion causing a nonlethal change such as transformation. There is evidence that may indicate that this does in fact occur.

Figure 3 shows the proportion of transformed colonies among $C_3H/10T\frac{1}{2}$ colonies surviving exposures to a range of X-ray doses. Fractionation reduced the transformation frequency at high doses and enhanced it at moderate doses; it appears that it had no effect at sufficiently low doses. This seemingly complicated behavior can be accounted for by the simple assumption that there is

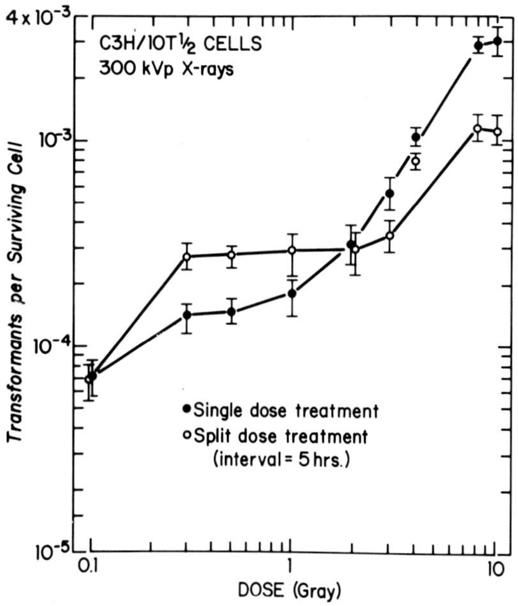

FIG. 3. Logarithmic representation of the transformation per surviving C3H/10T½ cells vs. X-ray dose for single and fractionated irradiations (R. C. Miller and E. J. Hall, personal communication 1979).

no interaction between the effects of the two-dose fractions. In this case the transformation frequency resulting from a dose D given in two fractions should be twice that following a single dose D/2. Thus, when the curve in this logarithmic representation has a slope equal to 1, fractionation has no effect, but the transformation frequency is doubled when the slope is zero and reduced when the slope reaches a value of about 2 for doses larger than a few gray. This elementary explanation would not be evident if one were not dealing with a homogenous cell population. The single-dose curve for a heterogenous population of transforming cells would be a superposition of single-dose curves of the shape shown in Figure 3 in which little or no plateau might be apparent. The latter is the case for the hamster embryo system studied by Borek and Hall (1973) (Figure 4), but when the effects of fractionation in the two systems are compared they seem to be identical (Figure 5).

Since it appears that the effects of temporal variations of dose can be explained by the shape of the dose-effect curve, the reasons for this shape must be of special interest. Microdosimetric considerations indicate that the initial linear rise in Figure 3 must be due to transformations initiated by single electrons traversing the cell (or more likely its nucleus), but there is no obvious reason why the slope of the curve must decrease before it rises even more steeply. A possible explanation is that transformations are due to lesions resulting from relatively rare combinations of sublesions. Other combinations that collectively are more probable can then be formed at moderate doses because additional electrons produce further sublesions that combine with those that would other-

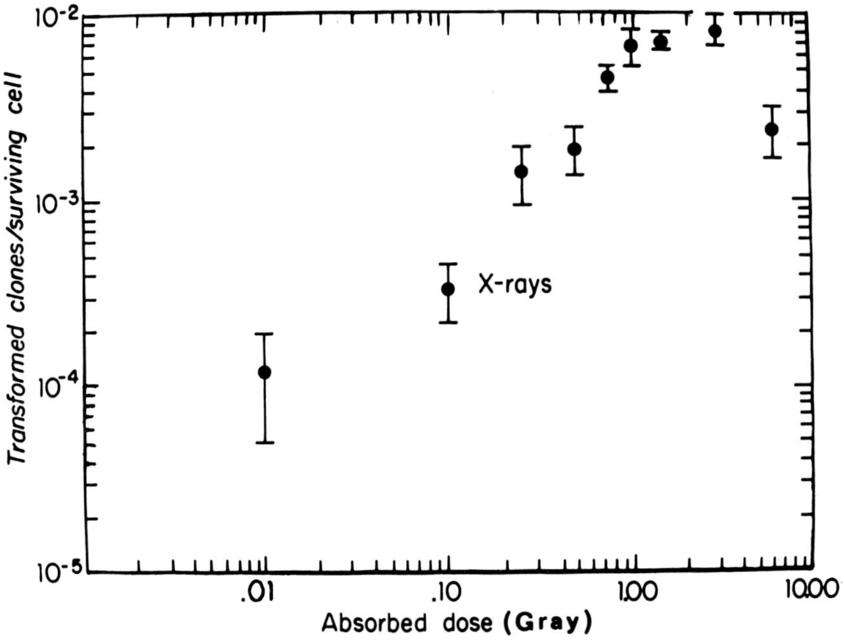

FIG. 4. Logarithmic representation of transformed clones vs. surviving cells of hamster embryo explants (replotted from data of Borek and Hall 1973).

wise have been partners in lesions resulting in transformations. These other lesions need not be lethal; they might merely cause undetected alterations. The second rise at higher doses is then due to a mechanism in which the two sublesions that can form the transformation lesion are produced by two electrons (rather

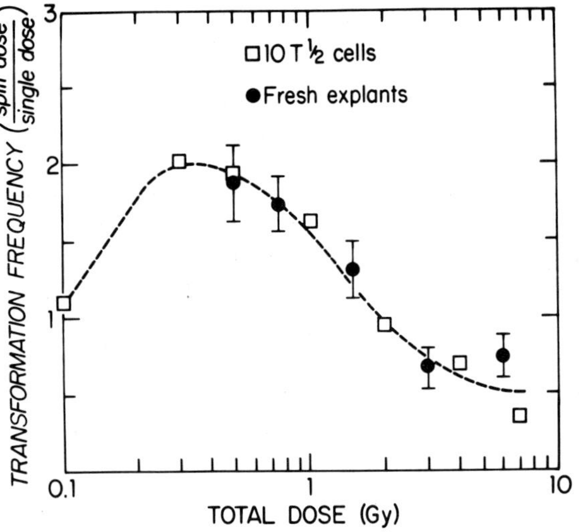

FIG. 5. Logarithmic representation of the ratio of transformation frequencies for single and split doses in 10T½ cells and fresh explants.

than one) with negligible production of competing sublesions. This is only a qualitative explanation and must, therefore, be regarded as an example of what I have termed "reasonable speculation."

THE SYSTEMIC RESPONSE

Although it is generally assumed that the process of radiation carcinogenesis is initiated by cell alteration, it is not known whether the cells involved in human carcinogenesis respond to radiation in the same manner as those subject to transformation experiments. As already stated, it must also be assumed that radiation carcinogenesis is a complex process and it follows that dose-effect curves for autonomous cells may have little relation to the curves relating the incidence of cancer to absorbed dose. A great variety of shapes of dose-response curves have in fact been observed in experimental radiation carcinogenesis and good examples of this are recent findings by Ullrich and his co-workers (Ullrich et al. 1976, 1977).

An additional significant finding is that the incidence of mammary neoplasms in the Sprague-Dawley rat does not increase linearly with neutron doses so low that only very few cells are traversed by neutron secondaries (Shellabarger et al. 1974). This tendency is also shown in some of the quoted results by Ullrich and co-workers. As shown in detailed consideration, this is proof that the carcinogenic process involves a collective cell response (Rossi and Kellerer 1972).

It should be noted, however, that the same analysis also discloses that the dependence of RBE on neutron dose for this evidently complicated process is the same as for a variety of end points for neutron energies of about 0.5 MeV as well as for "fission" neutrons. This relation is

$$\mathrm{RBE} \simeq 4.5\ (D_N/\mathrm{Gy})^{-1/2} \qquad (2)$$

where D_N is the absorbed dose of neutrons and the expression applies if D_N is between about 1 mGy and 1 Gy.

It seems likely that the reason for the ubiquitous applicability of this relation is that the dependence of RBE on absorbed dose is due to the basic biophysical process of dual radiation action. This precedes all subsequent biological processes that generally proceed in mechanisms that have little if any further dependence on radiation quality.

HUMAN CARCINOGENESIS

From the foregoing it might be expected that dose-effect curves for human carcinogenesis have a variety of shapes but that the dose-RBE dependence is the same as that for somatic effects in other higher organisms.

At this time the epidemiological observations made on A-bomb survivors in Japan constitute virtually the only usable data for RBE analysis. The Hiroshima bomb emitted a neutron fluence that was sufficiently intense to contribute roughly

10% of the absorbed dose in deep-lying organs, whereas the neutron component at Nagasaki was essentially negligible. However, even though the data from the two cities involve by far the largest cohorts studied in radiation epidemiology, the number of cases is limited, especially at Nagasaki. There, a somewhat smaller population was almost exclusively exposed to low-LET radiation with its relatively low biological potency.

For all types of leukemia collectively, the RBE increases with decreasing absorbed dose and the epidemiological data are consistent with a linear dose-effect relation for neutrons and a square relation for γ radiation (Rossi and Kellerer 1974). This was confirmed in a second analysis, which proceeded by a different method. This analysis also indicated that the numerical values of the RBE substantially agreed with equation 2 (Rossi and Mays 1978).

The induction in the two Japanese cities of other types of neoplasms has been too low to permit a corresponding analysis with reasonable statistical significance. An analysis is practicable, however, if malignant neoplasms are considered *in toto*. Figure 6, based on a recently published survey (Beebe et al. 1977), is a plot of the mortality per year as a function of kerma in Hiroshima and Nagasaki. Using available information on the relation between kerma and absorbed doses

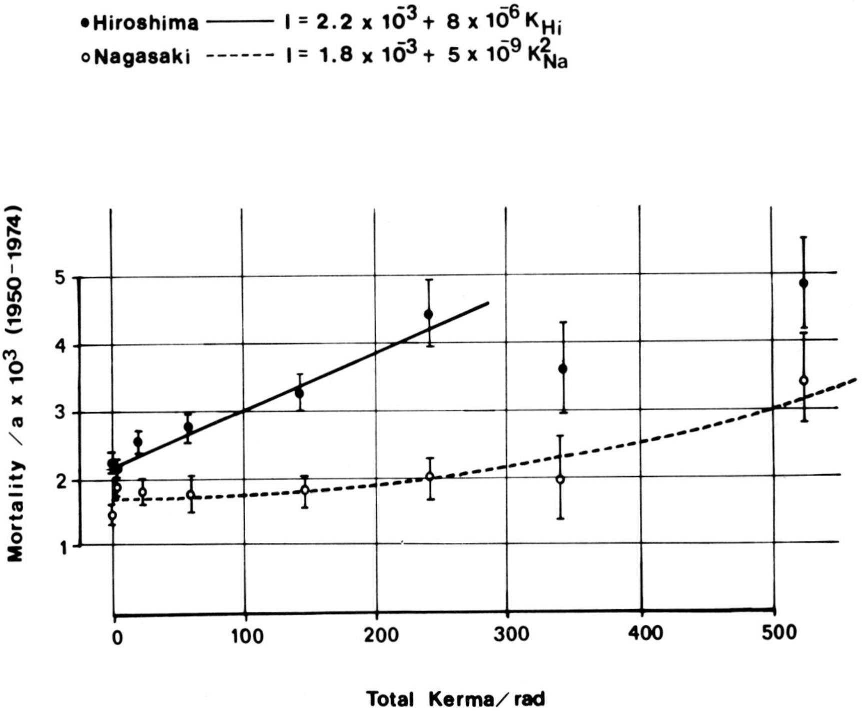

FIG. 6. Annual mortality from all malignant neoplasms in Hiroshima and Nagasaki (1950–1974) vs. total kerma. The limits correspond to ± one standard deviation.

for neutrons and gamma radiation (Jones 1977), one may estimate the annual mortality from all malignant neoplasms from 1950 to 1974 as a function of absorbed dose to the bone marrow, about $1.3 \times 10^{-2}(D_N/\text{Gy})$ for neutrons and $1.7 \times 10^{-4}(D_\gamma/\text{Gy})^2$ for gamma rays. These estimates are based on an analysis that ignores the two highest kerma points at Hiroshima. It seems probable that the mortality reaches a maximum at low neutron dose and that the concomitant much larger gamma dose causes a secondary peak at a larger (because of differences in RBE) kerma. The points in Figure 6 were obtained by simple division of pairs of numbers in the cited reference. A more refined statistical analysis might disclose a linear component at Nagasaki, where the zero-dose group appears to give a value below that indicated by the low-dose groups (the limits refer to ± one standard deviation). The zero-dose mortality also agrees with Japanese national statistics. On the other hand, the NIC (not in city) groups at both Nagasaki and Hiroshima, as well as the zero-dose group at Hiroshima, give values substantially equal to those observed for the low-dose groups at Nagasaki.

These uncertainties limit the reliability of the numerical risk estimates given above, and these estimates should not in any case be applied to low-LET radiation doses of about 10 mGy or less. It is also unjustified to assume that corresponding curves for specific solid cancers have the same shapes as shown in Figure 6.

CONCLUSIONS

Despite its statistical limitations the information contained in Figure 6 is of pragmatic, as well as fundamental, scientific interest. The pragmatic utility is that the relation between mortality due to cancer in general and dose of high-LET radiation is substantially linear at doses in the range of from roughly 10 mGy to 100 mGy, but the relation is nonlinear for doses of low-LET radiation that range from perhaps 50 mGy up to several gray. It follows that risk estimates for low-LET radiation obtained by linear extrapolation from high doses are too high. It also follows that while the hazard from low-LET radiations is thus likely to be less than given by most current risk estimates, the hazard of neutron radiation is greater than implied by a Q (quality factor) of 10.

The basic scientific conclusion is that data on human cancer are consistent in the theory of dual radiation action. Radiation carcinogenesis is therefore very likely to be a process initiated by lesions that in turn are due to the interaction of pairs of sublesions. It seems reasonable to speculate that these sublesions are double-strand breaks of DNA, but any further or more assured notions must await the availability of additional pertinent information.

ACKNOWLEDGMENTS

This investigation was supported by Grant Nos. CA 12536, CA 15307 to the Radiological Research Laboratory/Department of Radiology, and by Grant

No. CA 13696 to the Cancer Center/Institute of Cancer Research awarded by the National Cancer Institute, Department of Health, Education and Welfare, and by Contract EP-78-S-02-4733 from the Department of Energy.

REFERENCES

Awa, A. A. 1975. Chromosome aberrations in somatic cells. J. Radiat. Res. Suppl. 16:122–131.

Bateman, J. L., H. H. Rossi, A. M. Kellerer, C. V. Robinson, and V. P. Bond. 1972. Dose-dependence of fast neutron RBE for lens opacification in mice. Radiat. Res. 61:381–390.

Beebe, G., H. Kato, and C. Land, eds. 1977. Mortality Experience of Atomic Bomb Survivors 1950–74: Life Span Study 8. Radiation Effects Research Foundation, Technical Report, pp. 1–77.

Biola, M. T., R. LeGo, G. Ducatez, J. Dacher, and M. Bourguignon. 1971. Formation de chromosomes dicentriques dans les lymphocytes humains soumis in vitro a un flux de rayonnement mixte (gamms, neutrons), *in* Advances in Physical and Biological Radiation Detectors. International Atomic Energy Agency, Vienna, pp. 633–645.

Borek, C., and E. J. Hall. 1973. Transformation of mammalian cells in vitro by low doses of x-rays. Nature 243:450–453.

Cox, R., J. Thacker, and D. T. Goodhead. 1977. Inactivation and mutation of cultured mammalian cells by aluminium characteristic ultrasoft X-rays. II. Dose-responses of Chinese hamster and human diploid cells to aluminium X-rays and radiation of different LET. Int. J. Radiat. Biol. 31:561–576.

Hall, E. J., H. H. Rossi, A. M. Kellerer, L. Goodman, and S. Marino. 1973. Radiobiological studies with monoenergetic neutrons. Radiat. Res. 54:431–443.

Jones, T. D. 1977. CHORD operators for cell survival models and insult assessment to active bone marrow. Radiat. Res. 71:269–283.

Kellerer, A. M., and H. H. Rossi. 1972. The theory of dual radiation action. Curr. Top. Radiat. Res. Q. 8:85–158.

Kellerer, A. M., and H. H. Rossi. 1978. A generalized formulation of dual radiation action. Radiat. Res. 75:471–488.

Rossi, H. H. 1977. The effects of small doses of ionizing radiation: Fundamental biophysical characteristics. Radiat. Res. 71:1–8.

Rossi, H. H. 1979. Role of microdosimetry in radiobiology. Radiat. Environ. Biophys. (in press).

Rossi, H. H., R. Bird, R. D. Colvett, A. M. Kellerer, N. Rohrig, and Y. M. P. Lam. 1978. The molecular ion experiment, *in* Proceedings of the 6th International Symposium on Microdosimetry, J. Booz and H. Ebert, eds. Brussels, Euratom, pp. 937–947.

Rossi, H. H., and A. M. Kellerer. 1972. Radiation carcinogenesis at low doses. Science 175:200–202.

Rossi, H. H., and A. M. Kellerer. 1974. The validity of risk estimates of leukemia incidence based on Japanese data. Radiat. Res. 58:131–140.

Rossi, H. H., and C. W. Mays. 1978. Leukemia risk from neutrons. Health Phys. 34:353–360.

Shellabarger, C. J., R. C. Brown, A. R. Rao, J. P. Shanley, V. P. Bond, A. M. Kellerer, H. H. Rossi, L. J. Goodman, and R. E. Mills. 1974. Rat mammary carcinogenesis following neutron or x-irradiation, *in* Biological Effects of Neutron Irradiation (Symposium on the Effects of Neutron Irradiation upon Cell Function, Munich 1973).IAEA-S-179/28, 405–416.

Sparrow, A. H., A. G. Underbrink, and H. H. Rossi. 1972. Mutations induced in Tradescantia by millirad doses of X-rays and neutrons: Analysis of dose-response curves. Science 176:916–918.

Ullrich, R. L., M. C. Jernigan, G. E. Cosgrove, L. C. Satterfield, N. D Bowles, and J. B. Storer. 1976. The influence of dose and dose rate on the incidence of neoplastic disease in RFM mice after neutron irradiation. Radiat. Res. 68:115–131.

Ullrich, R. L., M. C. Jernigan, and J. B. Storer. 1977. Neutron carcinogenesis: Dose and dose-rate effects in BALB/c mice. Radiat. Res. 72:487–498.

Radiation Biology in Cancer Research, edited by
Raymond E. Meyn and H. Rodney Withers.
Raven Press, New York © 1980.

The Repair–Misrepair Model of Cell Survival

Cornelius A. Tobias, Eleanor A. Blakely, Frank Q. H. Ngo, and Tracy C. H. Yang

Department of Biophysics and Medical Physics and Biology and Medicine Division, Lawrence Berkeley Laboratory, University of California, Berkeley, California 94720

There is general agreement that the effects of ionizing radiation on living cells are the results of discrete quantum mechanical interactions and transitions, and that some of the most important consequences of these effects are the eventual development of lesions in genetic material. Furthermore, we are acutely aware that many living organisms, but particularly eukaryotic cells, possess enzymatic repair mechanisms that can in the course of time heal or alter the lesions in genetic material and thus profoundly modify the eventual results of radiation exposure at the cellular level. It should logically follow that models of cellular radiobiological phenomena should consider the structure of genetic material and physical radiation interactions with it, the radiation chemical consequences of initial energy transfer, and the time structure of enzymatic interactions.

In spite of such realizations, many of the quantitative models for cell survival are concerned almost exclusively with the physics and statistics of initial energy deposition events and attempt to correlate the eventual expression of biological effects directly with these events. Examples are the "target," "hit" (Lea 1955, Elkind and Sutton 1960, Wideröe 1966, Ehrenburg 1977), and the "dual action" theories (Jacobson 1957, Sinclair 1966, Neary 1965, Kellerer and Rossi 1972, 1978, Chadwick and Leenhouts 1973, 1978). A few models have incorporated time-dependent parameters (Kellerer and Hug 1963, Dienes 1966, Payne and Garrett 1975, Pohlit 1975, Garrett and Payne 1978, Braby and Roesch 1978, Niederer and Cunningham 1976).

In the repair-misrepair (RMR) model, we propose that the development of cellular biological effects from ionizing radiations has distinct and separable phases. For the present discussion, four phases are important.

1. The initial physical energy transfer and redistribution of energy by physical events.
2. Migration of the deposited energy and the establishment of long-lived molecular lesions as a result of radiation chemistry.

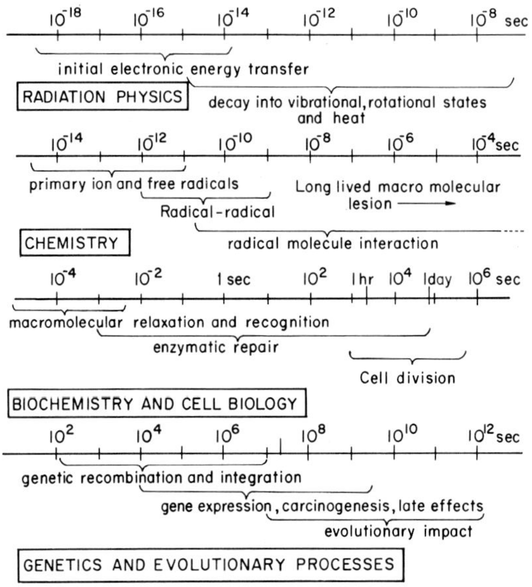

FIG. 1. Time sequence of the radiobiological events found with cell irradiation, from the initial electronic energy transfer through the late genetic effects. The physics events include time for heat transport.

3. Biochemical processes including repair or enhancement, coupled with progression of cells through various physiological states.
4. Genetic and evolutionary processes.

A timetable of events of this kind was shown at this conference by J. D. Chapman, and our version can be seen in Figure 1. A consideration of the above time sequence of events reveals that the first two basic processes can be separated from the last two.

Time Domains in Cell Inactivation by Ionizing Radiations

Radiation Physics

Briefly, the initial processes of radiation physics are comparable in their rate of occurrence to the time of passage by the ionizing particle or ray across an atom. Interactions begin at about 10^{-18} seconds. High-energy primary electronic exchanges occur in fast sequence; the initial local energy deposition events are then gradually thermalized. There is a set of steps involving reemission and reorganization of electronic energy levels. This is followed by energy transfer at the vibrational and rotational levels of molecules. The entire process including "thermalization" and heat flow is essentially complete in 10^{-8} seconds. An important role for radiation physics is then to quantitate the initial structure of energy deposition and the redistribution of energy in relation to molecular and cellular structures in living cells.

Radiation Chemistry

The radiation chemical phase overlaps the time sequence of radiation physics. This phase begins with the birth of highly reactive ion radicals and short-lived free radicals of water and organic molecules. The free radicals then react both with each other and with the macromolecular structure of the genetic apparatus and other cell organelles. The radical reactions are diffusion controlled; in the course of the chemical events the initial physical characteristics of the tracks of ionizing particles are gradually lost. Chemical radiation modifiers and sensitizers act during this phase. We know from many experts in radiology, chemistry, and biology that sensitizer action is important in the microsecond time domain. Recently, Shenoy et al. (1975) have shown that the oxygen-dependent damage in bacteria occurs in less than 100 microseconds. They failed to obtain an oxygen effect in mammalian cells when the oxygen was administered five milliseconds after the radiation exposure. We also have recent information that the modifying action of oxygen is limited by diffusion and that oxygen molecules can diffuse across one micrometer of cytoplasm in less than 10^{-3} seconds (Ling et al. 1978). From a biological point of view, the goal of radiation chemistry is the prediction and measurement of the yield of specific long-lived lesions in biologically important macromolecules such as DNA. Models for the radical chemistry phase were recently proposed by Magee (1979).

Radiation Biochemistry

Radiation biochemistry begins when long-lived macromolecular lesions have formed, for example in DNA. The essential steps are the recognition of lesions, mobilization of a sequence of specific enzymes to modify the lesions, energy-dependent resynthesis of DNA, and the reestablishment of the appropriate tertiary structures. Among the known types of macromolecular lesions are single- and double-strand breaks in DNA, stand-to-strand cross-links, base alteration and dimerization, and protein DNA cross-linkages. The rate of repair and the enzymatic sequences for each type of lesion are different; the time scale for repair can range from minutes to days. The rate of repair of DNA lesions in human cells was recently discussed by Cleaver (1978).

Cell Biology and Genetics

Biochemical repair processes occur simultaneously with the progression of normal cellular physiology, which is usually delayed as a consequence of injury. The expression of certain radiation effects such as lethality or mutations depends on progression through the cell cycle. Repairing genetic damage and the appearance of late effects, though initially coupled with biochemical repair processes, may continue through several generations of cells.

Separation of Dose-Dependent and Time-Dependent Parameters

One of the difficulties with modeling radiobiological phenomena has been that they are usually described as functions of two independent variables: dose and time. A general treatment of dose- and time-dependent processes has resulted in mathematical complexities. However, we shall demonstrate that for the purpose of modeling cellular phenomena these two variables can be separated. We must first consider the manner in which cells recognize the lesions induced by deleterious agents and the limitations imposed on radiobiological models by biological uncertainty.

Intracellular Recognition of Lesions

Our aim with the RMR model is to answer two questions: When do cells sense that they are being damaged? How do they respond to macromolecular injury? The simple answers are that a certain amount of time must elapse after the lesions are made before the cell can recognize them as lesions. After recognition, the cell's enzymatic machinery and energetics are mobilized to repair or resynthesize the essential molecules involved.

In order to estimate the time needed to recognize the damage, assume that the damage consists of discrete lesions in the DNA. Several types of lesions have been demonstrated experimentally, but we wish to take strand breaks as examples. Elkind has estimated in his discussion at this conference that a mean lethal dose of X rays produces about 10^3 single-strand breaks and perhaps 50 double-strand breaks in the DNA of a typical mammalian cell consisting of 10^9 base pairs. If this is the case, then on the average there are about 10^6 pairs of unbroken phosphate bonds between neighboring single-strand breaks and 2.4×10^7 pairs of unbroken bonds between double-strand breaks. The number of repair enzyme molecules available can be estimated to be about 10^6 per mammalian cell.

We can now make one of two assumptions for the process of recognition: either the repair enzyme complex can recognize damage at a distance, or it is necessary for the enzyme complex to be at the site of injury before damage can be recognized. In this latter case we visualize that a typical repair enzyme complex moves up and down DNA, continually testing its structure. We assume that the enzyme complex has a molecular weight of 10^5 daltons, that each enzyme complex has to test 10^3 base pairs, and that the process is diffusion limited. Calculations show that recognition of local damage may take on the average about one millisecond.

It is possible that the repair enzymes have a way of obtaining information about the occurrence of new lesions in DNA without first having to move to the actual site of the lesion. The enzyme (deployed adjacent to DNA, perhaps in nucleosomes) may sense the oscillations that are known to occur in DNA when it sustains local damage. For example, when a single-strand break occurs,

DNA removes the stress of supercoiling by uncoiling; when a double-strand break occurs, DNA snaps open. The relaxation time of bacteriophage DNA is of the order of magnitude of 10^{-4} second (Pritchard and O'Konski 1977). The relaxation time in mammalian cellular DNA might be longer than in phage DNA since much more DNA is involved in a more viscous milieu. We might assume that the oscillations caused by a strand break can be recognized in one relaxation time period, or 10^{-4} seconds. Figure 2 is a schematic drawing of this process.

Cells may be able to recognize various other types of lesions besides DNA lesions; membrane damage or significant concentrations of radiation products in the cytoplasm might also be recognized. However, reasonable calculations show that the timing for recognition and response to such events is probably not faster than the recognition of DNA nucleoprotein damage. The functional consequences of extranuclear damage are usually less serious for living cells than the consequences of a similar degree of nuclear damage.

Estimates for the rate of enzymatic repair can be made by considering the half-times for repair in mammalian cells (7 to 15 minutes for single-strand breaks and 80 to 90 minutes for double-strand breaks) (Ritter et al. 1977, Roots et al. 1979) and the number of repair enzyme molecules available. Based on

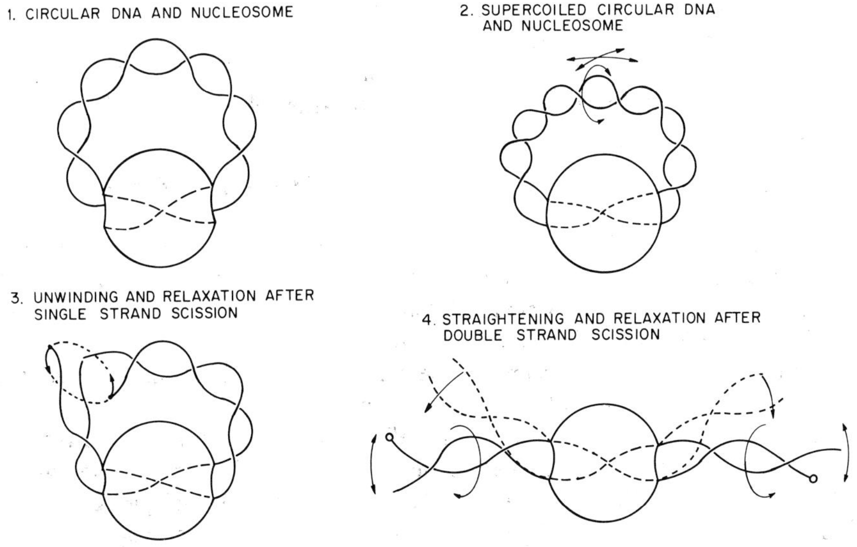

FIG. 2. The process of a DNA strand break: 1. Schematic view of DNA and nucleosome (based on nucleosomes in SV-40 virus). 2. The DNA is usually in supercoiled form. 3. The tension of supercoiling is released when a strand is broken. The unwinding from supercoiled form has characteristic relaxation times on the order of 10^{-4} seconds. 4. When a double-strand break is made, DNA unwinds and snaps open. The relaxation motions and change in coiling might be sensed by the production of stress in the macromolecular structure of nucleosomes.

the discussion above and on Figure 1 we can make three conclusions that have contributed to the development of the RMR model.

1. The enzymatic apparatus of cells is likely to spend at least 10^{-4} seconds recognizing a radiation-induced lesion in its DNA after that lesion has been established. Biochemical responses to lesions are unlikely to be significant in less than 10^{-3} seconds, but by this time the great majority of initial electronic physical energy transfers and radiation chemical transformations are complete.
2. The living cell can, however, recognize relatively long-lived macromolecular lesions in its own structure, particularly on the genetic material, DNA. It is very likely that cells cannot recognize specific radiations, e.g., X rays, neutrons, or heavy ions, as having distinctly separate properties because the specific interactions of these radiations occur too fast to be recognized by the cell. If two different deleterious physical or chemical agents produce the same kind of macromolecular lesion in DNA, it is likely the cell cannot distinguish between these two agents.
3. We may treat quantitative models of the biological action of ionizing radiations in two distinct and separate phases: physicochemical and biochemical-genetic. One aim of physicochemical experiments (and of modeling) should be to ascertain the yield per cell of each specific type of macromolecular lesion as this yield depends on dose, initial absorption events, track structure, and eventual chemical modification. Given the yield of macromolecular lesions, the second, biochemical phase of modeling is to establish how these lesions relate to the eventual expression of biological effects.

Limits of Available Information on the RMR Model: Biological Uncertainty

A salient feature of radiobiological phenomena is that the effects are expressed with a considerable time delay after an initial physical energy transfer. In the intervening time we are seriously limited in our knowledge about radiation-induced lesions and their relationship to the fate of the cells. Various names have been used to denote lesions at the early stages, such as "sublethal lesions," "potentially lethal lesions," "sublesions," or "prelesions."

Analysis of DNA extracted from cells does reveal an average number of specific lesions by molecular weight measurements, and it can also demonstrate repair by rejoining with the same technique. However, techniques do not exist to assay the integrity of DNA and its base sequences in a single living cell. Because of the very small dimensions of individual codons, any physical technique that we could use for examining the DNA in a living cell would, by necessity, cause new lesions in the DNA. For this reason, it seems prudent to assume for modeling purposes that the fate of a given radiation-induced macromolecular lesion is initially uncertain. Only after the cell attempts enzymatic repair will it be determined whether a given lesion will result in lethality or mutation or will be inconsequential. It seems logical that intracellular enzymatic structures of the cells have more information at the early stages of radiation injury than would the extracellular experimenter no matter what physical probe he uses.

Our model regards the fate of the early radiation-induced lesions in cells as uncertain. We will introduce probability factors to describe whether these lesions can be perfectly repaired or lead to lethality due to imperfect misrepair, which includes incomplete repair.

REPAIR-MISREPAIR MODEL

The model describes the yield of relevant macromolecular lesions per cell as a function of dose (D); the time-dependent (t) transformations of these lesions; and the time- and dose-dependent probabilities for survival (S), lethality (L), and mutation (M).

U stands for "uncommitted," which is what the lesions are before they are subject to enzymatic repair and modification. Lesions with somewhat similar properties are described in other quantitative models (Garrett and Payne 1978, Pohlit 1975, Powers 1962, Laurie et al. 1972).

We know that various kinds of deleterious agents produce various classes of U lesions, each of which might potentially produce a variety of expressed biological effects. For clarity, this is restricted to the discussion of a single class of U lesions. A given class of U lesions can be identified with a specific molecular lesion if the time rate of its biological transformations, as predicted by the model, agrees with the average rate of change measured by molecular techniques. Additional types of U lesions can be introduced in the model when a single type of lesion cannot account for all experimental data.

The general scheme of the RMR model is shown in Figure 3. The physical and chemical interactions, shown on the left side of the figure, are interesting only because they determine the dose-dependent yield U_0. The model itself is concerned with time-dependent transformations in the course of repair and with the probabilities that transformed states lead to the expression of biological effects.

We define R states (repair states) as the result of transformations of U states following enzymatic repair. R states are permanent in the sense that their presence may commit the cell to lethality, mutation, or survival. In the simplest form of the RMR model there are two R states: R_L is the yield of a linear repair process assumed to proceed as a monomolecular chemical reaction, and R_Q is the yield per cell of a repair process involving interaction between pairs of U lesions. R_Q is a "quadratic" repair process, and its rate is proportional to the square of the local density of U states. If the distribution of U lesions is uniform throughout the cell nucleus, then the rate of R_Q is proportional to $U(U-1) \simeq U^2$.

For a single, rapidly delivered dose of low linear energy transfer (LET) radia-

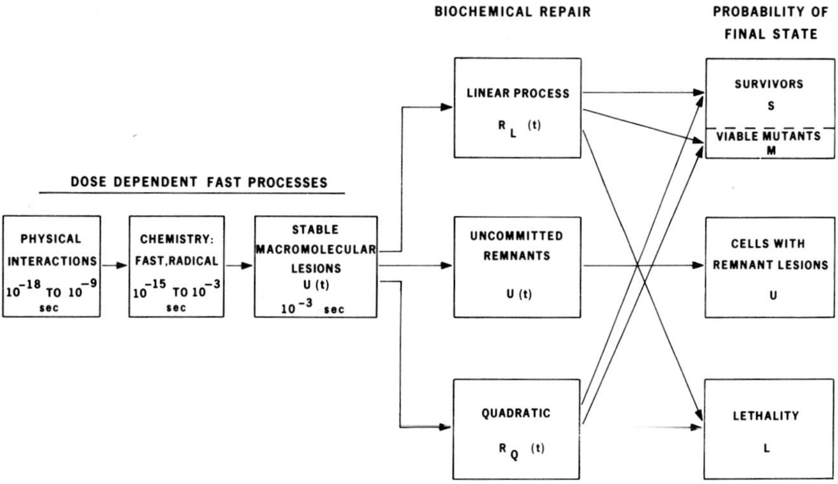

FIG. 3. Schematic drawing of the repair-misrepair model, showing the dose-dependent fast processes, different types of biochemical repair, and the probabilities of the final states of the cell.

tion, the time-dependent behavior of U is described by the first order quadratic differential equation:

$$dU/dt = -\lambda U(t) - kU^2(t) \tag{1}$$

In this equation λ is the coefficient of linear repair and k is the coefficient of quadratic repair. Integrating between limits 0 and t we find:

$$U(0) - U(t) = \int_0^t \lambda U\,(t)dt + \int_0^t k\,U^2\,(t)dt \tag{2}$$

With the definitions

$$R_L = \int_0^t \lambda U\,(t)dt$$

and

$$R_Q = \int_0^t k\,U^2\,(t)dt$$

we have

$$U\,(t) + R_L\,(t) + R_Q\,(t) = U(0) \tag{3}$$

Simple solutions exist for the decay of uncommitted U lesions and the growth of R states, with the assumption that for a specific cell type in a specific state, λ and k are constant and independent of time and dose. Let $U\,(0) = U_0$; $R_L(0) = R_Q\,(0) = 0$; $U\,(\infty) = 0$; and $\epsilon = \lambda/k$, the "repair ratio."

$$U = \frac{U_0 e^{-\lambda t}}{1 + \frac{U_0}{\epsilon}(1 - e^{-\lambda t})} \tag{4}$$

$$R_L(t) = \epsilon \ln \left[1 + \frac{U_0}{\epsilon}(1 - e^{-\lambda t}) \right] \tag{5}$$

$$R_Q(t) = \frac{U_0 \left(1 + \frac{U_0}{\epsilon}\right)(1 - e^{-\lambda t})}{1 + \frac{U_0}{\epsilon}(1 - e^{-\lambda t})} - \epsilon \ln \left[1 + \frac{U_0}{\epsilon}(1 - e^{-\lambda t}) \right] \tag{6}$$

Eurepair and Misrepair

A range of possible repair states occur after damage to cellular DNA. The cell may repair the damage accurately, making the DNA sequencing exactly like it was before radiation damage occurred. We call this type of true repair "eurepair." The other types of repair are variations of misrepair—ranging from viable mutants to alterations that eventually cause cell death. One form of misrepair is misreplication. The process of DNA synthesis and repair is often not completely accurate even in normal, unirradiated cells (Bernardi and Ninio 1978).

The states R_L and R_Q represent the products of biochemical repair. If U represents DNA strand breaks, for example, R_L and R_Q would be yields of reconstituted DNA with strand continuity, unless for some reason the repair is unsuccessful or incomplete. Let ϕ represent the probability that linear repair is eurepair, and δ the probability that quadratic repair is eurepair. The probabilities of misrepair will be represented by coefficients $1 - \phi$ and $1 - \delta$. It is obvious that incomplete repair is also misrepair.

Survival Probability S(t)

Equations 4 through 6 and the definitions found in Table 1 allow for the calculation of the probabilities of survival S(t) and lethality L(t). Assume that statistical variations in U, R_L, and R_Q are random. (Poisson statistics are used here; a more detailed discussion of statistical approaches is being prepared.) The probability of survival at time t clearly depends on the number of misrepaired lesions (R_{LM} and R_{QM}) and the number of unrepaired U lesions that a given cell can tolerate at cell division. Survival is usually measured as colony formation (reproductive integrity) after several cell divisions to limit the time variable such as $t < t_{max}$ where t_{max} is the time interval allowed for repair.

The genetic constitution of cells may also be a factor for survival. In order to describe the survival of higher ploidy or of binucleated cells, it may be necessary to consider additional constraints. Thus, survival probability should be considered separately for a variety of situations with specified constraints.

TABLE 1. *Definitions for the probabilities of eurepair and misrepair*

	Eurepair		Misrepair	
	Symbol	Probability	Symbol	Probability
Linear repair process				
R_L	R_{LE}	ϕ	R_{LM}	$1-\phi$
Quadratic repair process				
R_Q	R_{QE}	δ	R_{QM}	$1-\delta$

We have considered six different applications of the RMR model. These cases have been chosen to interpret a variety of types of radiobiological experiments. Each of these cases is briefly described below, and a detailed discussion for each will follow.

Case I: Assume that all linear repair is eurepair and that all quadratic repair results in lethal misrepair. This leads to the simplest RMR survival equation with two adjustable parameters: α, the yield of U lesions per rad, and ϵ, the repair ratio.

$$S = e^{-\alpha D}\left[1 + \frac{\alpha D}{\epsilon}\right]^{\epsilon} \tag{7}$$

There are interesting similarities and differences between this survival expression and the multitarget single-hit survival curves. Equation 7 is useful for fitting survival data from mammalian cells exposed to X rays.

As an illustration of Case I, an analysis of the X-ray survival of various repairless mutants of yeast cells will be made.

Case II: Assume that a fraction of linear repair is misrepair that causes more lethality than that found with Case I.

$$S = e^{-\alpha D}\left[1 + \frac{\alpha D}{\epsilon}\right]^{\epsilon\phi} \tag{8}$$

In addition to α and ϵ there is a third adjustable parameter, ϕ, which was defined in Table 1.

The survival probabilities of Case II are compared to the linear-quadratic survival equation. This form of survival equation is suitable for analysis of mammalian cell survival curves resulting from high-LET radiations, and the analysis of survival as a function of cell age.

Case III: For the application of the RMR model to split-dose and mixed modality exposures, we assume that a time interval τ separates two dose installments D_1 and D_2. The survival equation is:

$$S(D_1, D_2, \tau, t) = L(D_1, \tau) \cdot S(D_2, t-\tau) \tag{9}$$

At the end of period τ, remnant U lesions are added to the new lesions produced by dose D_2.

We will show an example of split-dose experiments with X rays on mammalian cells in the following discussion. Mixed radiation exposures may be analyzed in a similar manner. The analysis of split-dose experiments uses the concept of remnant U lesions. The relationship of remnant lesions to the initial slope of the survival curves will be discussed.

Case IV: For the calculation of mutation probabilities, we shall assume that a specific mutation corresponds to a specific kind of misrepair. The frequency of mutation induction as a function of dose is then, in a simple case, proportional to the amount of misrepair that is occurring while the cell recovers from a dose of radiation.

Case V: Repair processes that depend on the magnitude of administered dose may also be considered with the RMR model. The same dose of radiation that causes U lesions might either inactivate or enhance repair processes. Hence we obtain survival curves for repair inactivation by allowing the coefficient ϵ to be a decaying function of dose.

A second example allows ϵ to be advanced, that is to increase as an increasing function of dose. This results in survival curve shapes that have been described from experiments with bacteria and algae as "SOS repair."

Case VI: We will discuss survival and lethality from ionizing radiations that are delivered in protracted fashion at constant dose rate. Repair is occurring while the dose is still being administered, and the cells are left with accumulated U lesions at the time when radiation is stopped. The RMR model predicts certain types of dose rate effects for protracted doses.

We will now proceed with a detailed discussion of each of the six cases of the RMR model.

Case I of the RMR Model

Case I is based on the idea that linear repair, e.g., the rejoining of two adjacent broken ends of DNA in order to reconstitute the original unbroken piece, is always eurepair. Quadratic repair, which is the rejoining of pieces of DNA that did not originally belong together before the cell was exposed to radiation, is always misrepair, in fact misrepair causing a lethal effect.

If all R_L is eurepair, $\phi = 1$. If all R_Q is lethal misrepair, $\delta = 0$. If remnant U lesions are lethal, then at time (t) any cell that has R_Q lesions is committed to die and cells that have no R_Q or U lesions are committed to survive. According to Poisson statistics and based on equations 4 through 6

$$S(t) = \exp\left(-R_Q - U\right) = e^{-U_0}\left[1 + \frac{U_0}{\epsilon}(1 - e^{-\lambda t})\right]^{\epsilon} \qquad (10)$$

With the designation of:

$$\gamma(U_0, t) = \frac{\left(1 + \frac{U_0}{\epsilon}\right)(1 - e^{-\lambda t})}{1 + \frac{U_0}{\epsilon}(1 - e^{-\lambda t})} \tag{11}$$

$$L(t) = 1 - \exp(-R_Q) = e^{-\gamma U_0}\left[1 + \frac{U_0}{\epsilon}(1 - e^{-\lambda t})\right]^{\epsilon} \tag{12}$$

When $t \rightarrow \infty$ then $(S + L)_{t\rightarrow\infty} = 1$, which means that no U lesions are left; they have all been eurepaired or misrepaired. In Figure 4 the time dependence of the functions S(t) and L(t) are shown in the course of repair following a dose of radiation.

Dose Dependence of U and S

Dose does not explicitly enter into the survival probabilities (equations 10 through 12). However, U_0 is a function of dose. Generally, the dose-dependent initial yield of U lesions might be approximated by a power series that is subject to experimental verification at the molecular level:

$$U_0 = \sum_{i=1}^{i\,max} \alpha_i D^i \tag{13}$$

where α_i is a constant.

Without limiting the eventual applicability of the model, we shall restrict

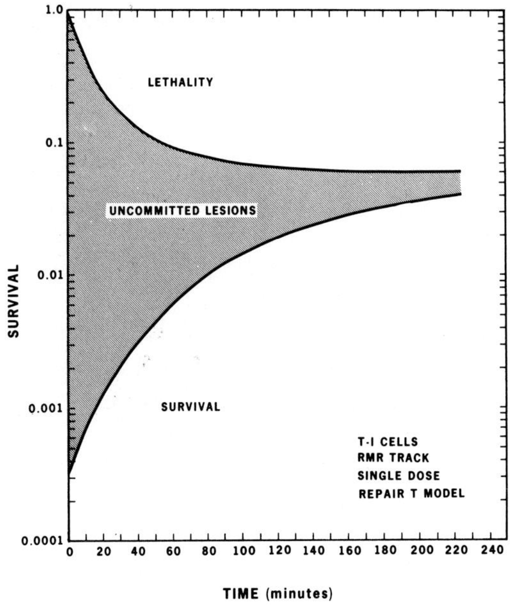

FIG. 4. Time dependence of the probabilities for survival (S) and lethality (1 − L(t)) as function of time elapsed after a single dose. The shape of the curves depends on the values of λ and k. The number of cells with uncommitted lesions decreases with time.

ourselves here to a discussion of single-strand DNA scissions and double-strand scissions produced by ionizing radiation. There is a variety of experiments available on the yield of strand breaks in microorganisms and in mammalian cells exposed to ionizing radiation. For example, in recent work on DNA of ϕX174 phage (Christensen et al. 1972, Hutchinson 1974, Corry and Cole 1973, Bonura et al. 1975, Sawada and Okada 1972, Veatch and Okada 1969) and on hamster cell DNA (Ritter et al. 1977), the yield of DNA strand breaks is proportional to dose. In phage, single- and double-strand breaks were measured separately, whereas in mammalian cells the sum of single- and double-strand breaks was measured. For both cases the yield of strand breaks was a linear function of dose:

$$U_0 = \alpha D \tag{14}$$

where α is the yield of strand breaks per cell per unit dose. The proportionality holds regardless of the particle or radiation used, whether it is X rays or carbon or argon ion beams; α does vary with beam quality, however.

There are some observations that tend to favor a quadratic relationship for the yield of strand breaks:

$$U_0 = \alpha_1 D + (\alpha_2 D^2) \tag{15}$$

where $\alpha_1, \alpha_2 =$ constants

For example, Dugle et al. (1976) observed this relationship for mammalian cell DNA with X rays at very high doses. Also, certain models, e.g., the dual action theory (Kellerer and Rossi 1972, 1978), propose relationships similar to equation 15.

The RMR model can use any of the forms of dose dependence for U_0 (equations 13 through 15) and this model may in fact be helpful in deciding which equations express the yield of U_0 lesions most accurately. In the present paper, we use only the linear relationship of equation 14; most biochemical evidence supports this choice.

The Rates of Repair: λ and k

For DNA strand scission, the rate of repair has usually been found experimentally to be proportional to the number of strand breaks present in the cell, leading to simple time-dependent exponential relationships for the decay of strand breaks. For X-ray-induced strand breaks measured by the alkaline sucrose method, the half-life for repair at 37°C is 7 to 15 minutes. Ritter et al. (1977) found that after heavy-ion irradiation about 50% of the breaks repaired with an 80-minute half-life, and up to 20% of the breaks remained unrepaired in a 12-hour time span. The slower rate of repair after heavy-ion exposures can be correlated with the increased incidence of double-strand scission. A variety of authors have demonstrated the repair of double-strand breaks (Hutchinson 1974, Corry and Cole 1973, Roots et al. 1979).

The term double-strand scission probably denotes a variety of lesions in mammalian cells that at some time following exposure to radiation reach a state in which both strands of DNA are severed. Most methods to assay the number of DNA strand breaks are indirect. There is also an indication that some double-strand breaks remain unrepaired even after 12 hours or more incubation of the damaged cells. Roots et al. (1979) have correlated the fraction of unrepaired breaks to the relative biological effectiveness (RBE) of heavy ions for inhibition of reproductive integrity in human kidney cells. However, we do not have direct evidence at present of whether the inability to repair causes death, or whether the cells that are dying have lost their ability to repair because of some other cause.

The increase in yield of DNA double-strand scissions appears to be intimately connected to the increased biological effectiveness of alpha particles and of accelerated heavy ions. Although most of the double-strand lesions are repaired, this repair is measured at the chemical level as an increase in molecular weight of DNA fragments. No chemical information is available on whether or not the entire coded message is preserved during the double-strand repair process, i.e., whether the repair is eurepair or misrepair. If we draw a parallel between the production and repair of double-strand breaks and chromosome breakage and repair, it becomes obvious that at the chromosomal level certain types of rejoinings, e.g., deletions or translocations, relate intimately to the survival or death of the cells and to the possible presence of mutants. It seems straightforward to assume that DNA double-strand scissions may often rejoin with DNA deletions and DNA rejoinings between two abnormal sets of broken DNA strands. Neary (1965) has theorized that abnormal chromosome rejoinings are proportional to the square of the dose.

The form of the RMR model, as given in equation 1, is patterned to fit the above ideas. The linear repair constant λ could represent the rate at which the broken strands of the same DNA molecule rejoin, and should be mostly eurepair unless the repair process for some reason cannot be completed. The rate constant k could represent DNA deletions and exchanges. The values of λ and k cannot be evaluated in a single survival experiment; this can be done, however, in split-dose experiments.

Practical Survival Equation for Case I Derived from Equations 10, 14, and 15

In Figure 4 we have plotted the time-dependent probabilities for cell survival, equations 10 and 12, for a single exposure to dose D which caused $U_0 = \alpha D$ lesions. On this figure we see that as the cells become committed to survive, or die due to the formation of R_{LE} or R_{QM} states, the probability of finding cells with U lesions diminishes. The actual observations of survival probability usually are made at a much later time after the cells have gone through several divisions and the cells that survived have formed colonies.

We now introduce the factor T as a time constraint that depends on the maximum time available for repair, t_{max}:

$$T = 1 - e^{-\lambda t_{max}} \tag{16}$$

t_{max} might be set as the time available from exposure to the first mitosis. In this case $T < 1$; alternatively, t_{max} might be the time of some other event in the cell cycle where repair ceases.

$$S = e^{-\alpha D}\left[1 + \frac{\alpha DT}{\epsilon}\right]^{\epsilon} \tag{17}$$

Usually, if $\lambda t_{max} \gg 1$, then $T \simeq 1$ and survival approximates equation 7.

$$S = e^{-\alpha D}\left[1 + \frac{\alpha D}{\epsilon}\right]^{\epsilon}$$

We find that this survival equation fits experimental data on lethal effects of low-LET radiation exceedingly well. For example, a survival curve for V-79 hamster cells exposed to X rays, obtained in our laboratory by Yang, is shown in Figure 5 together with an RMR survival curve fitted by nonlinear least squares method. In comparison, the linear multitarget (LMT) model (Elkind and Sutton 1960) and the repair saturation model (Green and Burki 1972) fit less well because of lack of fit in the "shoulder" region; the linear quadratic model deviates from the experimental data either in the shoulder region or at high doses.

Figure 6 demonstrates the manner in which the survival curves drawn from equation 10 or 7 vary when the repair parameters change. A hypothetical mammalian cell similar to V-79 hamster cells was modeled. The yield constant α remained the same for all the graphs. In Figure 6A, which uses equation 7, the repair constant ϵ was varied. When $\epsilon = 0$, an exponential survival was

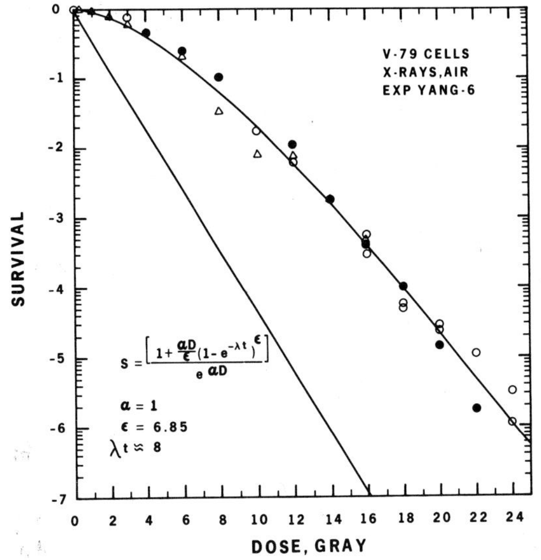

FIG. 5. Survival of V-79 cells exposed to X rays (taken from the work of T. Yang). The solid line through the experimental points is a fit by the RMR model. The exponential curve corresponds to $e^{-\alpha D}$.

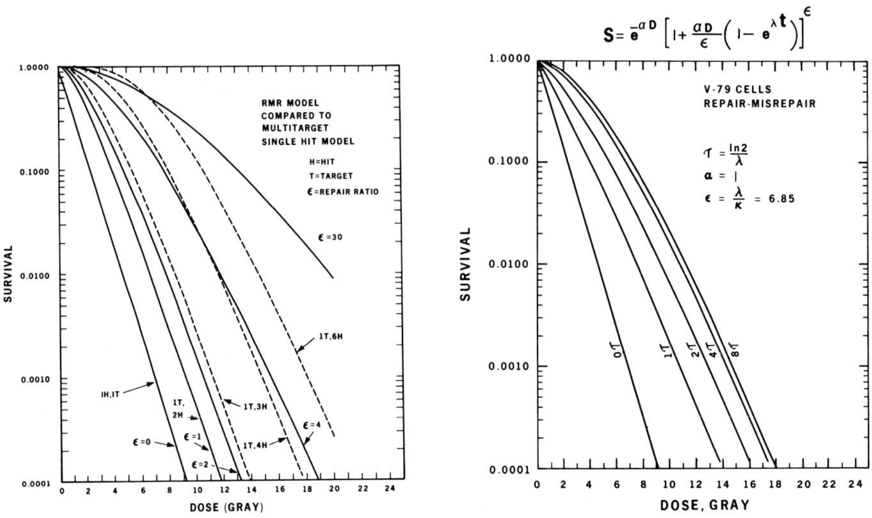

FIG. 6. A, Left, Theoretical RMR survival curves according to the equation

$$S = e^{-\alpha D}\left[1 + \frac{\alpha D}{\epsilon}\right]^{\epsilon}$$

are represented by solid lines. The value α was kept constant, and ϵ, the repair ratio, was varied from 0 to 30. When $\epsilon = 0$, there is no repair; when ϵ is large then linear eurepair is much more important than quadratic misrepair. The dotted lines represent the well-known single target, multihit survival equation:

$$S = e^{-\alpha D}\left[1 + D + \frac{(\alpha D)^2}{2!} + \cdot\ \cdot\ \cdot + \frac{(\alpha D)^{2}}{i!}\right]$$

The one-target, single-hit curve is identical with the RMR survival curve when $\epsilon = 0$. The one-target two-hit curve is identical with the RMR curve when $\epsilon = 1$. At higher hit numbers, there are significant differences between the two models both at low and at high dose levels. B, Right, The manner in which cells become committed to survive as repair proceeds in time. The coefficients α and ϵ were held constant, and the time available for repair was varied.

obtained; as ϵ is increased the survival curves have increasing shoulders and decreasing slopes at high dose levels.

Figure 6B allows variation of the time t_{max} of equation 10 while keeping α and ϵ constant. If the repair time is zero, the survival curve is exponential as in the case of $\epsilon = 0$. With increasing t_{max}, different survival curves are obtained with different initial slopes approximating the case of $t_{max} \rightarrow \infty$. This case corresponds to T = 1, which has zero initial slope.

Comparison with Conventional Single Target Multihit Theory

Conventional target theory with m number of hits and α inactivation constant gives the following survival equation:

$$S_{target}(m) = e^{-\alpha D} \sum_{i=0}^{m-1} \frac{(\alpha D)^i}{i!} \tag{18}$$

We can compare this directly with Case I if we expand equation 7 in the form of a power series. For the comparison, assuming $\epsilon = m - 1$ we obtain:

$$S_{RMR}(\epsilon = m - 1) = e^{-\alpha D} \sum_{i=0}^{m-1} \frac{(m-1)!}{i!\,(m-1-i)!} \left(\frac{\alpha D}{m-1}\right)^i \qquad (19)$$

Expressions 18 and 19 are similar except that the terms of 19 are smaller by the factor:

$$\frac{(m-1)!}{(m-1)^i\,(m-1-i)!}$$

When we deal with a single-hit survival curve $S = e^{-\alpha D}$, both expressions are the same. In the RMR model, we claim that $\epsilon = 0$; there is *no repair.* Both models agree that either a single U lesion or a single hit kills the cell.

A two-hit survival curve ($m = 2$) gives the same analytical form as $\epsilon = 1$. In this case, the multihit equation would claim that the cell could *always* tolerate one relevant lesion, never two or more. The RMR model states that $\epsilon = 1 = \lambda/k$; therefore, the coefficients for linear repair and quadratic misrepair are equal.

The dashed lines on Figure 6A are single-target, multiple-hit survival curves for the cases discussed above. Note that neither the constants D_q nor D_o of the target theory have a meaning for the RMR model; there is no "final" slope to measure D_o because the survival curves are continually bending. For the same reason, it is not valid to extrapolate the survival curve to zero dose in order to obtain the extrapolation number m.

Survival Curves of Cells with Genetic Defects in the Repair Mechanism

As an example of the use of the RMR method and to illustrate the validity of some of the concepts used, we have used the data of Ho and Mortimer (1973) on the X-ray survival curves of genetically tetraploid *Saccharomyces cerevisiae* (Ho 1976). These authors demonstrated that a mechanism for the lethal effect in these cells was the production of double-strand breaks in the nuclear DNA, but that in the wild type (+) of DNA, efficient repair mechanisms existed to repair double-strand breaks. A number of repair-deficient mutants were isolated; one of these (rad 52) was incorporated in the genome of five different yeast strains. Rad 52 is a repair-*deficient* gene, whereas the wild type gene can repair. The gene dose of rad 52 was 0, 1, 2, 3, or 4 in the five different mutants, whereas the wild type gene went from 4(+) to 0(+).*

---

* The genetic designations of the tetraploid strains of *S. cerevisiae* used are:

X3423 (X3406-1D × X3406-1A) *a/ a/ α/ α/ +/ +/ +/ +/*
XK11 (X3406-1D × X3443-6A) *a/ a/ α/ α/ +/ +/ +/rad 52/*
XK12 (X3406-1D × X3443-2B) *a/ a/ α/ α/ +/ +/rad 52/rad 52/*
XK13 (X3443-11B × X3443-6A) *a/ a/ α/ α/ +/ rad 52/rad 52/rad 52/*
X3452 (X3443-11B × X3443-2B) *a/ a/ α/ α/ rad 52/rad 52/rad 52/rad 52/*

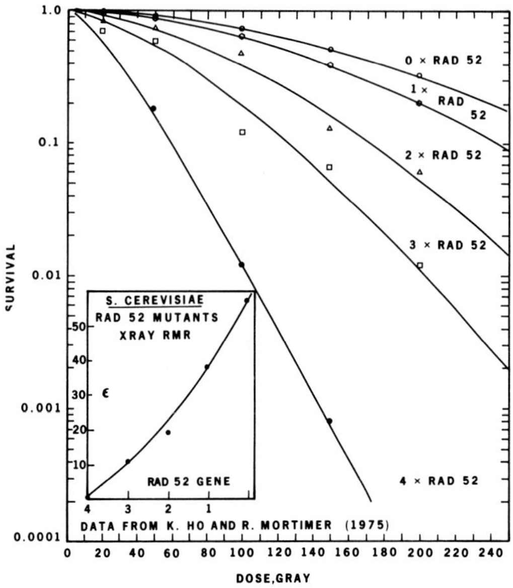

FIG. 7. Experimental survival curves for tetraploid yeast cells (from the work of Ho and Mortimer 1973), fitted by the RMR model. Rad 52 is a repairless gene. The survival curves vary with gene dose in a manner generally in agreement with the RMR model. With α and k fixed, the values for ϵ are given in the insert. If we assume that ϵ measures the repair rate and that ϵ is proportional to the repair enzymes, then it appears that the amount of repair enzyme available increases approximately proportionally to the gene dose of the wild type gene (+).

Figure 7 shows survival curves of the type of equation 7 fitted for each strain by the least squares method. The strain with four rad 52 genes had an almost pure exponential survival curve, showing very little repair. In the other strains, the yield parameter (α) was constant, whereas the repair parameter (ϵ) increased rapidly with gene dose. A plausible interpretation of these data is that the availability of the enzymes responsible for linear eurepair increased with dose to the wild type (+) gene, while intrinsic sensitivity of the genome for U lesions (α) and the rate of quadratic misrepair remained approximately constant. These experiments yielded only the value for $\epsilon = \lambda/k$; more could be learned about the values of λ and k in split-dose experiments (see Case III of the RMR model, which follows).

Case II of the RMR Model

It is necessary to extend the treatment of the RMR model to situations where linear repair is *not* always eurepair. For example, even though a repair enzyme may attach to a U_O lesion in a normal manner, it may be unable to complete repair. This would count as misrepair.

The analysis that follows is an obvious simplification of the very complex repair mechanisms that are known to occur in nature. Let $\phi \leq 1$ be the probability that linear repair (R_L) is eurepair (see Table 1).

We have shown elsewhere (Tobias et al. 1978) that in Case II, the RMR survival probability given in equation 7 can be modified to equation 8.

$$S = e^{-\alpha D}\left[1 + \frac{\alpha D}{\epsilon}\right]^{\epsilon\phi}$$

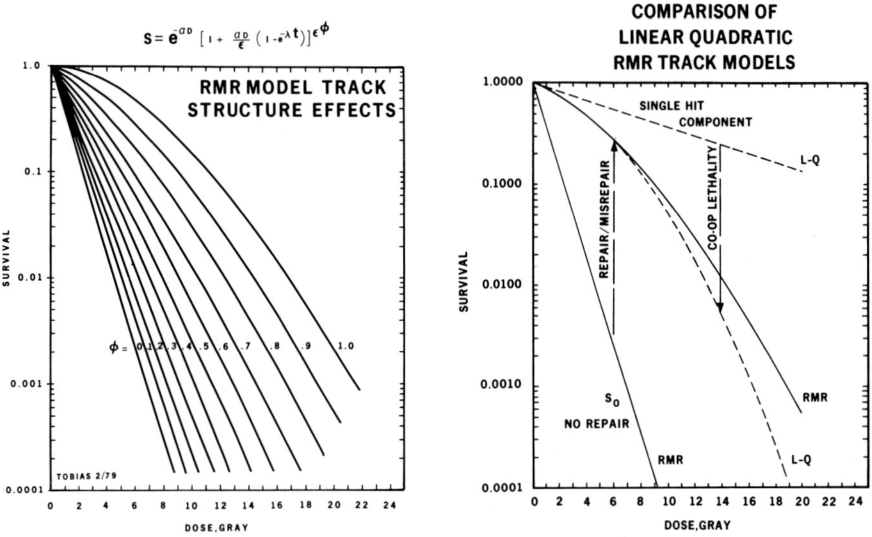

FIG. 8. A, Left, Theoretical survival curves and their dependence on the constant ϕ, which signifies the portion of linear eurepair. A single heavy ion produces several lesions along its track. The coefficient ϕ decreases as LET increases, and it measures the probability that all lesions produced in a single track are eurepaired. B, Right, Comparison of survival according to the repair-misrepair and linear-quadratic models. Continuous lines represent RMR curves, S_0 without repair and RMR with repair. The RMR curve is above the S_0 curve; repair helps survival. The linear quadratic survival (equation 21) is indicated by dashed lines. The straight line, which corresponds to the initial slope, is usually interpreted as being due to single-hit irreparable or irreversible lesions. The lower dashed curve indicates the cooperative lethality due to the quadratic term. If the initial slopes are the same then at very large doses the survival due to the linear quadratic model always dips below survival due to the RMR model.

An important consequence of equation 8 is that at low doses the survival curves have a finite negative slope;

$$\left(\frac{dS}{dD}\right)_0 = -\alpha(1-\phi) \tag{20}$$

In Figure 8A theoretical survival curves are plotted according to equation 8; the value ϵ is kept constant. When ϕ is zero, the survival is exponential. When $\phi = 1$, the curve is the same as that described by equation 7 and has zero initial slope. The family of curves with intermediate values of ϕ all had negative initial slopes; the slope gradually decreases as ϕ is increased.

We compared Case II to the linear quadratic survival equation (LQ) of Chadwick and Leenhouts (1973, 1978). Let x and y represent constant coefficients in the linear quadratic survival model expressed by the survival probability S_{LQ}:

$$S_{LQ} = e^{-xD - yD^2} \tag{21}$$

In Figure 8B we compared survival probabilities of equations 8 and 21, adjusted in such a manner that the initial slopes of the survival curves are the same.

Equating the initial slopes gives:

$$x = \alpha(1 - \phi) \tag{22}$$

The usual interpretation of x according to the LQ model is that it represents the yield of irreparable lesions per unit dose. The RMR model initially has no irreparable lesions, but in the course of time a fraction $(1 - \phi)$ of the initial U_0 lesions αD are misrepaired. The RMR model has a repair term:

$$\left[1 + \frac{\alpha D}{\epsilon}\right]^{\epsilon\phi} \geq 1 \tag{23}$$

which is greater than one if ϕ is positive, signifying repair of lesions.

On the other hand, the factor e^{-yD^2} of the LQ model may be regarded as a "potentiation term." Comparison of the two models in Figure 8B indicates that the survival due to the LQ model becomes progressively lower than the RMR survival at high doses. The slope of the RMR survival curve at large doses is always less steep than the slope of the RMR survival curve without the repair term, equation 23.

Equation 8 can also be used to describe enhancement of damage by changing the algebraic sign of ϕ from positive to negative. Enhancement corresponds to an *increase* in the number of U lesions over what has been initially produced by a dose of radiation. Enhancement may occur as a result of enzymatic action. RMR "enhancement" survival curves would lie well below the exponential curve S_0 of Figure 8B.

The Interpretation of Survival Data Obtained with Heavy Ions

We have analyzed survival data obtained from human kidney cells irradiated with a variety of accelerated heavy ions at the Bevalac accelerator (Blakely et al. 1979).

Control X-ray data were fitted to the RMR model equation 8; the value of ϕ was nearly 1. Data from high-LET radiations were then analyzed by nonlinear least square fitting of two constants: α and ϕ. We assumed that ϵ is the same regardless of particle velocity and LET, so that values obtained at low LET for ϵ were used to analyze the high-LET data.

We have used a neon beam of 425 MeV/amu nominal kinetic energy per nucleon. This beam has a useful range penetration of about 15 g/cm^2 in water. Survival curves were obtained along the Bragg ionization curve at eight different residual range values from 0.1 g/cm^2 to 12 g/cm^2 measured relative to the Bragg peak LET values ranged between 30 and 400 keV/μm. Figure 9 shows RMR fits to each of the survival curves. Note that the curves had less and less shoulder as the residual range was decreased and the LET increased.

In Figure 10 the RMR coefficients are analyzed as a function of the mean

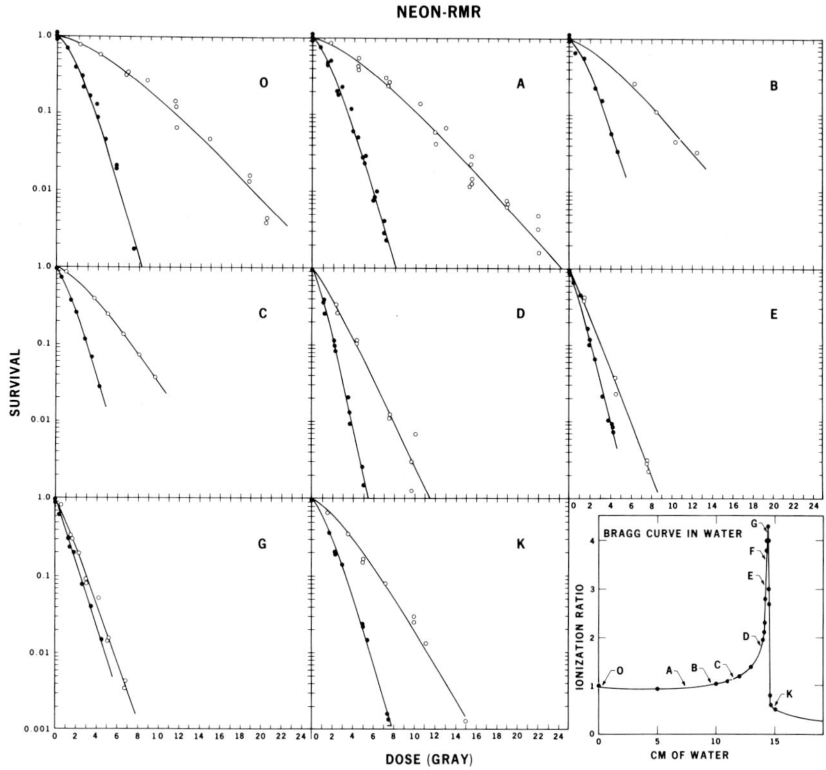

FIG. 9. Experimental survival curves for T-1 kidney cells in air and in nitrogen. The cells have been exposed to neon beams of various residual range values. The Bragg ionization curve at the bottom right panel shows the residual ranges, O through K, at which exposures were made. Solid squares indicate exposures in air; open squares indicate exposures in anoxic conditions. Note that the cells are more sensitive to neon particles near the Bragg peak and the oxygen effect is also reduced. The solid lines through the points are RMR least-squares fits.

LET_∞. The beam was contaminated at low residual range values by primary beam fragmentation. In Figure 10A the coefficient α, which measures the yield U_O lesions per cell, is plotted for exposures performed in air and those under hypoxic conditions.

If we consider the hypoxic conditions first, we see that α increases rapidly until it levels off above 100 keV/μm. In the presence of oxygen, however, there are only minor variations in the α coefficient, indicating that the yield of U_O lesions is nearly independent of LET in aerated conditions. At very high LET there is no significant difference between the yield of U_O lesions found in cells treated under either aerated or hypoxic conditions. Thus, the presence of oxygen during irradiation will significantly increase the initial yield of U_O lesions. It appears indeed that most, if not all, of the radiobiological oxygen effect relates

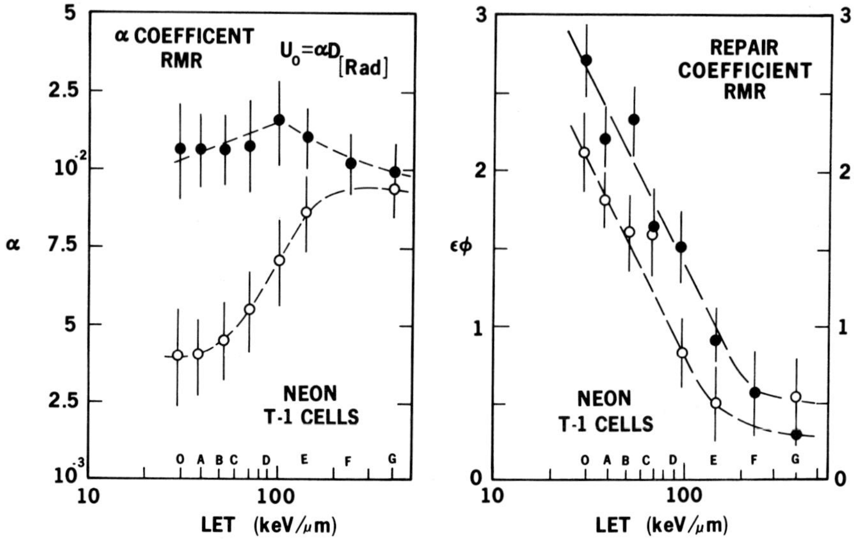

FIG. 10. A, Left, Values of α for the experiments in Figure 9: air, ●; nitrogen, ○. Since α is an indicator of U_0 lesions, we conclude that this yield increases with LET under anaerobic conditions. In air, however, the yield of U_0 lesions per unit dose is almost independent of LET. B, Right, Decrease in the exponent $\phi\epsilon$ of the survival equation as function of LET: air, ●; nitrogen, ○. The decrease of $\phi\epsilon$ indicates that there is much less eurepair at high LET than at low LET.

to the initial production of U_0 radiolesions during the early radiation physics and chemistry phases.

In Figure 10B the values of the exponent $\epsilon\phi$ are plotted as a function of LET. In air as well as under hypoxic conditions there is rapid decrease noted in $\epsilon\phi$ with increasing LET. Since ϵ is assumed to be constant, the measure of linear eurepair (ϕ) decreases rapidly with LET. We believe this is not because of a change in the value of λ or of k, but rather ϕ decreases because of the increased misrepair along individual ionizing particle tracks. The increased misrepair is caused by the physical closeness of U_0 lesions along the individual particle tracks. We visualize that a particle track with very high energy density has a high efficiency in producing DNA lesions wherever the expanding core of this track intersects DNA. Calculations show that 10 to 20 U_0 lesions per ionizing track are likely in mammalian cell nuclei when the LET is greater than 100 kev/μm. At this LET the core diameter of a high-speed heavy ion can be on the order of 10 to 20 nm (Chatterjee et al. 1973). This is the same order of magnitude as the size of nucleosomes; therefore, the damage might be extensive if a nucleosome is in the pathway of such a heavy-ion track.

We believe that the repair processes for heavy ion-induced radiolesions are quite similar to the repair processes for X-ray-induced lesions. The key question is whether or not all lesions made by a single track can be eurepaired; if only one of the lesions is misrepaired it may cause a lethal effect. Further analysis of this problem is in progress.

Applications to the Radiation Responses of Synchronous Cell Populations

It is well known that mammalian cells exhibit variations in radiation sensitivity during the cell division cycle. Although a good deal of empirical material is available, this problem has received relatively little analysis from the point of view of quantitative mechanisms. The RMR model can give some information on the variations in sensitivity for producing U_O lesions, and on changes in repair.

We have an experimental program to measure cell radiosensitivity in various stages of the cell cycle. The data given here should be regarded as preliminary because we are still finding a good deal of variation in both the α and ϵ values from experiment to experiment. This may be because the process of synchronization could interfere with the amount of the intracellular repair enzyme and with the chemical end groups that can modify radiosensitivity.

For X rays, the most sensitive part of the cell cycle is mitosis, and we find ϵ, the repair ratio, to be very small for mitotic cells ($\epsilon = 0.02$) (Figure 11A). However, the yield of U_O lesions (α) is comparable to and even somewhat lower than α at other cell phases. In the G and S phases, there is repair; much more in S ($\epsilon = 22.5$) than in G_1 ($\epsilon = 5.93$).

There is much less repair for any of the four cell phases irradiated with argon ions compared to the repair seen with X rays (Figure 11B). Survival curves obtained from cells irradiated with argon are almost purely exponential,

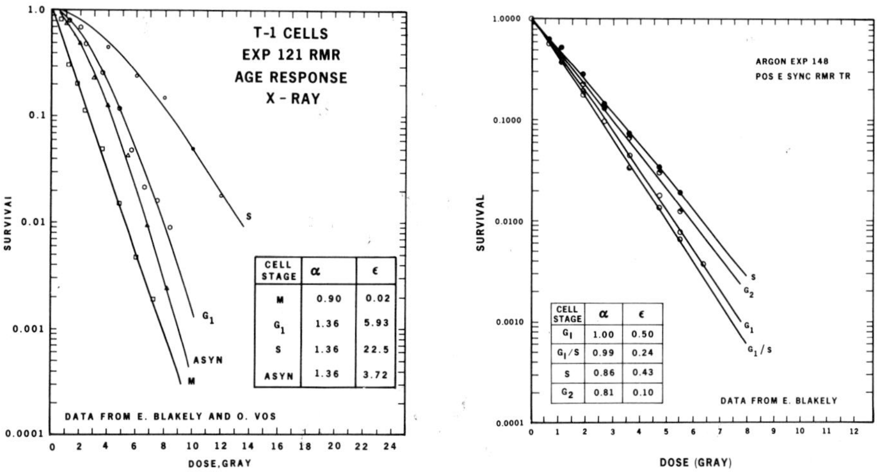

FIG. 11. A, Left, T-1 cell survival curves obtained with 220 kV X rays for cells synchronized by mitotic shake-off. The data are from Vos et al. (1966) and Blakely et al. (unpublished). B, Right, T-1 cell survival curves obtained with argon ions of 0.35-cm residual range in water.

and therefore it is more difficult to accurately determine the ϵ coefficient. However, ϵ is less than 1 for all of the cell cycle phases, and the yield of U lesions (α) are all within 10% of each other. Interestingly, Sasaki and Okada (1979) found the initial yield of strand breaks in mammalian cell DNA to be independent of the stage in the cell cycle.

We are only beginning to work on the synchrony problem. It appears that the rates of repair λ and k of equation 1 are both functions of cell age. It is likely that synchrony experiments will demonstrate a need for more detailed RMR models of the repair process than are given here.

Case III of the RMR Model: Split-Dose Experiments and Mixed Modalities

Equations 4, 5, and 6 are solutions of the basic RMR differential equation 1, and they give a detailed account of the time-dependent quantities of U and R states following a single dose of radiation.

Assume that a dose D_1 is given first and that this is followed by a second dose D_2 of the same radiation after a time interval τ. (All relevant quantities of the first and second exposure are denoted by subscripts 1 and 2, respectively.) We propose to deal with this problem by introducing the concept of a "remnant lesion."

The symbol for remnant lesions, U_R, stands for the quantity of uncommitted lesions present per cell at time τ.

$$U_R(\tau) = \frac{U_1(0)e^{-\lambda\tau}}{1 + \dfrac{U_1(0)}{\epsilon}(1 - e^{-\lambda\tau})} \tag{24}$$

If time scale t_2 begins with the second dose D_2 ($t = t_1 - \tau$), we can write a solution to equation 1 by prescribing new boundary values:

$$U_2(0) = U_R + \alpha D_2, \text{ where } U_2(\infty) = 0 \tag{25}$$

Analogous to equations 4, 5, and 6 we have:

$$U_2(t_2) = \frac{(U_R + \alpha D_2)\, e^{-\lambda t_2}}{1 + \dfrac{(U_R + \alpha D_2)}{\epsilon}(1 - e^{-\lambda t_2})} \tag{26}$$

$$R_L(t_2) = \epsilon \ln\left[1 + \frac{(U_R + \alpha D_2)(1 - e^{-\lambda t_2})}{\epsilon}\right] \tag{27}$$

$$R_Q(t_2) = \left[\frac{(U_R + \alpha D_2)\left(1 + \dfrac{U_R + \alpha D_2}{\epsilon}\right)(1 - e^{-\lambda t_2})}{1 + \dfrac{(U_R + \alpha D_2)}{\epsilon}(1 - e^{-\lambda t_2})}\right] \cdot$$

$$- \epsilon \ln\left[1 + \frac{(U_R + \alpha D_2)(1 - e^{-\lambda t_2})}{\epsilon}\right] \tag{28}$$

Calculation of the survival after two dose installments involves renormalization. We know that after time $t_1 = \tau$ of the first dose D_1 some cells are already committed to die since they have R_{QM} lesions. The number that still survive is $1 - L(\tau)$ of equation 12. The probability of survival after two doses, D_1 and D_2, separated by τ is:

$$S_{1,2} = [1 - L(\tau)]\, S(t_2) \tag{29}$$

$$S(D_1,D_2;\tau,t_2) = \exp\left[-U_R(\tau) - \alpha\{\gamma(\tau)D_1 + D_2\}\right] \cdot \left[\left(1 + \frac{D_1(1 - e^{-\lambda\tau})}{\epsilon}\right)\left(1 + \frac{(U_R + \alpha D_2)(1 - e^{-\lambda t_2})}{\epsilon}\right)\right]^{\epsilon} \tag{30}$$

where U_R is given in equation 24.

In Figure 12 we show a graphic analysis of the time dependence of the survival probabilities (S) and lethality probabilities (L) as functions of time. Three different conditions are compared: a single dose of 800 rad, a single dose of 400 rad, and two split doses of 400 rad each separated by a time interval τ. The area representing cells that have only U lesions is shaded. It is obvious from the figure that at τ there are cells with remnant lesions, and that the probability for U lesions increases stepwise with the addition of a second dose.

We have analyzed some split-dose X-ray experiments on V-79 hamster cells performed by Ngo et al. (1978). In this type of split-dose experiment it is necessary to evaluate the constants α and ϵ from single exposures in advance of the split-dose experiment. The usual experiment, shown in Figure 13, involves administering a preset single dose (D_1), and varying the size of the second

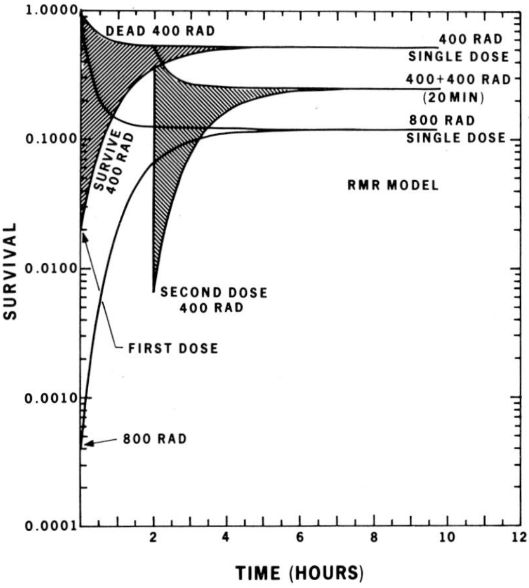

FIG. 12. The time dependence of the functions S and 1 − L are shown for a single dose of 400 rad, a single dose of 800 rad, and for split doses of 400 rad each separated by 20 minutes. The eventual observable survival levels are at large time values. Shaded areas corresponded to uncommitted lesions. The time rates of change of these curves depend on the coefficient λ (equation 25). For this example, λ was assumed to be about three times greater than is actually the case in V-79 cells.

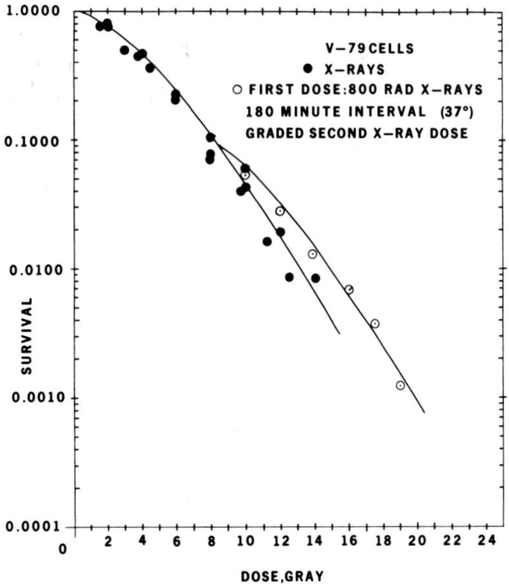

FIG. 13. Split-dose experiment on V-79 cells with X rays: experimental data and RMR theory. ●, single dose, X rays; ○, graded second doses; continuous lines, RMR model, following equation 30.

dose (D_2), keeping τ constant. When this is done, we can evaluate the remnant lesions U_R and also the value of λ, the time rate constant of linear repair. When the values of ϵ and λ are known, the values for $k = \lambda/\epsilon$ can then be calculated.

Mixed Radiations

The equations given for split-dose experiments can also be extended for mixed beams in special cases. One may initially ask the question whether or not two radiations can make uncommitted lesions of the same type for each other. This question can only be answered after extensive experimentation. The usual experimental design starts with a dose of radiation 1 (e.g., neon ion beam), which is followed by a series of doses of radiation 2 (e.g., X rays). A time interval τ is set between the two exposures. The situation can be described by equations similar to equations 24 through 30, except that $\epsilon_1 \neq \epsilon_2$ and $\alpha_1 \neq \alpha_2$. (We do not show the explicit equations in this paper.)

Figure 14 graphs data from a mixed radiation experiment. Experiments like these established that an exposure to heavy ions (carbon, neon, or argon) produces remnant U lesions for X rays and vice versa (Ngo et al. 1978). We suspect, however, that the interaction between two modalities, and also between split doses of the same modality, is more complex than a mere overlap of sublethal lesions. This is illustrated by the fact that experimentally large doses of high-LET neon or argon ions can under certain circumstances potentiate the effect of a second modality.

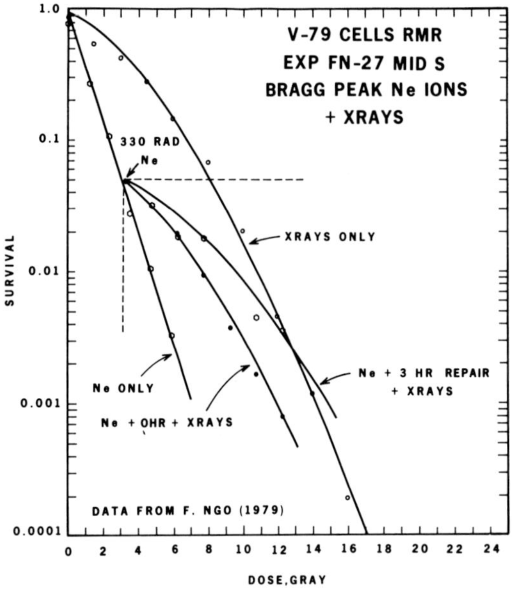

FIG. 14. Mixed beam experiments with V-79 cells. The curves shown are for X rays only, neon only, and neon followed by X rays. These experiments by F. Ngo et al. (unpublished) show that a previous dose of neon ions produces U lesions for X rays. After 330 rad neon dose the "remnant" lesions for X rays correspond to about 210 rad of X rays; after waiting three hours at room temperature the remnant lesions from 303 rad neon correspond to about 150 rad of X rays.

It appears quite likely that split dose and mixed modality exposures may uncover new repair mechanisms. It is of particular interest to extend this model to account for the effects from a variety of deleterious agents. Among the possible interactions are: (1) One modality makes entirely different molecular lesions than another modality and the repair mechanisms are also different; (2) One modality makes remnant lesions for another modality; (3) One modality may cause extranuclear effects (e.g., membrane damage), which may result in an alteration of the number of U lesions that can be produced by the other modality; and (4) Exposure to a deleterious agent can impair or potentiate the repair mechanisms for the other modality.

The RMR model might be suitable for further expansion because it separates the dose-dependent production of radiolesions from the time-dependent repair, and because it has a built-in "memory" for U_R lesions. Eventually the RMR model may be helpful in classifying a variety of lesions produced by a variety of agents.

The Role of Remnant Lesions and Their Relationship to the Initial Slope of the Survival Curves at Low Doses

We may generalize the role of remnant lesions, U_R, which is defined in equation 24. These should also be related to radiobiological experiments on mammalian cells when a single dose is delivered. It appears that remnant U_R lesions may be present not only as a consequence of a previous dose of radiation, but also because of other, nonspecific events in the life of the cell that are not precisely understood at present. Remnant lesions can lower the plating efficiency of cells, and they are also able to alter the initial slope of the survival curve. Although it is quite likely that most of the plating efficiency variations in radiobiological experiments are not related to U lesions for ionizing radiations, the RMR model might be useful in unraveling some of the causes of variations in plating efficiency.

Consequently, by the RMR model there are at least three possible reasons for a finite initial slope of the survival curve: (1) A limited time (t_{max}) is available for eurepair, usually the time to first mitosis (see equations 10 and 17); (2) Not all linear repair is eurepair—$\phi < 1$ (see equation 8); and (3) There are remnant U lesions from a previous dose of the same radiation, or a previous dose of another deleterious agent (see equations 24 and 30).

The problem of the initial slope of survival curves, and the initial slopes of mutation and transformation curves, is a serious one from the point of view of public health risk estimation. In some models, e.g., the dual action theory (Kellerer and Rossi 1972, 1978), a finite initial slope is firmly related to irreversible direct radiation injury. This is a rather different conclusion from that of the RMR model, where at least three different classes of phenomena can alter the initial slope and where the survival curve of cells depends on their recent history of exposure to a variety of deleterious agents. We hope that the RMR model can be used to design crucial experiments that would point to the most important factors in the causation of low-dose effects.

Case IV of the RMR Model: Mutations and Cell Transformation

There are several classes of mutations. The exact probability of producing a specific mutant or transformant depends on the number and types of changes that must occur in DNA to produce the specific new structure. It is possible, and even likely, that a number of cell generations and several crucial events must occur to complete a mutation or transformation process. There is not sufficient room in this paper to discuss all factors affecting mutation rates; however, an example is given of how the mutation rate might be calculated.

Assume that mutations derive from misrepaired R states. According to our definitions (Table 1), R_{LM} and R_{QM} are states with abnormal genetic structures. If a cell with such states survives, there is a chance that its progeny will be mutants. Assuming a constant probability, we give a formula for the probability M (D,t) that a survivor is a mutant, based on survival equation 8.

$$M(D,t) = 1 - \left[1 + \frac{\alpha D(1 - e^{-\lambda t})}{\epsilon}\right]^{-\Delta\phi\epsilon} \qquad (31)$$

Here $\Delta\phi/\phi$ is the fraction of linear repair that results in a mutant. At very low doses, we obtain a linear relationship between dose and mutation:

$$M(D,t) = \Delta\phi \cdot \alpha\,(1 - e^{-\lambda t})D \qquad (32)$$

whereas at high doses the mutant/survivor ratio is saturated. The mutant thus produced corresponds to a chromatid aberration, but whether or not it is expressed depends on further genetic developments. The absolute number of mutations produced is nonlinear and has a maximum. Equation 31 generally fits the shapes of mutation yields obtained in heavy-ion enhancement of virus-induced cell transformation studies by Yang et al. (1979).

Case V of the RMR Model: Radiation Effects on the Kinetics of the Repair Process

Inactivation of the Repair Mechanism

Because the dose- and time-dependent processes are handled separately, the RMR model is well suited for the study of the dose dependence of the kinetics of the repair process. In the survival expression equation 10 the coefficients λ, k, and ϵ were regarded as constants. For Case V we consider these coefficients to be dose-dependent quantities. We may regard λ as proportional to the available repair enzyme so that if the enzyme should be inactivated by a dose D of radiation, λ may change accordingly: $\lambda = \lambda$ (D). In Figure 15A we show theoretical examples of mammalian cell survival curves in which the repair process is inactivated. We assumed that the inactivation kinetics are linear, along with the coefficient ψ.

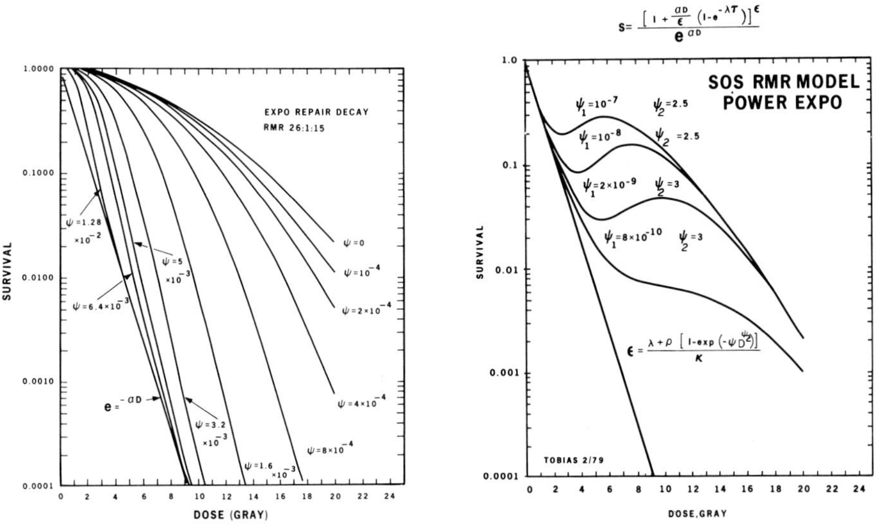

$$S = \frac{\left[1 + \frac{\alpha D}{\epsilon}\left(1 - e^{-\lambda T}\right)\right]^{\epsilon}}{e^{\alpha D}}$$

$$\epsilon = \frac{\lambda + \rho\left[1 - \exp\left(-\psi_1 D^{\psi_2}\right)\right]}{K}$$

FIG. 15. A, Left, Theoretical survival curves, assuming dose-dependent inactivation of the repair mechanism. It was assumed that α (from equation 16) was constant. The repair ratio, ϵ, is a function of the dose delivered according to equation 33. The values of ψ on the graph are given in units of rad^{-1}. When ψ is about 10^{-3}, the survival curves have a quasi-exponential portion. B, Right, SOS survival curves generated by the RMR model. To do this, we assumed a dose dependence of ϵ, as shown by the equations on the graph.

$$\epsilon(D) = \epsilon_o e^{-\psi D} \tag{33}$$

where $\epsilon_o = \epsilon$ at zero dose and ϵ (D) of equation 33 was substituted for ϵ in survival equation 7.

It is interesting to note that in Figure 15A the shape of the survival curves change from the usual continuously bending RMR form to shapes that behave like a simple exponential at medium high doses (8 to 16 Gy). This occurs when we assume that the repair mechanism is about ten times more resistant to ionizing radiation than the genome of the cells is to the production of U lesions. This type of dose-dependent behavior resembles a repair saturation curve such as proposed by Green and Burki (1972). However, at higher doses where the repair is completely inactivated the curves merge into the "repairless" curve $e^{-\alpha D}$. Thus, according to the RMR model, survival curves with shoulders and an exponential portion might be indicators of the presence of repair mechanisms that are damaged by a dose of radiation. It would also follow that repair would proceed more slowly after a large dose of radiation than after a small dose.

SOS Repair

The induction of repair by a dose of radiation is termed "SOS repair." Recently, new types of repair of UV-induced lethal lesions were described in bacteria (Radman 1975, Devoret 1978, Witkin 1976), in which a large dose of radiation induced a repair mechanism that was not present when a small dose of radiation was given. Howard and Cowie (1978) also showed SOS repair in algae. As far as we know, the genes controlling repair enzymes in mammalian cells are constitutive; however, more investigations appear necessary.

The RMR model is quite suitable for the study of rate processes involved in SOS repair. In Figure 15B we show survival probabilities for SOS repair when an appropriate function of dose is inserted in the ϵ parameter of equation 17.

Case VI of the RMR Model: Dose Rate Effects

In all of the preceding cases it was tacitly assumed that the dose was delivered so fast to the cells that exposure to radiation was complete before the repair processes were under way. In this section, we shall demonstrate that the RMR model can be used for modeling survival at low as well as high dose rates. At very high dose rates (e.g., $> 10^5$ Gy/min), the RMR model is probably not valid.

Assume a constant dose rate $\dot{D}$. Uncommitted lesions accrue at the constant rate $a(\dot{D})$ in accordance with equations 13 through 15. In our example, we choose $a(\dot{D}) = \alpha \dot{D}$ from equation 14. The differential equation for uncommitted lesions per cell analagous to equation 1 is

$$\frac{dU}{dt} = a - \lambda U - kU^2 \tag{34}$$

Integrating and using the quantities R_L and R_Q defined in equations 2 and 3 we have

$$U + R_L + R_Q = at \tag{35}$$

With boundary values of U(0) = 0 and U(∞) = U, we can solve for U(t):

$$U(t) = U_\infty \cdot \gamma_a(t) \tag{36}$$

where
$$U_\infty = -\frac{\epsilon}{2} + \left[\left(\frac{\epsilon}{2}\right)^2 + \left(\frac{a}{k}\right)\right]^{1/2}$$

$$\lambda_a = \lambda + 2kU_\infty;\ \epsilon_a = \frac{\lambda a}{k}$$

and
$$\gamma_a(t) = \frac{\left(1 - \dfrac{U_\infty}{\epsilon_a}\right)(1 - e^{-\lambda_a t})}{1 - \dfrac{U_\infty}{\epsilon_a}(1 - e^{-\lambda_a t})}$$

If we proceed in a manner similar to equations 1 through 6 and 17 we can calculate time-dependent probabilities of survival, S(a,t), and lethality, L(a,t):

$$S(a,t) = \left[1 - \frac{U_\infty}{\epsilon_a}(1 - e^{-\lambda_a t})\right]^{\epsilon} \exp(\lambda_a U_\infty - a)t \tag{37}$$

$$1 - L(a,t) = \left[1 - \frac{U_\infty}{\epsilon_a}(1 - e^{-\lambda_a t})\right]^{\epsilon} \exp[(\gamma_a U_\infty) + (\lambda_a U_\infty - a)t] \tag{38}$$

If radiation with a constant dose rate D is delivered for a time period τ, a remnant lesion of $\gamma_a(\tau)$ U will be present at the end of period τ. Repair will continue with the further passage of time.

$$S_t = S(a,\tau)\left[1 + \frac{\gamma_a(\tau)U_\infty}{\epsilon}(1 - e^{-\lambda(t-\tau)})\right]^{\epsilon} \cdot e^{-\gamma a(\tau)U_\infty} \tag{39}$$

where $S(a,\tau)$ is taken from equation 37.

In Figure 16 we have plotted the functions S(a,t) and S(t) for widely varying dose rates, but each with the same total dose. The survival depends on dose rate if the dose is delivered in about the same length of time as the half time for repair.

The example for a typical mammalian cell system assumes that the half-life for repair is longer than one hour. In this example, the dose can be delivered in any time interval, from 0 to about 10 minutes, without appreciably affecting survival.

Kaplan (1974) found "fast" repair processes in bacterial DNA which were

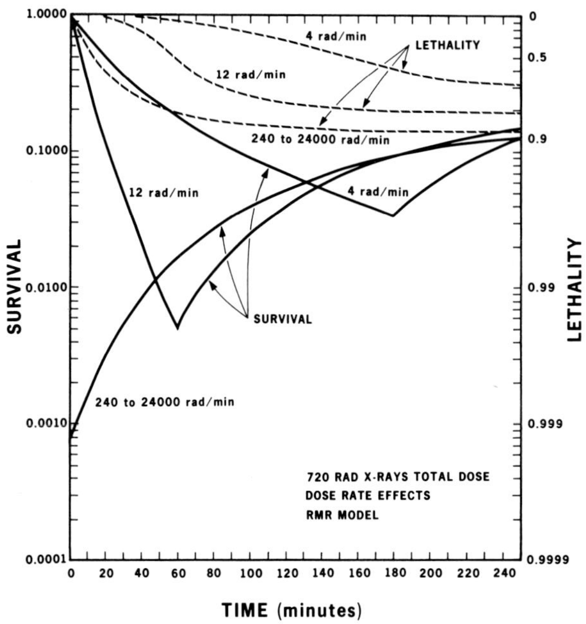

FIG. 16. Prediction of the RMR model for continuous dose rate exposure. S and (1 − L) curves are shown for 4 rad/min and 12 rad/min doses. Above 10,000 rad/min the model predicts no dose-rate effect. Note that the U lesions accumulate with time while the radiation is "on."

on the order of one minute, and a similar rate of repair process was found recently by Braby and Roesch (1978), in the algae chlamydomonas, which was irradiated continuously. The model described here is suitable for calculating survival for repair rates that are $\lambda \simeq 10^3$/sec or lower, but in mammalian cells we have no evidence at present for such high repair rates.

It is important for the validity of the RMR model that the rate-dependent events involving fast-radical chemistry do not significantly overlap in time with the enzymatic repair processes we discuss here. From the work of Epp et al. on bacteria (1968, Michaels et al. 1978), it appears that very high dose rates (high enough so that the entire dose is delivered in about 10^{-4} seconds) are necessary to modify the oxygen effect. Anoxic radiosensitivity, on the other hand, appears to be insensitive to high dose rates up to 10^{12} rad/minute. Todd et al. performed very high dose-rate experiments in electron beams with human kidney cells (1968). Their work, reinterpreted recently by Braby and Roesch (1978), is indicative of a small dose-rate effect above 10^{11} rad/minute. Such results essentially confirm the assumptions of our model: that dose- and time-dependent physical-chemical interactions are essentially over in less than 10^{-3} seconds.

Continuous Background Radiation and the Initial Slope of Survival Curves

Background radiation, according to equations 34 through 38, continuously delivers new U lesions to cells and, in spite of repair, some U lesions are always present in DNA under these conditions.

The background radiation caused by cosmic rays, and radioactivity found in the tissues and the environment, continuously produce lesions in DNA nucleoprotein. Misrepair of such lesions is also bound to occur, resulting in occasional lethal effects. As a result of the presence of remnant lesions from background radiation, the RMR model predicts that the initial dose-dependent slope of survival curves following acute low- or high-LET radiation is never exactly zero but always has a finite, albeit possibly very low, value.

SUMMARY

A new model is presented for cell survival, lethality, and mutation caused by ionizing radiations: the repair-misrepair model (RMR). We have shown that the fast events of physical energy transfer and of radiation chemistry are largely complete before the enzymes of a living cell can recognize relevant macromolecular processes and before biochemical repair processes are under way. This allows the model to be separated into dose-dependent and time-dependent processes; the shapes of the survival curves depend on both. Initial macromolecular lesions are regarded as uncommitted because the eventual fate of cells remains uncertain for some time after exposure. The enzymatic repair processes yield either eurepaired states or misrepaired states with altered structures. Survival is a result of competition between eurepair and misrepair. A fraction of misrepair leads to lethality; other misrepair fractions produce mutants. The general features of the misrepair process are analogous to chromosome rejoinings.

RMR survival kinetics have been applied to a variety of radiobiological processes including the analysis of repair-deficient mutants, the cell age response, the effects of accelerated heavy ions, split doses, survival from mixed modalities, and induction of mutations. The model provides a flexible framework for testing mechanisms for the biological effects of ionizing radiations and of other deleterious agents. Dose-rate effects have also been modeled. Work is in progress to adapt it to such processes as repair inactivation and SOS repair.

ACKNOWLEDGMENTS

The authors thank John Magee, Aloke Chatterjee, Ruth Roots, and E. Alpen for productive discussion, M. C. Pirruccello for editing, and L. E. Hawkins for typing the manuscript.

These studies were supported by the Office of Health and Environmental Research of the U.S. Department of Energy under contract No. W-7405-ENG-48, and the National Cancer Institute (Grant No. CA 15184).

LIST OF SYMBOLS

t = time

D = dose (usually delivered at t = 0); $\dot{D} = \frac{dD}{dt}$ = dose rate

$S(t)$ = probability of survival
$L(t)$ = probability of lethality
$M(t)$ = probability of mutation
U = number of uncommitted lesions/cell
$U(0)$ = U_0 initial number of U lesions
U_R = remnant U lesion (at time t)
R = repair states
  R_L = yield of a linear repair processes
  R_Q = yield of a repair process involving interaction between pairs of U lesions
λ = coefficient of linear repair
k = coefficient of quadratic repair
ϵ = repair ratio (λ/k)
ϕ = probability that linear repair = eurepair
δ = probability that quadratic repair = eurepair
α = yield of U lesions per unit dose
τ = time interval separating two dose installments
T = time constraint = $(1 - e^{-\lambda t})$
m = number of hits in conventional target theory
x,y = constant coefficients in the LQ model
$a(\dot{D})$ = rate of production of U lesions

REFERENCES

Bernardi, F., and J. Ninio. 1978. The accuracy of DNA replication. Biochimie 60:1083–1095.

Blakely, E. A., C. A. Tobias, T. C. Yang, K. C. Smith, and J. T. Lyman. 1979. Inactivation of human kidney cells by high-energy monoenergetic heavy-ion beams. Radiat. Res. (in press).

Bonura, T., C. P. Town, K. C. Smith, and H. S. Kaplan. 1975. The influence of oxygen on the yield of DNA double strand breaks in x-irradiated *Escherichia coli* K-12. Radiat. Res. 63:567–577.

Braby, L. A., and W. C. Roesch. 1978. Testing of dose-rate models with *Chlamydomonas reinhardi.* Radiat. Res. 76:259–270.

Chadwick, K. H., and H. P. Leenhouts. 1973. A molecular theory of cell survival. Phys. Med. Biol. 13:78–87.

Chadwick, K. H., and H. P. Leenhouts. 1978. The rejoining of DNA double strand breaks and a model for the formation of chromosomal rearrangements. Int. J. Radiat. Biol. 33:517–529.

Chatterjee, A., H. D. Maccabee, and C. A. Tobias. 1973. Radial cutoff LET and radial cutoff dose calculations for heavy charged particles in water. Radiat. Res. 54:479–494.

Christensen, R. C., C. A. Tobias, and W. D. Taylor. 1972. Heavy-ion induced single and double strand breaks in ϕX-174 replicative form DNA. Int. J. Radiat. Biol. 22:457–477.

Cleaver, J. E. 1978. DNA repair and its coupling to DNA replication in eukaryotic cells. Biochim. Biophys. Acta 516:480–516.

Corry, P. M., and A. Cole. 1973. Double strand rejoining in mammalian DNA. Nature 245:100–101.

Devoret, R. 1978. Inducible error-prone repair: One of the cellular responses to DNA damage. Biochimie 60:1135–1140.

Dienes, G. J. 1966. A kinetic model of biological radiation response. Radiat. Res. 28:183–202.

Dugle, D. L., C. J. Gillespie, and J. D. Chapman. 1976. DNA strand breaks, repair, and survival in X-irradiated mammalian cells. Proc. Natl. Acad. Sci. USA 73:809–812.

Ehrenberg, L. 1977. A note on the shape of shouldered dose-response curves. Int. J. Radiat. Biol. 31:503–506.
Elkind, M. M., and H. Sutton. 1960. Radiation response of mammalian cells grown in culture. I. Repair of x-ray damage in surviving Chinese hamster cells. Radiat. Res. 13:556–593.
Epp, E. R., H. Weiss, and A. Santomasso. 1968. The oxygen effect in bacterial cells irradiated with high intensity pulsed electrons. Radiat. Res. 34:320–325.
Garrett, W. R., and M. G. Payne. 1978. Applications of models for cell survival: The fixation time picture. Radiat. Res. 73:204–211.
Green, A. E. S., and J. Burki. 1972. A note on survival curves with shoulders. Radiat. Res. 60:536–540.
Ho, K. S. Y. 1976. Induction of dominant lethal damage and DNA double-strand breaks by x-ray in a radiosensitive strain of yeast *Saccharomyces cerevisiae.* Ph.D. Thesis, University of California, Berkeley.
Ho, K. S. Y., and R. K. Mortimer. 1973. Induction of dominant lethality by x rays in a radiosensitive strain of yeast. Mutat. Res. 20:45–51.
Howard, A., and F. G. Cowie. 1978. Induced resistance in *Closterium:* Indirect evidence for the induction of repair enzyme. Radiat. Res. 75:607–616.
Hutchinson, F. 1974. Relationships between some specific DNA lesions and some radiobiological effects in bacteria, *in* Physical Mechanisms in Radiation Biology, R. D. Cooper and R. W. Wood, eds. U.S. Atomic Energy Commission, Washington, D.C.
Jacobsen, B. S. 1957. Evidence for recovery from x-ray damage in chlamydomonas. Radiat. Res. 7:395–406.
Kaplan, H. S. 1974. Repair of x-ray damage to bacterial DNA and its inhibition by chemicals, *in* Advances in Chemical Radiosensitization. International Atomic Energy Agency, Vienna, pp. 123–142.
Kellerer, A., and U. Hug. 1963. Zur Kinetik der Strahlenwirkung. Biophysik 1:33–50.
Kellerer, A. M., and H. H. Rossi. 1972. The theory of dual radiation action. Curr. Top. Radiat. Res. Q. 8:85–158.
Kellerer, A. M., and H. H. Rossi. 1978. A generalized formulation of dual radiation action. Radiat. Res. 75:471–488.
Laurie, J., J. S. Orr, and C. J. Foster. 1972. Repair processes and cell survival. Br. J. Radiol. 45:362–368.
Lea, D. E. 1955. Action of Radiations on Living Cells. Cambridge University Press, London and New York.
Ling, C. C., H. B. Michaels, E. R. Epp, and E. C. Peterson. 1978. Oxygen diffusion into mammalian cells following ultrahigh dose rate irradiation and lifetime estimates of oxygen sensitive species. Radiat. Res. 76:522–532.
Magee, J. 1979. A radical diffusion model for cell survival (Abstract), *in* Proceedings, 6th International Congress of Radiation Research. Japanese Association for Radiation Research, Tokyo.
Michaels, H. B., E. R. Epp, C. C. Ling, and E. C. Peterson. 1978. Oxygen sensitization of CHO cells at ultrahigh dose rates: Prelude to oxygen diffusion studies. Radiat. Res. 76:510–521.
Neary, G. J. 1965. Chromosome aberrations and the theory of RBE. I. General considerations. Int. J. Radiat. Biol. 9:477–502.
Ngo, F. Q. H., E. A. Blakely, and C. A. Tobias. 1978. Does an exponential survival curve of irradiated mammalian cells imply no repair? (Abstract) Radiat. Res. 74:588.
Niederer, J., and J. R. Cunningham. 1976. The response of cells in culture to fractionated radiation: A theoretical approach. Phys. Med. Biol. 21:823–839.
Payne, M. G., and W. R. Garrett. 1975. Some relations between cell survival models having different inactivation mechanisms. Radiat. Res. 62:388–394.
Pohlit, W. 1975. The shape of dose-effect curves for diploid yeast cells irradiated with ionizing particles, *in* Proceedings of the Sixth L. H. Gray Conference, T. Alper, ed. John Wiley and Sons, New York.
Powers, E. L. 1962. Considerations of survival curves and target theory. Phys. Med. Biol. 7:3–28.
Pritchard, A. E., and C. T. O'Konski. 1977. Dynamics of superhelical DNA and its complexes with ethidium bromide from electro-optic relaxation measurements. Ann. N.Y. Acad. Sci. 30:159–169.
Radman, M. 1975. SOS repair hypothesis: Phenomenology of an inducible DNA repair which is accompanied by mutagenesis, *in* Molecular Mechanisms for Repair of DNA, P. Hanawalt and R. B. Setlow, eds. Plenum Press, New York, pp. 355–367.

Ritter, M. A., J. E. Cleaver, and C. A. Tobias. 1977. High-LET radiations induce a large proportion of nonrejoining DNA breaks. Nature 266:653–655.

Roots, R., T. C. H. Yang, L. Craise, E. A. Blakely, and C. A. Tobias. 1979. Impaired repair capacity of DNA breaks induced in mammalian cellular DNA by accelerated heavy ions. Radiat. Res. 78:38–49.

Sasaki, H., and S. Okada. 1979. Unequal segregation of nuclear materials in irradiated cultured mammalian cells (abstract), *in* Proceedings of the Sixth International Congress of Radiation Research. Japanese Association for Radiation Research, Tokyo.

Sawada, S., and S. Okada. 1972. Effects of BUdR-labelling on radiation-induced DNA breakage and subsequent rejoining in cultured mammalian cells. Int. J. Radiat. Biol. 21:599–602.

Shenoy, M. A., J. C. Asquith, G. E. Adams, B. D. Michael, and M. E. Watts. 1975. Time resolved oxygen effects in irradiated bacteria and mammalian cells: A rapid-mix study. Radiat. Res. 62:498–512.

Sinclair, W. S. 1966. The shape of radiation survival curves of mammalian cells cultured in vitro, *in* Biophysical Aspects of Radiation Quality. Technical Report Series 58. International Atomic Energy Agency, Vienna, pp. 21–43.

Tobias, C. A., E. A. Blakely, F. Q. H. Ngo, and A. Chatterjee. 1978. Repair-misrepair (RMR) model for the effects of single and fractionated dose of heavy accelerated ions. (Abstract) Radiat. Res. 74:589.

Todd, P., H. S. Winchell, J. M. Feola, and G. E. Jones. 1968. Pulsed high-intensity roentgen rays. Acta Radiol. 7:22–26.

Veatch, W., and S. Okada. 1969. Radiation-induced breaks of DNA in cultured mammalian cells. Biophys. J. 9:330–346.

Vos, O., H. A. E. M. Schenk, and D. Bootsma. 1966. Survival of excess thymidine synchronized cell populations in vitro after x-irradiation in various phases of the cell cycle. Int. J. Radiat. Biol. 11:495–503.

Wideröe, R. 1966. High-energy electron therapy and the two-component theory of radiation. Acta Radiol. 4:257–278.

Witkin, E. M. 1976. Ultraviolet mutagenesis and inducible DNA repair in *E. coli.* Bacteriol. Rev. 40:869–907.

Yang, T. C. H., E. A. Blakely, L. M. Craise, I. S. Madfes, C. Perez, and J. Howard. 1979. Cocarcinogenic effects of high-energy neon particles on the viral transformation of mouse C3H10T1/2 cells in vitro. Radiat. Res. (in press).

Radiation Biology in Cancer Research, edited by
Raymond E. Meyn and H. Rodney Withers.
Raven Press, New York © 1980.

Models of Radiation Inactivation and Mutagenesis

D. T. Goodhead

MRC Radiobiology Unit, Harwell, Didcot, OX11 ORD, England

Numerous models of radiation action have been developed in an attempt to understand the basic mechanisms by which radiation produces biological damage and change and for purely practical purposes (such as to better understand radiotherapy treatment schedules or to assist in setting acceptable limits in radiological protection). It is often difficult or impossible to make direct experimental measurements on the biological effectiveness of radiation at the very low doses or dose rates that may be of interest. For example, the frequency of biological effects may be too small at low doses, or the biological system may not be constant during a protracted low dose-rate experiment. Such information can then be sought only by extrapolation from higher doses or dose rates using additional assumptions to suggest an appropriate form of the extrapolation.

For some effects on single mammalian cells, probably including loss of clonogenic capacity (inactivation) and induction of mutation, it is valid to assume that the cell nucleus contains the radiosensitive elements of the cell. This assumption implies that the dose-response curve is linear below about 0.1 rad of γ rays for a typical mammalian cell nucleus of about 600 μm^3 volume because over this dose range very few cells will have received more than one energy deposition event; consequently these events must act independently of one another (Booz 1975). (This linear part of the curve may have zero slope, in which case there will be a threshold in the dose response.) Similarly, the dose-response curve should be linear for dose rates of less than about 0.1/T rad $hour^{-1}$ where T is the time in hours over which two separate radiation events in the same nucleus can together influence the response. Additional assumptions are required to make the extrapolation from these low doses to the lowest experimental points, which may well be at 10–100 rad. This can be done either on a purely empirical basis or within a theoretical framework using a model. Examples of the purely empirical approach are a simple linear extrapolation from the lowest measured point or extrapolation by a fitted curve of some simple mathematical form, such as a quadratic one ($\alpha D + \beta D^2$), which may have been observed to give good fits over the region of experimental data. The alternative is to select an equation based on the theoretical assumptions or hypotheses of a model, such as the quadratic equation either from the "theory of dual radiation action" (Kellerer and Rossi 1972) or from the "molecular theory of radiation action" (Chadwick and Leenhouts 1973).

SOME MODELS MUST BE REJECTED

Figure 1 shows a small selection of the many models that can be fitted to experimental data. Attempts to use statistical goodness of fit to determine the appropriate equation for general use have not been particularly successful, sometimes favoring one equation, sometimes another, and often leading to the conclusion that a variety of equations are statistically acceptable. Such statistical methods alone are not adequate to test the assumptions of the models, especially since an identical equation may be based on two totally contradictory sets of assumptions. Examples of this are the quadratic equation based on interaction of sublesions over micrometer distances by Kellerer and Rossi (1972), but over only nanometer distances by Leenhouts and Chadwick (1978), or the use of the multitarget equation based on the accumulation of sublesions by Bender and Gooch (1962) or on a depletion of repair capacity by Wideröe (1978).

Nevertheless, it is essential to discriminate between the different equations and models to understand the mechanisms of radiation action or to extrapolate

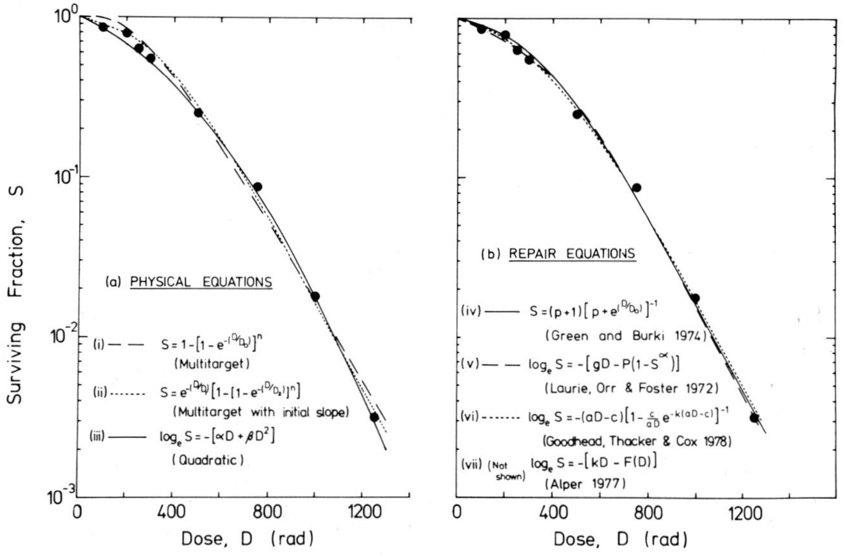

FIG. 1. Examples of models of radiation action fitted to observed results of the inactivation of Chinese hamster V79 cells after irradiation with γ rays (Cox et al. 1977b). (a) Selection of purely biophysical models based on the assumption of accumulation or interaction of "sublesions," namely (i) multitarget equation (Katz et al. 1971), (ii) multitarget with initial slope (Barendsen 1961, Bender and Gooch 1962; also used by Wideröe 1978 in a repair model), and (iii) quadratic exponent equation (Kellerer and Rossi 1972, Chadwick and Leenhouts 1973). (b) Selection of simple repair models based on the assumption of lesions being repairable by a dose-dependent or saturable repair system (iv) Green and Burki 1974, (v) Laurie et al. 1972 and (vi) Goodhead et al. 1978). Alper (1977) has not suggested a form of F(D), so no curve can be plotted. For the illustrative purposes of this figure most of the curves have been fitted by trial and error, so minor differences in curve shape should not be regarded as statistically significant.

TABLE 1. *Low dose extrapolation of curves of Figure 1*

MODEL EQUATION	EXAMPLE OF USE	KILLED FRACTION AT D=10 rad	SLOPE AS D→0 (rad^{-1})
PHYSICAL:			
(i) $S = 1-[1-e^{-(D/D_0)}]^n$ (Multitarget)	Katz et al. 1971	$6{\cdot}9 \times 10^{-7}$	0
(ii) $S = e^{-(D/D_1)}[1-[1-e^{-(D/D_2)}]^n]$ (Multitarget with initial slope)	Barendsen 1960; Bender & Gooch 1962; Wideröe 1978*	$1{\cdot}2 \times 10^{-2}$	$1{\cdot}2 \times 10^{-3}$
(iii) $\log_e S = -[\alpha D + \beta D^2]$ (Quadratic)	Kellerer & Rossi 1972; Chadwick & Leenhouts 1973	$1{\cdot}4 \times 10^{-2}$	$1{\cdot}4 \times 10^{-3}$
REPAIR:			
(iv) $S = (p+1)[p+e^{(D/D_0)}]^{-1}$	Green & Burki 1974	$7{\cdot}4 \times 10^{-3}$	$7{\cdot}3 \times 10^{-4}$
(v) $\log_e S = -[gD - P(1-S^{\alpha})]$	Laurie, Orr & Foster 1972	$1{\cdot}4 \times 10^{-2}$	$1{\cdot}4 \times 10^{-3}$
(vi) $\log_e S = -(aD-c)[1-\frac{c}{aD}e^{-k(aD-c)}]^{-1}$	Goodhead, Thacker & Cox 1978	$8{\cdot}7 \times 10^{-3}$	$8{\cdot}5 \times 10^{-4}$
(vii) $\log_e S = -[kD - F(D)]$	Alper 1977	'Any'	'Any'

*Interpreted as repair

experimental observations for practical purposes. Table 1 shows the extrapolation to low doses of the data of Figure 1. The various extrapolated values at 10 rad differ by a factor of 10^4, and the factor increases to infinity as the dose tends to zero. These will lead to gross differences in relative biological effectiveness (RBE) of other radiations if the γ-ray response is used as reference. The apparent agreement (within a factor of 2) between all but one of the models shown may be misleading since only the very simplest repair functions have been considered; in principle, almost any extrapolation is possible depending on the form of the dose-dependent repair function.

TESTING OF MODELS

Can we identify at least some of the models that are invalid for mammalian cell inactivation and mutation induction? We must seek critical experimental evidence that tests the assumptions or predictions of particular models or groups of models; this conventional application of the scientific method often makes most rapid progress when the results of the critical experiment are found to contradict a prediction and by implication the hypothesis on which it is based.

I shall confine this presentation to experimental evidence from our own laboratory on in vitro single-cell inactivation and mutation induction (induction of resistance to 6-thioguanine). Available evidence suggests that ionizing radiation induces the thioguanine-resistant phenotype in mammalian cells predominantly by causing fairly large structural changes to the DNA rather than by causing classical base-change mutations (Cox and Masson 1978, 1979, Thacker et al. 1979, Goodhead et al. 1979a) and that this mechanism may be fairly general for radiation-induced mutation in mammalian cells (Thacker et al. 1978). The present discussion emphasizes the inactivation experiments, since they usually have greater statistical accuracy than mutation experiments. However, the results of the mutation experiments often correlate well with those of inactivation (Thacker and Cox 1975, Munson and Goodhead 1977, Goodhead et al. 1979a) and are consistent with most of the present conclusions. Nevertheless, they should be regarded as less stringent tests of the hypotheses.

Identical Lesions for Mutation and Inactivation?

A hypothesis that can be tested readily is that the biochemical lesions responsible for mutation are of one type only and identical to those for inactivation, for example a double-strand break of the DNA molecule (Chadwick and Leenhouts 1976). This hypothesis leads to the prediction that the number of mutagenic lesions should be proportional to the number of lethal lesions for a given cell type irrespective of radiation quality. The average numbers of lesions of each type (mutagenic and lethal) per cell are measured experimentally by the induced

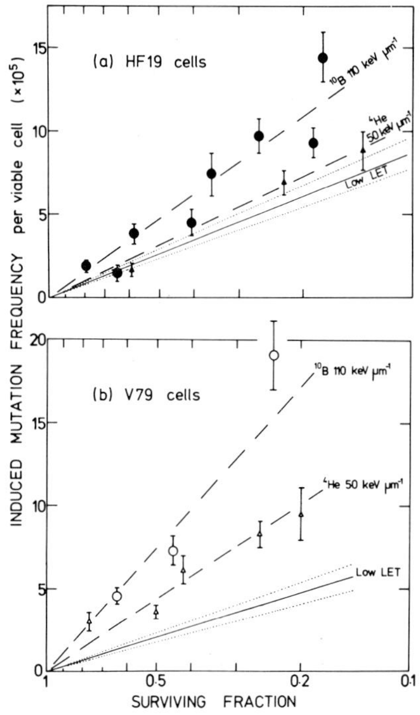

FIG. 2. Relationship between observed induced mutation frequency per viable cell and logarithm of cell survival after irradiation with low- and high-LET radiations (Goodhead et al. 1979a). The low-LET solid lines are weighted linear regression lines with 95% confidence limits (dotted) for (a) HF19 human fibroblasts (Cox and Masson 1976) and (b) V79 Chinese hamster cells (Thacker et al. 1977, Thacker and Cox 1975). For higher LET radiations straight lines have been drawn by eye through the data points for helium ions of 50 keV μm^{-1} (▲,△) and boron ions of 110 keV μm^{-1} (●,○) obtained by similar methods (Cox and Masson 1979, Thacker et al. 1979).

mutation frequency and by the negative logarithm of the surviving fraction of cells, respectively; so a plot of these experimental quantities should give the same straight line independent of radiation quality (Thacker and Cox 1975, Munson and Goodhead 1977). In direct contradiction to this, Figure 2 shows that radiations of higher linear energy transfer (LET) produce lines of steeper slope, implying relatively more mutagenic lesions (Cox et al. 1977a, Goodhead et al. 1979a). This conclusion does not appear to be altered by consideration of such complicating factors as the correlation of mutagenic and lethal lesions produced by individual high-LET tracks (Goodhead et al. 1979a).

Interaction of Sublesions?

Another hypothesis open to direct test is that of the relative roles of one-track and two-track action in the production of lesions, usually closely associated with the hypothesis of the need for interaction or accumulation of sublesions to produce lesions. This common hypothesis is illustrated schematically in Figure 3, where A and B are sublesions that interact to produce a lesion (which leads to the observable biological effect). A combination of the two modes of action can lead to a quadratic dose response of

$$\text{Effect} = \alpha D + \beta D^2$$

where the linear term is dominant for high-LET radiation and the dose-squared term is dominant at sufficiently high doses of low-LET radiation (e.g., Kellerer and Rossi 1972). A direct consequence of this hypothesis for mammalian cell inactivation and mutation is that the interaction distance must be of the order of one micrometer because the observed dose responses are visibly curved at doses of a few hundred rad of low-LET radiation, at which doses the average distance between tracks is of the order of micrometers.

Effectiveness of Short Tracks

The above hypothesis is tested directly by observing the effectiveness of radiations that produce only tracks much shorter than this interaction distance. Such tracks are provided by ultrasoft X rays (Figure 4). According to the hypothesis they should be very inefficient in one-track action, so the dose response should have only a very small initial slope and be dominated by the two-track term.

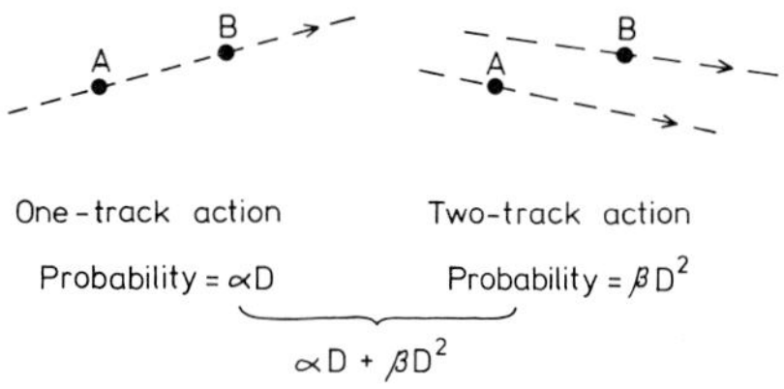

FIG. 3. Schematic diagram of one-track and two-track hypothesis of radiation action. It is often assumed that two sublesions (A and B) must interact to form a lesion with a dose-dependent probability of $\alpha D + \beta D^2$.

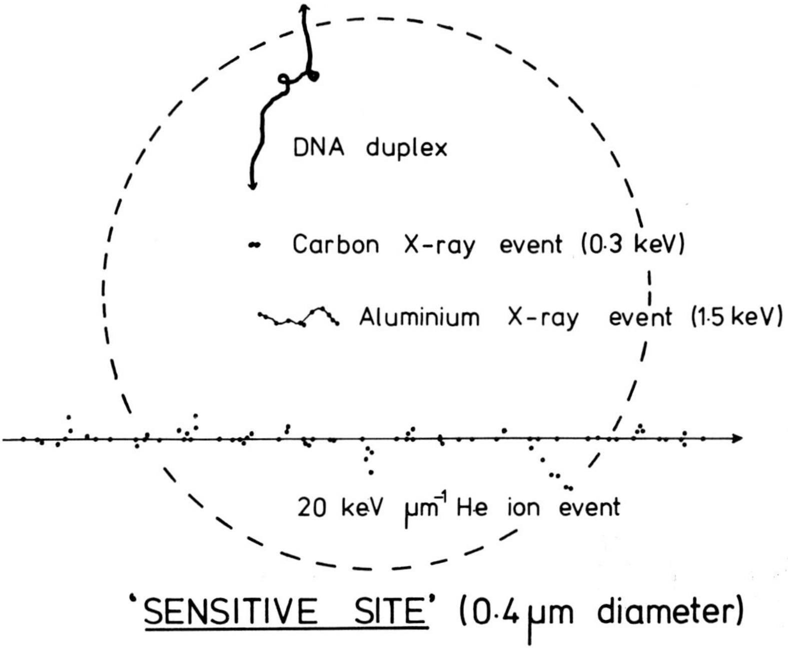

FIG. 4. Schematic diagram showing the short tracks (lengths ~70 nm and ~7 nm) produced by the absorption of aluminum K and carbon K ultrasoft X-rays, respectively. Each dot represents about six ionizations. Shown for comparison are a segment of a DNA double helix of diameter about 2 nm and a segment of the track of a fast helium ion of 20 keV μm^{-1}. The broken circle represents a sphere of diameter 400 nm, which is typical of the distance within which sublesions are assumed to interact within a sensitive site according to the theory of dual radiation action.

Their effectiveness should be less than or similar to that of ^{60}Co γ rays (Goodhead 1977, Goodhead et al. 1979b, Goodhead et al. 1978). The observed effect of these radiations for inactivation and mutation induction is quite contrary to these predictions; both aluminum and carbon ultrasoft X rays are seen to be much more effective than ^{60}Co γ rays (Figure 5).

We have shown elsewhere (Goodhead 1977, Goodhead et al. 1978, 1979b) in particular detail how the results disagree with the theory of dual radiation action in its original form (Kellerer and Rossi 1972) and its later generalized form (Kellerer and Rossi 1978), and a similar conclusion applies to other theories based on interaction or accumulation of sublethal damage. For example, the observations disagree strongly with the theory of Katz et al. (1971), which predicts that the ultrasoft X rays should act by one-track action (γ kill) only with zero initial slope and a purely dose-squared response. In addition the observed curvature of the dose-response curves for ultrasoft X rays (Figure 5) are in contradiction to the molecular theory of radiation action (Chadwick and Leenhouts 1973). This theory assumes that two-track action takes place

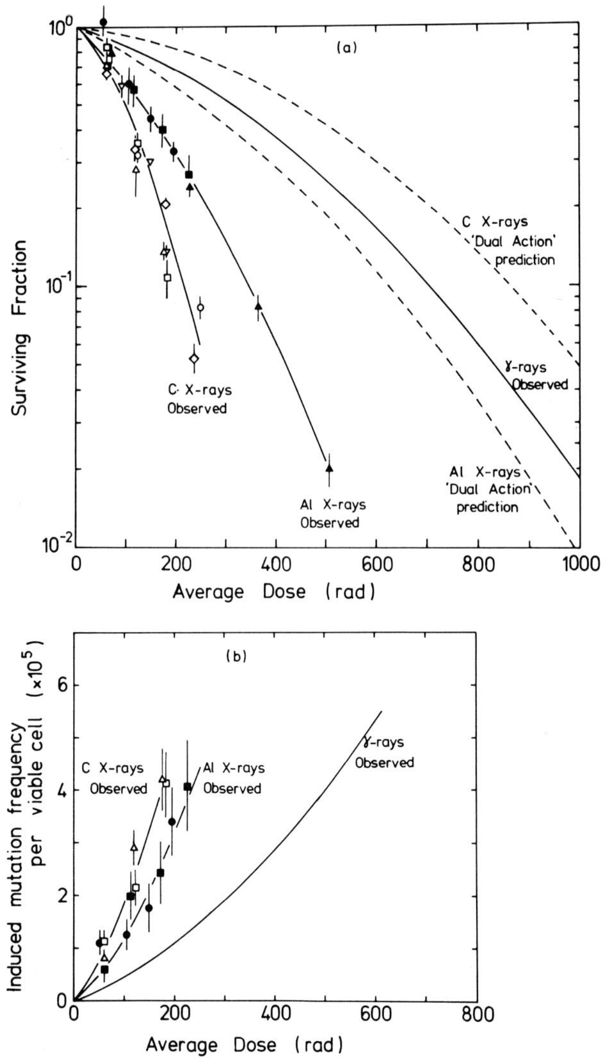

FIG. 5. Observed inactivation and mutation induction of V79 Chinese hamster cells after irradiation with 1.5 keV aluminum K ultrasoft X rays (solid symbols) (Cox et al. 1977b) and 0.3 keV carbon K ultrasoft X rays (open symbols) (Goodhead et al. 1979b) compared with the observed curves for γ-rays (Cox et al. 1977b). The two broken lines show the predicted curves for aluminum and carbon X rays according to the theory of dual radiation action (Kellerer and Rossi 1972). Details of the experimental methods, dosimetry, and average-dose calculations of the ultrasoft X rays are given by Goodhead and Thacker (1977), Cox et al. (1977b), and Goodhead et al. (1979b), and the predicted curves were calculated as described by Goodhead (1977) and Goodhead et al. (1979b).

over distances of about 6 nm, with the second track having a greatly enhanced effectiveness over the first (Leenhouts and Chadwick 1975). The probability of two tracks occurring this close together in a nucleus at a dose of a few hundred rad of carbon or aluminium X rays is $\ll 10^{-6}$, so there should be no curvature in the dose-response curve for these radiations. However, the observed responses (Figure 5) produce a curve similar to that of γ rays if the dose axes are scaled so that the curves correspond to one another.

The most direct explanation of the fact that the very short electron tracks from ultrasoft X rays are highly effective (Goodhead 1977, Goodhead et al. 1979b) is that the lesions themselves are determined directly by the energy deposited within single tracks, that is, within distances as small as $\lesssim$ 70 nm (deduced from aluminum X rays) or $\lesssim$ 7 nm (deduced from carbon X rays). Therefore, if sublesions are involved at all they must interact within $\lesssim$ 7 nm, and one-track action will be dominant for all conventional radiations.

Absence of a Shoulder

Further disagreement with the sublesion hypothesis arises from the fact that a shoulder on the dose-response curve for low-LET radiation is not a general characteristic of mammalian cells. For established cells, its presence varies through the cell cycle while for asynchronous early passage human diploid

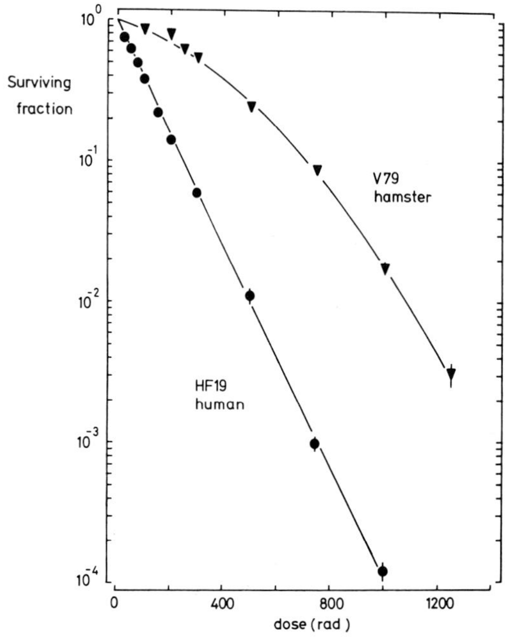

FIG. 6. Inactivation of HF19 human diploid fibroblasts after low-LET irradiation (250 kV X rays) compared with that of an established cell line (V79 Chinese hamster cell; γ rays) (Cox et al. 1977b).

fibroblasts there is no visible shoulder under conventional in vitro growth conditions (Figure 6) (Cox and Masson 1974, 1975, Weichselbaum et al. 1976, C. F. Arlett, personal communication). If the absence of a shoulder is explained by lack of susceptibility to interaction of sublethal damage, then these early passage cells should be more resistant than established cell lines to low-LET radiation; but they are observed not to be more resistant. If, instead, it is suggested that they are so sensitive that they are inactivated by what would be individual sublethal lesions for other cell types, then at higher LET there should be energy wastage and RBE values less than unity; in contradiction to this, the observed RBE values for human fibroblasts are substantially greater than unity and are similar to those of established cell lines (Cox et al. 1977a, Cox and Masson 1979, Goodhead et al. 1979a).

ALTERNATIVE HYPOTHESES

The above experimental evidence contradicts the hypothesis that the dominant mechanism of production of lesions is by the interaction of sublesions over large distances. Rejection of this hypothesis means rejection of one of the most common explanations for the curvature of low-LET dose-response curves and also for the shape of the relationship between RBE and LET. Are there reasonable alternative explanations? In principle, alternatives consistent with the above observations could be based upon the following hypotheses:

1. Lesions are produced by small amounts of energy in small distances so the dominant mode of production is by one-track action and the number of lesions produced is proportional to dose; the efficiency of production must depend on LET.
2. Some of these lesions can be acted on by a dose-dependent repair process whose efficiency decreases with increasing dose due to saturation.

Supporting Evidence

There is a variety of supporting evidence for such alternatives, including the following:

1. An approximate biophysical analysis of the variation of RBE for inactivation and for mutation with LET (Figure 7) suggests that there are two dominant types of lethal lesion and two dominant types of mutagenic lesion (Goodhead et al. 1979a). In this analysis one type of lethal and mutagenic lesion is produced efficiently at low LET by energy deposition of about 100 to 300 eV (3–9 ionizations) within a distance of $\sim$ 3 nm. The other type is produced efficiently at high LET and requires $\gtrsim$ 300 eV ($\gtrsim$ 10 ionizations) within $\sim$ 3 nm for a lethal lesion and $\gtrsim$ 600 eV for a mutagenic lesion. This analysis is based upon rather imprecise physical data of track structure at these small dimensions (Howard-Flanders 1958), but it is hoped that future computations will remove this

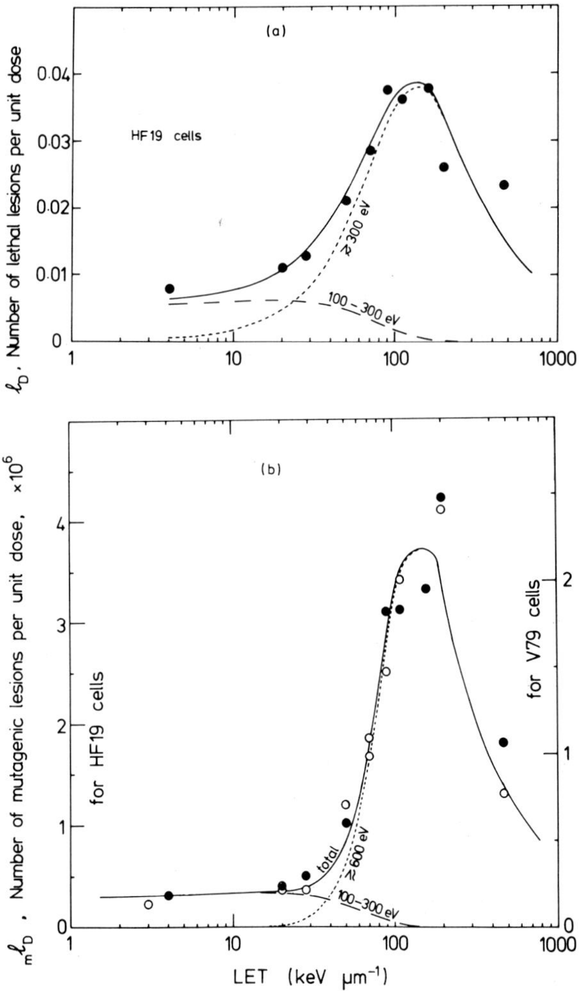

FIG. 7. Effectiveness of radiations of different LET (γ rays or 250-kV X rays and ^{4}He, ^{10}B and ^{14}N heavy ions) in producing (a) lethal (HF19 cells) and (b) mutagenic (HF19 and V79) lesions. Solid symbols are for HF19 human diploid fibroblasts (Cox and Masson 1979) and open symbols are for V79 hamster cells (Thacker et al. 1979). The solid lines show the total number of lesions calculated on a model that assumes lesions are produced by about 100 to 300 eV and $\gtrsim$ 300 eV (or $\gtrsim$ 600 eV in the case of mutation) in sites of about 3 nm (Goodhead et al. 1979a). The broken and dotted lines show the numbers of lesions of each type; the sum of them produces the solid lines. (The experimental and theoretical data have been expressed as numbers of lesions per cell per unit dose rather than as RBE because doing so takes into account correlation effects from small numbers of heavy ions, in particular the probability of a single track producing two or more lethal lesions in the same cell or of producing a mutagenic lesion that is not observable due to the same track independently killing the cell (Goodhead et al. 1979a). Where the dose responses are nonlinear, high-dose values have been used.)

imprecision and thereby enable support, modification, or rejection of these suggestions.

2. Other evidence also suggests that there are at least two different types of lethal lesion, one produced predominantly by low-LET and the other predominantly by high-LET radiation. It seems likely that the ultrasoft X rays produce low-LET type lesions (with great efficiency), as may be indicated by their inactivation dose-response curve (see Figure 5) for those cells for which a curved dose response is characteristic of low LET. This is supported by experiments with repair-deficient fibroblasts from patients with the radiation-sensitive syndrome ataxia telangiectasia. These cells show an approximately threefold increase in sensitivity (relative to fibroblasts from normal humans) for low-LET radiation (Cox et al. 1978), but preliminary experiments indicate that their increase in sensitivity is much smaller for high-LET damage (helium ions of 90 keV μm^{-1}) (Cox and Masson, personal communication). The observed increase in sensitivity for ultrasoft X rays (Figure 8) suggests that ultrasoft X rays produce low-, rather than high-, LET damage.

3. If the shoulder on the dose-response curve of established cell lines is due to the repair of lesions by a dose-dependent, saturable repair system, then the unshouldered response of freshly isolated human diploid fibroblasts could well be due to these cells having little or none of the repair system under conventional growth conditions. That they do not in all cases totally lack the system is implied by the fact that a shoulder can appear under various growth conditions when greater repair capacity may be available. A pronounced shoulder may appear when cell metabolism is altered after irradiation, either by holding in nutrient-deficient medium or by delaying the replating of plateau-phase cells

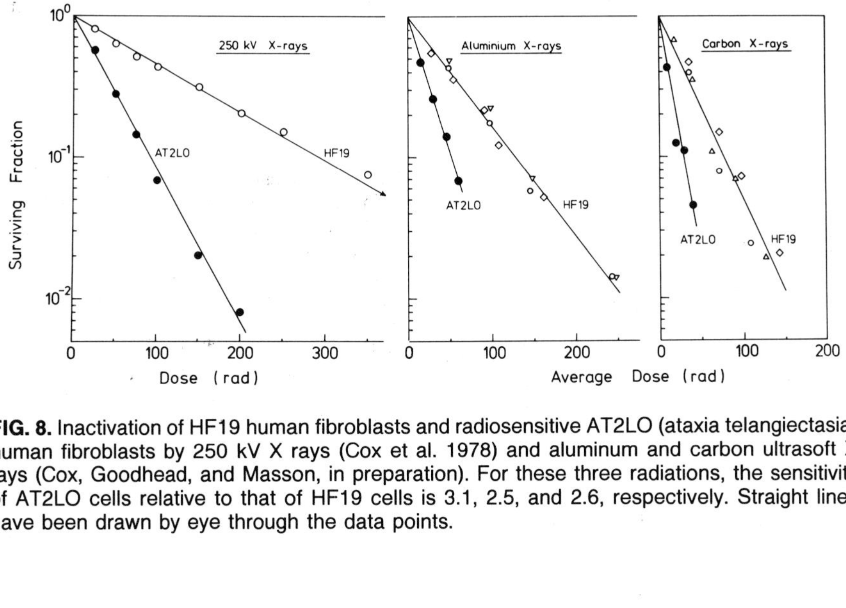

FIG. 8. Inactivation of HF19 human fibroblasts and radiosensitive AT2LO (ataxia telangiectasia) human fibroblasts by 250 kV X rays (Cox et al. 1978) and aluminum and carbon ultrasoft X rays (Cox, Goodhead, and Masson, in preparation). For these three radiations, the sensitivity of AT2LO cells relative to that of HF19 cells is 3.1, 2.5, and 2.6, respectively. Straight lines have been drawn by eye through the data points.

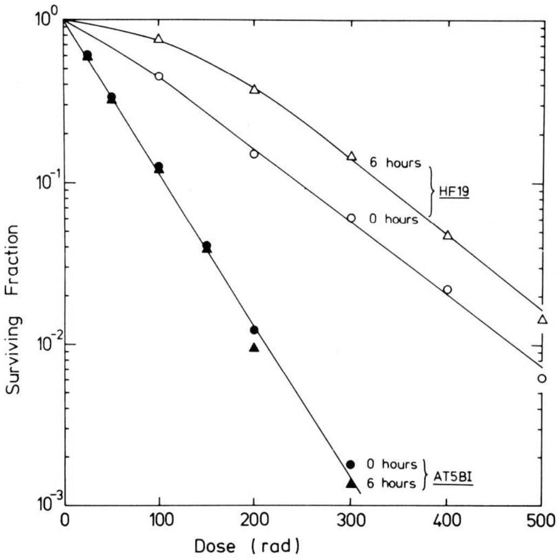

FIG. 9. Modification of survival curve of HF19 human fibroblasts by delaying replating after irradiation of plateau phase cells by zero hours (○) and 6 hours (△). Also shown are the results of a similar experiment with radiosensitive AT5BI ataxia telangiectasia human fibroblasts (Cox and Masson, personal communication). Lines have been drawn by eye.

(Figure 9) (Cox and Masson, personal communication). Slowly cycling senescent human cells can also show a small shoulder (Cox and Masson 1974).

4. Further association of the shoulder with repair processes is given by the repair-deficient ataxia telangiectasia fibroblasts. They show no evidence of a shoulder even under the above conditions of altered metabolism (Cox and Masson, personal communication) (Figure 9).

There are numerous possibilities whereby a repair function whose efficiency decreases with increasing dose could give shouldered dose-response curves. Some of the simplest of the many mathematical possibilities have been shown in Figure 1(b) and Table 1. For example, the saturable-repair model of equation vi assumes

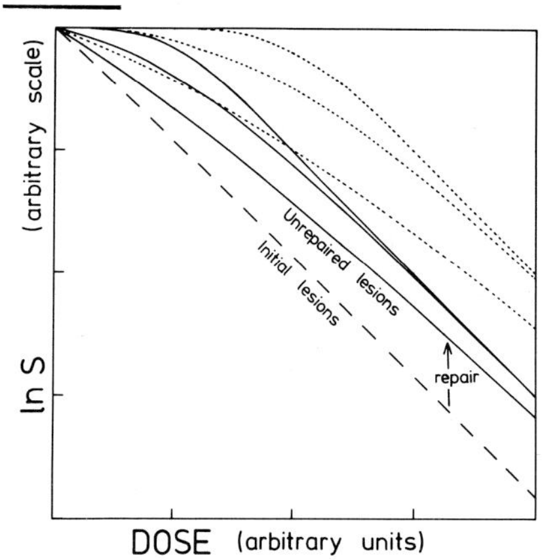

FIG. 10. A possible saturable-repair model of radiation action based on simple second-order kinetics (see text) (Goodhead et al. 1978). The straight diagonal broken line shows the numbers of lesions produced by the radiation, while the curved lines show the numbers that remain after repair. The two sets of curved lines (solid and dotted) correspond to different initial numbers of repair mediators, and the three lines in each set correspond to different total repair times (or reaction rates).

simply that the average number of lesions repaired in any small interval of time is proportional to the number of unrepaired lesions and the number of unused repair mediators (Goodhead et al. 1978). Figure 10 shows that a great variety of curve shapes can follow from such a hypothesis. The apparent goodness of fit of such an equation to biological data does not of itself verify the model since it is only one of many possibilities. Rather, it shows that this approach is feasible and worth further investigation.

Working Hypotheses—Small Lesions and Saturable-Repair Model

Table 2 summarizes the main experimental evidence considered above and its implications. Based on this and other evidence, we put forward the following working hypotheses (Goodhead et al. 1978).

1. Lethal and mutagenic lesions can be produced by
   (a) ~100 to 300 eV in ~3nm. This is the dominant mode of lesion production at low LET.
   (b) $\gtrsim$ 300 eV in ~3 nm (or $\gtrsim$600 eV for mutation). These are the dominant modes of lesion production at high LET.
2. Some (or all) of the lesions of type (a) above can be repaired by a repair process that becomes saturated with increasing dose.

We can speculate that the lesions are produced directly in DNA. In this case the estimated absolute probabilities per rad of occurrence of the above energy deposition events in the DNA of a single cell can be compared with the observed numbers of lethal lesions per rad with the following consequences. Only a very small proportion (possibly $\lesssim$1%) of energy events of type (a) above can lead to lethal lesions (i.e., the effects of the vast majority are either eliminated at all doses or are irrelevant to the survival of the cells). Some (or all) of the lesions produced are subject to the saturable repair process causing the shoulder. About 10% of the energy events of type (b) lead to lethal lesions (the effects of the remainder being eliminated at all doses or irrelevant) (Goodhead et al. 1979a). These lesions are not repairable by the saturable repair system.

We put forward the above working hypotheses to seek any critical experimental evidence against them or evidence that can discriminate for or against them relative to other hypotheses. There is also a need for experimental and theoretical evidence by which to limit the possible forms of the saturable repair function. It may well not follow the simple kinetics assumed by the repair models of Figure 1(b), in which case simple extrapolation to low doses cannot be made with confidence. On the one hand, if it could be shown that an appreciable proportion of the low-LET lesions are inherently unrepairable by this process, the extrapolation may be simplified. On the other hand, if it turned out that some of the repair capacity of the cell could be stimulated by small doses of radiation as is suggested by some recent experiments (D'Ambrosio and Setlow 1976, Sarasin and Hanawalt 1978, Leenhouts et al. 1978), the extrapolation

TABLE 2. *Some critical experimental evidence contradicting hypotheses of models*

Experimental observation	Effect studied (& cell type)	Reference	Contradicted hypothesis	Example of use	Implication
Mutation versus survival relationship	Inactivation & mutation (V79 & HF19)	1,8,9	Mutagenic & lethal lesions identical in all cases	Chadwick & Leenhouts (1976) (molecular theory of radiation action)	High LET produces relatively more mutagenic lesions.
Effectiveness of aluminum ultra-soft X-rays (1.5 keV)	Inactivation & mutation (V79 & HF19)	2,3,4	Long-range interaction or accumulation of sublesions Slow electrons damage 2-fold target	Kellerer & Rossi (1972) (theory of dual radiation action) Katz et al. (1971) (γ-kill) Burch (1970)	Most lesions are produced by one-track action of <1.5 keV within $\lesssim 70$ nm.
Effectiveness of carbon ultrasoft X rays (0.28 keV)	Inactivation & mutation (V79 & HF19)	4,5	All as for aluminum X rays above	All as for aluminum X rays above	All as above but <0.28 keV within $\lesssim 7$ nm.
Shoulder on ultra-soft X-ray dose-responses	Inactivation & mutation (V79)	2,3,5	Shoulder due to two-track interaction of sublesions over $\lesssim 6$ nm.	Leenhouts & Chadwick (1975)	Shoulder due to some other mechanism, e.g., dose-dependent repair of lesions.
Human fibroblasts: no shoulder at low LET, but RBE >1 at high LET	Inactivation & mutation (HF19)	2,3,6, 1,4,7,8	Shoulder on dose-response of other cell types due to two-track interaction of sublesions	Kellerer & Rossi (1972) (theory of dual radiation action) Katz et al. (1971) (γ-kill) Chadwick & Leenhouts (1973) (molecular theory of radiation action)	Even low-LET effect is by one-track action, & RBE increase with LET is due to track structure over nanometer distances.

References: 1. Goodhead et al. (1979a); 2. Cox et al. (1977b); 3. Goodhead (1977); 4. Goodhead et al. (1978); 5. Goodhead et al. (1979b); 6. Cox and Masson (1974); 7. Cox et al. (1977a); 8. Cox and Masson (1979); 9. Thacker et al. (1979).

would be very uncertain. In the latter case, RBE values at low doses are very difficult to predict unless the stimulation process is understood.

CONCLUSION

Experimental evidence from our laboratory has been used to discriminate between two broad alternative explanations of the cell-killing effect of radiation on mammalian cells, namely the production of sublesions that interact over large distances to produce the effect and the direct production of lesions some of which can be repaired by a saturable repair system. The evidence is contrary to the first explanation and is circumstantially in favor of the second. It is likely that similar conclusions apply to mutation induction in mammalian cells, although the mechanism is not always identical to that of inactivation. A set of working hypotheses has been put forward based on lesions being produced by small energy deposition within small distances, some of the lesions being repairable by the saturable repair system. It is hoped that these hypotheses will be tested against existing and future critical experimental evidence.

ACKNOWLEDGMENTS

I wish to thank my colleagues Dr. J. Thacker and Dr. R. Cox for experimental and theoretical collaboration in most of the work discussed in this paper.

REFERENCES

Alper, T. 1977. Elkind recovery and "sub-lethal damage": a misleading association. Br. J. Radiol. 50:459–467.

Barendsen, G. W. 1961. Damage to the reproductive capacity of human cells in tissue culture by ionizing radiations of different linear energy transfer, *in* The Initial Effects of Ionizing Radiations on Cells, R. J. C. Harris, ed. Academic Press, London, pp. 183–199.

Bender, M. A., and P. C. Gooch. 1962. The kinetics of X-ray survival of mammalian cells in vitro. Int. J. Radiat. Biol. 5:133–145.

Booz, J. 1975. Microdosimetric spectra and parameters of low LET radiations, *in* Proceedings of the Fifth Symposium on Microdosimetry, Verbania Pallanza, Italy, J. Booz, H. G. Ebert, and B. G. R. Smith, eds. Commission of the European Communities, EUR 5452, pp. 311–344.

Chadwick, K. H., and H. P. Leenhouts. 1973. A molecular theory of cell survival. Phys. Med. Biol. 18:78–87.

Chadwick, K. H., and H. P. Leenhouts. 1976. The correlation between mutation frequency and cell survival following different mutagenic treatments. Theor. Appl. Genet. 47:5–8.

Cox, R., G. P. Hosking, and J. Wilson. 1978. Ataxia telangiectasia: Evaluation of radiosensitivity in cultured skin fibroblasts as a diagnostic test. Arch. Dis. Child. 53:386–390.

Cox, R., and W. K. Masson. 1974. Changes in radiosensitivity during the in vitro growth of diploid human fibroblasts. Int. J. Radiat. Biol. 26:193–196.

Cox, R., and W. K. Masson. 1975. X-ray survival curves of cultured human diploid fibroblasts, *in* Proceedings of the Sixth L.H. Gray Conference, T. Alper, ed. Institute of Physics—John Wiley, London, pp. 217–222.

Cox, R., and W. K. Masson. 1976. X-ray induced mutation to 6-thioguanine resistance in cultured human diploid fibroblasts. Mutat. Res. 37:125–136.

Cox, R., and W. K. Masson. 1978. Do radiation-induced thioguanine resistant mutants of cultured mammalian cells arise from HGPRT gene mutation or X-chromosome rearrangement? Nature 276:629–630.

Cox, R., and W. K. Masson. 1979. Mutation and inactivation of cultured mammalian cells exposed to beams of accelerated heavy ions. III. Human diploid fibroblasts. Int. J. Radiat. Biol. (in press).

Cox, R., J. Thacker, D. T. Goodhead, and R. J. Munson. 1977a. Mutations and inactivation of mammalian cells by various ionising radiations. Nature 267:425–427.

Cox, R., J. Thacker, and D. T. Goodhead. 1977b. Inactivation and mutation of cultured mammalian cells by aluminium characteristic ultrasoft X-rays. II. Dose response of Chinese hamster and human diploid cells to aluminium X-rays and radiations of different LET. Int. J. Radiat. Biol. 31:561–576.

D'Ambrosio, S. M., and R. B. Setlow. 1976. Enhancement of postreplication repair in Chinese hamster cells. Proc. Natl. Acad. Sci. USA 73:2396–2400.

Goodhead, D. T. 1977. Inactivation and mutation of cultured mammalian cells by aluminium characteristic ultrasoft X-rays. III. Implications for theory of dual radiation action. Int. J. Radiat. Biol. 32:43–70.

Goodhead, D. T., R. J. Munson, J. Thacker, and R. Cox. 1979a. Mutation and inactivation of cultured mammalian cells exposed to beams of accelerated heavy ions. IV. Biophysical interpretation. Int. J. Radiat. Biol. (in press).

Goodhead, D. T., and J. Thacker. 1977. Inactivation and mutation of cultured mammalian cells by aluminium characteristic ultrasoft X-rays. I. Properties of aluminium X-rays and preliminary experiments with Chinese hamster cells. Int. J. Radiat. Biol. 31:541–559.

Goodhead, D. T., J. Thacker, and R. Cox. 1978. The conflict between the biological effects of ultrasoft X-rays and microdosimetric measurements and application, *in* Proceedings of the Sixth Symposium on Microdosimetry, Brussels, J. Booz and H. G. Ebert, eds. Commission of the European Communities, London, EUR 6064, pp. 829–843.

Goodhead, D. T., J. Thacker, and R. Cox. 1979b. Effectiveness of 0.3 keV carbon ultrasoft X-rays for the inactivation and mutation of cultured mammalian cells. Int. J. Radiat. Biol. (in press).

Green, A. E. S., and J. Burki. 1974. A note on survival curves with shoulders. Radiat. Res. 60:536–540.

Howard-Flanders, P. 1958. Physical and chemical mechanisms in the injury of cells by ionizing radiations. Adv. Biol. Med. Phys. 6:553–603.

Katz, R., B. Ackerson, M. Homayoonfar, and S. C. Sharma. 1971. Inactivation of cells by heavy ion bombardment. Radiat. Res. 47:402–425.

Kellerer, A. M., and H. H. Rossi. 1972. The theory of dual radiation action. Curr. Top. Radiat. Res. 8:85–158.

Kellerer, A. M., and H. H. Rossi. 1978. A generalized formulation of dual radiation action. Radiat. Res. 75:471–488.

Laurie, J., J. S. Orr, and C. J. Foster. 1972. Repair processes and cell survival. Br. J. Radiol. 45:362–368.

Leenhouts, H. P., and K. H. Chadwick. 1975. Stopping power and the radiobiological effect of electrons, gamma rays and ions, *in* Proceedings of the Fifth Symposium on Microdosimetry, Verbania Pallanza, Italy, J. Booz, H. G. Ebert, and B. G. R. Smith, eds. Commission of the European Communities, EUR 5452, pp. 289–308.

Leenhouts, H. P., and K. H. Chadwick. 1978. The crucial role of DNA double-strand breaks in cellular radiobiological effects. Adv. Radiat. Biol. 7:55–101.

Leenhouts, H. P., K. H. Chadwick, and M. J. Sijsma. 1978. The influence of radiation stimulated repair processes on the shape of dose effect curves, *in* Proceedings of the Sixth Symposium on Microdosimetry, Brussels, J. Booz and H. G. Ebert, eds. Commission of the European Communities, London, EUR 6064, pp. 1011–1021.

Munson, R. J., and D. T. Goodhead. 1977. The relation between mutation frequency and cell survival—a theoretical approach and an examination of experimental data for eukaryotes. Mutat. Res. 42:145–160.

Sarasin, A. R., and P. C. Hanawalt. 1978. Carcinogens enhance survival of UV-irradiated simian virus 40 in treated monkey kidney cells: Induction of a recovery pathway? Proc. Natl. Acad. Sci. USA 75:346–350.

Thacker, J., and R. Cox. 1975. Mutation induction and inactivation in mammalian cells exposed to ionising radiation. Nature 258:429–431.

Thacker, J., M. A. Stephens, and A. Stretch. 1978. Mutation to ouabain-resistance in Chinese hamster cells: Induction by ethyl methanesulphonate and lack of induction by ionising radiation. Mutat. Res. 51:255–270.

Thacker, J., A. Stretch, and M. A. Stephens. 1977. The induction of thioguanine-resistant mutants of Chinese hamster cells by γ-rays. Mutat. Res. 42:313–326.

Thacker, J., A. Stretch, and M. A. Stephens. 1979. Inactivation and mutation of cultured mammalian cells exposed to beams of accelerated ions. II. Chinese hamster cells. Int. J. Radiat. Biol. (in press).

Weichselbaum, R. R., J. Epstein, J. B. Little, and P. L. Kornblith. 1976. In vitro cellular radiosensitivity of human malignant tumors. Eur. J. Cancer 12:47–51.

Wideröe, R. 1978. A comparison of radiation effects on mammalian cells in vitro caused by X-rays, high energy neutrons and negative pions. Radiat. Environ. Biophys. 15:57–75.

Responses to Low Doses and Low Dose Rates

Radiation Biology in Cancer Research, edited by
Raymond E. Meyn and H. Rodney Withers.
Raven Press, New York © 1980.

Variations in Responses of Several Mammalian Cell Lines to Low Dose-Rate Irradiation

Joel S. Bedford, James B. Mitchell, and Michael H. Fox

Department of Radiology and Radiation Biology, Colorado State University, Fort Collins, Colorado 80523

The fundamental problem of cancer therapy is to produce (or increase) a differential lethal effect in tumors relative to damage produced in normal tissues. Studies in radiation biology have suggested a number of promising approaches, many of which are being tried in the clinic. However, there is one firmly established approach, presumably rooted in fundamental biological differences between tumors and normal tissues, which clearly influences the efficiency of tumor sterilization within the limits of normal tissue tolerance. That factor is the *manner in which the dose is administered in time.* Spreading the dose over a period of time generally results in greater damage to tumors at the level of normal tissue tolerance. A number of possible explanations have been offered for the time-dose effect on the therapeutic ratio. Among these are tumor reoxygenation and various differences in the kinetics of cell population growth in tumor and surrounding normal tissues. Mean generation times of one to five days for cycling cells in some human tumors make cell cycle redistribution during treatment a factor of potential importance. In an attempt to unravel one aspect of time-dose relationships as they influence cellular radiosensitivity, we have studied the effect of dose rate on the survival and life cycle of a number of different cell lines.

RESULTS AND DISCUSSION

Survival and Growth in Randomly Dividing Log-Phase Cultures

For single acute X- or gamma-ray exposures, differences in survival curves for various mammalian cell lines are relatively small, but these differences can become appreciable for irradiations at lower dose rates. Figure 1 shows a summary of dose-survival curves obtained for log-phase cultures of six different cell lines irradiated at a dose rate of 142 rad per minute, 154 rad per hour, or 37 rad per hour. The principal reason for the increasing difference in response with decreasing dose rates is that radiation-induced cell cycle perturbations are very different for the different cell lines.

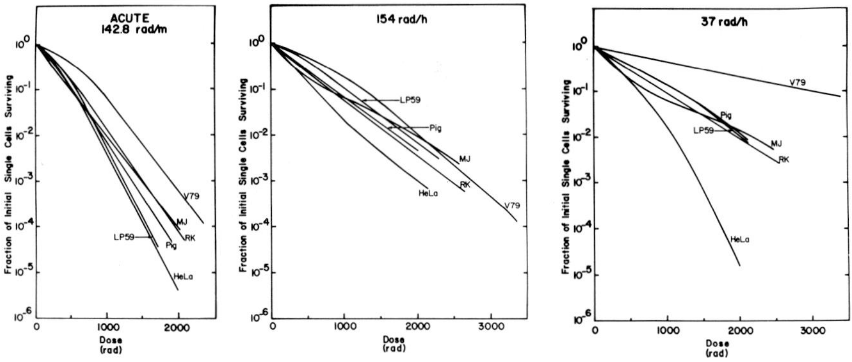

FIG. 1. A comparison of dose-survival curves for log-phase cultures of six different cell lines irradiated at different dose rates. Left panel, acute, 142.8 rad per minute. Middle panel, 154 rad per hour. Right panel, 37 rad per hour. The cell lines were Chinese hamster (V79), Indian muntjac (MJ), rat kangaroo (RK), PK15 pig kidney (Pig), mouse L-P59 (LP59), and human S3 HeLa (HeLa) (data replotted from Mitchell et al. 1979a).

The radiation exposures were started after the cells were in log-phase growth. The same multiplicity correction, using the average number of cells per colony at the start of the exposures, was applied to all the survival estimates. Survival, therefore, refers to the number of colonies eventually formed relative to the number of single cells present at the start of the exposures. In cases where appreciable cell division occurred during irradiation, as for V79 cells irradiated at 37 rad per hour, the results can be expressed several different ways. For example, it may be of greater interest to know the number of viable cells present at the end of an exposure relative to the number of viable cells at the start of the exposure. In the case of V79 cells irradiated at 37 rad per hour these survivals would be expected to differ by about 15% for a surviving fraction of 0.37 and by 6% for a surviving fraction of 0.10, by comparison with the survivals we have plotted. When extensive cell division occurs during exposure, the actual dose received by a cell is also subject to question, since several cell generations may have elapsed and cells present at the end of an exposure did not even exist as such when the exposure began.

An extreme example of the differences in cell population growth for two cell lines during irradiation at different dose rates is illustrated in Figure 2. A sevenfold difference in dose rate is required to produce approximately the same effect on cell population growth for the V79 cells in comparison with HeLa cells. Proliferation of V79 cells is not affected for at least three generations at a dose rate of 37 rad per hour, whereas HeLa cells do not proliferate at all at this dose rate.

A number of different properties of various cell lines were examined in an effort to determine whether a correlation existed between any of these and the critical dose rate necessary to stop cell population growth. Table 1 summarizes

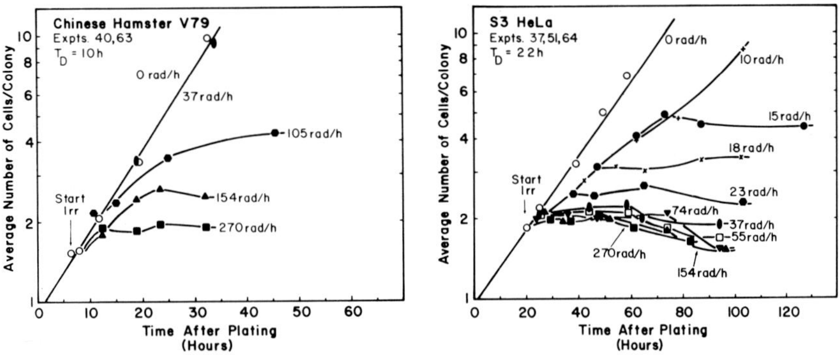

FIG. 2. The cell population growth of randomly dividing Chinese hamster V79 cells (left panel) and S3 HeLa cells (right panel) irradiated continuously at various dose rates. (Reproduced from Mitchell et al. 1979a, with permission of Academic Press.)

the doubling times, acute dose-survival curve parameters, acute dose division delays, and the dose rates necessary to stop cell population growth for seven cell lines. It appears that there is little or no correlation between the dose rate necessary to stop cell population growth and the doubling time for unirradiated cultures or the acute dose-survival curve parameters for these cell lines. Of course, this does not deny the possibility that the dose rate necessary to stop cell population growth may correlate with generation time for a given cell line under various conditions that alter proliferation rates. For these different cell lines, however, a rough correlation was apparent between the so-called critical dose rate and the acute dose division delay.

Radiation Effects on the Cell Cycle

Since progress through G_2 is particularly radiosensitive, dose rates that affect cell proliferation would be expected to result in a change in the life cycle distribution and the radiosensitivity of a cell population during exposure. Such changes have been observed during continuous irradiation (Bedford 1966, Bedford and Mitchell 1973, Kal et al. 1975, Szechter et al. 1978). Figure 3 shows the effect of continuous irradiation at a dose rate of 37 rad per hour on the life cycle of synchronized HeLa and V79 cells. Mitotic cells were obtained by mechanical harvest, and the average number of cells per colony, the tritiated thymidine flash labeling index, and the mitotic index were followed in each case. As expected from the cell population growth, V79 cells were virtually unaffected in their life cycle progress at this dose rate. By contrast, HeLa cells progressed through G_1 and S phases without appreciable delay but were temporarily arrested in G_2. The cells eventually entered a prolonged mitosis that few completed. Though the average number of cells per colony seems to indicate no cell division, colony size distributions and time-lapse cinemicrography have shown that as many as 15% to 20% of the cells may divide at this dose rate (Bedford and Mitchell

TABLE 1. *Properties of irradiated cell lines*

Cell Line	Doubling Time in F12 FC(10) (hrs at 37°C)	Acute Dose-Survival Curve		Estimated Minimum Dose Rate That:		Acute Dose Division Delay (min/rad)
		n	D_0	Immediately Stops Cell Pop. Growth	Allows < One Pop. Doubling	
S3 HeLa	22	2	175	30	18	2.04
Mouse LP59	22	6	141	74	34	1.98
Indian muntjac	28	1.5	215	28	12	9.12
Chinese hamster V79	10	6	215	~300	154	0.55
Mouse lymphoma L5178YS	10	1.5	50	4.8*	—	6 to 8†‡
Rat kangaroo	31	2.3	191	74	37	1.80
Pig kidney	26	2	200	270	74	0.76

* V. D. Courtenay 1969.
† W. L. Caldwell (personal communication).
‡ Ehmann et al. 1975.

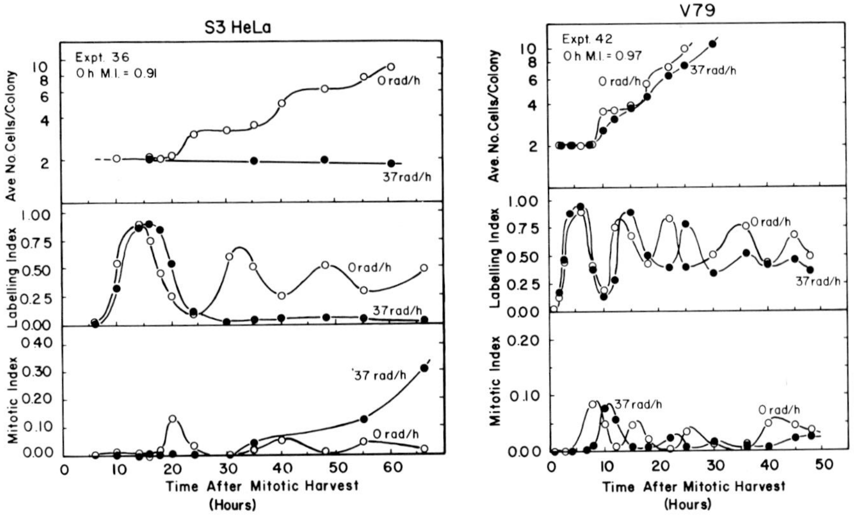

FIG. 3. The progress of synchronized S3 HeLa cells (left panel) and Chinese hamster V79 cells (right panel) through their life cycle during irradiation at 0 or 37 rad per hour. (Reproduced from Mitchell et al. 1979b, with permission of Academic Press.)

1977). A sevenfold increase in dose rate (270 rad per hour) is required to produce a similar effect on the life cycle of V79 cells.

In a previous paper (Bedford and Mitchell 1973) we examined the effect of continuous irradiation at 90 rad per hour on the life cycle of V79 cells, and, judging from the increase in average number of cells per colony, a severe inhibition of cell division occurred. Subsequently, we have carried out more detailed experiments over a wider range of dose rates and found that this higher dose rate of 270 rad per hour is required to produce an inhibition of cell division equivalent to that which we observe for HeLa cells irradiated at 37 rad per hour (Mitchell et al. 1979b). Even at this higher dose rate, however, V79 cells do not accumulate in mitosis to the same extent.

The life cycle of HeLa cells during continuous irradiation has also been examined in detail by flow microfluorometry (FMF). Cells initially synchronized in mitosis were irradiated continuously at dose rates of 0, 15, 34, and 74 rad per hour, and samples were taken at various intervals to determine the distribution of DNA content per cell following fixation and staining with chromomycin A_3. The FMF profiles are shown in Figure 4. For all these dose rates, cells progress through G_1 and S without appreciable delay, but a dramatic arrest in $G_2 + M$ is apparent for the 34 and 74 rad per hour irradiations. Virtually no cells divide at 74 rad per hour, but some cells succeed and reenter G_1 at 34 rad per hour. The relative number of cells in the G_1 peak for the later sampling times in the 34 rad per hour series indicates that approximately 17% of the cells were able to complete their first mitosis.

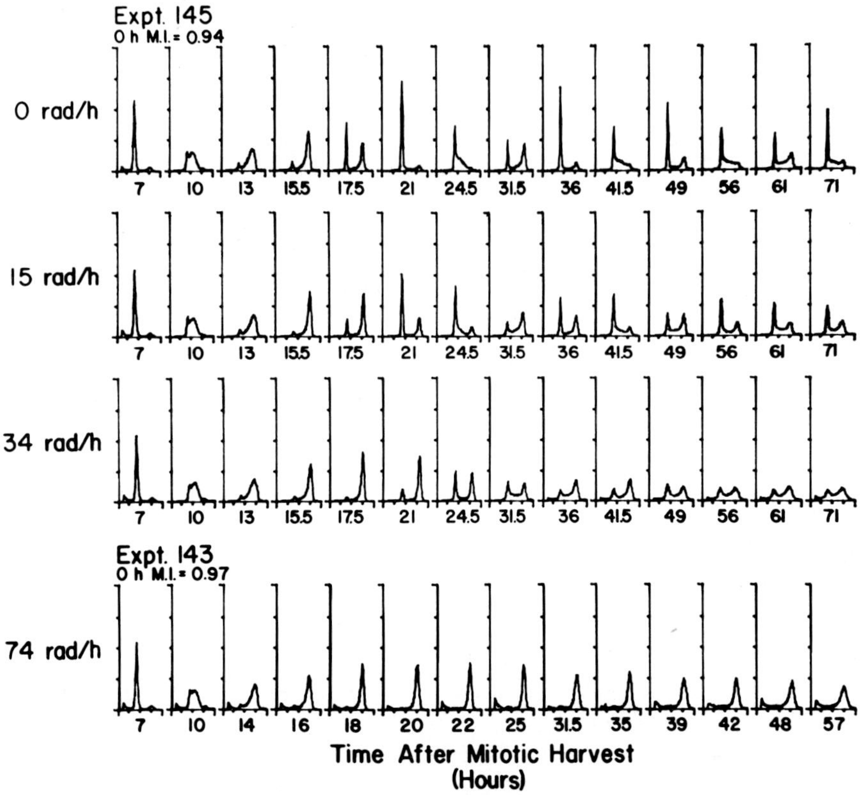

FIG. 4. The progress of synchronized S3 HeLa cells through their life cycle during irradiation, studied by flow microfluorometry (FMF). Each panel shows the distribution of cells in the population with respect to DNA content. The peaks to the left and right in each panel correspond to a G_1 and G_2 (or M) DNA content, respectively. The numbers beneath each panel refer to the time in hours after mitotic harvest when the samples were fixed for analysis. In each panel, the ordinate at full scale represents 4,000 cells and the abscissa at full scale, 100 channels.

As a result of cell cycle redistribution and division during exposure, cell survival at low dose rates and long exposure times are quite different from what would be expected if sublethal damage repair were the only factor determining the dose-rate effect. In fact, in some cases a reduction in dose rate can actually result in an increased rate of cell killing per unit dose. This is illustrated in Figure 5 for HeLa cells. With respect to the acute dose-survival curve, there is a clear loss in effect per unit dose for irradiation at 154 rad per hour, but further reductions in dose rate to 74, 55, and 37 rad per hour result in a progressive increase in effect per unit dose at the higher dose levels. Very little cell division occurs during exposure at these dose rates, but at 15 and 10 rad per hour G_2 progress is not severely delayed and appreciable cell division occurs. Thus, there is a further loss in effect per unit dose.

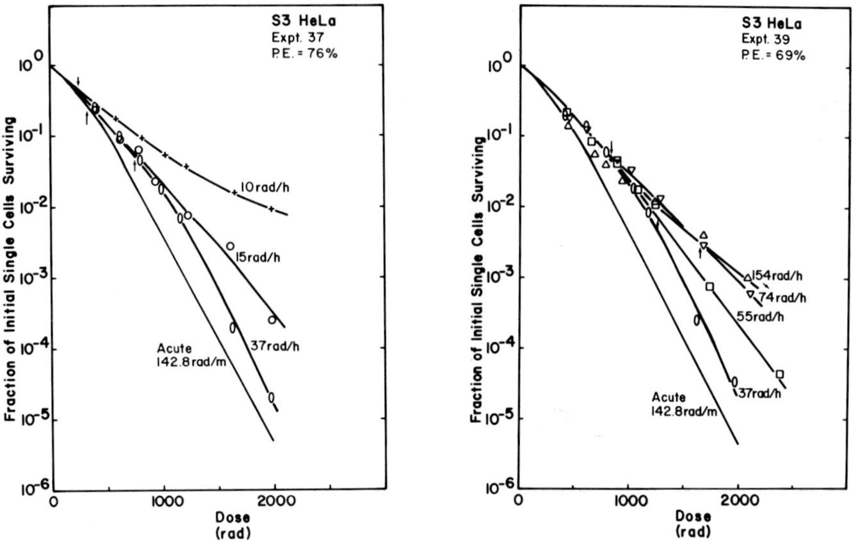

FIG. 5. Survival curves for log-phase S3 HeLa cells irradiated at different dose rates (cesium-137 gamma rays). The right panel illustrates the survival curves for cases in which cell division was not appreciable during exposures. The left panel illustrates survival curves for which, in some cases (10 rad per hour and 15 rad per hour), appreciable cell division did occur during exposures. Vertical arrows on the curves indicate the accumulated dose for each dose rate after a time equivalent to one generation time for unirradiated cultures. (Reproduced from Mitchell et al. 1979a, with permission of Academic Press.)

The inflection in effect as dose rates were reduced was most pronounced for HeLa cells, but it also occurred for other cell lines. This is illustrated in Figure 6, in which the total dose necessary to reduce the surviving fraction to 10^{-3} is plotted against the radiation dose rate for HeLa, V79, pig kidney, and mouse L-P59 cells. The pronounced inflection for HeLa cells is apparent, as is the lack of an inflection for the V79 cells. For the pig kidney and mouse L-P59 cells, there is very little change in the total dose necessary to produce this level of sterilization over the dose-rate range between about 50 and 150 rad per hour.

On the basis of these studies, we have concluded that dose-rate effects on cell killing reflect not only sublethal damage repair during irradiation, but also a gradual accumulation of cells into a radiosensitive G_2 phase for dose rates sufficient to prevent cell proliferation. It has been demonstrated by Griffiths and Tolmach (1976) that early to mid-G_2 HeLa cells, on the S phase side of the X-ray–induced G_2 arrest point, are as resistant as late S phase cells. This observation has been confirmed by Tomasovic, Spiro, and Dewey (personal communication 1978) using Chinese hamster ovary cells. Thus it might be argued that a gradual accumulation of cells in G_2 should result in the cell population becoming more resistant, instead of more sensitive as we have observed. However,

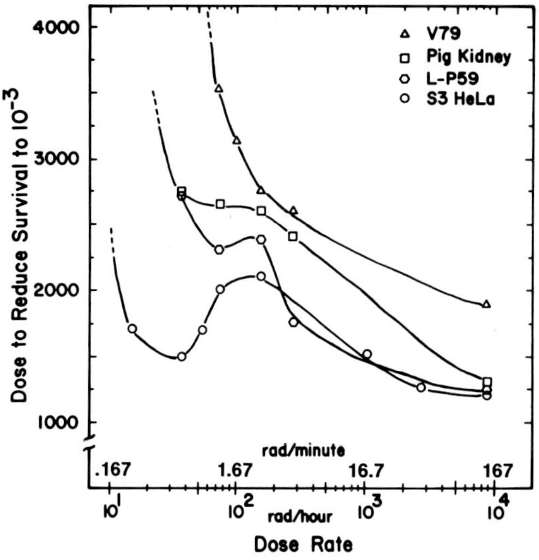

FIG. 6. The total dose necessary to reduce the fraction of cells surviving to 10^{-3} of the initial value as a function of the dose rates for log-phase V79 Chinese hamster, pig kidney, mouse L-P59, and S3 HeLa cells.

these observations are not as inconsistent as they seem. The experiments of Griffiths and Tolmach and of Tomasovic, Spiro, and Dewey indicate that early G_2 cells are resistant and undergo a very sharp transition to "mitotic-like" radiosensitivity as they enter or pass the X-ray–induced G_2 arrest point. The radiosensitivity at the G_2 arrest point is not known precisely. Also, these experiments do not refer to cells irradiated while suffering a G_2 arrest as would be the case for our continuous irradiation experiments. Further, Kal et al. (1975) and Szechter et al. (1978) have demonstrated that cells irradiated continuously at low dose rate show an increased sensitivity in their subsequent acute-dose response.

From the cell cycle analyses presented it is, of course, impossible to know whether cells that survive doses corresponding to surviving fractions less than about 10^{-2} derive from G_2-arrested cells or from a very small fraction that may be in other phases of the cell cycle. If a small proportion of slowly cycling cells or cells capable of division at 37 rad per hour were responsible for most or all of the colony-forming cells after the prolonged irradiations, then incubation with high specific activity ^{3}HTdR during irradiation should kill a significant proportion of these survivors. On the other hand, if the colonies arise from cells blocked in G_2 the addition of ^{3}HTdR should have little effect on the fraction of cells surviving. Several experiments of this nature (which will be reported in detail elsewhere) were carried out to determine the cell cycle origin of cells that survive exposures longer than one generation time at a dose rate of 34 rad per hour.

In one experiment, HeLa cells were synchronized in mitosis and two identical sets of cultures were irradiated continuously with ^{137}Cs gamma rays at 34 rad per hour for 45 hours. To one set, high specific activity ^{3}HTdR (5 μCi/

ml, 50 Ci/mmol) was added after 24 hours of exposure, by which time most of the cells would have progressed to G_2 and would be experiencing a G_2 arrest. No ^{3}HTdR was added to the other set. After continuing the gamma-ray exposure to 45 hours both sets of cultures were removed and rinsed three times with medium containing 5×10^{-5} M nonradioactive thymidine, fresh standard medium was added, and the cultures were incubated for colony formation. An average of 159 colonies per culture (surviving fraction $= 9.35 \times 10^{-4}$) were observed on the series to which no ^{3}HTdR was added, but there were no colonies on any of the plates that received ^{3}HTdR. From this, we have concluded that unless continuously irradiated G_2-blocked cells incorporated significant amounts of the ^{3}HTdR, cells that survive after long exposure at this dose rate must derive from a small proportion of cycling cells. This observation, however, does not contradict our conclusion that the increased rate of cell killing for longer exposure times at this dose rate results from an arrest of cells at a radiosensitive stage of the cell cycle.

Dose-Rate Effects in Plateau-Phase Cultures

Since the dose-rate effect is influenced by cell cycle redistribution and cell division in log-phase cultures, it seemed of interest to examine dose-rate effects in plateau-phase cultures in which cell cycle times are greatly lengthened and in which there is little or no net cell population growth. We therefore determined cell survival curves for fed and unfed plateau-phase cultures of the HeLa and V79 cells. The medium for fed cultures was changed daily before and during irradiation. The cell density in these cultures reached a steady-state value of about 4.8×10^5 and 1.0×10^6 cells per cm^2 for the HeLa and V79 cells, respectively. For the unfed plateau phase cultures, these values were approximately 1.4×10^5 cells per cm^2 for HeLa cultures and 1.8×10^5 cells per cm^2 for the V79 cultures. Average cell cycle transit times were estimated from percent labeled mitoses curves. For HeLa cultures, these times increased from approximately 22 hours to 25 hours for log-phase cultures, to 44 hours for unfed plateau cultures, and to 66 hours for fed plateau cultures. For the V79 cells, these values were estimated to be 9 hours, 22 hours, and 52 hours, respectively. For radiation exposures, there was 4 ml of medium in the T25 flasks containing the cultures. The survival for fed plateau-phase V79 cells following an acute dose of 2,000 rad was 0.0022, 0.0026, and 0.0035 for flasks containing 2, 4, or 8 ml of medium, respectively. Thus, with the 4-ml overlay of medium used in these experiments, respiration-induced hypoxia did not appear to be a problem. After each irradiation, the cultures were incubated for four hours at 37°C to allow for maximum repair of potentially lethal damage at each dose level, and the cultures were then trypsinized, counted, and plated to determine the colony-forming ability of single cells.

Figure 7 is a summary plot of the survival curves obtained for fed plateau-phase cultures of HeLa cells and V79 cells irradiated at 143 rad per minute,

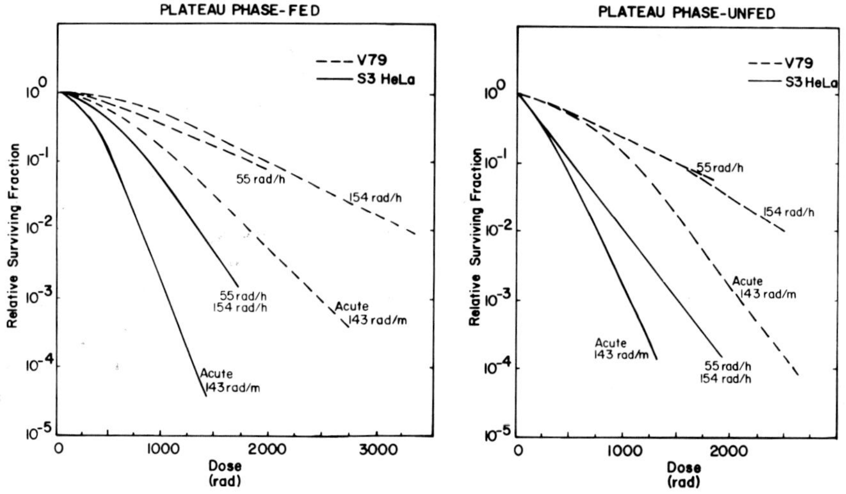

FIG. 7. A summary of survival curves at different dose rates for fed (left panel) and unfed (right panel) plateau phase cultures. The dashed lines refer to results for V79 Chinese hamster cells and the solid lines, for HeLa cells. (Redrawn from Mitchell et al. 1979c.)

154 rad per hour, and 55 rad per hour. The right panel illustrates the result for unfed plateau-phase cultures. Two features of these results are quite striking. First, there is a clear decrease in the rate of cell killing when the dose rate is reduced from 143 rad per minute to 154 rad per hour, but no further dose-rate effect was observed for a further reduction in dose rate to 55 rad per hour. Since the effect of cell cycle redistribution is either eliminated or plays a very minor role in determining the response, all of the sublethal damage that is capable of being repaired must be repaired during irradiation at dose rates less than 154 rad per hour under these conditions.

Appreciable dose-rate effects have been observed below 154 rad per hour when cell proliferation rates were greatly reduced or stopped by reducing the temperature (Szechter and Schwarz 1977). However, these authors, and others, have also shown that a reduction in temperature reduces the rate of repair of sublethal damage (Szechter and Schwarz 1977, Elkind et al. 1965). If repair rates are reduced, the lower limit to the dose-rate effect should also be reduced.

The second important feature of these results is that reducing the dose rate does not result in a magnification of the difference in responses for HeLa and V79 cells. Reducing the dose rate reduces the effect per unit dose to about the same extent for both cell lines. This contrasts with the result shown in Figure 1 for which log-phase cultures were used.

The results we have obtained using both log- and plateau-phase cells in culture tend to show very little loss in effect when dose rates are reduced through the range between about 150 and 50 rad per hour. This result is of interest in terms of the in vivo low dose-rate studies of Kal and Barendsen (1972), Hill and Bush (1973), Fu et al. (1975), and Tubiana and Boisgerie (1978). These workers have reported various cases in which little or no reduction in effect was observed when dose rates were lowered in this range. In some cases, an increase in effectiveness was observed for lower dose rates.

Several recent clinical reports indicate that superfractionation or low dose-rate teletherapy may yield results superior to those obtained by conventional dose fractionation (Suit 1977, Pierquin et al. 1978). Many radiotherapists claim that low dose-rate brachytherapy also yields superior results. Cell cycle redistribution may play an important role in such treatment modalities, in addition to tumor reoxygenation during treatment (Hill and Bush 1973) and the effective reduction in oxygen enhancement ratio that accompanies low dose-rate irradiation (e.g., Neary 1955, Hall et al. 1966, Bedford and Hall 1966, Hall and Fairchild 1970, Hall et al. 1974).

SUMMARY

As a result of these studies, we have concluded that:

1. Appreciable differences can exist from one cell line to another in response to low dose-rate irradiation.
2. The differences in response are governed to a large extent by differences in cell cycle perturbations during irradiation.
3. The differences depend perhaps as much on dose rate as they do on total dose.
4. Dose-rate effects differ, and they depend on whether the cell population contains a large fraction of rapidly cycling cells or a more inactive and slowly proliferating one.
5. These factors may be important for radiotherapy.

ACKNOWLEDGMENTS

This investigation was supported by Grant Numbers CA18023 and CA09236 awarded by the National Cancer Institute, Department of Health, Education and Welfare.

REFERENCES

Bedford, J. S. 1966. The influence of oxygen and dose-rate on the survival of cultured mammalian cells exposed to ionizing radiation. D. Phil. Thesis, Oxford University.

Bedford, J. S., and E. J. Hall. 1966. Threshold hypoxia: Its effect on the survival of mammalian cells irradiated at high and low dose-rates. Br. J. Radiol. 39:896–900.

Bedford, J. S., and J. B. Mitchell. 1973. Dose rate effects in synchronous mammalian cells in culture. Radiat. Res. 54:316–327.

Bedford, J. S., and J. B. Mitchell. 1977. Mitotic accumulation of HeLa cells during continuous irradiation: Observations using time-lapse cinemicrography. Radiat. Res. 70:173–186.

Courtenay, V. D. 1969. Radioresistant mutants of L5178Y cells. Radiat. Res. 38:186–203.

Ehmann, U. K., H. Nagasawa, D. F. Petersen, and J. T. Lett. 1974. Symptoms of x-ray damage to radiosensitive mouse leukemic cells: Asynchronous population. Radiat. Res. 60:453–472.

Elkind, M. M., H. Sutton-Gilbert, W. B. Moses, T. Alescio, and R. W. Swain. 1965. Radiation response of mammalian cells grown in culture. V. Temperature dependence of the repair of x-ray damage in surviving cells (aerobic and hypoxic). Radiat. Res. 25:359–376.

Fu, K. K., T. L. Phillips, L. J. Kane, and V. Smith. 1975. Tumor and normal tissue response to irradiation in vivo: Variation with decreasing dose rates. Radiology 114:709–716.

Griffiths, T. D., and L. J. Tolmach. 1976. Lethal response of HeLa cells to x-irradiation in the latter part of the generation cycle. Biophys. J. 16:303–318.

Hall, E. J., J. S. Bedford, and R. Oliver. 1966. Extreme hypoxia; its effect on the survival of mammalian cells irradiated at high and low dose-rates. Br. J. Radiol. 39:302–307.

Hall, E. J., and R. G. Fairchild. 1970. Radiobiological measurements with Californium-252. Br. J. Radiol. 43:263–266.

Hall, E. J., L. A. Roizin-Towle, and R. D. Colvett. 1974. RBE and OER determinations for radium and Californium-252. Radiology 110:699–704.

Hill, R. P., and R. S. Bush. 1973. The effect of continuous or fractionated irradiation on a murine sarcoma. Br. J. Radiol. 46:167–174.

Kal, H. B., and G. W. Barcndscn. 1972. Effccts of continuous irradiation at low dose rates on a rat rhabdomyosarcoma. Br. J. Radiol. 45:279–283.

Kal, H. B., G. W. Barendsen, R. Bakker-Van Homne, and H. Roelse. 1975. Increased radiosensitivity of rat rhabdomyosarcoma cells induced by protracted irradiation. Radiat. Res. 63:521–530.

Mitchell, J. B., J. S. Bedford, and S. M. Bailey. 1979a. Dose-rate effects in mammalian cells in culture. III. Comparison of cell killing and cell proliferation during continuous irradiation for six different cell lines. Radiat. Res. (in press).

Mitchell, J. B., J. S. Bedford, and S. M. Bailey. 1979b. Dose-rate effects on the life cycle and survival of S3 HeLa and V79 cells. Radiat. Res. (in press).

Mitchell, J. B., J. S. Bedford, and S. M. Bailey. 1979c. Dose-rate effects in plateau-phase cultures of S3 HeLa and V79 cells. Radiat. Res. (in press).

Neary, G. J. 1955. The dependence of the oxygen effect on the intensity of gamma irradiation in *Vicia faba, in* Progress in Radiobiology (Proceedings of the IV International Congress of Radiobiology, 1955). Oliver and Boyd, Edinburgh, pp. 355–362.

Pierquin, B. M., W. K. Mueller, and F. Baillet. 1978. Low dose rate irradiation of advanced head and neck cancer: present status. Int. J. Radiat. Oncol. Biol. Phys. 4:565–572.

Suit, H. D. 1977. Superfractionation. Int. J. Radiat. Oncol. Biol. Phys. 2:591–592.

Szechter, A., and G. Schwarz. 1977. Dose-rate effects, fractionation, and cell survival at lowered temperatures. Radiat. Res. 71:593–613.

Szechter, A., G. Schwarz, and J. K. Towne. 1978. Cell redistribution during continuous irradiations. (Abstract) Radiat. Res. 74:493.

Tubiana, M., and G. Boisgerie. 1978. Response of bone marrow and tumor cells to acute and protracted irradiation. (Abstract) Am. J. Roentgenol. 131:1107.

Radiation Biology in Cancer Research, edited by
Raymond E. Meyn and H. Rodney Withers.
Raven Press, New York © 1980.

Low Dose and Low Dose-Rate Effects on Cytogenetics

Antone L. Brooks

Inhalation Toxicology Research Institute, Lovelace Biomedical and Environmental Research Institute, Albuquerque, New Mexico 87115

The mechanisms by which ionizing radiation interact with living things to produce disease have been extensively studied. These studies have utilized biological models ranging in complexity from molecular changes to disease incidence in human populations. The responses of cells, tissues, and organs in these biological systems are dependent on radiation dose, dose rate, and quality. An adequate understanding of how changes in radiation exposure conditions alter the observed biological response is important in estimating the hazards associated with environmental radiation exposures.

The increase in chromosome aberration frequency induced by exposure to ionizing radiation is a sensitive, dose-related, biological change. Since the frequency of these aberrations changes as a function of radiation conditions, chromosome aberrations provide a useful measurement to link radiation dose and biological response.

A review of the classical dose-response relationships for chromosome aberrations following exposure to external low and high linear energy transfer (LET) radiation delivered at different radiation dose rates will be presented for background information. These biological generalizations for external radiation exposure have been tested in animals exposed to internally deposited radioactive material. The response of chromosomes to low dose rates similar to those observed from environmentally derived internal contamination have been measured and will be discussed. This will provide a framework to determine how well the results of research conducted at higher dose rates with external radiation predict the response to environmental radiation exposures.

To complete the cycle, a discussion of the limitations on the use of data derived from chromosome damage produced by internally deposited material and the use of chromosome aberrations for estimating the dose, dose distribution, or the level of activity deposited in the body will be given. This will provide some indication of how aberration frequency may or may not relate to risk in exposed populations or in people involved in occupational accidents.

EXTERNAL LOW-LET RADIATION

Dose-Response Relationships

After exposure of cells to acute low-LET radiation, the frequency of chromosome aberrations increases as a nonlinear function of radiation dose. This nonlinear dose–related increase in chromosome damage was observed by Sax (1939) and has since been described for many cell systems both in vitro (Lloyd et al. 1975, Bender and Barcinski 1969, Evans 1967) and in vivo (Brooks et al. 1971, Kelly and Brown 1965, McFee 1977). A variety of mathematical relationships have been fit to chromosome aberration data. The most useful model for describing both the physical and biological phenomena involved in producing chromosome aberrations is the quadratic equation described by Catcheside and coworkers (1946) as $Y = a + \alpha D + \beta D^2$, where Y represents the frequency of aberrations/cell, a is the background frequency of aberrations, α and β are constants, and D is dose in rads. This quadratic equation implies that some aberrations are produced by single ionizing events and increase linearly with dose, while other aberrations require the interaction of two separate ionizing events for their production and increase in number as the square of the radiation dose.

At a low total dose, only a few chromosomes are hit and, as a limited number of active sites are present in any given cell, little interaction between chromosomes results (Brewen 1963, Brooks et al. 1971, Liniecki et al. 1977). As the dose increases, more chromosomes are hit in a single cell and more interactions occur, resulting in an increased aberration frequency. This results in the chromosome aberration frequency increasing at a rate greater than linearly with dose. Thus, high total doses and dose rates produce aberrations by two mechanisms: single events and double events (Scott et al. 1970, Neary 1965).

Attempts have been made to define the shape of the dose-response relationship for chromosomes in human lymphocytes from exposures in the range of natural background (Pohl-Rüling et al. 1976, 1978). It was suggested that at doses ranging from background (170 mrad/year) to 800 mrad/year, the aberration frequency increased at a rate much higher than predicted from exposures in the 10–100 rad range. This implies that the biological damage per unit dose is underestimated in classical dose-response studies. Such an underestimate in damage may result in a risk value per rad that is too low following background levels of exposure. Interpretation of the results of this study is difficult because of problems associated with determining the dose to the blood lymphocytes that were scored for chromosome aberrations: The lymphocytes scored for chromosome aberrations were exposed to both internal and external sources and mixtures of both alpha and gamma irradiation. However, these studies have raised some interesting questions, and additional efforts to define the very low–dose portion of the dose-response curve may be useful in understanding mechanisms of chromosome damage and repair at environmental levels of exposure.

Dose-Rate Relationships

If low-LET radiation is fractionated or delivered at a low dose rate, the yield of chromosome and chromatid aberrations decreases (Brewen and Luippold 1971, Purrott and Reeder 1976a, 1976b, Brooks and McClellan 1969, Lloyd et al. 1975, Wolff 1972). This may be related to repair of chromosome damage, changes in cell cycle stage during exposure, or cell division with death of damaged cells during the radiation exposure. In cells that are not in proliferative stages during the radiation exposure, cell cycle stage sensitivity and cell division can be removed as variables and the rate of chromosome repair determined. As the dose rate decreases, the chromosome damage from one event can be repaired before additional damage can be produced in the same cell. This causes the interaction component, represented by the dose-squared constant of the quadratic dose-response curve, to decrease. At low dose rates, it should then be possible to study only those aberrations produced by a single ionizing event in the cell population. Experiments with low dose-rate radiation to nondividing cell populations have been conducted (Scott et al. 1970, Brewen and Luippold 1971, Lloyd et al. 1975, Wolff and Luippold 1955, Liniecki et al. 1977), and the repair rate and relationship between those aberrations produced by a single ionizing event and those that require multiple ionizations have been evaluated.

When the ratio of the constants alpha/beta equals 1 in the quadratic equation for chromosome damage produced at high dose rates, the dose is thought to be that at which the frequencies of aberrations produced by single- and double-track events are equal. This hypothesis has been tested by exposing the same cell system to high- and low-dose rates. The linear dose-response curve observed for the low dose rate had a slope similar to that predicted from the linear portion of the quadratic equation derived from high dose-rate exposure (Purrot and Reeder 1976b, Brewen and Luippold 1971, Brooks et al. 1971). It is of interest to determine the dose rate at which all the aberrations are the result of single events and the aberration frequency becomes completely independent of radiation dose rate. With internally deposited radioactive materials, experiments were conducted in which cells were exposed to low enough dose rates to produce damage that was the result of single events. The response of the chromosomes to these very low dose rates is included in the section on internal emitters.

EXTERNAL HIGH-LET RADIATION

Dose-Response Relationships

The number of chromosome aberrations following high-LET radiation, such as with neutrons and alpha particles, for the most part increases as a linear function of radiation dose (Bender and Gooch 1966, Lloyd et al. 1976a, 1978, Bender 1970). This is thought to be related to the large amount of energy deposited in the nucleus for each interaction or event. Neutron exposures not showing a linear increase in chromosome aberrations with dose have a very large alpha/beta ratio (Lloyd et al. 1976a). Thus, the nonlinear component is small, which implies that a small fraction of the chromosome aberrations produced by high-LET radiation is repaired and that perhaps even less repair of damage related to the ability of the cells to divide occurs. Because chromosome aberrations and cell killing are closely related (Carrano 1973, Carrano and Heddle 1973, Dewey et al. 1970), this lack of repair of chromosome damage following high-LET exposures would be an advantage in cancer therapy requiring cell killing.

Dose-Rate Effects

The frequency of chromosome aberrations after high-LET exposure does not, as would be predicted from the linear dose-response relationship observed after high dose rates, decrease as a function of decreasing dose rate or as the dose is fractionated. This again implies that single particles or ionization tracks produce most of the aberrations that result from neutron or alpha exposure and that these aberrations have a limited capability for repair. Thus, the frequency of neutron-induced aberrations is independent of radiation exposure conditions.

INTERNAL RADIATION EXPOSURE

Additional information is needed to utilize the data base from external radiation exposure as an indication of what would be predicted for environmental radiation exposures from internally deposited radioactive material. For internally deposited materials, the dose rates are very low, resulting in the dose being protracted over months and years in contrast with the minutes and hours for external exposures. Thus, exposure time, dose rate, and total dose must be known to relate biological response to exposure conditions. The distribution of the dose within the animal, organ, tissue, and cells may be very ununiform. This can also alter the responses, since the cells scored for chromosome damage may not be the ones exposed. Finally, the rate of cell division in the tissue under study is important in determining the response to low dose-rate exposures protracted over long periods of time. In dividing cell populations, the cells scored for cytogenetic analysis may not be from the first generation but may represent a mixed population that has received a wide range of total doses since the last cell division. This consideration is important since cells with certain types of chromosome aberrations die when they undergo cell division (Carrano 1973, Dewey et al. 1970, Carrano and Heddle 1973, Bedford et al. 1978).

Research at Lovelace Inhalation Toxicology Research Institute is directed towards quantitating chromosome damage from internally deposited radioactive isotopes that emit both high- and low-LET radiations. Our approach is to design experiments to investigate with internally deposited radioactive materials the variables of total dose, dose rate, cell division rate, and cells at risk. This research can then be related to the vast experience with external radiation exposure. We studied two basic systems, the bone marrow and the liver, which are two tissues with a wide range of cell turnover rates.

In bone marrow, many of the cell types divide with a total cell cycle time of less than 24 hours. Thus, the very low dose-rate exposure of cells in this system from internally deposited radioactive material represents exposure of many cell generations. By using bone marrow, we can investigate the influence of dose and dose rate on rapidly dividing cells.

Conversely, the cells in the liver divide at a slow rate. To determine the turnover rate of liver cells in the Chinese hamster, animals at 3, 4, 5, 7, 9, 12, and 15 days of age were injected with ^{14}C-thymidine to label the DNA of the liver cells. The rate of elimination of the ^{14}C-thymidine was measured to indicate cell turnover rate (Figure 1). The slope of the long-term component of the retention curve is an estimate of the average life span of the long-lived cells in the Chinese hamster liver, 1,200 days. With this long life span, the same cell can be exposed by an internal emitter at a very low dose rate over a period of years and chromosome damage accumulated. To score the damage, the liver cells can then be stimulated to divide by partial hepatectomy, allowed to pass through a single cell cycle, stopped at metaphase with colchicine at the first mitosis, and scored for chromosome aberrations. In this system, the cell acts as an integrator of radiation dose and the variable of cell division during exposure is minimized.

Effects of Internally Deposited Low-LET Radiation

Studies have been conducted with ^{90}Sr-^{90}Y, a bone-seeking isotope, and ^{144}Ce-^{144}Pr, a liver- and bone-seeking isotope. To understand the patterns of dose and dose rate in a cell population after deposition of these radioactive

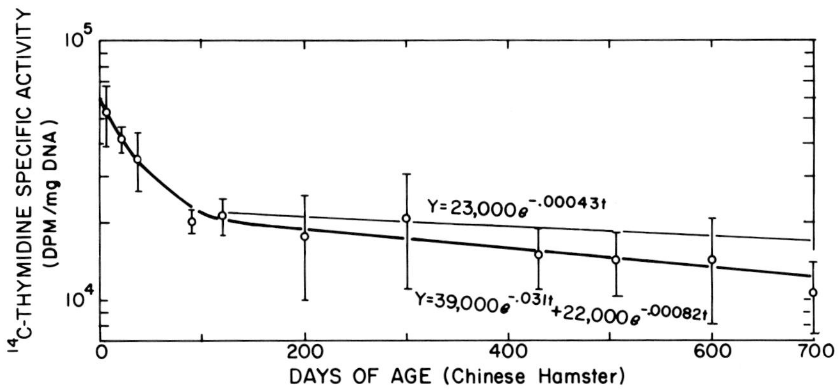

FIG. 1. The retention of ^{14}C-thymidine in Chinese hamster liver as a measure of cell turnover.

materials in the body, distribution and retention studies have been conducted (Brooks and McClellan 1969, Sturbaum et al. 1970). Animals were serially sacrificed and the retention pattern of the isotope determined in the tissue of interest. These patterns were integrated to derive the dose to the tissue scored for chromosome damage. The dose and dose rate were then related to the chromosome aberration frequency in that tissue. An example of the dose and dose-rate pattern derived from such studies is shown in Figure 2, in which dose and dose rate are plotted as a function of time after injection of ^{90}Sr-^{90}Y. Autoradiographic studies were done in connection with the retention distribution studies to evaluate the dose distribution within tissues. This was necessary to determine that the site of scoring cells was not different from the site of deposition of the isotope. For example, only if the dose to the liver and bone marrow cells is uniformly distributed can the average dose calculated be representative of the dose to the cells. And only if this holds can the abcrrations scored provide a realistic indication of the hazard from this exposure for late-occurring diseases.

Effect of Cell Division Rate

Use of liver and bone marrow permitted studies with cell populations with a wide range of division rates. Thus, the influence of cell division rate on chromosome aberration frequency could be determined following low dose-rate expo-

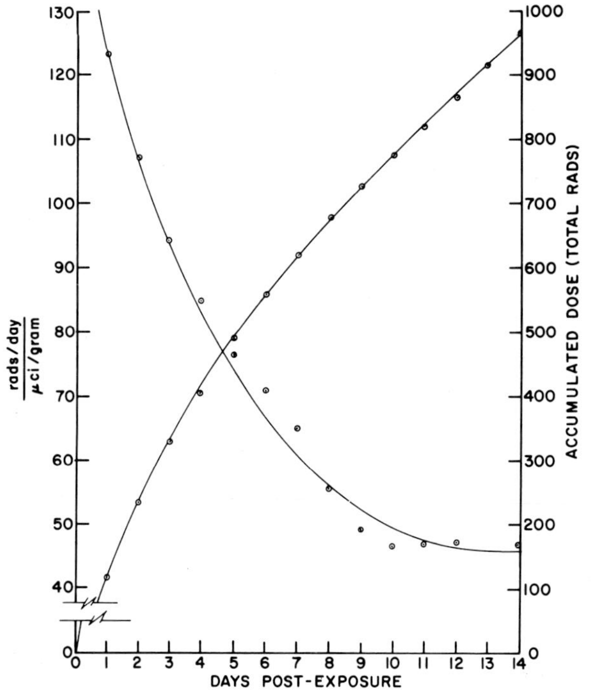

FIG. 2. The dose and dose-rate pattern for bone in Chinese hamsters injected with ^{90}Sr-^{90}Y.

sures. Bone marrow and liver cells were scored for chromosome damage over time after deposition of graded doses of ^{90}Sr-^{90}Y and ^{144}Ce-^{144}Pr.

Following ^{90}Sr-^{90}Y exposure, the types of aberrations in the bone marrow were mostly of the chromatid type with some symmetrical chromosome exchanges being found (Brooks and McClellan 1968, 1969). In fact, evidence for clones of abnormal cells with symmetrical chromosome exchanges were detected at longer exposure times. The frequency of chromosome aberrations in bone marrow is shown as a function of total dose in Figure 3. It indicates that the dose-response relationship was very dependent on the time of sampling. At longer times, the response of the chromosomes per unit dose was less than was observed at shorter times.

However, if chromatid and isochromatid deletions per cell were plotted against calculated dose rate on the day of sacrifice, there was a linear response with no effect of exposure time or total dose on the chromosome response. This indicated that the total dose accumulated per cell division was the important variable and that past radiation played little role in the frequency of these aberrations. If the types of aberrations produced were related to time of exposure, the type of abnormalities changed. At longer times, the stable symmetrical chromosome exchanges increased. Cells that contain these aberrations can survive cell division, form clones, and be maintained in the cell population. Survival

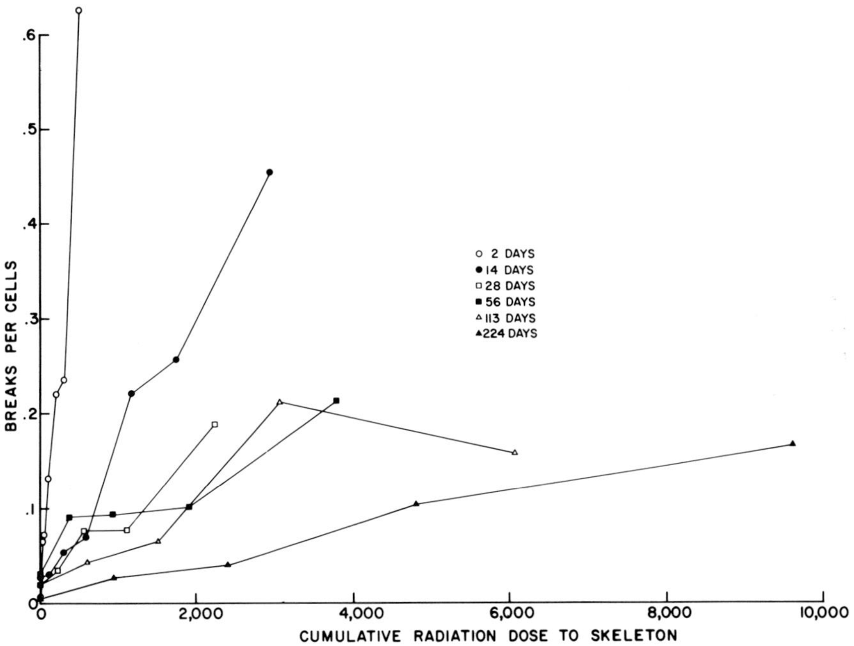

FIG. 3. The frequency of chromosome aberrations in Chinese hamster bone marrow sampled at a series of times after injection with ^{90}Sr-^{90}Y.

of cells with chromosome deletions or asymmetrical exchanges is limited, and these cells are eliminated from the population with time (Carrano 1973, Dewey et al. 1970).

The response of the chromosomes in the rapidly dividing, marrow cell population was compared to the response observed in slowly dividing liver cells following low dose-rate exposure to ^{144}Ce-^{144}Pr deposited in the liver. The response in liver cells increased linearly with dose and was independent of the time of exposure and dose rate (Figure 4) through 362 days (Brooks et al. 1972). The dose rate for these experiments ranged from 0.5 to 300 rads/day. This response illustrates that, for liver cells, 300 rads/day or 0.2 rads/minute (the highest rate used) was a dose rate at which the aberrations were produced by a single event. Reducing the dose rate to 3×10^{-4} rads/minute, the lowest rate used, did not further reduce the aberration yield in a slowly dividing cell population. When all the times of sacrifice and dose rates were combined, the aberration frequency increased as a linear function of dose with a slope of 3.1×10^{-4} aberrations/cell/rad. These data provide evidence that a lower limit exists in the response to low dose-rate exposure and that the effectiveness of low-LET gamma radiation delivered at environmental levels will not continue to decrease as dose rate decreases, but rather reaches a constant value that can be predicted. It was also of interest to determine whether the slope of the dose-response curve would be constant at lower total doses. As a check on the effect of total

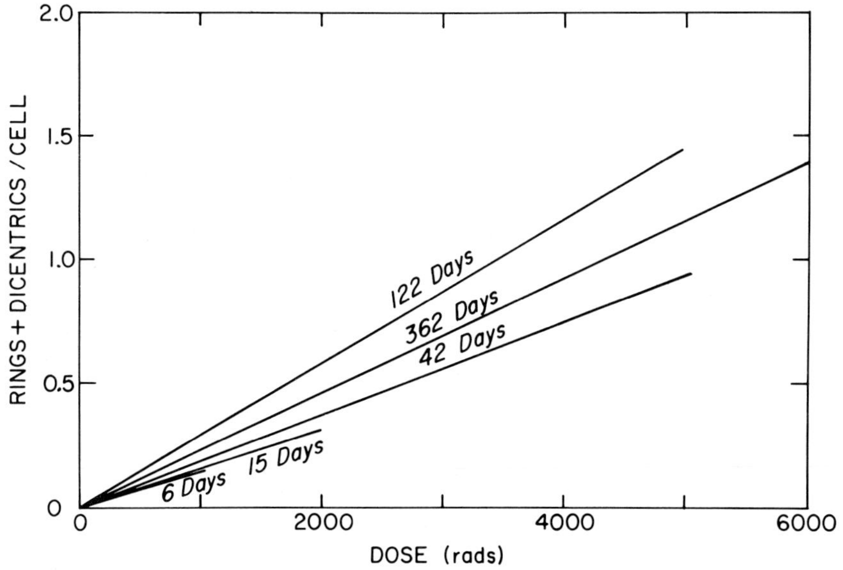

FIG. 4. The dose-response relationship between chromosome aberrations and ^{144}Ce-^{144}Pr. Animals were sacrificed at a series of times after injection. Cells were scored for chromosome damage 54 hours after partial hepatectomy.

dose, the data for the cells exposed to less than 1600 rads were analyzed separately; a curve of slope 2.9×10^{-4} aberrations/cell/rad was produced. This was very similar to the frequency of 3.1×10^{-4} aberrations/cell/rad for the total experiment.

For animals sacrificed at 722 days after injection with ^{144}Ce-^{144}Pr, the slope of the dose-response curve decreased to 1.8×10^{-4}. This decrease was thought to be related to the hyperplastic liver nodules observed in pathological sections. These are produced by cell division as the result of the aging process and may eliminate cells with chromosome aberrations from the cell population.

Effect of Internally Deposited High-LET Alpha Emitters

Animals were injected with ^{239}Pu, ^{238}Pu, and ^{241}Am in the citrate form to establish the relative hazards of alpha-emitting radioactive material in producing chromosome damage. The results will aid in the evaluation of health hazards from fuels and waste material from nuclear power production. The isotopes were retained in the bone and liver of the Chinese hamster with long effective half-lives and resulted in a large dose to the liver cells (Brooks 1975).

Chromosome aberrations in the liver cells produced by these isotopes also showed no dependence on time of exposure or dose rate (McKay et al. 1972, Brooks 1975). They increased linearly with dose as seen in Figure 5, with slopes of 4.8×10^{-3}, 3.8×10^{-3}, and 7.2×10^{-3} aberrations/cell/rad for ^{239}Pu, ^{238}Pu, and ^{241}Am, respectively. Data on chromosome aberrations from ^{252}Cf, acute ^{60}Co, protracted ^{60}Co, and ^{144}Ce are shown in Figure 5 for comparison. From these data, it is clear that the low dose-rate alpha irradiation was more effective by a factor of about 20 than low dose-rate beta or gamma exposure in producing chromosome damage. Since the chromosome response to acute ^{60}Co exposure changes as a function of radiation dose, it is very difficult to use acute ^{60}Co or X rays as baselines to derive a factor relating the hazards or effects produced by high- and low-LET radiations. Thus, for determining relative effectiveness of two radiations, they should both be delivered at low dose rates, which are more representative of environmental exposures. At these rates the response for both exposure types will be linear. Slopes can then be compared and standards set for both high- and low-LET radiation delivered at very low dose rates.

Relationship between Chromosome Aberration Frequency, Body Burden, and Dose in Humans

The frequency of chromosome aberrations has been measured in human beings exposed to radiation in work environments (Brandom et al. 1972, 1978, Hoegerman et al. 1976) and in people accidentally or medically exposed to external radiation (Dolphin et al. 1973, Bender and Gooch 1966, Lloyd et al. 1973) or to internally deposited radioactive material (Lisco and Conrad 1967, Schofield

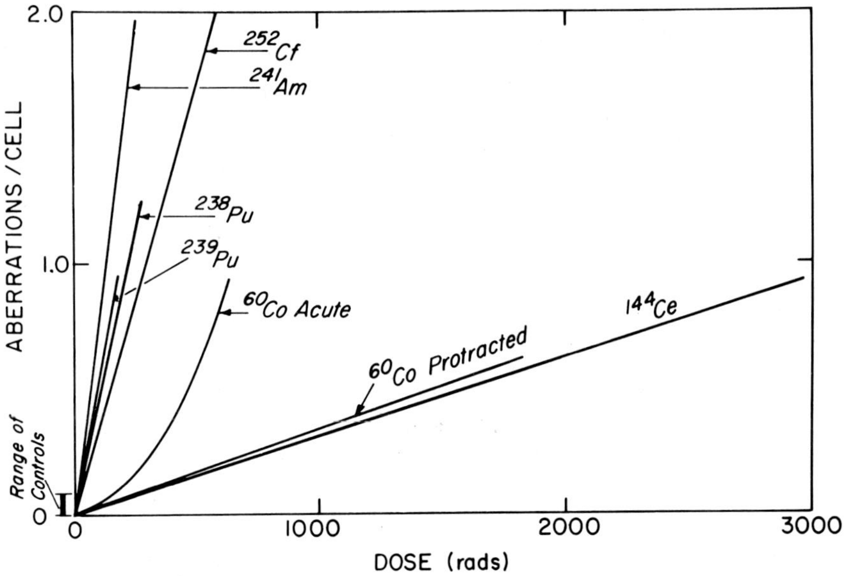

FIG. 5. Chromosome aberration frequency in the liver of the Chinese hamster following exposure to alpha particles from ^{239}Pu, ^{238}Pu, ^{241}Am, ^{252}Cf, beta particles from ^{144}Ce, or acute or chronic exposure to ^{60}Co gamma rays.

et al. 1974, Pohl-Rüling et al. 1978, Fischer et al. 1966, Lloyd et al. 1976b). There have been attempts to evaluate radiation dose or internal body burden of the isotopes and relate this to the hazard from the exposure by using the aberration frequencies observed. For external whole-body exposure, the biological dose derived from the aberration frequency was in good agreement with physical dosimetry (Dolphin et al. 1973, Bender and Gooch 1966, Kelly and Brown 1965).

For internally deposited radioactive material, the chromosome response, dose, and risk have a more complex relationship. First, the dose to the cells scored for chromosome damage may or may not be related to the dose to the tissues at risk for late-occurring disease such as cancer. For example, in monkeys that inhaled $^{239}PuO_2$ (LaBauve et al. 1974), the dose to the lungs and the biological response in the lungs were large and, in fact, caused early death in some of the animals, whereas the chromosome aberration frequency in the blood lymphocytes remained close to the background level. The same lack of relationship may exist in patients exposed to Thorotrast (Fischer et al. 1966), for whom the dose and risk is to the liver, or in patients or animals treated with ^{131}I (Lloyd et al. 1976b, Moore and Colvin 1968) for whom the dose and risk is to the thyroid and not to blood lymphocytes scored for chromosome damage.

Second, since many other agents in the environment, such as chemicals and viruses, produce chromosome damage (Fishbein 1976, Bartsch and Montesano

1975, Bartsch 1970), it is difficult to sample a worker population and relate the observed chromosome aberrations to a single environmental insult or agent. The chromosome aberration frequency is a reflection of the chemical, viral, or radiation dose or interaction with the blood lymphocytes and may or may not be related to dose, activity deposited, or the risk of development of tumors in other organs.

Finally, the sensitivity of the method is important to consider, since at low doses large numbers of cells must be scored. In addition, the dose to the blood lymphocytes is dependent on the turnover rate, exchangeable pool size (Field et al. 1972), and the blood flow rate and volume in the organ in which the activity is deposited. These and other difficulties have been reviewed (Sharpe 1969, Dolphin et al. 1973, Lloyd et al. 1973) and must be considered in relating aberration in blood lymphocytes of workers to their environmental exposures to radiation, especially that from internally deposited alpha-emitting radioactive materials.

SUMMARY

High dose-rate exposure to low-LET radiation produces a nonlinear dose-response relationship between chromosome aberrations and dose. Protraction of fractionation of low-LET radiation exposure decreases the magnitude of the chromosome response and results in a linear dose-response relationship. For high-LET radiation, the dose-response relationship for chromosome damage is linear regardless of dose rate or exposure conditions.

Extrapolation of the results from external radiation exposure to radioactive material deposited in the body is possible only if information is available on cells at risk and their rate of cell division. In rapidly dividing cell populations, the chromosome aberration frequency changes as a function of dose rate with little dependence on total dose. Chromatid aberrations are the primary aberration type scored in proliferating cell systems at early times after exposure. At later times, there is an increase in symmetrical chromosome exchanges since cells with these aberrations can survive multiple cell divisions.

Following low-LET exposure of cells with slow cell turnover rates, there is little change in the slopes of the dose-response curves for chromosome damage as a function of dose rate if the rate is between 0.2 and 3×10^{-4} rads/minute. The aberration frequency increases linearly with dose in this dose-rate range.

Protracted irradiation from alpha-emitting radionuclides also produces a linear dose-response relationship. There is no effect of dose rate or time of exposure on the dose-response relationships observed for alpha emitters. They are 15–20 times more effective per unit dose than protracted beta or gamma exposures in producing chromosome damage.

Chromosome aberration frequency is of little use as an indicator of internal contamination with alpha-emitting radionuclides unless the cell division rate, cells irradiated, and cells scored for damage are all well characterized. The aberration frequency in human blood lymphocytes is a reflection of the total chemical, viral, and radiation interaction with chromosomes. However, it may provide little information on radiation dose or associated risk for people exposed to internally deposited alpha-emitting radionuclides.

ACKNOWLEDGMENTS

This research was performed under U.S. Department of Energy Contract Number EY-76-C-04-1013 and was conducted in facilities accredited by the American Association for the Accreditation of Laboratory Animal Care.

REFERENCES

Bartsch, H. D. 1970. Virus-induced chromosomal alterations in mammals and man, *in* Chemical Mutagenesis in Mammals and Man, F. Vogel and G. Rohrorn, eds. Springer-Verlag, Berlin-New York, pp. 420–432.

Bartsch, H., and R. Montesano. 1975. Mutagenic and carcinogenic effects of vinyl chlorides. Mutat. Res. 32:93–114.

Bedford, J. S., J. B. Mitchell, H. G. Griggs, and M. A. Bender. 1978. Radiation-induced cellular reproductive death and chromosome aberrations. Radiat. Res. 76:573–586.

Bender, M. A. 1970. Neutron-induced genetic effects: A review. Radiat. Bot. 10:225–247.

Bender, M. A., and M. A. Barcinski. 1969. Kinetics of two-break aberration production by x-rays in human leukocytes. Cytogenetics 8:241–246.

Bender, M. A., and P. C. Gooch. 1966. Somatic chromosome aberrations induced by human whole-body irradiation: The "recuplex" criticality accident. Radiat. Res. 29:568–582.

Brandom, W. F., A. D. Bloom, P. G. Archer, V. E. Archer, R. W. Bistline, and G. Saccomanno. 1979. Somatic cell genetics of uranium miners and plutonium workers: A biological dose-response indicator, *in* International Symposium on the Late Biological Effects of Ionizing Radiation, 1978. IAEA-SM-224/310. International Atomic Energy Agency, Vienna, pp. 310–317.

Brandom, W. F., G. Saccomanno, V. E. Archer, P. G. Archer, and M. E. Coors. 1972. Chromosome aberrations in uranium miners occupationally exposed to ^{222}radon. Radiat. Res. 52:204–215.

Brewen, J. G. 1963. Dependence of frequency of x-ray-induced chromosome aberrations on dose rate in the Chinese hamster. Proc. Natl. Acad. Sci. USA 50:322–329.

Brewen, J. G., and H. E. Luippold. 1971. Radiation-induced human chromosome aberrations: in vitro dose rate studies. Mutat. Res. 12:305–314.

Brooks, A. L. 1975. Chromosome damage in liver cells from low dose rate alpha, beta, and gamma irradiation: Derivation of RBE. Science 190:1090–1092.

Brooks, A. L., and R. O. McClellan. 1968. Cytogenetic effects of strontium-90 on the bone marrow of the Chinese hamster. Nature 219:761–763.

Brooks, A. L., and R. O. McClellan. 1969. Chromosome aberrations and other effects produced by ^{90}Sr-^{90}Y in Chinese hamsters. Int. J. Radiat. Biol. 16:585–561.

Brooks, A. L., R. O. McClellan, and S. A. Benjamin. 1972. The effects of ^{144}Ce-^{144}Pr on the metaphase chromosomes of the Chinese hamster liver cells in vivo. Radiat. Res. 52:481–498.

Brooks, A. L., D. K. Mead, and R. F. Peters. 1971. Effect of chronic exposure to ^{60}Co on the frequency of metaphase chromosome aberrations in the liver cells of the Chinese hamster, (in vivo). Int. J. Radiat. Biol. 20:599–604.

Carrano, A. V. 1973. Chromosome aberrations and radiation-induced cell death. I. Transmission and survival parameters of aberrations. Mutat. Res. 17:341–353.

Carrano, A. V., and J. A. Heddle. 1973. The fate of chromosome aberrations. J. Theor. Biol. 38:289–304.

Catcheside, D. G., D. E. Lea, and J. M. Thoday. 1946. The production of chromosome structural changes in *Tradescantia micropores* in relation to dosage, intensity, and temperature. J. Genet. 47:137–145.

Dewey, W. C., S. C. Furman, and H. H. Miller. 1970. Comparison of lethality and chromosomal damage induced by x-rays in synchronized Chinese hamster cells in vitro. Radiat. Res. 43:561–581.

Dolphin, G. W., D. C. Lloyd, and R. J. Purrott. 1973. Chromosome aberration analysis as a dosimetric technique in radiological protection. Health Phys. 25:7–15.

Evans, H. T. 1967. Dose-response relations from in vitro studies, *in* Human Radiation Cytogenetics, H. J. Evans, W. M. Court Brown, and A. S. McLean, eds. John Wiley, New York, pp. 20–36.

Field, E. O., H. B. A. Sharpe, K. B. Dawson, V. Andersen, S. A. Killmann, and E. Weeke. 1972. Turnover rate of normal blood lymphocytes and exchangeable pool size in man, calculated from analysis of chromosomal aberrations sustained during extracorporeal irradiation of the blood. Blood 39:39–56.

Fischer, P., E. Golob, E. Kunze-Muhl, A. Ben Haim, R. A. Dudley, T. Mullner, R. M. Parr, and H. Vetter. 1966. Chromosome aberrations in peripheral blood cells in man following chronic irradiation from internal deposits of Thorotrast. Radiat. Res. 29:505–515.

Fishbein, L. 1976. Industrial mutagens and potential mutagens. I. Halogenated aliphatic derivatives. Mutat. Res. 32:267–307.

Hoegerman, S. F., H. T. Cummins, and J. F. Bronec. 1976. Chromosome breakage in lymphocytes from humans with body burdens of ^{226}Ra, *in* Radiation and the Lymphatic System. Energy Research and Development Agency Symposium Series 37, CONF-740930, pp. 113–119.

Kelly, S., and C. D. Brown. 1965. Chromosome aberrations as a biological dosimeter. Am. J. Pub. Health 55:1419–1423.

LaBauve, R. J., A. L. Brooks, J. A. Mewhinney, R. O. McClellan, and R. K. Jones. 1974. Cytogenetic changes in blood lymphocytes of the Rhesus monkey following acute inhalation exposure to polydisperse $^{239}PuO_2$. Radiat. Res. 59:207.

Liniecki, J., A. Bajerska, K. Wyszynska, and B. Cisowska. 1977. Gamma-radiation-induced chromosomal aberrations in human lymphocytes: Dose-rate effects in stimulated and non-stimulated cells. Mutat. Res. 43:291–304.

Lisco, H., and R. A. Conard. 1967. Chromosome studies on Marshall Islanders exposed to fallout radiation. Science 157:445–447.

Lloyd, D. C., R. J. Purrott, and G. W. Dolphin. 1973. Chromosome aberration dosimetry using human lymphocytes in simulated partial body irradiation. Phys. Med. Biol. 18:421–431.

Lloyd, D. C., R. J. Purrott, G. W. Dolphin, D. Bolten, A. Edwards, and M. J. Corp. 1975. The relationship between chromosome aberrations and low LET radiation dose to human lymphocytes. Int. J. Radiat. Biol. 28:75–90.

Lloyd, D. C., R. J. Purrott, G. W. Dolphin, and A. A. Edwards. 1976a. Chromosomal aberrations induced in human lymphocytes by neutron irradiation. Int. J. Radiat. Biol. 29:169–182.

Lloyd, D. C., R. J. Purrott, G. W. Dolphin, P. W. Horton, K. E. Halnan, J. S. Scott, and G. Mair. 1976b. A comparison of physical and cytogenetic estimates of radiation dose in patients treated with iodine-131 for thyroid carcinoma. Int. J. Radiat. Biol. 30:473–485.

Lloyd, D. C., R. J. Purrott, E. J. Reeder, A. A. Edwards, and G. W. Dolphin. 1978. Chromosome aberrations induced in human lymphocytes by radiation from ^{252}Cf. Int. J. Radiat. Biol. 34:177–186.

McFee, A. F. 1977. Chromosome aberrations in the leukocytes of pigs after half-body or whole body irradiation. Mutat. Res. 42:395–400.

McKay, L. R., A. L. Brooks, and R. O. McClellan. 1972. The retention, distribution, dose and cytogenetic effects of ^{241}Am citrate in the Chinese hamster. Health Phys. 22:633–640.

Moore, W., and M. Colvin. 1968. Persistence of chromosomal aberrations in Chinese hamster thyroid following administration of ^{131}I. J. Nucl. Med. 9:165–167.

Neary, G. J. 1965. The relation between the exponent of dose response for chromosome aberrations and the relative contribution of "two-track" and "one-track" processes. Mutat. Res. 2:242–246.

Pohl-Rüling, J., P. Fischer, and E. Pohl. 1976. Chromosome aberrations in the peripheral blood lymphocytes dependent on various dose levels of natural radioactivity, *in* Biological and Environmental Effects of Low-Level Radiation, Vol. II (Proceedings of a Symposium on Biological Effects of Low-Level Radiation Pertinent to Protection of Man and His Environment), IAEA SM-202/701. International Atomic Energy Agency, Vienna, pp. 317–324.

Pohl-Rüling, J., P. Fischer, and E. Pohl. 1978. The low-level shape of dose-response for chromosome aberrations, *in* Fundamental Symposium on the Late Biological Effects of Ionizing Radiation, 1978, IAEA SM-224/2403. International Atomic Energy Agency, Vienna, pp. 403–413.

Purrott, R. J., and E. Reeder. 1976a. Chromosome aberration yields in human lymphocytes induced by fractionated doses of x-radiation. Mutat. Res. 34:437–446.

Purrott, R. J., and E. Reeder. 1976b. The effect of changes in dose rate on the yield of chromosome aberrations in human lymphocytes exposed to gamma radiation. Mutat. Res. 35:437–444.

Sax, K. 1939. The time factor in x-ray production of chromosome aberrations. Proc. Natl. Acad. Sci. USA 25:226–233.

Schofield, G. B., H. Howells, F. Ward, J. C. Lynn, and G. W. Dolphin. 1974. Assessment and management of a plutonium contaminated wound case. Health Phys. 26:541–554.

Scott, D., H. B. A. Sharpe, A. L. Batchelor, H. J. Evans, and D. G. Papworth. 1970. Radiation-induced chromosome damage in human peripheral blood lymphocytes in vitro. II. RBE and dose-rate studies with ^{60}Co γ- and x-rays. Mutat. Res. 9:225–237.

Sharpe, H. B. A. 1969. Pitfalls in the use of chromosome aberration analysis for biological radiation dosimetry. Br. J. Radiol. 42:943–944.

Sturbaum, B., A. L. Brooks, and R. O. McClellan. 1970. Tissue distribution and dosimetry of ^{144}Ce in Chinese hamsters. Radiat. Res. 44:359–367.

Wolff, S. 1972. The repair of x-ray induced chromosome aberrations in stimulated and unstimulated human lymphocytes. Mutat. Res. 15:435–444.

Wolff, S., and H. E. Luippold. 1955. Metabolism and chromosome break rejoining. Science 122:231–232.

Radiation Biology in Cancer Research, edited by
Raymond E. Meyn and H. Rodney Withers.
Raven Press, New York © 1980.

Medical Radiation and Possible Adverse Effects on the Human Embryo

Mary Esther Gaulden and Robert C. Murry

Radiation Biology Section and Radiological Physics Section, Department of Radiology, The University of Texas Health Science Center at Dallas, Dallas, Texas 75235

The effects of high doses of radiation on development of the mammalian embryo are well established in experimental organisms (United Nations 1977). Data on several hundred human embryos, exposed mainly to therapeutic radiation (Murphy 1929, Dekaban 1968), lead to the conclusion that, except for time-scale, all mammalian embryos have in general the same radiation response. We wish to discuss here the effects of low doses of radiation (less than 10 rad), about which much less is known, from the viewpoints of a clinical radiation biologist and radiological physicist who are confronted with requests for advice on what to do about a human embryo or fetus exposed to diagnostic radiation. Limited human data force us to use basic knowledge of cell and radiation biology in reaching decisions.

When a woman has been irradiated and it is later determined that she is pregnant, the first question she and the physician ask is: Will the baby be normal? What they are really asking is: Will the baby be deformed at birth? The answer depends on the developmental stage at the time of exposure and the dose of radiation, and even with that information it is not always possible to give an unequivocal answer. The crucial question seldom asked is: If the baby has no malformations, will it be "normal"?

Data on experimental animals indicate that the period of development most susceptible to the induction of gross anomalies by radiation, down to exposures of 25 R (Russell 1957), is that period in the human which corresponds to the second to sixth weeks of gestation, i.e., the period of major organogenesis. The basic cellular effects, one or more of which can result in a gross developmental defect, are: cell killing, mitotic inhibition, altered cell migration and interaction, and chromosome or gene mutations affecting development of one or more organs. At doses of less than 10 rad the collective magnitude of these effects is probably not great enough to produce a gross developmental defect. Support for this is suggested by the limited data of Neumeister (1976), who has reported on 32 pregnancies in which the human embryo received less than 7 rad during the first six weeks of gestation. He was able to follow the children in 15 cases for one to three years and found no detectable abnormalities. Our experience confirms this. More significant is a large Finnish study in which Granroth (1979) has found significantly increased fetal X-ray exposure in children born with a central nervous system (CNS) defect, but the data do not yet permit determination of a cause-effect relationship.

At doses of less than 10 rad the most significant and probably the only cell effect in the embryo would be the induction of genetic effects, i.e., chromosome aberrations or gene mutations, or both, in some cells to produce a mosaic individual. We use the term "genetic effect" to denote one or more changes in the genetic material of either a somatic or germ cell. The consequences of somatic mosaicism, though real, would be expected to be subtle in that they would not be recognizable at birth. We present here the rationale for concern about low-dose effects on the embryo.

Data will be presented on (1) doses to the embryo from direct beam diagnostic radiography as well as scatter radiation from radiation therapy of tumors, (2) types of effects to be expected from low doses, (3) some data on low-dose effects observed in human embryos, (4) guidelines for terminating pregnancy after irradiation, and (5) the use of elective scheduling of abdominal diagnostic X-ray examinations in women to avoid irradiation of unsuspected conceptuses.

RADIATION DOSES

Before doses are discussed, one point merits special mention: partial body irradiation of the mother's pelvis is whole-body irradiation of the embryo. Whole-body irradiation presents a larger "target" and thereby increases the risk of cell effects, including carcinogenesis, which is now acknowledged as a significant effect of low doses on the exposed individual (United Nations 1977). The human embryo at the end of the period of major organogenesis, i.e., at the end of six weeks of development, is only about 2 cm long (crown to rump), so it will only be in the X-ray beam in those examinations that involve the mother's pelvic region. A 14 × 17 inch abdominal film of the mother later in pregnancy will, of course, expose the entire fetus.

It should also be noted that up to 99% of the photons in the entrance X-ray beam are absorbed by the patient; in other words, the X-ray image on the film or the fluoroscopic screen is produced by only a small fraction of the radiation entering the patient (Christensen et al. 1978). Those diagnostic procedures that require impingement of the beam on the uterus (in the pelvis), such as a barium enema, will of course result in a higher dose to the embryo than will a procedure, such as cholecystography, for which the beam is ideally collimated to the upper abdomen.

In Table 1 are listed the abdominal diagnostic procedures that would be expected to contribute to embryo dose. The median and maximum absorbed ovarian doses are taken from a survey of medical X-ray exposure of the United States population that was conducted in 1970 by the U.S. Public Health Service (Bureau of Radiological Health [BRH] Report 1976b); the doses from fluoroscopy

TABLE 1. *Representative abdominal diagnostic radiography doses*

	Ovarian Dose per Exam*		Per Radiograph, AP Projection		
Examination	Median mrad	Maximum mrad	Entrance Exposure mR†	Ovarian Dose mrad‡	Fetal Dose mrad‡
Cholecystography	27	1603	620	1.9	1.6
Upper gastrointestinal	91	1228	960	31	24
Intravenous or retrograde pyelogram	448	3069	590	120	156
Barium enema	574	9218	1320	268	350
Abdomen	144	1416	670	136	178
Pelvis	158	1587	610	126	173
Lumbar spine	608	2901	2180	410	545

* From Table 5–4, BRH Report, 1976b. Fluoroscopy doses not included.
† From Table 35 (all types of medical facilities), BRH Report, 1973.
‡ Calculated from entrance exposure by use of Tables 12a–12p of BRH Report, 1976a. A 2.5-mm Al HVL was assumed.

were not included in the survey results. In the cases we have investigated, the usual doses, including fluoroscopy, lie between the median and maximum ones of the survey. It is not uncommon for all of the first five procedures in Table 1 to be used on a patient complaining of abdominal pain, which would yield a total median dose of 1,284 mrad and a maximum dose of 16,534 mrad.

To determine how the gonadal doses compare with those of the fetus, we have calculated fetal dose per radiograph on the basis of entrance beam exposure to the mother. For consistency with the national survey data (BRH 1976b), BRH data (1976a) were used in the calculations given in Table 1. It can be seen that the higher dose procedures give fetal doses about 30% greater than the ovarian doses.

In Table 2 are presented the doses to the human embryo from scatter radiation from some of the more common radiation procedures for tumor therapy. We asked a radiotherapist to specify a representative treatment for each procedure listed. A dosimetry plan was prepared to deliver the prescribed dose, and a Farmer ionization chamber was inserted into a Rando-man phantom at the position of the uterus. The phantom was then set up for treatment with a ^{60}Co teletherapy unit, and the scattered dose rate in the uterus was determined. The doses in Table 2 were calculated by applying the dose rate to the treatment time in the dosimetry plan. Doses were also calculated from scatter-air ratios and found to be significantly lower than the measured ones. This is due to the fact that the phantom, like man, does not represent a homogeneous scatter material infinite in extent. We feel, therefore, that when a pregnant woman requires radiation therapy, the fetal dose should be measured in a phantom according to the exact prescribed treatment plan.

TABLE 2. *Scatter dose to human embryo from selected radiation therapy procedures as measured with phantom*

Tumor Site	Dose (Rad)
Brain	3
Breast (unilateral)	9
Neck (bilateral)	15
Lung (unilateral, hilar & supraclavicular)	20
Complete upper mantle (including diaphragm)	23

RADIATION EFFECTS

Development

As mentioned above, the available data on experimental animals and human beings indicate that there is a rather low probability that radiation doses of 10 rad or less delivered during the first six weeks of development will cause a *gross* developmental defect in human beings that would be detectable at birth. In this connection it is important to remember that in the absence of radiation, or any other known teratogenic agent or condition, every pregnancy carries a risk of some abnormality resulting from the background of genetic defects in the population and from congenital (not heritable) effects of unknown origin. These risks include approximately 2–3% of newborns who have a malformation plus an additional risk of approximately 3–5% of an abnormality that will be manifest later in life (Shepard 1976, United Nations 1977). In counseling patients it is advisable to bring this point to their attention.

To provide perspective on the early development of the human embryo, some events that occur during the first six weeks of human gestation are listed in Table 3. Several points are worthy of note. The fertilized egg is slow in dividing, so the second division is not completed until day 3 after fertilization. Implantation begins about a week after fertilization, and two germ layers, ectoderm and entoderm, are differentiated by day 8. A primitive placental circulation begins on day 11, which is before the mother will have missed her first menstrual period. By day 18 the ectoderm has formed the neural plate, which will give rise to the CNS and the peripheral nervous system. Major organogenesis is completed by day 42.

We propose that the embryo is most susceptible to the developmental effects of low doses of radiation during the first six weeks of development, and especially during the first two weeks. Our line of reasoning is as follows.

First, the frequency of one-hit chromosome aberrations and gene mutations bears a linear relation to low doses of radiation. Quantitative data on eukaryotic organisms show that at very low doses the most likely cell effect is a change in the genetic material of the one-hit type, such as a terminal chromosome deletion or the induction of a point (gene) mutation. A linear dose response for both types of these genetic effects has been demonstrated down to doses

TABLE 3. *Some events in early human development**

Day	Event
1	Fertilization
2	Two cells
3	Four cells; rapid division to 16 cells; enters uterus
4	Inner cell mass (embryo proper) forms
5	Blastocyst
6	Implantation begins
8	Establishment of ectoderm and entoderm germ layers
11	Primitive placental circulation
14	Prochordal plate formed; major organogenesis begins
	FIRST MISSED MENSTRUAL PERIOD
18	Neural plate
20	Primitive brain
21	Primordial germ cells visible in yolk sac wall near allantois
22	Heart begins to beat
27	Arm and leg buds
30	Lens vesicles, optical cups forming
36	Oral and nasal cavities confluent
42	Major organogenesis completed

* From Moore (1973)

as low as 0.25 rad (cf. Gaulden and Read 1978). In other words, there is no threshold dose, i.e., no absolutely safe dose, of radiation with respect to the induction of effects on DNA, so there is a finite probability that a low dose to an embryo may induce a genetic change in one or more cells. A viable chromosome aberration with a phenotypic effect, such as a deletion, mimics a dominant mutation in that it is expressed in the next generation when induced in a germ cell and is expressed in the exposed individual when induced in a somatic cell. Our emphasis in this discussion is on the somatic cell effects of a viable chromosome aberration.

Second, the induction of a viable chromosome aberration or mutation in one or more of a group of cells, such as an embryo, would result in a mosaic, i.e., an individual with some normal and some mutant cells. (A mosaic is derived from one zygote whereas a chimera results from the combination of cells from two or more zygotes.) Genetic mosaicism has been recognized in experimental organisms since the early part of this century (cf. Stern 1968). One of the first humans analyzed with modern cytogenetic techniques in 1959 was found to be a mosaic for the sex chromosomes: XXY/XX (Ford et al. 1959). Soon after that an entire fetus, approximately 70 days old, was analyzed cytogenetically and found to be a sex chromosome mosaic: XXY/XY (Klinger and Schwarzacher 1962). All human females, as a matter of fact, are mosaic for expression of the genes on the X chromosome in that one X is inactivated in every cell early in development, the inactivation being randomly the maternally derived X in some cells and the paternally derived one in others (Lyon 1972).

Radiation-induced mosaics for specific genes determining certain coat colors in mice were first produced by Russell and Major (1957) in 10¼-day-old embryos; at that developmental stage there are approximately 175 precursor cells (melanocytes). This so-called spot test has in recent years been increasingly used by Russell and others to detect in vivo chemical mutagenesis in somatic cells (Russell 1978a). Cytogenetic mosaics have been induced by X irradiation of 13-day-old rat embryos (Soukup et al. 1965); even though high doses were used, which were lethal to many cells, the chromosome aberration frequency was still above the spontaneous one five days later. Russell and Montgomery (1970) have shown that large autosomal deficiencies or deletions are viable in the mouse if they occur in no more than half of the somatic cells of an embryo. Obviously much more work is needed on low-dose induction of cytogenetic mosaics in the mouse embryo.

Genetic mosaicism is being used to an increasing extent as an experimental tool with which to investigate the fate of certain embryonic cells as well as the genetic control of patterns of development (Baker 1978, Gehring 1978, Russell 1978b). In addition, the use of chimeras, experimentally constructed by the injection of cells from one embryo into the blastocyst of another, has provided another method for studying gene influence on development (McLaren 1976). All of these investigations taken together present convincing evidence that a large part of development is controlled by genes.

There has been a remarkable increase in the number of reports of human cytogenetic mosaicism in the last few years. With the recent development of high-resolution Giemsa-banding techniques for human prometaphase chromosomes (Francke and Oliver 1978), it will be possible to detect very small aberrations, and this will permit finer "tuning" on mosaicism. There is, however, no doubt now that mosaicism is a part of the genetic variability in man as well as in other organisms and that its induction by irradiation of an embryo could have developmental consequences.

Third, one would not expect to be able to detect at birth the phenotypic effects of mosaicism for most chromosomal aberrations. Let us first look at the phenotypic effects of a small terminal deletion that is inherited, that is, a chromosome aberration that is in every cell of the individual. These have been described in the human for most of the chromosomes, but for our purposes the 18q- syndrome will suffice as an example. Children who inherit from one parent a chromosome 18 with approximately the terminal half of the long arm deleted are not grossly abnormal; in fact, some of them would not be recognized at birth as having any abnormalities, the only physical features being narrow external ear canals and tapering fingers (Schinzel et al. 1975). Gestation usually goes to term and the birth weight may be slightly low to normal. In other words, these children are not always diagnosed at birth, but the one striking and consistent abnormality in all of the patients described so far is mental retardation. If this syndrome can go undetected at the birth of a child who has the small chromosomal abnormality in every cell, it is easy to understand how a child who is mosaic for this abnormality would also be undiagnosed at birth. If a small degree of mental retardation were the only effect, the 18q- origin of the condition might escape notice at a later age.

Fourth, the earlier in embryogenesis a mutant cell is produced, the larger will be the number of cells derived from it in the fully developed individual. By the same token, the larger the number of mutant cells in an organ, the higher should be the probability of abnormal function or development. That the number of mutant cells in an individual determines the degree of abnormal development is suggested by a case investigated in our laboratory (unpublished). A child born with gross developmental defects (almost complete absence of brain, anomalous genitalia, no thumbs or big toes) was mosaic in 80% of his peripheral lymphocytes for a ring D chromosome. The formation of a ring involves the deletion of a small portion from each end of a chromosome with the subsequent union of the broken ends of the centric fragment. The father of this child was found to have the same ring D chromosome in 8% of his peripheral lymphocytes; he was normal physically and mentally as far as we could determine. What degree of mosaicism for this chromosome aberration existed in the other tissues of these individuals could not be determined. Verma et al. (1978) have reported a low level of mosaicism for a ring D (number 13) chromosome in lymphocytes and skin of a three-year-old child with only minimal physical deviations but with retarded psychomotor development. They point out the possible correlation between genetic imbalance at a critical developmental stage and the occurrence of clinical features.

There is now a substantial amount of data on mosaicism for trisomy 21 (three chromosomes 21 instead of the normal two), which is responsible for Down's syndrome in man. Attention has been focused on the lack of correlation between the number of mutant cells in the blood and in the skin (fibroblasts), as well as between the number of mutant cells in either of these tissues and the number or severity of the physical stigmata associated with the syndrome (Taysi et al. 1970). In fact, the simian palmar crease, so characteristic of non-mosaic Down's syndrome patients, can be absent in a mosaic who has severe mental retardation. Although mosaic Down's syndrome individuals show more variability in mental ability than do non-mosaic ones, in most of the cases with very small numbers of mutant lymphocytes or fibroblasts, there is subnormal mental ability (Johnson and Abelson 1969). In fact, some of the mosaics are not recognized as having Down's syndrome until they themselves have a child with the syndrome (Kaffe et al. 1974, Mikkelsen and Vestermark 1974).

It has been suggested (Ford 1967, Taylor 1968) that the lack of correlation between the number of mutant cells in the blood or skin or both and the extent of mental disability in mosaic Down's syndrome patients possibly results from in vivo selection for mutant cells in the brain and against them in the skin and blood. Such a selection could occur with cell division, the mutant cells being overgrown by the non-mutant ones. Cell division in the brain is completed by about age two years in the human (Dobbing and Sands 1973), whereas it continues throughout life in the skin and blood-forming tissues. It seems futile, however, to attempt to correlate the degree of mosaicism in skin and blood, which are derived from embryonic mesoderm, with the function (mental ability) of the CNS, which is derived from ectoderm, because even in the absence of selection one would not expect tissues derived from different germ layers to have the same degree of mosaicism unless the mutant cell was induced prior to differentiation of the germ layers. In spontaneous mosaics it is usually assumed that the mutant cell arises soon after cleavage, but this is not necessarily the case in those with small numbers of mutant cells. One fact does appear certain in most mosaic Down's syndrome patients: irrespective of the degree of cytogenetic mosaicism in testable tissues (Taysi et al. 1970), psychomotor abilities decline until stabilization is reached in adulthood (Kohn et al. 1970).

It is possible that a few mutant cells in the CNS may cause effects in this highly complex and interrelated cell system more readily than in other tissues. A review of cases of human mosaicism, which we have only touched on here, leads to the conclusion that the number of mutant cells in a given tissue at a critical developmental stage determines the degree of phenotypic effect as has been demonstrated in experimental organisms. Some indication of cytogenetic mosaicism in the CNS of the human might be obtained from examination of other tissues derived from the ectoderm, but they are not amenable to in vitro chromosome analysis (peripheral nervous system; sensory epithelia of ear, nose, and eye; hair and nails, mammary, pituitary, and subcutaneous glands; tooth enamel). On this point, therefore, we will have to rely on data from embryos and infants of experimental organisms.

Fifth, the organ system most likely to be affected by genetic mosaicism is the central nervous system. There are many genes controlling the development, structure, and function of the CNS. Although some CNS abnormalities in man are caused by a single gene, e.g., the dominant gene responsible for Huntington's chorea (onset usually in middle age), many aspects of the CNS are under polygenic control, e.g., mental ability and reading disability (Finucci et al. 1976). The large number of CNS genes are evidently distributed among all 23 pairs of chromosomes, because most inherited chromosome abnormalities of all types affect psychomotor function. Even some balanced chromosomal translocations and inversions, which were previously thought to be without effect in somatic cells, have been found to be associated with mental retardation in children (Funderburk et al. 1977), one possible explanation being that a break point occurs within a gene. If, therefore, a deletion is induced at any point in one chromosome in an embryonic neuroblast, there is a good probability that one or more CNS genes will be affected.

Is there evidence for mosaicism in human embryos exposed to very low doses of radiation and what is its relation to CNS effects? A few cases of cytogenetic mosaicism following diagnostic X-ray examinations in utero have been reported (cf. Gaulden 1974). For example, Sato (1966) found chromosome deletions and dicentrics in the chorionic fragments of an abortus induced five days after exposure at six weeks of gestation to 3.9 rad (barium meal and enema). Especially pertinent to our discussion is a child who was exposed at one week of development for "radiopelvimetry"; the dose was not estimated and indications for X-ray examination were not given (Lejeune et al. 1964). By two years of age its physical and mental growth were obviously abnormal; chromosome analyses revealed the presence of one to three abnormal (marker) chromosomes in all the peripheral lymphocytes and in 29% of the skin fibroblasts examined.

One cannot prove that the chromosome aberrations in these cases were induced by the radiation, but such argument would certainly not apply to the following case reported by Kucerova (1970). A fetus was exposed at 20–30 weeks of gestation to at least 200 rad of X rays during therapy to the mother for uterine cervix carcinoma. When delivered by Cesarian section at eight months the child was within normal limits for weight and length and showed no congenital malformations. However, she did not walk until 21 months or talk until 36 months. At 4½ years she had a mental age of 2 years. All types of chromosome aberrations were observed and in relatively large numbers over 4 years after birth.

This case demonstrates several points. The CNS, unlike all other organ systems, is sensitive to radiation throughout gestation, a fact known in humans since Murphy's studies in 1929, especially as regards the induction of microcephaly. Since inherited chromosome aberrations and gene mutations are known to affect the CNS, it stands to reason that the induction of mosaicism for these genetic effects in embryonic cells that are destined to form or are forming the CNS will have similar consequences but perhaps of less severity. This case also shows that a child who appears "normal" at birth, i.e., has no gross anomalies, can have severe neuromotor and mental deficiencies that only become evident with time. Hicks and D'Amato (1966), in reviewing radiation-induced CNS effects in the human embryo, reported on a child whose mother received approximately 1,800 R to the pelvis during the fourth to seventh months of pregnancy. The child was "normal" for a few years, but eventually had to be institutionalized for a "serious behaviour problem." With the recent increased research activity in behavioral genetics of man (Childs et al. 1976), more information in the future may become available on the embryonic effects of radiation on behavior as well as mental ability.

The prolonged sensitivity of the mammalian CNS to radiation is undoubtedly due to the fact that it is developing throughout gestation and beyond birth (Lemine 1974). In the human brain it has been found that a high rate of neuroblast division occurs during the 10th to 18th weeks of gestation; brain growth continues until about two years after birth, approximately ⅚ of it being in the postnatal period (Dobbing and Sands 1973). Neuroblasts have been demonstrated in several organisms to be unusually sensitive to radiation. Effects have been shown in the mouse embryo down to doses as low as 10 rad of X rays (Hicks and D'Amato 1966). In the grasshopper embryo neuroblast, mitotic inhibition and chromosome aberrations have been demonstrated after 1 rad of X rays (Gaulden and Read 1978). In cultures of human embryonic cells there is some evidence that "brain" cells in vitro are more sensitive to the induction of chromosome aberrations by radiation (down to 9 rad) than are lung cells (Böök et al. 1962).

Most of the research on radiation effects on the embryo has been focused on gross morphological criteria. Recent research in chemical teratogens, including medicinal compounds (Lewis et al. 1977), has pointed to more subtle effects or what has been called "behavioural teratology" (Werboff and Gottlieb 1963). Langman et al. (1975) have put the need for modern approaches to teratogen research in perspective: "The central nervous system produces special cell types throughout gestation and even after birth and since the cells for various CNS structures are produced at different times, a brief insult may damage some structures and spare others. Which structures sustain damage depends on the population of cells proliferating at the time of insult. . . . The data suggest that late gestation and early postnatal life are not periods of resistance to teratological agents. Rather, insults at these times result in alterations of anatomy and function that are difficult to observe with standard teratological evaluation techniques." This may also apply to very low doses of X rays in early development.

We must reorient our thinking and research on the embryonic effects of low doses of radiation in light of recent advances not only in current knowledge about the genetics, development, and differentiation of the CNS but also their interrelations. Anatomical and biochemical abnormalities can be worked out in experimental animals, but the demonstration of accompanying changes in function, such as cognitive ability and behavior, has seemed until recently to be limited and, therefore, of little value in extrapolating to man. This situation may be improving: for example, single gene behavioral mutants in the mouse (Sidman 1974) and memory mutants in *Drosophila* (Quinn et al. 1979) offer much promise in correlating structural and functional changes with specific mutations. As pointed out by Brent (1977), such correlations are vital to advancing our understanding of normal brain development and function and the effects of radiation on them.

Neoplasia

Space does not permit discussion of this subject in depth, but a few comments on genetic change and neoplasia are pertinent. Mulvihill (1977) has compiled a list of human genes known to involve susceptibility to neoplasia: it is impressive in number (143 genes), in the proportion of dominant genes (about half), and in the number of tissues affected (nearly all). Many tumors unquestionably bear some relation to mutation, which in some cases is inherited, in others is induced in a somatic cell, and in still others is both (Knudson 1977, 1979). A person with a tumor may be viewed as a mosaic in the sense that the tumor represents a clone of cells different from the other somatic cells, one reason being they have acquired among other things the heritable ability (from cell division to cell division) to continue growing without limitation.

Fetal exposure to low doses of radiation and carcinogenesis has generated much discussion (United Nations 1972, 1977). The two largest studies (Stewart et al. 1958, MacMahon 1962) concluded that diagnostic X-ray exposures in utero increase by approximately 50% the postnatal incidence of neoplasia, especially leukemia. This opinion met with great resistance that may be softening a little in that there is now more acceptance of the possibility that low doses may induce neoplasia. Contributing to this acceptance are probably the demonstration of heritable tumors and the recent data strongly suggesting that most carcinogens are also mutagens (Strong 1977, Burnet 1978). It follows that mutation may be an initiating event in radiation-induced carcinogenesis, and if so, there should be no threshold dose and the dose response should be linear at low doses. That this may be true is suggested by Stewart's (1971) calculation of a rough linear relation of relative neoplasia risk to the number of X-ray films in utero. Brown (1977) has examined data on tumor incidence in several populations of irradiated human children and adults and finds some evidence for a linear response at very low doses. In view of the various possible tumor-promoting factors, e.g., hormones, and the preventive factors, e.g., the immune system, one would not expect to find a strict linear dose response in a genetically heterogeneous population such as man.

If a tumor is initiated by mutation in one cell, all of its cells should be alike; there is supporting evidence for the clonal nature of most tumors (Nowell 1976). For example, data on tumors in women known to be heterozygous for two alleles of a gene that determines an enzyme structure and that is located on the X chromosome (only one X is functional in a somatic cell) have revealed genetic homogeneity, which means single-cell origin.

Most tumors have one or more chromosome aberrations, but individual tumors of a given type may not always have the same chromosome change (Harnden 1977). Chronic myelogenous leukemia (CML) is one neoplasm with an unusually consistent chromosome aberration. Cytogenetic analysis has been done on 569 patients with CML (Rowley 1978): in all cases the malignant cells had a translocation and in 95% of the cases it was a balanced one between the long arms of chromosomes 22 and 9. A cause-effect relation of the chromosomal change to the neoplasia is not established, but the circumstantial evidence is certainly strong. Another tumor of interest is retinoblastoma, which results from two mutations. In inherited cases, one mutation is inherited and the other occurs in a somatic cell; in sporadic cases both mutations are somatic (Knudson 1977, 1979). This tumor can also result from inheritance of a small interstitial deletion of a specific segment of the long arm of a chromosome 13, with the second mutation occurring somatically (Sparkes et al. 1979).

Any tumor involving gene mutation may be induced by irradiation of an embryo or fetus at any time during gestation. At low doses one would expect the frequency to be low. Many years of latency usually lie between the initiating carcinogenic event and the development of a detectable neoplasm. More epidemiological studies are needed that cover a large number of individuals over a long period of time. In the meantime, the available data have been taken to indicate that "the risk per unit absorbed dose of fatal induced malignancies by foetal irradiation may be in the region of 200–250 10^{-6} rad^{-1}" (United Nations 1977, Annex J). This is a value difficult for most patients to comprehend, so it is helpful to compare it with the risk of being killed in an automobile accident in their community. Although this risk seems small, our experience is that patients are more willing to take it with automobiles than with fetal irradiation, especially that which is unnecessary.

Conclusions

We would like to support Dr. Alper's plea that researchers not limit themselves to mammalian studies but that they use any cell or organism that is advantageous. We feel this is especially true for very low-dose radiation effects on the embryo. Examination of such effects in man will undoubtedly require sophisticated and large-scale epidemiological methods that involve behavioral genetics tools. Until sufficient data are available on the irradiated human embryo we offer the hypothesis of radiation-induced genetic mosaicism with CNS and neoplasia consequences. All the current evidence taken together makes it the most reasonable one with which to predict effects of low doses of ionizing radiation on the human embryo. The hypothesis is subject to test, and it is hoped that the above discussion will stimulate others to join in attempts to prove or disprove it.

DECISION GUIDELINES FOR IRRADIATED EMBRYOS OR FETUSES

When an embryo has been exposed to diagnostic radiation or scatter radiation from therapy, an estimate of the absorbed dose must be obtained; in some cases this may entail a repeat of the examination(s) with a phantom. Just because an embryo or fetus has been irradiated does not mean that therapeutic abortion is mandatory (Dalrymple and Baker 1977). Next, determination of the approximate date of conception should be made so that the gestational age at the time of radiation exposure can be estimated. For an embryo irradiated during the first six weeks of gestation we have developed a set of guidelines to help physicians and patients in trying to reach a decision on the difficult subject of termination. With the current paucity of knowledge about low-dose effects on experimental embryos, much less on human embryos, it is impossible to establish "rules." The guidelines given below are, therefore, at best rough ones, but we found that physicians want some idea of the limits of relative risks with which to make a judgment even if we cannot give them definitive risk levels at this time. The guidelines represent an expansion of those originally set forth by Hammer-Jacobsen (1959) and are based on the possible biological effects of low doses of radiation detailed above.

At a dose of less than 1 rad to the embryo, the risks do not, in our opinion, justify any action.

The total dose of most examinations in the cases referred to us lies between 1 and 5 rad. There is no way we can assure the patient of zero risks, but we feel that all factors taken together do not usually warrant any action at this dose level. One exception is when irradiation has occurred in the first week of gestation, during which time the risk is highest for induction of a mosaic with relatively large numbers of mutant cells. The physician should discuss with the patient and her husband the possible subtle effects of low doses of radiation as well as other factors that alone or together with radiation might contribute to a subtle or overt defect. These include the patient's age, her physical condition, current stage of pregnancy, medicinal agents taken during pregnancy, genetic defects on either side of the family, inherent risks in absence of radiation, etc. Sometimes young couples prefer not to take any "avoidable" risks no matter how small, radiation being considered such a risk, but the decision should be theirs.

At embryo doses of 5 to 10 rad the risks are naturally higher. We arbitrarily chose 5 rad as a point of increased hazards, because there have been reports of morphological anomalies in mouse and rat embryos given 5 rad of X rays (United Nations 1977, Annex J, Table 5). There is dispute about these data (Brent 1977, Mole 1979). Michel et al. (1977) have reported embryonic effects from 1 rad of X rays when given with lucanthone. The highly inbred nature of the experimental animals used may be a factor in the anomalies; the human by contrast is a relatively highly outbred species. Although the probability of the induction of a gross anomaly in a human embryo by 5 to 10 rad seems very low, the possibility must be kept in mind, and together with increased risks of mosaicism, it calls for more serious consideration of the radiation exposure than at lower doses.

At doses of 10 rad or more to an embryo during the first six weeks of development we recommend termination, the risks of mosaicism for one or more viable chromosome aberrations being significantly increased. In addition, the induction of a morphological anomaly at doses above 10 rad becomes a real possibility.

Missing from these guidelines are risk estimates at the different dose levels. There simply are not yet enough data on radiation-induced mosaics with which to make reasonable estimates. As noted above, the risks for neoplasia induction by diagnostic X-ray exposure in utero have been estimated (United Nations 1977, Mole 1979), but more work is needed on genetic and cytogenetic mosaic induction.

For embryos and fetuses older than six weeks the decision to do nothing after low doses and to abort after high ones appears simple, perhaps more so than is justified. Much more difficult are the cases with doses of approximately 20 to 100 rad, because the data on their CNS effects in the human are almost as scarce as are data on the very low doses. It is necessary "to feel one's way" in this situation. We are often asked about the usefulness of amniocentesis in such cases. That test is useless, even for exposures at less than 16 weeks of pregnancy (when it is usually performed), because if no chromosome aberrations are observed one cannot be confident none are present, and gene mutations cannot be detected.

Because many factors can affect embryonic development, a consultation on an irradiated pregnancy should be as thorough and searching as that performed for any other medical condition. Each case is unique and must be individually evaluated. An obstetrician helping a patient make a decision should have as much relevant information as the radiation biologist, radiological physicist, radiologist, and referring physician can supply.

ELECTIVE SCHEDULING OF DIAGNOSTIC RADIATION EXAMINATIONS

Therapeutic abortion is a most undesirable decision for a pregnant patient and for all concerned. We prefer not to be confronted with the need for considering it. In addition, it is ideal to avoid any dose of radiation to the embryo or fetus if possible. In 1970, therefore, Dr. Edward E. Christensen, a diagnostic radiologist in our department, designed and initiated a system at Parkland Memorial Hospital, Dallas, Texas, for elective scheduling of abdominal diagnostic X-ray examinations of fertile women. The system, which reduces radiation exposure of unsuspected conceptuses, was later extended to include nuclear medicine. "Fertile" females include all those between ages 12 and 45 (in 8,801 births in 1978 at Parkland Hospital, 20% of the mothers were under age 17).

Parkland Hospital is a 1,000-bed city-county hospital. Because of the workload—approximately 250,000 diagnostic radiographic examinations and 10,500 nuclear medicine procedures in 1978—the use of the so-called 10-day rule was deemed not practical for all examinations, so it was limited to the nuclear medicine procedures and to the higher dose radiographic exams, such as barium enema, pyelograms, etc. Examinations for patients beyond the first 10 days after beginning of their last menstrual period (LMP) are delayed and rescheduled when the next period begins. A 28-day rule was instituted for the lower dose exams, such as films of abdomen, sacrum, hips, etc.; only examinations for patients who are more than 28 days beyond LMP are delayed until menses begins. With these two "rules," higher doses are avoided if there is even a small probability of pregnancy, and no radiation is given if the probability of pregnancy is high.

Among the objections to the system voiced prior to its initiation were: it would be too much trouble for the radiology department staff; the referring physicians would consider it an interference; delaying of examinations would compromise patient care; it could not be used with the large number of emergency room patients. None of these objections has proved to be valid.

Approximately half of the requests for X-ray examinations at Parkland Hospital originate in the emergency room; a delayed examination there is usually out of the question. For emergency or hospitalized patients who are potentially pregnant or known to be pregnant, the examination is tailored by the radiologist

TABLE 4. *X-ray examinations, United States**

Exam	1964	1970
Radiographic		
# people	66,086,000	76,449,000
# exams	104,987,000	129,070,000
#/100 people	56.2	64.6
Fluoroscopic	7,800,000	—
Dental		
exams	46,000,000	68,000,000
# films	225,000,000	280,000,000

* From Bureau of Radiological Health Report (1973 and 1976b) and Dr. Larry Crabtree, personal communication

for the particular condition in consultation with the referring physician so that one or two films are made rather than a routine series.

How many women whose examinations are delayed do indeed prove to be pregnant? In a random sample of 500 potentially pregnant women in a five-month period, we found that the X-ray exams of 20% were delayed because of the 10-day rule. This represented only 4% of the total number of all patients being given the same exams in that time period; in other words, there was approximately one delayed exam/day. Of the women whose examinations were delayed, 5% proved to have been pregnant at the time of the original X-ray request. Our clinicians report that many patients and referring physicians decide the exam is unnecessary after it has been delayed; usually the patient becomes asymptomatic.

The extensive use of radiographs and nuclear medicine procedures at Parkland is not unique as can be seen from the national survey figures in Tables 4 and 5. It is our impression that a number of American radiologists are now using elective scheduling. We agree with Carmichael and Warrick (1978) that it is "an essential part of good radiological and radiographic practice." It also provides the simplest method for avoiding the medical and legal complications that may arise from irradiation during pregnancy. The key to the success of an elective scheduling system is the education of all personnel involved, including

TABLE 5. *Population exposure to medical radioisotopes**

Year	# Patient Doses
1959	400,000
1966	1,600,000
1971	4,000,000
1976	~17,000,000†

* Moeller (1971) and estimate for 1976
† Procedures

the referring physicians, on the reasons for and the methods of implementing it. We believe that the benefits are well worth the effort.

ACKNOWLEDGMENTS

We thank our colleagues Drs. Edward E. Christensen and Roberto Restrepo for their generous assistance with clinical details and for many helpful discussions, and Dianna Hallford for technical assistance.

REFERENCES

Baker, W. K. 1978. A genetic framework for Drosophila development. Annu. Rev. Genet. 12:451–470.

Böök, J. A., M. Fraccaro, K. Gredga, and J. Lindsten. 1962. Radiation induced chromosome aberrations in human foetal cells grown in vitro. Acta Genet. Med. Gemellol. (Roma) 11:356–387.

Brent, R. L. 1977. Radiations and other physical agents, *in* Handbook of Teratology, J. G. Wilson and F. C. Fraser, eds. Plenum Press, New York, pp. 153–223.

Brown, J. M. 1977. The shape of the dose-response curve for radiation carcinogenesis. Extrapolation to low doses. Radiat. Res. 71:34–50.

Bureau of Radiological Health Report. 1973. Population exposure to X rays, U.S. 1970. DHEW Publication (FDA) 73–8047.

Bureau of Radiological Health Report. 1976a. Organ doses in diagnostic radiology. DHEW Publication (FDA) 76–8030.

Bureau of Radiological Health Report. 1976b. Gonad doses and genetically significant dose from diagnostic radiology, U.S. 1964 and 1970. BRH Tech. Publ. FDA-76–8034.

Burnet, F. M. 1978. Cancer: somatic-genetic considerations. Adv. Cancer Res. 28:1–29.

Carmichael, J. H. E., and C. K. Warrick. 1978. The ten day rule—principles and practice. Br. J. Radiol. 51:843–846.

Childs, B., J. M. Finucci, M. S. Preston, and A. E. Pulver. 1976. Human behavior genetics. Adv. Hum. Genet. 7:57–97.

Christensen, E. E., T. S. Curry III, and J. E. Dowdey. 1978. An Introduction to the Physics of Diagnostic Radiology. 2nd ed. Lea and Febiger, Philadelphia, pp. 190–193.

Dalrymple, G. V., and M. L. Baker. 1977. Post-irradiation abortion: a slaughter of innocents? J. Arkansas Med. Soc. 73:474–476.

Dekaban, A. S. 1968. Abnormalities in children exposed to X-radiation during various stages of gestation: tentative timetable of radiation injury to the human fetus, Part I. J. Nucl. Med. 9:471–477.

Dobbing, J., and J. Sands. 1973. Quantitative growth and development of human brain. Arch. Dis. Child. 48:757–767.

Finucci, J. M., J. T. Guthrie, A. L. Childs, H. Abbey, and B. Childs. 1976. The genetics of specific reading disability. Ann. Hum. Genet. 40:1–23.

Ford, C. E. 1967. Discussion of mosaic mongols, *in* Mongolism (Ciba Foundation Study Group No. 25). Little, Brown and Co., Boston, pp. 71–76.

Ford, C. E., P. E. Polani, J. H. Briggs, and P. M. F. Bishop. 1959. A presumptive human XXY/XX mosaic. Nature 183:1030–1032.

Francke, U., and N. Oliver. 1978. Quantitative analysis of high-resolution trypsin-Giemsa bands on human prometaphase chromosomes. Hum. Genet. 45:137–165.

Funderburk, S. J., M. A. Spence, and R. S. Sparkes. 1977. Mental retardation associated with "balanced" chromosome rearrangements. Am. J. Hum. Genet. 29:136–141.

Gaulden, M. E. 1974. Possible effects of diagnostic X-rays on the human embryo and fetus. J. Arkansas Med. Soc. 70:424–435.

Gaulden, M. E., and C. B. Read. 1978. Linear dose-response of acentric chromosome fragments down to 1 R of X-rays in grasshopper neuroblasts, a potential mutagen-test system. Mutat. Res. 49:55–60.

Gehring, W. J., ed. 1978. Genetic Mosaics and Cell Differentiation. Springer-Verlag, New York.
Granroth, G. 1979. Defects of the central nervous system in Finland. IV. Association with diagnostic x-ray examinations. Am. J. Obstet. Gynecol. 133:191–194.
Hammer-Jacobsen, E. 1959. Therapeutic abortion on account of X-ray examination during pregnancy. Dan. Med. Bull. 6:113–122.
Harnden, D. G. 1977. Cytogenetics of human neoplasia, *in* Genetics of Human Cancer, J. J. Mulvihill, R. W. Miller, and J. F. Fraumeni, Jr., eds. Raven Press, New York, pp. 87–104.
Hicks, S. P., and C. J. D'Amato. 1966. Effects of ionizing radiation on mammalian development. Advances in Teratology 1:197–250.
Johnson, R. C., and R. B. Abelson. 1969. Intellectual, behavioral, and physical characteristics associated with trisomy, translocation, and mosaic types of Down's syndrome. Am. J. Ment. Defic. 73:852–855.
Kaffe, S., L. Y. F. Hsu, and K. Hirschhorn. 1974. Trisomy 21 mosaicism in a woman with two children with trisomy 21 Down's syndrome. J. Med. Genet. 11:378–381.
Klinger, H. P., and H. G. Schwarzacher. 1962. XY/XXY and sex chromatin positive cell distribution in a 60 mm human fetus. Cytogenetics 1:266–290.
Kohn, G., T. Taysi, T. E. Atkins, and W. J. Mellman. 1970. Mosaic mongolism. I. Clinical correlations. J. Pediatr. 76:874–879.
Knudson, A. G. 1977. Genetics and etiology of human cancer. Adv. Hum. Genet. 8:1–66.
Knudson, A. G. 1979. Hereditary cancer. JAMA 241:279.
Kucerova, M. 1970. Long-term cytogenetic and clinical control of a child following intrauterine irradiation. Acta Radiol. [Ther.] 9:353–361.
Langmen, J., W. Webster, and P. Rodier. 1975. Morphological and behavioural abnormalities caused by insults to the CNS in the perinatal period, *in* Teratology Trends and Applications, C. L. Berry and D. E. Poswillo, eds. Springer-Verlag, New York, pp. 182–200.
Lejeune, J., R. Berger, L. Archambault, R. Gorin, and R. Turpin. 1964. Mosaique chromosomique, probablement radioinduite in utero. C. R. Acad. Sci. [D] (Paris) 259:485–488.
Lemine, R. J. 1974. Embryology of the central nervous system, *in* Scientific Foundations of Paediatrics, J. A. Davis and J. Dobbing, eds. W. B. Saunders Co., Philadelphia, pp. 547–564.
Lewis, P. D., A. J. Patel, G. Béndek, and R. Balázs. 1977. Do drugs acting on the nervous system affect cell proliferation in the developing brain. Lancet 1:399–401.
Lyon, M. F. 1972. X-chromosome inactivation and developmental patterns in mammals. Biol. Rev. 47:1–35.
MacMahon, B. 1962. Prenatal X-ray exposure and childhood cancer. J. Natl. Cancer Inst. 28:1173–1191.
McLaren, A. 1976. Mammalian Chimaeras. Cambridge University Press, Cambridge, England.
Michel, C., H. Fritz-Niggli, H. Blattman, and I. Cordt-Riehle. 1977. Effects of low-dose irradiation with X-rays and pi-mesons on embryos of two different mouse strains. Strahlentherapie 153:674–681.
Mikkelsen, M., and S. Vestermark. 1974. Karyotype 45, XX, -21/46, XX, 21q- in an infant with symptoms of G-deletion syndrome I. J. Med. Genet. 11:389–393.
Moeller, D. W. 1971. Meeting radiological health manpower needs. Am. J. Public Health 61:1938–1946.
Mole, R. H. 1979. Radiation effects on pre-natal development and their radiological significance. Br. J. Radiol. 52:89–101.
Moore, K. L. 1973. The Developing Human. W. B. Saunders Co., Philadelphia.
Mulvihill, J. J. 1977. Genetic repertory of human neoplasia, *in* Genetics of Human Cancer, J. J. Mulvihill, R. W. Miller, and J. F. Fraumeni, Jr., eds. Raven Press, New York, pp. 137–143.
Murphy, D. P. 1929. The outcome of 625 pregnancies in women subjected to pelvic radium or roentgen irradiation. Am. J. Obstet. Gynecol. 18:179–187.
Neumeister, K. 1976. Problems arising from effects of low radiation doses in early pregnancy, *in* Proceedings of a Symposium on Biological Effects of Low-Level Radiation Pertinent to Protection of Man and His Environment, vol. II. International Atomic Energy Agency, Vienna, Publ. STI/PUB/409, pp. 261–270.
Nowell, P. C. 1976. The clonal evolution of tumor cell populations. Science 194:23–28.
Quinn, W. G., P. P. Sziber, and R. Booker. 1979. The *Drosophila* memory mutant *amnesiac.* Nature 277:212–214.
Rowley, J. D. 1978. Chromosomes in leukemia and lymphoma. Semin. Hematol. 15:301–319.
Russell, L. B. 1957. Effects of low doses of X-rays on embryonic development in the mouse. Proc. Soc. Exp. Biol. Med. 95:174–178.

Russell, L. B. 1978a. Somatic cells as indicators of germinal mutations in the mouse. Environ. Health Perspect. 24:113–116.

Russell, L. B., ed. 1978b. Genetic Mosaics and Chimeras in Mammals. Plenum Publishing Corp., New York.

Russell, L. B., and M. H. Major. 1957. Radiation-induced presumed somatic mutations in the house mouse. Genetics 42:161–175.

Russell, L. B., and C. S. Montgomery. 1970. Comparative studies on X-autosome translocations in the mouse. II. Inactivation of autosomal loci, segregation, and mapping of autosomal breakpoints in five T(X:1)'s. Genetics 64:281–312.

Sato, H. 1966. Chromosomes of irradiated embryos. Lancet 2:551.

Schinzel, A., K. Hayashi, and W. Schmid. 1975. Structural aberrations of chromosome 18. II. The 18q- syndrome. Report of three cases. Humangenet. 26:123–132.

Shepard, T. H. 1976. Catalogue of Teratogenic Agents, 2nd ed. Johns Hopkins University Press, Baltimore.

Sidman, R. L. 1974. Contact interaction among developing mammalian brain cells, *in* The Cell Surface in Development, A. A. Moscona, ed. John Wiley & Sons, New York, pp. 221–253.

Soukup, S. W., E. Takacs, and J. Warkany. 1965. Chromosome changes in rat embryos following X-irradiation. Cytogenctics 4:130–144.

Sparkes, R. S., H. Muller, and I. Klisak. 1979. Retinoblastoma with 13q- chromosomal deletion associated with maternal paracentric inversion of 13q. Science 203:1027–1029.

Stern, C. 1968. Genetic Mosaics and Other Essays. Harvard University Press, Cambridge, Massachusetts.

Stewart, A., J. Webb, and D. Hewitt. 1958. A survey of childhood malignancies. Br. Med. J. 1:1495–1508.

Stewart, A. 1971. Low dose radiation cancers in man. Adv. Cancer Res. 14:359–390.

Strong, L. C. 1977. Theories of pathogenesis: mutation and cancer, *in* Genetics of Human Cancer, J. J. Mulvihill, R. W. Miller, and J. F. Fraumeni, Jr., eds. Raven Press, New York, pp. 401–415.

Taylor, A. I. 1968. Cell selection in vivo in normal/G trisomic mosaics. Nature 219:1028–1030.

Taysi, K., G. Kohn, and W. J. Mellam. 1970. Mosaic mongolism. II. Cytogenetic studies. J. Pediatr. 76:880–885.

United Nations. 1972. A Report of the United Nations Scientific Committee on the Effects of Atomic Radiation to the General Assembly, with Annexes. Vol. II: Effects. United Nations Publication No. E.72.IX.18, New York.

United Nations. 1977. A Report of the United Nations Scientific Committee on the Effects of Atomic Radiation to the General Assembly, with Annexes. United Nations Publication No. E.77.IX.1, New York.

Verma, R. S., H. Dosik, I. H. Chowdhry, and R. C. Jhaveri. 1978. Ring chromosome-13 in a child with minor dysmorphic features—irregular phenotypic expression of ring-13 syndrome. Am. J. Dis. Child. 132:1018–1021.

Werboff, J., and J. Gottlieb. 1963. Drugs in pregnancy: behavioural teratology. Obstet. Gynecol. Surv. 18:420–423.

Radiation Biology in Cancer Research, edited by
Raymond E. Meyn and H. Rodney Withers.
Raven Press, New York © 1980.

Radiation Transformation In Vitro: Modification by Exposure to Tumor Promoters and Protease Inhibitors

Ann R. Kennedy and John B. Little

Department of Physiology, Harvard University School of Public Health, Boston, Massachusetts 02115

It has previously been shown that the induction of oncogenic transformation in vitro by radiation is a complex process which involves at least two steps: (1) the initiation and fixation of the transformed state (presumably an irreversible alteration of DNA) as a heritable cellular property, and (2) its subsequent phenotypic expression in terms of morphologically altered cells (Terzaghi and Little 1975, 1976, Little 1977). In an attempt to gain more information about the mechanisms of carcinogenesis, we have been examining the effects of various agents on the different steps in transformation in vitro. Of particular interest has been an examination of factors that influence the expression of X-ray damage, as these may be of particular importance following low-dose or low dose-rate exposure. One critical factor is postirradiation cell proliferation (Terzaghi and Little 1976). Another is the exposure to certain chemical promoting agents. We have previously shown that incubation with a nontoxic concentration of the classical phorbol ester promoting agent 12-0-tetradecanoyl-phorbol-13-acetate (TPA) during the expression period will greatly enhance X-ray transformation in mouse C3H 10T½ cells (Kennedy et al. 1978). We report here on further studies of the modification of X-ray transformation by TPA and on the effects of exposure to several inhibitors of proteolytic enzymes.

The concept of "two-stage" carcinogenesis (initiation-promotion) was initially derived from experiments on the induction of cancer in mouse skin by painting it with polycyclic hydrocarbon carcinogens (Berenblum 1975, Boutwell 1974, Scribner and Süss 1978). It was observed that the incidence of skin papillomas induced by a single low-dose application of an initiating carcinogen could be markedly enhanced if the area were subsequently treated with repeated applications of croton oil—an irritant that is noncarcinogenic by itself. The effect of croton oil was found to be due to the phorbol ester compounds it contains, one of the most active of which is TPA (Hecker 1971). TPA has subsequently been shown to be nonmutagenic (McCann et al. 1975, Trosko et al. 1977) and not to induce DNA damage or repair processes at doses at which it is biologically active as a promoting agent (personal communication, James Trosko). The fact that TPA will enhance transformation in vitro suggests that its effect is a general one, not specifically restricted to the mouse skin system.

Our interest in the effects of protease inhibitors on radiation transformation was stimulated by the reports that some of these agents could inhibit error-prone ("SOS") DNA repair and mutagenesis in bacteria (Meyn et al. 1977, Umezawa et al. 1977) as well as inhibit TPA-enhanced carcinogenesis in vivo (Troll et al. 1970, Hozumi et al. 1972). Thus, protease inhibitors might influence both the fixation and expression steps in in vitro transformation. It is our hypothesis that much of the transformation observed in rodent cell lines in culture may result from the action of an error-prone DNA repair process such as has been shown to exist in bacterial cells. Supporting evidence for such a mechanism was found in previous experiments in which an enhancement in the transformation frequency per viable cell was observed during the repair of potentially lethal radiation damage in 10T½ cells (Terzaghi and Little 1975, Little 1977). Protease inhibitors are thought to suppress error-prone repair in bacterial cells by inhibiting the proteolytic cleavage of specific protein repressors involved in the SOS repair process (Witkin 1976, Radman et al. 1977).

We have investigated the effects of three protease inhibitors—antipain, leupeptin, and soybean trypsin inhibitor (SBTI)—that specifically inhibit different proteolytic enzymes. The enzymes inhibited include: antipain—papain, trypsin, cathepsin A and B, and plasmin and plasminogen activator (Umezawa 1976, Aoyagi and Umezawa 1975, Zimmerman et al. 1978); leupeptin—papain, trypsin, plasmin, and cathepsin B (Umezawa 1972, Aoyagi and Umezawa 1975); and SBTI—trypsin, chymotrypsin (Kunitz 1947), and chick embryo Rous sarcoma TPA-induced plasminogen activator (personal communication, James Quigley). SBTI is of particular interest as it is a naturally occurring protease inhibitor found in the normal diet; it may thus play a role in the prevention of cancer in man (Troll et al. 1979). It has recently been found that carcinogenesis in three organ systems can be inhibited in animals kept on a soybean diet rich in protease inhibitors (Troll et al. 1979).

Antipain and leupeptin are small molecular weight (approximately 500–600 daltons) natural products isolated from actinomycetes. They have been characterized by Umezawa (1976) and Aoyagi and Umezawa (1975). Effective concentrations of these two inhibitors are nontoxic, either with or without radiation exposure, and have been shown to have no effect on growth, RNA, or protein synthesis in bacteria (Meyn et al. 1977) or mammalian cells (Weber et al. 1975, Hynes et al. 1975, Blumberg and Robbins 1975). SBTI is a high molecular weight (approximately 24,000 daltons), macromolecular type protease inhibitor (Kunitz 1947), also nontoxic to C3H 10T½ cells at the concentrations used. A preliminary report of our results has been published (Kennedy and Little 1978).

MATERIALS AND METHODS

We used a C3H mouse embryo–derived cell line (10T½ clone 8) isolated and characterized by Reznikoff et al. (1973a, 1973b) and adapted for studies of radiation transformation (Terzaghi and Little 1975, 1976, Little 1977). Stock cultures were maintained in 60-mm Petri dishes and were passaged by subculturing at a 1:20 dilution every seven days. The cells used were in passages 9–14. They were grown in a humidified 5% CO_2 atmosphere at 37°C in Eagle's basal medium supplemented with 10% heat-inactivated fetal calf serum and antibiotics. Cells were irradiated 24 hours after seeding of replicate 100-mm Petri dishes with 100–400 viable cells each. Irradiation was carried out at room temperature with a 100-kV Philips MG-100 industrial X-ray generator operating at 9.6 ma and yielding a dose rate to the cells of 78 rads/minute. Cloning efficiencies in each experimental group were determined from dishes seeded with a cell density one-fifth that of those for the transformation assay and terminated 10 days after irradiation. The protocol for X-ray transformation experiments for 10T½ cells is shown schematically in Figure 1. Methodological details for these experiments have been previously described (Terzaghi and Little 1975, 1976).

Incubation with TPA at a concentration of 0.1 μg/ml was begun 48 hours after irradiation and maintained for the entire six-week expression period as previously described (Kennedy et al. 1978); in the subculture experiments, treatment with TPA was begun immediately after subculturing. Protease inhibitor treatment was begun immediately after irradiation and continued for one day, five days, or the entire course of the transformation experiments (six weeks). TPA and protease inhibitors were suspended in fresh complete medium. Concentrations of protease inhibitors used were as follows: antipain, 600 μg/ml (1 mM) for one or five days or 50 μg/ml added continuously for six weeks; leupep-

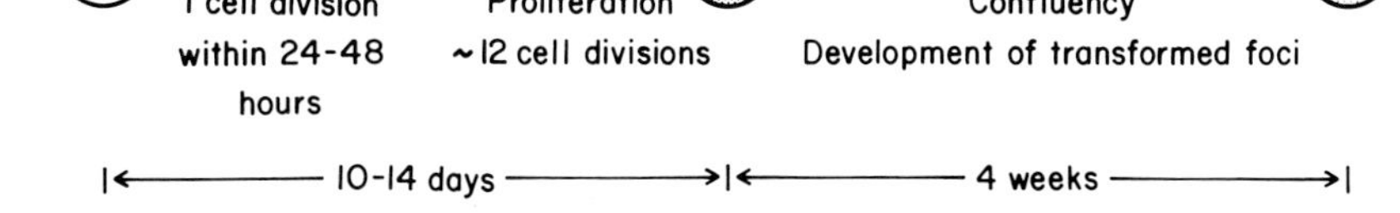

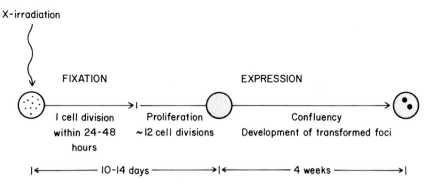

FIG. 1. Development of malignant transformation in vitro. Specific parameters shown are for 10T½ cells. About 300 viable cells are seeded in a 100-mm Petri dish (left), and irradiated. These cells proliferate until confluency is reached and cell division ceases 10–14 days later (center). Dense, piled up, transformed foci as are shown in Figure 1 appear overlying confluent monolayer 4–5 weeks later (right).

tin, 600 μg/ml (1.2 mM) for one or five days or 50 μg/ml added continuously; SBTI, 300 μg/ml for one day or 100 μg/ml added continuously. Stock solutions of 10 mg/ml of the protease inhibitors were prepared in distilled water, filtered through 0.2-μg Nalgene filter units, and kept at −80°C. The protease inhibitor solutions used in these experiments were prepared from the stock solutions each time an addition was to be made to the cultures. Antipain and leupeptin were obtained from Walter Troll, SBTI from Sigma.

A preliminary experiment was carried out with the A31-11 mouse BALB/3T3 cell line (Kakunaga 1973) to determine whether the suppressive effect of antipain was cell line specific. Stock cultures were passaged as for 10T½ cells. Cells were seeded 24 hours before irradiation at densities that yielded about 4,000 viable cells per 100-mm Petri dish. They were grown in Eagle's minimal essential medium supplemented with 10% heat-inactivated fetal calf serum and antibiotics. Transformed foci were scored at four weeks (Kakunaga 1973, Little, in press). Antipain was added at 50 μg/ml (0.08 mM) and TPA at 0.1 μg/ml.

RESULTS

In the classical mouse skin carcinogenesis experiments (Berenblum 1975, Boutwell 1974, Scribner and Süss 1978), repeated applications of phorbol ester promoting agents were equally effective whether exposure to the promoter began immediately or several months after treatment with the initiating carcinogen. One mechanism proposed for the action of these agents was that they stimulated cell proliferation, facilitating expression of the damage produced by the initiating agent. We have previously shown that the capacity for cell proliferation is critical for the phenotypic expression of X-ray transformation in 10T½ cells (Terzaghi and Little 1976, Little 1977); specifically, as is indicated in Figure 1, these cells must undergo on the average at least 12–13 rounds of cell division after irradiation in order for transformation to occur. In the present investigation, experiments were designed to examine these two questions in vitro: (1) Will TPA enhance oncogenic transformation if exposure to it is delayed for a finite period after X-irradiation; and (2) can the effect of TPA be related to a stimulation of cell proliferation? The results of such an experiment are shown in Table 1.

In the experiment shown in Table 1, cultures containing about 300 viable 10T½ cells were X-irradiated with 400 rads and returned to the incubator for 10–14 days until they reached confluency. Under these conditions, the cells will go through an average of 13 rounds of cell division before proliferation ceases as the cultures become density inhibited (Figure 1). The confluent cultures were then trypsinized, single-cell suspensions prepared, and the cells reseeded into new dishes at the original density of 300 cells per dish to allow another series of 13 cell divisions. After this reseeding at low density, the new dishes were returned to the incubator for six weeks and the appearance of transformed

TABLE 1. *Effect of TPA and reseeding at confluency on X-ray-induced transformation*

Group	Treatment	No. Transformed Foci (types 2 + 3) Predicted	Observed	Transformation Frequency (× 10^{-4})	No. Plates Containing Transformed Foci*
a	400 rads	—	6	15	5/16 = 31.3%
b	400 rads + reseeded	6.5†	5	11	5/20 = 25.0%
c	400 rads + reseeded + TPA	5.0‡	18	40	13/20 = 65.0%

* Tests of significance:
χ^2 on total number of plates containing transformants;
groups a vs. b, 5/16 vs. 5/20 $p > .05$
groups b vs. c, 5/20 vs. 13/20 $p < .001$
† Predicted figure based on the transformation frequency observed in group a and based on the assumptions described in the text.
‡ Calculated on the observed transformation seen in group b.

foci scored. As can be seen in groups a and b in Table 1, reseeding did not affect the ultimate transformation frequency. As increasing the number of postirradiation cell divisions from 13–26 did not enhance transformation, it appears unlikely that the effect of TPA can be explained primarily on the basis of such a mechanism. Incubation with TPA during the second series of divisions (after subculture), however, did significantly enhance the transformation frequency (Table 1, group c vs. a). Thus, delayed exposure to TPA was effective in facilitating expression of transformation, as was observed in the mouse skin experiments.

Table 2 shows the results of X-ray transformation experiments in which the cells were incubated with protease inhibitors and TPA for the entire six-week

TABLE 2. *Effect of continuous incubation with protease inhibitors (PI) on X-ray transformation*

Treatment	Transformation Frequency (× 10^{-4}) ± SE: Antipain	Leupeptin	SBTI*
600 rads	20.8 ± 2.8	20.8 ± 2.8	23.1
600 rads + PI	<3.1	<3.9	19.7
400 rads	7.5 ± 1.0	7.5 ± 1.0	9.7
400 rads + PI	—	—	12.5
400 rads + TPA	32.9 ± 2.2	32.9 ± 2.2	18.5
400 rads + TPA + PI	<4.1	18.6 ± 10.4	3.8

* Based on results of one experiment.

postirradiation expression period. As can be seen, continuous exposure to antipain and leupeptin significantly reduced the transformation frequency resulting from exposure to 600 rads alone; SBTI did not affect transformation induced by X rays alone. SBTI did, however, suppress X-ray transformation enhanced by TPA. Antipain completely suppressed TPA-enhanced radiation transformation, whereas additional experiments indicate that the suppressive effects of leupeptin on radiation transformation enhanced by TPA are not significant.

In order to determine whether the effect of protease inhibitors might be cell line specific, the effect of antipain on radiation transformation was examined in the A31-11 line of mouse 3T3 cells. The results of this experiment are shown in Table 3. Antipain suppressed radiation transformation as well as radiation transformation enhanced by TPA in these cells, indicating the effect is not related specifically to 10T½ cells.

The effects on X-ray transformation of short-term incubation of C3H 10T½ cells with protease inhibitors are shown in Table 4. The cells were incubated with the inhibitors beginning immediately after X irradiation. As can be seen in Table 4, only antipain significantly suppressed transformation when present for 24 hours; leupeptin and SBTI did not affect radiation transformation when present for only a 24-hour interval after irradiation. Leupeptin did, however, have a significant depressive effect on radiation transformation when present for five days.

The fact that antipain suppressed X-ray-induced transformation during the first 24 hours after irradiation suggests that it may have an effect on the fixation process of the initial transformational damage. As antipain has been shown to suppress error-prone DNA repair in bacterial cells (Meyn et al. 1977), its effect at early postirradiation times in 10T½ cells could be due to a similar mechanism. Experimental evidence that an error-prone repair process may play a role in X-ray transformation is shown in Figure 2. When X-irradiated 10T½ cells

TABLE 3. *Effect of continuous incubation with antipain on oncogenic transformation in 3T3 cells by X rays plus TPA*

Treatment	Total No. Cells at Risk	Total No. Transformed Foci	Transformation Frequency ($\times 10^{-4}$)
400 rads	60,500	66	10.9
400 rads + antipain	43,450	14	3.2
400 rads + TPA	52,800	91	17.2
400 rads + TPA + antipain	50,050	0	<0.2
100 rads + TPA	34,400	29	8.4
100 rads + TPA + antipain	33,500	0	<0.3

TABLE 4. *Effect of short-term incubation with protease inhibitors (PI) on X-ray transformation*

	Transformation Frequency (× 10^{-4}) ± SE*		
Treatment	Antipain	Leupeptin	SBTI
600 rads	20.7 ± 2.6	20.7 ± 2.6	23.1
600 rads + PI 1 day	5.6 ± 0.5	16.7 ± 3.3	18.3
600 rads + PI 5 days	<3.2 ± 0.7	7.8 ± 3.4	—

* Based on results of three experiments, except SBTI, which is one experiment. Incubation with protease inhibitors begun immediately after irradiation.

were allowed to repair potentially lethal damage by being held in a nonproliferative state for several hours after irradiation (similar to liquid holding recovery experiments in bacterial cells), an enhancement in transformation per surviving cell accompanied the enhancement that occurred in survival. One explanation for this phenomenon is that additional transformational lesions were actually induced in the DNA during repair of the initial X-ray damage.

DISCUSSION

It is clear from our results that both TPA and protease inhibitors given either alone or together can greatly influence the frequency of transformation induced by X rays. These results suggest that factors that influence the expression of the initial X-ray damage may largely determine the apparent X-ray-induced transformation frequency.

The reseeding experiments shown in Table 1 using radiation alone yield some information about the mechanisms of radiation transformation and its enhancement by TPA. If the potentially transformed cells in irradiated cultures were damaged in such a way that all of their progeny could eventually express the transformed phenotype, and these progeny proliferated at the same rate as those

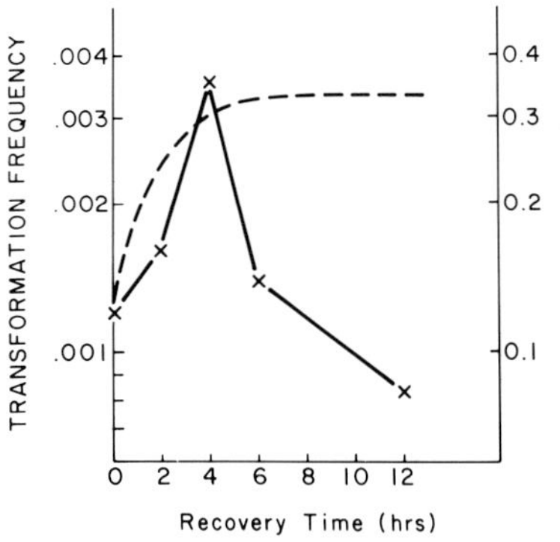

FIG. 2. Changes in transformation frequency during repair of potentially lethal X-ray damage. Cells were irradiated with 400 rads during stationary growth and allowed recovery intervals of 0–12 hours before subculture at low density to assay for transformation. Note that transformation actually increased during early recovery periods. Dashed line represents the change in survival observed in parallel experiments.

from untransformed cells, the ratio of potentially transformed cells to normal cells in the population at confluency should be the same as the ratio in the original treated plates. For example, if one out of 300 viable cells were potentially transformed in the initial culture, at confluency (about 2.5×10^6 total cells) both the normal and potentially transformed cells will have gone through about 13 rounds of cell divisions. This would result in approximately 8,000 progeny of the potentially transformed cell being present among 2.5×10^6 total cells or a ratio of one to 300. By subculturing to 300 viable cells again, on the average one of these potentially transformed progeny should be present, which would ultimately give rise to one transformed focus. The expected number of foci shown in Table 1 was calculated on the basis of this assumption. The fact that the observed number of foci in group b equal the expected number suggests that all of the progeny of the original potentially transformed cell were capable of giving rise to transformed foci (phenotype) in the radiation-alone group.

We have previously shown (Kennedy et al. 1978) that TPA added immediately after X-irradiation with 400 rads and maintained throughout the six-week expression period leads to about a threefold enhancement in transformation. The results in Table 1 show that when TPA was added only after subculture (group c), a similar enhancement occurred. In this subcultured group, exposure to TPA did not begin until at least 13 postirradiation cell divisions had already taken place. On the other hand, the additional proliferation that took place in group b did not appear to enhance transformation. We interpret these results to indicate (1) that, as in vivo, TPA is effective even when treatment begins a long time after exposure to the initiating agent, and (2) that the enhancement of the expression of X-ray transformation by TPA is not due to a simple stimulation of cell proliferation. This latter conclusion was also reached by Mondal et al. (1976) for TPA enhancement of methylcholanthrene-induced transformation in C3H 10T½ cells.

The results of the experiment shown in Table 1 argue against several hypotheses that have been proposed to explain the mechanism of TPA action. As pointed out above, TPA enhancement of X-ray transformation does not appear to occur merely by increasing cell proliferation. Furthermore, TPA enhancement does not appear to result simply from the conversion of premutational lesions in DNA to mutations, as the initiating event is self-replicating and transmitted to progeny cells, and exposure to TPA many generations later is equally effective in enhancing transformation.

Similarly, these experiments also argue against the effect being related to an effect of TPA on DNA repair. Promoting agents have been shown to inhibit excision repair (Cleaver and Painter 1975, Gaudin et al. 1972, Teebor et al. 1973), although this effect does not occur at TPA dose levels associated with biological activity as a promoting agent (personal communication, James Trosko). It has been reported that TPA has no effect on unscheduled DNA repair synthesis and postreplication repair in mammalian cells (Trosko et al.

1975). Despite these facts, it has been proposed that inhibition of repair of DNA damage may be responsible for promotion. Furthermore, it has been postulated that TPA could exert its promoting effect through an error-prone repair system that acts on long-lived DNA lesions (Troll et al. 1978). Although it is possible that some DNA lesions might not induce SOS repair activity unless TPA is present, it is unlikely that these lesions would still be present in significant numbers after 13 rounds of cell division. Finally, a common hypothesis concerning promoter action is that a latent tumor cell is formed by the initiating agent or that an initiated cell goes into a resting or G_0 state until acted upon by TPA (for discussion of this hypothesis, see Berenblum 1975). The subculture experiments shown in Table 1 speak strongly against this hypothesis, as the initiated cells are clearly replicating or they would be almost completely diluted out by the time of subculture. Many other hypotheses have been presented to explain the effects of promoting agents (Berenblum 1975, Boutwell 1974, Scribner and Süss 1978, Weinstein et al. 1979).

It is clear from our results using protease inhibitors that these substances greatly influence radiation transformation, and that this effect is not cell line specific. However, the three protease inhibitors we have studied appear to affect the transformation process in different ways. It can be seen in Table 2 that when antipain was added at 50 μg/ml throughout the six-week fixation and expression periods, it significantly suppressed radiation transformation as well as radiation transformation enhanced by TPA. Leupeptin significantly suppressed radiation transformation, but did not significantly affect radiation transformation enhanced by TPA. We are currently examining the effects of higher concentrations of leupeptin. SBTI had no effect on transformation induced by X rays alone, but did suppress transformation enhanced by TPA. These results suggest that there may be three different proteases involved in transformation induced by radiation. It is of interest that Kuroki and Drevon (personal communication) also found evidence for a role for several proteases in their experiments using protease inhibitors and chemically induced transformed clones of C3H 10T½ cells.

The results in Table 4 indicating that antipain present for only 24 hours after X-irradiation suppressed transformation suggests that some of the suppressive effect of antipain on radiation transformation may be due to an effect on a DNA repair process. It has been shown by Meyn et al. (1977) that antipain specifically suppresses error-prone (SOS) repair and mutagenesis in bacterial cells. The results in Figure 2 suggest that such a system may be involved in the induction of mutations leading to transformation in 10T½ cells. Interestingly, in experiments similar in design to that shown in Figure 2, a parallel enhancement was observed in the frequency of sister chromatid exchanges (SCE) induced during early recovery intervals (Nagasawa and Little, in press), presumably reflecting the activity of a molecular repair process. Antipain suppressed the induction of these SCE during the first four hours of recovery after X irradiation (Nagasawa and Little, in press). The results presented in Table 4 on the suppression of transformation by 24-hour exposure to antipain may thus in part reflect the suppression of an error-prone process for X-ray-induced DNA damage. Although antipain significantly depressed radiation transformation when present for one day after radiation, total inhibition of transformation occurred only when the inhibitor was present for the entire fixation and expression period (Table 2).

There is some additional evidence from the work of Borek et al. (in press) that the presence of antipain at early times is necessary for complete suppression of transformation. They found that antipain added immediately after radiation and continued for the whole expression period suppressed transformation, but suppression did not occur if antipain treatment was begun 24 hours after radiation exposure. The fact that leupeptin had no effect on radiation transformation when present for 24 hours following radiation exposure is consistent with its lack of ability to suppress error-prone repair in bacteria (personal communication, Walter Troll).

We postulate that another protease (besides the "repair" protease) is involved in the suppression of transformation by antipain. This protease is primarily involved in the expression phase of radiation transformation, and is affected by both antipain and leupeptin but not SBTI. We postulate that this protease is different from the one involved in the TPA enhancement of transformation. The protease involved with TPA action seems to be affected by antipain and SBTI. Endogenous proteases have been implicated previously in the mechanism of tumor promotion (Troll 1976). The induction of proteases by phorbol esters in vivo has been demonstrated by Troll et al. (1970) and Hozumi et al. (1972). These investigators also showed that tumor promotion by TPA could be suppressed by protease inhibitors (Troll et al. 1970, Hozumi et al. 1972). Plasminogen activator is a protease thought to play a central role in cell transformation (Rifkin et al. 1975). The induction of plasminogen activator activity by TPA has been demonstrated by Wigler and Weinstein (1976) and Luskotoff and Edgington (1977) in cultured cells. It is possible that plasminogen activator is the protease involved in the TPA promotion phase of radiation transformation.

Our studies on protease inhibitors give further evidence to support the hypothesis that complete carcinogens, which both initiate and promote, are different in their actions from the two-stage carcinogenesis scheme of an initiating agent and a promoting agent. Scribner and Süss (1978) have recently reviewed the evidence for the nonequivalence of complete carcinogenesis and the two-stage process in vivo. Our studies suggest that two different proteases are involved in the expression process that is part of complete carcinogenesis and the promotion brought about by TPA exposure.

These experiments do not directly address the question of the carcinogenic effects of low dose and low dose-rate irradiation. They indicate, however, that ancillary factors and agents may greatly influence the expression of transformation initiated by low doses of ionizing radiation, and may thus be of considerable importance in radiation carcinogenesis resulting from environmental exposures.

ACKNOWLEDGMENTS

We thank Dr. Walter Troll and Dr. I. Bernard Weinstein for helpful discussions concerning the mechanisms of action of TPA and protease inhibitors. We thank Dr. Walter Troll and the U.S.–Japan Cooperative Cancer Research Program for providing the antipain and leupeptin used in this study. We thank Gary Murphy for expert technical assistance. This research was supported by Grant No. CA-22704, awarded by the National Cancer Institute, Department of Health, Education and Welfare, and by Grant No. ES-00002, awarded by the National Institute of Environmental Health Sciences, Department of Health, Education and Welfare.

REFERENCES

Aoyagi, T., and H. Umezawa. 1975. Structures and activities of protease inhibitors of microbial orgin, *in* Proteases and Biological Control, E. Reich, D. B. Rifkin, and E. Shaw, eds. Cold Spring Harbor Laboratories, Cold Spring Harbor, New York, pp. 429–454.

Berenblum, I. 1975. Sequential aspects of chemical carcinogenesis: Skin, *in* Cancer: A Comprehensive Treatise, F. F. Becker, ed. Plenum Press, New York, pp. 323–344.

Blumberg, P. M., and R. W. Robbins. 1975. Relation of protease action on the cell surface to growth control and adhesion, *in* Proteases and Biological Control, E. Reich, D. B. Rifkin, and E. Shaw, eds. Cold Spring Harbor Laboratories, Cold Spring Harbor, New York, pp. 945–956.

Borek, C., R. Miller, C. Pain, and W. Troll. 1979. Conditions for the inhibiting and enhancing effect of the protease inhibitor antipain on X-ray induced neoplastic transformation in hamster and mouse cells. Proc. Natl. Acad Sci. USA (in press).

Boutwell, R. K. 1974. The function and mechanisms of promoters of carcinogenesis. CRC Critical Rev. Toxicol. 2:419–443.

Cleaver, J. E., and R. B. Painter. 1975. Absence of specificity of inhibition of DNA repair replication by DNA-binding agents, cocarcinogens and steroids in human cells. Cancer Res. 35:1773–1778.

Gaudin, D., R. S. Gregg, and K. L. Yielding. 1972. Inhibition of DNA repair by cocarcinogens. Biochem. Biophys. Res. Commun. 48:945–949.

Hecker, E. 1971. Isolation and characterization of the co-carcinogenic principles from croton oil. Methods in Cancer Research 6:439–484.

Hozumi, M., M. Ogawa, T. Sugimura, T. Takeuchi, and H. Umezawa. 1972. Inhibition of tumorigenesis in mouse skin by leupeptin, a protease inhibitor from actinomycetes. Cancer Res. 32:1725–1728.

Hynes, R. O., J. A. Wyke, J. M. Bye, K. C. Humphryes, and E. S. Pearlstein. 1975. Proteases involved in altering surface proteins during virus tranformation, *in* Proteases and Biological Control, E. Reich, D. B. Rifkin, and E. Shaw, eds. Cold Spring Harbor Laboratories, Cold Spring Harbor, New York, pp. 931–944.

Kakunaga, T. 1973. A quantitative system for assay of malignant transformation by chemical carcinogens using a clone derived from BALB/3T3. Int. J. Cancer 12:463–473.

Kennedy, A. R. and J. B. Little. 1978. Protease inhibitors suppress radiation-induced malignant transformation in vitro. Nature 276:825–826.

Kennedy, A. R., S. Mondal, C. Heidelberger, and J. B. Little. 1978. Enhancement of X-ray transformation by 12-0-tetradecanoyl-phorbol-13-acetate in a cloned line of C3H mouse embryo cells. Cancer Res. 38:439–443.

Kunitz, M. 1947. Crystalline soybean trypsin inhibitor. II. General properties. J. Gen. Physiol. 30:291–310.

Little, J. B. 1977. Radiation carcinogenesis in vitro: Implications for mechanisms, *in* Origins of Human Cancer, H. H. Hiatt, J. D. Watson, and J. A. Winston, eds. Cold Spring Harbor Conferences on Cell Proliferation, Vol. IV. Cold Spring Harbor Laboratory, Cold Spring Harbor, New York, pp. 923–939.

Little, J. B. 1979. Quantitative studies of radiation transformation with the A31-11 mouse BALB/3T3 cell line. Cancer Res. (in press).

Luskutoff, D. J., and T. S. Edgington. 1977. Synthesis of a fibronolytic activator and inhibitor by endothelial cells. Proc. Natl. Acad. Sci. USA 74:3903–3907.

McCann, J., E. Choi, E. Yamasaki, and B. N. Ames. 1975. Detection of carcinogens as mutagens in the Salmonella/microsome test: assay of 300 chemicals. Proc. Natl. Acad. Sci. USA 72:5135–5139.

Meyn, M. S., T. Rossman, and W. Troll. 1977. A protease inhibitor blocks SOS functions in *Escherichia coli:* Antipain prevents repressor inactivation, ultraviolet mutagenesis, and filamentous growth. Proc. Natl. Acad. Sci. USA 74:1152–1156.

Mondal, S., D. W. Brankow, and C. Heidelberger. 1976. Two-stage chemical oncogenesis in cultures of C3H/10T½ cells. Cancer Res. 36:2254–2260.

Nagasawa, H., and J. B. Little. 1979. Effect of tumor promoters, protease inhibitors, and repair processes on x-ray-induced sister chromatid exchanges in mouse cells. Proc. Natl. Acad. Sci. USA (in press).

Radman, M., G. Villani, S. Boiteux, M. Defais, P. Caillet-Fauquet, and S. Spadari. 1977. On the mechanism and genetic control of mutagenesis induced by carcinogenic mutagens, *in* Origins of Human Cancer, H. H. Hiatt, J. D. Watson, and J. A. Winston, eds. Cold Spring Harbor Laboratories, Cold Spring Harbor, New York, pp. 903–922.

Reznikoff, C. A., J. S. Bertram, D. W. Brankow, and C. Heidelberger. 1973a. Quantitative and qualitative studies of chemical transformation of cloned C3H mouse embryo cells sensitive to postconfluence inhibition of cell division. Cancer Res. 33:3239–3249.

Reznikoff, C. A., D. W. Brankow, and C. Heidelberger. 1973b. Establishment and characterization of a cloned line of C3H mouse embryo cells sensitive to postconfluence inhibition of cell division. Cancer Res. 33:3231–3238.

Rifkin, D. B., L. P. Beal, and E. Reich. 1975. Macromolecular determinants of plasminogen activator synthesis, *in* Proteases and Biological Control, E. Reich, D. B. Rifkin, and E. Shaw, eds. Cold Spring Harbor Laboratories, Cold Spring Harbor, New York, pp. 841–847.

Scribner, J. D., and R. Süss. 1978. Tumor initiation and promotion, *in* International Review of Experimental Pathology, G. W. Richter and M. A. Epstein, eds. Academic Press, New York, pp. 138–198.

Teebor, G. W., N. J. Duker, S. A. Puacan, and K. J. Zachary. 1973. Inhibition of thymine dimer excision by the phorbol ester, phorbol myristate acetate. Biochem. Biophys. Res. Commun. 50:66–70.

Terzaghi, M., and J. B. Little. 1975. Repair of potentially lethal radiation damage in mammalian cells is associated with enhancement of malignant transformation. Nature 253:548–549.

Terzaghi, M., and J. B. Little. 1976. X-radiation induced transformation in a C3H mouse embryo-derived cell line. Cancer Res. 36:1367–1374.

Troll, W. 1976. Blocking tumor promotion by protease inhibitors, *in* Fundamentals in Cancer Prevention, P. N. Magee et al. eds. University Park Press, Baltimore, pp. 41–55.

Troll, W. A., A. Klassen, and A. Janoff. 1970. Tumorigenesis in mouse skin: Inhibition of synthetic inhibitors of proteases. Science 169:1211–1213.

Troll, W., M. S. Meyn, and T. G. Rossman. 1978. Mechanisms of protease action in carcinogenesis, *in* Carcinogenesis, Vol. 2, Mechanisms of Tumor Promotion and Cocarcinogenesis, T. J. Slaga, A. Sivak, and R. K. Boutwell, eds. Raven Press, New York, pp. 301–312.

Troll, W., R. Weisner, S. Belman, and C. J. Shellabarger. 1979. Inhibition of carcinogenesis by feeding diets containing soybeans. Proc. Am. Assoc. Cancer Res. 20:265.

Trosko, J. E., C. Chang, L. P. Yotti, and E. H. Y. Chu. 1977. Effect of phorbol myristate acetate on the recovery of spontaneous and ultraviolet light–induced 6-thioguanine and ouabain-resistant Chinese hamster cells. Cancer Res. 37:188–193.

Trosko, J. E., J. D. Yager, G. T. Bowden, and F. R. Butcher. 1975. The effects of several croton oil constituents on two types of DNA repair and cyclic nucleotide levels in mammalian cells in vitro. Chem.-Biol. Interact. 11:191–205.

Umezawa, H. 1976. Structures and activities of protease inhibitors of microbial origin. Methods Enzymol. 45:678–695.

Umezawa, H., T. Matsushima, and T. Sugimura. 1977. Antimutagenic effect of elastatinal, a protease inhibitor from actinomycetes. Japan Acad. Ser. B 53:30–33.

Weber, M. J., A. H. Hale, and D. E. Roll. 1975. Role of protease activity in malignant transformation by Rous sarcoma virus, *in* Proteases and Biological Control, E. Reich, D. B. Rifkin, and E. Shaw, eds. Cold Spring Harbor Laboratories, Cold Spring Harbor, New York, pp. 915–930.

Weinstein, I. B., H. Yamasaki, M. Wigler, L. S. Lee, P. B. Fisher, A. Jeffrey, and D. Grunberger.

1979. Molecular and cellular events associated with the action of initiating carcinogens and tumor promoters, *in* Carcinogens: Identification and Mechanisms of Action (The University of Texas System Cancer Center 31st Annual Symposium on Cancer Research, 1978), A. C. Griffin and C. R. Shaw, eds. Raven Press, New York, pp. 399–418.

Wigler, M., and I. B. Weinstein. 1976. Tumor promoter induces plasminogen activator. Nature 259:232–233.

Witkin, E. M. 1976. Ultraviolet mutagenesis and inducible DNA repair in *Escherichia coli.* Bact. Rev. 40:869–907.

Zimmerman, M., J. P. Quigley, B. Ashe, C. Dorn, R. Goldfarb, and W. Troll. 1978. Direct fluorescent assay of urokinase and plasminogen activators of normal and malignant cells: Kinetics and inhibitor profiles. Proc. Natl. Acad. Sci. USA 75:750–753.

Radiation Biology in Cancer Research, edited by
Raymond E. Meyn and H. Rodney Withers.
Raven Press, New York © 1980.

Carcinogenesis in Mice after Low Doses and Dose Rates

R. L. Ullrich

Biology Division, Oak Ridge National Laboratory, Oak Ridge, Tennessee 37830

The relationship between radiation dose and its carcinogenic effectiveness and the influence of dose rate on that relationship have been the subject of much interest, debate, and research for many years. Estimates of risk from the low dose and low dose rate to which human beings are usually exposed must be derived by extrapolation from observations at high doses, based on assumptions about dose and dose-rate relationships.

In experimental animals, there are few instances in which the influence of dose and dose rate on tumor induction have been analyzed over a sufficiently wide dose range to adequately define the shape of the low linear energy transfer (LET) dose-response curve and to assess the influence of dose rate on this relationship. Most data on radiation carcinogenesis are in the dose range of 50–500 rad, and data at doses below 50 rad are limited. These data have been derived mainly from studies in the mouse and to a lesser extent in the rat. Even for mice and rats, however, data are available for only a few tumor types. Therefore, it is difficult to derive generalizations from the information available. This paper will discuss the influence of dose and dose rate for those experimental systems in which sufficient data are available, as well as the limitations of these data with regard to the mechanistic interpretations. A large portion of the information presented has been derived from a series of experiments on the influence of dose and dose rate on the induction of neoplastic disease that has been completed recently in our laboratory.

For the discussion on dose-rate effects, distinction must be made at the outset between studies of protraction and dose rate and studies that utilize fractionated exposures. Many of these latter studies have used relatively high doses per fraction, and the applicability of these studies to dose-rate effects is not known. Although fractionation studies can be useful for the understanding of carcinogenic mechanisms and eventually for understanding the basis for dose-rate effects, at the present time the relative importance of dose per fraction, fraction interval, total dose, and dose rate are not clear and fractionation experiments must be interpreted with caution. Protraction is used to describe exposures that are spread over a long period of the life span, during which susceptibility to the induction of tumors may change. Protraction of the period over which an animal is irradiated may alter the effect because of (1) true dose-rate influences, and (2) age-dependent changes in susceptibility to tumor induction. Because there have been no adjustments for changes in susceptibility with age, or in some instances for competing risk, these studies cannot be used for the assessment of dose-rate effects. The present discussion will be limited to those studies directly applicable to dose-rate effects. Studies using fractionated exposures or in which exposures have been protracted over a significant portion of the animal's lifetime will not be considered.

RADIATION LEUKEMOGENESIS

Much of the data on the form of the dose response for radiation-induced neoplasia, as well as for time-dose relationships for tumor induction, have been derived from studies on radiation leukemogenesis, in particular, the induction of myeloid leukemia. Upton (1961, Upton et al. 1958, 1964, 1966) has shown that male RFM mice are particularly sensitive to the induction of myeloid leukemia and has examined the dose response over a wide dose range of X rays or gamma rays, including doses as low as 25 rad. In studies over the dose range of 0–150 rad, the incidence appears to vary with the square of the dose, although the precise shape cannot be determined and a linear relationship cannot be excluded (Figure 1). A dose-squared relationship for myeloid leukemia has also been recently reported in male CBA mice by Major and Mole (1978) although the dose range was quite limited.

Under conditions of continuous exposure for 23 hours daily at low dose rates, Upton et al. (1970) observed that the yield of myeloid leukemia per rad was reduced as compared to the yield at similar doses delivered at high dose rates (Figure 1). This effect could generally be characterized as a diminution of the dose-squared component observed at high dose rates, resulting in a more linear response at the lower dose rates. Little change in age susceptibility to the induction of myeloid leukemia has been observed over the time interval used for the chronic radiation exposures and the differences appear to be due to true dose-rate effects.

Generalizations from these results on dose and dose-rate relationships are

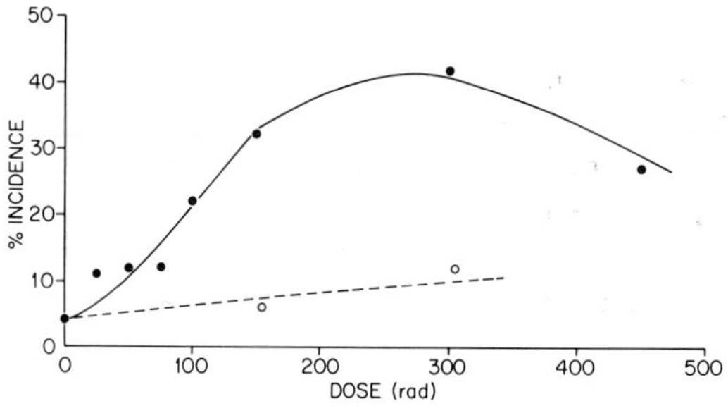

FIG. 1. Incidence of myeloid leukemia in male RF mice after single (●) or daily (○) X- or γ-ray irradiation (adapted from Upton et al. 1970).

complicated. It has been shown that females are less sensitive to the induction of myeloid leukemia and that sensitivity to induction can also vary with a number of host factors including genetic background, hormonal status, age, proliferative state of the bone marrow, and the conditions of the environment in which the animals are maintained (Upton et al. 1964, 1966). For example, animals housed in conventional animal facilities seem to be most susceptible, while animals maintained in a germfree or specific-pathogen-free (SPF) environment show a reduced sensitivity to the induction of myeloid leukemia but an increased sensitivity to the induction of thymic lymphoma (Walburg et al. 1968).

Recent studies in our laboratory show that in addition to a decreased sensitivity to induction of myeloid leukemia in SPF mice, the form of the dose response differed as well (Ullrich and Storer, unpublished observations). Rather than a dose-squared or linear dose-squared response as suggested by the earlier data in RFM and CBA mice, our data indicate a predominantly linear response over the 0–300 rad range (Figure 2). Although the reasons for this difference are not known, it is possible that the same factor which influences the sensitivity to radiation also influences the shape of the dose-response curve. This interpretation would suggest that the basis for the dose-squared component in earlier studies was related to host factors influencing induction or expression rather than factors related to microdosimetric considerations.

In view of the important role of host factors in the expression of myeloid leukemia (Upton et al. 1964, 1966), it is also unclear from available data whether the basis of the dose-rate effect reported by Upton is related to influences on factors involved in tumor expression (e.g., hormones, cell turnover) or, in fact, related to dose-rate effects on the initial events.

One of the most common forms of radiation-induced neoplasms of the lymphoreticular system in mice is thymic lymphoma. Although a number of investigators have intensively studied the induction of thymic lymphoma after radiation exposure, most have been concerned with modifying factors, the time course of the disease, or the sequence of events leading to its development rather than

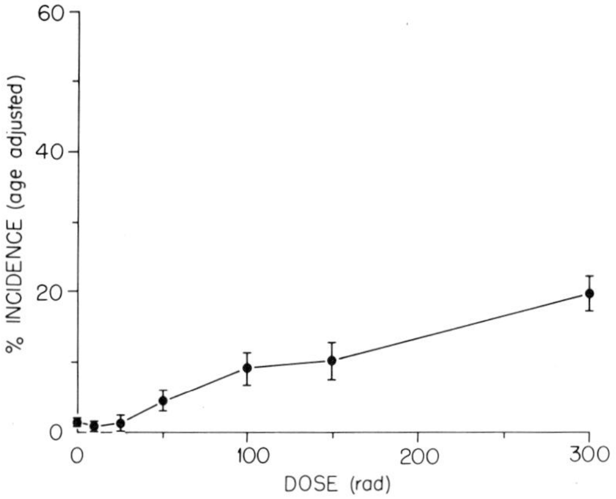

FIG. 2. Incidence of myeloid leukemia in male RFM mice after γ-ray irradiation.

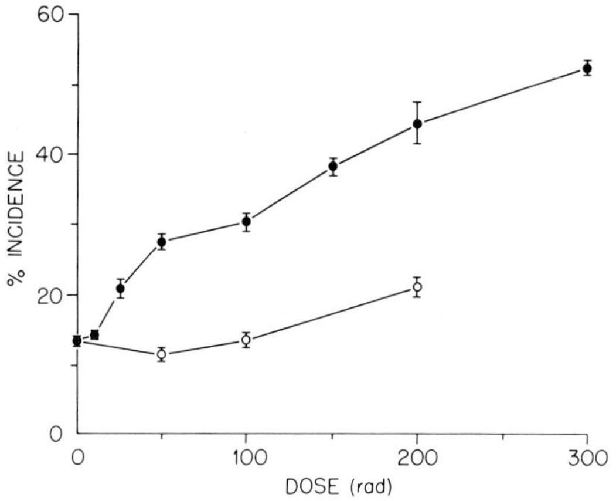

FIG. 3. Incidence of thymic lymphoma in female RFM mice after γ-ray irradiation at rates of 45 rad/minute (●) or 8.3 rad/day (○) (adapted from Ullrich and Storer 1978, with permission of the International Atomic Energy Agency).

the dose response (Kaplan 1964, 1967, Haran-Ghera 1973, Upton 1968). These studies have provided information that indicates that the mechanisms involved in induction after radiation exposure are quite complex and highly dependent on cell killing and the target cell–viral interactions that result. In view of this, it should not be surprising that dose and dose-rate relationships for this tumor are complex.

The most extensive information on dose and dose-rate relationships for thymic lymphoma are from data on female RFM mice (Ullrich and Storer, unpublished observations, 1978). After high dose-rate (45 rad/minute) irradiation over the dose range of 0–300 rad, no simple model (such as linear, dose-squared, or linear dose-squared) described the relationship over the entire dose range. The response appeared to consist of two components (Figure 3). Over the 0–25 rad range the incidence increased with the square of the dose, while over the 50–300 rad range a linear response is seen. Reducing the dose rate to 8.3 rad/day decreased the effectiveness of the irradiation so that the incidence of thymic lymphoma is lower in the low dose-rate group than in the high dose-rate group at all doses tested (Figure 3). At this lower dose rate, the response is best described by a linear dose-squared model with a shallow initial linear slope; linearity over the entire dose range can be rejected. Although these data are inconsistent with predictions based upon the theory of dual radiation action (Kellerer and Rossi 1975), considering the complexity of the proposed induction mechanism it would be unlikely that a model based upon actions within a single cell should apply in this instance.

SOLID TUMORS

Information on dose and dose-rate relationships for solid tumors is limited mainly to endocrine or endocrine-modifiable neoplasms, including ovarian, mammary, pituitary, and Harderian gland tumors.

A number of studies have shown that the incidence of ovarian tumors in RFM or BALB/c mice is greatly increased by doses of 50 rad (Upton et al. 1970, Ullrich and Storer 1978, Ullrich et al. 1976, Yuhas 1974). Most studies, in fact, indicate that the maximum tumor incidence is produced by doses in the range of 50–100 rad. It has also been observed repeatedly that the induction of ovarian tumors in these two mouse strains by radiation is highly dependent on the dose rate (Upton et al. 1970, Ullrich and Storer 1978, Yuhas 1974). Interpretation of dose-rate data is complicated by a decrease with age in the susceptibility of the mouse ovary to tumorigenesis. This protraction effect has been analyzed in connection with the influence of dose rate by Yuhas (1974), who found that approximately one third of the overall effect with protraction was attributable to age-dependent loss of susceptibility to ovarian tumorigenesis and two thirds was due to dose-rate effects (Figure 4). The basis for the observed dependency on dose rate remains to be fully elucidated. However, since the sequence of events leading to induction of ovarian tumors is believed to start with oocyte killing (Kaplan 1950, Bonser and Jull 1977), a reduction of tumorigenesis with low dose rates may be correlated, at least in part, with reduced cell killing of oocytes at low dose rates.

Relative to the dependency of ovarian tumorigenesis on cell killing, a comparison of the form of the dose-response relationships in RFM mice after ^{137}Cs gamma-ray irradiation delivered at 45 rad/minute or 8.3 rad/day indicates that

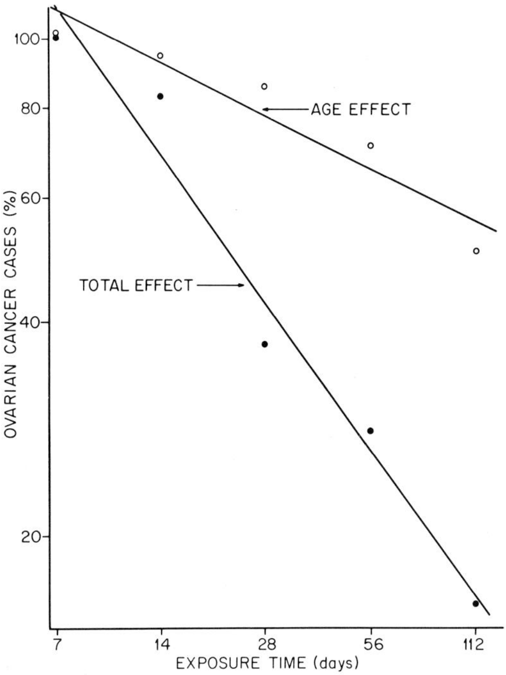

FIG. 4. Effect of exposure time (dose rate) on induction of ovarian tumors in BALB/c mice (adapted from Yuhas 1974).

both responses can be described by linear dose-squared models with a negative initial slope or by threshold models (Ullrich and Storer, unpublished observations, 1978) (Figure 5). In the absence of any biological basis for a negative initial slope and in view of the proposed mechanism for induction or expression of ovarian tumors, which most likely requires a degree of oocyte killing sufficient to start a sequence of events that leads to an alteration in the balance of the ovary-pituitary axis, altered hormonal effects on ovarian cells, and ultimately tumor formation (Kaplan 1950, Bonser and Jull 1977), it appears likely that a threshold exists for the induction of ovarian tumors. Using the high dose-rate data, the threshold model suggests that following the initial threshold the incidence of ovarian tumors increases with the square of the dose. Decreasing the dose rate alters the response in two ways: (1) the size of the threshold is increased from 12 rad to nearly 70 rad, and (2) the relationship between tumor incidence and dose following the threshold is linear at low dose rates rather than squared.

The incidence of mammary tumors in irradiated rats and mice is highly dependent upon strain, tumor type, and irradiation conditions (Shellabarger 1976). The most extensive dose-incidence data have been derived from studies using Sprague-Dawley rats, the majority of which develop mammary tumors spontaneously during the second year of life. In these animals, total body X irradiation or ^{60}Co irradiation at 1–2 months advances the onset of tumors so that the incidence of total tumors scored within 1 year after exposure increases as a linear function of the dose from 25–400 R with X rays and 16–250 R with ^{60}Co gamma rays (Bond et al. 1960, Shellabarger et al. 1969).

A number of aspects of this system complicate the interpretation of these data (Bond et al. 1960, 1964, Cronkite et al. 1960). First, the spectrum of tumor types is broad, including fibrosarcomas, fibroadenomas, and carcinomas. The dose-response relationship for these individuals tumor types is less clear. In general, a cut-off period of approximately 12 months is used in these studies

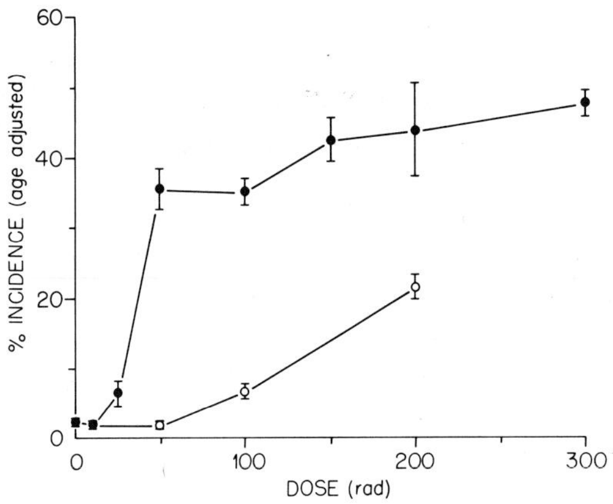

FIG. 5. Incidence of ovarian tumors in female RFM mice after γ-ray irradiation at rates of 45 rad/minute (●) or 8.3 rad/day (○) (adapted from Ullrich and Storer 1978, with permission of the International Atomic Energy Agency).

because beyond that point the control incidences begin to rise and near the end of the life span the total tumor incidence in control animals approaches that seen earlier in irradiated animals. This raises the serious question as to whether radiation is accelerating the normal process or truly inducing tumors. Under these circumstances, the interpretation of the dose-effect curve becomes unclear, and interpretations of the data in terms of mechanisms of radiation-induced tumorigenesis must be made with caution.

Although the dose response for mammary tumorigenesis in the Sprague-Dawley rat is reasonably well defined, there is very little information on the influence of dose rate. In some preliminary studies comparing the effectiveness of ^{60}Co gamma-ray irradiation at total doses of 88 R or 265 R delivered at 0.03 or 10 R/minute, Shellabarger and Brown (1972) observed that the yields of mammary neoplasms (i.e., adenocarcinomas and fibroadenomas) were similar at both dose rates. By examining adenocarcinomas and fibroadenomas separately, a small but significant dose-sparing effect was found only for the induction of mammary adenocarcinomas at a total dose of 265 R. In contrast to these data, fractionation of 400–500 R of X rays into as many as 32 exposures, delivered over a 16-week period, produced no apparent change in the total incidence compared to single doses, while an increase in the number of adenocarcinomas with fractionation was observed (Shellabarger et al. 1962, 1966). It is obvious that more systematic studies are needed to clarify the influence of dose rate in this system.

Information on dose response and dose-rate influence on mammary tumorigenesis in other strains of rats and in mice is less well characterized. Although sufficient data are not available to define the form of the dose-response curve, our recent studies comparing the yield per unit dose of mammary adenocarcinomas in female BALB/c mice after ^{137}Cs gamma-ray irradiation at dose rates of 45 rad/minute and 8.3 rad/day indicate a slight dose-sparing effect (Ullrich and Storer 1978, unpublished results). However, as with the data for the rat, interpretation of these data is also complicated. It is well known that the expression of mammary tumors in mice and rats is dependent upon a functional ovary. In view of the sensitivity of the BALB/c ovary to radiation and to dose-rate influences, it is quite likely that dose-rate effects on the ovary influence mammary tumorigenesis to different degrees at high and low dose rates. Because of these differing hormonal influences, information on the influence of dose rate on carcinogenic events within the mammary tissue itself cannot be derived.

Data on the dose response for radiation tumorigenesis and the influence of dose rate on the response to pituitary and Harderian gland tumors (Figure 6) in female RFM mice are available (Ullrich and Storer 1978, unpublished results). For these two tumor types, a comparison of the responses after high (45 rad/minute) or low (8.3 rad/day) dose-rate ^{137}Cs gamma-ray exposures indicates a linear dose-squared response for the high dose rate and a linear response for the low dose rate. Although the exact mathematical relationship varies for the two tumor types, the linear component of the dose response is similar for the high and low dose rate, suggesting that the primary influence of dose rate

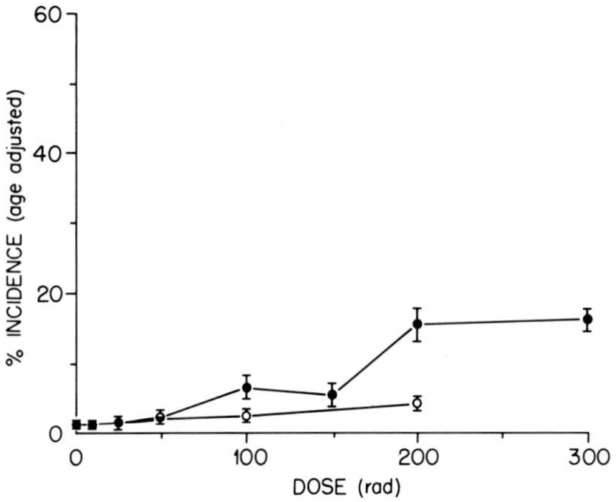

FIG. 6. Incidence of Harderian gland tumors in female RFM mice after γ-ray irradiation at rates of 45 rad/minute (●) or 8.3 rad/day (○) (adapted from Ullrich and Storer 1978, with permission of the International Atomic Energy Agency).

is to diminish the dose-squared component thereby resulting in a more linear response. Although these data are consistent with predictions based on the theory of dual radiation action, because of the known hormonal influences involved in the induction or expression of these two tumors (Furth et al. 1959, Fry et al. 1976), caution should be used in interpreting these data on the basis of this model. Specific experiments to examine whether the basis for the dose-rate effects is related to effects on the initial events or to effects on factors influencing tumor expression are required.

Information on other tumor types is inadequate to derive information on dose-effect relationships. Even for dose-rate influences, relatively little useful information is available other than that outlined above. Whole-body ^{137}Cs gamma-ray exposures delivered at the rate of 8.3 rad/day have been shown to be less effective on a per-rad basis in inducing lung adenocarcinomas in BALB/c mice than similar doses delivered at a rate of 45 rad/minute over the dose range of 0–200 rad (Ullrich and Storer 1978) (Figure 7). Since the tumors do not appear to be influenced by hormonal factors, these effects are most likely attributable to direct dose-rate influences on the lung. These data are complicated, however, by a study recently reported by Yuhas (in press). In this study, when

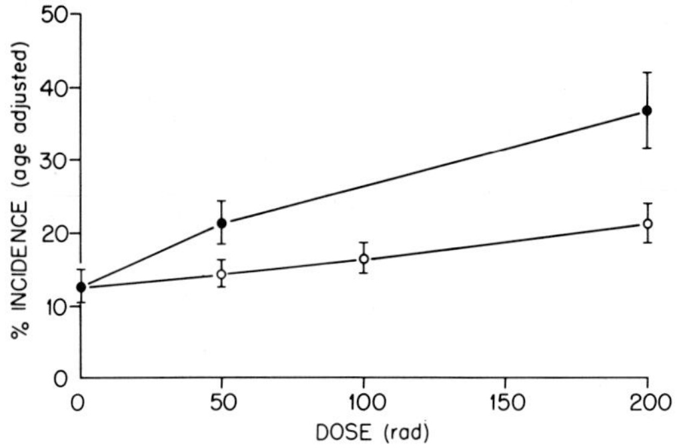

FIG. 7. Incidence of lung adenocarcinomas in female BALB/c mice after γ-ray irradiation at rates of 45 rad/minute (●) or 8.3 rad/day (○) (Ullrich and Storer, unpublished observations).

doses of 196 rad were delivered at varying dose rates from 1.75 to 112 rad/day, the incidence of lung adenocarcinomas in BALB/c mice increased with decreasing dose rate over the 112 to 14 rad/day dose rate range. At lower dose rates, the lung tumor incidence declined, and at rates of 1.75 and 3.5 rad/day the incidence was lower than that observed after high dose-rate irradiation (112 rad/day).

DISCUSSION

The results from the experimental systems reported here indicate that the dose-response curves for tumor induction in various tissues cannot be described by a single model. Furthermore, although the understanding of the mechanisms involved in different systems is incomplete, it is clear that very different mechanisms for induction are involved. For some tumors the mechanism of carcinogenesis may be mainly a result of direct effects on the target cell, perhaps involving one or more mutations. While induction may occur in many instances through such direct effects, the eventual expression of the tumor can be influenced by a variety of host factors including endocrine status, competence of the immune system, and kinetics of target and interacting cell populations. In other tumors, indirect effects may play a major role in the initiation or expression of tumors. Some of the hormone-modulated tumors would fall into this class. It is postulated that for ovarian tumors in mice, oocyte killing sufficient to alter the balance of the ovary-pituitary axis results in elevated gonadotropin levels that eventually lead to neoplastic development (Kaplan 1950, Bonser and Jull 1977). Cell killing also appears to play a role in the development of thymic lymphoma (Kaplan 1964, Upton 1968). Recent discussions of radiation carcinogenesis have generally described the dose response with a model that has an initial linear component with a quadratic component at higher doses (Kellerer and Rossi 1975, Brown 1977, Upton 1977). Such a model may be appropriate for tumors induced entirely by direct effects on the target cell, but it is unlikely that the complex interactions involved in tumor induction and expression by indirect effects are adequately described by such a model. In tumor induction involving a sequence of events in more than one tissue, the possibility of other models, including the possibility of a threshold, should be considered.

Information about the effects of dose rate are also complicated by the many types of radiation effects that may contribute to the carcinogenic process. The influence of dose rate may vary from one contributory factor to another and for each of these factors may vary with dose level; therefore, the total dose and dose rate can influence not only the extent of effects produced but also the nature of effects on target and interacting cell populations. In addition, the relative importance of the contributions made by the various initiating, promoting, and modifying mechanisms involved in the carcinogenic process may differ for different neoplasms. The basis for any observed dose-rate effect may range from influences of dose rate on events at the intercellular level to influences on mechanisms affecting tumor expression rather than induction. Unfortunately, there is neither a complete nor quantitative understanding of these mechanisms and their relative influences on any of the tumor types examined. As a result, we are dependent for the most part on empirical analysis of the available data.

Despite the complexities of the experimental systems and the lack of understanding of the types of mechanisms involved, in nearly every example the tumorigenic effectiveness per rad of low-LET radiation tends to decrease with decreasing dose rate. For some tumor types the differences may be small or may appear only with very low dose rates, while for others the dose-rate effects may be large.

In order to obtain general principles and to assess the general applicability of data on dose-rate influences, the most difficult problem is to obtain information on the basis for these effects. Of particular importance is the question of whether the degree to which dose-rate influences are a result of effects on mechanisms related to expression or rather a result of repair or recovery from the initial carcinogenic events. Related to this question is the extent to which latent carcinogenic effects may persist. The understanding of these questions has important implications not only when dose-rate effects and risk are being considered, but also when interactions with radiation and other carcinogens are considered. The demonstration of persistent lesions would suggest that exposures to multiple carcinogens at low doses or low dose rates, which themselves may be ineffective, could eventually combine to be tumorigenic.

ACKNOWLEDGMENTS

This research was sponsored by the Office of Health and Environmental Research, U.S. Department of Energy, under contract W-7405-eng-26 with the Union Carbide Corporation.

REFERENCES

Bond, V. P., E. P. Cronkite, S. W. Lippincott, and C. J. Shellabarger. 1960. Studies on radiation-induced mammary gland neoplasia in the rat. III. Relation of the neoplastic response to dose of total-body radiation. Radiat. Res. 12:276–289.

Bond, V. P., E. P. Cronkite, C. J. Shellabarger, and G. Aponte. 1964. Radiation-induced mammary gland neoplasia in the rat, *in* Mammalian Cytogenetics and Related Problems in Radiobiology, Proceedings of a Symposium held in Brazil, October 1972. Pergamon Press, New York, pp. 361–382.

Bonser, G. M., and J. W. Jull. 1977. Tumors of the ovary, *in* The Ovary, vol. II, Physiology, L. Zuckerman and B. J. Weir, eds. Academic Press, New York, pp. 129–184.

Brown, J. M. 1977. The shape of the dose-response curve for radiation carcinogenesis: Extrapolation to low doses. Radiat. Res. 71:34–50.

Cronkite, E. P., C. J. Shellabarger, V. P. Bond, and S. W. Lippincott. 1960. Studies on radiation-induced mammary gland neoplasia in the rat. I. The role of the ovary in the neoplastic response of the breast tissue to total- or partial-body x-irradiation. Radiat. Res. 12:81–93.

Fry, R. J. M., A. G. Garcia, K. H. Allen, A. Sullese, T. N. Tahmisian, L. S. Lombard, and E. J. Anisworth. 1976. The effect of pituitary isografts on radiation carcinogenesis in the mammary and Harderian glands of mice, *in* Biological Effects of Low-Level Radiation Pertinent to Man and His Environment, vol. I. International Atomic Energy Agency, Vienna, pp. 213–227.

Furth, J., N. Haran-Ghera, J. J. Curtis, and R. F. Buffett. 1959. Studies on the pathogenesis of neoplasms by ionizing radiation. I. Pituitary tumors. Cancer Res. 19:550–556.

Haran-Ghera, N. 1973. The role of immunity in radiation leukemogenesis, *in* Unifying Concepts of Leukemia, R. M. Dutcher and L. Chieco-Bianchi, eds. Karger, Basel, pp. 671–676.

Kaplan, H. S. 1950. Influence of ovarian function on incidence of radiation-induced ovarian tumors in mice. J. Natl. Cancer Inst. 11:125–132.

Kaplan, H. S. 1964. The role of radiation in experimental leukemogenesis. Natl. Cancer Inst. Monogr. 14:207–220.

Kaplan, H. S. 1967. On the natural history of the murine leukemias: Presidential address. Cancer Res. 27:1325–1340.

Kellerer, A. M., and H. H. Rossi. 1975. Biophysical aspects of radiation carcinogenesis, *in* Cancer: A Comprehensive Treatise. I. Etiology: Chemical and Physical Carcinogenesis, F. F. Becker, ed. Plenum Press, New York, pp. 405–439.

Major, I. R., and R. H. Mole. 1978. Myeloid leukemia in x-ray irradiated CBA mice. Nature 272:455–456.

Shellabarger, C. J. 1976. Modifying factors in rat mammary gland carcinogenesis, *in* Biology of Radiation Carcinogenesis, J. M. Yuhas, R. W. Tennant, and J. B. Regans, ed. Raven Press, New York, pp. 31–43.

Shellabarger, C. J., and R. D. Brown. 1972. Rat mammary neoplasia following ^{60}Co irradiation at 0.3R or 10R per minute. (Abstract) Radiat. Res. 51:493.

Shellabarger, C. J., V. P. Bond, and E. P. Cronkite. 1962. Studies on radiation-induced mammary gland neoplasia in the rat. VII. The effects of fractionation and protraction of sublethal total-body irradiation. Radiat. Res. 17:101–109.

Shellabarger, C. J., V. P. Bond, G. E. Aponte, and E. P. Cronkite. 1966. Results of fractionation and protraction of total-body radiation on rat mammary neoplasia. Cancer Res. 26:509–513.

Shellabarger, C. J., V. P. Bond, E. P. Cronkite, and G. E. Aponte. 1969. The relationship of dose of total-body ^{60}Co radiation to incidence of mammary neoplasia in female rats, *in* Radiation Induced Cancer, Proceedings of a Symposium on Radiation Induced Cancer. International Atomic Energy Agency, Athens, pp. 161–172.

Ullrich, R. L., and J. B. Storer. 1978. The influence of dose, dose rate, and radiation quality on radiation carcinogenesis and life shortening, *in* Late Biological Effects of Ionizing Radiation, vol. II. International Atomic Energy Agency, Vienna, pp. 95–113.

Ullrich, R. L., M. C. Jernigan, G. E. Cosgrove, L. C. Satterfield, N. D. Bowles, and J. B. Storer. 1976. The influence of dose and dose rate on the incidence of neoplastic disease in RFM mice after neutron irradiation. Radiat. Res. 68:115–131.

Upton, A. C. 1961. The dose-response relation in radiation-induced cancer. Cancer Res. 21:717–729.

Upton, A. C. 1968. The role of radiation in the etiology of leukemia, *in* Proceedings of the International Conference on Leukemia-Lymphoma, C. J. Zarafontis, ed. Lea and Febiger, Philadelphia, pp. 55–71.

Upton, A. C. 1977. Radiobiological effects of low doses: Implications for radiological protection. Radiat. Res. 71:51–74.

Upton, A. C., V. K. Jenkins, and J. W. Conklin. 1964. Myeloid leukemia in the mouse. Ann. N.Y. Acad. Sci. 114:189–202.

Upton, A. C., V. K. Jenkins, H. E. Walburg, Jr., R. L. Tyndall, J. W. Cronklin, and N. Wald. 1966. Observations on viral, chemical, and radiation-induced myeloid and lymphoid leukemias in RF mice. Natl. Cancer Inst. Monogr. 22:329–347.

Upton, A. C., M. L. Randolph, and J. W. Conklin. 1970. Late effects of fast neutrons and gamma rays in mice as influenced by the dose rate of irradiation: Induction of neoplasia. Radiat. Res. 41:467–491.

Upton, A. C., F. W. Wolf, J. Furth, and A. W. Kimball. 1958. A comparison of the induction of myeloid and lymphoid leukemias in x-irradiated RF mice. Cancer Res. 18:842–848.

Walburg, H. E., Jr., G. E. Cosgrove, and A. C. Upton. 1968. Influence of microbial environment on development of myeloid leukemia in x-irradiated RFM mice. Int. J. Cancer 3:150–154.

Yuhas, J. M. 1974. Recovery from radiation carcinogenic injury to the mouse ovary. Radiat. Res. 60:321–332.

Yuhas, J. M. 1979. Intrinsic and extrinsic variables affecting sensitivity to radiation carcinogenesis, *in* Radiosensitivity: Facts and Models, U.S./Italy Cooperative Seminar (in press).

Radiation Biology in Cancer Research, edited by
Raymond E. Meyn and H. Rodney Withers.
Raven Press, New York © 1980.

The Different Effects of Dose Rate on Radiation-Induced Mutation Frequency in Various Germ Cell Stages of the Mouse, and Their Implications for the Analysis of Tumorigenesis

W. L. Russell

Biology Division, Oak Ridge National Laboratory, Oak Ridge, Tennessee 37830

There is still a lively controversy over whether somatic mutation plays an important role in the initiation of cancer (Rubin 1976, Ames 1976). Some investigators feel that the weight of the evidence is against a mutational origin of malignant transformation. Others believe that most human cancer is initiated by somatic mutation resulting from DNA damage by radiations and chemicals. Still others avoid taking a polarized position. Thus, even though Mintz and Illmensee (1975) have provided strong evidence of an animal malignancy of nonmutational origin, Mintz (1978) proposes "that the critical *initiating* events in some cancers are nonmutational, whereas in some other cancers, they are mutational, genetic, or chromosomal changes."

The usefulness of the suggestions made in the present paper depends on the possibility that some radiation-induced cancers are, in fact, a result of mutational changes. What is proposed is that tumor induction by radiation be reexamined in the light of the extensive knowledge now available on mutation induction in mammals. Radiation-induced mutation in mice has turned out to be a far more complex process than was anticipated from work on lower organisms. Mutation frequency is markedly affected by physical factors, such as radiation dose rate, and by biological factors, such as sex and cell stage. The question therefore posed is whether tumor induction that is potentially mutational in origin might show similar responses to these factors. If some parallelisms are found, then the mutation data may be of use in predicting what to expect for tumor induction, for example, at low doses and low dose rates.

FACTORS AFFECTING MUTATION INDUCTION BY RADIATION IN THE MOUSE

The most extensive set of information on the factors affecting mutation induction by radiation in mammals has been obtained by the use of the specific-locus method in mice (Russell 1951). The method can detect a variety of mutational lesions ranging from small intralocus changes to deletions involving a few loci. The suggestions advanced here are based on the results of investigations with this method (Russell 1963, 1972, 1977, Searle 1974).

The effect of dose rate on mutation frequency in the male depends on germ cell stage. In spermatozoa, no statistically significant effect of dose rate has been detected. In the stem cell spermatogonia, however, a marked effect of dose rate has been repeatedly demonstrated. At low radiation dose rates, the mutation frequency is only 30% of what it is at high dose rates. The range of dose rates over which this effect occurs is from 90 R/minute to 0.8 R/minute. A higher dose rate of 1000 R/minute gave no significant increase in mutation frequency, and dose rates of 0.009 R/minute and as low as less than 0.001 R/minute showed no further reduction in mutation frequency over that obtained at 0.8 R/minute.

In females, the response to dose rate is quite different from that in males, but mutation frequency is again dependent on cell stage. In mature and maturing oocytes, over the range of dose rates tested from 1000 R/minute to 0.009 R/minute there is a continuous drop in mutation frequency, in contrast to the irreducible level reached in the male at 0.8 R/minute. At the lowest dose rate tested, the mutation frequency is extremely low. In four separate analyses of our data, combined with some fractionation data of others (Lyon and Phillips 1975), the mutation frequency from the low-level radiation ranged from $1/18$ to $1/46$ of that obtained with high single-dose acute irradiation (Russell 1977). Furthermore, only in the analysis giving the ratio of $1/18$ was the mutation frequency significantly higher than the spontaneous mutation rate. For oocytes in the immature, resting stage we made a surprising discovery. (Oocytes irradiated in this stage provide the offspring that are conceived more than six weeks after irradiation.) We found that mutation frequency seen in these offspring is not increased by irradiation of the mother. This is true not only for high doses of low dose-rate gamma irradiation, but also for as high a dose of acute X rays as can be given, 50 R, without seriously affecting fertility. Neutron irradiation also was ineffective.

In both males and females, large total doses of high dose-rate irradiation given in small fractions at appropriate intervals have a reduced effect, compared to that from single doses, which approaches the magnitude of reduction obtained with low dose-rate irradiation.

It is clear that mutational response to irradiation in mammalian germ cells is far from being a simple process. Depending on sex and cell stage, the result of lowering the dose rate can range from no effect at all, as in spermatozoa, through a moderate effect, as in spermatogonia, to an overwhelming effect, namely a reduction to a mutation frequency that is near the spontaneous level, as in mature and maturing oocytes. Exposure of immature, resting oocytes, even to acute irradiation, produces no mutations.

POSSIBLE BEARING OF THE MUTATION RESULTS ON TUMORIGENESIS

Having outlined this great variety of responses to mutation induction by radiation in mammalian germ cells, we can now turn to its possible bearing on tumorigenesis. In discussions by committees concerned with the cancer hazard from low-level radiation, it is sometimes suggested that, for cancers of presumed mutational origin, the risk from chronic radiation exposure may be only one-third the risk from acute exposure. This is based on our results on mutagenesis in spermatogonia. These results were the first ones reported, in our discovery of a dose-rate effect on mutation induction in mice, and are, perhaps, the most widely known. What I wish to emphasize here, however, is that this is only the response of spermatogonia, and that other cell types now investigated behave quite differently.

In considering the possible bearing of the mutation results on tumorigenesis, a distinction can be made between specific and general applications.

Specific Applications

A specific question that seems worth raising is whether certain somatic tissues or groups of tissues might, on the basis of their characteristics, be expected to have a mutagenic response—and, therefore, perhaps a tumorigenic response—to radiation that is similar to that in one or another of the germ cell stages. Unfortunately, we do not know the basis for the marked differences in response of the various germ cell stages. Plausible working hypotheses can, however, be advanced. It seems almost certain that the effect of dose rate must involve repair mechanisms. We have suggested that the failure of spermatozoa to show a dose-rate effect may be due to their extremely low metabolism and, therefore, a presumed low capacity for repair.

The apparent capacity for complete, or almost complete, repair at low dose rates in oocytes, in contrast to the more limited repair in spermatogonia, may be related to the fact that the spermatogonia are going through mitotic divisions during irradiation, whereas the oocytes are not.

With my limited knowledge of tumorigenesis, I hesitate to suggest specifically which tumors may be of mutational origin and which somatic tissues might be expected to behave like one or another of the germ cell stages. However, from the information now available, as summarized by Ullrich (1980, see pages 309 to 319, this volume), on the effect of dose rate on tumorigenesis, it would seem worthwhile to start hunting for parallelisms between the responses of certain tissues to tumor induction and the responses of particular germ cell stages to mutation induction.

For example, among tumors that could be of mutational origin and that show a dose-rate reduction factor of the order of three, are there similarities between the tissues involved and stem cell spermatogonia? Is the dose-rate effect limited to the same range of dose rates? For tumors that show a much larger dose-rate effect, is there any similarity between their tissues of origin and mature or maturing oocytes? Are the tissues that are highly resistant to tumor formation, even from high doses of acute irradiation, similar in any respect to immature, resting oocytes?

One must, of course, recognize some limitations in a comparison between mutations induced in somatic cells and mutations scored in the offspring of animals whose germ cells have been irradiated. Some types of genetic damage induced in germ cells are not transmitted to descendants, whereas some of these occurring in somatic tissues could conceivably initiate a tumor. One comparison has been made between radiation-induced mutations in somatic and germ cells for the same set of gene loci (Russell and Major 1957). The mutation frequencies were not significantly different, but the point estimate for the somatic cells was, as expected on the basis of the above consideration, somewhat higher than the frequency in the germ cells.

Another obvious difficulty in making the comparisons suggested here is the influence of secondary factors on tumor expression. Ullrich (1980, see pages 309 to 319, this volume) has rightly emphasized the problems inherent in trying to analyze the relative importance of the various initiating, promoting, and modifying mechanisms involved in a final observed dose-rate effect on tumor incidence.

In spite of the difficulties, an attempt to find specific parallelisms between mutagenesis and tumorigenesis seems worth making. If it turns out to be successful, it may prove useful, for example, in predicting the dose-rate response of a tumor that has not been investigated for the effect of this factor.

General Applications

Even if the search for specific applications proves fruitless, one can still ask what in general might be predicted for tumorigenesis that has a mutational initiating event.

In the past it has often been assumed that the frequency of radiation-induced tumors of mutational origin would be linearly related to dose over the whole dose range, and would be independent of dose rate. This assumption was based mainly on *Drosophila* mutation data. The mouse results summarized here lead to entirely different predictions, namely, that the effect of dose rate on tumors of mutational origin might range all the way from no effect, for some tumors, to an effect, for other tumors, so great that zero tumor incidence occurs below a certain dose rate. One might speculate further that since the absence of a dose-rate effect on mutation induction was found only in spermatozoa, and since spermatozoa are unique, highly specialized cells that have no obvious counterpart in somatic tissues, perhaps no tumors of mutational origin will have no dose-rate effect.

A linear response with dose would be predicted to occur only at low dose rates and only for some tumors. The dose response of other tumors might fall more rapidly than linearly even at low dose rates.

These predictions are certainly not in conflict with the summary of Ullrich (1980, see pages 309 to 319, this volume) of what has so far actually been determined for tumorigenesis: ". . . in nearly every example the tumorigenic effectiveness per rad of low-LET radiation tends to decrease with decreasing dose rate. For some tumor types the differences may be small or may appear only with very low dose rates, while for others the dose-rate effects may be large."

What I have tried to emphasize here is that the range of variability of tumor induction to dose rate is not necessarily dependent on the operation of secondary factors on tumor expression. The mutation data indicate that a whole range of effects is possible solely from the variability in response at the primary initiating mutational event. The secondary factors, of course, could be equally wide ranging in their effects, and would thus increase the likelihood that all tumors of mutational origin will show a dose-rate effect.

ACKNOWLEDGMENT

This research was sponsored by the Office of Health and Environmental Research, U.S. Department of Energy, under contract W-7405-eng-26 with the Union Carbide Corporation.

REFERENCES

Ames, B. N. 1976. Carcinogenicity tests. Science 191:241–245.

Lyon, M. F., and R. J. S. Phillips. 1975. Specific locus mutation rates after repeated small radiation doses to mouse oocytes. Mutat. Res. 30:375–382.

Mintz, B. 1978. Gene expression in neoplasia and differentiation, *in* The Harvey Lectures, Series 71. Academic Press, New York, pp. 193–246.

Mintz, B., and K. Illmensee. 1975. Normal genetically mosaic mice produced from malignant teratocarcinoma cells. Proc. Natl. Acad. Sci. USA 72:3585–3589.

Rubin, H. 1976. Carcinogenicity tests. Science 191:241.

Russell, W. L. 1951. X-ray-induced mutations in mice. Cold Spring Harbor Symp. Quant. Biol. 16:327–336.

Russell, W. L. 1963. The effect of radiation dose rate and fractionation on mutation in mice, *in* Repair from Genetic Radiation Damage, F. Sobels, ed. Pergamon Press, Oxford, pp. 205–217; 231–235.

Russell, W. L. 1972. The genetic effects of radiation, *in* Peaceful Uses of Atomic Energy, Vol. 13. International Atomic Energy Agency, Vienna, pp. 487–500.

Russell, W. L. 1977. Mutation frequencies in female mice and the estimation of genetic hazards of radiation in women. Proc. Natl. Acad. Sci. USA 74:3523–3527.

Russell, L. B., and M. H. Major. 1957. Radiation-induced presumed somatic mutations in the house mouse. Genetics 42:161–175.

Searle, A. G. 1974. Mutation induction in mice. Adv. Radiat. Biol. 4:131–207.

Ullrich, R. L. 1980. Carcinogenesis in mice after low doses and dose rates, *in* Radiation Biology in Cancer Research (The University of Texas System Cancer Center 32nd Annual Symposium on Fundamental Cancer Research, 1979), R. E. Meyn and H. R. Withers, eds. Raven Press, New York, pp. 309–319.

Radiation Biology in Cancer Research, edited by
Raymond E. Meyn and H. Rodney Withers.
Raven Press, New York © 1980.

X-ray-Induced Mutation Rates: A Target-Theory Estimate of Their Reduction In Vivo Owing to Selection and Repair

Henry I. Kohn

Department of Radiation Therapy, Harvard Medical School, and Shields Warren Radiation Laboratory of New England Deaconess Hospital, Boston, Massachusetts 02115

X-ray-induced mutations that are scored in vivo are the result of two groups of sequential processes—the initial "physical" ones involved in the absorption of energy and the induction of change in DNA, and the subsequent "biological" ones that lead from a changed gene to its recognized phenotypic expression. To what extent do these two groups independently contribute to setting the observed mutation rate?

METHODS

The sizes of a number of genes have been calculated from mutation-rate experiments by target theory. These calculated sizes have been compared to the "actual" size of a gene, known at least approximately from genetic theory. Do the two sizes match? Depending on the nature of the difference, inferences can be drawn relating to the independent contributions to the observed mutation rate of the physical and biological groups of reactions referred to above.

Target theory assumes that bigger molecules are more likely to be hit by a dose of ionizing radiation than smaller ones, and therefore should be more mutable. The calculation of size depends on the relation between the number of hits and the distribution of ion pairs (due to irradiation) in the exposed material. Such calculations have been successful in the study of enzymes and viruses (Lea 1946, Pollard et al. 1955).

The theoretical calculation of size, however, may be incorrect when applied to in vivo data (Hutchinson 1961). In this case, the mutation rate of a particular gene is calculated as "mutations per rad per viable mutant progeny examined," but (Kohn 1976) the frequency of observed hits (mutants) in these progeny is unlikely to equal the frequency of hits (mutations) in the parental germ cell since: (1) some hits will be repaired; (2) some hits will be carried in cells that die out or are selected against; (3) some hits will induce trivial mutations, that is, the mutation does not induce a significant change in phenotype although provided with an adequate opportunity to do so; and (4) some hits will not show because the mutant gene is quiet at the time of examination. These "deflating" factors will tend to make the target-theory estimate of gene size too small.

There are also "inflating" factors that tend to make the estimate too large: (1) in addition to direct hits, free radicals generated in nearby molecules (for example, water) may mutate the target gene; (2) since target is an operational term, its volume in fact might represent the aggregate volume of several separate genes, any one of which when hit can directly or indirectly induce the mutant phenotype.

Whether or not the calculated size of the gene, based on experimental data, will match the "actual size" of genetic theory will depend on the balance of the inflating and deflating factors. From the match or mismatch found, inferences

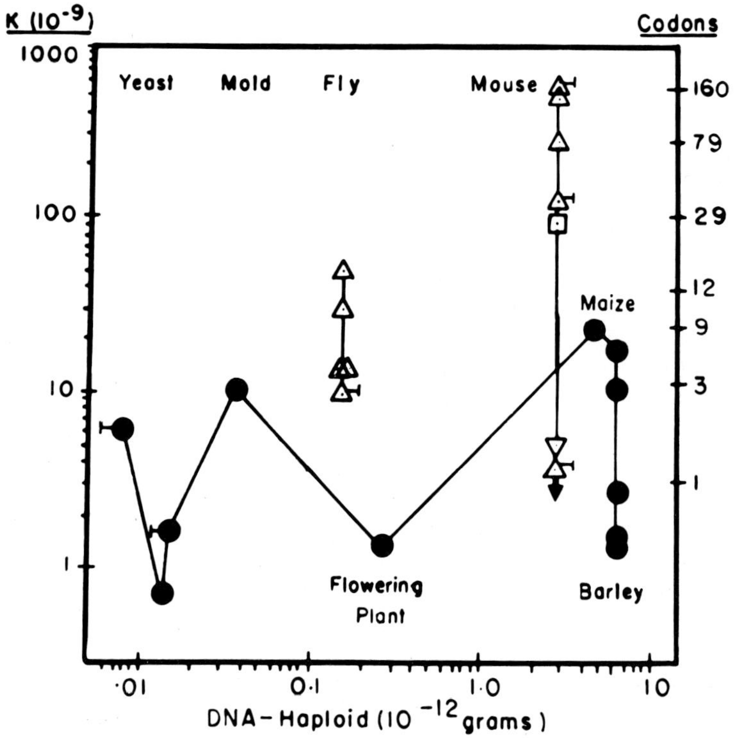

FIG. 1. K (mutations per gene per rad) and equivalent codons (estimated per target gene) plotted against test-organism haploid DNA for various X-irradiated genes and groups of genes. In the case of the mouse, three sets of genes were used: the 7-locus test, represented by triangles (plain, spermatogonia; hooked, mature oocytes; arrow and hooked, immature oocytes); the 6-locus test by a square (spermatogonia); the H-test by an inverted triangle (spermatogonia). The upper and lower points for mature oocytes were for 400 and 500 rads, respectively. The upper and lower spermatogonial points were for 1,000 rads (split, 1 day apart) and 300 rads, respectively.

can be drawn regarding the relative importance of the physical and biological groups of processes referred to above.

The data on which the calculations were made come from the review by Schalet and Sankaranarayanan (1976) and an additional report by Kohn and Melvold (1976). Eight diverse species are represented: *Schizosaccharomyces pombe* and *Saccharomyces cerevisiae* (yeast), *Neurospora crassa* (mold), *Drosophila melanogaster* (fly), *Arabidopsis thaliana* (flowering plant), *Zea mays* (maize), *Hordeum vulgare* (barley), and *Mus musculus* (mouse).

The mutation-rate constant K (mutations per locus per rad per observed viable progeny) was available for 16 specific genes or groups of genes that had been tested for specific locus, forward mutations. The dose-effect curves from which K was calculated were considered to be linear, or approximately so, except in the case of two high-dose points for mouse oocytes (Figure 1, uppermost two points for female) and the highest spermatogonial point. The latter were calculated as if linear (which minimized the estimated target size) for comparative purposes. All of the experimental data were obtained at relatively high dose rates (50 rads per minute or more).

Target size was estimated from K with a formula derived by Dertinger and Jung (1970): K times 5.8×10^{11} equals molecular weight. (In a previous analysis, I (Kohn 1976) employed Lea's (1946) factor, which is about twice as large.) The molecular weight was divided by 2,000 to obtain the equivalent number of codons.

RESULTS

The results are summarized in Figure 1, where the mutation rate *K* and the *equivalent number of codons per target* are plotted against *haploid DNA* for the species in question. The points for plants have been connected by straight lines. A survey of all of the data fails to suggest a dependency of *K* on *haploid DNA.*

Recalling that a typical protein would have 100–200 amino acids and thus be determined by 100–200 codons, it is striking to note in Figure 1 that (leaving aside the mouse data) 16 of 16 determinations fall below 15 codons, that 15 of the 16 fall below 10 codons, and that 10 of them fall below 3 codons. These absurdly small sizes indicate that a large fraction of the hit genes underwent repair or were selected against so that their mutant end-products did not appear in the progeny. Quantitatively the results indicate that more than 85% of the mutants were eliminated in the case of each gene, and that in more than half of them the elimination was greater than 97%. Evidently, the biological reactions that follow the initial physical ones play a "dominant" role in setting the mutation rate.

In the case of the mouse, the spread of the values of K for different genes and germ cells is more than 100-fold, a fact that has been emphasized by Russell (1973). The most dramatic difference is between the seven-locus test in immature oocytes (very low) and that in mature oocytes and in spermatogonia. Likewise, the H-test (histocompatibility genes) has a very low value of K. The two low values (about two codons) are not uniquely so, but are in line with the findings in other species with other genes (Figure 1). The spread of the determinations with the mouse viewed against the results with other species indicates that tests with additional loci are needed if we are to obtain a realistic picture of the "overall" X-ray-induced mutation rate of a mammalian genome.

ACKNOWLEDGMENTS

This investigation was supported by the National Institute of Allergy and Infectious Diseases and by Grant Number CA 12662, awarded by the National Cancer Institute, Department of Health, Education and Welfare.

REFERENCES

Dertinger, H., and H. Jung. 1970. Molecular Radiation Biology. Springer, New York.

Hutchinson, F. 1961. Molecular basis for action of ionizing radiation. Science 134:533–538.

Kohn, H. I. 1976. X-ray induced mutations, DNA and target theory. Nature 263:766–767.

Kohn, H. I., and R. W. Melvold. 1976. Divergent X-ray induced mutation rates in the mouse for *H* and "7-locus" groups of loci. Nature 259:209–210.

Lea, D. E. 1946. Actions of Radiations on Living Cells. Cambridge University Press, Cambridge.

Pollard, E. C., W. R. Guild, F. Hutchinson, and R. B. Setlow. 1955. The direct action of ionizing radiation on enzymes and antigens. Prog. Biophys. 5:72–108.

Russell, W. L. 1973. Mutagenesis in the mouse and its application to the estimation of the genetic hazards of radiation, *in* Advances in Radiation Research—Biology and Medicine, J. F. Duplan and A. Chapiro, eds. Gordon and Breach, New York, pp. 323–334.

Schalet, A. P., and K. Sankaranarayanan. 1976. Evaluation and re-evaluation of genetic radiation hazards in man. I. Interspecific comparison of estimates of mutation rates. Mutat. Res. 35:341–370.

Variations in Radiation Response

Radiation Biology in Cancer Research, edited by
Raymond E. Meyn and H. Rodney Withers.
Raven Press, New York © 1980.

Variations in Radiation Responses among Experimental Tumors

G. W. Barendsen

Radiobiological Institute of the Organization for Health Research TNO, Rijswijk, and Laboratory for Radiobiology, University of Amsterdam, Amsterdam, The Netherlands

In studies of responses of experimental tumors to various treatments, two aims can be distinguished: (1) elucidation of basic mechanisms that determine tumor growth and responses to treatments; (2) prediction of the effectiveness of specified treatments of cancer in man.

For the elucidation of basic mechanisms it is sometimes adequate to study a single type of response, e.g., growth delay, control probability, or cell survival, induced by only a single or a few doses, and to employ only a single type of tumor. However, for a detailed analysis of tumor responses to treatments with many fractions it is necessary to obtain data on a variety of end points including also cell kinetics, volume changes, and tumor bed effects or host responses. Moreover, in order to derive predictions for responses of tumors in man, it is necessary to study a variety of experimental tumors of different tissue of origin. Only a few animal tumors have as yet been studied in sufficient detail to allow an analysis of the influence of all factors determining responses to multiple dose fractions.

During the past ten years we have devoted considerable effort to the development of a number of experimental tumor systems, of which cells can be cultured in vitro and whereby these cells give rise to tumors upon inoculation in syngeneic animals (Barendsen et al. 1977).

Although results of various studies reported in the literature do not always agree quantitatively, it can be stated that among the many types of tumors in animals, as well as in man, significant differences have been demonstrated in responses to single and fractionated doses of ionizing radiation, not only for tumor growth delay and volume reduction, but also for the survival of cells and the doses required for local tumor control. These differences demonstrate that with respect to at least some of the many factors that determine responses to fractionated treatments, important diversity exists among various types of tumors.

The factors that determine responses of tumors to fractionated irradiation can be divided into four classes:

1. Radiation characteristics: dose, dose rate, dose fractionation schedule, radiation quality.
2. Responsiveness of tumor cells: intrinsic cellular radiosensitivity, accumulation of damage, cellular oxygenation conditions, cell cycle phase, nonproliferative phase of cells.
3. Responses of cells after irradiation: repair of sublethal damage, redistribution of cells in various phases of the cycle, reoxygenation of surviving hypoxic cells, repair of potentially lethal damage, recruitment of nonproliferating cells, repopulation by proliferation of surviving cells.
4. Retardation of tumor growth due to the influence of non-tumor-cell components: vascular supply, tumor bed influence, immunological and hormonal influences of the host.

The list of all these factors is of some interest in order to remind us of the complexity of the mechanisms involved in tumor growth and its radiation responses, but it is a much more difficult problem to quantitate the relative influence of these factors and to show to what extent they determine quantitatively the curability by fractionated treatments.

It is obviously impossible to discuss all these factors within the limits of this contribution, and in addition the available experimental data are insufficient to assess the influence of all relevant parameters and to provide a comprehensive review. Therefore, this presentation will be limited to a discussion of some of the factors for which quantitative data have been published.

DIFFERENCES IN RESPONSIVENESS OF TUMOR CELLS

Failure to obtain control of a tumor demonstrates that one or more tumor cells must have retained the capacity for unlimited proliferation. This implies that reproductive death of cells in tumors due to irradiation is of immediate consequence for the probability of local control, although host factors, e.g., immunologic defenses and hormonal influences, might play a part in some cases. It is therefore of interest to compare the radiosensitivities of cells from different types of tumors and to assess the influence of variations in intrinsic sensitivity of tumor cells on the doses required to obtain a given probability of local control.

In experiments that have been described in detail elsewhere, impairment of the clonogenic capacity of cells was measured as a function of the dose of 300-kV X rays and of 15-MeV neutrons for nine different cell lines in cultures derived from various types of tumors in rats and mice (Barendsen and Broerse 1977). The survival curves show a large variation in intrinsic radiosensitivity of the cells as well as in the relative biological effectiveness (RBE) of 15-MeV neutrons. A summary of doses required to obtain surviving fractions of cells of 0.5 and 0.1 is presented in Table 1. It can be concluded that the doses of X rays required to induce reproductive death in 50% of a cell population vary by a factor of 4, from 115 rad for L5178Y cells from a mouse lymphocytic

TABLE 1. *Doses of 300-kV X rays and of 15-MeV neutrons required to attain specified fractions of survival for cells from different types of animal tumors*

	Surviving Fraction of 0.5		Surviving Fraction of 0.1	
Origin of Cells	Dose X Rays / Dose Neutrons	= RBE	Dose X Rays / Dose Neutrons	= RBE
Rat skin basosquamous cell carcinoma RSC-1	400/190	= 2.1	980/520	= 1.9
Rat ureter squamous cell carcinoma RUC-1	270/120	= 2.3	760/375	= 2.0
RUC-2	450/150	= 3.0	1000/490	= 2.0
Rat mammary adenoma RMA-1	160/90	= 1.8	490/280	= 1.7
Rat mammary adeno-carcinoma RMA-2	300/120	= 2.5	780/380	= 2.0
Rat rhabdomyosarcoma R-1	210/100	= 2.1	560/290	= 1.9
Rat osteosarcoma ROS-1	240/110	= 2.2	660/330	= 2.0
Mouse lymphosarcoma MLS-1	120/65	= 1.9	390/220	= 1.8
Mouse lymphocytic leukemia L5178Y	115/55	= 2.1	290/175	= 1.7

leukemia, to 450 rad for RUC-2 cells derived from a rat ureter squamous cell carcinoma. With 15-MeV neutrons the variation in sensitivity is only slightly less, with differences larger than a factor of 3.

Consideration of the results presented in Table 1 shows that the radiosensitivity is largest for cells from a lymphosarcoma and from a lymphocytic leukemia, while the most resistant cells were derived from a squamous cell carcinoma. Cells from two sarcomas have an intermediate sensitivity, but it is likely that as more cell lines are investigated the distributions of sensitivities will overlap, i.e., it cannot be concluded that cells from all carcinomas are significantly more resistant than cells from all sarcomas.

The RBE values derived for 15-MeV neutrons show a significant variation from 3.0 to 1.8, i.e., they differ by a factor of at least 1.5. These differences become smaller with increasing dose.

It is important to point out that all of the cell lines for which data are given in Table 1 can be used to grow tumors in syngeneic mice or rats. It is evident that tumors grown from the resistant RUC-2 cells are expected to respond less to treatments with X rays as compared with tumors grown from the sensitive cells of a mouse lymphosarcoma. If such a lymphosarcoma is assumed to contain 10^9 clonogenic cells and, as shown in Table 1, a fraction of 0.1 survives after a dose of 390 rad of X rays, then a simple calculation shows that 10 dose fractions of 390 rad, for a total dose of 3,900 rad, would be required to control about 90% of the tumors. For the RUC-2 cells a similar calculation, assuming that only the intrinsic sensitivity of the cells determines this response, yields a total dose of 10,000 rad in 10 fractions for 90% control. Similar calculations for smaller doses per fraction based on data of the various cell lines yield differences up to a factor of about 4 with respect to total doses required for control of corresponding tumors. This implies that the large variations observed among responses of cells from different tumors are expected to play a very important part in causing differences among tumors in responses to fractionated treatments.

Variations in responsiveness of tumors to radiation can be caused by differences in intrinsic radiosensitivity among cells from various tumors, but the variations can be enhanced or decreased by other factors such as oxygenation conditions and proliferative status of the cells. An example of the influence of hypoxic cells is given in Figure 1, where survival curves for R-1 cells, irradiated in vitro and in vivo, are presented. Curve 1 is the survival curve measured with 300-kV X rays for R-1 cells in vitro (Barendsen and Broerse 1969). The actually measured data extend to 10^{-3}; the broken part is extrapolated. Curve 2 is the equivalent survival curve for 300 rad per fraction administered at intervals of four hours. Curves 3 and 4 represent the survival curves determined by preparation of cell suspensions from irradiated tumors, followed by clonogenic assay in vitro. Curve 3 represents data for single doses and curve 4 data for 300 rad of X rays per fraction with five fractions per week. It can be concluded

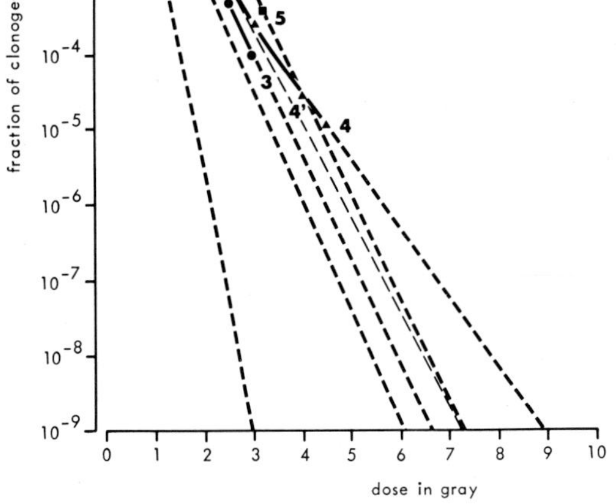

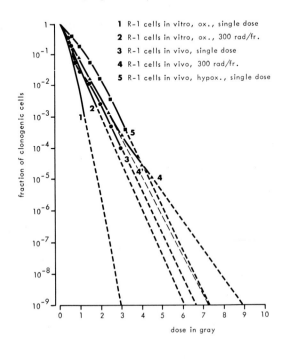

FIG. 1. Survival curves of R-1 cells, irradiated in vitro or in vivo and assayed for clone formation in vitro.

that for single doses the influence of the fraction of about 15% of hypoxic cells is very large (Barendsen and Broerse 1969, 1970, Barendsen 1974).

It is of interest to consider the doses required for 90% local control of small tumors containing 10^7 clonogenic cells. For this end point, doses required to obtain 10^{-8} survival can be estimated on the basis of the extrapolated curve to be larger by a factor of 2.2 than the dose expected from data on well-oxygenated cells in vitro. However, for fractionated treatments of tumors with 300 rad per fraction, the influence of hypoxic cells is diminished and the dose of 7,700 rad required to attain 10^{-8} survival is only a factor of 7,700/5,400 (1.4) larger than expected from data for cells in vitro (compare curves 4 and 2). As will be discussed later, the responses of these tumors to fractionated treatments are influenced by reoxygenation and repopulation.

It is of interest to note that this R-1 tumor contains approximately 60% noncycling cells. For noncycling cells cultured in vitro, it has been shown that they are not significantly more resistant to irradiation by single doses or fractionated doses than cycling cells. The data for survival of cells in vivo also do not indicate that noncycling cells are more resistant than cells in progress through the mitotic cycle. Furthermore, studies on the clonogenic fraction of cells in tumors at different time intervals between 0 and 96 hours after irradiation with doses of 1,000 and 2,000 rad of X rays have not shown a significant effect of repair of potentially lethal damage (Barendsen and Broerse 1969). For R-1 cells, in vitro repair of potentially lethal damage is also very small.

It can be concluded that with the R-1 sarcoma discrepancies between results obtained for survival of cells irradiated in vitro and in vivo are not observed and that the data obtained from studies of the clonogenic capacity of cells irradiated in tumors and analyzed by the dispersion technique are consistent with data obtained for local control. As will be discussed later, it is necessary, however, to obtain data on cell kinetics and changes in proliferation and about reoxygenation after single doses and in intervals between dose fractions in order to provide a complete description of responses to fractionated treatments. It is important to note further that such a good agreement between results obtained for cells in vitro and in vivo is not always observed. In Figure 2, single-dose cell survival curves are presented for RUC-2 cells from a moderately differentiated squamous cell carcinoma of a rat ureter. For comparison, corresponding data for the R-1 sarcoma are also presented. It is evident that, whereas the data for the R-1 tumor can be interpreted on the basis of the assumption that 15% of the cells are severely hypoxic, such a simple interpretation does not apply to the RUC-2 tumor. The change in slope of the survival curve for cells irradiated in RUC-2 tumors observed at high doses could be interpreted by assuming that about 3% of the cells are hypoxic. However, this would imply that at small doses, the in vivo curve should be close to the in vitro RUC-2 curve of oxygenated cells. In preliminary studies this is not observed and the possibility cannot be excluded that RUC-2 cells in vivo are more resistant than expected from in vitro data on cell survival. A high resistance has been observed

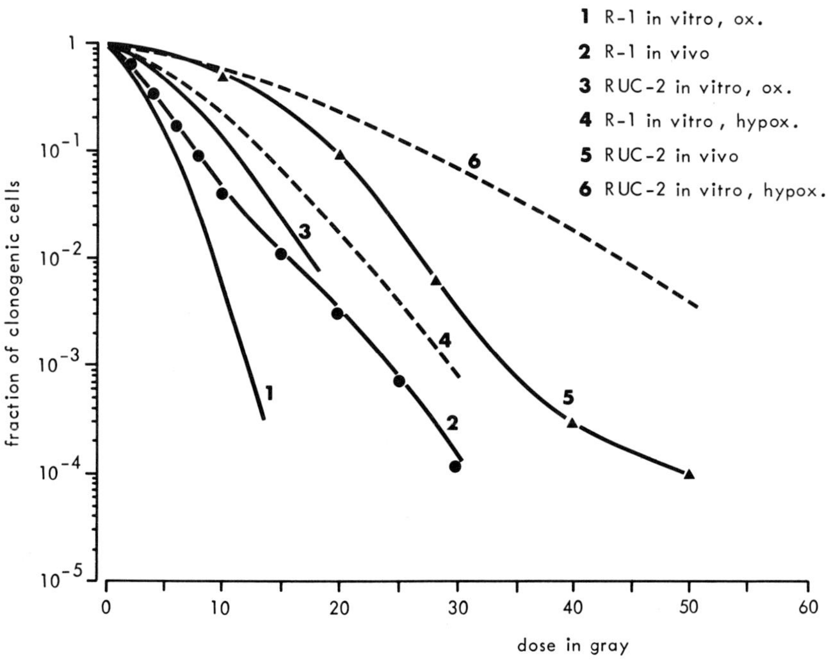

FIG. 2. Survival curves of R-1 cells (rat rhabdomyosarcoma) and RUC-2 cells (rat ureter squamous cell carcinoma) irradiated in vitro or in vivo and assayed for clone formation in vitro.

for cells in close contact to each other in spheroids, and such an increased resistance due to intercellular contact might play a role for RUC-2 cells (Durand and Sutherland 1975). It is also possible, however, that RUC-2 cells show extensive repair of potentially lethal damage. This repair must occur in a very short time because tumors have been excised and cell suspensions prepared within one hour after irradiation. Experiments with tumors irradiated in fully hypoxic conditions in dead animals and experiments with fast neutrons will be required to elucidate the nature of this high resistance of RUC-2 cells in vivo.

RESPONSES OF CELLS AFTER IRRADIATION

Responses of cells after irradiation can influence tumor responses by rendering subsequent doses either more or less effective at the cellular level or by causing more cells to repopulate tumors during intervals between and after treatments. It is evident that sublethal damage repair, redistribution of cells in various phases of the cycle, and reoxygenation of previously hypoxic cells belong to the first category, while repair of potentially lethal damage, recruitment of nonproliferating cells, and repopulation by proliferation of surviving cells belong to the second category.

Repair of sublethal damage has been shown to be a very common phenomenon, although exceptions associated with exponential survival curves, even for X rays, have been reported (Cox and Masson 1975). The relative influence of sublethal damage repair varies among cell types in vitro as well as in vivo. To provide a measure for the extent to which repair of sublethal damage can influence the dose required to attain tumor control, it is of interest to compare the initial slopes of survival curves at low doses with the final slopes at large doses. The initial slope represents the effectiveness of very small doses, where accumulation of sublethal damage plays a negligible part, while with very large single doses the maximum contribution of this accumulation of damage is obtained. In Table 2, values are presented for survival curve parameters for cultured cells from the different rat tumors discussed earlier. It is evident that the ratios of the final to the initial slopes vary from 2.3 to 5.7, i.e., by a factor of about 2.5. Similar values showing an even larger variation among cell types for cells of normal and malignant tissues have been discussed earlier by Elkind (1975).

Redistribution in the mitotic cycle of cells after irradiation can in principle cause large variations in responses to subsequent doses. Quantitative estimates of the influence of this factor for tumors irradiated with 200 rad per fraction at 24-hour intervals for several weeks have not been reported. Data are available, however, for effects of low dose-rate irradiations (Kal 1975). Differences in initial slopes of survival curves between S-phase cells and asynchronous cells by a factor of 3 have been deduced for V79 Chinese hamster cells in vitro (Elkind 1975, Durand and Sutherland 1975). However, the extent to which variations in responses among tumors are caused by this factor is difficult to estimate.

Reoxygenation has been demonstrated for a large number of experimental tumors, and it is likely to occur in many tumors in man. This process tends to decrease the influence of hypoxic cells and thus reduces possible differences among tumors due to hypoxia. For fractionated treatments with 300 rad per fraction, R-1 tumors were shown to reoxygenate to such an extent that the

TABLE 2. *Parameters of X-ray survival curves of cells in culture derived from different experimental tumors*

Type of Tumor*	D_0 (initial slope)	N (extrapolation number)	D_0' (slope at high doses)	D_0/D_0'
R-1	550	10	130	4.2
RUC-1	850	20	150	5.7
RUC-2	1200	20	220	5.5
ROS-1	550	4	165	3.3
RMS-1	450	10	110	4.1
MLS-1	280	5	120	2.3

* Notation refers to tumors described in Table 1.

fraction of hypoxic cells remained approximately constant at 15 to 30% throughout a three-week treatment (Barendsen and Broerse 1970). As a consequence, the survival curve 4, extrapolated according to 4′ in Figure 1, indicates that for fractionated doses of 300 rad of X rays per fraction, the effectiveness of this treatment is only a factor 1.2 less than expected on the basis of the in vitro survival curve 2 of Figure 1 for fractionated irradiations. This value can be compared with the value of 2.2 mentioned earlier for single doses, where hypoxic cells play a much larger role. Extensive experiments with hypoxic cell sensitizers for different tumors have also indicated that the influence of hypoxic cells is greatly reduced with fractionated treatments as compared with single doses (Denekamp and Fowler 1978, Denekamp and Stewart 1978). Thus, differences in radiosensitivity among tumors that result from the influence of hypoxic cells may be much smaller for fractionated treatments than for single doses (Barendsen 1973).

As a last factor repopulation by proliferation of surviving cells after irradiation and during a fractionated treatment will be discussed. In the R-1 sarcoma it has been demonstrated in extensive experiments on proliferation kinetics that as a result of the irradiation of these tumors with doses of 1,000 or 2,000 rad, an increased proliferation rate is observed (Hermens and Barendsen 1969, Barendsen and Broerse 1969). As a consequence the growth delay induced by a given dose is relatively small for the R-1 sarcoma, i.e., with the end point growth delay this tumor is relatively unresponsive although the corresponding in vitro survival curve indicates a relatively large cellular sensitivity. In Figure 3 it can be seen that the survival curve for R-1 cells in vitro is much steeper than the curve for RSC-1 cells from a rat skin squamous cell carcinoma, while with respect to growth delay RSC-1 tumors are much more responsive. Further

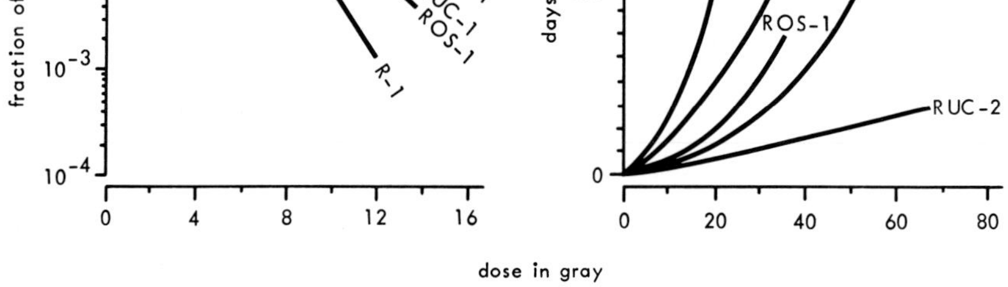

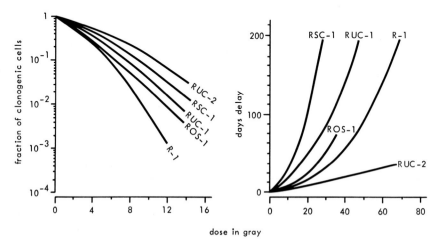

FIG. 3. Comparison of cell survival curves for different cell lines with growth delay curves for corresponding tumors in rats irradiated with single doses.

consideration of the differences among the cell survival curves and the growth delay curves of Figure 3 shows that the most resistant cells of RUC-2 produce also unresponsive tumors, but that the second most resistant type of cells in vitro, RSC-1, produce the most responsive tumors with respect to growth delay. These differences are presumably to a large extent due to differences in cell proliferation kinetics after irradiation.

Changes in proliferation kinetics have also been demonstrated during a three-week course of fractionated radiotherapy of the R-1 sarcoma (Barendsen and Broerse 1970, Tubiana 1973). Analysis of the data has shown that after the second week an increase in the rate of proliferation became a significant factor in determining the growth delay and the dose required for tumor eradication. This is shown by the differences between curves 4 and 4′ of Figure 1. Curve 4′ represents the extrapolation of data on survival of cells in irradiated tumors during the first two weeks of the treatment and curve 4 is extrapolated according to the last part of the three weeks of treatment. The difference between the doses that, according to these two curves, are required to attain a surviving fraction of 10^{-9} is equal to 8,900/7,300 (1.2). This result implies that the influence of this factor is approximately equal to that of the presence of hypoxic cells for this fractionated treatment.

RETARDATION OF TUMOR GROWTH BY HOST FACTORS

The host factors, mentioned in the first section of this paper, are undoubtedly important in tumor responses, but it is not possible to give quantitative estimates

TABLE 3. *Estimates of the relative influence of various characteristics on differences in responsiveness of tumors*

	Single Large Dose	Many Fractions
Radiation characteristics		
Dose fractionation	—	2–3
Dose rate	2–3	1–2
Radiation quality	1–2	1–2
Responsiveness of cells		
Intrinsic sensitivity	2–3	3–4
Accumulation of damage	1–2	2–3
Oxygenation conditions	2–3	1–2
Age distribution	1–2	1–2
Resting phase	—	—
Responses after a radiation dose		
Repair of sublethal damage	—	2–3
Redistribution in the cell cycle	1–2	1–2
Reoxygenation	—	1–2
Repair of potentially lethal damage	1–2	1–2
Recruitment of resting cells	—	—
Repopulation by proliferation	1–2	1–2

of their influence. It is well known that after large doses of radiation some tumors regrow more slowly than expected on the basis of their initial growth rate. This appears to occur in particular after the pretreatment volume has been attained again (Barendsen and Broerse 1970).

COMPARISON OF VARIOUS FACTORS

In Table 3, a summary is presented of estimates of the relative influence of the various parameters and characteristics discussed, expressed as the factors by which they can be expected to alter doses required to attain local control of tumors. The values are only presented to illustrate that some factors are considered more important than others. Finally, it can be concluded that depending on the combination of factors that can influence responses to treatments, very large differences must be expected and are indeed observed among different types of tumors with respect to various types of responses to fractionated irradiation.

REFERENCES

Barendsen, G. W. 1973. Quantitative biophysical aspects of responses of tumours and normal tissues to ionizing radiations. Curr. Top. Radiat. Res. Q. 9:101–108.

Barendsen, G. W. 1974. Characteristics of tumour responses to different radiations and the relative biological effectiveness of fast neutrons. Eur. J. Cancer 10:269–274.

Barendsen, G. W., and J. J. Broerse. 1969. Experimental radiotherapy of a rat rhabdomyosarcoma with 15 MeV neutrons and 300 kV X-rays. I. Effects of single exposures. Eur. J. Cancer 5:373–391.

Barendsen, G. W., and J. J. Broerse. 1970. Experimental radiotherapy of a rat rhabdomyosarcoma with 15 MeV neutrons and 300 kV X-rays. II. Effects of fractionated treatments, applied five times a week for several weeks. Eur. J. Cancer 6:89–109.

Barendsen, G. W., and J. J. Broerse. 1977. Differences in radiosensitivity of cells from various types of experimental tumors in relation to the RBE of 15 MeV neutrons. Int. J. Radiat. Oncol. Biol. Phys. 3:211–214.

Barendsen, G. W., H. C. Janse, B. F. Deys, and C. F. Hollander. 1977. Comparison of growth characteristics of experimental tumours and derived cell cultures. Cell Tissue Kinet. 10:469–475.

Cox, R., and W. K. Masson. 1975. X-ray survival curves of cultured human diploid fibroblasts, *in* Cell Survival after Low Dose of Radiation: Theoretical and Clinical Implications (The Sixth L. H. Gray Conference, 1974). The Institute of Physics, John Wiley & Sons, London, pp. 217–222.

Denekamp, J., and J. F. Fowler. 1978. Radiosensitization of solid tumors by nitroimidazoles. Int. J. Radiat. Oncol. Biol. Phys. 4:143–151.

Denekamp, J., and F. A. Stewart. 1978. Sensitization of mouse tumours using fractionated X-irradiation. Br. J. Cancer, 37:259–263.

Durand, R. E., and R. M. Sutherland. 1975. Intercellular contact: its influence on the D_q of mammalian survival curves, *in* Cell Survival after Low Dose of Radiation: Theoretical and Clinical Implications (The Sixth L. H. Gray Conference, 1974). The Institute of Physics, John Wiley & Sons, London, pp. 237–247.

Elkind, M. M. 1975. A summary and review of the conference, *in* Cell Survival after Low Dose of Radiation: Theoretical and Clinical Implications (The Sixth L. H. Gray Conference, 1974). The Institute of Physics, John Wiley & Sons, London, pp. 376–387.

Hermens, A. F., and G. W. Barendsen. 1969. Changes of cell proliferation characteristics in a rat rhabdomyosarcoma before and after X-irradiation. Eur. J. Cancer 5:173–189.

Kal, H. B. 1975. Effects of protracted gamma irradiation on cells from a rat rhabdomyosarcoma treated in vitro and in vivo, *in* Cell Survival after Low Dose of Radiation: Theoretical and Clinical Implications (The Sixth L. H. Gray Conference, 1974). The Institute of Physics, John Wiley & Sons, London, pp. 259–269.

Tubiana, M. 1973. Clinical data and radiobiological bases for radiotherapy. Curr. Top. Radiat. Res. Q. 9:109–118.

Radiation Biology in Cancer Research, edited by
Raymond E. Meyn and H. Rodney Withers.
Raven Press, New York © 1980.

Radiation Response of Human Tumor Cells In Vitro

Ralph R. Weichselbaum, John Nove, and John B. Little

Laboratory of Radiobiology, Harvard University School of Public Health, and Department of Radiation Therapy, Harvard Medical School, Boston, Massachusetts 02115

Confusion has arisen regarding the terms radioresponsive, radiocurable, and radiosensitive. Radioresponsive refers to the rate of tumor regression after a specific dose of radiation, whereas radiocurable designates whether a tumor remains locally controlled regardless of regression rate. Carcinoma of the prostate is relatively radiocurable but is not a radioresponsive tumor. Similarly, oat cell carcinoma of the lung is extremely radioresponsive; however, it is only moderately radiocurable, as one third of these patients have local treatment failures. "Radiosensitive" is strictly a laboratory term and refers to the response of individual cells to radiation. Radiosensitivity is measured in terms of the inverse of the slope (D_o) of the radiation survival curve.

Barranco et al. (1971) examined the in vitro radiation survival parameters of human melanoma cells and suggested that the lack of radiocurability of this tumor is based on a large extrapolation number ($\bar{n}$) of constituent cells of the tumor. Our group (Weichselbaum et al. 1976a, 1976b) studied the in vitro radiation survival curve parameters of two medulloblastomas (radiocurable), an osteosarcoma, and a glioblastoma (considered to be of low radiocurability). No differences were observed in radiation survival curve parameters among these tumors of varying radiocurability ($\bar{n}$ = 1.2–1.6; D_o = 130–145).

To further investigate the role of intrinsic radiosensitivity in human tumor radiotherapy, we have examined the in vitro radiation survival curve parameters of the following additional tumors: a second osteosarcoma, a hypernephroma, the melanoma reported by Barranco et al. (1971) (these tumors are considered nonradiocurable), a breast carcinoma, and a neuroblastoma (the last two are usually considered radiocurable). All experiments were performed on exponentially growing cell populations, and only those cells that had formed colonies of 50 or more cells were scored as survivors. The results are summarized in Table 1 (Weichselbaum et al., manuscript in preparation).

As can be seen in Table 1, no difference in survival curve parameters among tumors of varying radiocurability could be demonstrated in vitro. We agree with the findings of Barranco et al. that the radiosensitivity (D_o) of human melanoma is comparable to most mammalian cell lines as well as other tumors

TABLE 1. *Radiosensitivity of exponentially growing human tumor cell lines in vitro*

Cell Line	Tumor Type	D_0 (Exp.)	$\bar{n}$ (Exp.)
TX-4	Osteosarcoma	145 ± 7	1.8 ± 0.3
SAOS	Osteosarcoma	135 ± 9	2.2 ± 0.5
TX-7	Medulloblastoma	135 ± 14	1.5 ± 0.3
TX-14	Medulloblastoma	131 ± 8	1.6 ± 0.9
TX-13	Glioblastoma	143 ± 10	1.4 ± 0.5
MCF-7	Breast carcinoma	134 ± 13	1.3 ± 0.3
LAN-1	Neuroblastoma	149 ± 7	1.2 ± 0.2
MEL-H	Melanoma	150 ± 5	2.5 ± 0.4
PAS	Hypernephroma	131 ± 11	1.2 ± 0.4

we have studied, but we found an extrapolation number of only 2.5 (versus an extrapolation number of 40 reported by Barranco). This $\bar{n}$ was, however, the highest of all the lines studied. This discrepancy is attributable to differences in the surviving fraction at the 300–500 rad dose range; it remains unresolved. For the clinical radiotherapist, the extrapolation number ($\bar{n}$) may in some instances be more important than the radiosensitivity, since clinical doses are usually considered to be on the shoulder of the radiation survival curve. Weininger et al. (1978) also studied the in vitro survival parameters of three human melanoma lines. Their general conclusions are in agreement with ours. They specifically reported that a large extrapolation number was not characteristic of the melanomas they studied.

It should be pointed out that fibroblasts from patients with the autosomal recessive disease ataxia telangiectasia, associated with abnormal sensitivity to X rays in vivo, are abnormally sensitive to gamma radiation in vitro. This result is consistently reproducible between laboratories. We feel that if a marked intrinsic sensitivity or resistance of human tumor cells existed, it would also be detectable in vitro. In general, other investigators are in agreement that the D_o of the in vitro survival curve of human tumor cells is not a major factor in radiocurability (Smith et al. 1978). Also of interest is the observation that the ranges of D_o for the tumors studied in the present investigation were the same as those of a group of 11 normal human diploid fibroblasts (Weichselbaum et al., in press).

The enhanced survival that occurs when a radiation dose is split with an interval of several hours between fractions has been interpreted as being due to repair of sublethal damage induced by the first dose in cells that survive this dose (Elkind and Sutton 1959). These experiments when done in vitro are usually performed on exponentially growing cells. Since the extrapolation numbers of human tumor cells in vitro are relatively small, we thought perhaps the in vitro shoulder might underestimate the ability of tumor cells to perform sublethal damage repair. As is shown in Figure 1, however, split-dose experiments in two tumor cell lines show that the extrapolation number accurately reflects the amount of split-dose recovery measured in these human cells.

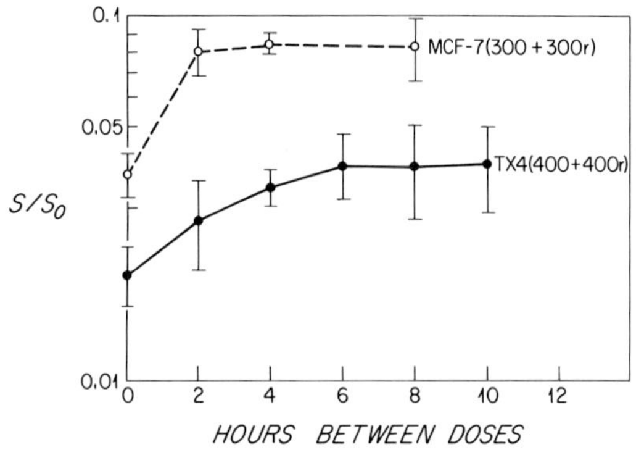

FIG. 1. Repair of sublethal damage by MCF-7 (○), a cell line derived from a carcinoma of the breast, and TX4 (●), derived from an osteosarcoma. Doses are as indicated.

Enhancement of survival after delay of subculture of cells irradiated in a density-inhibited state is referred to as potentially lethal damage repair (PLDR) and is analogous to liquid holding recovery in bacteria (Little 1969). Although PLDR has been demonstrated in animal tumors and in established cell lines, we wished to determine whether PLDR reflected a molecular repair phenomenon in human cells. Figures 2a and 2b show X-ray and ultraviolet (UV) PLDR recovery in normal diploid fibroblasts, fibroblasts from patients with ataxia telan-

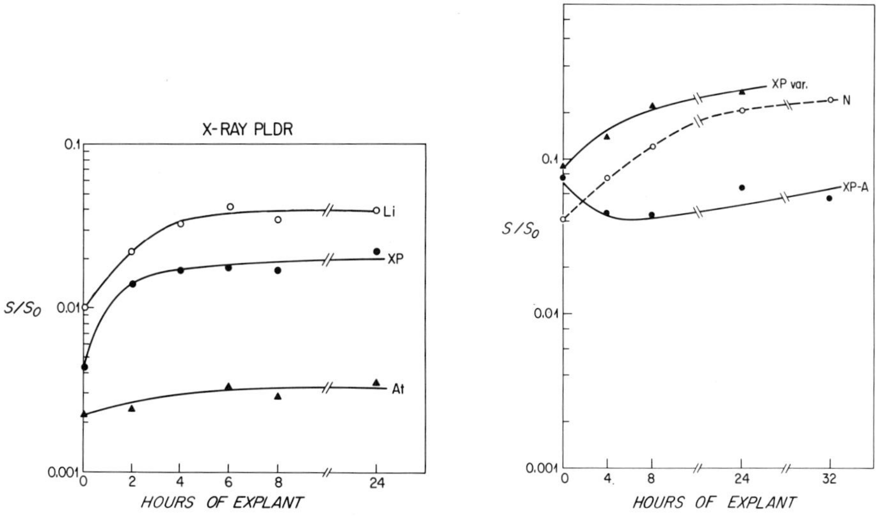

FIG. 2. A, Left, PLDR following X irradiation in three human diploid fibroblast strains: Li106 from a normal control (○); XP12BE from a complementation group A xeroderma pigmentosum patient (●); and CRL1343, from a patient with ataxia telangiectasia (▲). Doses used were 700 rad for the normal and XP strains and 350 rad for the AT strain. B, Right, PLDR following UV-irradiation in three human diploid fibroblast strains: Li106, from a normal control (○); XP4BE, from a xeroderma pigmentosum variant (▲); and XP12BE, from a complementation group A XP patient (●). Doses used were 12.0 J/m² for the normal strain, 6.0 J/m² for the XP variant, and 2.0 J/m² for the XPA strain.

giectasia (AT) (described above), and fibroblasts from patients with xeroderma pigmentosum (XP), an autosomal recessive disease associated with a sensitivity to the effects of UV light both in vivo and in vitro. This sensitivity has been shown to be due to a defect in the excision repair pathway. As can be seen, XP fibroblasts do not perform UV PLDR, and AT fibroblasts perform minimal X-ray PLDR as compared to normal controls. Fibroblasts from an XP variant proficient in excision repair were also proficient in UV PLDR, leading us to conclude that potentially lethal damage reflected a molecular repair phenomenon, likely excision repair, for the UV system and a molecular but still undetermined process for the gamma system (Weichselbaum et al. 1978). Since most cells in human tumors are nondividing, we believed in vitro density inhibition was a more accurate kinetic reflection of human solid tumors in vivo. Thus, we studied PLDR in tumors of varying radiocurability.

Experiments were performed with cells in a density-inhibited state. Medium was changed daily, following confluency, for three days to assure density inhibition (not nutritional depletion). Cells were irradiated and then subcultured at low cell density at various times from 0 to 24 hours after treatment to assay for colony formation. Only colonies of 50 or more cells were scored as survivors. Figure 3 shows PLDR experiments performed on LICH cells, human diploid fibroblasts, and TX4, a human osteosarcoma line. The osteosarcoma repairs significantly more potentially lethal X-ray damage than either of the other cell lines (Weichselbaum et al. 1977). Continuous labeling indices for LICH cells, osteosarcoma cells, and human diploid fibroblasts show that fewer cells are in the S phase in the last than in either LICH cells or osteosarcoma cells (Figure 4). This is an important finding, since PLDR has been shown to take place primarily in G_1 cells. Cultures with a large G_1 population might be expected to demonstrate more PLDR than cultures that are actively proliferating (Little and Hahn 1973). The increased capacity for osteosarcoma cells to repair PLD cannot be explained on the basis of the relative proportion of G_1 cells, and it

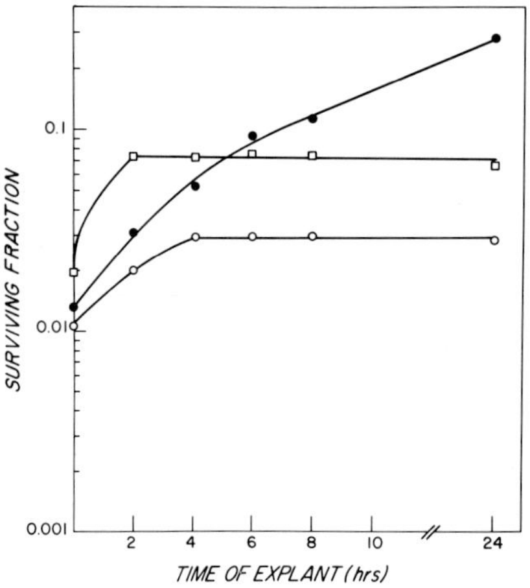

FIG. 3. PLDR following X irradiation in TX4, a human osteosarcoma (●); Li106, a normal human fibroblast strain (□); and LICH, an established human line of liver origin (○). Dose was 700 rad.

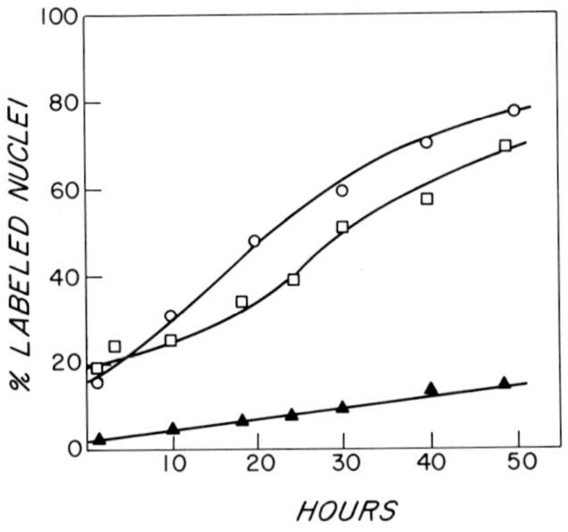

FIG. 4. Continuous labeling indices of plateau-phase cultures of Li106 (▲); TX4 (○); and LICH (□). ^{3}HTdR (0.5 μCi/ml) was added at 0 hr and samples fixed and prepared for autoradiography as a function of time.

appears that enhanced PLDR is characteristic of this osteosarcoma line. Figure 5 shows complete survival curves after 0 and 4 hours PLDR for this human osteosarcoma line. The D_o of the survival curve increased from 121 to 201 rad, significantly more than in the LICH cells (similar proliferation kinetics) in which the D_o increased from 122 to 162 rad (Little et al. 1973). Furthermore, significant PLDR occurred following doses as low as 200 rad, which are well within the clinical range. A D_o of 201 rad might render a tumor quite radioincurable.

Figures 6a and 6b show preliminary results of PLDR experiments for additional osteosarcomas, a melanoma, a hypernephroma, a breast carcinoma, and a neuroblastoma. As a group, tumors that are considered not radiocurable (osteosarcoma, melanoma, hypernephroma) appear to do significantly more PLDR than do radiocurable tumors (breast and neuroblastoma) that resemble ataxia telangiectasia in their ability to perform PLDR. It must be emphasized that

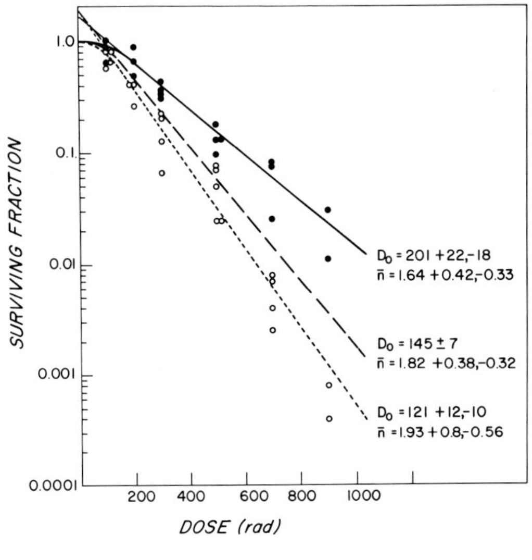

FIG. 5. PLDR survival curves for TX4. Cells were explanted immediately (○), or 4 hr after irradiation (●). Dashed line is a composite survival curve for exponentially growing cells (points not shown).

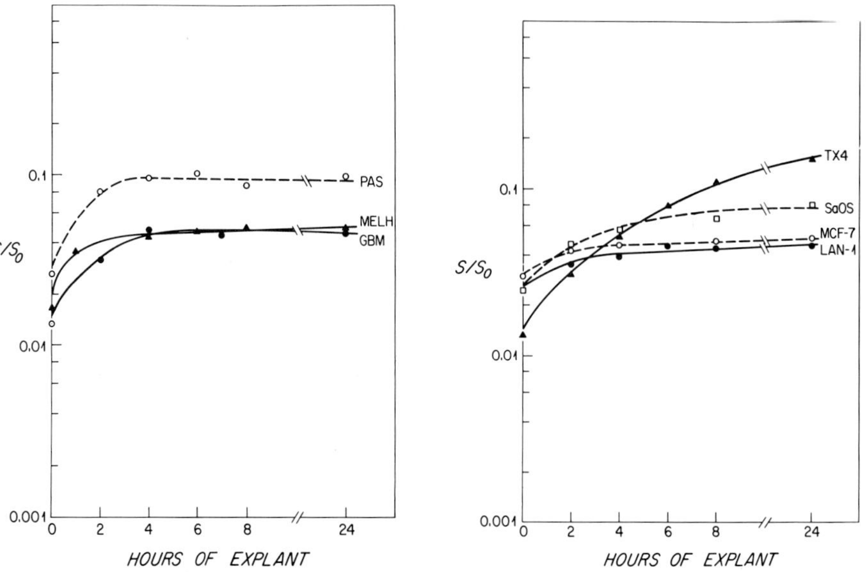

FIG. 6. PLDR following X irradiation in seven human tumor cell lines: A, Left, PAS, a hypernephroma (○); MELH, a melanoma (▲); GBM, a glioblastoma (●); B, Right, TX4, an osteosarcoma (▲); SaOS, an osteosarcoma (□); MCF-7, a carcinoma of the breast (○); and LAN-1, a neuroblastoma (●). Dose was 600 rads.

these findings are preliminary (Weichselbaum et al., manuscript in preparation). However, there is substantial evidence that, clinically, PLDR may be an important determinant of tumor curability (Little et al. 1973).

Plateau-phase cultures appear to be good in vitro models for in vivo tumor kinetics, since in vivo tumors have large populations of nondividing but potentially clonogenic cells. Previous investigations had suggested that high radiation doses were necessary to demonstrate PLDR (Little 1969, Little et al. 1973). An important preliminary finding in this report is that PLDR is demonstrated in all human tumor cells in clinical dose ranges and it appears that tumors that are relatively non-radiocurable as a group may do more PLDR than tumors that are radiocurable. Even a relatively low enhancement in survival (threefold to fourfold over 30 fractions) may make a significant difference in the ultimate surviving fraction and, therefore, radiocurability. These observations on the differential abilities of cells to repair PLD may explain some of the variations in local radiocurability of human tumors.

ACKNOWLEDGMENTS

We wish to acknowledge the assistance of Ann Schmit and Dr. Gary West. This work was supported by Grants No. CA 21848 and CA 11751 from the National Institutes of Health.

REFERENCES

Barranco, S. C., M. M. Romsdahl, and R. M. Humphrey. 1971. The radiation response of human malignant melanoma cells grown in vitro. Cancer Res. 31:830–833.

Elkind, M. M., and H. Sutton. 1959. X-ray damage and recovery in mammalian cells in culture. Nature 184:1293–1295.

Little, J. B. 1969. Repair of sublethal and potentially lethal radiation damage in plateau phase cultures of human cells. Nature 224:804–806.

Little, J. B., and G. M. Hahn. 1973. Life cycle dependence of radiation repair of potentially lethal damage. Int. J. Radiat. Biol. 23:401–407.

Little, J. B., G. M. Hahn, E. Frindel, and M. Tubiana. 1973. Repair of potentially lethal damage in vitro and in vivo. Radiology 106:689–694.

Smith, I. E., D. Courtenay, J. Mills, and M. J. Peckham. 1978. In vitro radiation response of cells from four human tumors propagated in immune suppressed mice. Cancer Res. 38:390–392.

Weichselbaum, R. R., A. J. Epstein, J. B. Little, and P. L. Kornblith. 1976a. The intrinsic radiosensitivity of human tumors of varying curability. Am. J. Roentgenol. 127:1027–1032.

Weichselbaum, R. R., A. J. Epstein, J. B. Little, and P. L. Kornblith. 1976b. In vitro cellular radiosensitivity of human malignant tumors. Eur. J. Cancer 12:47–51.

Weichselbaum, R. R., J. B. Little, and J. Nove. 1977. Response of human osteosarcoma in vitro to X-irradiation: Evidence for unusual cellular repair activity. Int. J. Radiat. Biol. 31:295–299.

Weichselbaum, R. R., J. Nove, and J. B. Little. 1978. Deficient repair of potentially lethal damage in ataxia telangiectasia and xeroderma pigmentosum fibroblasts. Nature 291:261–262.

Weichselbaum, R. R., J. Nove, and J. B. Little. 1979. X-ray sensitivity of 53 human diploid fibroblast cell strains from patients with characterized genetic disorders. Cancer Res. (in press).

Weininger, J., M. Guichard, A. M. Joly, E. P. Malaise, and B. Lachet. 1978. Radiosensitivity and growth parameters in vitro of three human melanoma strains. Int. J. Radiat. Biol. 34:285–290.

Radiation Biology in Cancer Research, edited by
Raymond E. Meyn and H. Rodney Withers.
Raven Press, New York © 1980.

Variations in Radiation Response of Tumor Subpopulations

David J. Grdina

Section of Experimental Radiotherapy, Department of Radiotherapy, The University of Texas System Cancer Center M. D. Anderson Hospital and Tumor Institute, Houston, Texas 77030

The success of radiation therapy for solid tumors growing in either human or animal hosts is dependent upon the successful destruction or sterilization of all the clonogenic tumor cells present. While the primary tumor mass is usually considered a single entity, it is actually a complex and heterogeneous cell system composed of neoplastic cells existing under a variety of physiological conditions. Included are proliferating tumor cells amply supplied with essential nutritional requirements for growth, nonproliferating but potentially clonogenic "G_0"-like tumor cells that under the proper conditions might be recruited to reenter a proliferative phase, and tumor cells that are about to lyse or die. Further complicating the situation are differences in oxygen tension and pH within the tumor to which the various classes of cells are exposed. Each of these factors can affect the radiation response of cells within the environment of the tumor and thus exert an effect on radiocurability of the tumor itself.

In this communication, the application of selected biophysical methods to separate and isolate subpopulations of cells from tumors growing in vivo is described. Emphasis is directed toward the separation of different functional classes of neoplastic cells from tumor systems rather than the separation of tumor from nontumor cell populations.

RATIONALE FOR CELL SEPARATION TECHNIQUES

Separation techniques for isolating unique subpopulations of cells were developed recently. The procedures are an extension of methods developed earlier in biochemistry and biophysics to fractionate and purify subcellular particles, macromolecules, and enzymes. The major limiting factors that have delayed progress in the development of cell separation techniques have been the difficulties encountered in preparing single-cell suspensions of viable cells and maintenance of cellular viability during and after the separation procedure. Analyses of these difficulties, along with the historical development of cell separation procedures, are presented in several excellent reviews (Shortman 1973, Harwood 1974, Pretlow et al. 1975, Castinpoolas and Griffith 1977).

Cell separation methods have been developed to exploit physical differences between morphologically similar cells, the rationale being that physical differences between cells also indicate functional differences. When this is true, purified subpopulations, each unique in a selected cell parameter, can be separated and isolated for study from complex and heterogeneous cell systems such as solid tumors. These purified subpopulations then become excellent substrates for studies involving the processes of cellular differentiation, clonogenicity, and malignancy. Two cell parameters that have been exploited with the most success in studies of tumor systems are density and size. Techniques used to exploit these parameters are density gradient centrifugation using Renografin (Grdina et al. 1975, 1976a) and centrifugal elutriation (Grdina et al. 1977, 1978b).

CELL SEPARATION BY DENSITY GRADIENT CENTRIFUGATION

Cells exhibit little or no variation in density as they traverse the cell cycle (Anderson et al. 1970). Rather, cells appear to undergo changes in density as a result of changes occurring in their environment. Cultured cells grown to plateau (i.e., stationary) phase as either fed or starved cultures become more dense than their exponentially growing counterparts (Bruns 1973, Grdina et al. 1974, Ng and Inch 1978). Likewise, the density of tumor cells growing in vivo can vary as a function of time following transplantation into host animals (Holczinger 1972). Because changes in cell density arise and add complexity to cell systems after environmental conditions become "severe," the resulting subpopulations, each with unique density, represent interesting target cell populations for radiobiological study.

A method of density gradient centrifugation with Renografin as the supporting medium has been used to separate at least five populations of clonogenic tumor cells from a murine fibrosarcoma (FSa). A detailed description of the tumor, Renografin, and the methods used to prepare suspensions and construct gradients are described elsewhere (Grdina 1976). This particular tumor was chosen for study because it can readily be made into a single-cell suspension. Cell viability is routinely greater than 95%, and the yield of viable cells is about 10^8 per gram of excised tumor tissue (Grdina et al. 1976). Renografin (methyl-glucamine N,N′-diacetyl-3, 5-diamino 2,4,6-triiodobenzoate; E. R. Squibb and Sons, New York, New York) was used in the construction of gradients because it is not cytotoxic and does not aggregate cells (Grdina et al. 1973). It has the additional advantages of having a high density and low viscosity and of being stable and inexpensive.

Figure 1 presents a representative density profile describing the separation of FSa cells in Renografin density gradients. The cell populations, designated B1 through B5, were collected at densities (g/cm^3) of 1.06 (B1); 1.08 (B2); 1.11 (B3); 1.14 (B4); and 1.17 (B5). Cloning efficiency (CE) was determined

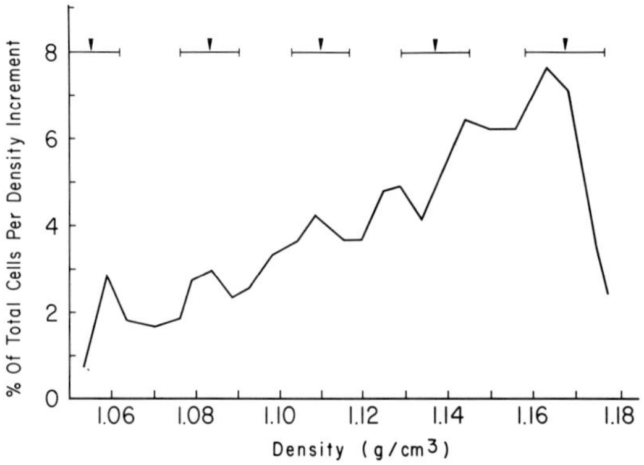

FIG. 1. A representative density profile of FSa cells from four tumors separated in a Renografin gradient. Since only selected bands were collected at times using a needle and syringe, symbols (⊢▼⊣) representing the mean density ± SD of five cell fractions collected by this method from 11 experiments are also presented at the top. (Redrawn, with permission of Academic Press Inc., from Grdina et al. 1975.)

by injecting cells i.v. into suitably prepared syngeneic mice, waiting two weeks, and then scoring the number of colonies formed in the lungs (Grdina et al. 1976). The CE of these separated populations varied significantly, with B2 being the most clonogenic (Table 1). Differences in CE were not attributable, however, to variations in either cell size or DNA content (Grdina et al. 1977).

The radiation response of selected density-separated populations was also characterized. Tumors in test animals were irradiated, excised, made into single-cell suspension, and separated in Renografin density gradients prior to assaying for clonogenic ability. Because large numbers of animals and a large volume of tumor tissue were needed, the experiments were limited to unseparated control (USC) cells, B2 cells ($\bar{\rho} = 1.08$), and B4 cells ($\bar{\rho} = 1.14$). These last two populations routinely contained about 11% and 25%, respectively, of the cells recovered from the entire gradient. Survival curves for USC, B2, and B4 cells from FSa

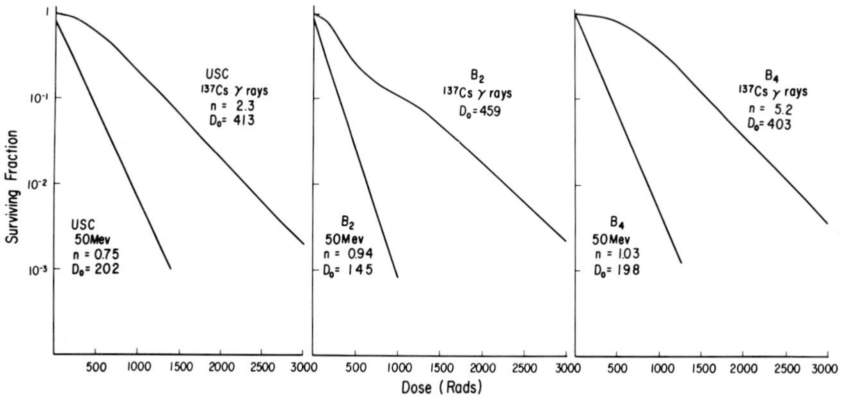

FIG. 2. Survival curves for FSa cells (USC, B2, and B4) irradiated in situ with either ^{137}Cs γ rays or E_d = 50 MeV (Be) neutrons. (Redrawn, with permission of The British Institute of Radiology, from Grdina et al. 1978a.)

TABLE 1. *Clonogenicity of tumor cells*

Population Tested	Expt. No.	Number of Tumor Cells Injected + 10^6 HIR*	Number of Lung Colonies† (10 mice/pt.: M ± SE)	Cloning Efficiency (CE) (%)	CE_B/CE_{USC}‡	P<
USC	1	1100	16.8 ± 1.8	1.53	1.0	—
	2	2980	21.7 ± 2.4	0.73	1.0	—
B1	1	1120	44.4 ± 2.5	3.95	2.6	0.001
	2	2010	38.8 ± 1.6	1.93	2.6	0.001
B2	1	980	52.2 ± 1.3	5.35	3.5	0.001
	2	2020	45.2 ± 2.3	2.24	3.1	0.001
B3	1	1170	34.4 ± 4.6	2.94	1.9	0.02
	2	4060	31.5 ± 1.3	0.78	1.1	0.5
B4	1	1280	36.8 ± 6.2	2.88	1.9	0.05
	2	4180	33.6 ± 3.6	0.80	1.1	0.5
B5	1	1250	13.4 ± 1.9	1.07	0.7	0.01
	2	7840	50.5 ± 2.2	0.64	0.9	0.5

Reprinted, with permission of Academic Press, Inc., from Grdina et al. 1975.

* HIR, heavily irradiated.

† WBI mice were used.

‡ To facilitate a comparison between experiments, the ratio of the CE of the unseparated control (USC) to that of each of the bands within the same experiment is presented.

tumors irradiated in situ with either ^{137}Cs γ rays or $E_d = 50$ MeV (Be) neutrons are presented in Figure 2 for comparison (Grdina et al. 1978a). The B2 cells, the most clonogenic of the separated populations, were the most sensitive to either low- or high-LET (linear energy transfer) radiation.

To determine the effect of hypoxia on radiation response, tumors were made acutely hypoxic by decapitating tumor-bearing animals 20 minutes before irradiation. The survival curves of B2 and B4 cells irradiated in either air-breathing or dead animals are presented in Figure 3 for comparison (Grdina et al. 1976). While conditions of acute hypoxia during irradiation had a marked effect on the radiation response of B2 cells, no change in the radiation exposure of the B4 population was observed. Because of this, it was concluded that the majority, if not all, of the cells collected at a density of 1.14 g/cm³ (i.e., B4) were chronically hypoxic in the tumor during the time of irradiation.

It is uncertain from these data, however, whether conditions of hypoxia directly induce the formation of denser cells. Presumably, cells that are chronically hypoxic in a tumor are also nutritionally starved. As described earlier, this latter condition is known to give rise to subpopulations of dense cells (Grdina et al. 1975). While the determining factor(s) leading to increased density is

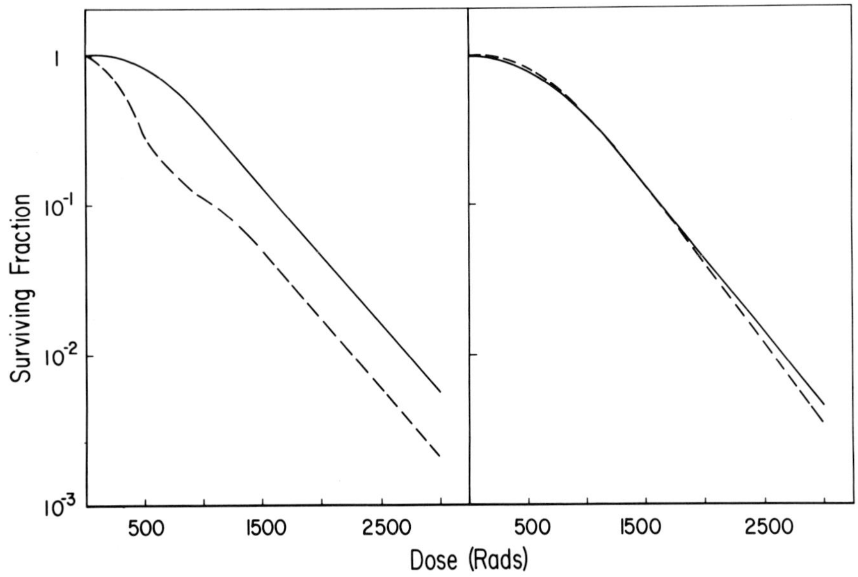

FIG. 3. Survival curves for B2 and B4 cells plotted together for comparison. Left panel: B2 irradiated under air breathing (---) or under acute hypoxic (———) conditions. Right panel: B4 irradiated under air breathing (---) or under acute hypoxic (———) conditions. (Redrawn, with permission of Academic Press, Inc., from Grdina et al. 1976a.)

unclear, density gradient centrifugation is a powerful technique for isolating chronically hypoxic cell populations from solid tumors.

CELL SEPARATION BY CENTRIFUGAL ELUTRIATION

Under uniform conditions of exponential growth, cell size increases steadily and exponentially with age during the division cycle (Anderson et al. 1969). Since the sedimentation rate of a cell is proportional to the two-thirds power of its volume, cell separation based on velocity sedimentation allows for the fractionation of an asychronous population of cells into subpopulations containing cells of similar size and DNA content.

The methods of velocity gradient centrifugation (Pretlow and Boone 1970) and velocity sedimentation at unit gravity (Peterson and Evans 1967, McDonald and Miller 1970) have been and are effective procedures for separating cells on the basis of their size and shape differences. Recently, however, a method employing centrifugal elutriation has been developed (Glick et al. 1971). Cells of different sizes separate in an elutriator rotor chamber (Beckman Instruments, Palo Alto, California) as a result of centripetally moving liquid carrying out of the centrifugal field only those cells with a sedimentation velocity below the equilibrium imposed by the opposition of *g* forces and velocity of the liquid. The relative advantages and disadvantages of centrifugal elutriation are discussed in detail elsewhere (Meistrich et al. 1977a).

Centrifugal elutriation has been used successfully to separate and synchronize murine L-P59 sarcoma and FSa tumor cells following growth in vitro (Meistrich et al. 1977b, Grdina et al. 1978b). It has not been effective, however, in suitably synchronizing cells derived directly from solid tumors growing in vivo (Grdina et al. 1977, Grdina et al. 1978b). This is due to the heterogeneity of the size of cells in the same phase of the division cycle in these tumors (Grdina et al. 1977). In contrast, the improved synchrony of tumor cells following growth in vitro is a reflection of the uniform culture conditions to which all the cells were exposed. As a result, differences in density between tumor cells are no longer observed and the relationship between cell size and DNA content becomes more apparent.

These results suggested that for centrifugal elutriation to be effective in synchronizing cells from an in vivo growing tumor, the tumor should be characterized as having a large growth fraction exposed to a relatively uniform environment. To approximate this situation, FSa cells were grown as pulmonary nodules (i.e., lung colonies) in C_3H mice. Animals were injected with viable FSa cells so as to give rise to between 100 and 150 pulmonary nodules. Thirteen days following injection, the animals were irradiated with ^{137}Cs γ rays at a dose rate of 1,000 rad/minute. Animals were then sacrificed, lungs containing tumor tissue removed, and a single-cell suspension made. Cell suspensions were prepared by mincing and trypsinization (Grdina et al. 1978b). Because FSa cells contain more DNA per cell than do normal murine cells, flow microfluorometry (FMF) was used not only to characterize the relative DNA contents of cells, but also to determine the proportion of normal cells in each suspension (Grdina et al. 1978a).

Approximately 3×10^8 tumor and lung cells were routinely separated by centrifugal elutriation. Because of the presence of a substantial number of normal cells in the original suspension, as well as in each of the elutriator fractions, it was important to analyze each population using FMF. Presented in Figure 4 is a representative sedimentation profile describing the separation of cells from pulmonary FSa nodules. The relative DNA distribution of cells recovered in each elutriator fraction is presented in Figure 5 for comparison.

After adjusting cell counts to reflect the proportion of tumor cells in each fraction, cells were injected i.v. into the lateral tail veins of suitably prepared animals to assay for clonogenic ability (Grdina et al. 1978c). After 13 days the animals were sacrificed, lungs removed and fixed, and colonies counted. The CE of the various irradiated cell populations were plotted as a function of elutriator-fraction number (Figure 6). The CE of these irradiated populations were compared to unirradiated controls obtained in a similar manner. The resulting surviving fractions, along with their corresponding age distributions, are summarized and presented in Figure 7 for comparison.

Populations most enriched with S phase cells were the most sensitive to irradiation in vivo. Similar results have been reported for FSa (Hunter et al. 1978) and mouse L-P59 cells (Dewey and Humphrey 1962) following irradiation in

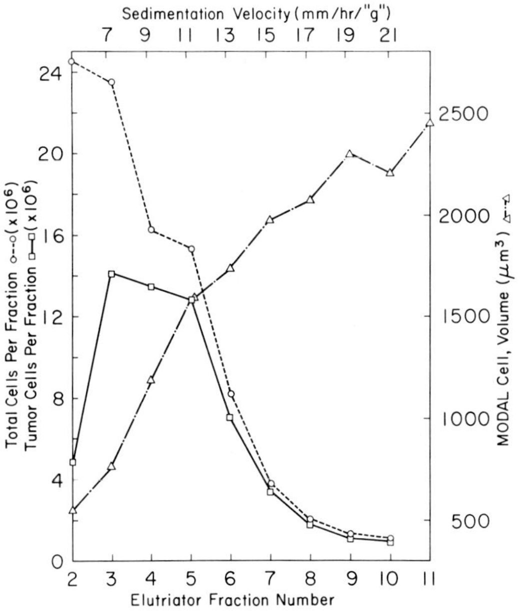

FIG. 4. Sedimentation profile of cells separated from FSa pulmonary nodules by centrifugal elutriation. Number of total cells (○) and number of FSa tumor cells (□) as determined by FMF analysis from each fraction are plotted as a function of fraction number and sedimentation velocity. Average volume of cells in each fraction (△) is calculated from modal channel number of Coulter volume distributions.

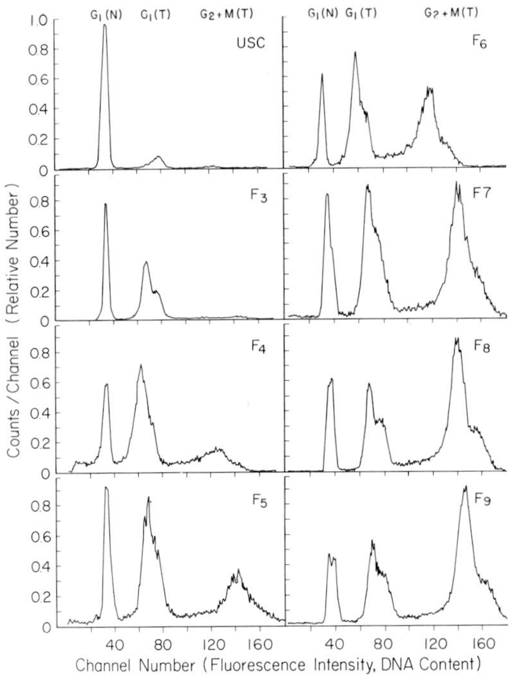

FIG. 5. FMF profiles describing the distributions of DNA content of an unseparated tumor cell suspension (USC) and cell populations separated from FSa pulmonary nodules (F_3-F_9) by centrifugal elutriation. Designations G_1(N), G_1(T), G_2 + M(T) refer to peaks representing G_1 phase normal diploid cells, G_1 phase tumor cells, and G_2 + M phase tumor cells, respectively.

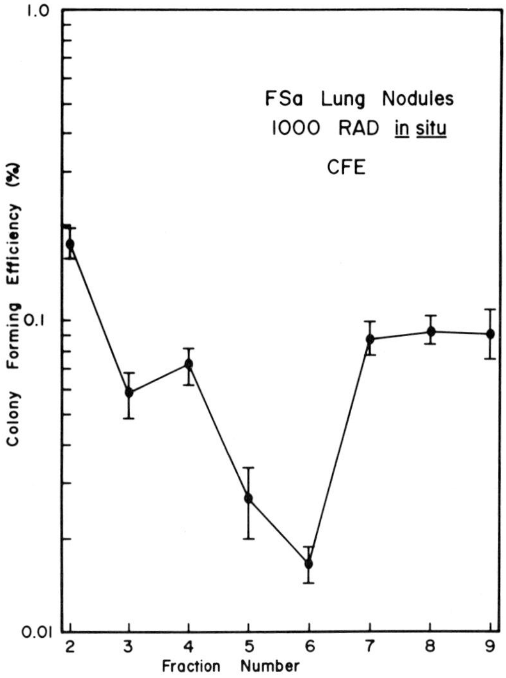

FIG. 6. Effect of 1,000 rad in situ on the colony-forming efficiency of cells in FSa lung nodules. Following irradiation, lungs were removed and cells were separated by centrifugal elutriation and injected into recipient mice. The colony-forming efficiency (CFE) of FSa cells are plotted as a function of elutriator fraction number. Vertical bar = SE.

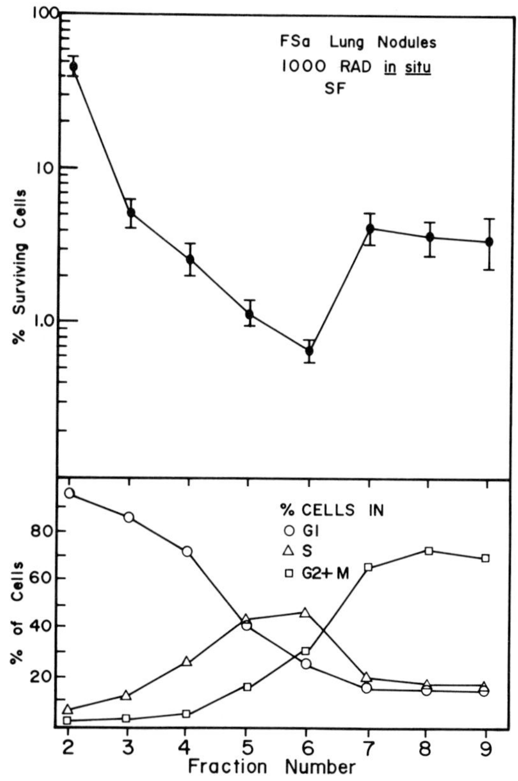

FIG. 7. The percentage of surviving FSa cells irradiated in situ in pulmonary nodules with 1,000 rad (top) and the percentage of cells distributed among the various cell cycle phases (bottom) are plotted as a function of elutriator fraction number. Vertical bar = SE.

vitro. Both of these cell lines are murine, malignant in C_3H mice, and heteroploid (~60–70 chromosomes).

Centrifugal elutriation has been demonstrated to be an effective method for separating and synchronizing tumor cells from pulmonary FSa nodules growing in mice. The method has been successfully applied to characterize the age response of FSa cells irradiated in vivo. In a similar manner, the effects of other deleterious agents that act on cells in specific phases of the cell cycle can be measured following in vivo treatment of tumor-bearing animals. Additionally, the effects of environment can be evaluated by comparing results from in vivo and in vitro treatments.

SUMMARY

Results presented here demonstrate the usefulness of cell separation methods in the study of tumor radiobiology. By exploiting the parameters of density or size or both, the heterogeneity of tumor systems can be significantly reduced. Populations of tumor cells can be not only isolated from normal cells, but also separated into functionally unique subpopulations. The choice of the appropriate method depends on the cell parameter to be exploited and the growth characteristics of the system studied. Relatively dense populations of cells appear following exposure to adverse environmental conditions. Consequently, separation by density gradient centrifugation is effective in isolating nutritionally starved or chronically hypoxic cells. These are excellent target populations in which to study the effects of in situ treatment with radiation, hyperthermia, or hypoxic cell sensitizers. Density, however, is invariant throughout the cell cycle. To isolate populations for the study of phase-specific problems, cell separation by velocity sedimentation is the preferred procedure. Centrifugal elutriation, for example, is an effective and rapid method with which to separate and synchronize cells on the basis of size following growth under relatively uniform environmental conditions. In this manner, it is possible to study and evaluate in a relatively short time the effectiveness of radiation, both single and fractionated doses, or chemotherapeutic agents in vivo against enriched populations of G_1, S, and G_2 tumor cells.

ACKNOWLEDGMENTS

This work was supported in part by NIH research grant number CA-18628, CA-06294, CA-11430, and CA-11138, awarded by the National Cancer Institute, Department of Health, Education and Welfare.

I gratefully acknowledge the help of Drs. M. L. Meistrich, L. J. Peters, and R. A. White. These studies were carried out in the laboratory of Dr. H. R. Withers, and I wish to thank him for his encouragement and support. Finally, I wish to acknowledge the excellent technical assistance of Ms. Nancy Hunter, Ms. Sandra Jones, Ms. Jean Jovonovich, and Mr. James Stutesman. I also wish to thank Ms. Rozanne Goddard for her assistance in the preparation of this manuscript.

REFERENCES

Anderson, E. C., G. I. Bell, D. F. Petersen, and R. A. Tobey. 1969. Cell growth and division. IV. Determination of volume growth rate and division probability. Biophys. J. 9:246–263.

Anderson, E. C., D. F. Petersen, and R. A. Tobey. 1970. Density invariance of cultured Chinese hamster cells with stage of the mitotic cycle. Biophys. J. 10:630–645.

Bruns, P. J. 1973. Cell density as a selective parameter in Tetrahymena. Exp. Cell Res. 79:120–126.

Castinpoolas, N., and A. L. Griffith. 1977. Preparative density gradient electrophoresis and velocity sedimentation at unit gravity of mammalian cells, *in* Methods of Cell Separation. Plenum Press, New York, pp. 1–23.

Dewey, W. C., and R. M. Humphrey. 1962. Relative radiosensitivity of different phases in the life cycle of L-P59 mouse fibroblasts and ascites tumor cells. Radiat. Res. 16:503–530.

Glick, D., V. E. Yuhas, and C. R. McEwen. 1971. Separation of mast cells by centrifugal elutriation. Exp. Cell Res. 65:23–27.

Grdina, D. J. 1976. Separation of clonogenic cells from stationary phase cultures and a murine fibrosarcoma by density gradient centrifugation. Methods Cell Biol. 14:213–228.

Grdina, D. J., I. Basic, S. Guzzino, and K. A. Mason. 1976. Radiation response of cell populations irradiated in situ and separated from a fibrosarcoma. Radiat. Res. 66:634–643.

Grdina, D. J., I. Basic, K. Mason, and H. R. Withers. 1975. Radiation response of clonogenic cell populations separated from a fibrosarcoma. Radiat. Res. 63:483–493.

Grdina, D. J., W. H. Hittelman, R. A. White, and M. L. Meistrich. 1977. The formation of lung colonies. An analysis based on cellular parameters of density, size, and DNA content. Br. J. Cancer 36:659–669.

Grdina, D. J., S. Linde, and K. Mason. 1978a. Response of selected tumor cell populations separated from a fibrosarcoma following irradiation in situ with fast neutrons. Br. J. Radiol. 51:291–301.

Grdina, D. J., M. L. Meistrich, and H. R. Withers. 1974. Separation of clonogenic cells from stationary phase cultures by density gradient centrifugation. Exp. Cell Res. 85:15–22.

Grdina, D. J., L. Milas, R. R. Hewitt, and H. R. Withers. 1973. Buoyant density separation of human blood cells in Renografin gradients. Exp. Cell Res. 81:250–254.

Grdina, D. J., L. J. Peters, S. Jones, and E. Chan. 1978b. Separation of cells from a murine fibrosarcoma on the basis of size. I. Relationship between cell size and age as modified by growth in vivo or in vitro. J. Natl. Cancer Inst. 61:209–214.

Grdina, D. J., L. J. Peters, S. Jones, and E. Chan. 1978c. Separation of cells from a murine fibrosarcoma on the basis of size. II. Differential effects of cell size and age on lung retention and colony formation in normal and pre-conditioned mice. J. Natl. Cancer Inst. 61:215–220.

Harwood, R. 1974. Cell separation by gradient centrifugation. Int. Rev. Cytol. 38:369–402.

Holczinger, L. 1972. Density distribution spectra of cell populations of Ehrlich ascites carcinoma in different stages of growth. Eur. J. Cancer 8:577–578.

Hunter, N., L. J. Peters, D. J. Grdina, R. A. White, and A. Bartel. 1978. Radiation sensitivity of murine fibrosarcoma cells separated by centrifugal elutriation and assayed in vivo. (Abstract) Radiat. Res. 74:532.

McDonald, H. R., and R. G. Miller. 1970. Synchronization of mouse L-cells by a velocity sedimentation technique. Biophys. J. 10:834–842.

Meistrich, M. L., R. E. Meyn, and B. Barlogie. 1977a. Synchronization of mouse L-P59 cells by centrifugal elutriation separation. Exp. Cell Res. 105:169–177.

Meistrich, M. L., D. J. Grdina, R. E. Meyn, and B. Barlogie. 1977b. Separation of mouse solid tumors by centrifugal elutriation. Cancer Res. 37:4291–4296.

Ng, C. E., and W. R. Inch. 1978. Comparison of the densities of clonogenic cells from EMT6 fibrosarcoma monolayer cultures, multicell spheroids, and solid tumors in Ficoll density gradients. J. Natl. Cancer Inst. 60:1017–1022.

Peterson, E. A., and W. H. Evans. 1967. Separation of bone marrow cells by sedimentation at unit gravity. Nature 214:824–826.

Pretlow, T. G., and C. W. Boone. 1970. Separation of malignant cells from transplantable rodent tumors. Exp. Mol. Pathol. 12:249–256.

Pretlow, T. G., E. E. Weir, and J. G. Zettergren. 1975. Problems connected with the separation of different kinds of cells. Int. Rev. Exp. Pathol. 14:91–204.

Shortman, K. 1973. Physical procedures for the separation of animal cells. Ann. Rev. Biophys. Bioeng. 7:93–130.

Kinetic Changes after Irradiation

Radiation Biology in Cancer Research, edited by
Raymond E. Meyn and H. Rodney Withers.

Apoptosis: Its Nature and Kinetic Role

John F. R. Kerr and Jeffrey Searle

Departments of Pathology, University of Queensland Medical School and Royal Brisbane Hospital, Herston, Brisbane, Queensland, Australia

Little attention has been devoted to defining the particular type of cell death that is produced in animals by radiation. Indeed, study of the mechanisms involved in death of cells has, in general, been relatively neglected in comparison with cell proliferation, and it is not widely appreciated that animal cells can die in two fundamentally different ways. The first, necrosis, is described in all textbooks of pathology and has long been recognized as being the outcome of irreversible injury to cells by agents such as toxins and ischemia. It is characterized by progressive swelling and degeneration of cellular components and disruption of ordered chemical activity (Trump and Ginn 1969). Our own observations and study of the published reports of others indicate that radiation, at least in doses of the order of those used in therapy, does *not* directly induce classical necrosis of cells. Rather, it results in their deletion by a structurally quite distinct but little known process, whose features suggest active self-destruction instead of degeneration. This latter phenomenon has been found to occur spontaneously in the tissues of healthy animals as well as under pathological conditions, and appears to play an opposite role to mitosis in cell population kinetics. It has been termed apoptosis (Kerr et al. 1972). (The word, like mitosis, comes from Greek, and means "falling off," as of petals from flowers or leaves from trees. The second "p" is silent.)

We shall first describe the sequence of morphological events in apoptosis and illustrate its general kinetic significance; it is, for example, implicated in tissue involution such as regression of the tadpole tail during metamorphosis and atrophy of endocrine-dependent mammalian organs after trophic hormone withdrawal, and it is a parameter in both the steady state kinetics of normal adult tissues and the growth of neoplasms. We shall then consider the induction of apoptosis in certain cell populations by radiation, cancer chemotherapeutic drugs, and hyperthermia, and speculate about the implications of the concept of apoptosis for radiation biology.

THE MORPHOLOGY OF APOPTOSIS

The sequence of morphological events in apoptosis is illustrated diagrammatically in Figure 1. Detailed documentation of the description that follows may be found in the paper by Kerr et al. (1972).

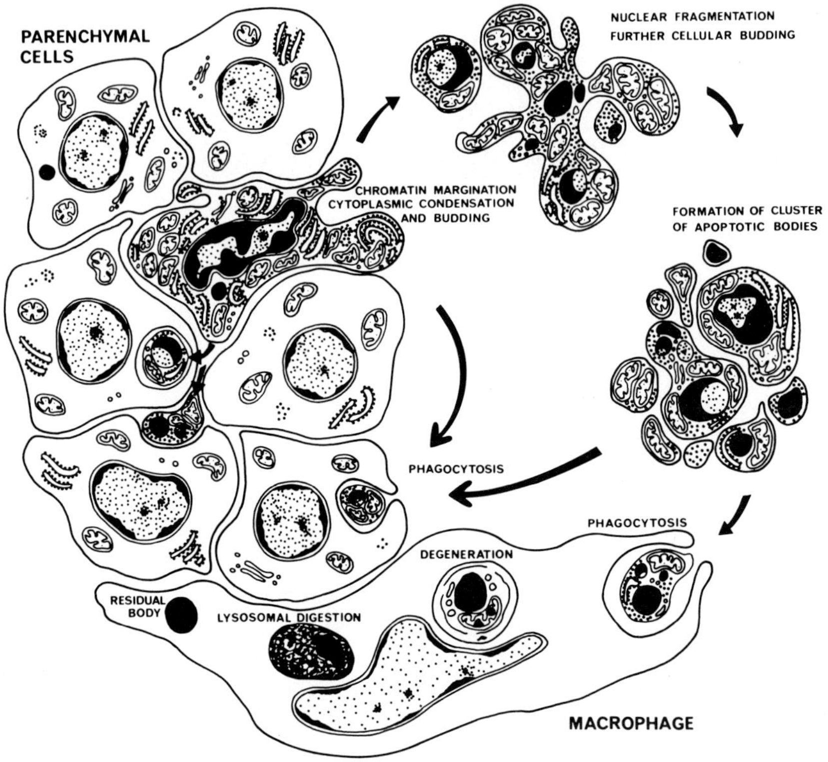

FIG. 1. Morphology of apoptosis. Top left, earliest recognizable stage. Top and right, later stages depicted without surrounding cells. Center left, movement of one apoptotic body along intercellular space and phagocytosis of another. Bottom, sequence of changes in apoptotic bodies that follows their phagocytosis. (Reproduced from Kerr and Searle 1974, with permission of Australian Cancer Society.)

Apoptosis takes place in two discrete phases. In the first, cells condense and bud to produce many membrane-enclosed apoptotic bodies. In the second, these bodies are phagocytosed and digested by nearby tissue cells; the phagocytosis may fail to take place when apoptosis affects cells dispersed in fluid, as in suspension cultures and ascites tumors. The same stereotyped sequence occurs in normal tissues and neoplasms, in mammals and amphibia, and in the embryo and adult.

Apoptosis characteristically affects scattered individual cells. In the earliest stage at present recognizable with the electron microscope, the chromatin has aggregated near the nuclear membrane, the cytoplasm has become condensed, microvilli (if originally present) have disappeared, and blunt protuberances have formed at the cell surface (Figures 1–3A). In solid tissues, affected cells separate from their neighbors at this stage, and the desmosomal attachments of affected epithelial cells break down. The closely packed organelles retain their structural integrity. The overall compaction of the cytoplasm is sometimes associated with the development of lucent vacuoles (Figure 3a). Continuation of cytoplasmic condensation and focal cell surface protrusion is then accompanied by breaking up of the nucleus into a number of fragments, which contain a variable admixture of dense and loosely textured chromatin and which are often but not always surrounded by membranes (Figures 1, 3B, 4). Finally, the surface protuberances pinch off with plasma membrane sealing at the sites of separation to produce membrane-enclosed apoptotic bodies of varying sizes: some of these contain nuclear fragments in addition to cytoplasm, others only closely packed, well-preserved organelles (Figures 1, 5–8).

When apoptosis affects cells in tissues, the resulting apoptotic bodies are squeezed along the intercellular spaces (Figures 1, 5–8) and are either shed from epithelial surfaces or rapidly phagocytosed by nearby cells. There is no associated inflammation with the exudation of specialized phagocytes as is evoked by the occurrence of necrosis, and the apoptotic bodies may be taken up by epithelial (Figures 9B, 10) and neoplastic cells (Figure 5) as well as by tissue macrophages (Figure 1). Inside phagosomes they undergo degeneration and

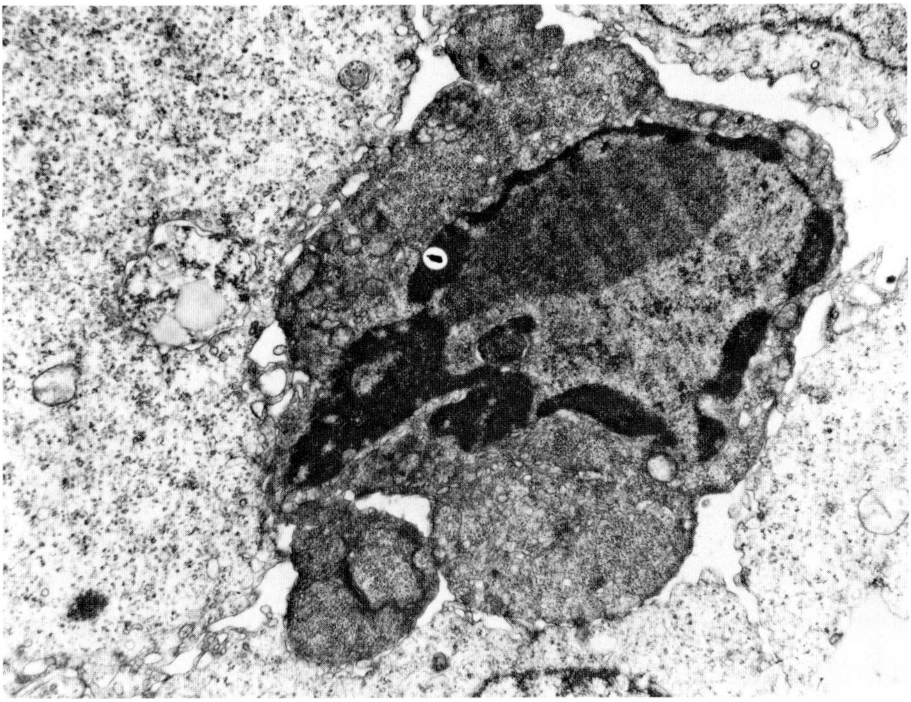

FIG. 2. Spontaneous apoptosis of neoplastic cell in Crocker mouse sarcoma 180 growing as subcutaneous nodule. Earliest stage of process recognizable with electron microscope. ×9,100. (Reproduced from Searle et al. 1975, with permission of *Journal of Pathology*.)

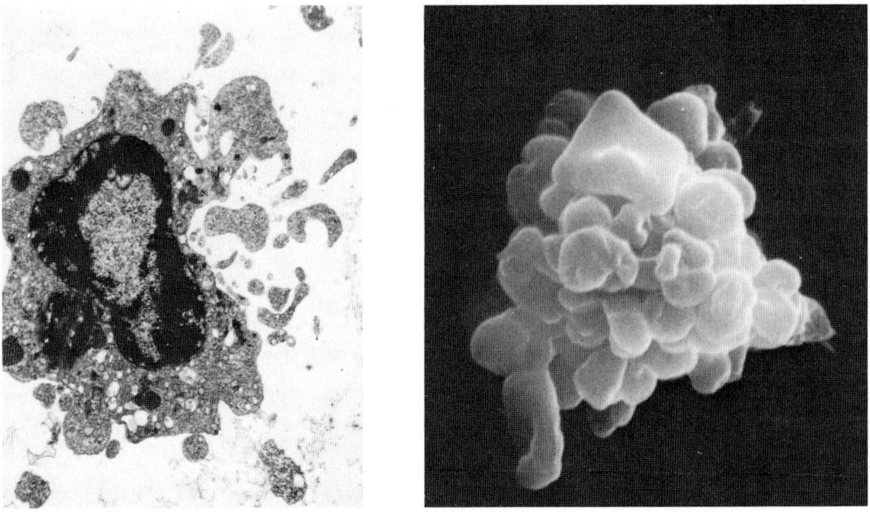

FIG. 3. Apoptosis of cells in sarcoma 180, growing as ascites tumor, 18 hours after intraperitoneal injection of actinomycin D in dose of 300 μg/kg body weight. A, left; Transmission electron micrograph. ×5,000. B, right; Scanning electron micrograph of cell proven by prior light microscopy to have fragmented nucleus. ×4,100.

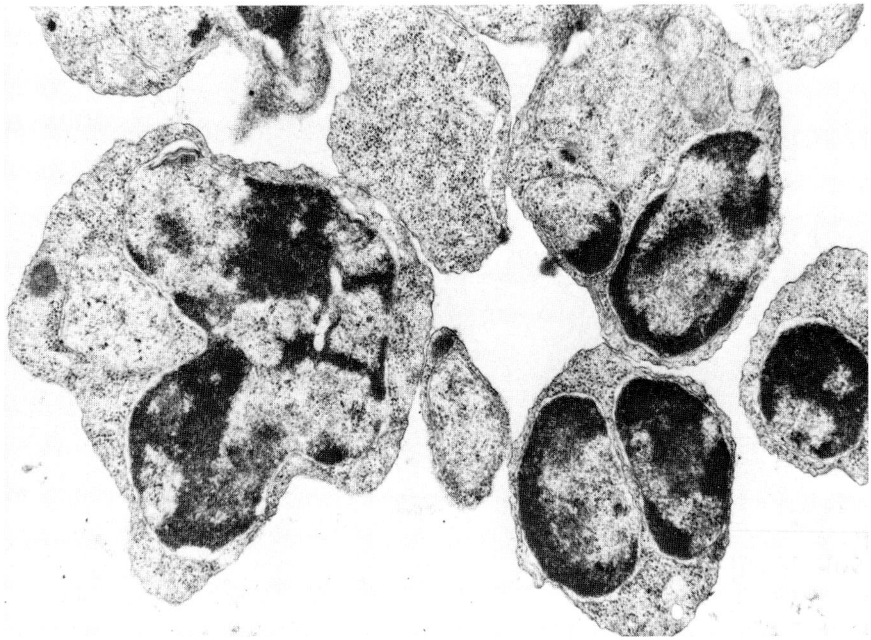

FIG. 4. Section through surface protuberances of ascites tumor cell similar to that illustrated in Figure 3B. Note multiple nuclear fragments. ×14,000.

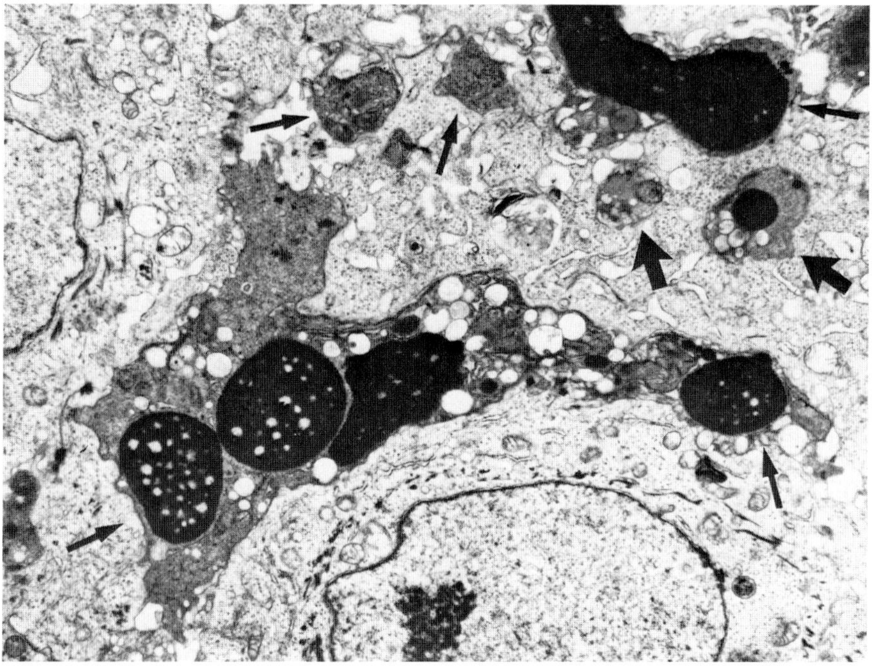

FIG. 5. Spontaneous apoptosis in squamous cell carcinoma of human cervix uteri. Some apoptotic bodies (thin arrows) are still extracellular. Others (thick arrows) have probably been phagocytosed by tumor cells. While serial sectioning would be required to prove this, the ability of squamous carcinoma cells to phagocytose apoptotic bodies has been unequivocally established by demonstration of digestion of bodies within phagolysosomes in their cytoplasm (Searle et al. 1973). ×5,700. (Reproduced from Searle et al. 1973, with permission of *Pathology*.)

are quickly degraded by lysosomal enzymes (Figures 1, 11, 12). Within a few hours, all that remains of a deleted cell is a small amount of compacted indigestible material in the lysosomes of its former neighbors.

It is the first phase of apoptosis that is unique. The second is merely a mopping-up operation, and is the only stage at which lysosomal enzymes are likely to participate. It is emphasized that it is the lysosomes of the phagocyte, not the cell being deleted, that are involved. The mechanism of the cellular condensation and budding is unknown, but presumably involves extrusion of water and rearrangement of cytoskeletal elements. There may be a change in the plasma membrane to allow the phagocytosis by nearby cells. The whole process of condensation and budding to form apoptotic bodies has been observed by phase contrast microscopy to take only a few minutes (Russell et al. 1972, Sanderson 1976). The conversion of a cell into a large number of membrane-enclosed fragments clearly results in a marked increase in the surface-to-volume ratio; estimates of the degree of condensation suggest that it is enough to account for this increase without new membrane synthesis. While the organelles of apoptotic bodies seen in the intercellular spaces of tissues always appear well preserved, apoptotic bodies formed in ascites tumors and suspension cell cultures, where phagocytosis is rare, soon undergo secondary swelling with membrane rupture (Searle et al. 1975, Don et al. 1977), a process that resembles necrosis. Apoptotic bodies, once formed, thus seem unable to maintain their integrity for long periods. Their phagocytosis in tissues prior to this secondary degeneration may explain the absence of inflammation.

Apoptosis is well suited to a role in cell kinetics. The cells surrounding those being deleted merely close ranks, and there is no tissue disorganization such as follows necrosis. The products of digestion of the deleted cells are presumably reused.

At the light microscope level, apoptosis is inconspicuous (Figures 9A, 13). The larger bodies appear as ovoid or spherical globules, but the smaller ones cannot be detected unless they contain at least one dense nuclear fragment. Further, the rapid phagocytosis and digestion of apoptotic bodies in tissues

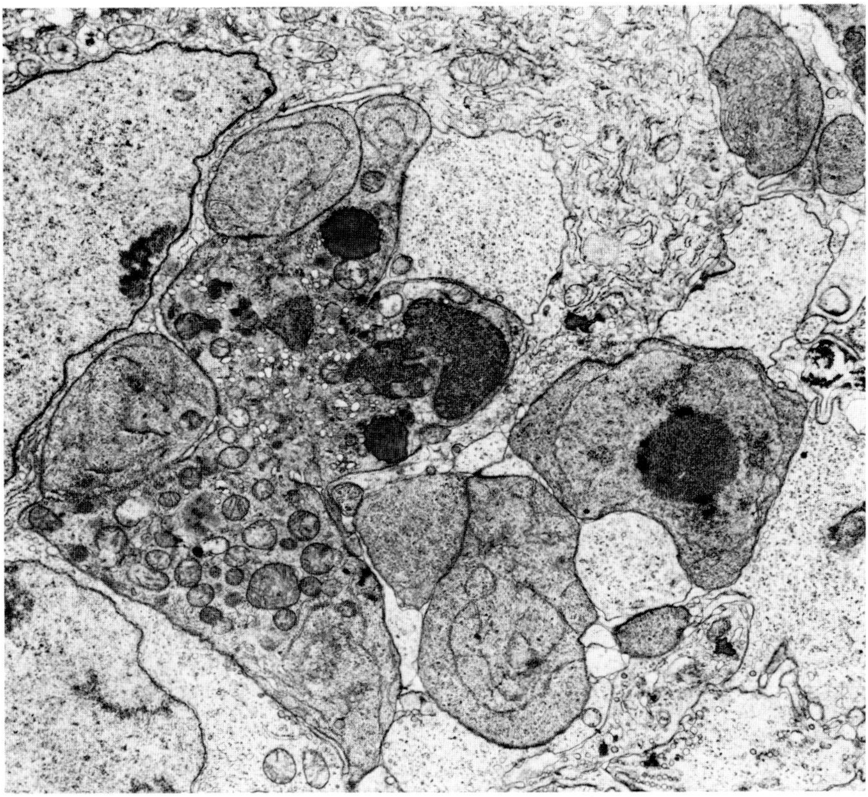

FIG. 6. Cluster of extracellular apoptotic bodies in intramuscular nodule of sarcoma 180 treated with 1,000 rads X rays six hours previously. ×7,200.

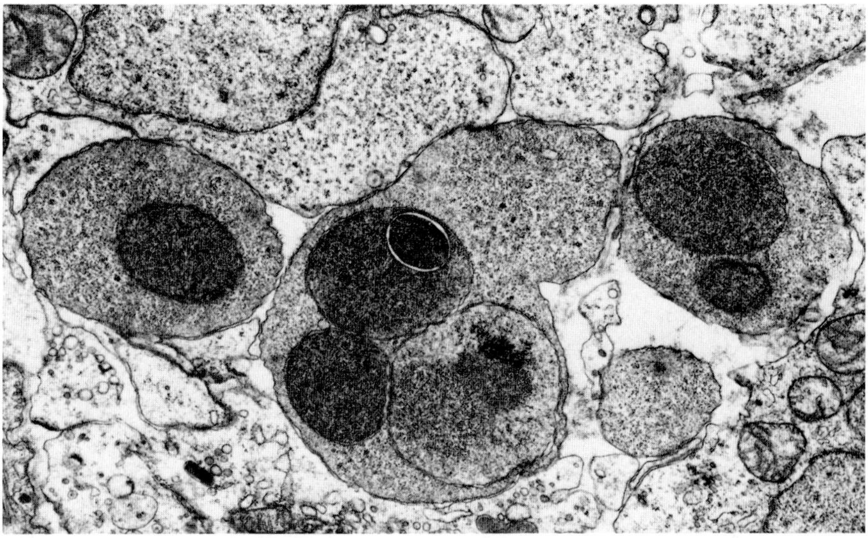

FIG. 7. Same tumor as illustrated in Figure 6. Four small apoptotic bodies, three containing nuclear fragments, lie in intercellular space. ×12,000.

mean that histological evidence of cell deletion by apoptosis soon disappears (Wyllie 1974). The process is also hard to quantify with the electron microscope because of its patchy distribution and the difficulty of distinguishing between autophagic vacuoles and phagocytosed apoptotic bodies that lack nuclear components (Figure 9B).

It should be noted that the histological changes classically designated as nuclear pyknosis (compaction) and karyorrhexis (fragmentation) occur in both necrosis and apoptosis (Kerr et al. 1972, Searle et al. 1975). It is thus not surprising that apoptosis, including that induced by radiation, should sometimes have been referred to as necrosis (Strange and Murphree 1972), in spite of the distinctive cytoplasmic changes.

Finally, something should be said about nomenclature. Apoptosis is not the only name that has been given to this process. In the phase contrast microscopic studies cited above, in which cell death produced by specifically allergized T lymphocytes was being described, the distinctive cellular budding to form spherical globules was referred to as popcorn-type cytolysis (Russell et al. 1972) and zeiosis (Sanderson 1976), respectively; that the authors were indeed observing what we have suggested be called apoptosis has been amply confirmed by subsequent electron microscopy of T cell–mediated immune killing of cells (Don et al. 1977, Liepins et al. 1977, Sanderson and Glauert 1977). Russell et al. did not relate their popcorn-type cytolysis to already known modes of cell death. The term apoptosis, which superseded our previous morphologically descriptive term shrinkage necrosis (Kerr 1965, 1971), was proposed to emphasize the

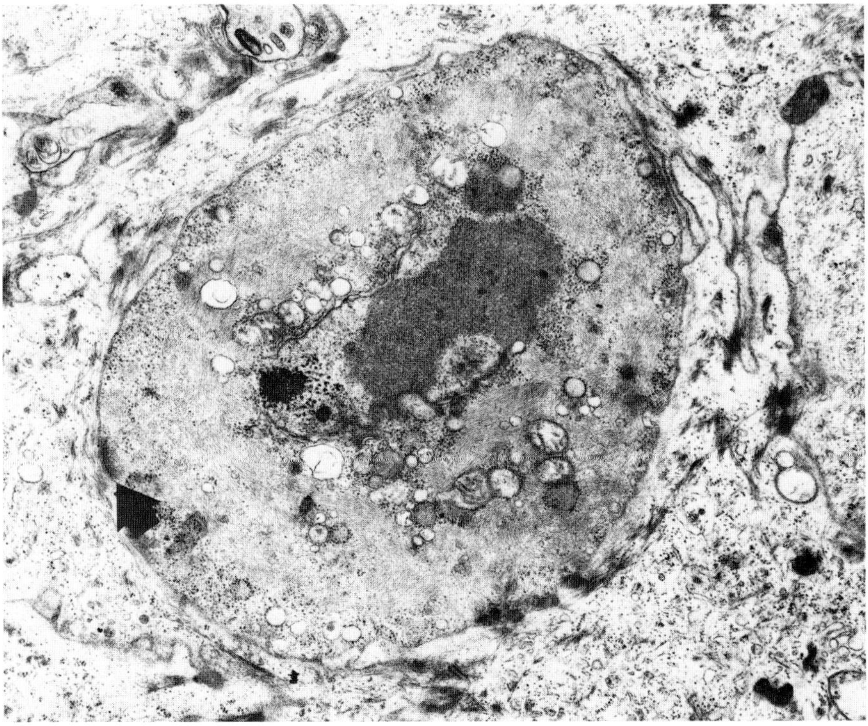

FIG. 8. Apoptotic body containing nuclear fragment lying in tortuous intercellular space in mouse epidermis irradiated with ultraviolet light of wave length 230–430 nm four hours previously. Dose at skin surface 6×10^{17} photons/cm^2. Note hemidesmosome (arrow) in apoptotic body, and close packing of its tonofilaments. ×15,700.

phenomenon's widely ranging occurrence and general kinetic significance (Kerr et al. 1972). Zeiosis was originally used by Costero and Pomerat (1951) to designate the *reversible* formation of focal protuberances at the cell surface, and others have continued to use the term in this way (Godman et al. 1975); its use for cell death is thus likely to create confusion. In recording the spontaneous occurrence of two distinct types of cell death in a reticulum cell sarcoma, Weinberger and Banfield (1965) referred to what we call apoptosis as necrobiosis. However, this term has also been used for such diverse processes as incipient necrosis (Müller 1955) and "degeneration" of collagen in the skin disease necrobiosis lipoidica (Lever 1961), and is so ambiguous that its use would be best discontinued altogether. Extrusion subdivision with the production of minisegregants (Johnson et al. 1975, Mullinger and Johnson 1976) appears to be the same as apoptosis (Johnson et al. 1978). Lastly, apoptotic bodies found in certain specific tissues were given names by light microscopists, often prior to the advent of electron microscopy, without full appreciation of their basic nature (Kerr et al. 1972). Such names include karyolytic bodies in gut crypts after X irradiation

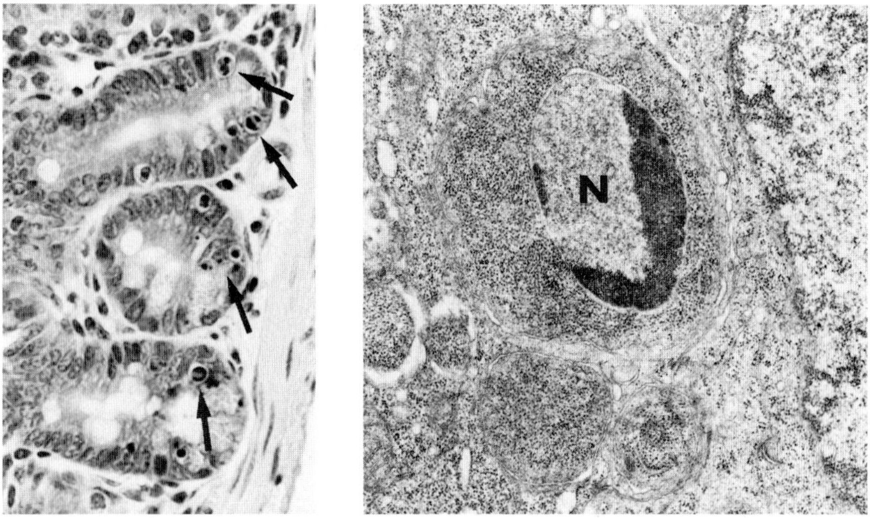

FIG. 9. Mouse small intestinal crypts four hours after treatment with 400 rads X rays. A, left, Numerous apoptotic bodies (arrows) lie in epithelium near crypt bases. Hematoxylin and eosin. ×480. B, right, An apoptotic body with typical nuclear fragment (N) has been phagocytosed by crypt epithelial cell. Membrane-enclosed masses of condensed cytoplasm without nuclear fragments are probably also phagocytosed apoptotic bodies, but possibility of their being autophagic vacuoles cannot be excluded. ×16,000.

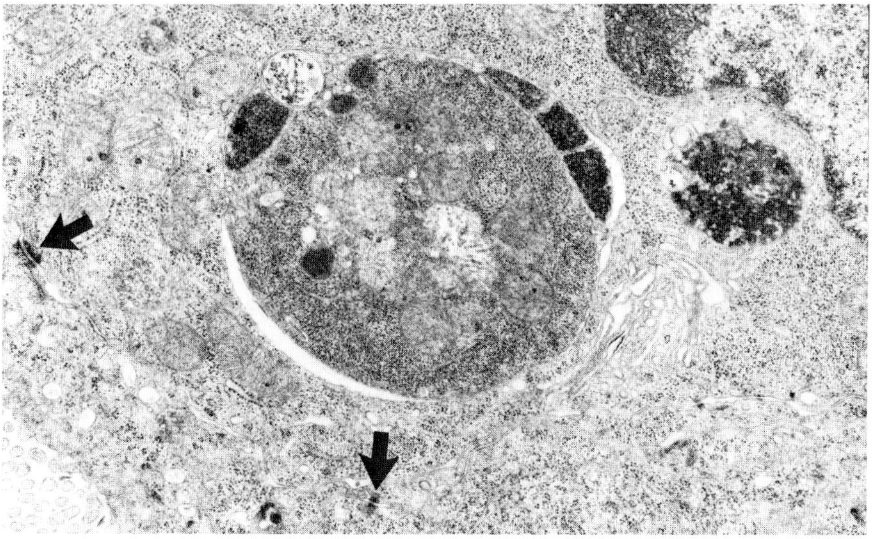

FIG. 10. Same tissue as illustrated in Figure 9. Apoptotic body containing small nuclear fragments lies within phagosome in crypt cell identified as epithelial by presence of desmosomes (arrows). Lysosomal residual body, probably derived from phagocytosed apoptotic body, also present. ×17,000.

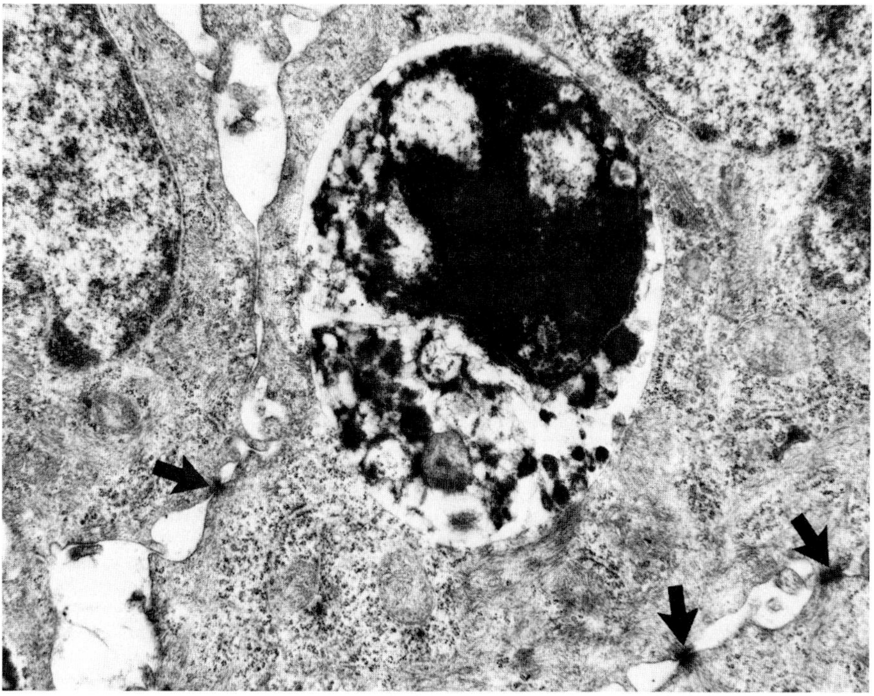

FIG. 11. Partly degraded apoptotic body with recognizable nuclear fragment in epidermis of tail of metamorphosing tadpole. Cell in which it lies identified as epithelial by presence of tonofilaments and desmosomes (arrows). Electron microscopic histochemistry shows presence of lysosomal enzymes in phagosomes containing such partly degraded bodies (Kerr and Searle 1972a). ×11,900. (Reproduced from Kerr et al. 1974, with permission of Company of Biologists Limited.)

(Montagna and Wilson 1955), Councilman bodies in the liver, and tingible bodies in lymph node germinal centers.

THE ROLE OF APOPTOSIS IN CELL POPULATION KINETICS

A corollary of the occurrence of cell division in many of the tissues of normal adult mammals (Cameron 1971) is that cells must also be lost. While it is clear that continual cell migration accounts for loss from rapidly renewing populations such as the bone marrow, it is uncertain whether this applies to slowly renewing populations, and it is often suggested that cells regularly die in organs such as the liver. Nevertheless, little has been written about the mechanism of the postulated death. Prolonged search of histological sections of normal liver discloses an occasional cell undergoing apoptosis (Kerr 1965, 1971), whereas necrosis is never seen. It is certain that apoptosis occurs spontaneously in various slowly renewing adult tissues, but difficulty in accurately quantifying it has so far precluded meaningful assessment of the balance between cell loss by this means and mitosis.

Endocrine-dependent tissues, which are capable of undergoing gross but reversible increase and decrease in size under hormonal control, clearly sometimes lose enormous numbers of cells during the phase of regression. Here the kinetic role of apoptosis has been well documented with the electron microscope. For example, lowering blood corticotrophin levels in mammals results in enhancement of apoptosis in the adrenal cortex (Wyllie et al. 1973), castration greatly enhances apoptosis in prostatic epithelium (Figure 13A) (Kerr and Searle 1973), an occurrence that can be completely prevented by androgen administration, and apoptosis is prominent in the human premenstrual endometrium (Hopwood and Levison 1976). Estrogens increase apoptosis as well as inhibit mitosis in the prostates of animals with intact testes (Kerr and Searle 1973), and glucocorticoids induce destruction of lymphoid cells by apoptosis (Kerr et al. 1972, Robertson et al. 1978).

Certain animal tissues normally undergo complete and irreversible regression, a dramatic example being loss of the anuran tadpole tail during metamorphosis. Here again apoptosis accounts for the cell deletion (Figure 11) (Kerr et al. 1974), the cells of the regressing tail continually cannibalizing their neighbors. The same process seems to be involved in insect metamorphosis (Goldsmith 1966).

Focal, precisely controlled, and often massive cell death has long been known to play a part in the normal development of vertebrate embryos (Glücksmann

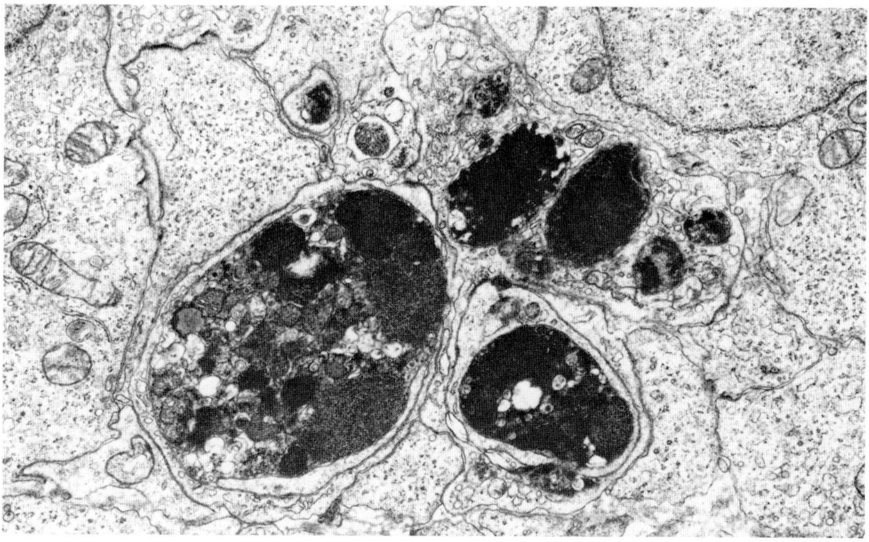

FIG. 12. Intramuscular nodule of sarcoma 180 treated with 1,000 rads X rays six hours previously. One phagocytosed partly degraded apoptotic body contains recognizable mitochondria and nuclear fragments. Others are being reduced to lysosomal residual bodies. ×9,700.

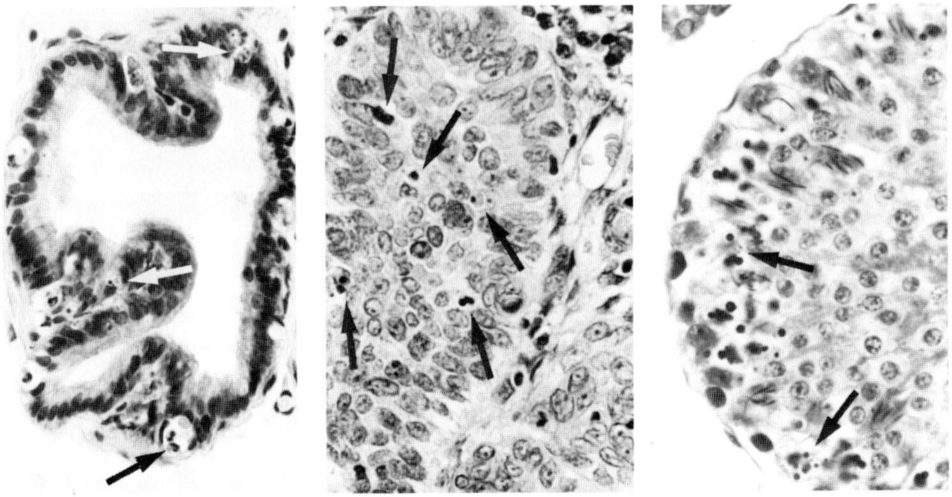

FIG. 13. Histological appearances of apoptosis (arrows). Hematoxylin and eosin. ×380. A, left, Rat prostate three days after castration. B, center, Untreated basal cell carcinoma of human skin. C, right, Rat testis seven hours after heating to 43°C for 15 minutes.

1951, Saunders 1966). It is responsible for the elimination of phylogenetic vestiges such as the rudimentary gills of mammals, and for various morphogenetic processes as the separation of digits. This "programmed cell death" of the embryologists has, however, been largely overlooked by those working on adult tissues. Realization that published descriptions of its morphology (Glücksmann 1951, Bellairs 1961, Farbman 1968, Manasek 1969, Webster and Gross 1970, Mottet and Hammar 1972) conformed exactly with the cell death observed in healthy adult animals provided a crucial strut in the formulation of the apoptosis concept (Kerr et al. 1972).

So far, we have considered the kinetic role of apoptosis under normal and paraphysiological conditions. Of particular interest in cancer research is its spontaneous occurrence in most, if not all, growing neoplasms (Cooper et al. 1975), and its possible involvement in the regression of neoplasms that follows successful nonsurgical therapy.

During the past decade there has been growing awareness of the major importance of spontaneous cell loss as a parameter in neoplastic growth (Iversen 1967, Steel 1967, Laird 1969, Clifton and Yatvin 1970, Lala 1971, Cooper et al. 1975). While a number of processes, including exfoliation, migration, and necrosis, may be implicated in this loss, there is no doubt that apoptosis often accounts for much of it (Figures 2, 5, 13B) (Kerr and Searle 1972b, Wyllie 1974, Cooper et al. 1975, Bird et al. 1976). In certain tumors, such as basal cell carcinomas of human skin, the extent of spontaneous apoptosis is manifestly so great in histological sections that it must grossly retard growth (Kerr and Searle 1972b); it is present in early tumors with only small clusters of neoplastic cells (Figure 13B) as well as in large tumors that have developed focal ischemic necrosis. There is a need for semiquantitative study of the contribution of apoptosis to total cell loss calculated from other measured kinetic parameters in a variety of different tumors.

It is noteworthy that Laird, in 1969, drew an analogy between the occurrence of morphologically demonstrable cell death during normal embryonic development and the spontaneous cell loss known from kinetic studies to take place in neoplasms, and suggested that programmed cell death might be a universal characteristic of metazoan growth, which is retained in variable degree in neoplasms along with other features of normal growth and differentiation. Our morphological studies are in accord with this prediction, but have shown that the type of cell death involved also occurs in normal static and diminishing cell populations.

That nonsurgical therapy of neoplasms might tip the balance between mitosis and apoptosis is an attractive hypothesis. Apoptosis has been shown to be implicated in regression of at least one endocrine-dependent animal tumor after hormone withdrawal (Kerr et al. 1972), and the finding that apoptosis can be induced by cell-mediated immune attack on cells, both in vitro (Don et al. 1977) and in vivo (Searle et al. 1977), raises the possibility that apoptosis might be produced in neoplasms by immunological means. The induction of apoptosis by other agents used in tumor therapy will be considered in the next section.

THE INDUCTION OF APOPTOSIS BY RADIATION, CANCER CHEMOTHERAPEUTIC DRUGS, AND HYPERTHERMIA

As was stated in the introduction to this paper, the cell death that is produced by various forms of radiation has rarely been compared morphologically with types of cell death known to occur under other circumstances. Examination of figures published in the literature and our own electron microscopy suggest that doses of radiation in the therapeutic range may induce apoptosis exclusively. For example, gross enhancement of apoptosis in the epithelium of intestinal crypts follows exposure to X rays (Figures 9, 10) (Montagna and Wilson 1955, Hugon and Borgers 1966, Potten 1977, Potten et al. 1978), beta (Cheng and Leblond 1974) and gamma rays (Potten et al. 1978), and protons (Ghidoni and Campbell 1969); extensive apoptosis has been figured in X- and gamma-irradiated lymphoid tissues (Jordan 1967, Lucas and Peakman 1969, Aleksandrova and Guljaev 1976), and in X-irradiated salivary glands (Pratt and Sodicoff 1972) and embryonic tissues (Strange and Murphree 1972); the "dyskeratotic" or "sunburn" cells found in the skin after ultraviolet irradiation (Wilgram et al. 1970) represent apoptosis of epidermal epithelium (Figure 8) (Olson and Everett 1975); X irradiation of a transplantable mouse tumor results in increase in apoptosis but not necrosis in the tumor nodules (Searle 1975a). In electron micrographs of such irradiated tumors (Figures 6, 7) it is, of course, not possible to tell whether apoptosis of an individual cell is spontaneous or radiation induced.

In 1970, Webster and Gross pointed out that the cell death that is produced in chick embryos by a number of cancer chemotherapeutic drugs is ultrastructurally identical with that occurring spontaneously in the embryos in areas such as the presumptive interdigital clefts. Subsequently we studied the ultrastructural features of the death produced by drugs of this group in rapidly renewing adult cell populations such as the epithelium of small intestinal crypts (Searle et al. 1975), the germinal centers of lymphoid follicles (Searle et al. 1975), and the spermatogonia of testicular seminiferous tubules (Harrison 1975), as well as in a mouse ascites tumor (Figures 3, 4) (Searle et al. 1975), and not surprisingly found that this is also apoptosis; we consider (Searle 1975b, Searle et al. 1975) that electron micrographs published by others (Verbin et al. 1972, 1973, Krishan and Frei 1975) confirm that these drugs induce apoptosis, not necrosis, of proliferating cells. By contrast, conventional toxins like carbon tetrachloride, which often preferentially affects cells with highly specialized biochemical functions but with low turnover rates such as hepatocytes and renal tubular epithelium, classically cause necrosis.

The possibility that shrinkage of tumors following therapy with hyperthermia might be associated with enhanced apoptosis has been little investigated. However, heating normal adult rat testes to 43°C for 15 minutes has been shown to be followed within six hours by a massive wave of ultrastructurally typical apoptosis involving B type spermatogonia and certain types of spermatocytes (Figure 13c) (Harrison 1975), the susceptibility of individual seminiferous tubules being rigidly related to their stage in the spermatogenic cycle. Further, an electron microscopic study of cell death induced by brief hyperthermia in the brains of 21-day guinea pig fetuses (Wanner et al. 1976) shows what we consider to be typical apoptosis. These findings clearly suggest that a study of heated tumors with the morphological characteristics of apoptosis in mind would be well worthwhile.

IMPLICATIONS OF THE CONCEPT OF APOPTOSIS FOR RADIATION BIOLOGY

The relative lack of interest in the morphology of radiation-induced cell death probably stems from the conventional view of cell death as a degenerative phenomenon. In classical necrosis, morphological changes indicative of death do indeed merely represent the inevitable dissolution of structure that ensues when an injured cell is unable to maintain a steady state in its vital homeostatic activities (Trump and Ginn 1969). It is the specific site of attack of the injurious agent and the earliest functional disturbances that are of real concern, rather than the postmortem degenerative changes discernible with the microscope.

However, the features of apoptosis indicate that it is not a form of cell degeneration. On the contrary, evidence summarized in the next paragraph suggests an intrinsically programmed process of active self-destruction. This obviously has implications for radiation biology.

The marked cellular condensation that characterizes apoptosis seems, a priori, likely to require the expenditure of energy; injuries that lead to impairment of energy production in cells are usually followed by swelling (Trump and Ginn 1969). Whitfield et al. (1968) have shown that pyknosis of nuclei of thymocytes produced by exposure to cortisol or radiation is dependent on initiation of a respiration-linked reaction. Lieberman et al. (1970) have shown that the occurrence of cell death in intestinal crypt epithelium after X irradiation can be prevented by inhibitors of protein synthesis. Johnson et al. (1978) have found that a proportion of minisegregants, which appear to be the same as apoptotic bodies, are capable of synthesizing nucleic acids and protein. Lastly, the very speed with which the condensation and budding are effected, the whole process of formation of a cluster of apoptotic bodies taking only a few minutes, is difficult to reconcile with passive degeneration. Thus, although the story is still very fragmentary, there is little doubt about the active nature of the first phase of apoptosis, and this carries the implication that the stereotyped sequence is genetically programmed. Regulation of the occurrence of apoptosis in embryonic and adult tissues by natural hormones that also regulate growth and differentiation (Saunders 1966, Kerr et al. 1972) is consistent with this contention.

Why does radiation in moderate doses trigger active cellular self-destruction rather than the degenerative phenomenon of necrosis? A clue may be provided by the effect of hyperthermia on the testis. Increasing the temperature at which spermatogenesis takes place increases the rate of genetic mutation in prospective gametes (Cowles 1965); the normal maintenance of the testes at a temperature below that of the rest of the body in adult scrotal mammals presumably minimizes mutation, and raising their temperature above a critical level produces aspermia. It makes biological sense for prospective gametes that have suffered serious heat-induced mutations to undergo self-destruction by apoptosis; transient infertility is better from the evolutionary point of view than the production of offspring carrying gross genetic abnormalities. Likewise, self-destruction of *somatic* cells that have developed serious unrepairable genetic abnormalities as a result of exposure to radiation or cancer chemotherapeutic drugs might have survival value for the animal in which the cells are located if they belong to populations programmed for continual replication. Neoplastic cells might retain this characteristic of normal proliferating cells in variable degree.

Finally, it is possible that an understanding of the general controls of initiation and inhibition of apoptosis at the molecular level may help define the critical targets and ultimate effector mechanisms in radiation-induced cell death.

ACKNOWLEDGMENTS

This work was supported by the National Health and Medical Research Council of Australia and the Queensland Cancer Fund. We are grateful to Christopher Potten and Terence Allen of the Paterson Laboratories, Christopher Bishop of the University of Queensland, and Robert Bourne of the Queensland Radium Institute for their help.

REFERENCES

Aleksandrova, S. E., and V. A. Guljaev. 1976. Die Dynamik der ultrastrukturellen Veränderungen des Kernes und des Chondrioms im Prozeβ des Thymozytentodes nach Bestrahlung. Radiobiol. Radiother. (Berl.) 17:597–605.

Bellairs, R. 1961. Cell death in chick embryos as studied by electron microscopy. J. Anat. 95:54–60.

Bird, C. C., A. H. Wyllie, and A. R. Currie. 1976. Ageing in tumours, *in* Scientific Foundations of Oncology, T. Symington and R. L. Carter, eds. William Heinemann, London, pp. 52–61.

Cameron, I. L. 1971. Cell proliferation and renewal in the mammalian body, *in* Cellular and Molecular Renewal in the Mammalian Body, I. L. Cameron and J. D. Thrasher, eds. Academic Press, New York and London, pp. 45–85.

Cheng, H., and C. P. Leblond. 1974. Origin, differentiation and renewal of the four main epithelial cell types in the mouse small intestine. V. Unitarian theory of the origin of the four epithelial cell types. Am. J. Anat. 141:537–562.

Clifton, K. H., and M. B. Yatvin. 1970. Cell population growth and cell loss in the MTG-B mouse mammary carcinoma. Cancer Res. 30:658–664.

Cooper, E. H., A. J. Bedford, and T. E. Kenny. 1975. Cell death in normal and malignant tissues. Adv. Cancer Res. 21:59–120.

Costero, I., and C. M. Pomerat. 1951. Cultivation of neurons from the adult human cerebral and cerebellar cortex. Am. J. Anat. 89:405–467.

Cowles, R. B. 1965. Hyperthermia, aspermia, mutation rates and evolution. Q. Rev. Biol. 40:341–367.

Don, M. M., G. Ablett, C. J. Bishop, P. G. Bundesen, K. J. Donald, J. Searle, and J. F. R. Kerr. 1977. Death of cells by apoptosis following attachment of specifically allergised lymphocytes in vitro. Aust. J. Exp. Biol. Med. Sci. 55:407–417.

Farbman, A. I. 1968. Electron microscope study of palate fusion in mouse embryos. Dev. Biol. 18:93–116.

Ghidoni, J. J., and M. M. Campbell. 1969. Karyolytic bodies. Giant lysosomes in the jejunum of proton-irradiated rhesus monkeys. Arch. Pathol. 88:480–488.

Glücksmann, A. 1951. Cell deaths in normal vertebrate ontogeny. Biol. Rev. 26:59–86.

Godman, G. C., A. F. Miranda, A. D. Deitch, and S. W. Tanenbaum. 1975. Action of cytochalasin D on cells of established lines. III. Zeiosis and movements at the cell surface. J. Cell Biol. 64:644–667.

Goldsmith, M. 1966. The anatomy of cell death. J. Cell Biol. 31:41A.

Harrison, M. W. 1975. Apoptosis in rapidly proliferating normal cell populations. Thesis presented for degree of Bachelor of Medical Science, University of Queensland.

Hopwood, D., and D. A. Levison. 1976. Atrophy and apoptosis in the cyclical human endometrium. J. Pathol. 119:159–166.

Hugon, J., and M. Borgers. 1966. Ultrastructural and cytochemical studies on karyolytic bodies in the epithelium of the duodenal crypts of whole body X-irradiated mice. Lab. Invest. 15:1528–1543.

Iversen, O. H. 1967. Kinetics of cellular proliferation and cell loss in human carcinomas. A discussion of methods available for *in vivo* studies. Eur. J. Cancer 3:389–394.

Johnson, R. T., A. M. Mullinger, and C. S. Downes. 1978. Human minisegregant cells. Methods Cell Biol. 20:255–314.

Johnson, R. T., A. M. Mullinger, and R. J. Skaer. 1975. Perturbation of mammalian cell division: human mini segregants derived from mitotic cells. Proc. R. Soc. Lond. (Biol.) 189:591–602.

Jordan, S. W. 1967. Ultrastructural studies of spleen after whole body irradiation of mice. Exp. Mol. Pathol. 6:156–171.

Kerr, J. F. R. 1965. A histochemical study of hypertrophy and ischaemic injury of rat liver with special reference to changes in lysosomes. J. Pathol. Bacteriol. 90:419–435.

Kerr, J. F. R. 1971. Shrinkage necrosis: a distinct mode of cellular death. J. Pathol. 105:13–20.

Kerr, J. F. R., B. Harmon, and J. Searle. 1974. An electron-microscope study of cell deletion in the anuran tadpole tail during spontaneous metamorphosis with special reference to apoptosis of striated muscle fibres. J. Cell Sci. 14:571–585.

Kerr, J. F. R., and J. Searle. 1972a. The digestion of cellular fragments within phagolysosomes in carcinoma cells. J. Pathol. 108:55–58.

Kerr, J. F. R., and J. Searle. 1972b. A suggested explanation for the paradoxically slow growth rate of basal-cell carcinomas that contain numerous mitotic figures. J. Pathol. 107:41–44.

Kerr, J. F. R., and J. Searle. 1973. Deletion of cells by apoptosis during castration-induced involution of the rat prostate. Virchows Archiv (Cell Pathol.) 13:87–102.

Kerr, J. F. R., and J. Searle. 1974. Apoptosis: nature of the process and its relevance to cancer. Cancer Forum 3:54–59.

Kerr, J. F. R., A. H. Wyllie, and A. R. Currie. 1972. Apoptosis: a basic biological phenomenon with wide-ranging implications in tissue kinetics. Br. J. Cancer 26:239–257.

Krishan, A., and E. Frei. 1975. Morphological basis for the cytolytic effect of vinblastine and vincristine on cultured human leukemic lymphoblasts. Cancer Res. 35:497–501.

Laird, A. K. 1969. Dynamics of growth in tumors and in normal organisms, *in* Human Tumor Cell Kinetics (National Cancer Institute Monograph 30), S. Perry, ed. National Cancer Institute, Bethesda, pp. 15–28.

Lala, P. K. 1971. Studies on tumor cell population kinetics, *in* Methods in Cancer Research, H. Busch, ed., vol. 6. Academic Press, New York and London, pp. 3–95.

Lever, W. F. 1961. Histopathology of the Skin, ed 3. Pitman, London, pp. 337–339.

Lieberman, M. W., R. S. Verbin, M. Landay, H. Liang, E. Farber, T.-N. Lee, and R. Starr. 1970. A probable role for protein synthesis in intestinal epithelial cell damage induced in vivo by cytosine arabinoside, nitrogen mustard, or x-irradiation. Cancer Res. 30:942–951.

Liepins, A., R. B. Faanes, J. Lifter, Y. S. Choi, and E. de Harven. 1977. Ultrastructural changes during T-lymphocyte-mediated cytolysis. Cell. Immunol. 28:109–124.

Lucas, D. R., and E. M. Peakman. 1969. Ultrastructural changes in lymphocytes in lymph-nodes, spleen and thymus after sublethal and supralethal doses of X-rays. J. Pathol. 99:163–169.

Manasek, F. J. 1969. Myocardial cell death in the embryonic chick ventricle. J. Embryol. Exp. Morphol. 21:271–284.

Montagna, W., and J. W. Wilson. 1955. A cytologic study of the intestinal epithelium of the mouse after total body X irradiation. J. Natl. Cancer Inst. 15:1703–1736.

Mottet, N. K., and S. P. Hammar. 1972. Ribosome crystals in necrotising cells from the posterior necrotic zone of the developing chick limb. J. Cell Sci. 11:403–414.

Müller, E. 1955. Der Zelltod, *in* Handbuch der Allgemeinen Pathologie, F. Büchner, E. Letterer, and F. Roulet, eds., vol. 2, part 1. Springer-Verlag, Berlin, Göttingen, Heidelberg, pp. 613–679.

Mullinger, A. M., and R. T. Johnson. 1976. Perturbation of mammalian cell division. III. The topography and kinetics of extrusion subdivision. J. Cell Sci. 22:243–285.

Olson, R. L., and M. E. Everett. 1975. Epidermal apoptosis: cell deletion by phagocytosis. J. Cutan. Pathol. 2:53–57.

Potten, C. S. 1977. Extreme sensitivity of some intestinal crypt cells to X and γ irradiation. Nature 269:518–521.

Potten, C. S., S. E. Al-Barwari, and J. Searle. 1978. Differential radiation response amongst proliferating epithelial cells. Cell Tissue Kinet. 11:149–160.

Pratt, N. E., and M. Sodicoff. 1972. Ultrastructural injury following X-irradiation of rat parotid gland acinar cells. Arch. Oral Biol. 17:1177–1186.

Robertson, A. M. G., C. C. Bird, A. W. Waddell, and A. R. Currie. 1978. Morphological aspects of glucocorticoid-induced cell death in human lymphoblastoid cells. J. Pathol. 126:181–187.

Russell, S. W., W. Rosenau, and J. C. Lee. 1972. Cytolysis induced by human lymphotoxin. Cinemicrographic and electron microscopic observations. Am. J. Pathol. 69:103–118.

Sanderson, C. J. 1976. The mechanism of T cell mediated cytotoxicity. II. Morphological studies of cell death by time-lapse microcinematography. Proc. R. Soc. Lond. (Biol.) 192:241–255.

Sanderson, C. J., and A. M. Glauert. 1977. The mechanism of T cell mediated cytotoxicity. V. Morphological studies by electron microscopy. Proc. R. Soc. Lond. (Biol.) 198:315–323.

Saunders, J. W. 1966. Death in embryonic systems. Science 154:604–612.

Searle, J. 1975a. Cell death by apoptosis. Thesis presented for degree of Doctor of Medicine, University of Queensland.

Searle, J. 1975b. Correspondence. Cancer Res. 35:2900–2901.

Searle, J., D. J. Collins, B. Harmon, and J. F. R. Kerr. 1973. The spontaneous occurrence of apoptosis in squamous carcinomas of the uterine cervix. Pathology 5:163–169.

Searle, J., J. F. R. Kerr, C. Battersby, W. S. Egerton, G. Balderson, and W. Burnett. 1977. An electron microscopic study of the mode of donor cell death in unmodified rejection of pig liver allografts. Aust. J. Exp. Biol. Med. Sci. 55:401–406.

Searle, J., T. A. Lawson, P. J. Abbott, B. Harmon, and J. F. R. Kerr. 1975. An electron-microscope study of the mode of cell death induced by cancer-chemotherapeutic agents in populations of proliferating normal and neoplastic cells. J. Pathol. 116:129–138.

Steel, G. G. 1967. Cell loss as a factor in the growth rate of human tumours. Eur. J. Cancer 3:381–387.

Strange, J. R., and R. L. Murphree. 1972. Exposure-rate response in the prenatally irradiated rat: Effects of 100R on day 11 of gestation to the developing eye. Radiat. Res. 51:674–684.

Trump, B. F., and F. L. Ginn. 1969. The pathogenesis of subcellular reaction to lethal injury, *in* Examples of Descriptive and Functional Morphology (Methods and Achievements in Experimental Pathology), E. Bajusz and G. Jasmin, eds., vol. 4. S. Karger, Basel and New York, pp. 1–29.

Verbin, R. S., G. Diluiso, and E. Farber. 1973. Protective effects of cycloheximide against 1-β-D arabinosylcytosine-induced intestinal lesions. Cancer Res. 33:2086–2093.

Verbin, R. S., G. Diluiso, H. Liang, and E. Farber. 1972. The effects of cytosine arabinoside on proliferating epithelial cells. Cancer Res. 32:1476–1488.

Wanner, R. A., M. J. Edwards, and R. G. Wright. 1976. The effect of hyperthermia on the neuroepithelium of the 21-day guinea-pig foetus: histologic and ultrastructural study. J. Pathol. 118:235–244.

Webster, D. A., and J. Gross. 1970. Studies on possible mechanisms of programmed cell death in the chick embryo. Dev. Biol. 22:157–184.

Weinberger, M. A., and W. G. Banfield. 1965. Fine structure of a transplantable reticulum cell sarcoma. I. Light and electron microscopy of viable and necrotic tumor cells. J. Natl. Cancer Inst. 34:459–479.

Whitfield, J. F., A. D. Perris, and T. Youdale. 1968. Destruction of the nuclear morphology of thymic lymphocytes by the corticosteroid cortisol. Exp. Cell Res. 52:349–362.

Wilgram, G. F., R. L. Kidd, W. S. Krawczyk, and P. L. Cole. 1970. Sunburn effect on keratinosomes. A report with special note on ultraviolet-induced dyskeratosis. Arch. Dermatol. 101:505–519.

Wyllie, A. H. 1974. Death in normal and neoplastic cells. J. Clin. Pathol. 27(suppl. 7): 35–42.

Wyllie, A. H., J. F. R. Kerr, I. A. M. Macaskill, and A. R. Currie. 1973. Adrenocortical cell deletion: the role of ACTH. J. Pathol. 111:85–94.

Radiation Biology in Cancer Research, edited by
Raymond E. Meyn and H. Rodney Withers.
Raven Press, New York © 1980.

Regeneration of Tumors after Cytotoxic Treatment

T. C. Stephens and G. G. Steel

Radiotherapy Research Unit, Divisions of Biophysics and Radiotherapy, Institute of Cancer Research, Sutton, Surrey, England

In 1967, Howard Skipper proposed a simple model to describe the relationship between the tumor volume response and the repopulation kinetics of the tumor cells that survived a cytotoxic treatment (Skipper 1967). The main features of the model were that any chemotherapeutic treatment was assumed to produce a constant reduction in clonogenic cell survival (constant log kill hypothesis) and that growth and regrowth of tumor cells were assumed to be exponential at a constant rate. Although the repopulation kinetics of surviving tumor cells in regenerating tumors has not been studied extensively in the last decade, recent investigations in this laboratory indicate the inadequacy of such a simple model in the radiotherapy and chemotherapy of solid tumors.

A MODEL OF TUMOR REGENERATION

Following subcurative cytotoxic treatment, solid tumors usually undergo a volume response, which is depicted in Figure 1. Soon after treatment (Rx), the tumor begins to shrink as the tumor cells that have died as a result of treatment lyse and are resorbed. R. H. Thomlinson has recently shown that the rate of tumor resorption is apparently characteristic for various types of solid tumors. With some treatments, tumors may continue to increase in size for a few days after treatment as a result of the posttreatment division of "doomed" cells or because there is a transient edema in the tumor and surrounding tissues. Concurrent with these events, surviving clonogenic tumor cells begin to proliferate and repopulate the tumor, eventually resulting in volume regrowth. Thus, it is the competing processes of resorption and repopulation that are the major factors determining the volume response curve of a tumor.

Clonogenic tumor cells only occupy part of the tumor mass, the remainder consisting of blood vessels, necrotic tissue, infiltrating host cells, and nonclonogenic tumor cells. If the fraction of tumor cells that are clonogenic is small, there may be a disproportionately large tumor volume response following treatment due to the loss of nonclonogenic decendents.

It has often been assumed that clonogenic tumor cell repopulation occurs

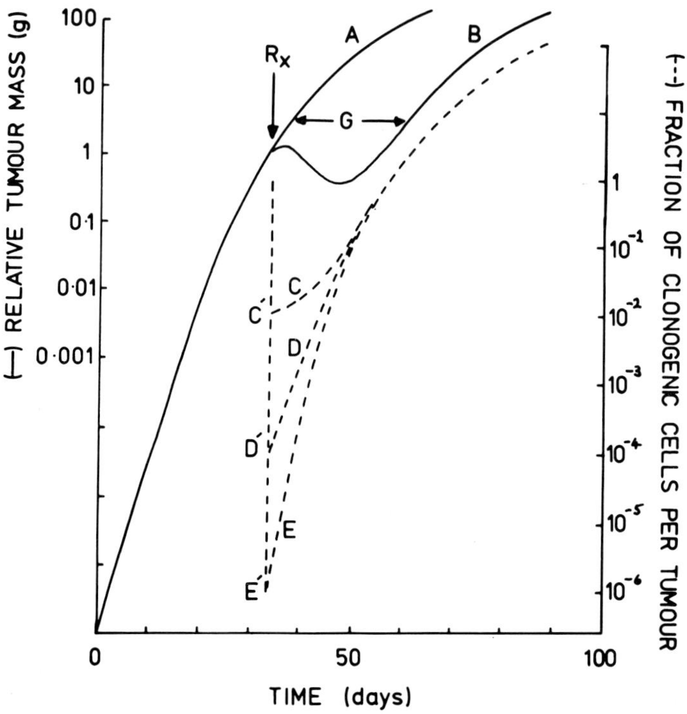

FIG. 1. A representation of the behavior of the regeneration of a solid tumor following a cytotoxic treatment, Rx. Curve A is the complete volume growth curve of the tumor from a few cells. Curve B is the volume response of the tumor following treatment. After treatment the tumor starts to shrink as dead tissue is resorbed, but later repopulation by surviving clonogenic tumor cells leads to volume regrowth. The time displacement G, between curves A and B, is the growth delay. Curves C, D, and E represent hypothetical repopulation curves for clonogenic tumor cells that survive treatment.

at a constant rate, irrespective of the level of tumor cell depopulation that has been achieved, and independent of the cytotoxic agent used to treat the tumor. Furthermore, it is often supposed that the repopulating cells in regenerating tumors follow the same growth curve as the original growth of the tumor from a few cells (Figure 1, curve D). This assumption is made in spite of the fact that repopulating cells may be exposed to a very different microenvironment from that which prevails in an untreated tumor containing the same number of viable cells. The tumor cells in regenerating tumors may be completely surrounded by cells that have been killed by treatment and are undergoing lysis and resorption. In addition, in regenerating tumors there will be a preexisting blood capillary network that may be more than adequate to support tumor regrowth up to the treatment size. On the other hand, any treatment-induced vascular damage may limit the regrowth potential of surviving tumor cells. Thus, there seems to be no *a priori* reason why repopulating clonogenic tumor cells should obey the same mathematical function that defines the untreated growth of the tumor, and, as shown in Figure 1, a given level of growth delay could derive from a smaller extent of cell killing (C′) coupled with slow repopulation (curve C) or a much greater extent of cell killing (E′) followed by very rapid repopulation (curve E).

REPOPULATION MAY DEPEND ON DIFFERENT AGENTS AND ON THE EXTENT OF CELL KILL

Clonogenic tumor cell repopulation in regenerating tumors may be assessed by two methods, the comparison of initial tumor cell survival with growth delay and the sequential measurement of tumor cell survival after treatment. From the relationship between the level of cell survival shortly after treatment and the ensuing growth delay it is only possible to derive an average rate of tumor cell repopulation. However, the sequential measurement of cell survival at various times after treatment allows changes in the repopulation rate during tumor regeneration to be detected.

We have used both methods outlined above to study the cellular repopulation patterns in two experimental mouse tumors. In Figure 2, the median time for Lewis lung tumors to grow to four times their treatment volume ($T_{4\times}$) is plotted against the surviving fraction after local irradiation. The same relationship was found when the mice breathed air during irradiation, when the tumors were made hypoxic by clamping the blood supply, and when the mice were treated with the hypoxic cell sensitizer misonidazole, although the dose-response curves (not shown) differed for each treatment. Growth delay was not linearly related to log survival, as would be expected if repopulation occurred at a constant rate independent of the level of cell kill. The observed growth delay per decade of cell killing became greater as the total extent of cell killing increased. While this could reflect a gradual decrease in the average repopulation rate with increasing radiation dose, the interpretation of the data when there is little cell killing may be difficult. Tumors did not shrink below treatment size with the lower radiation doses used in this study, so growth delay was estimated as the difference between the $T_{4\times}$ following treatment and the $T_{4\times}$ without treatment. It is a basic assumption that the tumor growth at $T_{4\times}$ will only reflect surviving tumor cell repopulation. However, when the $T_{4\times}$ is small (for one decade of cell killing it was only five days) the tumor volume at $T_{4\times}$ may be influenced by factors other than tumor cell repopulation. For instance, there may have been very little resorption of killed cells, many "doomed" cells may have undergone post-treatment divisions, and edema may still be present.

Sequential cell survival assays performed on the tumors of mice irradiated with several doses of ^{60}Co γ rays while breathing air show that the actual repopulation doubling time of clonogenic cells varied from about 1 to 1½ days, apparently decreasing slightly as the extent of tumor cell killing increased (Figure 3). However, there was also a gradual lengthening of the lag period before the clonogenic tumor cell repopulation rate reached its maximum (Stephens et al. 1978). This clearly demonstrates that repopulation rates derived from the comparison of cell survival and growth delay at best give only a crude average for the whole repopulation period.

The repopulation patterns observed in our study agree well with those reported by Barendsen and Broerse (1969), who used the rat R1-rhabdomyosarcoma and sequential cell survival assays to study repopulation after X and neutron irradiation. They found a lag phase before the clonogenic tumor cells began to repopulate, but the maximum repopulation rates were not significantly different for the two doses of X rays used.

As a further cautionary note, McNally (1973) found different relationships between cell survival and tumor growth delay in the rat RIB_5C tumor irradiated while the animals were breathing air, while the animals were breathing oxygen, or while the tumors were made hypoxic by clamping the blood supply. Much greater growth delay for a given level of cell kill was achieved when animals were irradiated while breathing air than with the other two treatments. Although this result is apparently consistent with differences in the rates of tumor cell

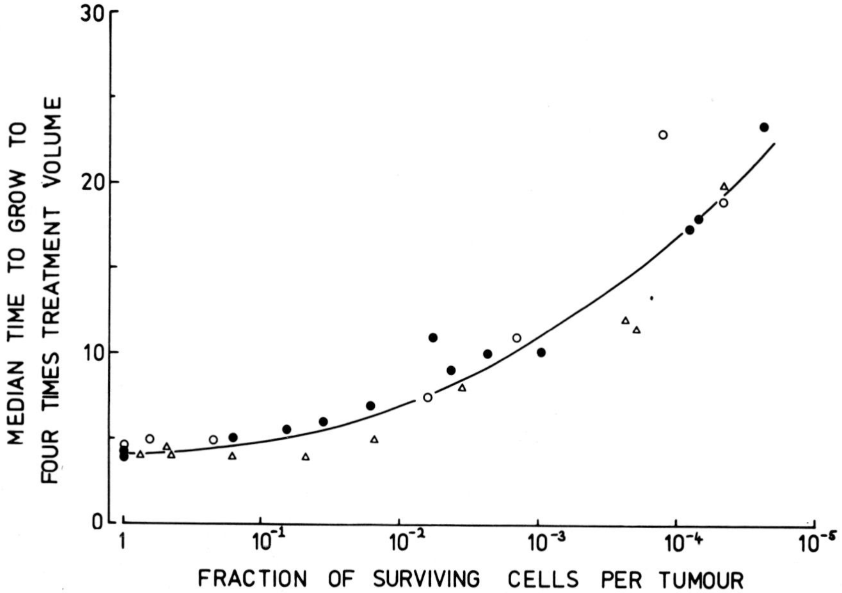

FIG. 2. Relationship between tumor volume response and clonogenic tumor cell survival in the Lewis lung carcinoma treated with ^{60}Co γ radiation under various conditions. Groups of mice bearing i.m. tumors of about 0.15 g were irradiated locally either while breathing air (●), while the tumor was made anoxic by clamping (△), or 30 minutes after an i.p. injection of misonidazole (1 mg/g body weight) (○). One day later two mice from each group were killed, their tumors dissected out and trypsinized, and the fraction of surviving cells per tumor determined by an in vitro clonogenic cell assay. The median time for the remaining tumors in each group to grow to four times the treatment volume ($T_{4\times}$) was determined from caliper measurements of the tumors.

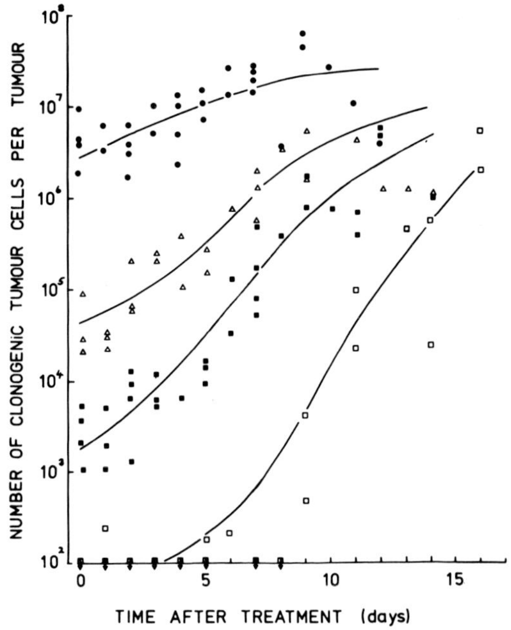

FIG. 3. Tumor cell repopulation in the Lewis lung carcinoma following irradiation. Intramuscular tumors of approximately 0.15 g were irradiated locally with various doses of ^{60}Co γ rays in air-breathing mice. Treatments are (△) 15 Gy, (■) 25 Gy, (□) 35 Gy, and (●) untreated control. At approximately daily intervals after treatment, pairs of mice were killed and cell survival measured by an in vitro clonogenic cell assay.

repopulation, direct measurements of cell survival at various times after the three treatments showed that the surviving fraction returned to unity at about the same time in each case. McNally suggested this discrepancy may arise if the clonogenic cell assay does not reliably indicate the radiosensitivity of the hypoxic cells in the tumor he used.

Although there was no difference in the relationship between cell survival and growth delay for three different conditions of irradiation of the Lewis lung carcinoma, we have observed differences with various drugs in the B16 melanoma (Stephens and Peacock 1977, Peacock and Stephens 1978). The growth delay for a given level of cell survival was much greater with cyclophosphamide than with either 1-(2-chloroethyl)-3-cyclohexyl-1-nitrosourea (CCNU) or melphalan (Figure 4). Sequential cell survival assays showed that differences in the repopulation rate could account for these results. The clonogenic cell repopulation rate was nearly twice as fast after treatment with CCNU as with cyclophosphamide (Figure 5, panel A), and after cyclophosphamide there may have been a lag period before repopulation accelerated to its maximum rate. Repopulation following melphalan treatment was also very rapid and was not significantly different from two levels of cell kill (Figure 5, panel B). This is consistent with the relationship between cell survival and growth delay for melphalan shown in Figure 4, for the approximate linearity of this curve suggests that the clonogenic cell repopulation rate probably did not vary significantly for surviving fractions down to 10^{-3}. In a rat brain tumor treated with BCNU, Rosenblum et al. (1976) found that the repopulation rate *decreased* with increasing cell kill, so

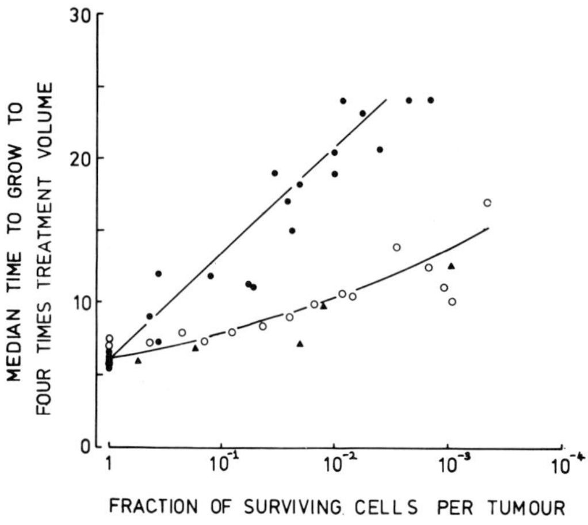

FIG. 4. Relationship between tumor volume response and tumor cell survival in the B16 melanoma treated with cyclophosphamide (●), CCNU (○), or melphalan (▲). Groups of mice were treated when their s.c. tumors weighed approximately 0.1 g. The other experimental details were the same as in Figure 2.

it seems that there is no general rule regarding repopulation rate and cell survival level in solid tumors.

THE ESTIMATION OF CELL SURVIVAL FROM REGROWTH CURVES IS HAZARDOUS

Several workers have attempted to estimate cell survival from the extent of growth delay induced by treatment (Lloyd 1975, Griswold 1975, Alfieri and Hahn 1978). The assumption has usually been that cellular repopulation after treatment occurs at the same rate as the growth of an equal-sized population of untreated cells. Both Lloyd and Griswold used the B16 melanoma and the drugs that were used in our studies cited above. Lloyd compared his estimates of cell survival from growth delay with direct bioassays of cell survival and found that the estimated cell kill due to treatment with CCNU or 1-(2-chloroethyl)-3-(4-methylcyclohexyl)-1-nitrosourea (MeCCNU) was lower than was obtained by direct bioassay. Griswold's estimates of cell kill due to MeCCNU were less than those found by Blackett and co-workers (1975) using a colony assay, but his estimated cell kill with cyclophosphamide was greater than we have observed (Stephens and Peacock 1977). These differences can be explained if the repopulation rates that were assumed in order to derive cell survival from growth delay data underestimated the actual rate after CCNU or MeCCNU but overestimated the rate after cyclophosphamide treatment.

HOST CELL REPOPULATION OF TUMORS

It is well established that many tumors are infiltrated by host cells (Evans 1972, Underwood 1974), and it has been shown that host macrophages can, under certain circumstances, be cytotoxic to some types of tumor cells in vitro

(Stewart and Beetham 1978). However, very little work has been done on the host cell complement of regenerating tumors.

In a recent study of tumor cell repopulation in irradiated Lewis lung carcinomas, Stephens and co-workers (1978) did notice that cell suspensions prepared from untreated tumors contained about 10% of small nucleated cells and more than half of these were cells of the monocyte-macrophage series. As the tumors grew, the proportion of small host cells remained roughly constant (Figure 6, panel A). In contrast, shortly after 25 Gy of local irradiation the proportion of host cells increased, and for a few days comprised about 60% of the cells in the regenerating tumor (Figure 6, panel B).

We also found that two types of colonies of distinctive morphology were obtained in the in vitro cell survival assay that was used in this study. Compact colonies were shown to consist of tumor cells, while diffuse colonies were found to be composed of macrophages. The cells giving rise to diffuse colonies were given the name *macrophage progenitors.* They required colony-stimulating factor in order to proliferate in vitro, and recent unpublished studies indicate that they comprise about 6% of the small cells present in untreated tumors. Following

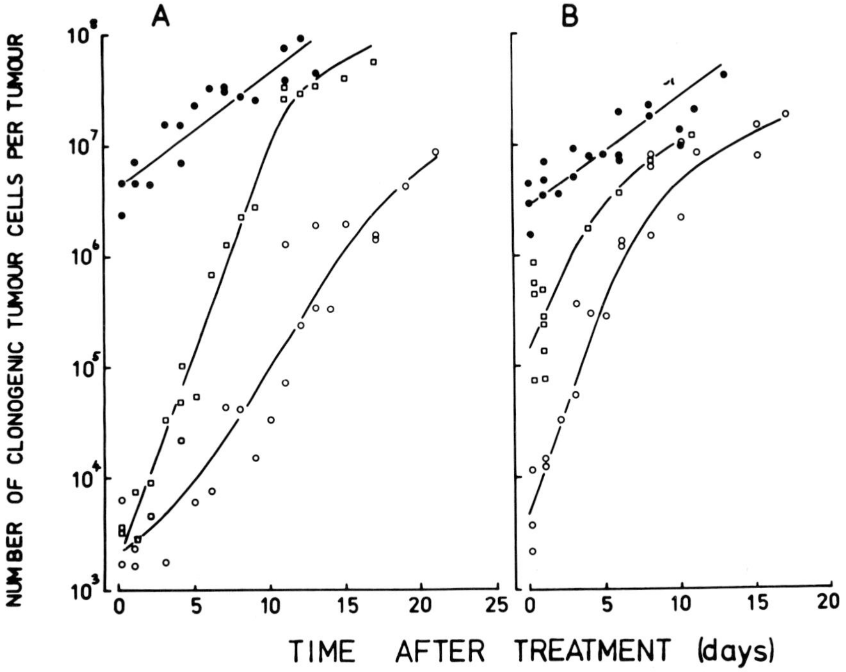

FIG. 5. Tumor cell repopulation in the B16 melanoma following drug treatment. Groups of mice bearing s.c. tumors of approximately 0.1 g were treated with various cytotoxic drugs, and sequential cell survival assays were performed as described in Figure 3. Panel A; untreated control (●), cyclophosphamide 300 mg/kg (○), CCNU 20 mg/kg (□). Panel B; untreated control (●), melphalan 7.5 mg/kg (□), melphalan 15 mg/kg (○).

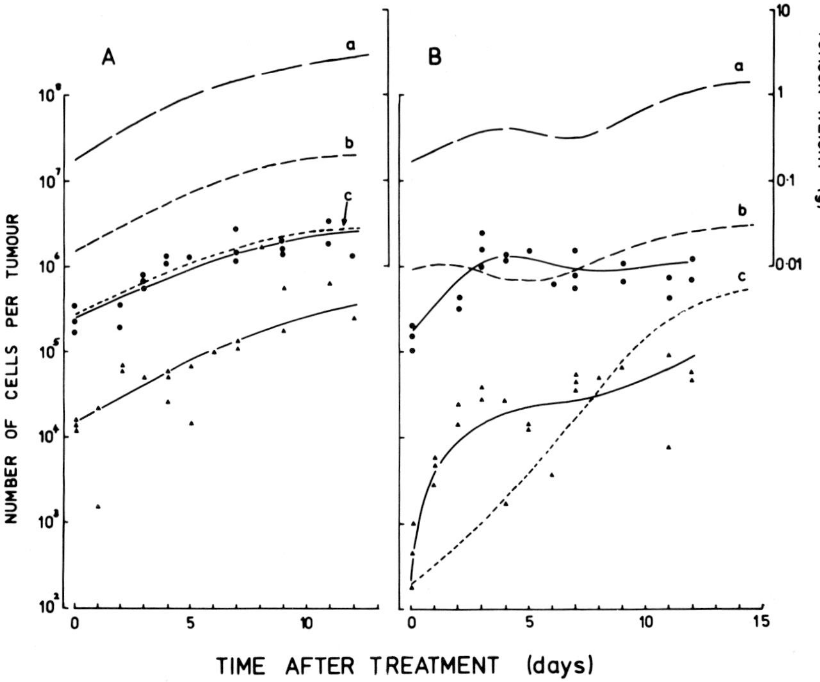

FIG. 6. Repopulation of Lewis lung tumor by cells of host origin. The changes in total host cell numbers per tumor (●) and clonogenic macrophage progenitors per tumor (▲) in untreated tumors (Panel A) and tumors regenerating after local treatment with 25 Gy of ^{60}Co γ rays given at time zero (Panel B) are shown. For comparison in each panel, curve a indicates tumor volume response, curve b is the change in the number of tumor cells per tumor, and curve c indicates the number of clonogenic tumor cells per tumor.

25 Gy of γ irradiation (Figure 6, panel B) the survival of macrophage progenitors was initially depressed by about two decades but recovered very quickly reaching pretreatment levels by about two days. This recovery was much quicker than repopulation by clonogenic tumor cells and did not occur if the mice were irradiated to the whole body. This implies that macrophage progenitors rapidly migrate into irradiated tumors from the unirradiated host.

Although the effect of host cells on tumor cell repopulation in regenerating tumors has yet to be demonstrated, these studies stress the importance of establishing the identity of the cells present in colonies grown in vitro and draw attention to the part played by host cells within regenerating tumors.

METASTASIS FROM REGENERATING TUMORS

We have recently begun to examine the patterns of metastasis from Lewis lung tumors regenerating after cytotoxic treatment. There are two main aims: (1) to investigate the relationship between the clonogenic cell content of tumors and their metastatic behavior and (2) to establish whether certain cytotoxic agents enhance the number of natural metastases seeded in the lung.

In a preliminary experiment we investigated the kinetics of metastasis from untreated tumors by sterilizing the tumors with 50 Gy of γ rays at various times after implantation and counting the number of macroscopic lung metastases present 14 days later. Following the implantation of 10^2 to 10^5 cells, metastasis always began when the tumor reached 0.1 to 0.2 g, and thereafter the number increased steadily for irradiation at larger tumor sizes. There was no change in the time of appearance or number of metastases when tumor-bearing mice were given 15 Gy of γ irradiation locally to the thorax just before the primary implant began to metastasize.

In the regeneration studies, primary implants into the leg were allowed to grow until they had just begun to metastasize. The primary implants were then given a test irradiation (Rx, Figure 7) of 15 or 25 Gy to induce regression and regrowth, and between one and three hours later the lungs were irradiated to destroy the tumor cells that had already seeded in the lungs. The primary implants were then given a sterilizing dose of γ rays at various times after the test dose, and the number of lung metastases were counted 14 days later. Treatment with 15 and 25 Gy of γ irradiation produced about three and six days of growth delay, respectively (Figure 7, panel A), but the seeding of metastases was delayed by about four and eight days, respectively (Figure 7, panel B).

The clonogenic cell repopulation pattern was also measured in this experiment and was identical to that shown in Figure 3. At the time of Rx, untreated tumors contained about 3×10^6 clonogenic tumor cells, while at the initiation of metastasis following the test treatment Rx, the tumors contained about 3×10^5 clonogenic cells. The reason treated tumors began to remetastasize with fewer clonogenic tumor cells present than in untreated tumors is not clear. This preliminary experiment suggests that it is not due to enhancement of natural lung metastasis by the thoracic irradiation, an effect that has been demonstrated when lung metastases are created artificially by i.v. injection of tumor cells (van Putten et al. 1975, Carmel and Brown 1977, Steel and Adams 1977). It may, however, reflect the presence of a preexisting blood capillary network in regenerating tumors. Metastasis through a preexisting tumor vascular system that may have been damaged by treatment might be more rapid than through the blood vessels of an untreated tumor.

SUMMARY

The regenerative response of solid tumors has often been treated in terms of a simple model in which repopulating tumor cells are assumed to proliferate at a constant rate irrespective of the level of cell killing achieved and independent of the cytotoxic agent used to treat the tumor. There is now evidence that, although the rate of regression of tumors may be independent of the cytotoxic agent used, the rate of tumor cell repopulation can vary both with the intensity

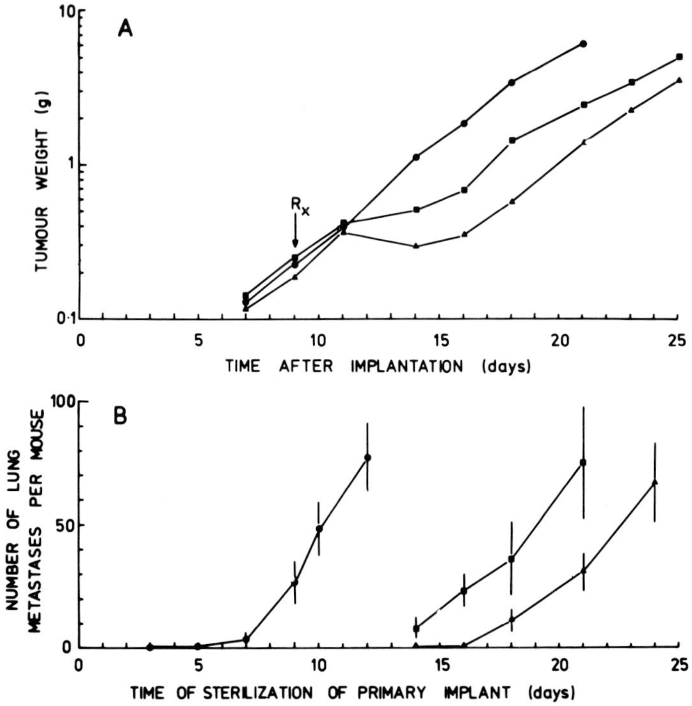

FIG. 7. Relationship between the volume growth (Panel A) and metastatic behavior (Panel B) of i.m. Lewis lung tumors, untreated (●) or following local irradiation (Rx) with 15 Gy (■) or 25 Gy (▲) of ^{60}Co γ rays. The curves in Panel B show the accumulation of lung metastases during tumor growth and regrowth after Rx. They were derived by sterilizing the primary tumors with 50 Gy of γ rays at various times after implantation and then counting the number of macroscopic nodules in the lungs 14 days later (the time required for a single cell to produce a countable nodule). The groups of mice receiving treatment (Rx) to the primary tumor were also given 15 Gy of thoracic irradiation shortly after treatment so that any metastases seeded before treatment (Rx) was given would be eradicated. Thus, only metastases formed during tumor regrowth were detected.

and type of treatment. This fact leads to the clinically important conclusion that there is no simple relationship between level of cell survival produced by treatment and the tumor growth delay that ensues.

Information is accumulating on the behavior during tumor regression and regrowth of host cells that may make up a substantial part of the cell population in tumors. However, the biological effects on tumor cell survival and repopulation of the coexisting population of host cells is not clear.

In this laboratory we have recently started to examine the metastatic behavior of regenerating tumors and have shown that the initiation of metastatic spread from irradiated tumors begins with fewer clonogenic tumor cells than in untreated tumors.

ACKNOWLEDGMENTS

We thank Miss K. Adams and Mr. J. H. Peacock for their assistance in carrying out the experiments described, and we are grateful for the support and encouragement of Professor M. J. Peckham.

REFERENCES

Alfieri, A. A., and E. W. Hahn. 1978. An in situ method for estimating cell survival in a solid tumor. Cancer Res. 38:3006–3011.

Barendsen, G. W., and J. J. Broerse. 1969. Experimental radiotherapy of a rat rhabdomyosarcoma with 15 MeV neutrons and 300 kV X-rays. Eur. J. Cancer 5:373–391.

Blackett, N. M., V. D. Courtenay, and S. M. Mayer. 1975. Differential sensitivity of colony-forming cells of hemopoietic tissue, Lewis lung carcinoma, and B16 melanoma to three nitrosoureas. Cancer Chemother. Rep. 59 (pt. 1):929–933.

Carmel, R. J. and J. M. Brown. 1977. The effect of cyclophosphamide and other drugs on the incidence of pulmonary metastases in mice. Cancer Res. 37:145–151.

Evans, R. 1972. Macrophages in syngeneic animal tumors. Transplantation 14:468–473.

Griswold, D. P. 1975. The potential of murine tumor models in surgical adjuvant chemotherapy. Cancer Chemother. Rep. 59 (pt. 2):187–204.

Lloyd, H. H. 1975. Estimation of tumor cell kill from Gompertz growth curves. Cancer Chemother. Rep. 59 (pt. 1):267–277.

McNally, N. J. 1973. A comparison of the effects of radiation on tumour growth delay and cell survival. The effect of oxygen. Br. J. Radiol. 46:450–455.

Peacock, J. H., and T. C. Stephens. 1978. Influence of anaesthetics on tumour-cell kill and repopulation in B16 melanoma treated with melphalan. Br. J. Cancer 38:725–731.

Rosenblum, M. L., K. D. Knebel, D. A. Vasquez, and C. B. Wilson. 1976. In vivo clonogenic tumor cell kinetics following 1,3-bis(2-chloroethyl)-1-nitrosourea brain tumor therapy. Cancer Res. 36:3718–3725.

Skipper, H. E. 1967. Kinetic consideration associated with therapy of solid tumors, *in* The Proliferation and Spread of Neoplastic Cells (The University of Texas System Cancer Center 21st Annual Symposium on Fundamental Cancer Research). Williams & Wilkins Co., Baltimore, pp 213–233.

Steel, G. G., and K. Adams. 1977. Enhancement by cytotoxic agents of artificial pulmonary metastasis. Br. J. Cancer 36:653–657.

Stephens, T. C., G. A. Currie, and J. H. Peacock. 1978. Repopulation of γ-irradiated Lewis lung carcinoma by malignant cells and host macrophage progenitors. Br. J. Cancer 38:573–582.

Stephens, T. C., and J. H. Peacock. 1977. Tumour volume response, initial cell kill and cellular repopulation in B16 melanoma treated with cyclophosphamide and 1-(2-chloroethyl)-3-cyclohexyl-1-nitrosourea. Br. J. Cancer 36:313–321.

Stewart, C. C., and K. L. Beetham. 1978. Cytocidal activity and proliferative ability of macrophages infiltrating the EMT6 tumor. Int. J. Cancer 22:152–159.

Underwood, J. C. E. 1974. Lymphoreticular infiltration in human tumours: prognostic and biological implications: A review. Br. J. Cancer 30:538–548.

van Putten, L. M., L. K. J. Kram, M. H. C. van Dierendonk, T. Smink, and M. Fusy. 1975. Enhancement by drugs of metastatic lung nodule formation after intravenous tumour cell injection. Int. J. Cancer 15:588–595.

Radiation Biology in Cancer Research, edited by
Raymond E. Meyn and H. Rodney Withers.
Raven Press, New York © 1980.

Evidence for the Recruitment of Noncycling Clonogenic Tumor Cells

Robert F. Kallman, C. A. Combs, Allan J. Franko, Bryan M. Furlong, Scott D. Kelley, Hannah L. Kemper, Rupert G. Miller, Diane Rapacchietta, David Schoenfeld, and Masaji Takahashi

Department of Radiology, Stanford University School of Medicine, Stanford, California 94305

It is well known that cells that survive a "treatment" undergo changes in their distribution with respect to phases of the generation cycle. Indeed, this *redistribution* (or reassortment) constitutes one of the four R's of radiotherapy (Withers 1975). In this sense, the term redistribution refers *primarily* to cells actively traversing the cell cycle, namely, P cells. In addition to P cells, however, significant numbers of cells may be in a nonproliferating state, i.e., they may be "resting" in a G_1-like or, occasionally perhaps, a G_2-like state. These are designated as quiescent, or Q, cells. Because such Q cells usually constitute the majority of tumor cells in solid tumors (Mendelsohn 1969, Frindel et al. 1968, Young and De Vita 1970) and because their reentry into the cell cycle as a consequence of therapeutic perturbation is a significant event, especially as this may determine the effectiveness of subsequent therapeutic intervention, it is essential that we examine and understand this special kind of redistribution. Indeed, the redistribution of Q cells into the P compartment deserves a separate "R": *recruitment* (Gabutti et al. 1969, Lampkin et al. 1971, Saunders and Mauer 1969). Insofar as Q cells retain clonogenic capability, the need to incorporate Q cells and their purported recruitment into realistic models of tumors is obvious. If we can demonstrate that Q cells can be recruited into P in experimental tumors, it must then be determined exactly when such recruited cells (Q → P) start cycling and the rate of recruitment or the number of cells recruited per hour per unit dose.

While the growth fraction is a major determinant of tumor sensitivity, it is also necessary to know the rate of cell loss (Steel 1968) in both unperturbed and perturbed tumors. In both situations, the extent to which clonogenic tumor cells are lost must be taken into account, and it is essential to know whether cell loss involves P cells, Q cells, or both.

Most methods that have been developed for kinetic analysis of tumors are limited strictly to unperturbed cell populations. They are not applicable to tumors that have been treated with radiation or with chemotherapeutic drugs. Cells that are "killed" by these modalities do not die instantaneously. Rather, they may continue metabolizing and cycling for one to several posttreatment cycles (Szczepanski and Trott 1975, Hurwitz and Tolmach 1969). Because classical cell kinetics methods, notably the percent labeled mitosis technique, can be applied only with fixed cells and tissues, it is impossible to determine the clonogenic potential of cells that may or may not incorporate radioactive label. It is necessary, therefore, to employ a method that can disclose not only whether cells are cycling or noncycling but also whether they have retained clonogenic potential.

Although many models of cell cycle kinetics in tumors (e.g., Dethlefsen et al. 1977, Steel 1977) recognize P and Q cell compartments and show the transition, P ⟶ Q, most do not show this as a reversible state transition. The model shown in Figure 1 does allow for such a transition, and its features, briefly, are as follows: physical measurement of a tumor gives us its actual volume, and its virtual volume is the volume to which it would have grown if there were no cell loss. Within the actual volume, the tumor is composed of three major classes of cells. P cells are those that are actively cycling and have clonogenic potential. Upon division, each P cell may form two P cells, two Q cells, or one of each. (We should also allow for the transition of a P to a Q cell without cell division.) Q cells may be of two kinds: (1) undifferentiated and noncycling or (2) maturing/differentiated. Figure 1 shows these as Q_1 and Q_2. Although Q cells in tumors are frequently referred to as G_0 cells, this usage is not strictly correct. The term "G_0" was originally introduced in 1963 (Lajtha

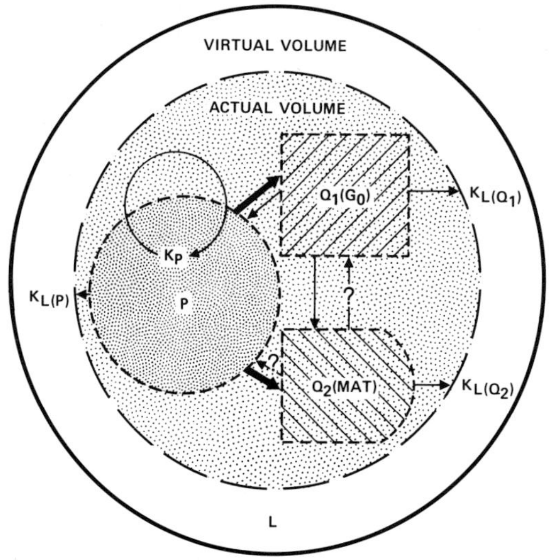

$$\phi = \frac{K_{L(P)} + K_{L(Q_1)} + K_{L(Q_2)}}{K_P}$$

FIG. 1. Model of tumor cell cycle kinetics.

1963) and is still used to designate cells in certain normal tissues, such as the liver, intestinal crypts, and bone marrow, that are noncycling, that contain the amount of DNA characteristic of pre-DNA-synthetic cells, and that can be induced to begin proliferating again upon appropriate specific stimulation. In experimental tumors, which are generally very anaplastic, most noncycling cells, i.e., Q_1 cells, probably cease proliferating because nonspecific environmental deficiencies prevent them from expressing their capacity for proliferation. In this sense, they are analogous to plateau-phase cells in vitro. In normal tissues, such cells are correctly termed G_0 and may be regarded functionally as quiescent stem cells. Some noncycling cells in well-differentiated tumors may be withheld from proliferation because they have not escaped completely from the homeostatic control mechanisms regulating the tissue of origin, and these constitute the Q_2, or *maturing*, compartment (*cf.* Bresciani and Nervi 1977). The model in Figure 1 suggests that the preferred movement is from P to Q (either Q_1 or Q_2). The lighter arrow from Q_1 to P constitutes recruitment; movement from Q_2 to P would also constitute recruitment, but it is doubtful that a maturing, i.e., differentiating, cell can dedifferentiate in this manner. Also, the Q_2 compartment can be fed from the Q_1 compartment, and it is similarly unlikely that Q_2 cells can dedifferentiate to reenter Q_1. Cell loss is designated as L, and this model suggests that cells can enter L from any of the three compartments visualized. This is in contrast to the "pipeline" model, wherein cells are born in P, either stay there or move to Q, and are lost only from Q. Thus, Figure 1 suggests that the cell loss factor, ϕ, is determined by the rate of cell loss from all three compartments, divided by the rate of cell production within the P compartment.

Two experimental tumor models have been used to investigate the recruitment of clonogenic Q cells in experimental solid tumors. The Rijswijk group (Barendsen et al. 1973, Hermens and Barendsen 1967, 1969, 1977, 1978) has utilized the R-1 rhabdomyosarcoma in rats for this purpose, and we have utilized the EMT6 mouse tumor. The approach used in these experiments is conceptually simple: one wishes to ascertain by autoradiographic examination of colonies growing in vitro whether such colonies were derived from cycling or noncycling progenitor cells when those progenitors resided in the solid in vivo tumor.

MATERIALS AND METHODS

Experiments were done with EMT6 tumors (Rockwell et al. 1972) in syngeneic BALB/c mice. As shown in Figure 2, tumors are grown by inoculating 10^5 EMT6 cells intradermally into the flanks of adult mice. At 10 to 14 days after inoculation, when the tumors have grown to a mean diameter of approximately 7 mm, the mice are given ^{3}HTdR (tritiated thymidine) (6.7 Ci/mM) by intraperitoneal injection; a standard course consists of five such injections, one every six hours over a 24-hour period, at a dose of 1 μCi/g/injection or 5 μCi/g over the 24-hour period. This schedule was chosen in terms of the EMT6 in

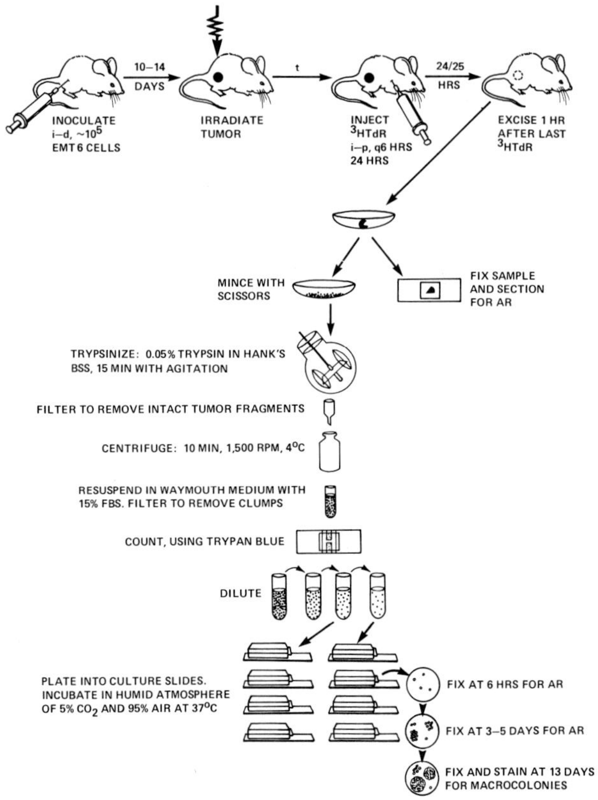

FIG. 2. Schematic representation of techniques used in experiments designed to identify clonogenic P and Q cells in solid EMT6 tumors.

vivo growth kinetic parameters determined by the percent labeled mitosis technique (Rockwell et al. 1972): T_C, 20.7 ± 7.1 hours; T_{G_1}, 7.3 ± 10.9 hours; T_S, 11.9 ± 3.6 hours; T_{G_2}, 1.9 ± 0.7 hours; growth fraction, 0.48; T_{pot}, 32 hours; T_D, 84 hours; K_P, 0.021; K_L, 0.013; and ϕ, 0.61.

The standard semicontinuous-labeling schedule just described was designed to ensure that all cycling cells are labeled and that noncycling cells remain unlabeled. Theoretically, there are two sources of unlabeled P cells under these conditions: (1) cells that have S phases less than six hours long, and (2) cells in which the sum $T_{G_2} + T_M + T_{G_1}$ is greater than 24 hours. Nonlabeling owing to *short* T_S would be expected in less than 1% of a population in which T_S is normally distributed with $\mu = 11.9$ and $\sigma = 3.6$ hours. Although less than 5% of a population of P cells with a log normal distribution of T_C would have cell cycle times greater than 24 hours + T_S, some of these slowly cycling cells would be in S phase during the 24-hour injection period. Correcting for this leads to the expectation that less than 1% of such a population of P cells would remain unlabeled because of *long* cycle times.

For experiments in which the semicontinuous labeling just described was to be compared with continuous labeling, ^{3}HTdR was infused continuously via a catheter fixed to a 27-gauge hypodermic needle inserted into the lateral tail vein and then taped in place to the tail.

One hour after the completion of thymidine administration, mice are sacrificed and their tumors carefully removed and freed of extraneous skin and connective tissue. The tumors are then converted to single-cell suspensions by the steps outlined in Figure 2; recently, the trypsin used for digestion has been replaced by a mixture of pronase (0.05%), collagenase (0.02%), and DNAase (0.02%). With this mixture, digestion is carried out at 37°C for 30 minutes with agitation. The pellet of cells produced during centrifugation is resuspended in Waymouth's medium containing 15% fetal bovine serum and is filtered to remove undigested clumps of cells. Viable tumor cells are counted using a hemacytometer under phase contrast illumination after the addition of an equal volume of trypan blue, 0.05%, to aid in the discrimination of dead cells.

A typical suspension contains a wide range of cell sizes and shapes, including red blood cells, white cells, and tissue and tumor cells as large as approximately 15 μm in diameter. In the critical step of cell enumeration, the putative tumor cells scored are those with diameters of at least approximately 8 μm. Cells are always counted differentially, with separate counts registered of large "viable" tumor cells (greater than 12 μm diameter), small "viable" tumor cells (8 to 12 μm diameter), large stained ("dead") tumor cells, and small stained cells; cells smaller than 8 μm are not scored. Cell concentrations are adjusted by serial dilution, and aliquots are plated onto culture slides (Lab-Tek chamber slides, single chamber model), which are incubated at 37°C in a humid atmosphere of 5% CO_2 and 95% air. As noted in Figure 2, samples of solid tumors prior to suspension are fixed, imbedded, and sectioned for tissue autoradiography, and representative culture slides are fixed at six hours and three to five days of incubation for cell autoradiography. Slides are dipped into NTB-2 emulsion, drained, exposed at 4°C in the dark for five weeks, developed with D-19, and stained with Giemsa stain.

For autoradiographic analyses throughout this report, we refer to clusters of cells as *colonies.* We recognize, however, that a preferable term is *foci,* for designating a group of cells as a focus does not imply that it is, or has the capacity to constitute, a colony. Thus, we use interchangeably the terms *focus* and *colony,* emphasizing that we use the latter mainly because it is so firmly rooted in related areas of study.

For the roughly four years that this project has been under way, autoradiographic scoring has been done manually by a standardized method. Each slide bearing cell or colony autoradiographs is scanned under low magnification (×100) according to a regular raster pattern. As soon as a cluster of cells comprising a colony is seen, the colony is centered in the microscope field and the X,Y coordinates are read from the vernier scales of the mechanical stage. Magnification is increased (to ×630 or ×1000), oil is added between slide and objective lens, and grains are scored using bright-field illumination. Upon completion of scoring of a given colony, the scan is resumed under low magnification along the same path until the next colony is encountered, at which time the procedure is repeated. Scoring in this manner usually requires 10 hours or more per colony slide.

Because of the slow rate of accumulation of data under these conditions, we have been developing a semiautomated instrument, which is described in Figure 3. Development of this instrument has just been completed, and it is used as follows: autoradiograph slides are placed on the motorized stepping stage of the microscope. The stepping stage controller is driven by the PDP11/04 computer or directly by joystick. The video microscope image can be displayed directly on the monitor or can be sliced by adjusting thresholds and related parameters in order to obtain an image suitable for digitization. This system digitizes a processed image and stores it in memory. The digitized image can be viewed directly or can be processed automatically to obtain counts of grains per cell.

The instrument is operated sequentially in two modes: (1) scanning/mapping and (2) grain counting. The slide is scanned by a trained operator who observes a low magnification image as it moves across a TV monitor. When colonies and/or cells appear, the operator stops the stage motion, engages the joystick, and brings a cross-hair cursor over each cell individually, as illustrated in Figure 4. Function keys allow the grouping of cells into sequentially numbered colonies. Each cell's position is thereby mapped with X,Y coordinates. As each cell is mapped, a symbol appears at the position of the cross-hairs, to preclude redun-

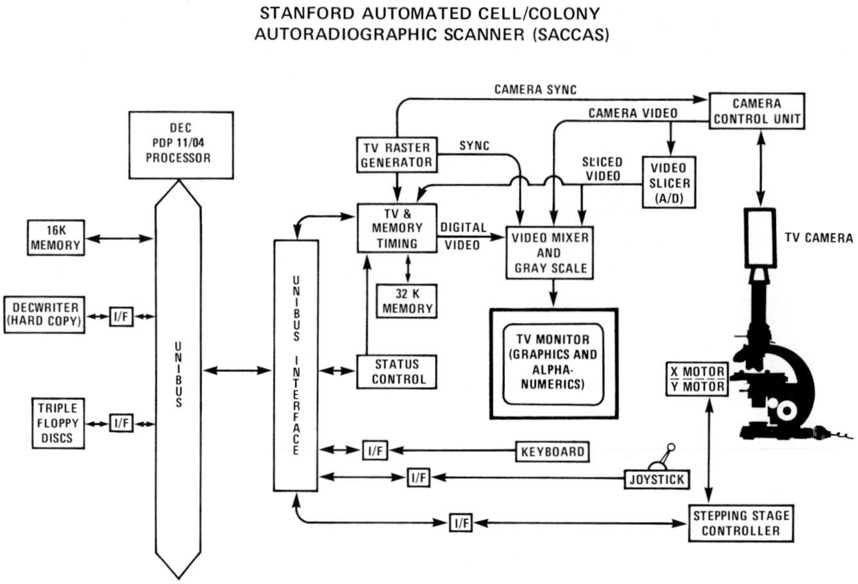

FIG. 3. Principal components and configuration of instrument for analyzing autoradiographs.

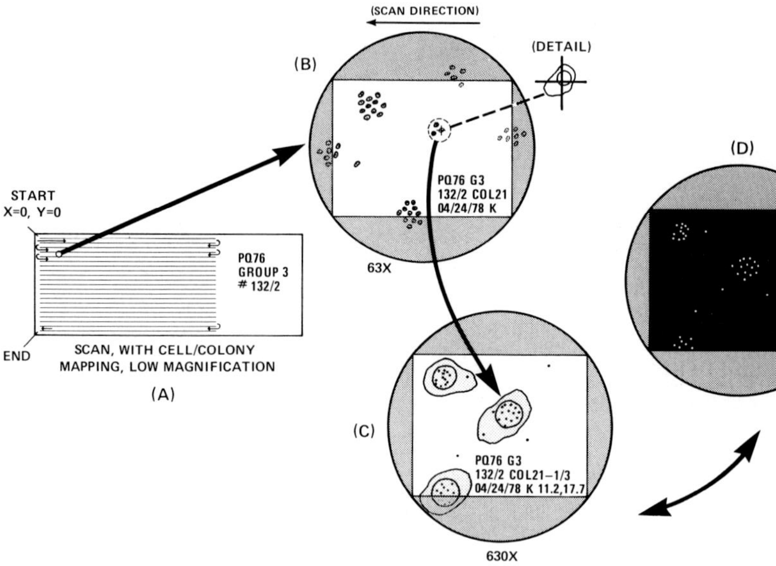

FIG. 4. Operational sequence used with SACCAS (cf. Figure 3).

dant scoring. When all cells in a monitor field are mapped, the scan is continued until the entire slide is mapped. The slide is then returned to reference point 0,0, and the magnification is increased. A function key is depressed, which causes the computer to position the slide at the first cell of the first colony mapped. Under bright-field observation, the nuclear boundary is traced with the cursor; dark-field conditions are then substituted for bright-field, the dark-field image is digitized, and the bright grains over the nucleus (Figure 4D) are counted automatically. Illumination is then switched back to bright field; the operator instructs the computer to drive the stage to the position of the next cell in the colony, its grains are scored in the same manner, and the procedure is repeated until the entire slide has been examined. Grain counts of cells are combined by the computer into colony grain counts, and these data are analyzed statistically by the computer for the determination of labeling indices.

RESULTS AND DISCUSSION

Rationale for and Pitfalls Associated with Microcolony Scoring

The universal criterion of clonogenicity is the formation of a macroscopic colony containing at least 50, but usually several hundred, cells. However, owing to the continuous dilution of label each time the cells divide, it is impossible to detect label autoradiographically in large colonies. Grain counting must therefore be performed in microcolonies, in which the number of grains incorporated by a typical progenitor cell has not been halved more than a few times by successive cell divisions. After preliminary experiments had shown that a suitable distribution of microcolony sizes was reached after incubation of unperturbed EMT6 cell suspensions for three days, this incubation time was adopted as a standard for subsequent experiments. In addition to the difficulties associated with tritium-grain count dilution, allowing colonies to grow for times appreciably longer than three days caused the total number of cells to be so large as to increase to an unreasonable degree the time required for scoring. Even though it was established that the incorporation of tritium, as ^{3}HTdR, did not compromise the viability and clonogenicity of cells (see below), it is to be expected that the probability of successful maturation of three-day microcolonies to, say, 13-day macrocolonies will be correlated with microcolony size: smaller colonies might be expected to abort more frequently than larger colonies. Since the investigation of Q-cell recruitment depends on microcolony-labeling indices and since microcolonies at three days range from two to perhaps 30 cells, it was essential to investigate abortion probability as a function of microcolony size.

Preliminary experiments to obtain this kind of information were done by mapping the positions of all colonies in a number of culture slides incubated for three days and then returning the slides to the incubator in order that colonies might continue growing for an additional seven days. Tissue culture medium was then drained from each slide, the cells were fixed and stained, and the original mapping coordinates (mechanical stage vernier scale settings) were used to relocate each colony position. In a few cases, overlap of a given position by a nearby large colony prevented determining whether a small microcolony had matured to a macrocolony, and these few colonies were disregarded in the data analysis.

These experiments were done with suspensions of cells prepared from untreated EMT6 tumors and from tumors that had been irradiated with 600 rads one hour prior to removal and preparation of single-cell suspensions. Table 1 shows that relatively few (about one fourth) of the unirradiated cells that have not divided within three days will start dividing subsequently and mature to a macrocolony. This occurrence is even rarer in irradiated single cells. The majority of two- to three-cell foci derived from unirradiated cells failed to mature to microcolonies, whereas four- to five-cell foci have an abortion rate of close to 0.5. Foci of between six and 11 cells have relatively low abortion rates (0.06 to 0.22), and all foci containing more than 11 cells matured to microcolonies. The abortion rates for foci derived from cells exposed to 600 rads differed primarily in that they tended to be slightly higher for 6- to 11-cell foci, and there was an occasional abortion above that focus size.

The extent to which our autoradiographic data are influenced by scoring at three days was further evaluated by comparing three-day with four-day statistics. In the data shown in Table 2, approximately one fourth of the three-day foci contained at least 12 cells, and this increased to 64% at four days. Despite the somewhat higher probabilities of abortion for the smallest foci, there were

TABLE 1. *Rates of abortion of 3-day foci (microcolonies), where abortion is defined as the failure to mature to a macrocolony*

Cells/Focus	Unirradiated		600 Rad*	
1	21/29†	0.72‡	13/14†	0.93‡
2–3	11/15	0.73	28/37	0.76
4–5	13/27	0.48	20/24	0.45
6–7	1/17	0.06	15/40	0.38
8–9	6/27	0.22	14/43	0.33
10–11	1/9	0.11	7/29	0.24
12–13	0/11	0	1/24	0.04
>13	0/110	0	0/139	0

* Tumors irradiated in situ one hour before excision for suspension and plating.
† Foci that did not mature to macrocolonies/foci observed.
‡ Abortion rate, the quotient from the column immediately to the left.

no appreciable differences between labeling indices for small and large foci in cultures that had been incubated for either three or four days. Neither was there a consistent increase or decrease in labeling index (LI) for the several focus sizes shown in the table. As expected, the plating efficiency based on three-day microcolonies was slightly higher than that for four-day colonies, and both were higher than the 20.4% based upon macrocolony counts.

Similar comparisons can be made for foci grown from tumor cells irradiated in situ three days before the start of labeling (or four days before sacrifice, tumor dispersion, and plating) (Tables 3 and 4). These data resemble those for unirradiated cells in that there is no consistent trend for LI to rise or fall

TABLE 2. *Comparison of three-day and four-day LI (col.) for foci grown from unirradiated EMT6 tumor cells*

	3-day		4-day	
Cells/Focus	N (%)*	LI (col.)	N (%)*	LI (col.)
2	40 (15)	0.89	12 (10)	0.87
3–5	76 (28)	0.76	12 (10)	0.94
6–11	88 (33)	0.86	20 (16)	0.94
12–22	66 (24)	0.83	31 (25)	0.83
>22	1 (0.4)	—	47 (39)	0.90
2–95	271	0.83	122	0.89
P.E.† 20.4%	34.5%		31.5%	

* The number of foci and percent of the total foci observed.
† Plating efficiencies based on: 13-day macrocolonies in cultures that were grown from aliquots of the suspensions fixed at shorter times for microcolony autoradiography; 3-day foci containing 2–95 cells; and 4-day foci containing 2–95 cells, in order.

TABLE 3. *Comparison of three-day and four-day LI (col.) for foci grown from tumor cells irradiated (300 rad) in situ four days before excision, suspension, and plating (excised one hour after 24 hours of semicontinuous labeling with $^{3}HTdR$)*

	3-day		4-day	
Cells/Focus	N (%)*	LI	N (%)*	LI
2	26 (15)	0.82	38 (12)	0.80
3–5	61 (34)	0.79	64 (20)	0.75
6–11	54 (31)	0.74	79 (25)	0.86
12–22	35 (20)	0.80	60 (19)	0.75
>22	1 (1)	—	77 (24)	0.88
2–95	177	0.78	318	0.81
P.E.† 17.5%	29.3%		35.1%	

* The number of foci and percent of the total foci observed.

† Plating efficiencies based on: 13-day macrocolonies in cultures that were grown from aliquots of the suspensions fixed at shorter times for microcolony autoradiography; 3-day foci containing 2–95 cells; and 4-day foci containing 2–95 cells, in order.

between three and four days of incubation. However, the overall LI may be seen to decrease with increasing radiation dose, either for three-day or four-day data. The expected decline in macrocolony plating efficiency as a function of radiation dose is also apparent. Interestingly, however, the plating efficiency (P.E.) based on four-day colony counts is higher for both irradiated groups than it is for three-day colonies.

The validity of these experiments depends heavily on the assumption that the cell suspension technique yields a representative sample of the P and Q

TABLE 4. *Comparison of three-day and four-day LI (col.) for foci grown from tumor cells irradiated (600 rad) in situ four days before excision, suspension, and plating (excised one hour after 24 hours of semicontinuous labeling with $^{3}HTdR$)*

	3-day		4-day	
Cells/Focus	N (%)*	LI	N (%)*	LI
2	73 (28)	0.85	42 (13)	0.67
3–5	93 (36)	0.60	70 (21)	0.66
6–11	60 (23)	0.83	83 (25)	0.71
12–22	29 (11)	0.84	72 (22)	0.78
>22	6 (3)	—	63 (19)	0.78
2–95	261	0.75	330	0.72
P.E.† 13.2%	26.1%		33.1%	

* The number of foci and percent of the total foci observed.

† Plating efficiencies based on: 13-day macrocolonies in cultures that were grown from aliquots of the suspensions fixed at shorter times for microcolony autoradiography; 3-day foci containing 2–95 cells; and 4-day foci containing 2–95 cells, in order.

cells from the solid tumor. If, for example, the enzymatic digestion procedure were preferentially to digest Q cells, or perhaps P cells, the autoradiographic data would be suspect. There is no evidence, as shown in Table 5, that such artifacts are produced. It must be borne in mind that the LI shown in this table were obtained by different statistical methods (see below). For unirradiated tumors, there was no significant difference in cell-labeling index (designated as LI[cells]) between tumor sections (2nd column) and suspended/plated cells (3rd column); furthermore, the colony-labeling index (designated as LI[col.]) data agreed satisfactorily for the two methods of statistical analysis of three-day data, and the four-day LI(col.) was only slightly different. Comparisons between these same statistics for cells from irradiated tumors lead to the same conclusions. Thus, there appear to be no artifacts associated with the preparative procedures that are basic to this experimental program.

In all autoradiographic preparations in which the presence of photographic grains must be attributed to the presence of radioisotopic label duly incorporated into cellular molecules, it is essential that one account for background grains. In most conventional autoradiographic preparations, *labeled* cells usually contain numerous grains, and it is relatively simple to discriminate these from cells labeled only with *background* grains. If, in contrast, one must decide whether a given progenitor cell was labeled by its having given rise to n descendant cells, the number of grains per descendant cell will decrease as n increases. As tritium toxicity will increase with the amount incorporated per cell, one cannot seek to ensure large numbers of grains in descendant cells simply by intentionally very heavily labeling the progenitor cells. For these and related reasons, the correction for background labeling is especially important in the determination of LI(col.) in our cultured cells. In the case of the R-1 tumor, Barendsen et al. (1973) reported that the level of background label obeyed the Poisson distribution. With such a distribution, it is relatively simple to establish expected background grain counts for colonies of different sizes and thus to decide if a given colony is "labeled" or "unlabeled."

Careful analyses of the grain counts of microcolonies derived from EMT6

TABLE 5. *Labeling indices computed according to three different statistical methods*

	LI(cell)		LI(col.)		
	Tumor	6 Hours	3 Days		4 Days
	(Stillström)	(Stillström)	(Schoenfeld)	(Miller)	(Miller)
Unirradiated	0.87	0.84	0.85	0.83	0.89
300 rad (-4 days)	0.76	0.75	0.81	0.78	0.81
600 rad (-4 days)	0.69	0.64	0.76	0.75	0.72

* Stillström, J. 1963. Grain count corrections in autoradiography Inter. J. Appl. Radiat. Isotopes 14:113–118.

progenitor cells in the present experiments, however, established that background did not obey this Poisson distribution. Rather, it seemed to be described by a negative binomial distribution. The computation of labeling indices after correction for background according to this distribution is described separately (Schoenfeld and Kallman 1978).

Because the method of Schoenfeld (Schoenfeld and Kallman 1978) is complex in its assumptions and computations and because accurate manual grain counting is tedious and excessively time consuming, a simpler method has been devised by R. Miller. This method is presented elsewhere in detail (Miller 1978) and is described briefly.

In the Miller approach, instead of counting the number of grains in a colony, the number of cells in the colony with one or more grains is counted. For example, in a colony of size six there might be four cells with one or more grains and two cells with no grains whatsoever, so the number four is recorded. For each colony size, the numbers of colonies having no cells with one or more grains, one cell with one or more grains, two cells with one or more grains, etc., are tabulated. The model assumption underlying the background correction is that each cell in a colony that has incorporated radioactive label has a probability, *l*, of exhibiting one or more grains, and each cell in a colony without incorporated label has a probability, b, of exhibiting one or more grains. For microcolonies up to about size 16 in these experiments, the probability, *l*, for a labeled cell is very high (*viz.*, 0.95 to 1.00), whereas for unlabeled cells the probability, b, is relatively small (*viz.*, 0.20 to 0.30). For a colony of size n the number i of cells exhibiting one or more grains has a binomial probability $\binom{n}{i} l^i(1-l)^{n-i}$, if the progenitor cell incorporated label. There is a different binomial probability $\binom{n}{i} b^i(1-b)^{n-i}$ for unlabeled colonies. The probability, b, for background grains is estimated from control experiments on cells not exposed to label by computing the average proportion of cells with one or more grains over all colony sizes.

For cultures exposed to label the colonies will consist of two types—those derived from a progenitor cell that had incorporated label and those derived from a cell that had not. Thus, the observed numbers of cells with one or more grains for any colony size are distributed as a mixture of two binomial distributions. There could be various methods for estimating *l* and the mixing proportion L = LI(col.), but the method of moments was selected for its simplicity and computational ease. For a fixed colony size, n, let

m_1 = average number of cells in the colony with one or more grains, and

m_2 = average squared number of cells in the colony with one or more grains.

Then, because the colonies exposed to label are a mixture of two populations, the following first two moment equations hold:

$$\frac{m_1}{n} = L\,l + (1 - L)b,$$

$$\frac{m_2 - m_1}{n(n-1)} = L\,l^2 + (1 - L)b^2.$$

The solution to these two equations is:

$$\hat{l} = \frac{\left[\dfrac{m_2 - m_1}{n(n-1)}\right] - b\dfrac{m_1}{n}}{\dfrac{m_1}{n} - b},$$

$$\hat{L} = \frac{\dfrac{m_1}{n} - b}{\hat{l} - b}.$$

After L has been estimated for each colony size, a weighted average over all the colony sizes is computed. This is the labeling index for the experiment. The weights used in the average are the numbers of colonies scored for the different colony sizes.

As shown in Table 5, there is good agreement between the LI(col.) for the methods of both Schoenfeld and Miller. Despite their being based on quite different grain counting procedures and probability models, these two methods give remarkably similar estimates of the LI. Although each method makes different assumptions and some of these assumptions are weaker than others, the fact that the two methods agree so well empirically bolsters one's faith in both methods, and it would seem that indeed the "weak" assumptions are considerably strengthened by this agreement.

In order to determine the lifetime of clonogenic Q cells, the three-day LI(col.) was determined as a function of duration of exposure of tumor cells in situ to ^{3}HTdR. In the experiments illustrated in Figure 5, LI(col.) was approximately 83% when ^{3}HTdR was administered over a 24-hour period. This increased to approximately 98% with three days of semicontinuous labeling. In this same experiment, ^{3}HTdR was also administered continuously by intravenous infusion, and the LI(col.) statistics did not differ for semicontinuous compared with continuous labeling, as summarized in Table 6. Interestingly, these results are in remarkably close agreement with those obtained several years ago for LI(cell) (Rockwell et al. 1972). From these data, it may be concluded that clonogenic Q cells have a maximum lifetime of approximately three days. After this time, all of the Q cells that were present at the start of the labeling period are lost. Although these data are inconclusive as evidence for the mechanism of cell loss from tumors, they are entirely consistent with deductions from independent experiments using the ^{125}IUdR method for studying the rate of cell loss (Franko

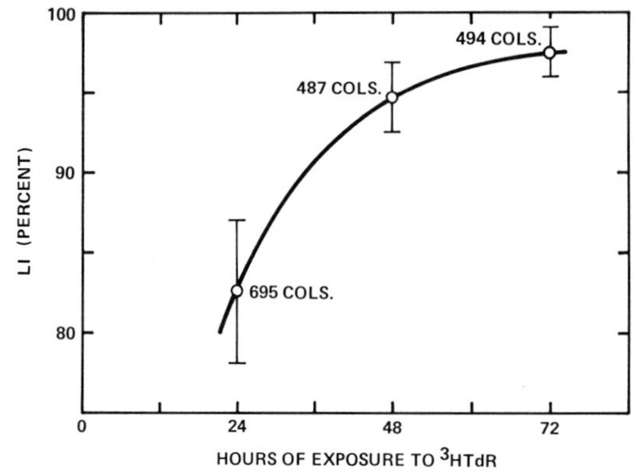

FIG. 5. Colony-labeling index (three-day microcolonies) as a function of duration of semicontinuous ^{3}HTdR labeling of EMT6 solid tumors.

and Kallman, in press). These experiments show that in some tumors, notably EMT6, it is essential that measured ^{125}IUdR tumor levels be corrected for influx of host cells labeled in sites other than the tumor itself, and after such correction, tumor IUdR declines, starting shortly after labeling. These findings are in accord with a cell loss model wherein P cells are lost, presumably randomly, at the same time as Q cells.

With this experimental system, the effect of various treatments upon the recruitment of Q cells into P may be investigated in two ways: (1) by first labeling all P cells and then administering treatment, or (2) by treating the tumor first and then labeling afterward. In both cases, labeling is carried out over a sufficiently long time (24 hours in these experiments) to ensure that all P cells have been allowed to incorporate label, and in both cases, treatment and labeling may be separated by variable lengths of time. The first approach has been used by Hermens and Barendsen (1977), who have reported that P cells move into Q at the rate of less than 1% per hour. They suggest that this recruitment may be a factor in explaining the repopulation by clonogenic cells observed in R-1 tumors after irradiation with 800 rads.

In experiments reported in a discussion at this symposium two years ago

TABLE 6. *Labeling indices (%) of three-day microcolonies from tumor cells labeled continuously (iv) or semicontinuously (ip) with the same amount (5 μCi/g body weight) of ^{3}HTdR*

Frequency	Duration	LI (col.)*	n†
Continuous	24 hours	82.6 ± 2.6	205
Every 3 hours		83.8 ± 2.7	193
Every 6 hours		81.4 ± 2.3	297
Continuous	48 hours	95.2 ± 1.4	238
Every 6 hours		94.2 ± 1.5	249

* Mean ± standard deviation.
† Number of colonies scored.

(Kallman 1977), we showed the results of experiments in which EMT6 tumors were irradiated either with 300 or 600 rads and were then labeled in situ for 24 hours at increasing times after irradiation. Grain scoring was performed on autoradiographs of culture slides that had been incubated for three days, and LI(col.) was computed either for all cells, for small colonies, or for large colonies. If (a) radiation were to trigger the recruitment of Q cells into P, (b) label were made available to all tumor cells after irradiation, and (c) the radiation survival probabilities of P and Q cells did not differ, i.e., there is no selective sensitivity or resistance of either class of cells, then the labeling index of surviving clonogenic cells would be expected to increase. As seen in Figures 6 and 7, when tumors were irradiated with 300 or 600 rads and were labeled, say, for the 24-hour period starting immediately after irradiation, LI(col.) decreased promptly and markedly. It decreased still further when labeling was started one day after irradiation and the cells were suspended and plated 24 hours later. When still more time was allowed to elapse between irradiation and labeling, LI(col.) appeared to return toward the control level. Only in the case of 600 rads, however, did LI(col.) reach the control level, and it exceeded the unirradiated level when there were five days between irradiation and the start of labeling (six days before suspension/plating). This could be interpreted as evidence of net Q cell recruitment, recruitment which only becomes apparent between approximately five and six days after irradiation. The upward direction of the LI(col.) curve starting at two to three days after both doses also is consistent with recruitment, insofar as the labeling index should rise as more and more noncycling cells are brought into cycle. Unfortunately, it is impossible to determine whether the upward excursions of these curves starting at two–three days were caused by recruitment or by rapid repopulation by surviving P cells, i.e., a speeding-up of the cycles of those cells already cycling. The latter interpretation is reasonable in the light of reports of accelerated cell cycle times in irradiated tumors (Van Peperzeel 1972, Hermens 1973), but it must

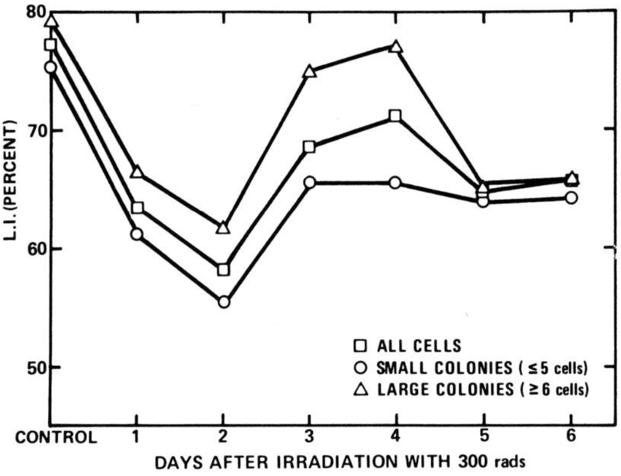

FIG. 6. Colony-labeling index (three-day microcolonies) as a function of time between irradiation of EMT6 tumors with 300 rads and tumor excision, suspension, and plating for cell culture. All tumors were labeled with ^{3}HTdR semicontinuously for 24 hours prior to the *days after irradiation* (abscissa); LI(col.) of unirradiated control tumors shown on ordinate axis.

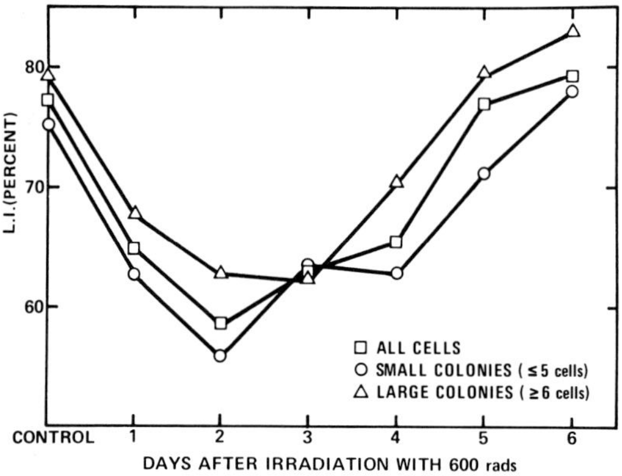

FIG. 7. Colony-labeling index (three-day microcolonies) as a function of time between irradiation of EMT6 tumors with 600 rads and tumor excision, suspension, and plating for cell culture. All tumors were labeled with ^{3}HTdR semicontinuously for 24 hours prior to the *days after irradiation* (abscissa); LI(col.) of unirradiated control tumors shown on ordinate axis.

be recognized that several others have reported slowing of the cell cycle in irradiated tumors (reviewed by Kallman and Rockwell 1977).

The initial drop in the curve of LI(col.) suggests that the clonogenic P cells of this tumor are more radiosensitive than the clonogenic Q cells. There is little or no direct experimental evidence bearing on the important question of differential P vs. Q cell radiosensitivity. To the extent that Q cells are in reality cells with very long G_1 phases, however, it could be predicted from the age response functions of cycling cells in general (Terasima and Tolmach 1963, Elkind and Whitmore 1967) and EMT6 cells in particular (Rockwell and Kallman 1974) that Q cells are more radioresistant. Such differential radiosensitivity between clonogenic P and Q cells is of great practical importance. Indeed this becomes more crucial in cancer therapy if Q cell recruitment is stimulated or enhanced by the therapy itself.

A differential age response of P and Q cells would mask recruitment as recruitment is revealed by the technique we have adopted. Although we cannot draw firm conclusions about whether Q cells are recruited in irradiated tumors, the evidence at hand is certainly consistent with the induction or acceleration of recruitment by irradiation. When the differential age response is fully defined, it should be possible to confirm this and to determine whether recruitment starts promptly or after an appreciable time delay. The fact that the curves in Figures 6 and 7 start ascending at approximately two days does not preclude that recruitment might start much sooner, perhaps even at time zero. During the first two days, the cells that incorporate label may be predominantly surviving P cells, and only later is there a preponderance of labeled Q → P colony-forming cells. It is obvious that additional experiments must be completed before firm general conclusions can be drawn.

CONCLUSIONS

The experiments described in this report show that (1) Q cells have the capacity to form colonies under the in vitro conditions used; (2) irradiation causes an immediate preferential depletion of P cells, owing presumably to their greater radiosensitivity; and (3) the rise in Q-cell-derived colonies starting at two–three days and overshooting at approximately five–six days after 600 rads suggests that Q cells are recruited into P in vitro, but other and quite different mechanisms could account for some of these same findings.

ACKNOWLEDGMENTS

This investigation was supported by Grant Numbers CA-03353, CA-10372 and CA-20527, awarded by the National Cancer Institute, Department of Health, Education and Welfare.

REFERENCES

Barendsen, G. W., H. Roelse, A. F. Hermens, H. T. Madhuizen, H. A. van Peperzeel, and D. H. Rutgers. 1973. Clonogenic capacity of proliferating and non-proliferating cells of a transplantable rat rhabdomyosarcoma in relation to its radiosensitivity. J. Natl. Cancer Inst. 51:1521–1527.

Bresciani, F., and C. Nervi. 1977. Growth kinetics in human squamous cell carcinoma, *in* Growth Kinetics and Biochemical Regulation of Normal and Malignant Cells (The University of Texas System Cancer Center 29th Annual Symposium on Fundamental Cancer Research), B. Drewinko and R. M. Humphrey, eds. Williams and Wilkins Co., Baltimore, pp. 643–661.

Dethlefsen, L. A., J. Sorenson, and J. Snively. 1977. Cell loss from three established lines of the C3H mouse mammary tumor: A comparison of the ^{125}I-UdR and the ^{3}H-TdR autoradiographic methods. Cell Tissue Kinet. 10:447–459.

Elkind, M. M., and G. F. Whitmore. 1967. The Radiobiology of Cultured Mammalian Cells. Gordon and Breach, New York.

Franko, A. J., and R. F. Kallman. 1979. Cell loss and influx of labeled host cells in EMT6 and RIF-1 tumors using ^{125}IUdR release. Cell Tissue Kinet. (in press).

Frindel, E., E. Malaise, and M. Tubiana. 1968. Cell proliferation kinetics in five human solid tumors. Cancer 22:611–620.

Gabutti, V., A. Pileri, R. P. Tarocco, F. Gavosto, and E. H. Cooper. 1969. Proliferative potential of out-of-cycle leukaemic cells. Nature 224:375–376.

Hermens, A. F. 1973. Variations in the cell kinetics and the growth rate in an experimental tumour during natural growth and after irradiation. Publ. No. 835, Radiobiological Institute TNO, Rijswijk (ZH), The Netherlands.

Hermens, A. F., and G. W. Barendsen. 1967. Cellular proliferation patterns in an experimental rhabdomyosarcoma in the rat. Eur. J. Cancer 3:361–369.

Hermens, A. F., and G. W. Barendsen. 1969. Changes of cell proliferation characteristics in a rat rhabdomyoscarcoma before and after X-irradiation. Eur. J. Cancer 5:173–189.

Hermens, A. F., and G. W. Barendsen. 1977. Effects of ionizing radiation on the growth kinetics of tumors, *in* Growth Kinetics and Biochemical Regulation of Normal and Malignant Cells (The University of Texas System Cancer Center 29th Annual Symposium on Fundamental Cancer Research), B. Drewinko and R. M. Humphrey, eds. Williams and Wilkins Co., Baltimore, pp. 531–545.

Hermens, A. F., and G. W. Barendsen. 1978. The proliferative status and clonogenic capacity of tumour cells in a transplantable rhabdomyosarcoma of the rat before and after irradiation with 800 rad of X-rays. Cell Tissue Kinet. 11:83–100.

Hurwitz, C., and L. J. Tolmach. 1969. Time lapse cinemicrographic studies of X-irradiated HeLa S3 cells. I. Cell progression and cell disintegration. Biophys. J. 9:607–633.

Kallman, R. F. 1977. Discussion, after paper presented by A. F. Hermens, *in* Growth Kinetics and Biochemical Regulation of Normal and Malignant Cells (The University of Texas System Cancer Center 29th Annual Symposium on Fundamental Cancer Research), B. Drewinko and R. M. Humphrey, eds. Williams and Wilkins Co., Baltimore, pp. 545–546.

Kallman, R. F., and S. Rockwell. 1977. Effects of radiation on animal tumor models, *in* Cancer:

A Comprehensive Treatise, F. F. Becker, ed., Vol. 6. Plenum Press, New York, pp. 225–279.

Lajtha, L. G. 1963. On the concept of the cell cycle. J. Cell. Comp. Physiol. (Suppl. 1) 62:143–144.

Lampkin, B. C., T. Nagao, and A. M. Mauer. 1971. Synchronization and recruitment in acute leukemia. J. Clin. Invest. 50:2204–2214.

Mendelsohn, M. L. 1969. Cell cycle kinetics and mitotically linked chemotherapy. Cancer Res. 29:2390–2393.

Miller, R. G., Jr. 1978. Estimation problems in microcolony autoradiography, *in* Biostatistics Casebook, Vol. II (Technical Report No. 36, PHS Grant R01 GM21215, Division of Biostatistics, Stanford University), pp. 111–135.

Rockwell, S., and R. F. Kallman. 1974. Cyclic radiation-induced variations in cellular radiosensitivity in a mouse mammary tumor. Radiat. Res. 57:132–147.

Rockwell, S. C., R. F. Kallman, and L. F. Fajardo. 1972. Characteristics of a serially transplanted mouse mammary tumor and its tissue-culture-adapted derivative. J. Natl. Cancer Inst. 49:735–749.

Saunders, E. F., and A. M. Mauer. 1969. Reentry of non-dividing leukemic cells into a proliferative phase of acute childhood leukemia. J. Clin. Invest. 48:1299–1305.

Schoenfeld, D., and R. F. Kallman. 1978. Technical Report No. 442, Sidney Farber Cancer Institute, Boston.

Steel, G. G. 1968. Cell loss from experimental tumours. Cell Tissue Kinet. 1:193–207.

Steel, G. G. 1977. Growth kinetics of Tumours. Clarendon Press, Oxford.

Stillström, J. 1963. Grain count corrections in autoradiography. Int. J. Appl. Radiat. Isotopes 14:113–118.

Szczepanski, L., and K. R. Trott. 1975. Post-irradiation proliferation kinetics of a serially transplanted murine adenocarcinoma. Br. J. Radiol. 48:200–208.

Terasima, T., and L. J. Tolmach. 1963. Variations in several responses of HeLa cells to x-irradiation during the division cycle. Biophys. J. 3:11–33.

Van Peperzeel, H. A. 1972. Effects of single doses of radiation on lung metastases in man and experimental animals. Eur. J. Cancer 8:665–675.

Withers, H. R. 1975. The four R's of radiotherapy. Adv. Radiat. Biol. 5:241–271.

Young, R. C., and V. T. DeVita. 1970. Cell cycle characteristics of human solid tumors. Cell Tissue Kinet. 3:285–290.

Radiation Biology in Cancer Research, edited by
Raymond E. Meyn and H. Rodney Withers.
Raven Press, New York © 1980.

In Quest of the Quaint Quiescent Cells

Lyle A. Dethlefsen

Section of Radiation and Tumor Biology, Department of Radiology, University of Utah Medical Center, Salt Lake City, Utah 84132

Quiescent cells, for the purposes of this paper, are operationally defined as those cells that are not in active proliferation during the course of time measurements are obtained. Normally, the measurements are obtained by ^{3}HTdR autoradiography. Many other terms (e.g., resting, dormant; Epifanova 1977) have been used to distinguish this population of cells from the actively proliferating cells, and depending on the context, the semantics has led to disagreement about the nature and extent of such populations both in vitro and in vivo. Thus, in an attempt to avoid ambiguity about solid tumors, I consider the term "quiescent" to include all cells out of cycle irrespective of the reason. This is in distinct contrast to my concept of the G_0 state, which, as originally defined by Lajtha (1963, also see Quastler 1963), is confined to viable cells that are out of cycle under normal physiological conditions (i.e., not nutrient deprivation per se) and can be recruited into active proliferation by a proper stimulus. The best examples of these cells come from normal tissues in vivo (liver, salvary gland, etc.); however, the stimulus does not have to be physiological (Baserga 1976). In passing, there is also value in pointing out that even though the stimulus frequently is considered a positive signal, it can just as well be removal of a negative signal, an endogenous mitotic inhibitor (See Lazzio et al. 1975).

For convenience, the proliferating and quiescent populations have frequently been labeled P and Q and I will do so throughout this manuscript; however, the initial use of the terms by Cairnie et al. (1965) was related to cell populations in the rat intestinal crypts. Thus, their Q designation probably reflects a state more analogous to the G_0 state than I intend for the solid tumors discussed here. To reiterate, in this manuscript, the Q population includes all cells out of cycle irrespective of the underlying mechanisms, and within this heterogenous population of cells there are cells out of cycle because of nutrient deprivation. There may also be G_0 cells, and these G_0 cells may also be heterogenous in nature (Epifanova and Terskikh 1969).

The presence of Q cells in experimental solid tumors has been known for over 15 years (Mendelsohn 1962a, 1962b), and since 1962 their presence has been documented extensively in both experimental and human solid tumors as well as in experimental ascitic tumors and experimental and human leukemias (Steel 1977). Thus, the relevance of these cells for the understanding of solid tumor biology cannot be disputed. Also, the relevance of the Q cells to chemobiology and cancer chemotherapy has been extensively documented (e.g., the recruitment of cells from Q to P enhances the cytotoxicity of cell cycle phase–specific agents; see the review of Valeriote and van Putten 1975). In contrast, even though Q cells appear to be strongly relevant, theoretically, to radiobiology and radiation oncology, this relevance in vivo has not been documented in detail. For example, Hermens and Barendsen (1969, 1978) have shown that the clonogenic population in rat rhabdomyosarcoma fluctuates dramatically following X irradiation, and Barendsen et al. (1973) have reported that the P and Q cells in this rhabdomyosarcoma have equal clonogenicity in vitro. However, the inherent radiosensitivity of the Q cells versus the P cells in various cell cycle phases has not been documented in vivo. For example, does the inherent cellular radiosensitivity of the Q cells have any importance at all for tumor eradication? I do not believe anyone knows. The Q cells may well have their own hierarchy of radiosensitivities independent of, but at times equal to or greater than, that of the P cells (Mendelsohn 1967), but this certainly needs to be confirmed.

An exception to this lack of substantiation for Q cell relevance in vivo is that fraction of Q cells that are quiescent because they are hypoxic (to be discussed in detail later). The oxygen effect and its relevance are extensively documented in the radiation biology literature, and the concept has been actively pursued in clinical radiation oncology (Gray et al. 1953). To quote Kaplan (1974): "This thesis was so seductively plausible as to have become almost an article of faith for many investigators. . . . "; however, as suggested by both Hall (1967) and Kaplan (1974), the hypoxic fraction may not be relevant to cure in many situations in the clinic because of the fractionated protocols used and reoxygenation. Irrespective of the eventual outcome of this hypothesis for clinical radiation oncology, one must still consider the hypoxic cell fraction relevant to radiation and tumor biology since it is one of the components of the Q population. In this regard, hypoxic Q cells will also take on increasing importance for chemobiology as new drugs selectively cytotoxic for hypoxic cells are studied in detail; e.g., electronic-affinic radiosensitizers (Brown 1977, Pettersen 1978, Wong et al. 1978).

In spite of the intense interest and extensive work in this area over the last 15 years, little is known about the nature of the Q cells in solid tumors. The lack of progress is due primarily to the complexity of the in vivo situation. The P versus Q classification is a convenient, simple bookkeeping device, whereas in reality solid tumors are made up of populations of cells that are heterogeneous in clonogenicity, cellular kinetics, karyotype, oxygen status, immunological properties, and chemo- and radioresponsiveness (Steel 1972, Tannock 1972, Grdina et al. 1975, 1977, Dexter et al. 1978, Heppner et al. 1978). Also, since the autoradiographic observations are static measurements, one may readily fall into the trap of assuming that these measurements actually reflect a static situation in the tumor. Thus, in an attempt to make progress, we frequently ignore the considerable data that indicate that tumors represent dynamic situations and that the P and Q cells are most probably moving back and forth between compartments as well as shifting among different states of "P-ness" and "Q-ness." For example, the motility of tumor cells is well documented (Abercrombie and Ambrose 1962; Pontén 1975, Gershman et al. 1978) as is the highly kinetic status of blood flow (Reinhold 1971, Gross et al. 1974, Swabb et al. 1974, Endrich et al. 1979). Thus, the finding of a Q cell at the instant of observation may indicate little about its status a short time before or what might have happened later if the system had not been perturbed by the measurement. Perhaps this complexity at the cellular level can be accommodated by a global statistical approach as is now done in ^{3}HTdR autoradiography; however, I contend that eventually we are going to have to acknowledge the dynamic state of solid tumors when doing experiments.

In spite of this pessimistic introduction and what at times will be a refrain of "we don't know," I actually am optimistic for the future and will conclude by discussing new methodologies that will help us answer many questions about Q cell biology. Without a doubt the initial experiments will have to be done in appropriate in vitro models. Thus, I will first review what little we do know or can reasonably hypothesize about the nature of the Q cell in experimental solid tumors, then review the nature of the in vitro models in relation to their appropriateness for the in vivo situation, and finally outline parameters that may have importance and the newer techniques that, when used in various combinations, offer exciting new probes for the identification and study of the Q cell.

THE NATURE OF THE Q CELLS IN SOLID TUMORS

Hypoxic Q Cells

A review of the literature indicates that very little is actually known about the physiological, biophysical, or biochemical nature of the Q cells in experimental solid tumors. The most substantial data, both qualitatively and quantitatively, concern the Q cells that are apparently quiescent because of oxygen deprivation. This fraction, as currently estimated, may constitute less than half of the total Q cell population (van Putten et al. 1977). These observations are circumstantial since one currently can neither measure the cellular pO_2 in vivo nor directly relate this variable to the cycling versus noncycling cells. However, the circumstantial evidence is strong and the in vitro data tend to support the hypothesis that, depending on the oxygen concentration, the hypoxic cells can be quiescent but clonogenic when reoxygenated.

Tannock and co-workers (Tannock 1968, 1970, 1976, Tannock and Steel 1970, Tannock and Howes 1973), using a murine tumor that tends to grow as longitudinal cords with axial blood vessels, have demonstrated that: (a) Morphologically viable cells form a ring around the blood vessels while the morphologically dead and dying cells are predominantly found at a radius of 85–100 μm and greater. (b) The centrally located viable cells are pushed (or migrate) toward the necrotic regions as cells proliferate. (c) The radial axis for the viable cord is in generally good agreement with oxygen diffusion calculations; thus, cell death appears to correlate with severe chronic oxygen deprivation. (d) The mitotic index and ^{3}HTdR labeling index are highest centrally and decrease with increasing radial distance from the blood vessels, while the median cell cycle time shows little variation; thus, the growth fraction decreases toward the necrotic areas. (e) Tumors growing in hypoxic mice have cords with smaller radii and lower labeling indices. (f) X irradiation with single large doses (2,000–3,000 rad) tends to produce a higher degenerate index centrally, indicating a relatively higher radiosensitivity there than closer to the necrotic regions. Bateman and Steel (1978) have also published circumstantial evidence that suggests that the cell cycle age distribution for hypoxic cells is similar to that in the oxygenated cells in the Lewis lung tumors, and Hermens and Barendsen (1978) have reported evidence from the rat R-1 rhabdomyosarcomas that suggests that the clonogenic Q cell population contains a larger proportion of hypoxic cells than the clonogenic P cell population.

The in vitro studies of Koch et al. (1973a) indicate that for V-79 S-171 Chinese hamster fibroblasts, there is little effect on the population doubling time when the oxygen tension drops from 1,000 to 500 ppm; however, at 200 ppm, the doubling time increases appreciably and the cell population appears to double two or three times before entering plateau-phase growth. At 24 ppm of O_2, there was only one population doubling before entry into the plateau phase. Although the data were not shown, the authors state that clonogenicity remains high in the plateau phase for at least four days under extreme hypoxic conditions. Koch et al. (1973b) have also shown that the cells in hypoxic growth phase ($<$ 25 ppm O_2) were arrested in G_1; however, when reoxygenated, they did not respond like synchronous (mitotic selection) G_1 cells. Bedford and Mitchell (1974) have also studied the effects of hypoxia on the growth of CHL-F Chinese hamster cells in vitro and reported that at 70 ppm O_2, the cells go into a plateau phase of growth with a lower labeling index, a lengthened life cycle, a decrease in the proportion of cells in S phase, and a G_1 phase that is lengthened disproportionately to the rest of the cell cycle. The absolute plating efficiency also drops dramatically during 80 hours of culture at 70 ppm O_2. In a more recent paper, Born et al. (1976) reported that the survival of B-14-FAF 28 Chinese hamster cells decreased significantly after short periods ($<$ 1 day) of incubation under hypoxic conditions (O_2 concentration not measured but assumed to be $\simeq$ 0.1 μmol/liter). They also reported a significant increase in the S-phase duration with a G_1-phase prolongation.

The apparent discrepancy between in vivo and in vitro data on hypoxia's effect on the cell cycle could be due to the lack of resolution of the in vivo assays as well as the inability to measure cellular oxygen tension. However, it seems equally likely that the discrepancy is at least partially due to the heterogenous nature of the in vivo hypoxic cells. Nevertheless, the different in vitro experiments indicate the importance of other factors such as cell type and medium (environmental factors) as well as the varying oxygen tension. Tannock (1972) has addressed this complex issue for in vivo tumors. Briefly stated, the assumption of a two-compartment model of well-oxygenated versus anoxic cells for solid tumors in vivo is unrealistic and may lead to misleading conclusions.

Cell Density and/or Size

The separation of heterogenous populations of mammalian cells based on density by centrifugation in linear density gradients was described in 1970 (Pertoft 1970, Pretlow and Boone 1970) and on size by centrifugal elutriation in 1971 (Glick et al. 1971) and Grdina et al. (1974, 1975, 1976, 1977) have used the former technique to extensively characterize the FSa murine fibrosarcoma cells growing in vivo. Using a modified version of the lung-colony assay of Hill and Bush (1969) along with flow cytometric analysis (Steinkamp et al. 1974) and the premature chromosome condensation assay of Hittelman and Rao (1974), they have demonstrated that: (a) five distinct bands ranging in density from $\simeq$ 1.06 to 1.17 g/cm^3 can be identified; (b) the lighter density bands tend to show a larger median cell volume; however, there is considerable overlap across the density bands; (c) cells from the less dense bands tend to show a higher clonogenic efficiency (CE) than the more dense cells; (d) the CE apparently is independent of cell size and cell cycle stage; (e) the more dense bands contain a higher percentage of cells with S and G_2 DNA content, and they tend to have more chromosomal damage; (f) the more dense cells appear to be more radioresistant when irradiated in situ, as assayed by the lung-colony procedure; and (g) this resistance appears to be due to a chronic state of hypoxia.

Two reports present data that are in apparent conflict with these conclusions. The first is an initial study using centrifugal elutriation and flow cytometric analysis of FSa cells prelabeled with ^{3}HTdR in vivo whose data suggest, based on the state of Q-ness or P-ness, that the separations based on size were probably not unique (Meistrich et al. 1977). However, in subsequent studies Sigdestad and Grdina (1978), utilizing linear density gradients and a different ^{3}HTdR labeling procedure, have shown that the less dense, more clonogenic cells have a higher labeling index (L.I.) (25–30%) than the more dense, less clonogenic cells (10–15%). Thus, cell density may be a more sensitive variable than cell size for these particular studies. In the second report with apparently conflicting data, centrifugal elutriation of FSa cells was also done (Suzuki et al. 1977), and the data suggest that the CE is both cell-size dependent and cell-cycle dependent; however, the lung-colony assays were not done in preconditioned mice. Thus, environmental factors are apparently critical in this assay, since in two subsequent papers by Grdina et al. (1978a, 1978b) the data indicate no dependence of CE on cell size or cell cycle stage when the mice were preconditioned. These reports are consistent with the studies discussed earlier.

Interestingly, when the elutriation and flow cytometric studies were done on cells 48 hours after incubation in vitro rather than immediately after dispersal, then the elutriation-synchronization was considerably more effective. This suggests that, in contrast to the situation in vitro, there may be considerable size heterogeneity within any one phase of the cell cycle when cells are growing as solid tumors in situ.

Nuclear Morphometry/Chromatin Configuration

Analytical and quantitative cytology is certainly not a new field (see Wied 1966), but only in more recent studies has this approach been used in a systematic way to study nuclear morphometry ("in situ" chromatin)—as a function of both cell cycle stage (Kendall et al. 1977) and the transition of WI-38 cells from a resting (arrested) to a proliferating state (Nicolini et al. 1977a)—and to relate this information to alterations in the physical-chemical properties of isolated chromatin. Briefly, single cells are spread on a slide, Feulgen stained, and scanned by automated instrumentation utilizing densitometer recordings and image analysis. The integrated optical density (IOD; DNA content), area, projection, and perimeter of each nuclear image are measured, and from these the average optical density (AOD), form factor, and mean bound path are computed. Nicolini et al. (1977c) have used this approach in conjunction with ^{3}HTdR autoradiography and flow cytometry for analyzing B16 melanoma cells growing in C-57 mice. The data show that all three approaches gave similar estimates for the proportion of quiescent cells as well as those in the various cell cycle phases (i.e., G_1, S, and $G_2 + M$). The data also suggest that geometric and densitometric parameters may discriminate between proliferating G_1 cells and Q cells with a G_1 DNA content; however, these interesting preliminary observations need to be confirmed. In this regard Potmesil and Goldfeder (1971) have reported that nucleolar morphology may also be an index for distinguishing proliferating from nonproliferating cells in three transplantable mouse mammary carcinomas.

Cellular RNA Content

It has long been established that Q cells have a lower RNA content than P cells (Cooper 1971) and that total RNA content as well as rate of RNA synthesis varies significantly as the cells traverse the cell cycle (Terasima and Tolmach 1963; Thilly et al. 1977). Also, a recent preliminary report by Watson and Chambers (1977) presented data that suggested that the RNA content of highly clonogenic in vivo EMT6 tumor cells was appreciably higher than in those cells with low clonogenic potential. These authors used differential acridine orange staining procedures (Darzynkiewicz et al. 1975, Traganos et al. 1977) and reported that the cells with the lowest RNA content (red fluorescence, wavelength unspecified) are in the process of disintegration, those with the highest RNA content have the highest in vitro plating efficiency, and those with intermediate RNA content have a low plating efficiency. Unfortunately, the red fluorescence was not quantitated against a biochemical determination of RNA content, and the proportions of Q and P cells were not documented; thus, these data can only be considered suggestive at this time.

G_0 Cells in Solid Tumors

Do G_0 cells exist in solid tumors? Theory, intuition, and circumstantial data suggest that there should (or may) be some G_0 cells in the Q population of experimental solid tumors, but we do not actually know. Experiments have been done to show recruitment of Q cells into P (Valeriote and van Putten 1975), but we do not know if the Q cells are all quiescent because of nutrient deprivation or if some of the Q cells are analogous to resting lymphocytes waiting to be stimulated by phytohemagglutinin or to hepatocytes before partial hepatectomy (Cooper 1971). The situation is best expressed by a quote from Steel and Lamerton (1969):

> Among the various uncertainties, one which has not received a great deal of attention is how long a cell in a growing tumor might be expected to remain in a state of suspended proliferation. If the environmental factor concerned is hypoxia, it is unlikely that the cell could remain for long without dying, or without being reoxygenated, particularly in a treated tumor. The same reasoning could apply to other nutritional deficiencies, and the assumption that tumor cells can spend long periods in a G_0 state may well be justified. Theory has gone as far as it can on this question of the G_0 cells in tumors, and we need a great deal more experimental work on the difficult problem of distinguishing G_0 cells from sterile cells or those which may be proceeding slowly through the cycle.

Unfortunately, the same situation holds ten years later, indicating the complexity of the problem.

Summary on Q-Cell Biology

The above reports all demonstrate the heterogeneity of in situ solid tumor cells in their pO_2, density, size, nuclear morphometry, clonogenicity, and RNA content. They also show that for many of these parameters there is considerable variability within the cell cycle itself, and the values for P cells may well overlap the mean values for Q cells (See Table 1 for a summary). Thus, one is frequently looking for subtle effects in vivo with complex assays. The resolution of the assays may be poor, and the probability for large replication errors is high. Table 2 documents this complexity by indicating some apparent conflicting data from the literature. As indicated, the work of Barendsen et al. (1973) suggests that the P and Q cells have equal clonogenicity. However, the studies of Sigdestad

TABLE 1. *Variables analyzed relative to cellular P- and Q-ness in solid tumors*

Variable	P	vs.	Q
	(variables are a continuum)*		
Environmental status			
Oxygen	↑		↓
Other nutrients	↑ (?)		↓ (?)
Catabolic products	↓ (?)		↑ (?)
G_o	—		?‡
Clonogenicity	= (?)		= (?)
Cell density	↓		↑
Cell size	↑		↓
Nuclear morphometry†	G_1 cells		G_o + Q cells
IOD	=		=
AOD	↑		↓
Form factor	↑		↓
Mean bound path	↑		↓
Horizontal projection	↓		↑

* The direction of the arrows indicates the relative measures (e.g., the P cells have a higher concentration of oxygen than the Q cells). ?, no data or conflicting, uncertain data.
† See text and Nicolini et al. (1977a, 1977c) for details.
‡ If any G_o cells, most likely a small percentage.

and Grdina (1978, 1979) suggest that P and Q cells can be separated by density, and the Q cells have a lower CE and are probably chronically hypoxic in situ. I do not believe that these data conflict. For example, the in vivo ^{3}HTdR method gives an estimate for the total Q compartment, and the in vitro clonogenic

TABLE 2. *Apparent conflicting data*

	P	vs.	Q
1) Clonogenicity* (in vitro)	=		=
Cell density†	↓ (↑ C.E.) (↑ L.I.)		↑ (↓ C.E.) (↓ L.I.) (chronically hypoxic ?)
2) Hypoxia (cell-cycle age distribution)			
In vivo‡	=		= (?)
In vitro§	⟶//		G_1 block

* Barendsen et al. (1973).
† Sigdestad and Grdina (1978).
‡ Tannock (1968, 1976) and Bateman and Steel (1978).
§ Koch et al. (1973b) and Bedford and Mitchell (1974).
// This implies a long G_1 block of the P cells; thus they become Q cells.
For other symbols, see Table 1.

assay, if properly done, reflects this entire population. In contrast, differential centrifugation allows one to select a partially purified fraction from the Q compartment, and these cells were probably quiescent and chronically hypoxic in situ. Based on the reports of Koch et al. (1973a, 1973b), many of these are most likely permanently nonclonogenic. The same argument probably applies to the apparently disparate cell cycle results from the in vivo hypoxia studies and the in vitro hypoxia studies. By their very nature, the in vivo studies are more global and thus the analysis may miss some subtle effects of hypoxia. Of course, the cellular states of hypoxia are also much more heterogenous in vivo than the well-maintained and carefully monitored in vitro experiments; thus, the mean or modal response in vivo may be quite different from the average response of a homogenous population in vitro.

The G_0 status of Q cells in solid tumors is certainly debatable at this time; however, my contention is that the data on solid tumors discussed above indicate that the majority of the Q cells are quiescent because of the local milieu (i.e., nutrient deprivation or excessive catabolic products or both). Conversely, the proportion of G_0 cells is probably vanishingly small, if not zero. Thus, if we concentrate on G_0 cells in our studies, I suspect we will be looking for the proverbial needle in the haystack. Given the above assumptions, I believe that the quantitative interpretations in all the above reports must be taken with a liberal dose of skepticism; however, the qualitative interpretations are most likely valid and certainly suggest hypotheses for further detailed experimentation. Since the variables studied do show marked heterogeneity in vivo, the assays are complex, and a G_0 state in solid tumors has not been identified conclusively, the most prudent course would be to first demonstrate what qualitative and quantitative differences exist in an appropriate in vitro model system. This is certainly not an original idea of mine; however, I do feel that the in vitro models in use have not been critically evaluated as they relate to: (1) the various biochemical, biophysical, and physiological parameters that can be measured accurately for both P and Q cells, and, more importantly, (2) the relationship of these specific variables to the heterogeneity of the Q cells from solid tumors in vivo.

IN VITRO STUDIES

Models

The liver following partial hepatectomy and the isoproterenol-stimulated salivary gland are two good examples of in vivo tissues whose G_0–G_1 transitions have been studied extensively (See Baserga 1976, for a review of these and other tissues). There are obvious limitations to in vivo studies; thus, many in vitro models have been used in an attempt to more rigorously define the molecular biology of this transition, the so-called growth control. Among these, the resting versus phytohemagglutinin-stimulated lymphocyte is an old, popular, and proba-

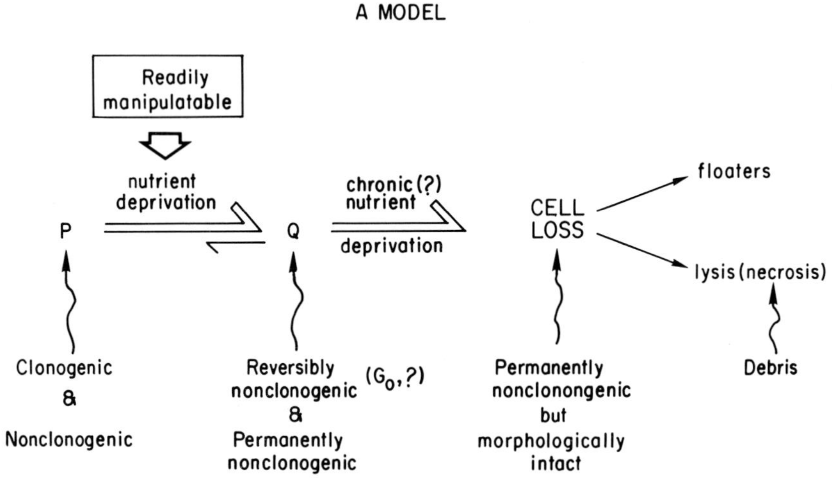

FIG. 1. A suggested in vitro model with the minimum number of compartments and subcompartments required to reasonably represent an in vivo solid tumor. The proliferating (P) and quiescent cells (Q) are subdivided into clonogenic and nonclonogenic cells with the latter being either reversible or permanent. The question mark after G_0 indicates the uncertainty of how many, if any, true G_0 cells exist in solid tumors. Cell loss represents cells irreversibly moving toward death and lysis.

bly valid model; however, more recent studies have utilized the stationary-phase cultures of 3T3 and WI-38 cells (Baserga 1976). The rationale for the model is that when subjected to nonpermissive growth conditions (serum deprivation), these cells enter an apparent G_0 state more easily than transformed (possibly malignant) cells (Schiaffonati and Baserga 1977). Under such conditions, apparently normal cells will remain viable for extended periods of time, which may be analogous to the in vivo tissue situation, while the transformed cells begin dying (lysing) quite quickly. Insofar as this distinction is valid, the stationary-phase cultures of so-called normal cells probably do represent a reasonable model for studying the G_0–G_1 transition; however, if the Q cells in solid tumors are quiescent predominately because of nutrient deprivation and few or no cells exist in the G_0 state, and I believe this to be the case, then other in vitro models using malignant or malignant-acting cells would be more appropriate (see Figure 1).

Plateau-Phase Cultures

As a first approximation, the in vitro model's nutrient status should be readily manipulatable, including oxygen concentration, to achieve a heterogenous mixture of hypoxic, P, Q, dead, clonogenic, and viable but nonclonogenic cells. Hahn and Little (1972) have described in detail the status of two cell lines, Chinese hamster (HA2) and Chang human liver cells (LICH), that they have used in radio- and chemobiology studies. The cultures are fed daily and upon confluence may be kept viable almost indefinitely; moreover, these cultures are a cell-renewal system wherein cell loss (floaters and lysis) is balanced by proliferation in the monolayer. The authors suggest that for the HA2 line, the limitation in cell density is caused by nutrient deficiency due to diffusion limitation rather than the release of a specific growth inhibitor. The validity of this observation is not crucial since both mechanisms probably are (or could be) functioning in vivo and the significance of each contribution to the overall effect is simply not understood. In contrast, unfed cultures enter the plateau phase of growth at a much lower cell density, and this cell density is maintained for only four to six days with the resultant death of the entire population. The significant changes in cell number–dependent cell cycle parameters in the fed cultures include an increase in the S-phase duration, a lower growth fraction, and cell loss. There are also changes in cellular radiation sensitivity and repair of radiation damage.

Ross and Sinclair (1972) used the V79-S171 subline of Chinese hamster cells with either a minimal medium (EM-15) or an enriched medium (HUT-15), and allowed the cells to reach a stationary phase due to nutrient deficiency. They found, in contrast to Hahn and Little (1972), that cells were arrested in both G_2 and G_1, with the proportion and rate of accumulation in G_1 depending on the medium employed. Also, a fraction of the cells synthesized DNA in the stationary phase, but these could not be stimulated to renewed growth by the introduction of fresh medium.

Mauro et al. (1974a) have also described a plateau-phase system for V79-735-B-(SSL) Chinese hamster cells and HeLa S3 cells, each growing in two different media. They did not routinely use fed cultures but did supplement the medium with 15% fetal calf serum, and reported different quantitative results for the different lines and within the lines when grown in different media. The Chinese hamster cell system was characterized extensively for further antineoplastic studies (Mauro et al. 1974b) and the following compartments were identified: attached cells, both P and Q as well as clonogenic and nonclonogenic cells, and floating cells (nonclonogenic). They reported an increase in the median cell cycle time and an S-phase prolongation. Another interesting observation was that the clonogenic cells seemed to include only those that were not synthesizing DNA at the time of trypsinization.

The different quantitative results in these various reports indicate the importance of both cell type and environment; however, the qualitative conclusion of Mauro et al. (1974a) that the overall kinetic pattern for the plateau-phase cultures resembles the in vivo cell renewal systems is in agreement with Hahn and Little (1972). Thus, one or more of these cell lines in exponential growth versus fed and unfed and early and late plateau-phase cultures should serve as good working models for in vivo solid tumors.

The isoleucine-deficient suspension-culture approach of Tobey and Ley (1970) has also been used for growth-control studies, and because of ease of handling and abundance of P and Q cells, this model may seem an attractive alternative for the monolayer cultures described above; however, this specific nutrient-deprivation method apparently arrests cells in a state analogous to G_0 (Sander and Pardee 1972, Yen and Pardee 1978). Thus, it is probably not an ideal in vitro model for in vivo solid tumors.

Multicell Spheroids

Any in vitro model will most likely have some disadvantages and in some ways be inappropriate to the situation in vivo; thus, a good alternative in vitro model for the plateau-phase monolayer culture is the multicellular spheroid model of cells growing in suspension (Sutherland and Durand 1976). The Chinese hamster V79-121b cells were used initially, but now many other cells have been adapted to spheroid growth and are used effectively for radio- and chemobiology studies. As the spheroids grow, they develop a Q cell compartment as well as a central compartment of dead and dying cells. With increasing size, both the central necrosis and the Q compartment increase; thus, the growth fraction decreases but the cell cycle transit time remains constant (Durand 1976). Also, it appears that many, if not all, of the Q cells are quiescent because of hypoxia; however, a fraction of the Q cells do remain clonogenic (Sutherland and Durand 1976), and the Q cells tend to be smaller and denser than the P cells (Durand 1975). In a recent report (Franko and Sutherland 1978), the rate of death of acutely and chronically hypoxic cells was discussed in depth. All spheroid cells were made hypoxic (< 100 ppm O_2). The resulting rate of cell death was biphasic, with 15% dying in the first six hours. Also, the chronically hypoxic inner cells were more sensitive than the previously oxic outer cells, of which 5% were still surviving after as long as six days. Changes in the medium's glucose concentration did not affect survival of the outer cells. The data indicate that the inner cells die at a much faster rate than cells in monolayer, whereas the resistance of the outer cells is equal to or greater than the monolayer cells. These reports demonstrate the heterogeneity in size, density, clonogenicity, proliferative status, and hypoxic status of cells when growing in spheroids. Thus, the spheroids and plateau-phase monolayer cultures should complement each other, as they both function as convenient models for identifying distinguishing features of the P and Q cells of in vivo solid tumors.

Discriminatory Variables for P-ness and Q-ness

The variables already discussed and the methods used on a few selected in vivo tumors are obviously first candidates for work with the in vitro models; i.e., cellular hypoxia, cell size, cell density, nuclear morphometry, and cellular RNA content. Amazingly little is actually known about these variables for the in vitro models presented above, especially about how these variables from in vitro cells correlate with the same variables for the same cells growing in vivo. However, a limited number of studies have used the same methodologies to

TABLE 3. *Possible determinants for P-ness vs. Q-ness*

Parameter	P	vs.	Q
	(relative measures)*		
Cell size†	↑		↓
Cell density†	↓		↑
Clonogenicity†	= or ↑ (?)		= or ↓ (?)
Cell surface	rough		smooth
Membrane transport	↑		↓
Chromatin structure and function†	↑		↓
Chromatin template activity	↑		↓
RNA and protein content†	↑		↓
Macromolecular synthesis	↑		↓
Macromolecular turnover	↓		↑
Anabolic enzyme activity	↑		↓
Catabolic enzyme activity	↓		↑

* As Epifanova (1977) points out, considerable variability is noted for the P cells as they traverse the cell cycle; however, principal differences may be found by comparing the means. The direction of the arrows indicates the difference (e.g., the P cells tend to be larger in size than the Q cells).

† Some work has been done on in vivo solid tumors.

analyze the same cells growing in vitro and in vivo (e.g., Grdina et al. 1978a,b), and in my opinion, this is the best approach for making advances in this area.

Surprisingly little attention has been paid to either pH or tonicity by investigators in this field even though these variables are relevant to growth and affect cellular response to external perturbations (Raaphorst and Kruuv 1977a, 1977b, Raaphorst et al. 1977, Gerweck 1977, Henle and Dethlefsen 1979). Also, the cyclic nucleotides and polyamines tend to be ignored even though their role, albeit confusing at times, is of much interest to investigators in growth control (Friedman 1976, Jänne et al. 1978). I do not drop these caveats to further obscure an already complex issue but instead to help maintain a broader perspective of the problem. Progress can be made if studies are first focused on the variables already discussed and then complemented by new extended studies based on the literature on the molecular biology of the reported G_0–G_1 transition. Such an approach should serve as a reasonable feed-back loop for attempts at sorting out the P-ness and Q-ness of in vivo solid tumor cells. Thus, listing the variables that do seem to discriminate between P and Q cells seems prudent (Table 3). These specific variables will not be discussed here because: (a) they have been reviewed recently (Epifanova 1977, Pardee et al. 1978); (b) there is obvious overlap with some variables already discussed; and (c) others will be briefly discussed later under Methods.

Methods

Although seven general classifications of methods are listed in Table 4, my intent is not to describe each in detail but only to alert the reader to these

TABLE 4. *Methods*

1. Histology and ^{3}HTdR autoradiography (light microscopy, transmission and scanning E.M.)
2. Clonogenic assays (in vitro and in vivo)
3. Differential centrifugation (linear density gradient, elutriation)
4. Biochemical and molecular procedures (RNA synthesis, HnRNA, mRNA, etc.; protein synthesis: histones, nonhistone chromosomal proteins; enzyme activity assays; membrane transport)
5. Chromatin structure and function (nuclear morphometry, circular dichroism, ethidium bromide binding)
6. Flow cytometry and sorting (differential acridine orange staining, BUdR-mithramycin technique, dual DNA-specific staining, viable staining (e.g., 33342 Hoechst), and sorting)
7. Data analyses (analytical interpretations, i.e., modeling)

approaches. The first four will not be discussed at all because numbers 1, 2, and 4 (histology and ^{3}HTdR autoradiography, clonogenic assays, and biochemical and molecular procedures) are older, reasonably well understood, and extensively referenced in the general literature as well as in the specific reviews discussed in this report, and the third area (differential centrifugation) has already been discussed by Grdina (1980, see pages 353 to 363, this volume) earlier in this monograph and by me in this specific review.

Chromatin Structure and Function

Nuclear morphometry has been discussed and referenced earlier; however, the circular dichroism and ethidium bromide assays have not, to my knowledge, been used in studies on chromatin from in vivo solid tumors. The circular dichroism of chromatin is complex. One can detect at least six bands, and the molecular ellipticities are based on concentrations of the nucleotide bases and protein residues as well as the helical configuration of DNA and proteins (Simpson and Sober 1970). Also, the number of available binding sites (fluorescence intensity) of ethidium bromide is apparently related to the conformational heterogeneity of DNA in chromatin (Lawrence and Daune 1976). In spite of this complexity, Nicolini et al. (1975a) have used both assays on HeLa cells and reported that mid-S-phase chromatin shows an increase in maximum ellipticity (250- to 300-nm region) and an increased ability to bind ethidium bromide compared to mitotic chromatin. G_1 chromatin has intermediate values. Nicolini et al. (1975b, 1977b) also reported evidence that these end points do demonstrate appreciable quantitative changes in the chromatin of resting versus stimulated WI-38 cells. Thus, it is possible that these assays will complement the nuclear morphometry assays for studies on the Q cells of the in vitro models.

Flow Cytometry (FCM) and Sorting

The utility and efficacy of FCM analysis and sorting have been extensively documented for studies on many aspects of cell biology, and specifically for studies on cellular kinetics in vitro and in vivo (Arndt-Jovin and Jovin 1978, Gray et al. 1979). This section will serve to identify several more, unique approaches that utilize FCM and have exciting potential for studies of Q cell biology.

Darzynkiewicz et al. (1977) used the metachromatic dye acridine orange to stain RNA and DNA differentially (green fluorescence relates primarily to intercalation into double-stranded DNA, while the red fluorescence relates primarily to dye stacking on single-stranded RNA; Melamed et al. 1977) and to stain nuclear chromatin after removal of the cellular RNA with RNase pretreatment. They reported that unstimulated lymphocytes (G_0 state) could be distinguished from the proliferating lymphocytes in G_1 and that mitotic chromatin stained differently from interphase chromatin. Ashihara et al. (1978) used a similar approach in analyzing AF8 cells that were either quiescent (serum deprivation; perhaps G_0) or synchronized in early G_1 (mitotic detachment) and reported fluorescent data compatible with the concept that the two cell populations are in different physiological states. However, they also reported evidence suggesting that the changes in red fluorescence (> 600 nm) may not be strictly related to quantitative changes in RNA content. Regardless of the specific biophysical mechanism(s) for the changes in red fluorescence, this approach should be useful for the further analysis of Q-cell biology. In a related approach, with differential fluorochromasia, Swartzendruber (1977a, 1977b) has shown that the incorporation of 5-bromodeoxyuridine in place of thymidine increases the fluorescent intensity of mithramycin-stained cells. Thus, at least for the simpler in vitro systems, this approach may help distinguish between the P and Q cells and merits some attention in relation to the Q cell problem in solid tumors.

An alternative approach for studying chromatin configuration, in the sense of DNA accessibility in chromatin, results from studies of energy transfer between pairs of DNA-specific fluorescent stains (Brodie et al. 1975, Sahar and Latt 1978, Langlois and Jensen 1979). Hoechst 33258, which binds tightly to DNA, for example, can be used as an energy donor for ethidium bromide, since the fluorescence emission spectrum of the former extensively overlaps with the absorption spectrum of the latter. Thus, if two bound molecules are close together and the donor is excited, there is a high probability of energy transfer to ethidium bromide, resulting in a more intense fluorescence. The biophysical interaction in cells can be extremely complex (Langlois and Jensen 1979); however, if experiments are done carefully, information can be obtained on the amount and distribution of stain molecules in the DNA, and from these data inferences can be made concerning chromatin configuration. Thus, the different chromatin configurations that apparently exist in cells in the P and Q state may possibly be analyzed by such an approach.

One of the past limitations of FCM analysis has been the need to fix the cells for the quantitative determination of DNA content. This precluded most biochemical assays on sorted cells as well as the clonogenic assays. However, in 1977 Arndt-Jovin and Jovin reported that Hoechst 33342 could be used to quantitatively stain DNA in viable, unfixed cells, and two recent preliminary reports have confirmed this observation for in vitro studies. In these studies, the cells were treated with a chemotherapeutic agent or irradiation and the cell-age redistributions and clonogenicity of cells sorted from specific cell cycle phases were determined (Gray and George 1978, Pallavicini et al. 1979). This general approach should be a most welcome complement to the various methods discussed above.

Data Analyses

Terming data analyses a method may seem presumptious; however, this heading does not refer only to biostatistical analyses. Each of the first seven general areas listed in Table 4 are complex and the use of two or more combined methods will result, at a minimum, in a three-dimensional array of data. A complex analytical problem can result, with two levels of analysis being critical. Cell cycle kinetics and FCM serve as good examples for the points to be made. Extracting reliable quantitative information from FCM histograms is complex, but it has been done by several authors (e.g., Dean and Jett 1974). The resulting data are helpful for interpretations; however, if these data are to be compared to cellular kinetic data (e.g., ^{3}HTdR incorporation and autoradiography), then another level of complexity is introduced. Usualiy, visual interpretations or intuition are insufficient; thus, modeling becomes an important second step for data interpretation (e.g., Dethlefsen et al. 1976, 1979). The models, obviously, should be consistent with the current state of knowledge, but the assumptions behind the models, which are research tools, are not as important as the fact that they are invaluable aids for data interpretation and the design of further experiments.

Summary of Methods

What is apparent is the fact that no one of the general approaches will by itself give significant new insight into the nature of Q cells either in vitro or in vivo. A judicious combination of two, three, or more must be utilized because of the complexity of the problem. Also, all indications suggest that frequently there are subtle quantitative differences between P and Q cells, instead of quantum jumps or easily identifiable qualitative differences.

SUMMARY

This review is intended to point out the complexity of the quest for the quiescent cells in solid tumors and specifically the heterogeneity that apparently exists within the Q cell population as well as the overlap of variables between the P and Q cells in solid tumors in vivo. In a way, this is meant to keep us humble; however, in no way is it intended to demoralize. In contrast, I believe the immediate future holds considerably more promise for significant advances than did the recent past. The considerable data from studies on the G_0-G_1 transition in vitro will serve as functional guidelines, even though the results from these models are most likely not directly applicable to in vivo solid tumors. Also, the several new approaches in analytical and quantitative cytology, which are being refined and exploited, will continue to stimulate new studies and the development of other new techniques. What is apparent, at least to me, is that appropriate in vitro models should be used along with a combination of several sophisticated biophysical-cytological techniques in the attack on the biology of Q cells in solid tumors. I feel confident that the diligent application of such an approach will result in significant progress and help delineate the relevance of solid tumor Q cells to experimentalists and our friends in clinical oncology.

ACKNOWLEDGMENTS

This investigation was supported by Grants CA 14165 and CA 22188, awarded by the National Cancer Institute, Department of Health, Education and Welfare.

A special thanks to Drs. Joe W. Gray and Stephen P. Tomasovic for the critique and careful reading of this manuscript, and to Sherrie Stewart for the typing.

REFERENCES

Abercrombie, M., and E. J. Ambrose. 1962. The surface properties of cancer cells: A review. Cancer Res. 22:525–548.

Arndt-Jovin, D. J., and T. M. Jovin. 1977. Analysis and sorting of living cells according to deoxyribonucleic acid content. J. Histochem. Cytochem. 25:585–589.

Arndt-Jovin, D. J., and T. M. Jovin. 1978. Automated cell sorting with flow systems. Ann. Rev. Biophys. Bioeng. 7:527–558.

Ashihara, T., F. Traganos, R. Baserga, and Z. Darzynkiewicz. 1978. A comparison of cell cycle-related changes in postmitotic and quiescent AF8 cells as measured by cytofluorometry after acridine orange staining. Cancer Res. 38:2514–2518.

Barendsen, G. W., H. Roelse, A. F. Hermens, H. T. Madhuizen, H. A. van Peperzeel, and D. H. Rutgers. 1973. Clonogenic capacity of proliferating and nonproliferating cells of a transplantable rat rhabdomyosarcoma in relation to its radiosensitivity. J. Natl. Cancer Inst. 51:1521–1526.

Baserga, R. 1976. The prereplicative phase of G_0 cells, *in* Multiplication and Division in Mammalian Cells, The Biochemistry of Disease, Vol. 6. Marcel Dekker, New York, pp. 53–77.

Bateman, A. E., and G. G. Steel. 1978. The proliferative state of clonogenic cells in the Lewis lung tumour after treatment with cytotoxic agents. Cell Tissue Kinet. 11:445–454.

Bedford, J. S., and J. B. Mitchell. 1974. The effect of hypoxia on the growth and radiation response of mammalian cells in culture. Br. J. Radiol. 47:687–696.

Born, R., O. Hug, and K.-R. Trott. 1976. The effect of prolonged hypoxia on growth and viability of Chinese hamster cells. Int. J. Radiat. Oncol. Biol. Phys. 1:687–697.

Brodie, S., J. Giron, and S. L. Latt. 1975. Estimation of accessibility of DNA in chromatin from fluorescence measurements of electronic excitation energy transfer. Nature 235:470–471.

Brown, J. M. 1977. Cytotoxic effects of the hypoxic cell radiosensitizer Ro 7–0582 to tumor cells in vivo. Radiat. Res. 72:469–486.

Cairnie, A. B., L. F. Lamerton, and G. G. Steel. 1965. Cell proliferation studies in the intestinal epithelium of the rat. II. Theoretical aspects. Exp. Cell Res. 39:539–553.

Cooper, H. L. 1971. Biochemical alterations accompanying initiation of growth in resting cells, *in* The Cell Cycle and Cancer, R. Baserga, ed. Marcel Dekker, New York, pp. 197–226.

Darzynkiewicz, Z., F. Traganos, T. Sharpless, and M. Melamed. 1975. Conformation of RNA in situ as studied by acridine orange staining and automated cytophotometry. Exptl. Cell Res. 95:143–153.

Darzynkiewicz, Z., F. Traganos, T. K. Sharpless, and M. R. Melamed. 1977. Cell cycle-related changes in nuclear chromatin of stimulated lymphocytes as measured by flow cytometry. Cancer Res. 37:4635–4640.

Dean, P. N., and J. H. Jett. 1974. Mathematical analysis of DNA distributions derived from flow microfluorometry. J. Cell Biol. 60:523–527.

Dethlefsen, L. A., J. W. Gray, Y. S. George, and S. Johnson. 1976. Flow cytometric analysis of the perturbed cellular kinetics of solid tumors: problems and promises, *in* Pulse-Cytophotometry (Second International Symposium). European Press, Ghent, Belgium, pp. 188–200.

Dethlefsen, L. A., R. M. Riley, and J. L. Roti Roti. 1979. Flow cytometric analysis of adriamycin-perturbed mouse mammary tumors. J. Histochem. Cytochem. 27:463–469.

Dexter, D. L., H. M. Kowalski, B. A. Blazar, Z. Fligiel, R. Vogel, and G. H. Heppner. 1978. Heterogeneity of tumor cells from a single mouse mammary tumor. Cancer Res. 38:3174–3181.

Durand, R. E. 1975. Isolation of cell subpopulations from in vitro tumor models according to sedimentation velocity. Cancer Res. 35:1295–1300.

Durand, R. E. 1976. Cell cycle kinetics in an in vitro tumor model. Cell Tissue Kinet. 9:403–412.

Endrich, B., M. Intaglietta, H. S. Reinhold, and J. F. Gross. 1979. Hemodynamic characteristics in microcirculatory blood channels during early tumor growth. Cancer Res. 39:17–23.

Epifanova, O. I. 1977. Mechanisms underlying the differential sensitivity of proliferating and resting cells to external factors. Int. Rev. Cytol. (Suppl. 5), pp. 303–335.

Epifanova, O. I., and V. V. Terskikh. 1969. On the resting periods in the cell life cycle. Cell Tissue Kinet. 2:75–93.

Franko, A. J., and R. M. Sutherland. 1978. Rate of death of hypoxic cells in multicell spheroids. Radiat. Res. 76:561–572.

Friedman, D. L. 1976. Role of cyclic nucleotides in cell growth and differentiation. Physiol. Rev. 56:652–708.

Gershman, H., W. Katzin, and R. T. Cook. 1978. Mobility of cells from solid tumors. Int. J. Cancer 21:309–316.

Gerweck, L. E. 1977. Modification of cell lethality at elevated temperatures. The pH effect. Radiat. Res. 70:224–235.

Glick, D., D. von Redlich, E. T. Juhos, and C. R. McEwen. 1971. Separation of mast cells by centrifugal elutriation. Exp. Cell Res. 65:23–26.

Gray, J. W., P. M. Dean, and M. L. Mendelsohn. 1979. Quantitative cell cycle analysis: Flow cytometry and sorting, *in* Flow Cytogenetics and Sorting, M. Melamed, P. Mullaney, and M. L. Mendelsohn, eds. John Wiley & Sons, New York (in press).

Gray, J. W., and Y. George. 1978. The response of Lewis lung tumor cells to ara-C and hydroxyurea (Abstract). Proc. Am. Assoc. Cancer Res. 19:205.

Gray, L. H., A. D. Conger, M. Ebert, S. Hornsey, and O. C. A. Scott. 1953. The concentration of oxygen dissolved in tissues at the time of irradiation as a factor in radiotherapy. Br. J. Radiol. 26:638–648.

Grdina, D. J. 1980. Variation in radiation response of tumor subpopulations, *in* Radiation Biology in Cancer Research (The University of Texas System Cancer Center 32nd Annual Symposium on Fundamental Cancer Research, 1979), Raven Press, New York, pp. 353–363.

Grdina, D. J., I. Basic, K. A. Mason, and H. R. Withers. 1975. Radiation response of clonogenic cell populations separated from a fibrosarcoma. Radiat. Res. 63:483–493.

Grdina, D. J., I. Basic, S. Guzzino, and K. A. Mason. 1976. Radiation response of cell populations irradiated in situ and separated from a fibrosarcoma. Radiat. Res. 66:634–643.

Grdina, D. J., W. N. Hittelman, R. A. White, and M. L. Meistrich. 1977. Relevance of density, size, and DNA content of tumour cells to the lung colony assay. Br. J. Cancer 36:659–669.

Grdina, D. J., L. Milas, K. A. Mason, and H. R. Withers. 1974. Separation of cells from a fibrosarcoma in renograffin density gradients. J. Natl. Cancer Inst. 52:253–257.

Grdina, D. J., L. J. Peters, S. Jones, and E. Chan. 1978a. Separation of cells from a murine fibrosarcoma on the basis of size. I. Relationship between cell size and age as modified by growth in vivo or in vitro. J. Natl. Cancer Inst. 61:209–214.

Grdina, D. J., L. J. Peters, S. Jones, and E. Chan. 1978b. Separation of cells from a murine fibrosarcoma on the basis of size. II. Differential effects of cell size and age on lung retention and colony formation in normal and preconditioned mice. J. Natl. Cancer Inst. 61:215–220.

Gross, J. F., M. Intaglietta, and B. W. Zweifach. 1974. Network model of pulsatile hemodynamics in the microcirculation of the rabbit omentum. Am. J. Physiol. 226:1117–1123.

Hahn, G. M., and J. B. Little. 1972. Plateau-phase cultures of mammalian cells: An in vitro model for human cancer. Curr. Top. Radiat. Res. 8:39–83.

Hall, E. J. 1967. The oxygen effects: Pertinent or irrelevant to clinical radiotherapy. Br. J. Radiol. 40:874–875.

Henle, K. J., and L. A. Dethlefsen. 1979. Sensitization to hyperthermia (45°C) of normal and thermotolerant CHO cells by anisotonic media. Int. J. Radiat. Biol. (in press).

Heppner, G. H., D. L. Dexter, T. De Nucci, F. R. Miller, and P. Calabresi. 1978. Heterogeneity in drug sensitivity among tumor cell subpopulations of a single mammary tumor. Cancer Res. 38:3758–3763.

Hermens, A. F., and G. W. Barendsen. 1969. Changes of cell proliferation characteristics in a rat rhabdomyosarcoma before and after x-irradiation. Eur. J. Cancer 5:173–189.

Hermens, A. F., and G. W. Barendsen. 1978. The proliferative status and clonogenic capacity of tumour cells in a transplantable rhabdomyosarcoma of the rat before and after irradiation with 800 rad of x-rays. Cell Tissue Kinet. 11:83–100.

Hill, R. P., and R. S. Bush. 1969. A lung colony assay to determine the radiosensitivity of the cells of a solid tumor. Int. J. Radiat. Biol. 15:435–444.

Hittelman, W. N., and P. N. Rao. 1974. Premature chromosome condensation. I. Visualization of x-ray-induced chromosome damage in interphase cells. Mutat. Res. 23:251–258.

Jänne, J., H. Pösö, and A. Raina. 1978. Polyamines in rapid growth and cancer. Biochim. Biophys. Acta 473:241–293.

Kaplan, H. S. 1974. On the relative importance of hypoxic cells for the radiotherapy of human tumours. Eur. J. Cancer 10:275–280.

Kendall, F., R. Swenson, T. Borun, J. Rowinski, and C. Nicolini. 1977. Nuclear morphometry during the cell cycle. Science 196:1106–1109.

Koch, C. J., J. Kruuv, and H. E. Frey. 1973a. The effect of hypoxia on the generation time of mammalian cells. Radiat. Res. 53:43–48.

Koch, C. J., J. Kruuv, H. E. Frey, and R. A. Snyder. 1973b. Plateau phase in growth induced by hypoxia. Int. J. Radiat. Biol. 23:67–74.

Lajtha, L. G. 1963. On the concepts of the cell cycle. Cell Comp. Physiol. 62:143–145.

Langlois, R. G., and R. H. Jensen. 1979. Interactions between pairs of DNA-specific fluorescent stains bound to mammalian cells. J. Histochem. Cytochem. 27:72–79.

Lawrence, J., and M. Daune. 1976. Ethidium bromide as a probe of conformational heterogeneity of DNA in chromatin. The role of histone H_1. Biochemistry 15:3301–3307.

Lazzio, B. B., C. B. Lazzio, E. G. Bamberger, and S. V. Lair. 1975. Regulators of cell division: Endogenous mitotic inhibitors of mammalian cells. Int. Rev. Cytol. 42:1–47.

Mauro, F., B. Falpo, G. Briganti, R. Elli, and G. Zupi. 1974a. Effects of antineoplastic drugs on plateau-phase cultures of mammalian cells. I. Description of the plateau-phase system. J. Natl. Cancer Inst. 52:705–713.

Mauro, F., B. Falpo, G. Briganti, R. Elli, and G. Zupi. 1974b. Effects of antineoplastic drugs on plateau-phase cultures of mammalian cells. II. Bleomycin and hydroxyurea. J. Natl. Cancer Inst. 52:715–722.

Meistrich, M. L., D. J. Grdina, R. E. Meyn, and B. Barlogie. 1977. Separation of cells from mouse solid tumors by centrifugal elutriation. Cancer Res. 37:4291–4296.

Melamed, M. R., Z. Darzynkiewicz, F. Traganos, and T. Sharpless. 1977. Cytology automation by flow cytometry. Cancer Res. 37:2806–2812.

Mendelsohn, M. L. 1962a. Chronic infusion of tritiated thymidine into mice with tumors. Science 135:213–215.

Mendelsohn, M. L. 1962b. Autoradiographic analysis of cell proliferation in spontaneous breast cancer of C3H mouse. III. Growth fraction. J. Natl. Cancer Inst. 28:1015–1029.

Mendelsohn, M. L. 1967. Cell cloning experiments as models for radiotherapy: A critical appraisal. Natl. Cancer Inst. Monogr. 24:157–167.

Nicolini, C., K. Ajiro, T. W. Borun, and R. Baserga. 1975a. Chromatin changes during the cell cycle of HeLa cells. J. Biol. Chem. 250:3381–3385.

Nicolini, C., S. Ng, and R. Baserga. 1975b. Effect of chromosomal proteins extractable with low concentrations of NaCl on chromatin structure of resting and proliferating cells. Proc. Natl. Acad. Sci. USA 72:2361–2365.

Nicolini, C., W. Giaretti, C. DeSaive, and F. Kendall. 1977a. The G_0-G_1 transition of WI38 cells. II. Geometric and densitometric texture analyses. Exp. Cell Res. 106:119–125.

Nicolini, C., F. Kendall, R. Baserga, C. DeSaive, B. Clarkson, and J. Fried. 1977b. The G_0-G_1 transition of WI38 cells. I. Laser flow microfluorimetric studies. Exp. Cell Res. 106:111–118.

Nicolini, C. A., W. A. Linden, S. Zietz, and C. T. Wu. 1977c. Identification of nonproliferating cells in melanoma B16 tumour. Nature 270:607–609.

Pallavicini, M. G., M. E. Lalande, R. G. Miller, and R. P. Hill. 1979. Cell cycle phase specific clonogenicity of KHT tumor cells following in vivo irradiation. (Abstract) Cell Tissue Kinet. (in press).

Pardee, A. B., R. Dubrow, J. L. Hamlin, and R. F. Kletzien. 1978. Animal cell cycle. Ann. Rev. Biochem. 47:715–750.

Pertoft, H. 1970. Separation of cells from a mast cell tumor on density gradients of colloidal silica. J. Natl. Cancer Inst. 44:1251–1256.

Pettersen, E. O. 1978. Radiosensitizing and toxic effects of the 2-nitroimidazole Ro-07-0582 in different phases of the cell cycle of extremely hypoxic human cells in vitro. Radiat. Res. 73:180–191.

Pontén, I. 1975. Contact inhibition, *in* Cancer, A Comprehensive Treatise, Vol. 4, Biology of Tumors: Surfaces, Immunology and Comparative Pathology, F. F. Becker, ed. Plenum Press, New York, pp. 55–100.

Potmesil, M., and A. Goldfeder. 1971. Nucleolar morphology, nucleic acid synthesis, and growth rates of experimental tumors. Cancer Res. 31:789–797.

Pretlow, T. G., and C. W. Boone. 1970. Separation of malignant cells from transplantable rodent tumors. Exp. Mol. Pathol. 12:249–256.

Quastler, H. 1963. The analysis of cell population kinetics, *in* Cell Proliferation (A Guiness Symposium, University of Dublin, Trinity College, 1962). L. F. Lamerton and R. J. M. Fry, eds. F. A. Davis Co., Philadelphia, pp. 18–34.

Raaphorst, G. P., H. E. Frey, and J. Kruuv. 1977. Effects of salt solutions on radiosensitivity of mammalian cells. III. Treatment with hypertonic solutions. Int. J. Radiat. Biol. 32:109–126.

Raaphorst, G. P., and J. Kruuv. 1977a. Effect of salt solutions on radiosensitivity of mammalian cells. I. Specific ion effects. Int. J. Radiat. Biol. 32:71–88.

Raaphorst, G. P., and J. Kruuv. 1977b. Effect of salt solutions on radiosensitivity of mammalian cells. II. Treatment with hypotonic solutions. Int. J. Radiat. Biol. 32:89–101.

Reinhold, H. S. 1971. Improved microcirculation in irradiated tumours. Eur. J. Cancer 7:273–280.

Ross, D. W., and W. K. Sinclair. 1972. Cell cycle compartment analysis of Chinese hamster cells in stationary phase cultures. Cell Tissue Kinet. 5:1–14.

Sahar, E., and S. L. Latt. 1978. Enhancement of banding patterns in human metaphase chromosomes by energy transfer. Proc. Natl. Acad. Sci. USA 75:5650–5654.

Sander, G., and A. B. Pardee. 1972. Transport changes in synchronously growing CHO and L cells. J. Cell. Physiol. 80:267–272.

Schiaffonati, L., and R. Baserga. 1977. Different survival of normal and transformed cells exposed to nutritional conditions nonpermissive for growth. Cancer Res. 37:541–545.

Sigdestad, C. P., and D. J. Grdina. 1978. Separation of murine fibrosarcoma cells by density centrifugation following tritiated thymidine labelling. (Abstract) Radiat. Res. 74:542.

Simpson, R. T., and H. A. Sober. 1970. Circular dichroism of calf liver nucleohistone. Biochemistry 9:3103–3109.

Steel, G. G. 1972. The cell cycle in tumours: An examination of data gained by the technique of labelled mitosis. Cell Tissue Kinet. 5:87–100.

Steel, G. G. 1977. Growth Kinetics of Tumours. Clarendon Press, Oxford, pp. 144–216.

Steel, G. G., and L. F. Lamerton. 1969. Cell population kinetics and chemotherapy. Natl. Cancer Inst. Monogr. 30:29–50.

Steinkamp, J. A., A. Romero, P. K. Horan, and H. A. Crissman. 1974. Multiparameter analysis and sorting of mammalian cells. Exp. Cell Res. 84:15–23.

Sutherland, R. M., and R. E. Durand. 1976. Radiation response of multicell spheroids—an in vitro tumor model. Curr. Top. Radiat. Res. 11:87–139.

Suzuki, N., M. Frapart, D. J. Grdina, M. L. Meistrich, and H. R. Withers. 1977. Cell cycle dependency of metastatic lung colony formation. Cancer Res. 37:3690–3693.

Swabb, E. A., J. Wei, and P. M. Gullino. 1974. Diffusion and convection in normal and neoplastic tissue. Cancer Res. 34:2814–2822.

Swartzendruber, D. E. 1977a. Microfluorometric analysis of cellular DNA following incorporation of bromodeoxyuridine. J. Cell. Physiol. 90:445–454.

Swartzendruber, D. E. 1977b. A bromodeoxyuridine (BUdR)-mithramycin technique for detecting cycling and non-cycling cells by flow microfluorometry. Exp. Cell Res. 109:439–443.

Tannock, I. F. 1968. The relation between cell proliferation and the vascular system in a transplanted mouse mammary tumor. Br. J. Cancer 22:258–273.

Tannock, I. F. 1970. Effects of pO_2 on cell proliferation kinetics, *in* Time and Dose Relationships in Radiation Biology as Applied to Radiotherapy. BNL-50203 (C-57), pp. 215–224.

Tannock, I. F. 1972. Oxygen diffusion and the distribution of cellular radiosensitivity in tumours. Br. J. Radiol. 45:515–524.

Tannock, I. 1976. Oxygen distribution in tumours: Influence on cell proliferation and implications for tumour therapy. Adv. Exp. Med. Biol. 75:597–603.

Tannock, I., and A. Howes. 1973. The response of viable tumor cords to a single dose of radiation. Radiat. Res. 55:477–486.

Tannock, I. F., and G. G. Steel. 1970. Tumor growth and cell kinetics in chronically hypoxic animals. J. Natl. Cancer Inst. 45:123–133.

Terasima, T., and L. J. Tolmach. 1963. Growth and nucleic acid synthesis in synchronously dividing populations of HeLa cells. Exp. Cell Res. 30:344–362.

Thilly, W. G., D. I. Arkin, and G. N. Wogan. 1977. Nucleic acid content of HeLa S3 cells during the cell cycle: Variations between cycles. Cell Tissue Kinet. 10:81–88.

Tobey, R. A., and K. D. Ley. 1970. Regulation of initiation of DNA synthesis in Chinese hamster cells. I. Production of stable, reversible G_1-arrested populations in suspension culture. J. Cell Biol. 46:151–157.

Traganos, F., Z. Darzynkiewicz, T. Sharpless, and M. R. Melamed. 1977. Simultaneous staining of ribonucleic and deoxyribonucleic acid in unfixed cells using acridine orange in a flow cytofluorometric system. J. Histochem. Cytochem. 25:46–56.

Valeriote, F., and L. van Putten. 1975. Proliferation-dependent cytotoxicity of anticancer agents: A review. Cancer Res. 35:2619–2630.

van Putten, L. M., J. Keizer, and R. F. Evenwel. 1977. Resting cells and cancer biology, *in* Growth Kinetics and Biochemical Regulation of Normal and Malignant Cells (The University of Texas System Cancer Center 29th Annual Symposium on Fundamental Cancer Research, 1976). Williams & Wilkins Co., Baltimore, pp. 91–98.

Watson, J. V., and S. H. Chambers. 1977. Fluorescence discrimination between diploid cells on their RNA content: A possible distinction between clonogenic and non-clonogenic cells. Br. J. Cancer 36:592–600.

Wied, G. L., ed. 1966. Introduction to Quantitative Cytochemistry. Vols. 1 & 2. Academic Press, New York.

Wong, T. W., G. F. Whitmore, and S. Gulyas. 1978. Studies on the toxicity and radiosensitizing ability of misonidazole under conditions of prolonged incubation. Radiat. Res. 75:541–555.

Yen, A., and A. B. Pardee. 1978. Arrested states produced by isoleucine deprivation and their relationship to the low serum produced arrested state in Swiss 3T3 cells. Exp. Cell Res. 114:389–395.

Tissue Responses to Radiation

Radiation Biology in Cancer Research, edited by
Raymond E. Meyn and H. Rodney Withers.
Raven Press, New York © 1980.

The Pathobiology of Late Effects of Irradiation

H. Rodney Withers, Lester J. Peters, and H. Dieter Kogelnik*

*Section of Experimental Radiotherapy, Department of Radiotherapy, The University of Texas System Cancer Center M. D. Anderson Hospital and Tumor Institute, Houston, Texas 77030; and *Strahlentherapeutische Klinik, Allgemeines Krankenhaus der Stadt Wien, Vienna, Austria*

Contrary to the general belief that acute effects of irradiation result from killing of parenchymal cells and late effects from vascular injury, we propose that both types of effect may result directly from radiation-induced parenchymal or stromal cell depletion or both. The rate of development of overt injury depends upon the rate at which the cells of the tissue divide: acute injury occurs in rapidly proliferating tissues, late injury in tissues which turn over slowly. Thus, a late effect is analogous to an acute effect but delayed in its expression by the slow rate at which the target cells turn over.

GENERAL CHARACTERISTICS OF LATE EFFECTS

Genetic, teratogenic, or carcinogenic effects are not discussed in this paper, which is restricted to a consideration of structural and functional changes that develop in normal tissues irradiated coincidentally during radiotherapy for cancer. The rate of development of acute and late effects varies from tissue to tissue, reflecting a broad spectrum of turnover kinetics. It is conventional to consider changes developing two to three months or longer after completion of radiotherapy to be "late." They occur in such slowly proliferating tissues as lung, liver, heart, nervous system, cartilage, bone, muscle, and connective tissues.

THE VASCULAR THEORY OF LATE EFFECTS

There is no general agreement about what vascular lesion is thought to lead to late radiation injury. It may be predominantly an endarteritis affecting small arteries and arterioles (White 1976, Casarett 1964), an increased "reactivity" and segmental narrowing of small arterioles (Lindop et al. 1970), swelling and degeneration of endothelial cells in small vessels, especially capillaries (Casarett 1964, Stearner et al. in press, Phillips et al. 1972), with or without microthrombi (Fajardo and Stewart 1971), or increased permeability of capillary walls leading to an interstitial transudate and, ultimately, fibrosis (Jolles and Harrison 1966, Ullrich and Casarett 1977, Law and Thomlinson 1978), with parenchymal degeneration from an increase in the histohematic connective tissue barrier (Casarett 1972). Obviously, all these processes could occur and contribute to late injury, although, in fact, none has been conclusively demonstrated to be the cause of changes of the type seen in the various slowly responding tissues after irradiation.

The vascular theory of late radiation injury probably had its origin in the consistent observation of endarteritis and telangiectasia in tissues severely injured by radiation. The theory has been so widely accepted that the various other vascular changes observed after radiation have subsequently been assumed to also play a role in the etiology of late effects. However, capillary structure and function have not been shown to be grossly or permanently compromised by radiation and have commonly returned to normal by the time late effects develop (Phillips 1966, Glatstein 1973, Stearner and Christian 1978, Lindop et al. 1970, Moustafa and Hopewell 1979, Hopewell 1980, see pages 449–459, this volume), while the subintimal proliferation and other endarteritic and medial wall changes in arteries and arterioles become prominent only after the late changes are well advanced. The arteriolar changes resemble those found in other atrophic or involutional processes, except for the slow loss of medial muscle (Stearner and Christian 1978, Hirst et al. 1979), presumably from a direct cytocidal effect of radiation on these slowly dividing cells.

GENERAL REASONS OPPOSING A VASCULAR ORIGIN OF LATE INJURY

There is no universal "late effect" syndrome: rather, the manifestations of late radiation injury are diverse and develop at different rates in different tissues. Although they depend to some extent on dose, some examples of approximate latent intervals between irradiation and clinical appearance of late effects in man are 2–4 months for pneumonitis, 6–24 months for myelitis, 12 months to several years for nephritis or associated hypertension, 1–5 years for blindness, and 6 months to many years for progressive "fibrosis" in subcutaneous and other connective tissues.

This wide range in the times at which late injury appears is consistent with a large variation in the proliferation kinetics of the "target" cells. There are also large differences in the proliferation kinetics of the vascular endothelium in various tissues (Engerman et al. 1967, Tannock and Hayashi 1972) and the spread in latent intervals to the appearance of late effects could be consistent with these cells being the target cells. However, as mentioned above, endothelial injury usually appears appreciably earlier than the parenchymal injury, and there seems no reason to consider that it is the slowly proliferating endothelial cells, rather than the slowly proliferating parenchymal cells, which determine the parenchymal response. Types of vascular injury other than endothelial cell degeneration would not be expected to show a consistently different response rate from tissue to tissue, and hence the most plausible explanation for the variability in latent intervals probably lies in the variability of parenchymal proliferation patterns.

Dose Effect Relationships

There are wide differences in the doses required for late injury in various tissues. For functional end points the volume of the organ irradiated is critical, but even when the whole of the critical tissue is included, there is considerable variation in the dose that gives an acceptably low probability of injury. For example, when treatment is given in daily 200-rad fractions, approximate tolerable doses for whole-organ irradiation are kidneys, 2,000 rads; lungs, 1,500 rads; heart, 3,500 rads; brain, 5,000 rads; bladder and prostate, 6,500–7,000 rads. These differences could be easily explained in terms of varying initial numbers of "target" cells in the different tissues or variations in their radiosensitivity, whereas it is difficult to implicate blood vessel damage as the universal basis for effects requiring such a wide range of doses.

Differences in $RBE_{n/\gamma}$

If all late effects were due to vascular injury, the $RBE_{n/\gamma}$ would be constant for all such effects. Multifraction experiments show that when size of dose per fraction is accounted for, the RBE is, in fact, fairly constant for the few late effects investigated—except for myelitis, for which it is significantly higher (Hussey et al. 1980, see pages 471–488, this volume)—a difference that is unlikely to be accounted for solely on the basis of increased absorption of neutrons in the central nervous system.

SPECIFIC LATE INJURIES ILLUSTRATING THAT BLOOD VESSEL INJURY IS PROBABLY NOT AN ETIOLOGICAL FACTOR

Spinal Cord and Peripheral Nerves

The typical lesion in radiation-induced spinal cord injury is demyelination of the long tracts with little or no demonstrable primary change in the unmyelinated neurons of the gray matter (see Figure 2A of Hussey 1980, see pages 471–488, this volume, van der Kogel 1977, 1980, see pages 461–470, this volume, Phillips and Buschke 1969). The vascular supply to the spinal cord is from anterior and posterior spinal arteries, and some areas of gray and white matter are supplied by the same terminal arterioles. With a common blood supply, it is unlikely that vascular injury would so consistently spare gray matter and produce, as the initial lesion, only demyelination of nerve tracts in the white matter. Furthermore, the arteries and arterioles show no appreciable abnormalities at the time the demyelination develops, and since the motor and sensory changes begin insidiously and progress over weeks or months, capillary microthrombosis, although possible, is also an unlikely etiological factor.

The oligodendrocytes, which produce the myelin sheaths for the nerve fibers of the white matter, are a slowly dividing population of cells, and radiation-induced impairment of their reproductive integrity would result in a slow depletion of their number and a consequent slow demyelination. Thus, the oligodendrocytes are a more logical "target" for radiation myelitis than are blood vessels.

Radiation in sufficiently high doses may also cause demyelination of peripheral nerves (Cheng and Schulz 1975), and it seems reasonable to assume that depletion of the myelin-producing Schwann cells, rather than vascular injury, is the basis for radiation neuropathy. In the treatment of supraclavicular lymph nodes, the brachial plexus and the stellate (sympathetic) ganglion receive essentially the same dose, and yet in reports of brachial plexus injury, there is no mention of Horner's syndrome, which would result from injury to the unmyelinated sympathetic nerves from the stellate ganglion. Likewise, injury to the sympathetic nerves supplying the iris is not mentioned in reports of optic nerve injury (Cheng and Schulz 1975, Shukovsky and Fletcher 1972), even though it would be obvious if present. The difference in susceptibility to radiation injury appears, therefore, to be related to the presence or absence of a myelin sheath around the nerve, which is consistent with our suggestion that the biological basis for radiation neuropathy is depletion of Schwann cells, not vascular injury.

Retina

The radiation response of the retina is interesting because this tissue has but one arterial supply. Shukovsky and Fletcher (1972) reported on a series of patients whose retinae were irradiated during treatment of squamous carcinomas of the ethmoid sinuses and nasal cavities. Nine of 15 long-surviving patients who had received large doses to one or both retinae lost vision—in a total of 12 eyes. In only 2 eyes was there a sudden acute loss of vision due to retinal artery thrombosis, and these episodes occurred at 9 and 15 months after irradiation. In 7 eyes, blindness resulted from slow retinal atrophy occurring between 2 and 3½ years after therapy. In another group of 3 eyes, optic nerve degeneration led to a slow deterioration of vision over several months between 4 and 5 years after therapy. In these last two groups, the vessels were reported to be normal. In other studies (Wara et al. 1979, Fitzgerald et al., in press) exudates and hemorrhages were seen, but these could be the result of retinal atrophy rather than the cause.

Therefore, vascular injury in the form of occlusion of the single afferent artery is an uncommon cause of blindness, it occurs earlier than optic nerve and retinal atrophy, and would be an even less frequent problem if there were collateral arteries.

Kidney

The kidney may provide another example of a late effect that results from parenchymal cell depletion rather than vascular injury. Glatstein (1973) measured the uptake of ^{86}Rb by mouse kidneys as a means of determining blood flow through the organ after exposure to single doses of X rays of up to 1,900 rad. After 1,900 rad, the clearance of ^{86}Rb per gram of tissue began decreasing from about 90% of control values at two months to about 50% between three and eight months. After 1,500 rad, ^{86}Rb extraction per gram of kidney decreased to 75% of control values at two months, and persisted essentially unchanged until the end of the experiment 12 months after irradiation.

At first glance, these data support the generally accepted concept that radiation-induced vascular injury leads to ischemic sclerosis. However, as Glatstein points out, parenchymal cell depletion could lead to decreased clearance of ^{86}Rb even if blood flow were normal. The correction he made for the reduced weight of the irradiated organ by expressing the results as extraction per gram of kidney reduced the error in the measurement of renal blood flow, but may not have eliminated it. Weight loss after irradiation is not uniform throughout the kidney, but results primarily from depopulation of proximal tubules (Figure 1). Thus, the decrease in parenchymal cells extracting ^{86}Rb is proportionately greater than the decrease in total renal weight. The 40% to 50% decrease in

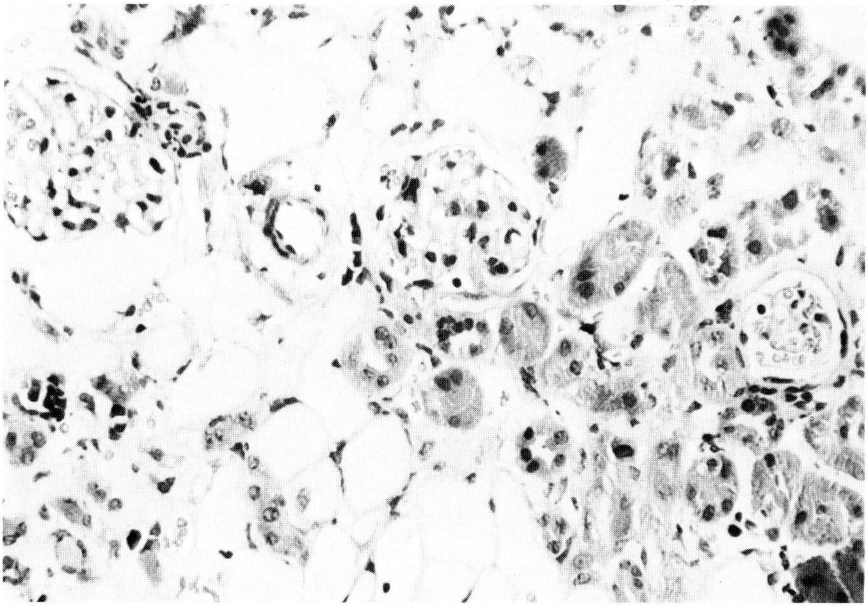

FIG. 1. Photomicrograph of mouse kidney 12 months after a single dose of 1,900 rad. Note extensive loss of tubule epithelium but normal glomeruli and blood vessels.

^{86}Rb extraction per gram of kidney after 1,900 rad and the 25% decrease after 1,500 rad may reflect merely the selective loss of cells from proximal tubules and not a reduction of blood flow. However, even if blood flow were reduced to 60% of controls, extensive necrosis of parenchymal tubule cells would not result. After single doses of 1,900 rad, extensive necrosis of proximal tubule cells does occur and is easily seen in histological sections, not at the time of the initial drop in ^{86}Rb clearance, but months later. Thus, alterations in blood flow, if they do in fact occur, are not simultaneous with extensive proximal tubule degeneration, and it is reasonable to consider them as largely independent phenomena. As in other tissues, and as illustrated in Figure 1, arteriolar changes are not obvious early in the course of parenchymal degeneration, and capillary changes in the glomerulus are minimal (Wachholz and Casarett 1970).

Connective Tissues

The most common late effect in normal tissues following radiotherapy is so-called fibrosis, which develops in most tissues. It is commonly seen in the dermis. The change is more an atrophic contraction than fibrosis. Histologically, the connective tissues become replaced by dense, hypocellular collagen. In severe cases, necrotic ulcers may develop and heal slowly, or not at all, primarily because the dense collagen does not contract.

Traditionally, these changes have been ascribed to ischemia resulting from vascular injury, but during their development, there is relatively little change in blood flow (Hopewell and Young 1978, Moustafa and Hopewell 1979), and the histological picture is not that typical of ischemia. Recently, it has also been postulated that the changes may result from organization of fibrin that collects in interstitial tissues as a result of the increased permeability of irradiated capillaries (Law and Thomlinson 1978). However, late contraction is also seen after prolonged irradiation at dose rates low enough that a significant increase in vascular permeability is unlikely (Scallon et al. 1969). Another postulated mechanism is dysfunction of irradiated fibrocytes (Rantanen 1973) leading to an imbalance of collagen production and resorption. We agree that the lesion probably results from such an imbalance, but suggest an alternative, or possibly supplementary hypothesis, viz. that a slow loss of fibroblasts following irradiation could lead to incomplete resorption of collagen already deposited by the fibroblasts before their death. This collagen, which is no longer in a kinetic steady state of secretion and resorption, would slowly undergo cross-linking and aging, forming the hard contracted plaques seen years later in heavily irradiated tissues. This dense acellular collagen may be regarded as an exoskeleton of the long-departed fibroblasts and is analogous to sequestrated bone devoid of osteocytes. The poor healing of necrotic ulcers probably also reflects the lack of viable fibroblasts necessary for the contraction phase of wound healing.

Heart

Irradiation of human or rabbit heart may lead to exudative pericarditis and some myocardial fibrosis beginning about two months after treatment (Fajardo and Stewart 1971). This is preceded by swelling of capillary endothelial cells with resultant narrowing and occasional microthrombi in the lumen. Fajardo and Stewart (1971) found that the ratio of capillaries to myocardial fibers in rabbits was later reduced, but Fajardo and Brown (1973) and Stearner et al. (in press) found no such reduction in irradiated mouse hearts. It has been assumed that in rabbits and man vascular changes underlie cardiac injury, which is primarily a pericarditis. In mice, the pericardium is unaffected, but thrombi develop, attached to the endocardium (Fajardo and Brown 1973), presumably a sequel of endocardial injury. These investigators noted that the level of plasminogen activator is high in rat pericardium (it has not been measured in mice), but low in rabbits, and they postulated that the fibrinous pericardial lesion in rabbits results from an absence or low level of fibrinolysins. Thus, an alternative hypothesis for the pathogenesis of the pericardial injury seen in man and rabbits, which is as plausible as the present notion that capillary injury is the underlying cause, is that the lethally injured mesothelial cells of the pericardial sac are lost slowly, and the nonspecific fibrinous exudate that follows is not resorbed because of the low level of plasminogen.

Blood Vessels

Endothelial cells proliferate slowly (Tannock and Hayashi 1972, Engerman et al. 1967) and exhibit radiation survival characteristics similar to those of other cells (Van den Brenk 1972, Reinhold and Buisman 1973, Fike and Gillette 1978). Radiation should cause slow depletion of these cells, and the resulting interference with vascular competence will depend on many factors, including dose, regeneration of survivors, and migration into the injured vessels of cells from outside the exposed area.

Blood vessels appear to have a great capacity to recover from radiation injury. For example, when malignant cells are implanted into heavily irradiated subcutaneous tissue or lungs, they can grow into large tumors (Hewitt and Blake 1968, Hill and Bush 1969, Clifton and Jirtle 1975, Urano and Suit 1971), requiring an increase in the local vasculature by several orders of magnitude. Another example of the apparent resistance of blood vessels to radiation injury is the lack of effect on structure or function of the small bowel after large doses have been delivered to its mesenteric vessels (Hirst et al. 1979). Large arteries and veins have been generally acknowledged to show little detectable effect of irradiation.

The human liver provides an exception to our hypothesis: when the whole organ is exposed to moderately high doses (e.g., 2,500 to 3,000 rad in 10 fractions), thromboses develop in the centrilobular veins leading to portal hypertension (Ingold et al. 1965, Wharton et al. 1973).

Cataracts

Radiation accelerates the rate at which cataracts develop in the lens of the eye. This late effect could hardly result from vascular injury because there are no blood vessels in the lens.

CONCLUSION

Although it has been widely accepted that vascular injury leads to late damage to parenchyma and stroma, the mechanism by which the observed vascular changes lead to tissue damage is not obvious and the time sequence of the respective changes is not usually consistent with a cause-effect relationship. It seems more likely that the late changes in blood vessels follow radiation-induced parenchymal or stromal depletion and are, in large part, the result, not the cause, of the late injury. The early changes in blood vessel endothelium are transitory and usually precede the late effect by months or even years and have resolved by the time late injuries become apparent. The radiation-induced loss of arteriolar smooth muscle may precede or follow parenchymal injury, suggesting that this late effect is independent of others and is not their cause. The various characteristics of late radiation injury can be more easily understood, and more profitably investigated, if one assumes that their pathobiological basis is depletion of parenchymal or stromal cells or both.

REFERENCES

Casarett, G. W. 1964. Similarities and contrasts between radiation and time pathology, *in* Advances in Gerontological Research, B. Strehler, ed. Academic Press, New York, pp. 109–163.

Casarett, G. W. 1972. Aging, *in* Radiation Effect and Tolerance of Normal Tissues (Basic Concepts in Radiation Pathology), Vol. 6, Frontiers of Radiation Therapy and Oncology, J. Vaeth, ed. S. Karger, New York, pp. 479–485.

Cheng, V. S. T., and M. D. Schulz. 1975. Unilateral hypoglossal nerve atrophy as a late complication of radiation therapy of head and neck carcinoma: A report of four cases and a review of the literature on peripheral and cranial nerve damages after radiation therapy. Cancer 35:1537–1544.

Clifton, K. H., and R. Jirtle. 1975. Mammary carcinoma cell population growth in preirradiated and unirradiated transplant sites. Viable tumor growth, vascularity and the tumor bed effect. Radiology 117:459–465.

Engerman, R. L., D. Pfaffenbach, and D. M. Davis. 1967. Cell turnover of capillaries. Lab. Invest. 17:738–743.

Fajardo, L. F., and J. M. Brown. 1973. Cardiac mural thrombi caused by radiation. Radiat. Res. 55:387–389.

Fajardo, L. F., and J. R. Stewart. 1971. Capillary injury preceding radiation-induced myocardial fibrosis. Radiology 101:425–433.

Fike, J. R., and E. L. Gillette. 1978. ^{60}Co and negative pi-meson irradiation of microvasculature. Int. J. Radiat. Oncol. Biol. Phys. 4:825–828.

Fitzgerald, C. R., J. Hird, K. E. Ellingwood, and R. R. Million. 1979. Ocular complications of radiotherapy for malignant tumors of the paranasal sinuses and orbit. Arch. Ophthalmol. (in press).

Glatstein, E. 1973. Alterations in rubidium-86 extraction in normal mouse tissues after irradiation. An estimate of long-term blood flow changes in kidney, lung, liver, skin and muscle. Radiat. Res. 53:88–101.

Hewitt, H. B., and E. R. Blake. 1968. The growth of transplanted murine tumours in pre-irradiated sites. Br. J. Radiol. 22:808–824.

Hill, R. B., and R. S. Bush. 1969. A lung colony assay to determine the radiosensitivity of the cells of a solid tumor. Int. J. Radiat. Biol. 15:435–444.

Hirst, D. G., J. Denekamp, and E. L. Travis. 1979. The response of mesenteric vessels to irradiation Radiat. Res. 77:259–275.

Hopewell, J. W. 1980. The importance of vascular damage in the development of late radiation effects in normal tissues, *in* Radiation Biology in Cancer Research (The University of Texas System Cancer Center 32nd Annual Symposium on Fundamental Cancer Research), R. E. Meyn and H. R. Withers, eds. Raven Press, New York, pp. 449–459.

Hopewell, J. W., and C. M. A. Young. 1978. Changes in the microcirculation of normal tissues after irradiation. Int. J. Radiat. Oncol. Biol. Phys. 4:53–58.

Hussey, D. H., C. A. Gleiser, J. H. Jardine, G. L. Raulston, and H. R. Withers. 1980. Acute and late normal tissue effects of 50 $MeV_{d \rightarrow Be}$ neutrons, *in* Radiation Biology in Cancer Research (The University of Texas System Cancer Center 32nd Annual Symposium on Fundamental Cancer Research), R. E. Meyn and H. R. Withers, eds. Raven Press, New York, pp. 471–488.

Ingold, J. A., G. B. Reed, H. S. Kaplan, and M. A. Bagshaw. 1965. Radiation hepatitis. Am. J. Roentgenol. 93:200–208.

Jolles, B., and R. G. Harrison. 1966. Enzymic processes and vascular changes in the skin radiation reaction. Br. J. Radiol. 39:12–18.

Law, M. P., and R. H. Thomlinson. 1978. Vascular permeability in the ears of rats after X-irradiation. Br. J. Radiol. 51:895–904.

Lindop, P. J., A. Jones, and A. Bakowska. 1970. The effect of 14-MeV electrons in the blood vessels of the mouse ear lobe, *in* Time and Dose Relationships in Radiation Biology as Applied to Radiotherapy, V. P. Bond, H. D. Suit, and V. Marcial, eds. Brookhaven National Laboratory Report BNL-5032 (C-57), pp. 174–180.

Moustafa, H. T., and J. W. Hopewell. 1979. Blood flow clearance changes in pig skin after single doses of X rays. Br. J. Radiol. 52:138–144.

Phillips, T. L. 1966. An ultrastructural study of the development of radiation injury in the lung. Radiology 87:49–54.

Phillips, T. L., S. Benak, and G. Ross. 1972. Ultrastructural and cellular effects of ionizing radiation, *in* Frontiers of Radiation Therapy and Oncology, Vol. 6, J. Vaeth, ed. S. Karger, New York, pp. 21–43.

Phillips, T. L., and F. Buschke. 1969. Radiation tolerance of the thoracic spinal cord. Am. J. Roentgenol. 105:659–664.

Rantanen, J. 1973. Radiation injury of connective tissue. A biochemical investigation with experimental granuloma. Acta. Radiol. Suppl. 330:1–92.

Reinhold, H. S., and G. H. Buisman. 1973. Radiosensitivity of capillary endothelium. Br. J. Radiol. 46:54–57.

Scallon, J. E., C. A. Sondhaus, S. Snyder, B. H. Feder, and W. G. Gunn. 1969. Permanent interstitial therapy using low energy and long half-life radiation sources. Am. J. Roentgenol. 105:157–164.

Shukovsky, L. J., and G. H. Fletcher. 1972. Retinal and optic nerve complications in a high dose irradiation technique of ethmoid sinus and nasal cavity. Radiology 104:629–634.

Stearner, S. P., and E. J. B. Christian. 1978. Long-term vascular effects of ionizing radiations in the mouse: Capillary blood flow. Radiat. Res. 73:553–567.

Stearner, S. P., V. V. Yang, and R. L. Devine. 1979. Cardiac injury in the aged mouse: Comparative ultrastructural effects of fission spectrum neutrons and gamma-rays. Radiat. Res. 78:429–447.

Tannock, I., and S. H. Hayashi. 1972. The proliferation of capillary endothelial cells. Cancer Res. 32:77–82.

Ullrich, R. L., and G. W. Casarett. 1977. Interrelationship between the early inflammatory response and subsequent fibrosis after radiation exposure. Radiat. Res. 72:107–121.

Urano, M., and H. D. Suit. 1971. Experimental evaluation of tumor bed effect of C_3H mouse mammary carcinoma and for C_3H mouse fibrosarcoma. Radiat. Res. 45:41–49.

Van den Brenk, H. A. S. 1972. Macrocolony assay for measurement of reparative angiogenesis after X-irradiation. Int. J. Radiat. Biol. 21:607–611.

van der Kogel, A. J. 1977. Radiation tolerance of the rat spinal cord: Time-dose relationships. Radiology 122:505–509.

van der Kogel, A. J. 1980. Mechanisms of late radiation injury in the spinal cord, *in* Radiation Biology in Cancer Research (The University of Texas System Cancer Center 32nd Annual Symposium on Fundamental Cancer Research), R. E. Meyn and H. R. Withers, eds. Raven Press, New York, pp. 461–470.

Wachholz, B. W., and G. W. Casarett. 1970. Radiation hypertension and nephrosclerosis. Radiat. Res. 41:39–56.

Wara, W. M., A. R. Irvine, R. E. Neger, E. L. Howes, Jr., and T. L. Phillips. 1979. Radiation retinopathy. Int. J. Radiat. Oncol. Biol. Phys. 5:81–83.

Wharton, J. T., L. Delclos, S. Gallager, and J. P. Smith. 1973. Radiation hepatitis induced by abdominal irradiation with the cobalt 60 moving strip technique. Am. J. Roentgenol. 117:73–80.

White, D. C. 1976. The histopathologic basis for functional decrements in late radiation injury in diverse organs. Cancer 37:1126–1143.

Radiation Biology in Cancer Research, edited by
Raymond E. Meyn and H. Rodney Withers.
Raven Press, New York © 1980.

The Importance of Vascular Damage in the Development of Late Radiation Effects in Normal Tissues

J. W. Hopewell

Churchill Hospital Research Institute, University of Oxford, Headington, Oxford, England 0X3 7LJ

In current radiotherapeutic practice, it is invariably the risk of serious late radiation damage to normal tissues that is dose limiting when attempting to improve local tumor control rates. If this important clinical problem is to be totally avoided or even dealt with, more information is required about the pathogenesis of these late radiation sequelae. The recognition of early indicators of severe late effects would also be of some advantage. One important hypothesis is that late radiation damage is preceded by and is consequently a function of radiation effects on the vascular system. For such a hypothesis to be valid, evidence of similar time-related vascular changes would have to be shown in a number of normal tissues. For both functional and morphological end points some of this information is now available, and it has added considerable weight to the view that vascular effects are of primary importance in determining the late radiation tolerance of normal tissues.

STUDIES IN PIG SKIN

The main evidence in support of the above hypothesis has been provided by the results of a series of investigations carried out in pig skin. The skin's accessibility makes it a convenient tissue model in which to study vascular changes with a range of complementary assay methods. In addition, the similarities in the microvasculature of pig and human dermis (Donovan 1975) allow the results obtained in the pig to be directly extrapolated to man, with some degree of certainty.

Direct visual observations (skin scoring) of radiation-induced changes in the skin have provided a guide to the initial vascular changes in the dermis. The system used assessed the degree of erythema, epithelial desquamation, and dermal necrosis separately. Arbitrary numerical scores were then attached to these biological reactions according to their severity (Table 1).

A separate analysis of the data for erythema alone has demonstrated two

TABLE 1. *Skin reactions in pig skin and associated numerical scores*

Erythema	Desquamation
Minimal ~ 1	Dry ~ 3
Moderate ~ 2	Moist < ½ field ~ 4
Bright red ~ 3	Moist > ½ field ~ 5
	Dermal necrosis
Mauve reaction ~ 2.75	< ½ field ~ 5.5
Dusky reaction ~ 3.5	> ½ field ~ 6.5

distinct waves of reaction in the first 16 weeks after irradiation. This applied to skin treated with both single and fractionated doses of radiation, although the relative severity of each wave was modified by fractionation (Hopewell et al. 1978). In the example cited for treatment with three large fractions in six days, the time-related changes in the erythema response are plotted for individual skin sites (Figure 1). In the first wave, a maximum bright red erythema was

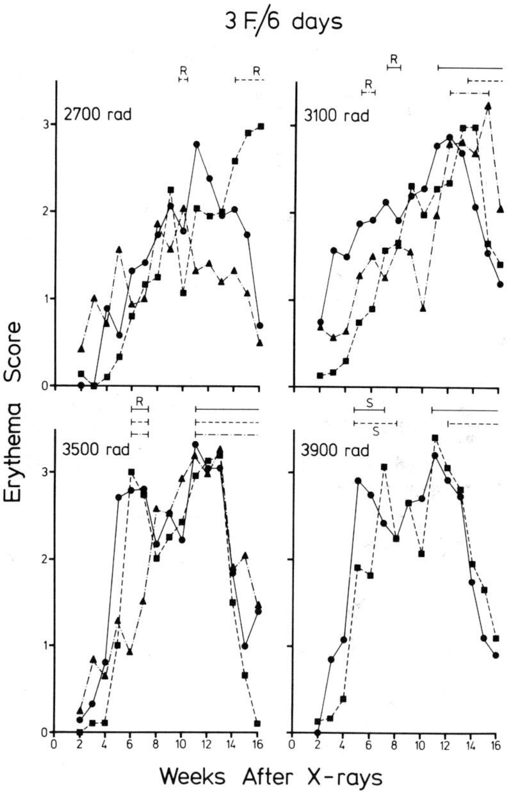

FIG. 1. Time-related changes in the intensity of the erythema reaction in individual skin fields, on three separate pigs, irradiated with a range of total doses given in three fractions/six days. The duration of any associated epithelial reactions and dermal necrosis for individual fields are indicated by the bars (R ~ dry desquamation; S ~ moist desquamation < ½ field; no letter, ~ dermal necrosis).

reached four to six weeks after treatment, and this may be associated with dry or moist desquamation. Epidermal reactions were transient if the dose required to produce these effects was not greatly exceeded. In the second wave, the maximum reaction, which reaches a peak 10–14 weeks after irradiation, was exemplified by a dusky or mauve appearance to the skin, reactions characteristic of ischemia. When tolerance doses were exceeded dermal necrosis resulted. The appearance of dermal necrosis without previous moist desquamation, reported here for three fractions/six days, was not a characteristic of skin fields treated with multiple small fractions in which transient moist desquamation in the first wave was not followed by dermal necrosis (Hopewell et al. 1978).

These direct visual observations and the conclusions drawn from them have been complemented by dermal blood flow measurements. The technique, which measures the local clearance rate of intradermally injected ^{99m}Tc, as the pertechnetate ion (Moustafa and Hopewell 1976), provides parameters associated with the nutrient blood flow to both the papillary (fast exponent) and reticular (slow exponent) dermis (Young and Hopewell 1979). In skin irradiated with single doses of X rays, two waves of modification in isotope clearance from

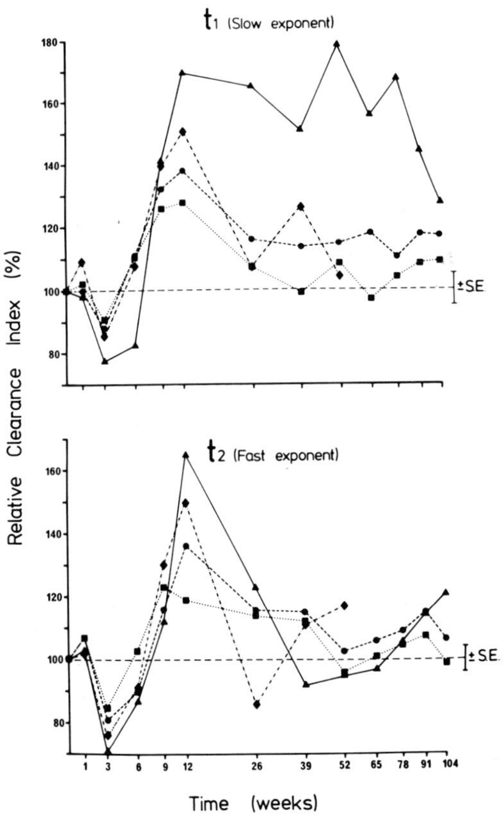

FIG. 2. Time- and dose-related changes in the relative clearance index (half clearance time irradiated/half clearance time control (%)) for the fast (t_2) and slow (t_1) exponent of isotope clearance after the intradermal injection of a small volume (0.03 ml) of tracer into pig skin. ± SE, standard error of the mean for pretreatment values in normal pig skin; ■, 800 rad; ●, 1,800 rad; ◆, 2,070 rad; ▲, 2,340 rad. (Reproduced from Moustafa and Hopewell 1979a, with permission of *British Journal of Radiology*.)

the dermis were observed. These changes were dose related and applied equally to the fast and slow exponents of isotope clearance (Figure 2).

In the immediate period after irradiation, isotope clearance times were shorter than in control skin, suggesting an increase in blood flow. This coincides with the first wave of erythema, and it is suggested that these changes represent a secondary inflammatory response of the dermal vasculature to epidermal damage. After 12 weeks, isotope clearance times are significantly longer than in control skin, indicating a net reduction in vascular function. This is in keeping with the skin reactions at this time, which had a bluish tinge characteristic of ischemia.

The impairment in vascular function was transient, isotope clearance parameters returning to normal over the period 12–26 weeks, the same period over which late tissue atrophy develops (Hopewell and Young 1978, Hopewell et al. 1979). Late tissue atrophy, which was measured by comparing the linear dimensions of previously delineated irradiated and control fields on the same animal (relative linear field contraction), was constant over the period 6–12 months after irradiation (Hopewell and Young 1978, Hopewell et al. 1979). The suggestion that the reduction in vascular function at 12 weeks is the major contributory factor in the subsequent development of late atrophy is supported by the good correlation between them (Figure 3).

Qualitative and quantitative histological studies in the dermis have provided additional evidence of vascular changes that are in keeping with the functional findings. A major histological finding 10–12 weeks after irradiation was the appearance of arterioles totally or partially occluded by endothelial cells. Most of these lesions were found in the vessels of the deep dermal plexus (Figure 4), although occluded vessels were found at all levels of the dermis and in the underlying fat. These occlusive changes were, as might be expected, associated with a significant reduction in the density of blood vessels at all levels of the pig dermis 12 weeks after irradiation (Young 1978). Like changes in the functional parameters, this reduction in vascular density was transient, and in mea-

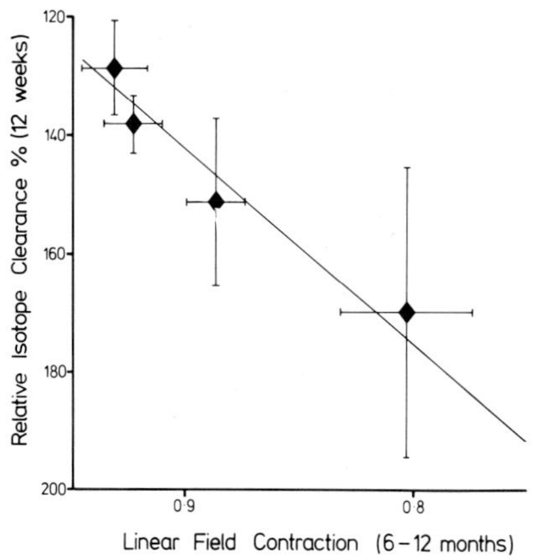

FIG. 3. Correlation between the reduction in blood flow in the dermis (t_1, slow exponent) 12 weeks after irradiation with single doses and the subsequent severity of relative linear field contraction at 6–12 months. Correlation coefficient, 0.93. (Reproduced from Hopewell et al. 1979, with permission of *Radiology*.)

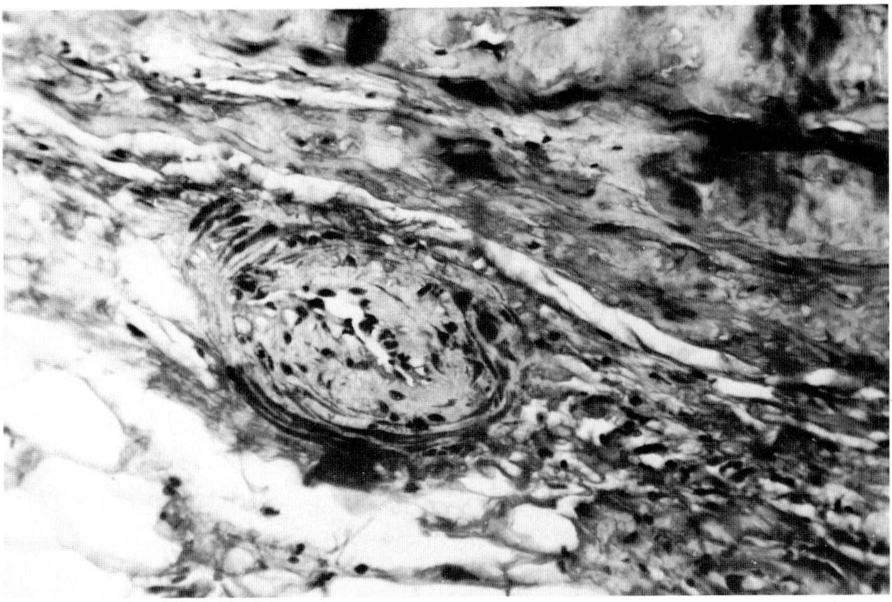

FIG. 4. Arteriole in the deep dermal plexus of pig skin 11 weeks after exposure to 3,500 rad (three fractions/six days). The vessel is nearly totally occluded by endothelial cells (Mallory ×300).

surements undertaken in pig skin 12 and 24 months after irradiation the vascular density was normal. At this time period after irradiation, the dermis and the subcutaneous fat were reduced in thickness (Figure 5), an observation consistent with parenchymal atrophy. Viewed superficially, these areas of irradiated skin are similar in appearance to the areas of induration reported in human skin after photon (Gauwerky and Langheim 1978) or neutron irradiation (Stone 1948).

Despite the apparent normality of the regional blood supply at 12 and 24 months after irradiation, the response of this vascular system to the additional stress of surgical trauma was similar to that found 12 weeks after irradiation,

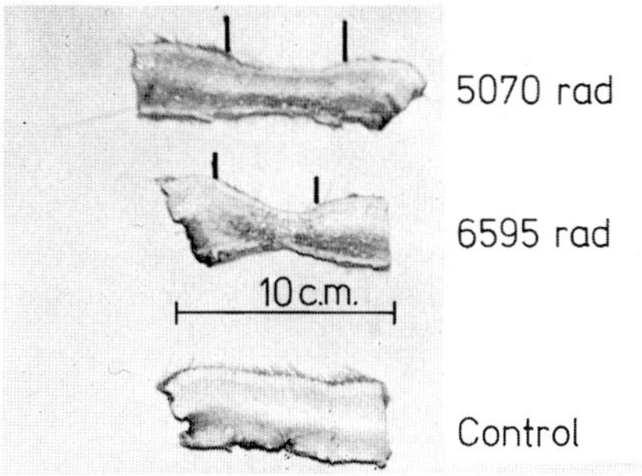

FIG. 5. Slices of pig skin two years after exposure to radiation given as 14 fractions/39 days. The thickness of the dermis and subcutaneous fat is reduced when compared with normal skin taken from an equivalent site on the opposite flank of the same pig.

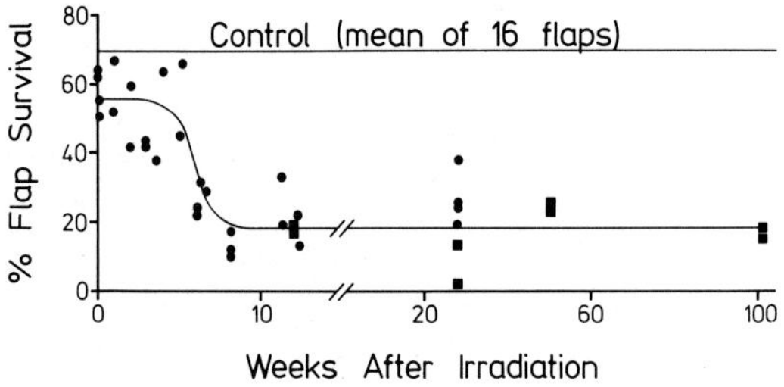

FIG. 6. Time-related changes in the survival of skin flaps raised from irradiated skin (1,800 rad, single dose) in the pig. The graph is redrawn from results presented by Wiernik et al. 1974 (●) with the addition of some of our new data (■).

when vascular impairment was maximal (Figure 6). Surgical stress was achieved by raising the whole of the irradiated area as a conventional single-pedicle skin flap based on the ventral margin of the treated area (Patterson 1968, Wiernik et al. 1974). The eventual surviving length of the flap, one week after operation, was used as a measure of the functional capacity of the inherent blood supply under stress.

This finding is consistent with current radiotherapeutic teaching, which suggests that late radiation necrosis, which arises many years after treatment, may occur as a consequence of additional traumatic stress. Necrosis of the skin following excessive exposure to wind and sun and bone necrosis after tooth extraction are frequently quoted examples.

STUDIES IN OTHER NORMAL TISSUES

Many of the gross morphological, functional, and histological evidence quoted above for the existence of vascular damage in the dermis of pig skin after irradiation and the role played in the development of generalized late tissue damage have also been reported in a number of other normal tissues. The time course for these changes also shows a remarkable similarity in different normal tissues.

In the oral mucosa of rhesus monkeys (Jardine et al. 1975), an anatomical site in many ways comparable with the skin, gross evidence of two waves of reaction was again noted. The acute reaction was characterized by an erythema and an associated mucositis, while the second wave was associated with oromucosal necrosis. The first sign of necrosis was a blanching of the mucosa, a feature often associated with pig skin just prior to dermal breakdown. The time course for the onset of necrosis in both systems was similar, as was their response to modifications in fractionation, thereby suggesting a similar vascular mechanism.

On the basis of the results of functional studies, two waves of response have also been reported in the irradiated hamster cheek pouch (Hopewell 1975). Here ^{51}Cr-labeled red blood cells were used to measure time- and dose-related changes in the blood volume of the cheek pouch (Figure 7). The results show an early rise in the relative blood volume of cheek pouches irradiated with 2,000 and 3,000 rad. This modification in vascular volume occurs in association with a wave of epithelial desquamation, a finding that supports the view that

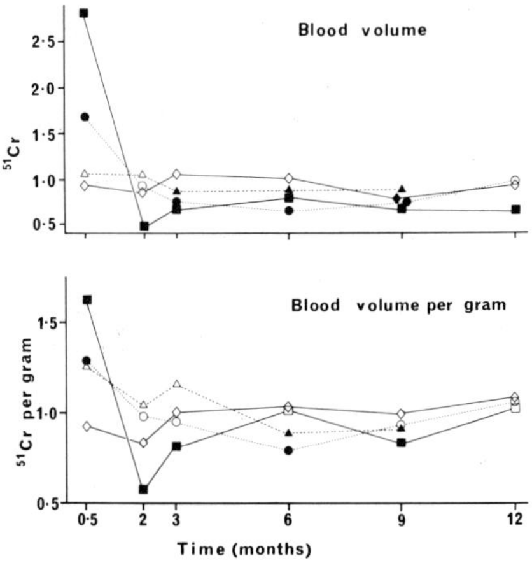

FIG. 7. Time- and dose-related changes in the red cell volume of the irradiated hamster cheek pouch. Results are expressed relative to control values obtained from control pouches in the same animals with respect to the whole pouch and per gram of tissue. △, 500 rad; ◇, 1,000 rad; ○, 2,000 rad; □, 3,000 rad. Modifications that exceeded the usually accepted levels of significance ($P < 0.05$) are indicated by the solid symbols (Adapted from Hopewell 1975).

acute improvements in vascular function are related to a hyperemic response of the vasculature to primary epithelial cell killing. By two to three months after irradiation, the blood volume of the pouch was reduced. Here again the effect was transient, the blood volume per unit mass of tissue having returned to normal by 12 months. This phenomenon was again related to parenchymal atrophy, the weights of irradiated pouches being significantly less than those of controls as a late consequence of irradiation (Hopewell 1975).

The reduction in vascular volume two to three months after irradiation with doses of 2,000 rad and above was associated with the appearance of focal occlusive changes in vessels (Figure 8). Preliminary quantitative morphological studies would suggest a reduction in the capillary network at this time (Gunn, personal communication). The histological information as to the nature of these occlusive changes has still to be established, although careful histological studies have produced evidence of occlusive endothelial changes in the vasculature of the brain (Hopewell 1974), heart (Fajardo and Steward 1971), and kidney (Figure 9). Suggestions (Casarett 1964) that these vascular lesions are the cause of the marked tubular atrophy that is produced in the kidney three to six months after irradiation have still to be verified.

Further indirect evidence as to the transient nature of the late impairment in the vasculature, which is such a feature of the functional studies, has been provided by investigations into the "tumor bed" effect. Experiments involving the transplantation of an epithelial tumor into the hamster cheek pouch (Hopewell and Young 1978) or the subcutaneous implant of a mouse melanoma (Kummermehr and Holler 1978) at various times after irradiation of the host vasculature revealed periods when the growth of the transplanted tumor was reduced. These coincided with times when it was anticipated that the density of the host vasculature was reduced. A comparable explanation was offered for the transient dose-dependent reduction in the ability of the bone marrow to support colony-forming units (Nelson et al. 1977). The time period over which these

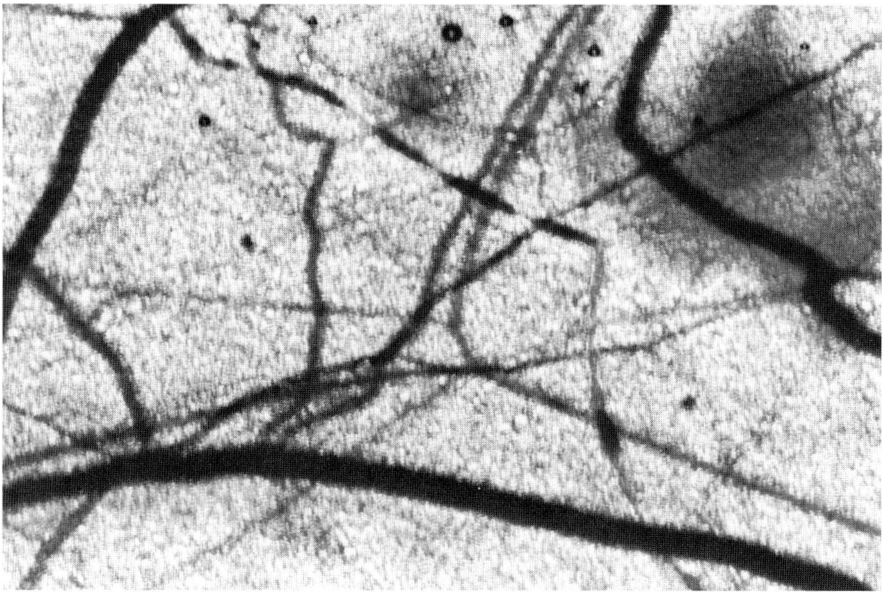

FIG. 8. Localized constrictions in a small arteriole in the cheek pouch of a hamster, three months after local irradiation with 2,500 rad. Magnification ×10. (Photograph supplied by Dr. Yvonne Gunn.)

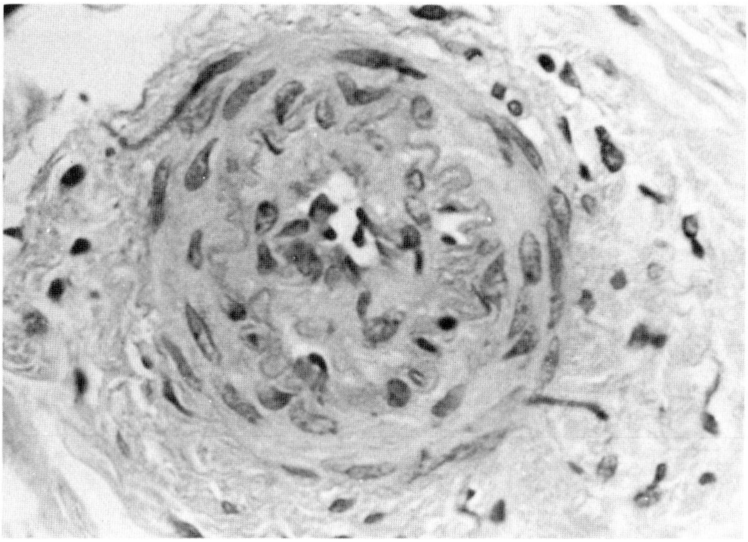

FIG. 9. Arteriole in the renal medulla of a pig kidney two months after local irradiation with a single dose of 1,050 rad. The vessel is occluded by endothelial cells (H & E, ×700).

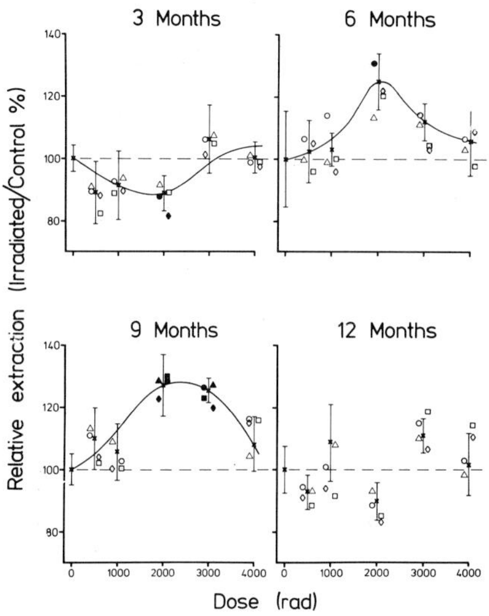

FIG. 10. Dose-related changes in the relative extraction of iodoantipyrine in different brain regions 3–12 months after local irradiation. Error bars indicate the standard error of the mean for the whole brain. Solid symbols show significant modifications ($P < 0.05$) in the different brain regions when compared with age-matched controls. X, whole brain; △, cerebral hemispheres; ○, mid-brain; □, cerebellum; ◇, brain stem. (Reproduced from Moustafa and Hopewell 1979b, with permission of the *British Journal of Radiology.*)

changes were found, three to six months after irradiation, was similar to those reported in other systems. Secondary degeneration of the bone marrow after local irradiation with 2,000 rad has also been reported after a similar time period, a phenomenon again seen as an effect on the vascular system (Sams 1965).

Functional studies, utilizing the extraction of iodoantipyrine, have suggested a small reduction in cerebral blood flow in the brain three months after a dose of 2,000 rad (Moustafa and Hopewell 1979b). However, unlike other normal tissues the tracer extraction ratio did not immediately return to normal, nor was there any evidence of tissue atrophy. Measurements of cortical thickness indicated no modification by irradiation (Keyeux et al. 1976). In the period after irradiation, there was, surprisingly, a transient increase in the tracer extraction ratio after doses of 2,000 and 3,000 rad (Figure 10); this phenomenon is now believed to be associated with a regulatory response to tissue hypoxia (Moustafa and Hopewell 1979b). This regulatory capacity of the brain circulation has also been demonstrated in rats breathing increasing concentrations of carbon dioxide (Keyeux and Ochrymowicz-Bemelman 1978). In the irradiated brains the effect was greatest in the mid-brain.

While the final link between this transient increase in iodoantipyrine extraction and the appearance of characteristic focal vascular lesions in the brain (Hopewell and Wright 1970, Hopewell 1979) is unknown, it is possible to speculate that the lesions are a result of the random failure of the vasoregulatory reaction. The greater oxygen requirement of the mid-brain as compared with the cerebral hemispheres could account for the more marked functional effects noted in this region and for the subsequent appearance of a greater number of vascular lesions in the mid-brain (Reinhold and Hopewell 1979). The time course for the appearance of these late vascular lesions was comparable with the appearance of similar lesions in the spinal cord (van der Kogel 1980, see pages 461 to 470, this volume).

The lack of any functional vascular changes after a dose of 4,000 rad is in keeping with previous histological findings (Hopewell and Wright 1970). With these higher doses radiation damage would appear to be mediated through the neuroglia both in the brain (Hopewell and Wright 1970, Hopewell 1979) and spinal cord (van der Kogel 1979). The final tolerance dose is, however, limited by damage to the vascular system.

CONCLUSIONS

Through a multidisciplinary approach, evidence of a reduction in vascularity was found in a number of normal tissues after a latent period of approximately three months. This impairment in vascularity was transient; it returned to normal as late tissue damage developed. The only exception appeared to be the brain, where a vasoregulatory response may subsequently be involved.

Histological studies in pig skin and other normal tissues have demonstrated focal occlusive changes in arterioles, and it is suggested that this may be the primary vascular lesion responsible for the development of late normal tissue damage.

ACKNOWLEDGMENTS

I am endebted to my colleagues, Drs. Foster, Gunn, Moustafa, Patterson, Young, and Wiernik, for allowing me to reproduce their own and our joint data, to Mr. John Wilkinson for preparing the illustrations, and to Miss Carol Evans for typing the manuscript. Many of the investigations involving the examination of radiation effects on pig skin and hamster cheek pouch are supported by the Cancer Research Campaign to whom the author gratefully acknowledges personal support. The cerebral blood flow studies formed part of a multidisciplinary research project carried out as part of the program of the European Late Effects Project Group (EULEP).

REFERENCES

Casarett, G. W. 1964. Pathology of single intravenous doses of polonium. Radiat. Res. Suppl. 5:246–321.

Donovan, W. E. 1975. Experimental models in skin flap research. *in* Skin Flaps, W. C. Grabb and M. B. Myers, eds. Little Brown and Co., Boston, pp. 11–20.

Fajardo, L. F., and J. R. Steward. 1971. Capillary injury preceding radiation induced myocardial fibrosis. Radiology 101:429–433.

Gauwerky, F., and F. Langheim. 1978. Der zeitfaktor bei der strahleninduzierten subkutanen fibrose. Strahlentherapie 154:608–616.

Hopewell, J. W. 1974. The late vascular effects of radiation. Br. J. Radiol. 47:157–158.

Hopewell, J. W. 1975. Early and late changes in the functional vascularity of the hamster cheek pouch after local x-irradiation. Radiat. Res. 63:157–165.

Hopewell, J. W. 1979. Late radiation damage to the central nervous system—A radiobiological interpretation. Neuropath. Appl. Neurobiol. 5:329–343.

Hopewell, J. W., J. L. Foster, Y. Gunn, H. F. Moustafa, T. J. S. Patterson, G. Wiernik, and C. M. A. Young. 1978. Role of vascular damage in the development of late radiation effects in the skin, *in* Late Biological Effects of Ionising Radiation, vol. 1. IAEA, Vienna, pp. 483–492.

Hopewell, J. W., J. L. Foster, C. M. A. Young, and G. Wiernik. 1979. Late radiation damage to pig skin. The effects of overall treatment time and number of fractions. Radiology 130:783–788.

Hopewell, J. W., and E. A. Wright. 1970. The nature of latent cerebral irradiation damage and its modification by hypertension. Br. J. Radiol. 43:161–167.

Hopewell, J. W. and C. M. A. Young. 1978. Changes in the microcirculation of normal tissues after irradiation. Int. J. Radiat. Oncol. Biol. Phys. 4:53–58.

Jardine, J. H., D. H. Hussey, D. D. Boyd, G. L. Raulston, and T. J. Davidson. 1975. Acute and late effects of 16- and 50-$MeV_{d \rightarrow Be}$ neutrons on the oral mucosa of rhesus monkeys. Radiology 117:185–191.

Keyeux, A., and D. Ochrymowicz-Bemelmans. 1978. Late response of the cerebral circulation to x-irradiation of the brain in the rat, *in* Late Biological Effects of Ionizing Radiation, vol 2. IAEA, Vienna, pp. 251–260.

Keyeux, A., H. S. Reinhold, J. W. Hopewell, G. B. Gerber, W. Calvo, and J. R. Maisin. 1976. A multi disciplinary approach to radiation late effects in the brain circulatory system: First results. Bibl. Anat. 15:326–330.

Kummermehr, J., and E. Holler. 1978. The effects of tumour bed irradiation on growth and histogenesis of Harding-Passey melanoma transplants. Proceedings of 14th Annual Meeting of the European Society of Radiation Biology (Julich).

Moustafa, H. F., and J. W. Hopewell. 1976. The evaluation of an isotopic clearance technique for the measurement of skin blood flow in the pig. Microvas. Res. 11:147–153.

Moustafa, H. F., and J. W. Hopewell. 1979a. Blood flow clearance changes in pig skin after single doses of x-rays. Br. J. Radiol. 52:138–144.

Moustafa, H. F., and J. W. Hopewell. 1979b. Late functional changes in the vasculature of the rat brain after local x-irradiation. Br. J. Radiol. (in press).

Nelson, D. F., J. T. Chaffey, and S. Hellman. 1977. Late effects of x-irradiation on the ability of mouse bone marrow to support hematopoiesis. Int. J. Radiat. Oncol. Biol. Phys. 2:39–45.

Patterson, T. J. S. 1968. The survival of skin flaps in the pig. Brit. J. Plastic Surg. 21:113–117.

Reinhold, H. S. and Hopewell, J. W. 1979. Late changes in the architecture of blood vessels of the rat brain after irradiation. Bri. J. Radiol. (in press).

Sams, A. 1965. The long term effects of 2000 R of x-rays on the bone marrow of the mouse tibia. Br. J. Radiol. 38:914–919.

Stone, R. S. 1948. Neutron therapy and specific ionization. Am. J. Roentgenol. 59:771–785.

van der Kogel, A. J. 1980. Mechanisms of late radiation injury in the spinal cord, *in* Radiation Biology in Cancer Research (The University of Texas System Cancer Center 32nd Annual Symposium on Fundamental Cancer Research), R. E. Meyn and H. R. Withers, eds. Raven Press, New York, pp. 461–470.

Wiernik, G., T. J. S. Patterson, and R. J. Berry. 1974. The effect of fractionated dose-patterns of x-irradiation on the survival of experimental skin flaps in the pig. Br. J. Radiol. 47:343–345.

Young, C. M. A. 1978. Functional and morphological changes in the dermis of pig skin following surgery and x-irradiation. D. Phil. Thesis, University of Oxford, United Kingdom.

Young, C. M. A., and J. W. Hopewell. 1979. Evaluation of an isotopic clearance technique in pig skin: Correlation of functional and histological changes. Microvas. Res. (in press).

Radiation Biology in Cancer Research, edited by
Raymond E. Meyn and H. Rodney Withers.
Raven Press, New York © 1980.

Mechanisms of Late Radiation Injury in the Spinal Cord

A. J. van der Kogel

Radiobiological Institute of the Organization for Health Research TNO, Rijswijk, The Netherlands

In studies concerning the mechanisms of development of radiation damage in various tissues, especially late effects, a fundamental problem is the assessment of the type of cell that is primarily involved. For the central nervous system, two main hypotheses have been advanced to account for the pathogenesis of radiation encephalopathy or myelopathy. According to one hypothesis, damage to the glial cell population is regarded as the primary stage in development of the lesions, while the other hypothesis considers the vascular system to be the primary target.

In most reports on human radiation myelopathy, it was concluded that the lesions were of vascular origin, while a minority of the authors regarded the vascular lesions as secondary to white matter necrosis (reviewed by Jellinger and Sturm 1971). The latter suggestion was confirmed by a number of experimental studies, leading to thc conclusion that white matter necrosis as observed in radiation myelopathy is primarily due to damage to the oligodendroglial cells (Carsten and Zeman 1966, Bradley et al. 1977). However, most of the reports have neglected the possibility that the histopathological changes might depend on the dose level. Most authors used only a single dose, which was twice as high as the threshold for the induction of paralysis. Hopewell and Wright (1970), in a study on the rat brain, used various doses and suggested that the sudden death occurring after 2,000–3,000 rad was due to vascular damage, while glial cell depletion led to white matter necrosis after doses of 3,000–4,000 rad.

It will be shown here that different pathological mechanisms are involved in the induction of paralysis, depending on the dose and time after irradiation. The suggestion will be made that different cell types are involved as target cells, and the implications for the radiation tolerance of the spinal cord and the time-dose-fractionation relationships will be discussed.

MATERIALS AND METHODS

Male WAG/Rij rats, 12–14 weeks of age, were irradiated locally in the region of the cervical spinal cord (region C_5-T_2) with 300-kV X rays. Irradiations and dosimetry were carried out as described previously (van der Kogel and Barendsen 1974). For experiments employing up to 10 daily fractions, the rats were anesthetized with Nembutal (60 mg/kg). In multifractionation experiments with up to 60 fractions (two fractions/day with an interval of six hours), inhalation anesthesia with 1% enflurane/99% O_2 was used. Using this method, the anesthetic state is reduced to 10 minutes as compared to about two hours with Nembutal.

The rats were weighed and examined every one to two weeks, but this was extended to daily examination when they showed signs of disease or neurological damage. The development of paralysis was scored by observing movements of a rat inside its cage or on a smooth surface and by simple tests for reflexes and muscular strength. Various grades of increasing severity have been distinguished (van der Kogel 1977). Most lesions were progressive, although the rate of development was dose dependent and differed for the various syndromes to be described. The rats were observed until definite myelopathy had developed and were subsequently killed by perfusion fixation with phosphate-buffered formalin while under ether anesthesia. The spinal cords of all rats were examined histologically.

RESULTS

Histopathological Observations

White matter demyelination and necrosis

After irradiation with single doses in the range of 2,000–4,000 rad, the latent period decreases with increasing dose from about 200 days to 120 days. In rats killed after definite symptoms of paralysis had developed, the predominant lesion was demyelination and necrosis of the white matter (Figure 1). The lesions were irregularly distributed over the white matter of the irradiated segments, with a predominance in the lateral and ventral columns. Around most necrotic areas there was a slight glial reaction, with hypertrophied astrocytes. In some cases, cysts filled with lipid macrophages ("foam cells") were observed.

Late vascular damage

After irradiation with doses below the level for induction of white matter necrosis, paralysis may develop after very long latent periods up to the end of the normal lifespan of about two years. Lesions consisted of vascular damage and were markedly different from the white matter necrosis type of lesion. The vascular damage varied from an increased capillary density and telangiectasis (dilated capillaries or venules) in neurologically normal rats to a massive hemorrhage in acutely paralyzed rats. Between these extreme stages, various grades

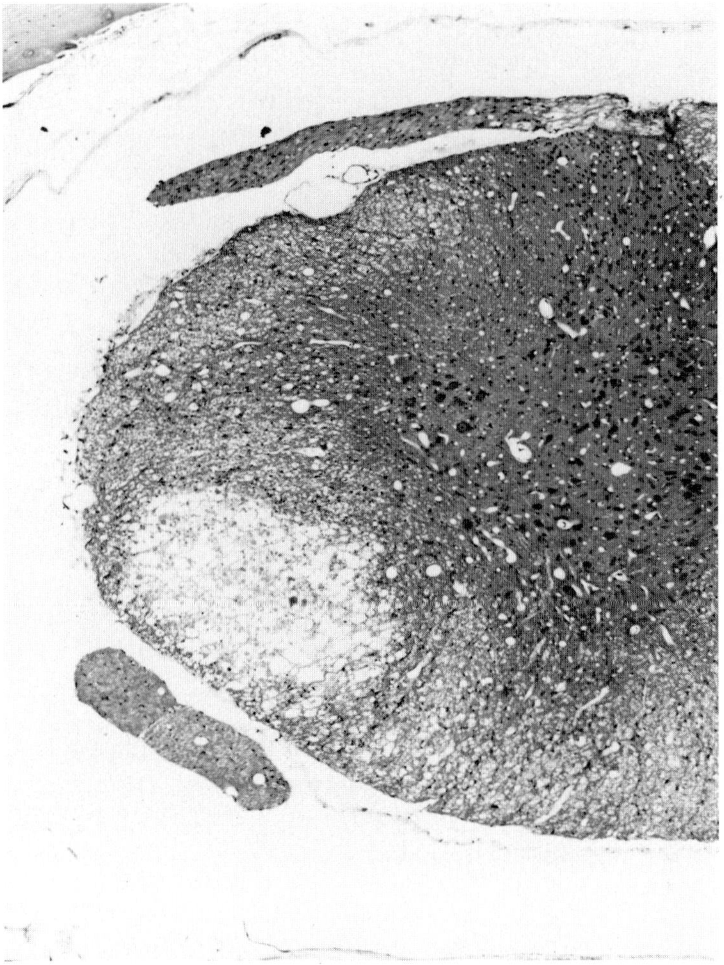

FIG. 1. White matter necrosis. 164 days after 2 × 1,500 rad (1-week interval) (×80).

of lesions were observed, and three main categories of damage were arbitrarily designated:

I. Increased capillary density, in both white and gray matter, with some dilated capillaries or venules (Figure 2). In some cases, microthrombi and petechial hemorrhages in combination with moderate gliosis are present. The reactive cells are mainly hypertrophic astrocytes and small round macrophages, which are probably microglial cells. Lesions of this type are observed in rats that do not show paralytic symptoms and are killed at the approximate end of their lifespan.

II. Extensive telangiectases with focal hemorrhage and fibrin deposits (Figure

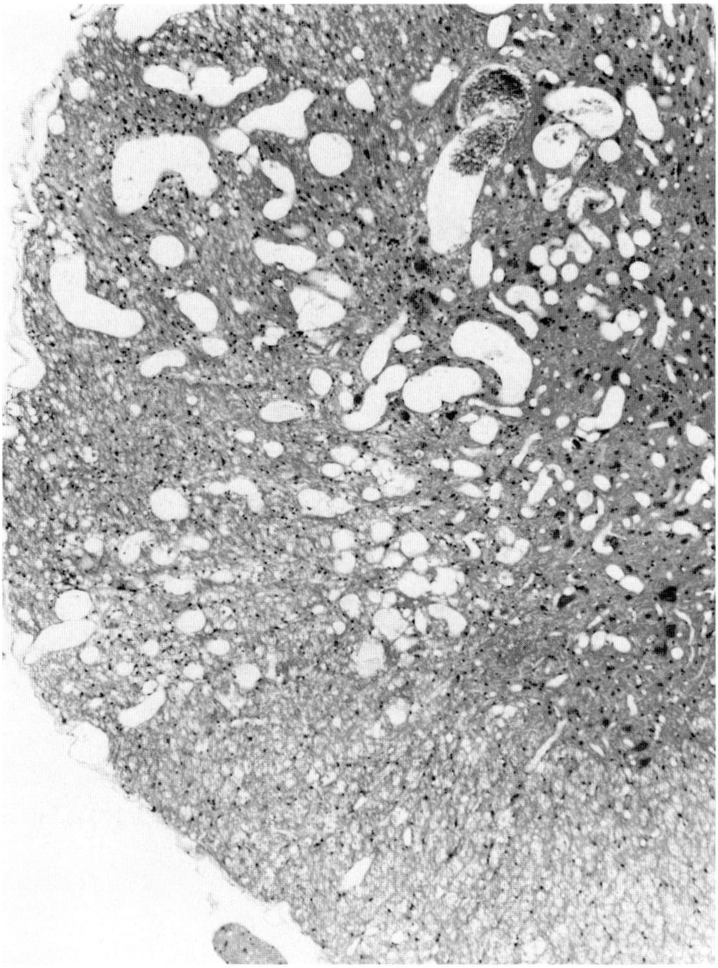

FIG. 2. Increased capillary density and telangiectasis in white and gray matter. 443 days after 1,850 rad (×80).

3). In addition to a strong reaction by endogenous cell types (astrocytes, microglial cells) or hematogenous monocytic infiltrates, inflammation with neutrophilic leukocytes is observed, especially in cases with extensive fibrin deposits and hemorrhages. In areas with a strong cellular reaction, other cell types of interest are monstrous astrocytes with large nuclei and macrophages laden with erythrocytes or hemosiderin pigment. The lesions of this type represent a slowly progressing, chronic process of vascular degeneration. The distribution varies from small foci to involvement of a number of cord segments over a large area. This is reflected in a large variability in neurological damage, some rats showing no

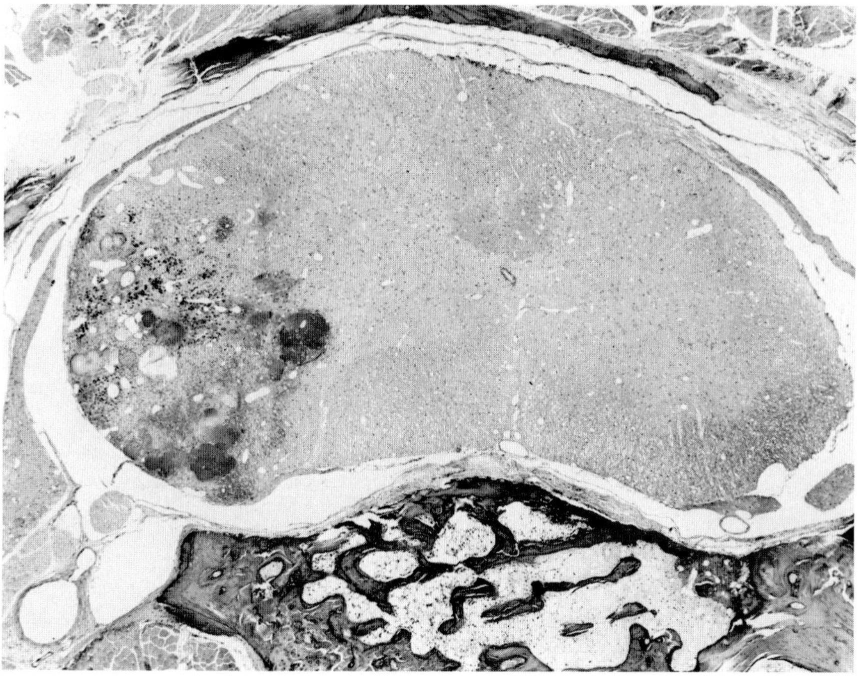

FIG. 3. Chronic vascular damage with telangiectasis, numerous thrombi, and strong inflammatory response. 325 days after 2 × 1,200 rad (1-day interval) (×32).

symptoms at all, and others having a slight or severe paralysis that is unilateral in most cases.

III. Instead of the slowly progressing, chronic lesions, a large, acutely developing hemorrhage extending over a number of cord segments (Figure 4). This leads to sudden paralysis due to invasion and compression of the surrounding tissue. Depending on the time course of the process, a cellular reaction is absent or moderate.

For those animals developing paralysis at one year or later after irradiation, it is especially important to determine the histological type of damage, because the neurological symptoms may also be due to invasion and compression of the cord by tumors induced in the surrounding irradiated vertebrae and muscular tissue.

Time-Dose-Isoeffect Relationships

In addition to their histological characteristics, the two main categories of damage (white matter necrosis and vascular damage) can be distinguished on the basis of the time of development, with the time threshold at about 7 months

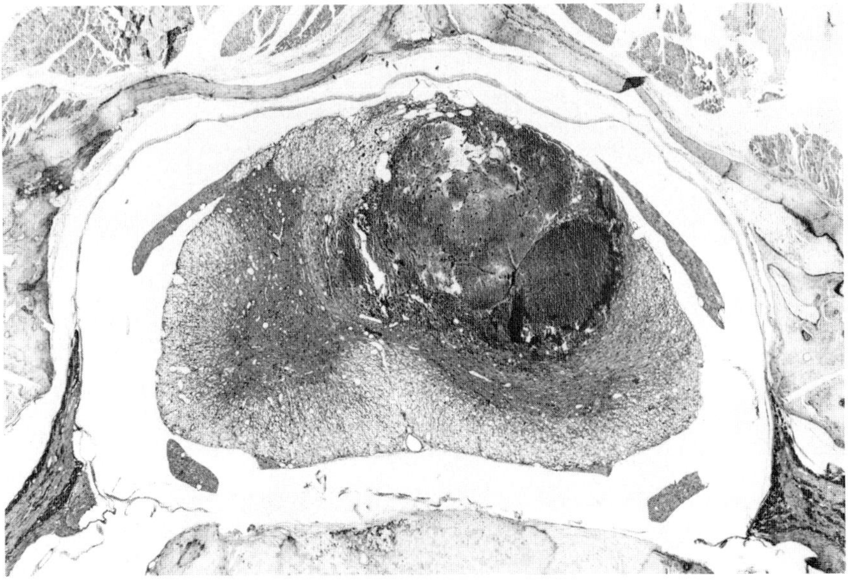

FIG. 4. Acute hemorrhage causing compression of the surrounding tissue. 221 days after 10 × 500 rad (×32).

after irradiation. Thus, by separation in time, dose-response curves can be constructed, representing a specific lesion, presumably correlated with damage to a specific target cell.

Dose-response curves for single-dose and two-fraction experiments are shown in Figure 5. Some of the curves were constructed by computer probit analysis and are accompanied by the 95% confidence intervals, but sigmoid curves were fitted by eye to the data, which did not fulfill the requirements for probit analysis. The curves in A represent paralysis induction due to white matter necrosis, which occurs within seven months after irradiation, while the curves in B represent the induction of neurological symptoms due to vascular damage. From the dose-response curves, isoeffect doses can be derived, e.g., the ED_{50}, a dose at which 50% of a group of rats develop paralysis or a histologically defined specific type of damage. In Figure 6, ED_{50} values for the induction of paralysis due to white matter necrosis are plotted versus the number of fractions of equal magnitude on a double logarithmic scale (curve 1). A straight line is fitted by least square analysis for 1–10 fractions, with a slope of 0.46 (correlation coefficient 0.996). At the highest doses applied in the experiments with 30 and 60 fractions (8.000 and 10.800 rad, respectively), no paralysis was observed at less than one year postirradiation. For the vascular damage, it was only for single-dose and two-fraction experiments that complete dose-response curves could be constructed so that ED_{50} values could be obtained. For the experiments with 5,

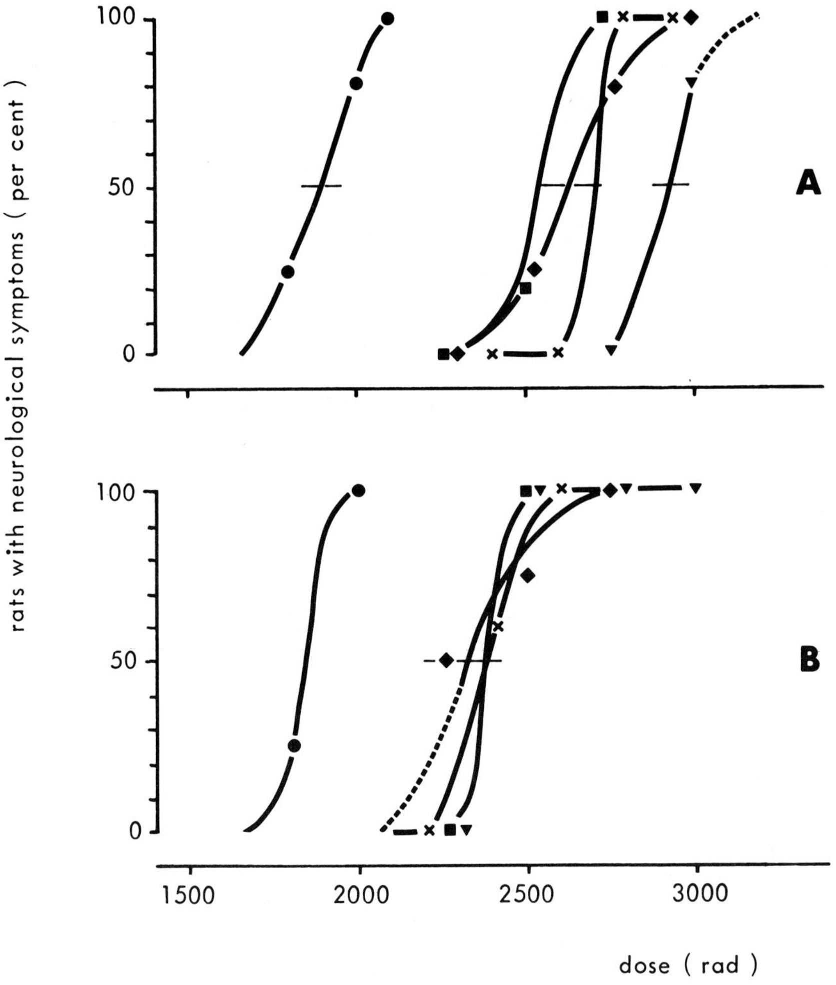

FIG. 5. Dose-response curves for two types of lesions in the cervical spinal cord. A, White matter necrosis (latent period < 7 months). B, Vascular damage (latent period 8–18 months). ●, single dose; **X**, 2 equal fractions—1-day interval; ■, 1-week interval, 2 fractions (1st fraction = 1,500 rad); ◆, 8-week interval, 2 fractions (1st fraction = 1,500 rad); ▼, 16-week interval, 2 fractions (1st fraction = 1,500 rad).

10, and 60 fractions, threshold doses were approximated by using the mean values between the highest dose at which no vascular damage was observed for at least one year postirradiation and the lowest dose at which vascular damage was observed. A straight line is fitted for the ED_{50} values and the approximate threshold dose; it has a slope of 0.42 (correlation coefficient 0.999). Curve 2 thus represents the minimal tolerance level, below which no rats of the WAG/Rij strain are expected to develop any type of damage during their

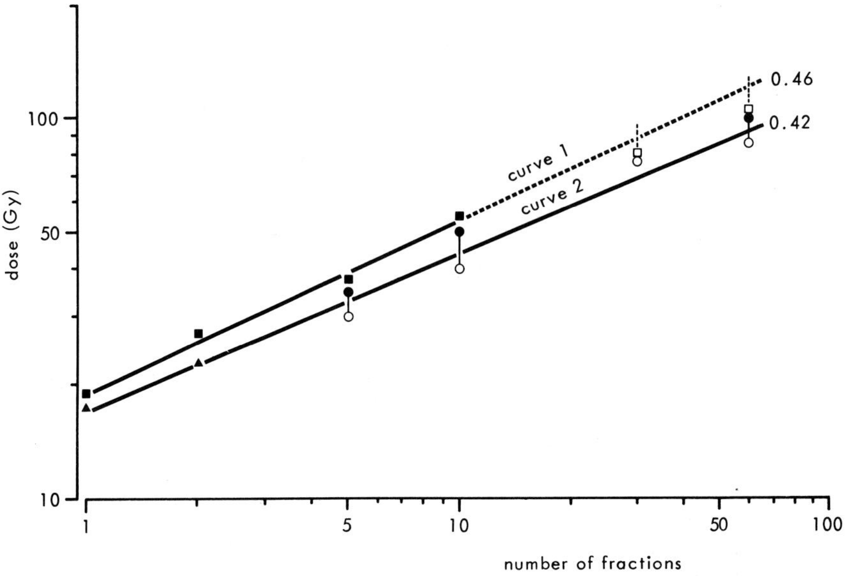

FIG. 6. Dose-fractionation relationships for different types of lesions in the cervical spinal cord. Curve 1, White matter necrosis. ■, ED_{50}; □, Highest dose used, no white matter necrosis observed. Curve 2, vascular damage. ▲, ED_{50}; ●, lowest dose at which vascular damage is observed; O, highest dose with no vascular damage.

lifespan. These values are found to be lower than the ED_{50} data for white matter necrosis.

The influence of the overall time has been estimated in split-dose experiments with various time intervals (Figure 5). With intervals of up to 56 days, there is no significant difference in the ED_{50} for white matter necrosis. However, in the experiment with an interval of 112 days, significant additional recovery equivalent to a dose (ED_{50}) of about 300 rad was observed. This increase in ED_{50} is not seen for the induction of late vascular damage.

DISCUSSION

Irradiation of the rat cervical spinal cord can lead to at least two specific syndromes, depending on dose and time after irradiation. The early type is characterized by a rapid progression of symptoms and a latent period inversely related to the dose. This type of damage, and its predominance in the white matter, has been described for both the spinal cord and the brain of rats (Innes and Carsten 1962, Hopewell and Wright 1975), rabbits (Berg and Lindgren 1958), and monkeys (Innes and Carsten 1962), and is associated with doses above the therapeutic range. The pathogenesis suggests the presence of a primarily demyelinating lesion, for which the oligodendrocyte is regarded as the affected cell type (van der Kogel 1979). This is supported by recent observations of several authors on the pathogenesis and the kinetics of this type of damage (Hubbard 1977, Mastaglia et al. 1976).

The late vascular damage develops after irradiation with doses in the range of clinical radiotherapy. A general aspect of these lesions is an increased density of capillaries and the occurrence of telangiectasia. The presence of petechial hemorrhages might be regarded as an initial stage in the development of the more severe chronic vascular damage or the acute hemorrhages. Because the lesions apparently start in the capillaries, the endothelial cell is the most probable target.

The slowly progressing chronic lesions probably represent the most common type of late radiation damage in the human spinal cord, but four other syndromes described in humans are also observed in rats (van der Kogel 1979). This suggests the presence of common mechanisms for the induction of damage, which also emphasizes the suitability of the rat spinal cord as an animal model for predictions in man.

The dose-response curves shown in Figure 5 for the two types of damage may be assumed to represent damage in different types of target cells, namely the oligodendrocyte and the endothelial cells. The capacity for repair of subeffective damage is determined by comparing the ED_{50} of the single-dose experiment with the two-fraction experiment (one-day interval). The D_2–D_1 value is about 750 rad for the white matter necrosis and about 550 rad for the late vascular damage. This suggests that the two types of target cells have different repair capacities. That the lower D_2–D_1 value does not necessarily predict a strongly reduced repair capacity with multiple small fractions is reflected by the slope of 0.42 of the curve for vascular damage as compared with 0.46 for white matter necrosis (Figure 6).

In addition to the great capacity for repair of subeffective damage, additional time-dependent recovery is observed in the two-fraction experiment with an interval of 16 weeks and equivalent to a dose of about 300 rad. This type of recovery appears to be involved in only the white matter necrosis syndrome, indicating another difference between the two types of target cells.

TABLE 1. *Characteristics of radiation-induced lesions in the cervical spinal cord*

	White matter necrosis	Vascular damage
Dose range (single doses)	2,000 — 4,000 rad	$<$ 2,000 rad
Latent period	inversely related to the dose (210 → 120 days)	variable, 220–550 days
Possible target cell	oligodendrocyte	endothelium
Repair (D_2–D_1)	750 rad	550 rad
Repopulation	starting after about 8 weeks	not observed for intervals $\leq$ 16 weeks

The delayed onset of the time-dependent recovery suggests that repopulation, and not a process of slow repair that starts immediately upon irradiation, takes place (Field and Hornsey 1977). This is supported by observations on the rat lumbar spinal cord (White and Hornsey 1978) showing a delay of between 15 and 32 days before the start of additional recovery. The characteristics of the two syndromes observed in the cervical cord are summarized in Table 1.

It can be concluded that at least two mechanisms are involved in the development of radiation myelopathy and that these differ in repair characteristics and time-dose relationships.

REFERENCES

Berg, N. O., and M. Lindgren. 1958. Time-dose relationship and morphology of delayed radiation lesions of the brain in rabbits. Acta Radiol., Suppl. 167:1–118.

Bradley, W. G., J. D. Fewings, W. J. K. Cumming, and R. M. Harrison. 1977. Delayed myeloradiculopathy produced by spinal X-irradiation in the rat. J. Neurol. Sci. 31:63–82.

Carsten, A. L., and W. Zeman. 1966. The control of variables in radiopathologic studies on mammalian nervous tissue. Int. J. Radiat. Biol. 10:65–74.

Field, S. B., and S. Hornsey. 1977. Repair in normal tissues and the possible relevance to radiotherapy. Strahlentherapie 153:371–379.

Hopewell, J. W., and E. A. Wright. 1970. The nature of latent cerebral irradiation damage and its modification by hypertension. Br. J. Radiol. 43:161–167.

Hopewell, J. W., and E. A. Wright. 1975. The effects of dose and field size on late radiation damage to the rat spinal cord. Int. J. Radiat. Biol. 28:325–333.

Hubbard, B. M. 1977. Late effects of ionising radiation on the central nervous system of the rat. Ph.D. Thesis, University of Oxford.

Innes, J. R. M., and A. Carsten. 1962. A demyelinating or malacic myelopathy and myodegeneration—Delayed effect of localized X-irradiation in experimental rats and monkeys, *in* Response of the Nervous System to Ionizing Radiation, T. J. Haley and R. S. Snider, eds. Academic Press, New York, pp. 233–247.

Jellinger, K., and K. W. Sturm. 1971. Delayed radiation myelopathy in man. J. Neurol. Sci. 14:389–408.

Mastaglia, F. L., W. I. McDonald, J. V. Watson, and K. Yogendran. 1976. Effects of X-radiation on the spinal cord: An experimental study of the morphological changes in central nerve fibres. Brain 99:101–122.

van der Kogel, A. J. 1977. Radiation tolerance of the rat spinal cord: time-dose relationships. Radiology 122:505–509.

van der Kogel, A. J. 1979. Late effects of radiation on the spinal cord. Dose-effect relationships and pathogenesis. Ph.D. Thesis, University of Amsterdam. Publication of the Radiobiological Institute TNO, Rijswijk, The Netherlands.

van der Kogel, A. J., and G. W. Barendsen, 1974. Late effects of spinal cord irradiation with 300 kV X-rays and 15 MeV neutrons. Br. J. Radiol. 47:393–398.

White, A., and S. Hornsey. 1978. Radiation damage to the rat spinal cord: the effect of single and fractionated doses of X-rays. Br. J. Radiol. 51:515–523.

Radiation Biology in Cancer Research, edited by
Raymond E. Meyn and H. Rodney Withers.
Raven Press, New York © 1980.

Acute and Late Normal Tissue Effects of 50 $MeV_{d \rightarrow Be}$ Neutrons

David H. Hussey, Chester A. Gleiser,* John H. Jardine,* Gilbert L. Raulston,* and H. Rodney Withers

*Department of Radiotherapy and *Section of Experimental Animals, The University of Texas System Cancer Center M. D. Anderson Hospital and Tumor Institute, Houston, Texas 77030*

In October 1972, M. D. Anderson Hospital initiated a pilot study of fast-neutron therapy using the Texas A&M variable energy cyclotron (TAMVEC). During the early phase of this program, the acute and late effects of fast neutrons were investigated by irradiating a variety of organ systems in large animals using dosage schedules similar to those that have been employed clinically. The organs included: (1) pig skin (Withers et al. 1977, 1978), (2) rhesus monkey oral mucosa (Jardine et al. 1975), (3) rhesus monkey spinal cord (Jardine et al., in press), and (4) rhesus monkey kidney (Raulston et al. 1978).

This paper is a review of the large-animal normal-tissue studies. The specific objectives are: (1) to summarize the results in terms of the relative biological effectiveness (RBE) of 50 $MeV_{d \rightarrow Be}$ neutrons and the tolerance of normal tissues to fractionated neutron irradiation; and (2) to compare the experimental results with the clinical observations in the pilot study at TAMVEC. The RBE for acute and late effects are discussed in some detail since clinical impressions in pilot studies (Hussey et al. 1975, Ornitz et al., in press) have indicated that the late effects of neutrons are more severe than would be predicted on the basis of the acute reactions observed.

Dosimetric Considerations

The animals were irradiated with ^{60}Co γ rays at M. D. Anderson Hospital or Texas A&M University, or with neutrons using TAMVEC. Neutrons were produced by bombarding a thick beryllium target with 50 MeV deuterons (50 $MeV_{d \rightarrow Be}$). This beam has depth-dose properties similar to those of 4 MV X rays. The depth of maximum dose (D_{max}) is 1.0 cm, and the beam is attenuated to 50% of D_{max} at 13.8 cm (10 × 10 cm field).

The γ-ray doses were determined by standard dosimetry techniques. The neutron doses were measured with a 0.1 cm^3 tissue-equivalent ionization chamber immersed in tissue-equivalent liquid ($\rho = 1.07$). The neutron beam doses have been expressed in rads including both the neutron and gamma components ($rad_{n\gamma}$). The source-skin distances were 80 cm with ^{60}Co and 140 cm with neutrons. The dose rates differed with each study, but ranged from 120 to 180 rad/min with ^{60}Co and from 60 to 90 $rad_{n\gamma}$/min with neutrons.

Pig Skin

Twenty-four miniature pigs each weighing approximately 40 pounds were used to evaluate the acute and late effects on skin. Four fields measuring 8 × 8 cm were tattooed on each animal. Three fields were irradiated, and the fourth served as a control to compensate for growth during the experiment. The portals were irradiated with: (1) ^{60}Co, twice weekly, (2) ^{60}Co, five times weekly, (3) neutrons, twice weekly, or (4) neutrons, four times weekly. The doses are outlined in Table 1. The total duration of treatment was 6½ weeks. The neutron portals were bolused with 10 mm of lucite and the ^{60}Co portals with 5 mm of lucite in order to eliminate skin sparing.

The acute responses were scored on a scale of 0 to 10 based on a subjective assessment of severity (Withers et al. 1977). A score of 5 indicated a patchy, moist desquamation. The acute reactions were evaluated by two to five observers at least twice weekly during the course of treatment and for 11 days following completion of irradiation.

The late responses were quantitated by the degree of contraction of the irradiated fields relative to the control fields (Withers et al. 1977). A subjective scoring system based on epidermal atrophy, epilation, induration, and ulceration was also used, and it gave results similar to those obtained with measurements of contraction. The late reactions were scored twice monthly for eight months following completion of irradiation.

Results

The results of the pig skin experiment are summarized in Figure 1. The acute reactions are shown as dashed lines, and the late reactions, i.e., contraction of the irradiated fields, as solid lines. For both neutron schedules and the ^{60}Co twice-weekly fractionation schedule, the acute reactions were almost imperceptible (erythema to dry, coarse scaliness).

The late effects of neutrons were severe relative to the acute effects. After even the highest neutron dose, there was minimal acute skin reaction, but the later contraction was more severe than that observed in either of the ^{60}Co γ-ray groups.

The dose response for late effects with neutrons was independent of the number of fractions (over a range of two to four fractions per week, 75 to 200 $rad_{n\gamma}$ per fraction). This means that when changing treatment schedules from twice weekly to four times weekly one can expect to see the same late reactions if the same total dose and overall time are employed.

TABLE 1. *Doses per fraction, total doses, and number of fields treated with γ rays or neutrons given in different numbers of fractions per week for 6½ weeks*

Cobalt-60						Neutrons					
Twice Weekly			5 Times Weekly			Twice Weekly			4 Times Weekly		
Dose/Fx (rad)	Fields	Total Dose (rad)	Dose/Fx (rad)	Fields	Total Dose (rad)	Dose/Fx ($rad_{n\gamma}$)	Fields	Total Dose ($rad_{n\gamma}$)	Dose/Fx ($rad_{n\gamma}$)	Fields	Total Dose ($rad_{n\gamma}$)
360	4	4680	180	4	5760	150	2	1950	75	2	1950
385	4	5015	200	4	6400	160	3	2080	80	3	2080
410	4	5330	220	4	7040	170	3	2210	85	3	2210
435	5	5655	240	5	7680	180	3	2340	90	3	2340
460	4	5980	260	4	8320	190	3	2470	95	3	2470
						200	2	2600			

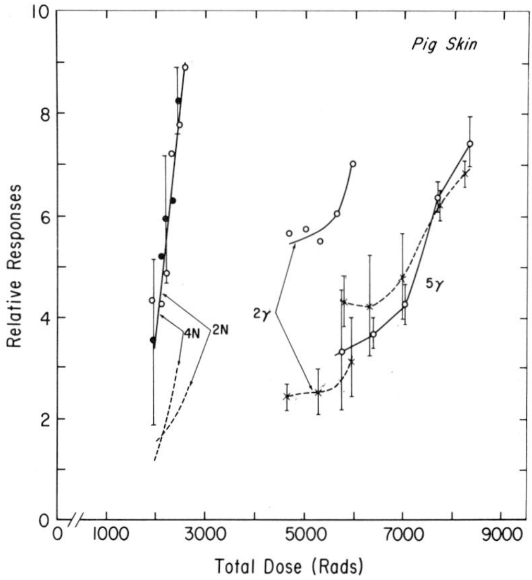

FIG. 1. Dose-response relationships for pig skin. Contraction of pig skin five to eight months after irradiation is shown as solid lines, and the acute response as dashed lines. In only the five-fraction-per-week γ-ray group do the acute and late responses parallel one another. In the two-fraction-per-week ^{60}Co animals, and both neutron groups, the late effects are relatively more severe than the acute effects. The methods used for quantitating responses are described in the text. The errors shown are a representative sample of the standard errors. (Reproduced from Withers et al. 1977, with permission of Pergamon Press.)

The late effects of ^{60}Co γ rays given twice weekly were more severe than would be predicted on the basis of comparisons with the acute and late reactions of five-times-weekly ^{60}Co irradiation. None of the acute reactions from 13 fractions of 360 to 460 rad were as severe as those resulting from 32 fractions of 180 to 260 rad. Yet the contraction measured eight months later was, in most cases, more severe in the animals irradiated with only 13 fractions. A more complete analysis (Withers et al. 1977) shows that the slope of the isoeffect line (over a range of 180 to 460 rad per fraction) for acute response was less than would have been predicted by the Ellis NSD formula (Ellis 1968), whereas the slope of the isoeffect line for late response was greater than would have been predicted by the NSD formula.

Relative Biological Effectiveness

A direct comparison of the data in Figure 1 indicates that the RBE for late effects (skin contraction) with twice-weekly irradiation (~ 400 rad ^{60}Co fractions) lies in the range of 2.2 to 2.6. The RBE for late effects for five-times-weekly fractionation cannot be determined by a direct comparison of the data in Figure 1 since the neutrons were delivered only four times weekly. However, since the dose-response relationship for late injury from neutrons is the same for four-times-weekly fractionation (26 fractions) as for twice-weekly fractionation (13 fractions), it is reasonable to assume that it will also be the same for five-times-weekly fractionation (32 fractions). If this is correct, the RBE for late effects for five-times-weekly fractionation (~200 rad ^{60}Co fractions) lies in the range of 3.1 to 3.4, since a dose of 2,080 $rad_{n\gamma}$ with neutrons (two or four times weekly) produced the same amount of late skin contraction as 6,400 to 7,040 rad with ^{60}Co γ rays (five times weekly).

Since the acute reactions with neutrons were too mild, only the upper limits of the RBE for acute effects could be determined. The RBE for acute effects for twice-weekly fractionation (~400 rad ^{60}Co fractions) was <2.0, since a dose of 4,680 rad in 13 fractions with ^{60}Co γ rays produced a more severe reaction than 2,340 $rad_{n\gamma}$ in 13 fractions with neutrons. Similarly, since 5,760 rad in 32 fractions with ^{60}Co produced a more severe reaction than 2,340 $rad_{n\gamma}$ in 26 fractions with neutrons, the RBE for acute effects for five-times-weekly fractionation (~200 rad ^{60}Co fractions) was <2.5. Since the acute responses to neutrons are not necessarily independent of a change from five to four fractions per week, this estimate could change slightly if neutrons were given five times weekly.

Rhesus Monkey Oral Mucosa

Thirty mature female rhesus monkeys *(Macaca mulatta)* were used to evaluate the acute and late effects on oral mucosa. The animals were assigned to six groups of five each and irradiated with ^{60}Co γ rays or 50 $MeV_{d \rightarrow Be}$ neutrons according to the treatment schedules listed in Table 2. The dosage schedules for groups I-III were designed to deliver a dose approximately equivalent to 6,500 rad in 32 fractions in 6½ weeks with ^{60}Co γ rays. Equivalent doses had been determined by matching acute mucosal reactions in human patients treated through parallel opposing portals covering the oropharynx and faucial arch (Jardine et al. 1975). Three dose levels were used for the animals treated

TABLE 2. *Incidence of oromucosal necrosis with mandibular exposure in rhesus monkeys*

Group	Modality	Dosage Schedule	Oromucosal Necrosis
I	^{60}Co	5525 rad/13Fx/6½ weeks* (425 rad—twice weekly)	5/5
II	^{60}Co	6500 rad/32 Fx/6½ weeks* (203 rad—5x weekly)	0/5
III	Neutrons	2215 $rad_{n\gamma}$/13 Fx/6½ weeks* (170 $rad_{n\gamma}$—twice weekly)	3/5
IV	Neutrons	2000 $rad_{n\gamma}$/32 Fx/6½ weeks (62.5 $rad_{n\gamma}$—5x weekly)	0/5
V	Neutrons	2200 $rad_{n\gamma}$/32 Fx/6½ weeks (69 $rad_{n\gamma}$—5x weekly)	3/5
VI	Neutrons	2400 $rad_{n\gamma}$/32 Fx/6½ weeks (75 $rad_{n\gamma}$—5x weekly)	5/5

* The dosage schedule in groups I–III produced equivalent acute oropharyngeal mucosal reactions in patients treated at M. D. Anderson Hospital or TAMVEC.

with neutrons five times weekly (groups IV-VI) because the equivalent doses for acute mucosal reactions had not been tested for five-times-weekly fractionation. The total duration of treatment was 6½ weeks.

The monkeys were irradiated through a single right 4½ × 7 cm portal encompassing the lower one half of the oral cavity and oropharynx, the horizontal ramus of the mandible, and the submandibular triangle. The radiation doses were assessed at the axis midplane at the level of the first molar tooth.

The oral cavity was examined and photographed weekly during treatment and for six months posttreatment, and at least once a month for an additional two years. The animals were sacrificed when death secondary to oromucosal necrosis seemed imminent.

Results

The rhesus monkeys all developed equivalent acute mucosal reactions—a mild erythema. None developed the studded-to-confluent fibrinous mucositis that would be expected in human beings at this dose level. Consequently, an RBE for acute mucosal effects could not be determined.

The late effects with neutrons twice weekly were more severe than would be expected on the basis of a comparison of the acute and late reactions in human beings—using five-times-weekly ^{60}Co irradiation as a guideline. The dosage schedules used for groups II and III had produced the same acute oropharyngeal mucosal reactions in patients. Yet the incidence of oromucosal necrosis in the monkeys irradiated with neutrons (group III) was greater than that observed in the animals irradiated with ^{60}Co five times weekly (group II) (Table 2).

A fractionation effect for late injury was not observed with neutrons. The same incidence of necrosis (3/5) was observed when 2,215 $rad_{n\gamma}$ was delivered twice weekly (group III) as when 2,200 $rad_{n\gamma}$ was delivered five times weekly (group V). This observation is similar to that noted in the pig skin study.

A significant fractionation effect for late injury was observed with ^{60}Co γ rays. None of the five animals treated to a dose of 6,500 rad five times weekly (group II; NSD = 1,861 rets) developed oromucosal necrosis, whereas all five monkeys irradiated to a dose of 5,525 rad twice weekly (group I; NSD = 1,968 rets) developed oromucosal necrosis. The dosage schedules for groups I and II had produced equivalent acute mucosal reactions in patients.

Relative Biological Effectiveness

The RBE for late effects for twice-weekly fractionation (~400 rad ^{60}Co fractions) was <2.5, since 2,215 $rad_{n\gamma}$ in 13 fractions with neutrons resulted in a lower incidence of oromucosal necrosis than 5,525 rad in 13 fractions with ^{60}Co. The RBE for late effects for five-times-weekly fractionation (~200 rad ^{60}Co fractions) was >3.0 since 2,200 $rad_{n\gamma}$ in 32 fractions with neutrons produced a greater incidence of oromucosal necrosis than 6,500 rad in 32 fractions with ^{60}Co.

Rhesus Monkey Spinal Cord

The spinal cord study was undertaken because radiation myelitis had been observed clinically in pilot studies of fast-neutron therapy at other institutions (Catterall 1978, Laramore et al., in press). Twenty adult male rhesus monkeys were assigned to four groups of five each and irradiated with ^{60}Co γ rays or 50 $MeV_{d \to Be}$ neutrons using dosage schedules selected to permit an investigation of RBE values of 3.0 to 4.1 (Table 3). The total duration of treatment was 4½ weeks.

The treatments were delivered through single posterior 8 × 5 cm portals encompassing the spinal cord from C_1 to T_2. The radiation doses were assessed at the midplane of the cervical spinal cord as determined using radiographs obtained perpendicular to the treatment beam.

Neurologic function was evaluated by daily observation of behavior pattern, gait, and balance, and the mobility, tactile response, and prehensile response of the upper and lower extremities. The animals irradiated with neutrons (groups I, II) were sacrificed for postmortem examination when death due to spinal cord injury was imminent, or at twelve months in monkeys not developing neurologic signs. Motion pictures were obtained before the animals were sacrificed in order to document the presence or absence of gross neurologic abnormalities. The animals irradiated with ^{60}Co γ rays (groups III, IV) were irradiated more recently and have been followed only eight months.

Results

A significant dose-response relationship was observed in the animals irradiated with neutrons (Table 3). Whereas none of the monkeys in group I (1,300 $rad_{n\gamma}$)

TABLE 3. *Incidence of radiation myelitis in rhesus monkeys (clinical findings)*

Group	Modality	Spinal Cord Dose	Radiation Myelitis
I	Neutrons	1300 $rad_{n\gamma}$/9 Fx/4½ weeks (144 $rad_{n\gamma}$—twice weekly)	0/5
II	Neutrons	1550 $rad_{n\gamma}$/9 Fx/4½ weeks (172 $rad_{n\gamma}$—twice weekly)	5/5
III	^{60}Co	4620 rad/22 Fx/4½ weeks (210 rad—5x weekly)	0/5*
IV	^{60}Co	5390 rad/22 Fx/4½ weeks (245 rad—5x weekly)	0/5*

* Preliminary results; the animals in groups III and IV have been followed only 8 months.

showed clinical evidence of neurologic dysfunction, all five animals in group II (1,550 $rad_{n\gamma}$) developed paralysis. The earliest signs of neurologic deficit occurred following a latent period of four and a half to eight months. Over the next one and a half to three months, the animals developed quadriparesis that progressed to the point that the animals were unable to feed themselves and death due to debility was imminent. The animals were killed at this stage—six to eleven months after the start of irradiation.

Histopathologically, the lesions were most marked in the animals in group II (1,550 $rad_{n\gamma}$), all of whom exhibited severe malacia and moderate to severe demyelination of the lateral and posterior funiculi (Figure 2). The lesions in group I (1,300 $rad_{n\gamma}$) were minimal by comparison. There was almost complete absence of malacia in these animals and the demyelination was much less pronounced.

The monkeys irradiated with ^{60}Co γ rays (groups III, IV) have thus far developed no signs of neurologic dysfunction, although they have been followed for only eight months. One animal in each of these groups will be sacrificed at twelve months so that the histologic findings can be compared with those observed in the animals treated with neutrons. The remaining animals will be followed at least 24 months in order to determine the risk of neurologic injury appearing in the second year.

Relative Biological Effectiveness

If the late response of spinal cord to neutron irradiation is independent of fractionation (over a range of two to five fractions per week, 70 to 170 $rad_{n\gamma}$ per fraction), as was observed for late injury in pig skin and rhesus monkey oral mucosa, an RBE for late effects can be determined. The preliminary data indicate that the RBE for five-times-weekly fractionation (~200 rad ^{60}Co fractions) is >3.5 since all five monkeys irradiated to a dose of 1,550 $rad_{n\gamma}$/9 Fx/4½ weeks with neutrons developed significant quadriparesis, whereas none of the five animals irradiated to a dose of 5,390 rad/22 Fx/4½ weeks with ^{60}Co have developed signs of neurologic dysfunction.

Rhesus Monkey Kidney

Twenty adult rhesus monkeys were used to compare the renal effects of ^{60}Co γ rays and 50 $MeV_{d\rightarrow Be}$ neutrons. After all animals were determined to have normal renal function, unilateral nephrectomies were performed. Postnephrectomy baseline renal function studies were obtained. Six weeks after nephrectomy, the animals were irradiated according to the dosage schedules listed in Table 4. The treatments were delivered twice weekly over a period of four weeks.

The monkeys were irradiated through single left posterior oblique portals. The radiation doses were assessed at the center of the kidneys as determined using excretory urograms obtained perpendicular to the treatment beam. The

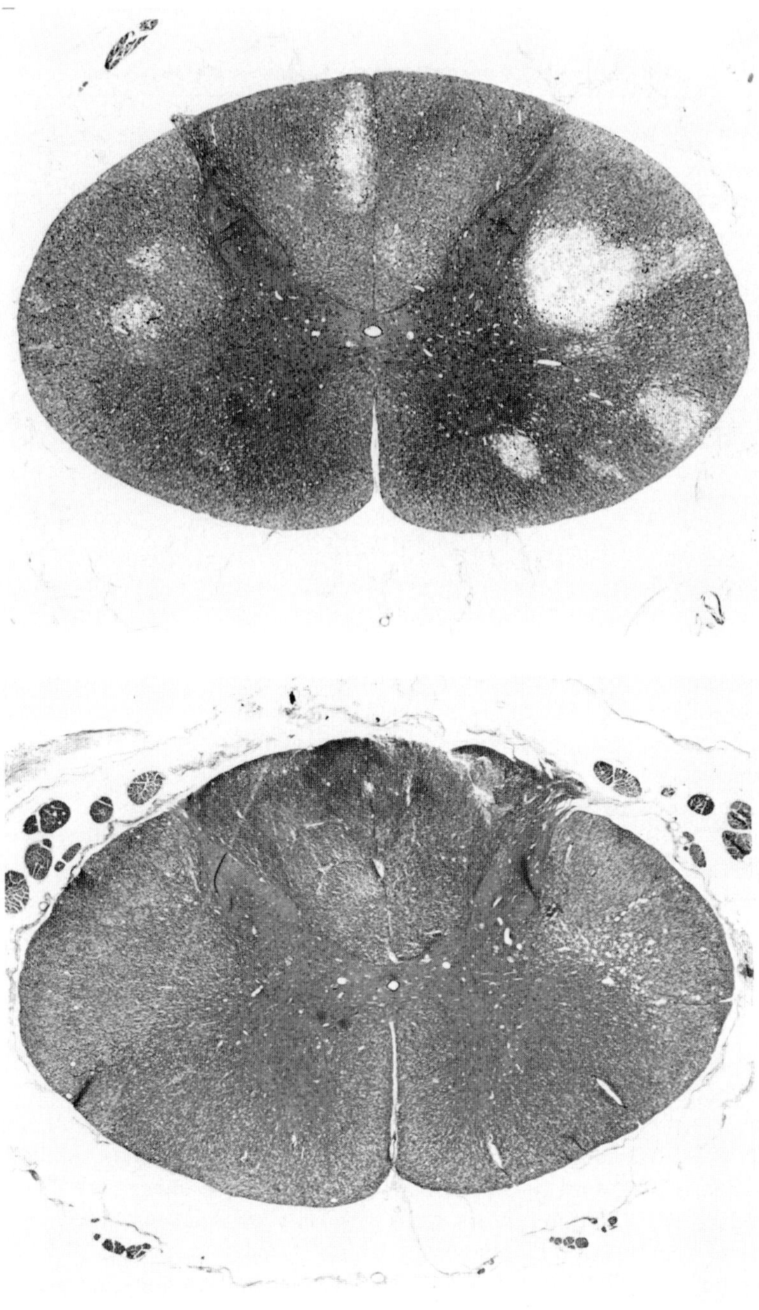

FIG. 2. A (top), Photomicrograph (×10) of a cervical spinal cord six months after irradiation to a dose of 1,550 $rad_{n\gamma}$ with 50 $MeV_{d \rightarrow Be}$ neutrons, demonstrating multiple discrete areas of encephalomalacia and demyelination in the lateral and posterior funiculi. B (bottom), Photomicrograph (×10) of a cervical spinal cord twelve months after irradiation to a dose of 1,300 $rad_{n\gamma}$ with 50 $MeV_{d \rightarrow Be}$ neutrons. There is minimal demyelination of the lateral funiculi and no malacia.

TABLE 4. *Incidence of fatal radiation nephritis in rhesus monkeys*

Group	Modality	Kidney Dose	Dead of Radiation Nephritis
I	Neutrons	960 $rad_{n\gamma}$/8 Fx/4 weeks (120 $rad_{n\gamma}$—twice weekly)	0/5
II	Neutrons	1080 $rad_{n\gamma}$/8 Fx/4 weeks (135 $rad_{n\gamma}$—twice weekly)	5/5
III	^{60}Co	2350 rad/8 Fx/4 weeks (294 rad—twice weekly)	0/5
IV	^{60}Co	2700 rad/8 Fx/4 weeks (338 rad—twice weekly)	0/4*

* The animals in group IV developed moderately abnormal renal function tests (average BUN > 50 mg%; average creatinine > 2.0 mg%) but survived.

monkeys irradiated with neutrons were treated through 9 × 6 cm portals, and those irradiated with ^{60}Co were treated through portals measuring 7½ × 5½ to 9 × 6 cm.

Follow-up studies were performed every 30 days for the first year and less frequently thereafter. The follow-up evaluation included a complete blood count, serum enzymes, urinalysis, blood urea nitrogen (BUN), serum creatinine, effective renal plasma flow (ERPF), and blood pressure determinations. The ERPF was determined by a modification of the ^{131}I-labeled iodohippurate method described by Blaufox et al. (1963). The monkeys in groups I-III have been followed for three years or until death; those in group IV were treated more recently and have been followed only 20 months. One monkey in group IV that died of an anesthetic complication during the second week of irradiation has been excluded from the analysis.

Results

A significant dose-response relationship was observed in the groups treated with neutrons (Table 4). All five animals treated with the higher dose of neutrons (group II—1,080 $rad_{n\gamma}$) died of radiation nephritis three to six months after irradiation, whereas all of the animals treated with the lower dose (group I—960 $rad_{n\gamma}$) are alive three years later with only minimal elevations in serum creatinine and BUN levels (Figure 3). None of the animals treated with ^{60}Co γ rays died of radiation nephritis, although those treated with the higher dose (group IV—2,700 rad) have developed moderately abnormal renal function tests (average BUN >50 mg/100 ml, average creatinine >2.0 mg/100 ml).

Histopathologic correlation was obtained in the five animals in group II that died of radiation nephritis and in one animal in group III that died of unrelated causes 15 months after treatment (Figure 4A). The histopathologic changes

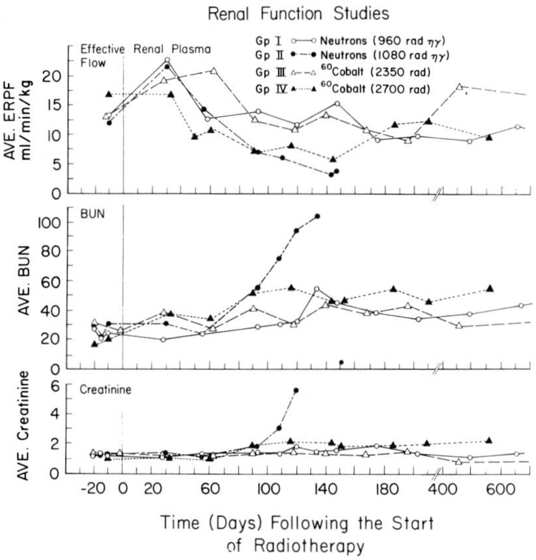

FIG. 3. Chronological plots of effective renal plasma flow (ERPF), blood urea nitrogen (BUN), and serum creatinine.

that followed neutron irradiation (group II—1,080 $rad_{n\gamma}$) were similar to those reported for radiation nephritis secondary to X- and γ-ray irradiation. The damage was characterized by hemorrhage, necrosis, and hyaline obliteration of the glomeruli, thickening of Bowman's capsule, tubular degeneration and atrophy, diffuse interstitial fibrosis, and fibrinoid necrosis of the arterioles. The lesions in the animal that died following irradiation with 2,350 rad of ^{60}Co γ rays (group III) were minimal by comparison (Figure 4B).

Relative Biological Effectiveness

The data indicate that the RBE for radiation nephritis (~325 rad ^{60}Co fractions) is approximately 2.5 to 2.8. The lower limit is well established since all five animals irradiated with 1,080 $rad_{n\gamma}$ with neutrons died of radiation nephritis, whereas none of the five animals irradiated with 2,700 rad with ^{60}Co died of radiation nephritis. The upper limit is less certain, although the animals irradiated with 2,700 rad with ^{60}Co have demonstrated more severe renal impairment than those irradiated with 960 $rad_{n\gamma}$ with neutrons.

COMPARISON OF RBEs

The RBEs for pig skin, rhesus monkey oral mucosa, rhesus monkey spinal cord, and rhesus monkey kidney are summarized in Table 5. With the exception of the RBEs for radiation myelitis in rhesus monkeys, the RBEs for late effects in these experiments were consistent with the values of 3.1 to 3.4 (~200 rad ^{60}Co fractions) and 2.2 to 2.6 (~400 rad ^{60}Co fractions) observed for pig skin.

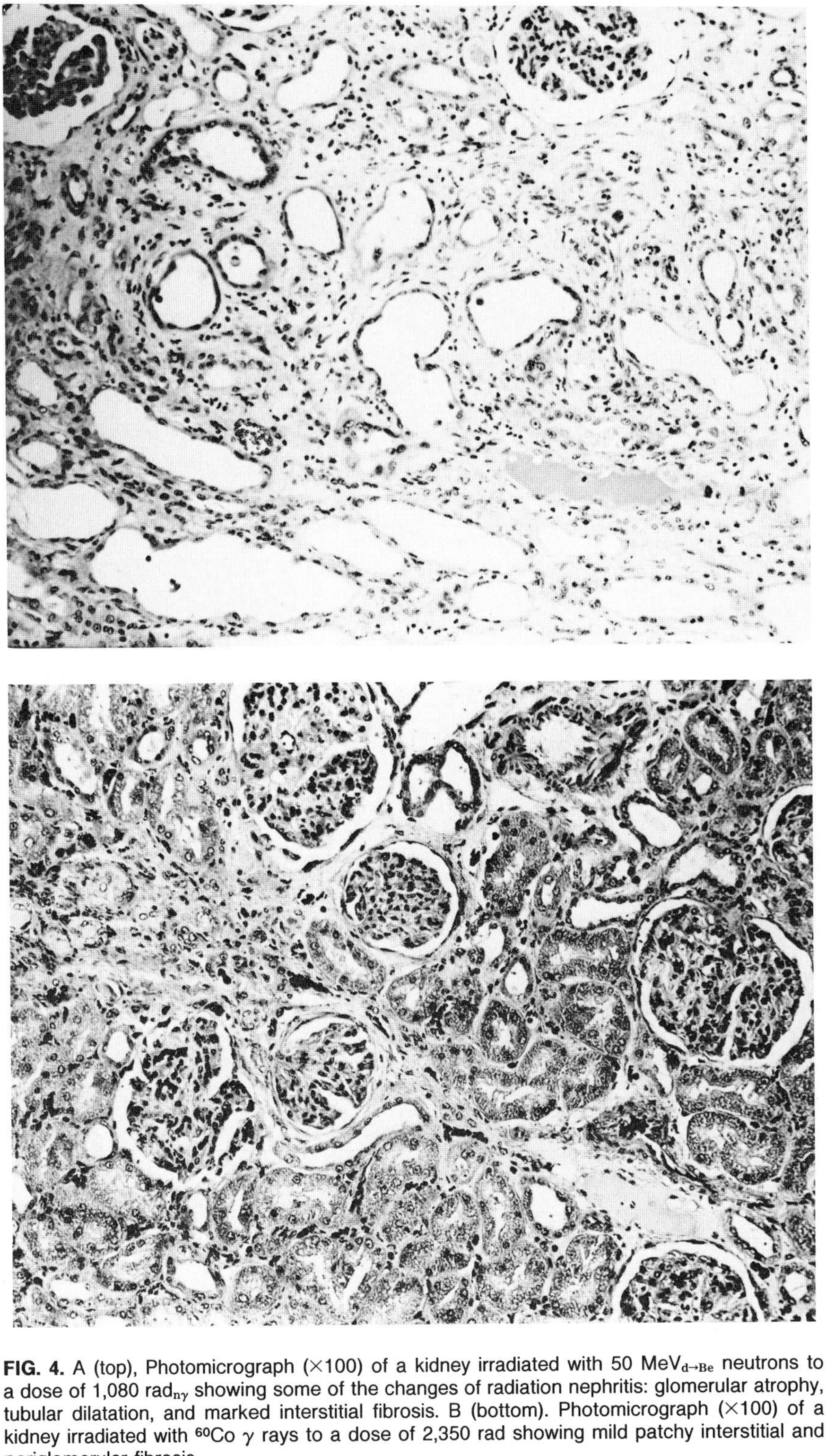

FIG. 4. A (top), Photomicrograph (×100) of a kidney irradiated with 50 $MeV_{d \rightarrow Be}$ neutrons to a dose of 1,080 $rad_{n\gamma}$ showing some of the changes of radiation nephritis: glomerular atrophy, tubular dilatation, and marked interstitial fibrosis. B (bottom). Photomicrograph (×100) of a kidney irradiated with ^{60}Co γ rays to a dose of 2,350 rad showing mild patchy interstitial and periglomerular fibrosis.

TABLE 5. *Summary of the RBEs observed in the normal tissue studies*

Organ/End Point	Photon Fraction Size ~200 rad	~325 rad	~400 rad
Skin			
Acute reaction in pig skin	<2.5		<2.0
Fibrosis in pig skin	3.1–3.4		2.2–2.6
Oral mucosa			
Necrosis in rhesus monkeys	>3.0		<2.5
Spinal cord			
Myelitis in rhesus monkeys	>3.5*		
Kidney			
Nephritis in rhesus monkeys		2.5–2.8	

* Preliminary data; the animals irradiated with ^{60}Co γ rays have been followed only 8 months.

The preliminary results of the spinal cord study indicate that the RBE for radiation myelitis is significantly greater than that observed for other tissues, since all five animals irradiated to a dose of 1,550 $\text{rad}_{n\gamma}$ with neutrons developed quadriparesis and none of the five animals irradiated to a dose of 5,390 rad with ^{60}Co have developed neurologic abnormalities.

The RBEs for acute effects in pig skin were significantly less than the RBEs for late effects. This was true for both the two- and five-times-weekly ^{60}Co fractionation schedules. For example, the RBE for acute skin reaction with five-times-weekly fractionation (~200 rad ^{60}Co fractions) was <2.5, a value significantly less than the RBE of 3.1 to 3.4 for late skin contraction.

TOLERANCE OF NORMAL TISSUES

The results of the large animal studies are listed in Table 6 in terms of the tolerance of the normal tissues to the late effects of 50 $\text{MeV}_{d\rightarrow Be}$. In this essay, tolerance is defined as the threshold radiation dose for a degree of injury in animals that would be unacceptable in patients. Radiation tolerance will depend on the site and volume irradiated and the incidence and severity of injury acceptable to the radiotherapist in a given clinical situation. It is influenced by a variety of patient factors and probably varies with different species.

The tolerance dose for fibrosis in pig skin is approximately 2,200 $\text{rad}_{n\gamma}$ in 6½ weeks. This produced ~25% contraction of the area of the irradiated field. This amount of fibrosis is probably acceptable for small boost fields. A dose of 2,340 $\text{rad}_{n\gamma}$ in 6½ weeks produced ~50% contraction, an unacceptable result in most clinical situations.

The limits of tolerance of the oral mucosa of rhesus monkeys is approximately 2,100 $\text{rad}_{n\gamma}$ in 6½ weeks. Whereas none of the five monkeys irradiated to a

TABLE 6. *Tolerance of normal tissues to late effects of 50 MeV$_{d \rightarrow Be}$ neutrons*

Organ	Tolerance Dose*	Observation
Skin	2200 rad$_{n\gamma}$/6½ wks	2200 rad$_{n\gamma}$ produced ~25% contraction of irradiated pigskin. 2340 rad$_{n\gamma}$ produced ~50% contraction.
Oropharyngeal mucosa	2100 rad$_{n\gamma}$/6½ wks	2200 rad$_{n\gamma}$ produced necrosis in 6 of 10 rhesus monkeys. 2000 rad$_{n\gamma}$ produced no necrosis.
Spinal cord	1300 rad$_{n\gamma}$/4½ wks	1550 rad$_{n\gamma}$ produced paralysis in 5 of 5 rhesus monkeys. 1300 rad$_{n\gamma}$ produced subtle histopathologic changes but no neurologic deficit.
Kidney	960 rad$_{n\gamma}$/4 wks	1080 rad$_{n\gamma}$ produced fatal nephritis in 5 of 5 rhesus monkeys. 960 rad$_{n\gamma}$ produced minimal physiological changes, but no death.

* Approximate values—tolerance will depend on the site and volume irradiated, possible variation with different species, and the level of damage acceptable to the radiotherapist.

dose of 2,000 rad$_{n\gamma}$ developed oromucosal necrosis, six of ten treated to a dose of 2,200 rad$_{n\gamma}$ developed necrosis.

The threshold for spinal cord injury in rhesus monkeys is approximately 1,300 rad$_{n\gamma}$ in 4½ weeks, since none of five animals irradiated with this dose developed signs of neurologic dysfunction, and all five animals irradiated to a dose of 1,550 rad$_{n\gamma}$ developed quadriparesis. Even 1,300 rad$_{n\gamma}$ probably carries some risk of spinal cord injury in view of the subtle histological changes that were observed in the animals irradiated with this dosage schedule (Figure 2B). One should remember that the spinal cord may receive a significant dose of scattered radiation from adjacent fields with neutrons after the cord is removed from the primary beam.

The tolerance dose for rhesus monkey kidney is approximately 960 rad$_{n\gamma}$ in four weeks, since a dose of 1,080 rad$_{n\gamma}$ resulted in fatal radiation nephritis in five monkeys. All five monkeys irradiated with 960 rad$_{n\gamma}$ have survived three years with minimal impairment of renal function.

CLINICAL EXPERIENCE WITH NEUTRONS

The normal tissue tolerances observed in the large animal studies correlate well with the clinical experience at TAMVEC. The dose relationships for major complications in the pilot study are listed in Figure 5A. The clinical material includes all patients who developed complications in areas treated entirely with neutrons. Abdominal complications were excluded since these might be expected to appear at lower doses than complications in the head and neck, breast, or soft tissues. The majority of the complications have developed in patients who received tissue doses greater than 2,120 $rad_{n\gamma}$ in 6–7½ weeks.

The dose relationships for local control of bulky head and neck tumors are plotted in Figure 5B. The best separation between local tumor control and failure occurred at a dose level of 2,010 $rad_{n\gamma}$ in 6–7½ weeks. These data indicate that the range of acceptable neutron doses is narrow—for this patient population with locally advanced cancers.

DISCUSSION

The observation that the late sequelae of fast neutron irradiation are greater than would be predicted on the basis of the acute reactions has been noted in clinical neutron therapy pilot studies (Hussey et al. 1975, Ornitz et al. 1978). These clinical impressions are quantitatively supported by the pig skin and rhesus monkey oromucosal experiments in which the late effects of neutrons were greater than would be expected on the basis of the acute reactions observed. The majority of neutron therapy pilot studies have used two- or three-times-weekly fractionation. The point of reference has been clinical experience with five-times-weekly fractionation with X and γ rays.

The disparity between the acute and late reactions observed clinically might be due in part to the large-dose, limited-fractionation treatment schedules employed in the pilot studies, since the pig skin experiment showed a dissociation of acute and late effects with twice-weekly fractionation with ^{60}Co γ rays (Figure 1). However, in pig skin there is dissociation of the acute and late responses with neutrons even with the more-fractionated four-times-weekly treatment

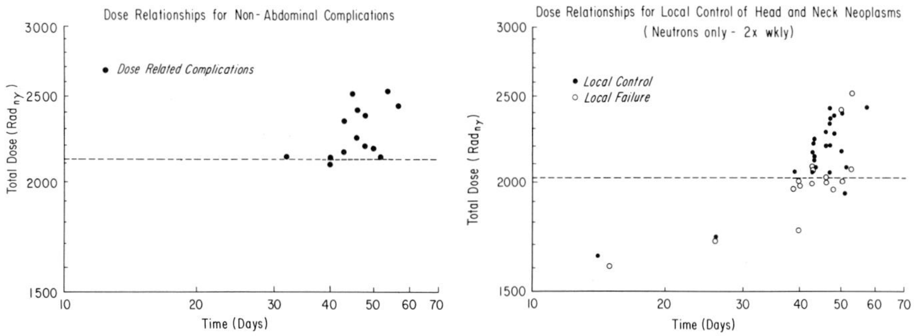

FIG. 5. A (left), Dose relationship for complications in patients treated entirely with neutrons at TAMVEC. The doses shown are the tissue doses in the region of injury. Bowel complications have been excluded since these might be expected to occur at lower doses. The majority of the complications occurred with tissue doses greater than 2,120 $rad_{n\gamma}$. B (right), Dose relationships for local control of advanced head and neck cancers treated entirely with neutrons. Fifteen patients with massive tumors who were deemed incurable prior to the start of treatment were excluded. The local control rate is significantly greater with tumor doses greater than 2,010 $rad_{n\gamma}$.

schedules. As a result, the RBEs for late effects in pig skin are greater than the RBEs for acute effects for both two- and five-times-weekly fractionation (Table 5).

The high RBE for late contraction in pig skin may be partly due to tissue inhomogeneity, since a greater neutron dose is deposited in the fatty subcutaneous tissue. Bewley (1963) has calculated that fat absorbs approximately 18% more energy with 16-$MeV_{d \rightarrow Be}$ neutrons than muscle tissue because of its greater hydrogen content. This differential absorption is slightly less important with 50 $MeV_{d \rightarrow Be}$ neutrons because relatively more energy is deposited by interactions with carbon and oxygen. Nevertheless, this factor could lead to increased subcutaneous fibrosis with either neutron beam.

Tissue inhomogeneity may also contribute to the high RBE for spinal cord since the central nervous system has a greater hydrogen content than muscle. However, tissue inhomogeneity does not explain why the RBE for spinal cord is significantly greater than the RBE for skin contraction since subcutaneous tissue contains even more hydrogen than the central nervous system, or why the RBE for late injury in oral mucosa, which is relatively devoid of fat, is as great as that observed for subcutaneous fibrosis.

The dissociation of the RBEs for acute and late effects of neutron irradiation is probably mainly due to differences in the way neutrons and gamma rays interact with the proliferative tissues responsible for acute effects and the nonproliferative (or slowly proliferating) tissues responsible for late effects. Numerous radiobiology studies have shown a diminished capacity for intracellular repair of radiation damage following neutron irradiation (Hall et al. 1975, Gragg et al. 1977, Field 1977), and cell cycle fluctuations in radiosensitivity are significantly less with neutrons than with γ rays (Gragg et al. 1978). There may also be differences in the proliferative response, e.g., division delay or stimulus for repopulation following neutron and gamma irradiation; but this area has not been studied extensively.

Differences in Repair of Sublethal Injury

Most cell lines have a diminished capacity to accumulate and repair sublethal injury following neutron irradiation compared to gamma irradiation (Hall et al. 1975, Gragg et al. 1977). As a result, neutron cell survival curves usually have little or no shoulder. If the shoulder of the γ-ray survival curve for the nonproliferative tissues responsible for late effects is broader than the shoulder of the γ-ray survival curve for the proliferative tissues responsible for acute reactions, and both are effectively eliminated on the neutron survival curves, then the RBE for late effects would be greater than the RBE for acute effects.

Differences in Slow Repair or Repair of Potentially Lethal Damage

Field (1977), noting an increase in the LD_{50} for lung damage as a function of the time interval between two fractions of gamma irradiation, has postulated a "slow repair" process, analagous to the repair of sublethal damage, but approximately 100 times slower. This long-term recovery was not observed with neutrons.

Recovery from potentially lethal damage occurs over a period of a few hours in cells irradiated in vitro when the postirradiation conditions are suboptimal for growth. Such conditions include maintaining cells in plateau phase, holding cells at suboptimal temperatures, or incubating cells in balanced salt solutions. Repair of potentially lethal damage occurs following gamma irradiation but has not been observed following neutron irradiation (Gragg et al. 1977).

If slow repair or repair of potentially lethal damage occurs in nonproliferative tissues but not in proliferative tissues, and does not occur, or is reduced, in both tissues after exposure to neutrons, then the RBE for late effects for protracted irradiation would be increased relative to the RBE for acute effects.

Differences in Cell-Cycle Variation in Radiosensitivity

By giving the same total dose of gamma radiation in a greater number of fractions, there may be relatively greater cell killing in proliferative tissues responsible for acute effects than in nonproliferative tissues responsible for late effects, because of progression of the surviving cells into sensitive phases of the cell cycle between dose fractions. There is significantly less cell cycle variation in radiosensitivity with neutrons than with γ rays (Gragg et al. 1978). Consequently, the net sensitization of proliferative tissues would be less with neutrons than with γ rays, resulting in a lower RBE for acute effects with fractionated treatment schedules.

CONCLUSIONS

1. The late effects of neutron irradiation in pig skin and rhesus monkey oral mucosa are severe relative to the acute effects, using five-times-weekly fractionation with ^{60}Co γ rays for comparison.

2. The dose response for late effects of neutron irradiation in pig skin and rhesus monkey oral mucosa appears to be independent of weekly fractionation—over a range of two to five fractions per week, 70 to 200 $rad_{n\gamma}$ per fraction.

3. The RBEs for late effects in pig skin, rhesus monkey oral mucosa, and rhesus monkey kidney are consistent with one another, whereas the RBEs for late effects in rhesus monkey spinal cord appear to be greater than that observed for other tissues. The RBEs for acute effects in pig skin are significantly less than the RBEs for late effects.

4. The tolerance doses for 50 $MeV_{d\rightarrow Be}$ neutrons in large animals are ~2,200 $rad_{n\gamma}$/6½ weeks for skin contraction, ~2,100 $rad_{n\gamma}$/6½ weeks for oromucosal necrosis, ~1,300 $rad_{n\gamma}$/4½ weeks for radiation myelitis, and ~960 $rad_{n\gamma}$/4 weeks for radiation nephritis. These values correlate well with the clinical observations at TAMVEC, since the majority of complications have occurred following tissue doses in excess of 2,120 $rad_{n\gamma}$.

ACKNOWLEDGMENT

This investigation was supported in part by Grant CA 12542, awarded by the National Cancer Institute, Department of Health, Education and Welfare.

REFERENCES

Bewley, D. K. 1963. Pre-therapeutic experiments with the fast neutron beam from the Medical Research Council cyclotron. II. Physical aspects of the fast neutron beam. Br. J. Radiol. 36:81–88.

Blaufox, M. D., H. G. W. Frohmuller, J. C. Campbell, D. C. Utz, A. L. Orvis, and C. A. Owen, Jr. 1963. A simplified method of estimating renal function with iodohippurate ^{131}I. J. Surg. Res. 3:122–125.

Catterall, M. 1978. Observations on the reactions of normal malignant tissues to a standard dose of neutrons. Third Meeting on Fundamental Practical Aspects of Fast Neutrons and Other High LET Particles in Clinical Radiotherapy, Sept 13–15, 1978, The Hague, The Netherlands.

Ellis, F. 1968. The relationship of biological effect to dose-time fractionation factors in radiotherapy. Curr. Top. Radiat. Res. 4:382.

Field, S. B. 1977. Early and late normal tissue damage after fast neutrons. Int. J. Radiat. Oncol. Biol. Phys. 3:203–210.

Gragg, R. L., R. M. Humphrey, and R. E. Meyn. 1977. The response of Chinese hamster overy cells to fast-neutron radiotherapy beams. II. Sublethal and potentially lethal damage recovery capabilities. Radiat. Res. 71:461–470.

Gragg, R. L., R. M. Humphrey, H. D. Thomas, and R. E. Meyn. 1978. The response of Chinese hamster ovary cells to fast neutron radiotherapy beams. III. Variations in RBE with position in the cell cycle. Radiat. Res. 76:283–291.

Hall, E. J., L. Roizin-Towle, and F. H. Attix. 1975. Radiobiological studies with cyclotron-produced neutrons currently used for radiotherapy. Int. J. Radiat. Oncol. Biol. Phys. 1:33–40.

Hussey, D. H., G. H. Fletcher, and J. B. Caderao. 1975. A preliminary report of the MDAH-TAMVEC neutron therapy pilot study, *in* Proceedings of the 5th International Congress of Radiation Research. Academic Press, New York, pp. 1106–1117.

Jardine, J. H., D. H. Hussey, D. D. Boyd, G. L. Raulston, and T. J. Davidson. 1975. Acute and late effects of 16- and 50-$MeV_{d \rightarrow Be}$ neutrons on the oral mucosa of rhesus monkeys. Radiology 117:185–191.

Jardine, J. H., D. H. Hussey, G. L. Raulston, C. A. Gleiser, K. N. Gray, J. I. Huchton, and P. R. Almond. 1979. The effects of 50 $MeV_{d \rightarrow Be}$ neutron irradiation on rhesus monkey cervical spinal cord. Int. J. Radiat. Oncol. Biol. Phys. (in press).

Laramore, G. E., J. C. Blasko, T. W. Griffin, M. T. Groudine, and R. G. Parker. 1979. Fast neutron teletherapy for advanced carcinomas of the oropharynx. Int. J. Radiat. Oncol. Biol. Phys. (in press).

Ornitz, R. D., E. Bradley, K. Mossman, F. Fender, M. C. Schell, and C. C. Rogers. 1979. Preliminary clinical observation of early and late normal tissue injury in patients receiving fast neutron radiation. Int. J. Radiat. Oncol. Biol. Phys. (in press).

Raulston, G. L., K. N. Gray, C. A. Gleiser, J. H. Jardine, B. L. Flow, J. I. Huchton, K. R. Bennett, and D. H. Hussey. 1978. A comparison of the effects of 50 $MeV_{d \rightarrow Be}$ neutron and cobalt-60 irradiation of the kidneys of rhesus monkeys. Radiology 128:245–249.

Withers, H. R., B. L. Flow, J. I. Huchton, D. H. Hussey, J. H. Jardine, K. A. Mason, G. L. Raulston, and J. B. Smathers. 1977. Effect of dose fractionation on early and late skin responses to γ-rays and neutrons. Int. J. Radiat. Oncol. Biol. Phys. 3:227–233.

Withers, H. R., H. D. Thames, Jr., D. H. Hussey, B. L. Flow, and K. A. Mason. 1978. Relative biological effectiveness (RBE) of 50 MV(Be) neutrons for acute and late skin injury. Int. J. Radiat. Oncol. Biol. Phys. 4:603–608.

Radiation Biology in Cancer Research, edited by
Raymond E. Meyn and H. Rodney Withers.
Raven Press, New York © 1980.

Slow Repair and Residual Injury

Shirley Hornsey and Stanley B. Field

*Cyclotron Unit, Medical Research Council, Hammersmith Hospital,
London W12 OHS, England*

SLOW REPAIR

When a dose of X or γ rays is split into two or more fractions with the dose fractions separated by a few hours, the total dose must be increased to produce the same level of tissue damage. This is due to recovery from sublethal damage between fractions, first described by Elkind and Sutton (1959). A similar phenomenon sometimes occurs between dose fractions separated by much longer periods of time, over days and weeks, and this has been termed "slow repair" (Field et al. 1976). The Elkind recovery from sublethal damage that occurs over the first few hours following radiation is responsible for the fractionation-dependent factor (N factor) while slow repair may be associated with the time factor (T) in the relationship between total dose, number of fractions (N), and overall treatment time (T) such as that described by Ellis (1969), i.e., total dose $\sim\alpha\ N^{\alpha}T^{\beta}$.

Slow Repair in Lung

Irradiation of the whole thorax of mice leads to death between 40 and 180 days postirradiation from radiation pneumonitis (Phillips and Margolis 1972). When two or more doses of X rays were given, with intervals between doses of one hour up to five weeks, Elkind repair of sublethal damage occurred in the first few hours followed by a slower repair with a half time of about ten days. The term slow repair was first used to describe this latter repair observed in fractionation studies on the lungs of experimental mice (Field et al. 1976). While the repair of sublethal damage after neutron irradiation was less than after X-ray irradiation, *no* slow repair was observed after neutrons (Field and Hornsey 1977). In Figure 1, the isoeffect curves ($LD_{50/180}$) for two or five fractions of X rays given over 24 hours to five weeks are shown. The slopes of these isoeffect curves (.06 to .07) give the exponent of T in the Ellis formula. Also shown is the isoeffect curve for two doses of neutrons.

It is not known which are the primary target cells in the lung whose damage leads to radiation pneumonitis. Phillips (1966) from electron microscopic studies has suggested that these are the capillary endothelial cells, while Van den Brenk (1971) suggested the type II alveolar cells that produce alveolar surfactant, a

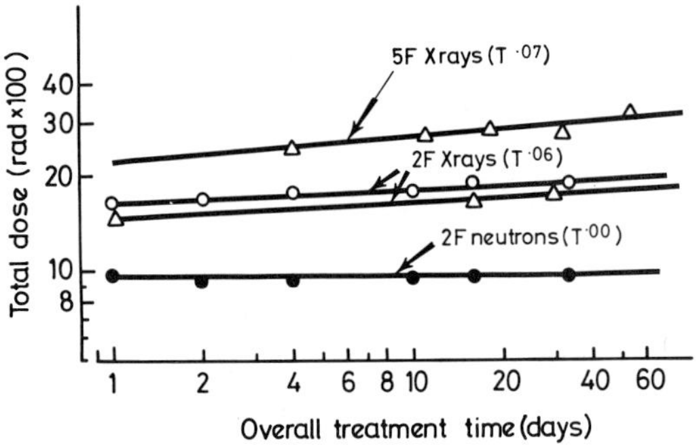

FIG. 1. Isoeffect curves relating total dose and overall treatment time for radiation pneumonitis in mouse lung irradiated with 250-kV X rays or fast neutrons.

phospholipid involved in the control of surface tension, are involved. Recent observations on the reduction of type II alveolar cells after irradiation suggest they could be the primary target (Ahier and Field, manuscript in preparation).

The normal turnover of cells in the lung is extremely low, and it was therefore considered that slow repair in the lung could not be explained by cell turnover. Also, the shape of the isoeffect curve (with a half-value of about 10 days) is inconsistent with homeostatic repopulation's being a major contributor to slow repair, because repopulation would have to start rapidly after irradiation. In addition, the proportion of alveolar cells labeled with tritiated thymidine does not rise above normal levels until about 10 days postirradiation, by which time half the slow repair has already taken place. In Figure 2, the isoeffect curve, plotted on a linear scale, is shown in relation to the labeling index for the alveolar type II cells and other alveolar cells, including the capillary endothelial cells (Coulter, Ahier, and Field, manuscript in preparation). It can also be seen in Figure 2 that the labeling index for the alveolar cells with time after X rays is the same as after neutrons, although no slow repair is observed after neutrons. These results suggest that slow repair in the lung cannot be accounted for by repopulation.

What Is Slow Repair?

If slow repair occurs before any repopulation takes place, we must conclude that, like Elkind repair, it occurs as an intracellular biochemical repair process. It could result from a genuinely slow reaction or because the associated enzyme (or enzymes) is in low concentration or because the repair is dependent on radiation-induced enzymes. Such a repair system has been postulated for *Chlamydomonas* by Hillová and Drášil (1967), in which radiation decreases the sensitivity, i.e., increases D_o, to further irradiation, and for *Oedogonium* by Horsley and Laszlo (1971), in which a larger value of D_q is obtained on a second dose-survival curve. The increase in D_o or D_q was prevented by inhibition of protein synthesis (Horsley and Laszlo 1971), which led Bryant (1975) to suggest an inducible enzyme system was involved. It is not clear from his results,

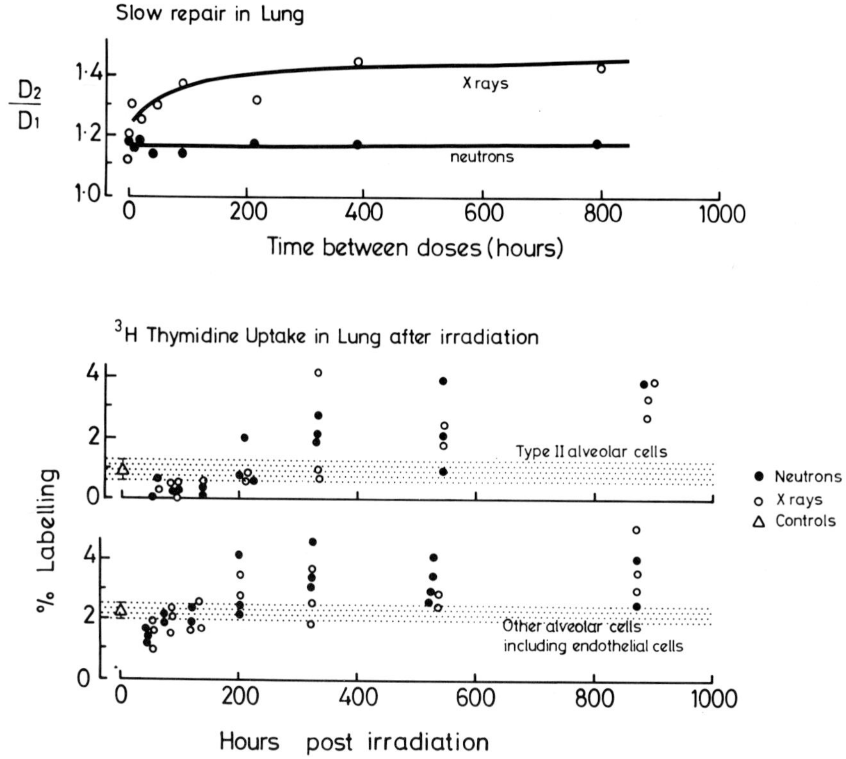

FIG. 2. The relationship between slow repair in the irradiated mouse lung and the postirradiation uptake of ^{3}H-thymidine into alveolar cells. The incorporation of ^{3}H-thymidine into alveolar cells occurs after most of the slow repair has occurred (by permission from Coultas, Ahier, and Field).

however, to what extent this repair system is interlinked with recovery from sublethal damage. The increase in D_0 in *Chlamydomonas* can, however, be induced by irradiation with helium nuclei of a high linear energy transfer (LET) (Davies et al. 1969) and is inhibited by ultraviolet light production of thymine dimers (Bryant and Parker 1977). An increase in D_0 for a second dose X-ray survival curve similar to that observed in *Chlamydomonas* was observed by Lockhart et al. (1961) in a HeLa S3 cell line, although this change in D_0 does not occur in many cell lines under the normal conditions of culture.

Slow Repair and PLD

Slow repair occurs during the time between two irradiations, while repair of potentially lethal damage (PLD) occurs with irradiation and some other subsequent stimulus, for example, plating tumor cells into in vitro culture or stimulating growth in situ. Reinhold and Buisman (1975) and Van den Brenk et al. (1974), stimulating growth of the vascular system at various times after X irradiation, showed repair of damage occurring over a time course similar to that observed for slow repair. It is unlikely that this repair in the capillaries is due to cell repopulation as the time course of repopulation is much slower. For example, Fajardo and Stewart (1973) showed repopulation, after 2,000 rad X rays to the heart, in the myocardial blood vessels of the rabbit starting between 15 and 20 days postirradiation. The repair observed by Reinhold and Buisman was half completed by 10 days. The labeling index in the capillary endothelial cells of myocardial vessels is among the highest observed in blood vessels, so this time course is most probably among the shortest for repopulation in vessels. Tannock and Hayashi (1972) found no increase in labeling of endothelial cells after 2,000 rad and 4,000 rad to the leg of the mouse while Hirst and Hobson (1978) found only a small increase in labeling of endothelial cells in the small arteries of the mouse mesentery three weeks after treatment with 4,500 rad, from 1% in control animals to 3% in irradiated ones.

Is Slow Repair a Generality?

Slow repair will be of greatest importance in tissues with a low cell turnover. It does not, however, occur in all tissues with a low cell turnover. While there is considerable recovery from sublethal damage in the spinal cord (van der Kogel 1977a, White and Hornsey 1978) there appears to be *no* slow repair in this tissue. In the CFHB strain of rats used at Hammersmith, no increase in the ED_{50} dose for radiation myelopathy was observed with X-ray doses separated by 1 to 20 days, but as the overall treatment time was increased further, the ED_{50} increased. This pattern of increase in ED_{50} dose only when the overall treatment time exceeded 20 days was the same for neutron irradiations (Figure 3) (White and Hornsey, manuscript in preparation). Preliminary observations on thymidine uptake by the glial cells in the cord indicate that the increase in ED_{50} after 20 days is associated with cellular repopulation (Hornsey and White, manuscript in preparation). In another strain of rats, the increase in ED_{50}

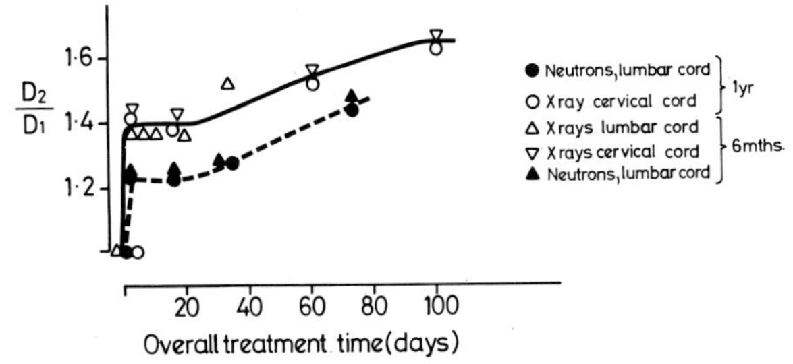

FIG. 3. Time-dependent repair in the CFHB rat spinal cord after X-ray or neutron irradiation.

occurs at a later overall treatment time, in excess of 80 days (van der Kogel 1977b), indicating that the timing of repopulation may be strain-dependent.

It appears, therefore, that slow repair may occur in some tissues, but not in others.

RESIDUAL INJURY

It is generally accepted that re-treatment of previously irradiated sites in patients must be done with caution as any residual damage in the tissues will reduce tolerance to subsequent treatment. "Residual injury" after irradiation has been observed in a number of tissues in different ways. One of the earlier demonstrations was by Weinbren and his co-workers (1960) in irradiated rat liver. They demonstrated latent chromosome abnormalities at times up to one year after irradiation, in cells stimulated into division in the anterior lobe by ligating the peduncle of the posterior lobes. Curtis (1967) subsequently showed that such abnormalities decreased in number with time after irradiation, but that the reduction with time was less after neutron irradiation of the liver than after X irradiation. More recently, Savage and Bigger (1978) have shown persistent chromosome changes in fibroblasts grown from subcutaneous tissue biopsies taken from radiotherapy patients treated to near tolerance levels. Symmetrical chromosome changes were observed up to 60 years after treatment.

Persistent injury may also be shown by the response to a second radiation treatment. Clearly, if repair continues for several months following irradiation there will be a decrease in residual injury as a result. Orton and Ellis (1973) suggested this could be allowed for by modifying the partial tolerance, according to $T^{.11}$ (from the Ellis formula), for up to 100 days.

Residual injury can be measured by giving a priming dose or course of dose fractions and measuring the reduction in a second dose required to produce a given response, compared with treating unirradiated age-related controls. Brown and Probert (1975) found 35–40% of the first course of treatment was "remembered" for late damage when the second treatment was given six months after the first. They also tested Ellis's prediction and found that the results were reasonably well predicted by the Ellis formula (Brown and Probert 1973). However, the remembered or residual damage for the acute skin response was much less than for late deformity, with only 10–15% remaining. This latter observation on residual injury in cells responsible for early acute response in skin is in agreement with observations by Denekamp (1975) and Field and Law (1976), who also found that six to eight months after a first treatment of either single or fractionated doses of X rays only about 10% of the first dose was remembered. Using necrosis in mouse tails, Hendry (1978) also showed about 10% residual damage between six weeks and ten months after X irradiation, but this was increased to about 35% residual damage after a second course or after up to six courses of irradiation given at six weekly intervals.

The early skin reaction is due primarily to damage to the basal epithelial cells, which, when stimulated by radiation injury, proliferate rapidly when cell death occurs (Denekamp 1973). Presumably, cells carrying any residual injury are mostly lost or swamped during this rapid proliferation and the tissue is restored to near normal. In this rapidly proliferating tissue, slow repair will therefore be of minor importance and indeed may be terminated by cell division (Field et al. 1976). Other rapidly proliferating tissues, e.g., hemopoietic tissues, also show little long-term residual injury (Porteous and Lajtha 1966, Ainsworth and Leong 1966, Hawes et al. 1966). Human skin is also able to repair epithelial damage almost fully (Hunter and Stewart 1977).

Irradiation damage in tissue with a slow cell turnover will not be shed or swamped so readily by cell proliferation, and the damage can therefore remain latent for much longer periods. It is in these tissues that slow repair may be of importance as the major means of repairing damage during the course of fractionated treatments or between split course treatments. It may also be the mechanism by which residual damage is repaired. In tissues that show an acute radiation response owing to damage to rapidly proliferating cells and a late response owing to damage to cells with a low cell turnover, we might expect a change in the relationship between early and late response because the rate of tissue repair due to rapid proliferation will be greater than that due to slow repair. The experimental observations on rodent skin (feet or ears) are equivocal. Denekamp (1977), using mouse feet, and Field and Law (1976), using mouse feet and ears, found the relationship between early and late damage remains the same for a wide range of fractionated treatments and re-treatments. However, Brown and Probert (1975) in treatment schedules of two ten-fraction courses of X rays found that residual damage for late effects (deformity) at one to ten months after the first treatment was much greater than residual damage for acute skin response, implying that early and late damage are disassociated when many fractions are given. The topic is by no means fully resolved.

In a recent series of experiments, Field, Marston, and Tompkins (manuscript in preparation) measured residual injury in mouse feet by re-treatment with either X rays or neutrons six months after priming treatments with X rays or neutrons. As in the earlier experiments of Brown and Probert (1975) two different end points were used, skin reactions and late deformity. Three priming doses of X rays or neutrons were given, which were estimated to be equivalent by using the measured relative biological effectiveness (RBE) for early skin damage in this system derived from previous experiments. The percentage of residual injury was measured for the acute skin reaction seven to 40 days after the second treatment and for foot deformities six months after the second treatment. The results show: (1) The relationship between acute and late reaction for non-pretreated feet was the same for test doses of X rays or neutrons (Figures 4 and 5). (2) The relationship between acute and late reaction changes if pretreatment is given with either X rays or neutrons so that a greater late deformity for a given early acute reaction is obtained, as was observed by Brown and Probert (1975) (Figures 4 and 5). (3) No significant difference was observed

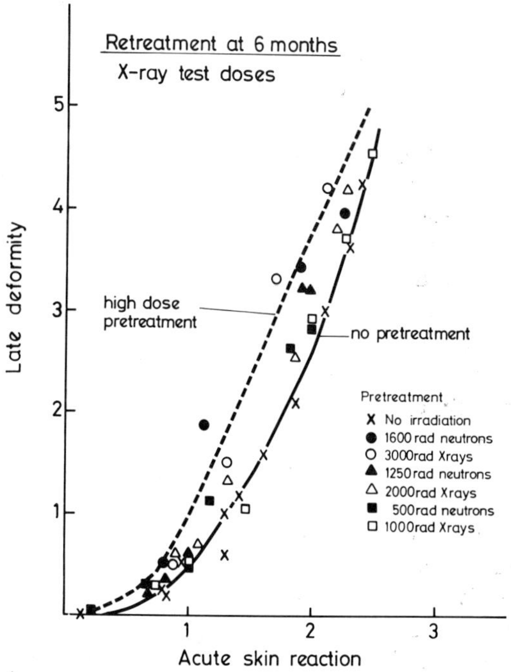

FIG. 4. The relationship between early acute skin reaction and late deformity in mouse feet given test doses of X rays six months after a pretreatment with either X rays or neutrons.

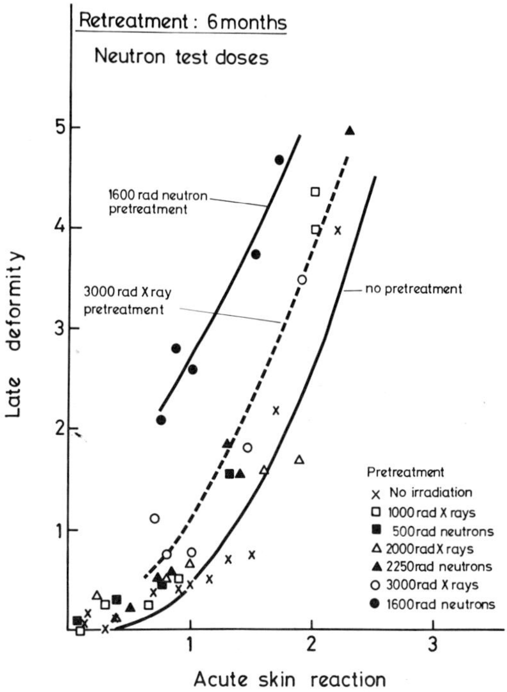

FIG. 5. The relationship between early acute skin reactions and late deformity in mouse feet given test doses of neutrons six months after a pretreatment with either X rays or neutrons.

between priming treatment with X rays or neutrons except when the test dose was given with neutrons. A much higher level of late deformity was observed for a given level of early acute damage after a priming dose of 1,600-rad neutrons than after a priming dose of 3,000-rad X rays (Figure 5). (4) The residual injury was not significantly different for early or late damage after pre-irradiation with X rays or for early acute damage after neutrons, but was significantly greater for late damage after *pretreatment* with neutrons (Figure 6). This is in agreement with the findings of Hendry et al. (1977), who observed more residual damage in mouse tails after neutron irradiation (25%) than after X irradiation (10%). (5) No significant difference in residual injury was generally observed by testing with neutrons or X rays except perhaps in late damage after neutron priming treatment, in which the residual injury appeared to be greater for all three test doses after neutron irradiation (Figure 6). This slight inconsistency between results derived from the early versus late damage analysis (Figures 4 and 5) and the estimate of residual injury (Figure 6) arises because the latter depends on control values while the former are independent of them. Both analyses indicate that after high, near-tolerance doses of irradiation there is more late damage than early residual damage, particularly so after high doses of neutrons. The sensitivity of the tissue to re-treatment with neutrons after a previous high dose of neutrons may be due to some hypoxia in the tissue at the time of re-treatment exacerbated by ischemia induced by the priming dose; while the hypoxia would protect against X-ray re-treatment, the protection

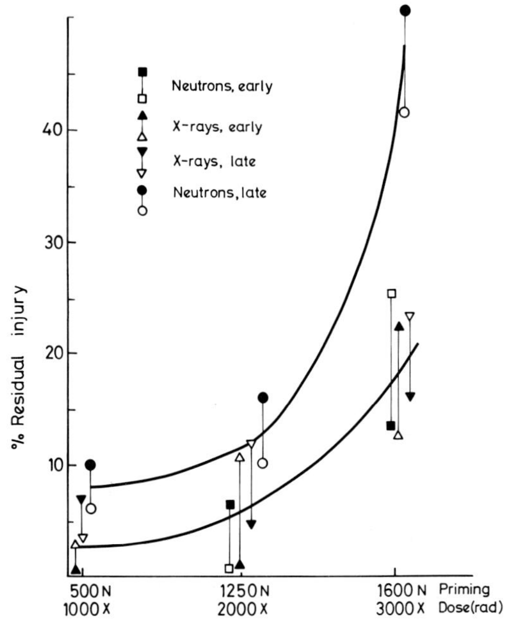

FIG. 6. The percentage of residual damage six months after priming doses of X rays or neutrons is shown for both early acute skin damage and for late deformity in mouse feet.

against neutron re-treatment would be less. This indicates that the RBE for re-treatment of previously heavily irradiated sites may be raised by hypoxia.

RBE

A slow repair process that occurs after X irradiation but not after neutron irradiation, as observed in lung damage, would lead us to expect more residual damage after neutron than X irradiation. This was observed by Curtis (1967) in liver. It indicates that re-treatment after neutron irradiation would be more difficult than after X-ray irradiation. Hendry et al. (1977), estimating necrosis in mouse tails re-treated six months after irradiation, found about 10% residual damage from the first X-ray treatment but about 25% remained after neutron treatment. We can estimate that the RBE for this residual injury is about 20% higher than for acute effects. This agrees with the RBE values observed by Field (1977) in a protracted schedule of skin irradiation in weekly fractions given over six months. The RBE values observed were also about 20% higher

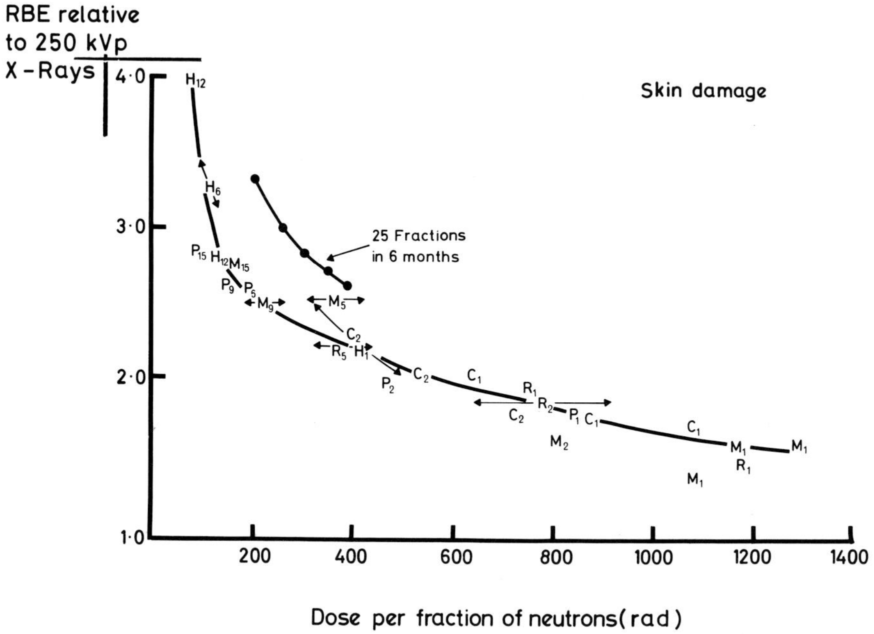

FIG. 7. The RBE for very protracted treatment times (weekly fractions over six months) of X rays or neutrons to mouse feet shown in relation to the RBE curve for skin damage in mice, rats, pigs, and man for treatments given over one day to one month. (Reproduced from Field, 1977, with permission of Pergamon Press.)

than those observed for skin reactions after normal fractionation regimes given in only a few weeks (Figure 7).

Estimates of residual damage can be made from split-dose experiments on lung, which indicate that one day after a priming dose of X rays or neutrons about 40% residual damage remains; over the next four to five weeks this is reduced to 28% after X-ray irradiation but remains at about 40% with neutrons. This means that the RBE for a fractionated course of treatment is increased by differences in slow repair between X rays and neutrons as well as by differences in the repair of sublethal damage.

While 40% residual damage remains one day after neutron irradiation of the spinal cord, only 30% remains after X-ray irradiation, reflecting the large amount of Elkind recovery after X-ray treatment. However, approximately 10% of damage is shed between 30 and 120 days after both X-ray and neutron irradiation, leaving ~ 20% residual X-ray damage and 30% residual neutron damage. Thus, the RBE for cord damage does not change significantly for protracted irradiation compared with acute irradiations over a few days. For lung irradiation the RBE increases by about 14% if the overall treatment time is increased from one day to four to five weeks. We may conclude that where repair of residual damage after irradiation is due to cell repopulation, as in many early acute types of tissue damage and some late damage such as of the spinal cord, extending the overall treatment time will have no effect on RBE. However, where repair of residual damage is due to slow repair, as in lung damage and possibly late damage in skin, the RBE will increase with prolongation of treatment.

REFERENCES

Ainsworth, E. J., and G. F. Leong. 1966. Recovery from radiation injury in dogs as evaluated by the split-dose technique. Radiat. Res. 29:131–142.

Brown, M. J., and J. C. Probert. 1973. Long term recovery of connective tissue after irradiation. Radiology 108:205–207.

Brown, J. M., and J. C. Probert. 1975. Early and late radiation changes following a second course of irradiation. Radiology 115:711–716.

Bryant, P. E. 1975. Decrease in sensitivity of cells after split-dose recovery: Evidence for the involvement of protein synthesis. Int. J. Radiat. Biol. 27:95–102.

Bryant, P. E., and J. Parker. 1977. Evidence for location of the site of accumulation of sub-lethal damage in *Chlamydomonas*. Int. J. Radiat. Biol. 32:237–246.

Curtis, H. J. 1967. Biological mechanisms of delayed radiation damage in mammals. Curr. Top. Radiat. Res. 3:139–174.

Davies, R. D., P. D. Holt, and D. G. Papworth. 1969. The survival curves of haploid and diploid *Chlamydomonas reinhardii* exposed to radiation of different LET. Int. J. Radiat. Biol. 15:75–87.

Denekamp, J. 1973. Changes in the rate of repopulation during multifraction irradiation of mouse skin. Br. J. Radiol. 46:381–387.

Denekamp, J. 1975. Residual radiation damage in mouse skin 5–8 months after irradiation. Radiology 115:191–195.

Denekamp, J. 1977. Early and late radiation reactions in mouse feet. Br. J. Cancer 36:322–329.

Elkind, M. M., and H. Sutton. 1959. X-ray damage and recovery in mammlian cells grown in culture. Nature 184:1293–1295.

Ellis, F. 1969. Dose, time and fractionation: a clinical hypothesis. Clin. Radiol. 20:1–7.

Fajardo, L. F., and J. R. Stewart. 1973. Pathogenesis of radiation-induced myocardial fibrosis. Lab. Invest. 29:244–247.

Field, S. B. 1977. Early and late normal tissue damage after fast neutrons. Int. J. Radiat. Oncol. Biol. Phys. 3:203–210.

Field, S. B., and S. Hornsey. 1977. Slow repair after X rays and fast neutrons. Br. J. Radiol. 50:600–601.

Field, S. B., S. Hornsey, and Y. Kutsutani. 1976. Effects of fractionted irradiation on mouse lung and a phenomenon of slow repair. Br. J. Radiol. 49:700–707.

Field, S. B., and M. P. Law. 1976. The relationship between early and late radiation damage in rodent skin. Int. J. Radiat. Biol. 30:557–564.

Hawes, C., A. Howard, and L. H. Gray. 1966. Induction of chromosome structural damage in Ehrlich ascites tumour cells. Mutat. Res. 3:79–89.

Hendry, J. H. 1978. The tolerance of mouse tails to necrosis after repeated irradiation with X rays. Br. J. Radiol. 51:808–813.

Hendry, J. H., I. Rosenberg, D. Greene, and J. G. Stewart. 1977. Re-irradiation of rat tails to necrosis at six months after treatment with a "tolerance" dose of X rays or neutrons. Br. J. Radiol. 50:567–572.

Hillová, J., and V. Drášil. 1967. The inhibitory effect of iodoacetamide on recovery from sublethal damage in *Chlamydomonas reinhardii.* Int. J. Radiat. Biol. 12:201–208.

Hirst, D. G., and B. Hobson. 1978. Repair of radiation damage in the blood vessels of the mouse mesentery and small intestine. Int. J. Radiat. Biol. 34:566–567.

Horsley, R. J., and A. Laszlo. 1971. Unexpected additional recovery following a first X-ray dose to a synchronous cell culture. Int. J. Radiat. Biol. 20:593–596.

Hunter, R. D., and J. G. Stewart. 1977. The tolerance to re-irradiation of heavily irradiated human skin. Br. J. Radiol. 50:573–575.

Lockhart, R. Z., M. M. Elkind, and W. B. Moses. 1961. Radiation response of mammalian cells grown in culture. II. Survival and recovery characteristics of several subcultures of HeLa S3 cells after X-irradiation. J. Natl. Cancer Inst. 27:1393–1404.

Orton, C. G., and F. Ellis. 1973. A simplification in the use of NSD concept in practical radiotherapy. Br. J. Radiol. 46:529–537.

Phillips, T. L. 1966. An ultrastructual study of the development of radiation injury in the lung. Radiology 87:49–54.

Phillips, T. L., and L. Margolis. 1972. Radiation pathology and the clinical response of lung and esophagus. Frontiers of Radiation Therapy and Oncology 6:254–273.

Porteous, D. D., and L. G. Lajtha. 1966. On stem cell recovery after irradiation. Br. J. Haematol. 12:177–188.

Reinhold, H. S., and G. H. Buisman. 1975. Repair of radiation damage to capillary endothelium. Br. J. Radiol. 48:727–731.

Savage, J. R. K., and T. R. L. Bigger. 1978. Aberration distribution and chromosomally marked clones in X-irradiated skin, *in* Mutagen-Induced Chromosome Damage in Man. H. J. Evans and D. C. Lloyd, eds. Edinburgh University Press, Edinburgh.

Tannock, I. F., and S. Hayashi. 1972. The proliferation of capillary endothelial cells. Cancer Res. 32:77–82.

Van den Brenk, H. A. S. 1971. Radiation effects on the pulmonary system, *in* Pathology of Irradiation, C. C. Berdjis, ed. Williams & Wilkins Co., Baltimore, pp. 569–596.

Van den Brenk, H. A. S., C. Sharpington, C. Orton, and M. Stone. 1974. Effects of X-radiation on growth and function of the repair blastema (granulation tissue). II. Measurements of angiogenesis in the Selye pouch in the rat. Int. J. Radiat. Biol. 25:277–289.

van der Kogel, A. J. 1977a. Radiation tolerance of the rat spinal cord: Time dose relationships. Radiology 122:505–509.

van der Kogel, A. J. 1977b. Radiation tolerance of the spinal cord: Dependence on fractionation and extended overall times, *in* Radiobiological Research and Radiotherapy, Vol. 1. International Atomic Energy Agency, Vienna, pp. 83–90.

Weinbren, K., W. Fitschen, and M. Cohen. 1960. The unmasking by regeneration of latent irradiation effects in the rat liver. Br. J. Radiol. 33:419–425.

White, A., and S. Hornsey. 1978. Radiation damage to the rat spinal cord: The effect of single and fractionated doses of X rays. Br. J. Radiol. 51:515–523.

Radiation Biology in Cancer Research, edited by
Raymond E. Meyn and H. Rodney Withers.
Raven Press, New York © 1980.

Quantitative Studies of the Radiobiology of Hormone-Responsive Normal Cell Populations

Kelly H. Clifton

Departments of Human Oncology and Radiology, Wisconsin Clinical Cancer Center, University of Wisconsin Medical School, Madison, Wisconsin 53792

Quantitative studies of the acute and late carcinogenic effects of irradiation on mammary and thyroid epithelial cells were undertaken in our laboratories because of the importance of these two radiogenic cancers in humans (Hempelmann 1968, BIER Report 1972) and in response to the following considerations:

1. Quantitative radiation dose–cell survival response data were available for but a few normal cell types and were restricted to those with prominent stem cell compartments and high proliferative rates (Elkind and Whitmore 1967, Withers 1967, Withers and Elkind 1969, Emery et al. 1970, Till and McCulloch 1970, Withers and Mason 1974).
2. Little was known concerning the relationship, if any, between acute cell death, intracellular repair processes, and late overt neoplasia.
3. The effects of physiological state—e.g., cell cycle stage, functional condition, cell age—at the time of radiation exposure and during the tumor latent period had not been totally unraveled, nor had the roles of physical and chemical factors such as LET and oxygen.

In order to gain information in these areas, we set out to develop systems that would meet the following criteria:

1. Known numbers of normal cells of a specific radiogenic neoplasm-susceptible type could be irradiated under controlled conditions in vivo or in vitro.
2. The post-irradiation cell survival could be estimated.
3. Abscopal radiation effects could be minimized or avoided.
4. Physical and physiological factors could be modified during exposure, and the latter before exposure and during the tumor latent period.
5. And, ultimately, the risk of neoplasia per rad per initial cell and per surviving cell could be estimated under well-defined conditions.

Two systems that approximate these criteria have been developed—the rat mammary epithelium and rat thyroid epithelium. These systems were chosen for initial study because of their susceptibility to radiogenic neoplasia and hormonal manipulation. For example, Shellabarger (1971) had reported induction of tumors in pieces of rat mammary tissue irradiated in vitro and autografted, and the hormonal potentiation of mammary carcinogenesis was well established (Yokoro et al. 1961, Furth 1973, Clifton and Sridharan 1975, Shellabarger 1976). Radiogenic thyroid neoplasia had been demonstrated (Malone 1975) and monodispersed thyroid cells successfully grafted (Kerkof and Chaikoff 1966). The techniques we have developed or adapted for these tissues appear, however, to be applicable to other cell types and to other physical or chemical carcinogenic or toxic insults.

In the following, the techniques are described, and the experience to date with mammary and thyroid epithelial cells is summarized, including questions raised as well as problems answered. The findings are discussed in relation to carcinogenesis studies. It is to be emphasized that the results from our laboratory represent the labors of many, including especially Drs. Michael Gould and Peter Mahler (mammary work), Dr. John Crowley (statistics), and Robert DeMott and Timothy Mulcahy (thyroid work).

METHODOLOGY AND RATIONALE

Cell Transplantation

Inbred Fischer, or occasionally W/Fu, rats have been used throughout. The procedures have been described in detail elsewhere and are summarized for thyroid cells in Figure 1.

Glands are removed from donor rats surgically while alive or after killing, minced, and enzymatically dissociated by incubation with collagenase followed

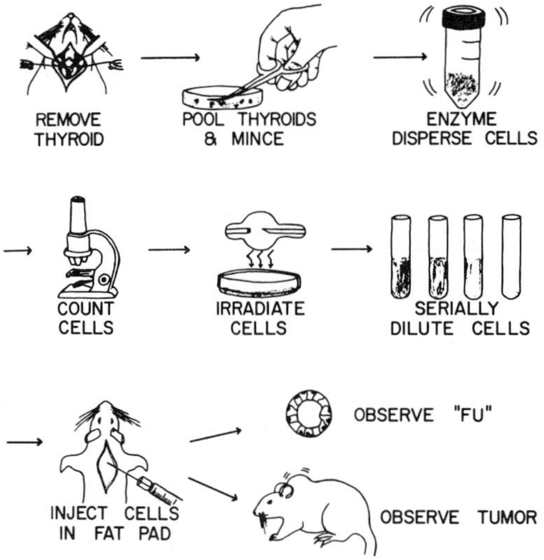

FIG. 1. Flow chart for preparation and transplantation of known numbers of monodispersed thyroid cells for survival assay and carcinogenesis studies. See text for details.

by pronase. The cell suspension is passed through a filter with ~ 50 μm pore size, and DNase is added before or after filtration to avoid DNA-mediated cell clumping.

The concentration of monodispersed "viable" cells is determined with a phase microscope and hemocytometer, and serial dilutions are prepared. Known cell numbers are then inoculated into fat pads of recipient animals. In the case of mammary cells, we have employed the interscapular white fat pad overlying the brown fat, which is free of mammary tissue. Thyroid cells have been inoculated in as many as five sites per recipient—the interscapular fat pad and two sites in each inguinal mammary pad. Fat pads have long been recognized as excellent sites for transplantation of mammary tissue (DeOme et al. 1959) and are better than other sites tested for thyroid cell grafts (Clifton et al. 1978).

Radiation

The glands may be irradiated in vivo before preparation of suspensions or in vitro thereafter. This feature adds important flexibility. In the studies reviewed here, we have employed 250 kvp X rays (Gould and Clifton 1977, DeMott 1978, DeMott et al., 1979) and ^{137}Cs gamma rays, and ~ 2 MeV modified fission neutrons (Clifton et al. 1975a, Clifton and Crowley 1978).

Cell Survival Assays

These assays are based on the capacity of inocula of monodispersed cells to give rise to glandular structures in syngeneic recipients. For ease of identification of these structures, it is frequently desirable to hormonally stimulate the grafted cells for three to four weeks to induce proliferation and secretion. After a given period—three to six weeks for mammary cells, usually four weeks for thyroid cells—the grafted fat pads are removed, fixed, stained, defatted, and stored in mineral oil. Glandular structures are readily identifiable by inspection with a dissecting microscope and may be confirmed by routine sectioning and microscopy (Gould et al. 1977, Clifton et al. 1978).

Successful grafting of mammary cells is indicated by one or more AU, or alveolar units, comprising a sphere of mammary epithelium and accompanying myoepithelial cells surrounding a milk-filled lumen and in turn surrounded by a basement membrane (Figure 2) (Clifton et al. 1976, Gould et al. 1977). Similarly, successful thyroid grafts comprise one or more FU, or follicular units, consisting of thyroid cells in normal arrangement around a colloid-distended lumen (Figure 3) (Clifton et al. 1978). The "AD50" for mammary cells and the "FD50" for thyroid cells, that cell number which results in one or more AU or FU in 50% of the graft sites in analogy to the tumor cell TD50 of Hewitt and Wilson (1959), is calculated according to the relationship derived by Porter et al. (1973). The probability, p, of one or more AU or FU per graft site is:

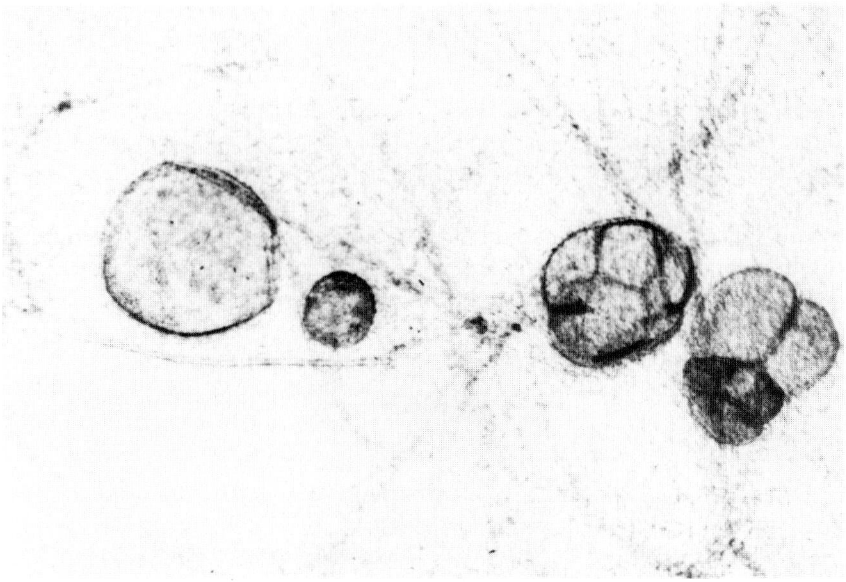

FIG. 2. Whole mount of a cluster of AU (alveolar units) that developed from inoculation of monodispersed mammary cells into the interscapular fat pad of a rat also grafted with MtH-secreting pituitary tumor strain MtT-F4. Note septa in some AU. Stained with hematoxylin and cleared (105X). Reproduced from Clifton et al. (1976) with permission of the International Atomic Energy Agency, Vienna.

$$p = 1 - e^{-M} \qquad \text{(Eq. A)}$$

and

$$\log M = \log K + S \log Z \qquad \text{(Eq. B)}$$

where M is the number of AU- or FU-forming cells in the inoculum, K is the fraction of cells capable of AU or FU formation, S is a constant that determines the slope of the relationship, and Z is the total number of cells inoculated. The fraction of cells surviving a given radiation exposure is then (AD50 or FD50 control)/(AD50 or FD50 irradiated) (Gould and Clifton 1977, DeMott et al. 1979) (Figure 4).

Hormonal Manipulation

Transplantable mammotropic pituitary tumors (MtT) have been used as indwelling sources of hormones in both mammary cell survival assays and carcinogenesis studies (Clifton et al. 1975b, Clifton and Crowley 1978, Gould et al. 1978). Alternatively, MtH (mammotropic hormone, prolactin) has also been more modestly elevated by grafts of anterior pituitary tissue beneath the kidney capsule (Clifton et al. 1975b). Finally, we are currently testing a technique in

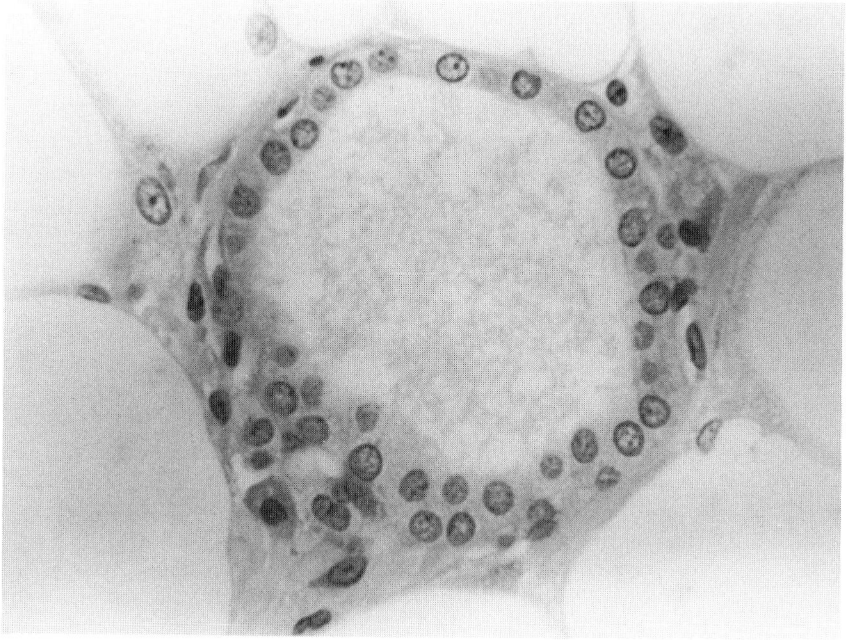

FIG. 3. Photomicrograph of a section through FU—a follicular structure present at the graft site four weeks after inoculation of monodispersed thyroid cells into an inguinal mammary fat pad of a hypothyroid recipient rat. Periodic acid–Schiff, hematoxylin stain (550X).

which anterior pituitary tissue is grafted in the spleen adjacent to a silastic capsule containing estrogen. Under these circumstances, we expect estrogenic stimulation of MtH release, most of which should reach the peripheral circulation. In contrast, most or all of the estrogen would be expected to be inactivated

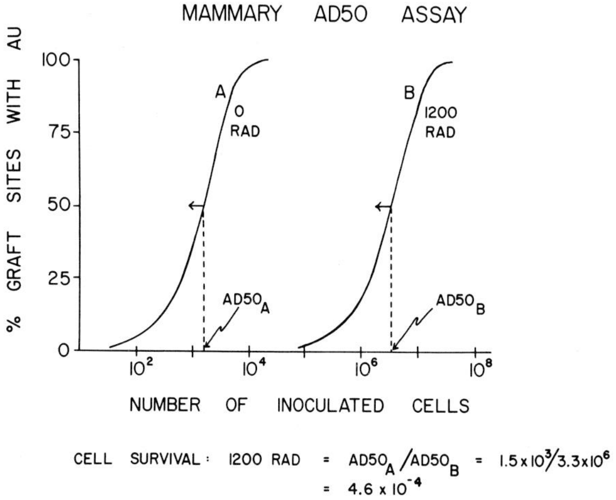

FIG. 4. Graphical representation of the effect of irradiation of the mammary gland on AU formation from inocula of monodispersed mammary cells and calculation of AD50 values and cell survival.

during passage through the liver and would not reach the peripheral circulation. In thyroid cell survival assays, elevated thyrotropin has been induced in recipient rats by thyroidectomy coupled with a low-iodine diet (Clifton et al. 1978, DeMott et al. 1979). Comparable specific physiological stimulation might be employed with cells of other types. Furthermore, specific hormonal inhibition of proliferation and/or differentiative function may be employed. For example, we have inhibited differentiation for milk secretion by induction of glucocorticoid deficiency in mammary carcinogenesis studies (Clifton et al. 1975b, Clifton and Crowley 1978, Gould and Clifton 1978a).

Carcinogenesis Experiments

To achieve the aims of the program it was necessary to establish conditions by which the efficiency of radiogenic neoplasia induction was maximized. Toward this goal, we have investigated the mammary tumor response of female rats to whole-body irradiation with fission neutrons (10–100 rads) or gamma rays (50–600 rads) and with a variety of hormonal conditions (Clifton et al. 1975b, Clifton and Crowley 1978, Gould et al. 1978). Maximum carcinoma incidence has been observed thus far in animals irradiated with 100-rad neutrons or 500-rad gamma rays, followed by exposure to elevated MtH in combination with glucocorticoid deficiency (Clifton and Crowley 1978). These conditions are being employed in current studies aimed at carcinoma induction from known numbers of grafted irradiated mammary cells.

As discussed below, thyroid epithelial cells may be better suited to the quantitative study of carcinogenesis in vitro. Indeed, carcinoma incidence exceeded 40% in grafts of thyroid cells irradiated with 500-rad X rays in vitro and grafted in thyroidectomized recipients maintained on low-iodine diets (DeMott 1978).

PROGRESS AND PROBLEMS

Nature of Mammary AU and Thyroid FU

The mammary AU that develop from inoculated monodispersed cells are normal insofar as light and electron microscopy have yet revealed (Gould et al. 1977). AU contain both epithelium and myoepithelium normally oriented within a basement membrane. When stimulated by MtH, the epithelial cells secrete, and the secretory products accumulate and distend the central lumen. With time, AU develop septa in apparent formation of alveolar clusters (see Figure 2), and duct-like buds are occasionally observed. Furthermore, three or more months after inoculation of large cell numbers, complete mammary structures have been found. And finally, both morphological observations of events soon after grafting and the relationship between the number of cells per inoculation site and the number of sites that develop AU are consistent with the conclusion that an AU can arise from a single cell (Gould et al. 1977, Gould and Clifton 1977).

At any given time AU-forming cells represent a small percentage of the total cell number, and their concentration varies with hormonal status. For example, the AD50 of cells from mammary glands of untreated virgin females was 1.5×10^3. In contrast, 8.2×10^3 cells from MtH-stimulated, highly proliferative glands were required to produce AU in 50% of the graft sites (Gould and Clifton 1978a).

Is there a mammary gland stem cell? Or are all or most mammary cells capable of AU formation at one or more stages in their functional-reproductive cycles? We do not know. Experiments designed to increase the efficiency of AU formation by cell separation with density gradient centrifugation have thus far failed (Gould and Clifton 1978b). However, efforts to perfect this system are continuing.

Are AU-forming cells the same as those from which carcinomas ultimately arise? Again, we do not know, but we suspect future experiments will show them to be so.

The thyroid cell transplantation system differs in a number of ways from the mammary system, some of which are advantageous for quantitative study. The FD50 is remarkably low; under optimal conditions, only 55 cells from immature male donors grafted in thyroid-deficient recipients (DeMott et al. 1979). Light and electron microscopy, as well as parameters of function, indicate that FU comprise physiologically normal cells. For example, they concentrate radioiodide and support the body growth of thyroidectomized recipients given normal diets (Clifton et al. 1978). The morphological response to differing hormonal conditions is nearly identical to that of normal thyroid. An added advantage is that assays may be designed such that several cell inoculation sites may be used in the same animal. We now routinely make five grafts per recipient rat.

A confusing problem is, however, that the cell dose–FU response relationship deviates from that expected for a one, two, three or more cell FU-origin model, although it is closest to one. A slope (S) (Eq. B) of one would be consistent with a single-cell origin of an FU. If more than one cell were required, the value of S would be greater than one. Our calculations indicate an S value of ~ 0.5 (DeMott et al. 1979). This suggests that the efficiency of FU formation per cell decreases as the number of cells in the inoculum increases. This relationship is not, however, altered by radiation exposure of the grafted cells and thus does not influence the generation of dose-survival curves.

The FD50 is heavily dependent on the hormonal status of the recipient rat. For example, when the recipients are thyroidectomized and placed on an iodine-deficient diet on the day of grafting, the FD50 is less than 100 cells. In contrast, when the recipient rats are euthyroid, the FD50 is approximately ten times greater (DeMott 1978).

Studies with ^{3}H-thymidine-labeled cells indicate that this difference is not attributable to loss of the inoculated cells from the graft site during the first five days after inoculations (Mulcahy and Clifton, unpublished data). Rather, the evidence suggests that the inoculated cells remain at the graft site in euthyroid animals but do not form recognizable glandular structures. If the euthyroid host is thyroidectomized four weeks after grafting and the graft sites inspected four weeks after surgery, the FD50 decreases to near that in initially hypothyroid hosts.

A related problem is the fate of the other cell types, i.e., parathyroid and medullary "C" cells, which are contained in our monodispersed thyroid suspensions. Immunohistochemical studies have demonstrated that C cells are present in small number in parafollicular location in the thyroid structures that develop from monodispersed cell inocula (Mulcahy and Clifton, unpublished data). Whether these are cells that persisted from the inocula without division or are the product of active proliferation is unknown. Furthermore, we do not have clear evidence of functional parathyroid tissue in the grafts, although parathyroid is included in the initial tissue minces. This may be related in part to the fact that in the rat, parathyroid tissue is not completely removed by surgical thyroidectomy, and there thus may be little stimulation of any grafted cells.

Acute Radiation Sensitivity

It is of interest that hormonal stimulation of the mammary cells before and during irradiation had no effect on the survival of rat mammary cells. The survival parameters of cells irradiated in vivo and removed immediately for AD50 assay were statistically indistinguishable if the donors were grafted with MtH-secreting tumors or were untreated virgins—and this despite the fivefold difference in unirradiated AD50 values between mammary cells from these two sources. The combined D_0 valuc was 129 rad with an extrapolation number, n, of 5 as calculated from the multitarget–single hit model (Gould and Clifton 1977, 1978a).

This relatively low extrapolation number is comparable to those reported for mammalian cells irradiated and assayed in culture and for normal marrow cells and tumor cells removed for transplant assay soon after exposure (Elkind and Whitmore 1967). It is, however, lower than those calculated from split-dose studies of skin and intestinal epithelial stem cells irradiated and assayed in situ (Withers 1967, Withers and Elkind 1969, Emery et al. 1970, Withers and Mason 1974). When 24 hours were allowed to elapse between irradiation of the donor rats and removal of the mammary tissue for AD50 assay, the D_0 value was unchanged but the extrapolation number was increased to 17 (Figure 5) (Gould and Clifton 1979). Cell-labeling studies indicated that this increase was not attributable to cell proliferation or cell loss. Rather, we believe it represents an increase in intracellular repair which is dependent on retention of the irradiated cells in their normal tissue arrangement, and have thus suggested that it be termed "in situ repair" (ISR) (Gould and Clifton 1979). Whether

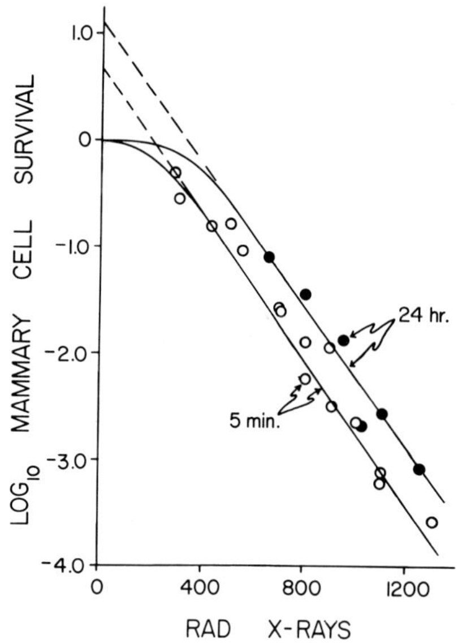

FIG. 5. Comparison of mammary cell survival when glands were removed from donor rats for survival assay five minutes after irradiation vs. 24 hours after exposure. Slopes and D_o values were insignificantly different. Note shift to right of "24 hour" curve indicating increased intracellular repair, termed ISR (in situ repair). Redrawn from Gould and Clifton (1979).

this tissue environment–dependent, increased intracellular repair (ISR) occurs in other cell types is as yet unknown.

Thyroid epithelial cells are significantly more radioresistant than mammary cells. The D_0 of thyroid cells irradiated with X rays in vitro was 192 rad, and the n value was 4 (Figure 6) (DeMott et al., in press). Recent data indicate that the parameters will be very similar for cells irradiated in vivo and immediately removed for assay (Mulcahy and Clifton, unpublished observations). We plan to investigate the possibility of ISR in thyroid in the near future.

Carcinogenesis

In parallel studies, we investigated the hormonal conditions for maximal expression of neoplastic change in the mammary tissue of the inbred Fischer and W/Fu rat strains. Building on extensive previous work, particularly that of Shellabarger and associates (Shellabarger 1976) and Furth and associates (Furth 1973), these studies have revealed or confirmed:

1. That MtH markedly potentiates mammary neoplasia and influences the nature of the resultant neoplasms, i.e., carcinoma vs. fibroadenoma (Clifton et al. 1975a, Clifton and Crowley 1978, Gould et al. 1978).
2. That glucocorticoid deficiency further potentiates radiogenic carcinoma induction (Clifton et al. 1975b, Clifton and Crowley 1978).
3. That an animal that has had one carcinoma is at a greater carcinoma risk than one that has not previously developed a tumor (Clifton and Crowley 1978, Gould et al. 1978).

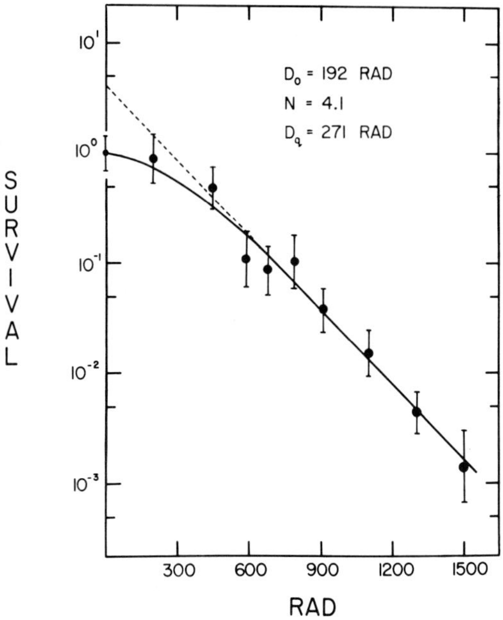

FIG. 6. Survival of thyroid cells irradiated with X rays in vitro under well-oxygenated conditions. Reproduced from DeMott et al. (1979) with permission of Academic Press, Inc.

4. That over the dose ranges of 10–100 rad fission spectrum neutrons and 50–500 rad ^{137}Cs gamma rays, the dose-carcinoma response approximates linearity. The neutron/gamma relative biological effectiveness (RBE) is ~4 (Clifton and Crowley 1978).

Consideration of the mammary cell survival relationship illustrates that the apparent linearity is probably fortuitous. For example, assuming a D_0 value of 127 rad and n value of 17 (Gould and Clifton 1979), one would expect surviving mammary cell fractions of 1.000 and 0.995 after low linear energy transfer (LET) radiation doses of 50 and 167 rad, respectively. After 500 rad, the surviving fraction would be reduced to 0.285. Thus, assuming potential carcinoma cells are as radiosensitive as normal cells, virtually all potential carcinoma cells would survive the acute effects of exposure at the two lower doses, but only about one fourth would survive at 500 rad. These results suggest that there may be a difference in the mechanism and/or sites of the lethal and neoplasm-inducing events. The incidence of carcinomas in known numbers of grafted irradiated cells is under investigation.

As noted above, the thyroid system may have advantages for the quantitative study of carcinogenesis. In two preliminary experiments, carcinomas were observed in grafts of known numbers of cells that had been irradiated in vitro and grafted in immature thyroidectomized recipients maintained on iodine-deficient diets. In one such experiment, carcinomas were observed in 44% of the grafts of 3×10^5 cells that had been irradiated with 500 rad (Table 1) (DeMott 1978).

TABLE 1. *Neoplasia in grafts of irradiated and unirradiated monodispersed thyroid cells in hypothyroid hosts*

				Neoplasms	
No. Inoculated Cell*	Radiation Dose (X rays)†	No. Surviving Cell‡	No. rats	No. Carcinomas	No. Adenomas
3.0×10^5	500 rad	7.9×10^4	16	7	9
6.0×10^4	0	6.0×10^4	13	2	10
3.0×10^5	0	3.0×10^5	6	0	6

* "Viable" cells as determined by phase microscopy; one inoculation site per rat.
† Irradiated in aerated suspension in vitro.
‡ Surviving fraction calculated assuming $D_0 = 192$ rad and $n = 4$ (DeMott et al., in press).

From the transplantation vs. FU relationship (Eq. B above), assuming an FD50 of 55 unirradiated cells (DeMott et al. 1979), it can be calculated that there is one FU-forming unit per 114 cells and that 38 unirradiated cells would yield one or more FU in 44% of the graft sites. If one further assumes, for purpose of discussion, that potential carcinoma cells follow the same transplantation relationship as normal cells, then the inocula contained one potential carcinoma cell per $(114/38) \times 3 \times 10^5$ inoculated cells, or 1 potential carcinoma cell per 9×10^5 initial cells.

Now, thyroid cell survival at 500 rad is 26% (see Figure 6). Thus, a 44% carcinoma incidence represents 1 carcinoma per 2.3×10^5 *surviving cells.* Finally, if one assumes that carcinomas arise from FU-forming cells, the incidence is 1 carcinoma per 8×10^3 initial FU-forming units, or 1 carcinoma per 2×10^3 surviving FU formers.

These calculations are presented only for the purpose of illustration. Experiments aimed at testing the assumptions noted are in progress.

CONCLUSION

The techniques and early results presented here offer the hope that a number of fundamental questions in the radiobiology and carcinogenesis of mammary and thyroid epithelia may be amenable to quantitative study. Furthermore, the procedures may be utilizable in the study of a variety of other problems in differentiation, cell-cell interaction, cytolethality, and carcinogenesis.

Finally, these techniques, which are adaptations and refinements of methods in common use in tumor radiobiology and endocrinology, appear applicable to other tissues. For example, R. Jirtle and G. Michalopoulos of Duke University have found that hepatocytes are similarly transplantable with high efficiency (personal communication). In our laboratories, we have encouraging early results with prostate cells and adrenocortical subcapsular cells.

The relationship between endocrinology and cancer research has been long and valuable. We hope this association of the two with cellular radiobiology will be of further profit.

ACKNOWLEDGMENTS

This work has been supported by Research Grant RO1 CA13881, Program Grant PO1 CA19278, and Comprehensive Cancer Center Grant P30 CA14520 from the National Cancer Institute, Department of Health, Education and Welfare, and by Research Grant PDT 86 from the American Cancer Society. We are indebted to Jane Barnes, Joan Eggert, Joan Mitchen, Dyan Nagle, and Laura Nettleton for excellent technical assistance.

REFERENCES

Advisory Committee on Biological Effects of Ionizing Radiation. 1972. Somatic effects of ionizing radiation, *in* BEIR Report: The Effects on Populations of Exposure to Low Levels of Ionizing Radiation. National Academy of Sciences, Washington, D.C., pp. 120–125.

Chen, K. Y., and H. R. Withers. 1972. Survival characteristics of stem cells of gastric mucosa exposed to localized γ-radiation in C3H mice. Int. J. Radiat. Biol. 21:521–534.

Clifton, K. H., and J. J. Crowley. 1978. Effects of radiation type and dose and the role of glucocorticoids, gonadectomy, and thyroidectomy in mammary tumor induction in mammotropin-secreting pituitary tumor-grafted rats. Cancer Res. 38:1507–1513.

Clifton, K. H., R. K. DeMott, R. T. Mulcahy, and M. N. Gould. 1978. Thyroid gland formation from inocula of monodispersed cells: Early results on quantitation, function, neoplasia, and radiation effects. Int. J. Radiat. Oncol. 4:987–990.

Clifton, K. H., E. B. Douple, and B. N. Sridharan. 1975a. Effects of grafts of single anterior pituitary glands on the incidence and type of mammary neoplasm in neutron- or γ-irradiated Fischer female rats. Cancer Res. 36:3732–3735.

Clifton, K. H., and B. N. Sridharan. 1975. Endocrine factors and tumor growth, *in* Cancer: A Comprehensive Treatise, F. F. Becker, ed. Vol. 3. Plenum Press, New York, pp. 249–285.

Clifton, K. H., B. N. Sridharan, and E. B. Douple. 1975b. Mammary carcinogenesis enhancing effect of adrenalectomy in irradiated rats with pituitary tumor MtT-F4. J. Natl. Cancer Inst. 55:485–487.

Clifton, K. H., B. N. Sridharan, and M. N. Gould. 1976. Risk of mammary oncogenesis from exposure to neutrons or gamma rays, experimental methodology and early findings, *in* Biological and Environmental Effects of Low-Level Radiation, Vol. 1. International Atomic Energy Agency, Vienna, pp. 205–212.

DeMott, R. K. 1978. The radiation response of the rat thyroid gland: Transplantation, cell survival and neoplastic development in grafts of monodispersed cells. M. S. Thesis in Radiological Science. University of Wisconsin, Madison, Wisconsin.

DeMott, R. K., R. T. Mulcahy, and K. H. Clifton. 1979. The survival of thyroid cells following irradiation: A directly generated single-dose survival curve. Radiation Res. 77:395–403.

DeOme, K. B., L. J. Faulkin, Jr., H. A. Bern, and P. B. Blair. 1959. Development of mammary tumors from hyperplastic alveolar nodules transplanted into gland-free mammary fat pads of female C3H mice. Cancer Res. 19:515–520.

Elkind, M. M., and G. F. Whitmore. 1967. The Radiobiology of Cultured Mammalian Cells. Gordon and Breach, New York, pp. 85–102.

Emery, E. W., J. Denekamp, and M. M. Ball. 1970. Survival of mouse skin epithelial cells following single and divided doses of X-rays. Radiat. Res. 41:450–466.

Furth, J. 1973. The role of prolactin in mammary carcinogenesis, *in* Human Prolactin, Proceedings of the International Symposium on Human Prolactin, J. L. Pasteels and C. Robyn, eds. American Elsevier, New York, pp. 233–248.

Gould, M. N., and K. H. Clifton. 1977. The survival of mammary cells following irradiation *in vivo:* A directly generated single dose survival curve. Radiat. Res. 72:343–352.

Gould, M. N., and K. H. Clifton. 1978a. The survival of rat mammary gland cells following irradiation *in vivo* under different endocrinological conditions. Int. J. Radiat. Oncol. 4:629–632.

Gould, M. N., and K. H. Clifton. 1978b. The quantitative transplantation of subpopulations of rat mammary gland cells separated on renograffin gradients. Exp. Cell Res. 114:451–454.

Gould, M. N. and K. H. Clifton. 1979. Evidence for a unique *in situ* component of the repair of radiation damage. Radiat. Res. 77:149–155.
Gould, M. N., W. F. Biel, and K. H. Clifton. 1977. Morphological and quantitative studies of gland formation from inocula of monodispersed rat mammary gland cells. Exp. Cell Res. 107:405–416.
Gould, M. N., K. H. Clifton, and J. Crowley. 1978. Effects of endocrinological conditions associated with acute versus chronic lactation on the incidence of mammary carcinomas in irradiated rats. J. Natl. Cancer Inst. 60:469–471.
Hemplemann, L. H. 1968. Risk of thyroid neoplasms after irradiation in childhood. Science 160:159–163.
Hewitt, H. B., and C. W. Wilson. 1959. A survival curve for mammalian cells irradiated *in vivo*. Nature 183:1060–1061.
Kerkof, P. R., and I. L. Chaikoff. 1966. Follicular reorganization and I^{131} utilization by cultured rat thyroid cells implanted into thyroidectomized rats. Endocrinology 7:1177–1188.
Malone, J. F. 1975. The radiation biology of the thyroid. Curr. Top. Radiat. Res. 10:265–368.
Porter, E. H., H. B. Hewitt, and E. R. Blake. 1973. The transplantation kinetics of tumor cells. Br. J. Cancer 27:55–62.
Shellabarger, C. J. 1971. Induction of mammary neoplasia after *in vitro* exposure to X-rays. Proc. Soc. Exp. Biol. Med. 136:1103–1106.
Shellabarger, C. J. 1976. Modifying factors in rat mammary gland carcinogenesis, *in* Biology of Radiation Carcinogenesis, J. M. Yuhas, R. W. Tennant, and J. D. Regan, eds. New York, Raven Press, pp. 31–43.
Till, J. E., and E. A. McCulloch. 1970. Early repair processes in marrow cells irradiated and proliferating *in vivo*. Radiat. Res. 41:450–466.
Withers, H. R. 1967. Recovery and repopulation *in vivo* by mouse skin epithelial cells during fractionated irradiation. Radiat. Res. 32:227–239.
Withers, H. R., and M. M. Elkind. 1969. Radiosensitivity and fractionation response of crypt cell of mouse jejunum. Radiat. Res. 38:598–613.
Withers, H. R., and K. A. Mason. 1974. The kinetics of recovery in irradiated colonic mucosa of the mouse. Cancer 34:896–903.
Yokoro, K., J. Furth, and N. Haran-Ghera. 1961. Induction of mammotropic pituitary tumors by X-rays in rats and mice: The role of mammotropes in development of mammary tumors. Cancer Res. 21:167–192.

Radiation Biology in Cancer Research, edited by
Raymond E. Meyn and H. Rodney Withers.
Raven Press, New York © 1980.

Effect of Lung Irradiation on Metastases: Radiobiological Studies and Clinical Correlations

Lester J. Peters, Kathryn A. Mason, and H. Rodney Withers

Section of Experimental Radiotherapy, Department of Radiotherapy, The University of Texas System Cancer Center M. D. Anderson Hospital and Tumor Institute, Houston, Texas 77030

Over 40 years ago, the first anecdotal reports of metastases localized to heavily irradiated skin appeared (Schürch 1935, Schwarz 1935). More recently, Dao and Kovaric (1962) reported on a series of 354 breast cancer patients that showed a higher incidence of ipsilateral lung and skin metastases in a selected group of 103 patients who received postoperative radiotherapy following mastectomy than in the remainder treated by surgery alone. However, this series was heavily biased in favor of patients treated by radical mastectomy alone, and the conclusions of the study could not be confirmed in a subsequent review of 674 breast cancer patients reported by Chu et al. (1967). Nonetheless, the occasional localization of metastases within irradiated tissues is a clinical reality, and a similar phenomenon has been amply demonstrated in a variety of animal tumor models (for review, see Von Essen and Stjernswärd 1978). Several investigators have documented an increase in the lung colony forming efficiency of intravenously injected tumor cells (Dao and Yogo 1967, Fisher and Fisher 1969, Brown 1973, Withers and Milas 1973, Van den Brenk et al. 1973, Thompson 1974) or of tumor cells shed spontaneously from autochonous or transplanted "primary" tumors (Owen and Bostock 1973, Van den Brenk et al. 1973, Peters 1974) following local irradiation of the lungs (usually with large single doses). However, the mechanism of the effect has remained elusive (Peters et al. 1978) and its relevance to clinical radiotherapy uncertain. In this paper, we describe experiments undertaken to define the radiobiological characteristics of the effect, including considerations of dose rate, fractionation, and linear energy transfer (LET), and we present data illustrating the importance of the timing of tumor cell dissemination to the lungs during a course of fractionated treatment as a determinant of the metastatic risk. From these experiments, certain potentially hazardous clinical practices are defined. However, in the absence of uncontrolled disease outside the irradiated volume, the risk of enhanced tumor growth in irradiated tissues appears to be low. A review of the M. D. Anderson experience with management of breast cancer by radiotherapy alone or by surgery plus postoperative radiotherapy failed to indicate any excess of metastases in the irradiated ipsilateral lung.

MATERIALS AND METHODS

Mice and Tumor

C_3Hf/Kam mice bred in the specific pathogen-free colony maintained in the Section of Experimental Radiotherapy at M. D. Anderson Hospital were used in all experiments. Mice of both sexes, aged 10–12 weeks at the initiation of the experiments, were used. The animals were housed five/cage and were maintained on a sterilized pellet diet.

The tumor used in these experiments is a methlycholanthrene-induced fibrosarcoma (FSa) that is demonstrably immunogenic in its syngeneic hosts (Suit and Kastelan 1970). However, the immunogenicity of the tumor is not a consideration in the phenomena to be described (Peters et al. 1978). Source material for these experiments was derived from fifth generation isotransplants of tumor brei stored in liquid nitrogen. The method used for preparing single-cell suspensions by mechanical mincing and trypsinization has been described previously (Milas et al. 1974).

Lung Colony Assay

Single-cell suspensions of FSa, diluted to contain $1\text{–}4 \times 10^4$ viable cells in 0.25 ml of Hsu's medium supplemented with 5% fetal calf serum, were injected intravenously into appropriately treated mice. Fourteen days later, the mice were killed, their lungs were removed and fixed in Bouin's solution, and the number of macroscopic tumor nodules was counted as originally described by Hill and Bush (1969).

Irradiation

For most experiments, a double-headed small-animal ^{137}Cs irradiator was used, at a dose rate of 968 rad/minute. For immobilization during irradiation, mice were anesthetized with pentobarbital, 65 mg/kg i.p. The portal for "local thoracic irradiation" (LTI) was 3 cm diameter, extending from the lower costal margin to the mid-neck of the mouse. Dosimetry, determined by TLD measurements, was homogeneous to $\pm 3\%$ of the nominal dose (Hranitzky et al. 1973). When required, the dose rate could be reduced to 180 rad/minute by insertion of 2.4 HVL lead attenuators.

Low dose-rate irradiations (15.1 rad/minute) were accomplished using a Theratron 80 ^{60}Co teletherapy machine at a source to mid-mouse distance of 168 cm. The beam was collimated with tapered lead blocks 10 cm thick standing on a 1-in. thick lucite plate immediately above the mice. These blocks defined a 3.0-cm wide exposure zone covering the thoracic regions of eight mice in two parallel rows. A water phantom containing 7.5 cm water was placed directly above the shielding blocks to attenuate the beam to the desired dose rate. The dose to the protected regions of the mice 1 cm under the shielding blocks was ~8% of the dose to the thorax. Anesthesia for these irradiations was provided by a mixture of pentobarbital, 50 mg/kg, and phenobarbital, 50 mg/kg, both given i.p.

Neutron irradiation was performed at the Texas A & M University Variable Energy Cyclotron (TAMVEC) using the 50 $MeV_{d \rightarrow Be}$ fast-neutron beam. A special collimator was constructed, consisting of six steel pipes of 3-cm inner diameter, embedded in a 50 × 20 × 20 cm block of water-extended polyethylene and iron filings (Oliver and Moore 1970). With the collimator properly aligned (using lasers), and the face of the collimator at a distance of 140 cm from the neutron source, the axes of the steel pipes projected exactly to the target. Mice, anesthetized with pentobarbital 65 mg/kg, were positioned for irradiation on a half-inch thick perspex plate, which was positively located on the collimator face with studs. The dose rate (neutron and γ) at the center of each of the six irradiation portals was 78 rad/min ± 3%. Photographic densitometry was used to determine the penumbral characteristics of the beam, and it indicated that a 2.8-cm diameter field was encompassed by the 95% isodose contour.

RESULTS

Time Course of Lung Colony Enhancement

Following a single dose of 1,000-rad local thoracic irradiation using ^{137}Cs γ rays at 968 rad/minute, 10^4 FSa tumor cells were injected at times varying from six hours to 28 days. The results, plotted in Figure 1, show a maximum effect when tumor cells were injected one to two days after irradiation, with a rapid fall-off to a plateau of about half the maximum effect occurring by four days. This residual level of effect persisted for at least one month. At the time of maximal enhancement of lung colony forming efficiency, the histological appearance of the lungs was completely normal (Figure 2).

Dose-Response Relationships for γ Rays and Neutrons

The number of lung colonies resulting from a standard injection of 10^4 FSa cells, given i.v. 24 hours after various doses of ^{137}Cs γ rays to the thorax, are plotted in Figure 3, upper panel. This shows an essentially linear dose response up to 1,750 rad, the maximum single dose that could be given without causing weight loss and inanition due to acute esophagitis.

An almost identical single dose-response relationship was obtained with 50 $MeV_{d \rightarrow Be}$ fast-neutron irradiation (Figure 4, upper panel), although in these experiments the upper dose limit was 1,250 rad.

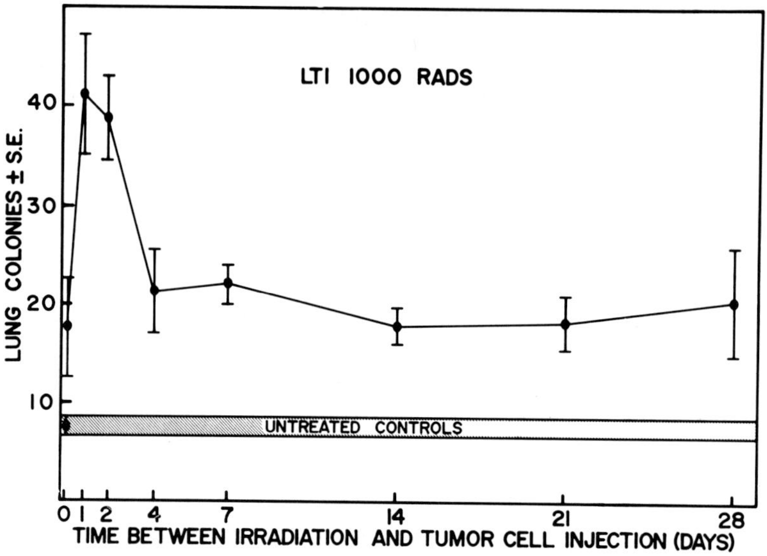

FIG. 1. Effect of the time interval between irradiation of the thorax and i.v. injection of tumor cells on the number of lung colonies formed. Mice were injected with 2 x 10^4 FSa cells at times ranging from three hours to 28 days after a single dose to the thorax of 1,000 rad γ rays. Maximum enhancement of lung colony forming efficiency occurred when tumor cells were injected one to two days after irradiation.

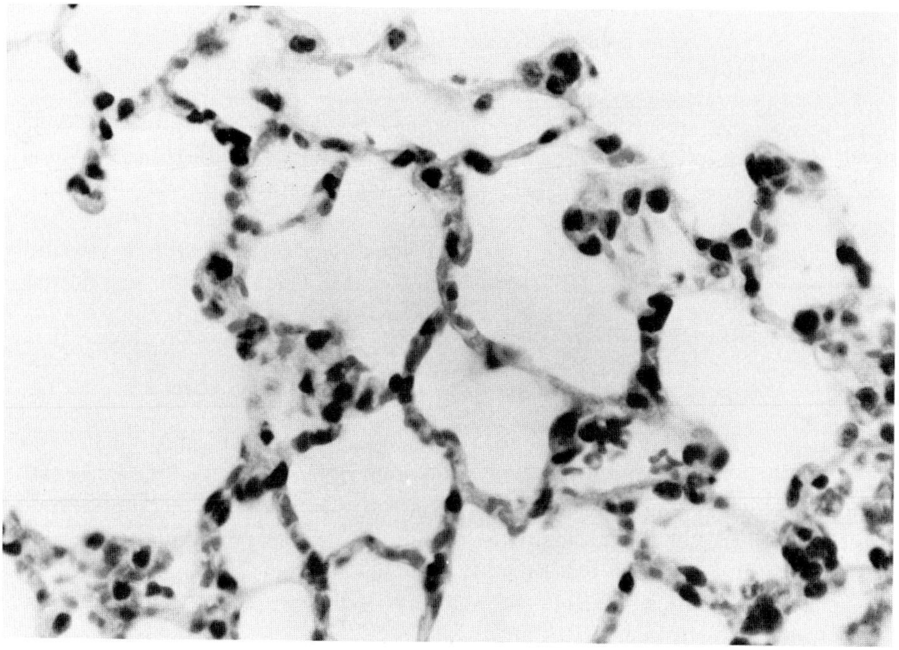

FIG. 2. Histological appearance of the lung of a mouse one day after thoracic irradiation with 1,000 rad ^{137}Cs γ rays. No morphological changes from normal can be distinguished by light or electron microscopy (H & E, ×320).

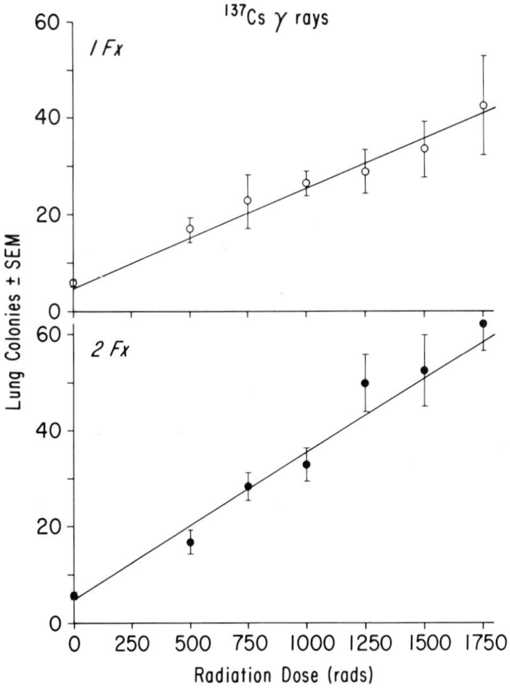

FIG. 3. Dose-response relationships for enhancement of lung colony forming efficiency by ^{137}Cs γ rays. The upper panel shows the response to single doses of radiation given to the thorax 24 hours before i.v. injection of 2×10^4 FSa tumor cells. The lower panel shows the response to two equal doses separated by four hours. Surprisingly, the slope of the dose-response curve for split-dose irradiation is significantly steeper than that for single doses.

Since very little difference in lung colony forming efficiency was observed after irradiation with equal single doses of neutrons or γ rays, approximate $RBE_{n/\gamma}$ values for doses in the range of 250–1,250 rad were estimated as the ratio of lung colonies at each dose of neutrons and γ rays, respectively. These RBE values ranged from 1.3 at 250 rad γ rays to 1.0 at 1,250 rad, almost exactly coinciding with the RBE previously obtained in this laboratory for hemopoietic stem cell survival, assayed by the endogenous spleen colony technique (Kogelnik and Withers, unpublished), but well below the RBE reported for other normal tissue end points (Field 1977).

Split-Dose Experiments

Dose-response relationships were also obtained for both ^{137}Cs γ rays and fast neutrons, using two equal-sized doses delivered four hours apart (Figures 3 and 4, lower panels). Tumor cells were injected i.v. 24 hours after the second radiation exposure. Unexpectedly, these experiments showed that split-dose irradiation with both γ rays and neutrons resulted in *greater* enhancement of lung colony forming efficiency than did single exposures to the same total doses. The difference in slope of the single and split-dose response curves achieved statistical significance with γ irradiation ($p < .01$), but not with neutrons

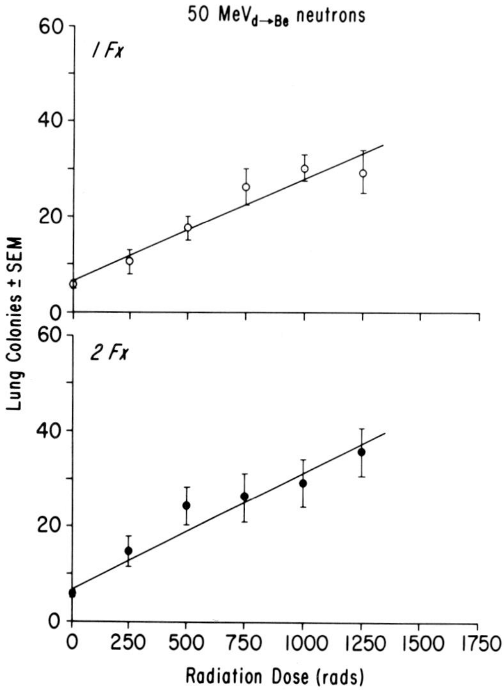

FIG. 4. Dose-response relationships for enhancement of lung colony forming efficiency by 50 MeV$_{d \rightarrow Be}$ neutrons. The upper panel depicts the single-dose response and the lower panel the split-dose response, as in Figure 2. Although the slope of the split-dose curve is again steeper, the difference in these experiments did not reach statistical significance, possibly because the maximum tolerable dose with neutrons was lower. The $RBE_{n/\gamma}$ for lung colony enhancement is low, ranging, for single doses, from 1.3 at 250 rad γ rays to 1.0 at 1,250 rad.

($p > .05$). Because of the greater split-dose enhancement seen with γ rays, estimates of $RBE_{n/\gamma}$ values in the 250–1,250 γ-ray rad range for divided doses were <1.0.

Influencc of Interfraction Interval

The preceding split-dose experiments were performed using a fixed four-hour interfraction interval. To determine whether the effect observed was dependent on a particular interval, further experiments were carried out, using a standard split dose of 900 + 900 rad ^{137}Cs γ radiation with interfraction intervals (Δt) of from 1 to 24 hours. These high doses were chosen to maximize the chance of resolving a split-dose enhancement. In all cases, the tumor cells were injected 24 hours after the second radiation dose. The results, plotted in Figure 5, show that at all times tested, divided doses were more effective than the single dose ($\Delta t = 0$).

Effect of Dose Rate

Mice were irradiated to 1,000 or 1,800 rad LTI using three widely differing dose rates: 968 rad/minute and 180 rad/minute of ^{137}Cs γ rays, and 15.1 rad/minute of ^{60}Co γ rays. Since the low dose-rate exposure required prolonged

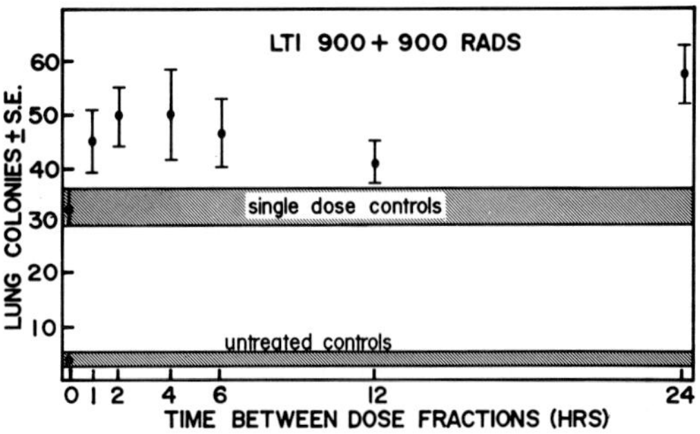

FIG. 5. Effect of duration of interfraction interval on split-dose response. Mice received two doses of 900 rads γ radiation to the thorax at intervals ranging from one to 24 hours. FSa tumor cells (2×10^4) were injected i.v. 24 hours after the second dose of irradiation. Irrespective of the interfraction interval, the lung colony forming efficiency was higher after split-dose irradiation than after a single dose of 1,800 rad.

anesthesia (> two hours), all mice in these experiments received the pentobarbital plus phenobarbital mixture described above. The results, presented in Table 1, indicate no difference in effect between the three dose rates at 1,000 rad, but significantly increased effectiveness of low dose-rate irradiation when the total dose was increased to 1,800 rads. This paradoxical effect of dose rate is consistent with the enhanced effect of dose fractionation previously noted.

Effect of Multifractionated Irradiation and of Seeding of Lungs During Treatment

In an attempt to simulate more closely the clinical situation, mice received LTI using daily fractions of 200 rads ^{137}Cs γ rays to a total dose of 2,000 rad. Before, at varying times during, or after irradiation, 10^4 tumor cells were injected i.v., and the number of lung colonies was scored either 14 days after completion of irradiation or 14 days after tumor cell injection when all the radiotherapy was delivered first. Obviously, in the experiments in which tumor cells were injected during treatment, any enhancement of lung colony forming efficiency due to the conditioning effect of the lung irradiation would be partially or totally offset by radiation killing of tumor cells lodged in the lungs. The results of these experiments, shown in Figure 6, indicate that for significant enhancement of lung colony formation to occur seeding must occur during the latter half of the course of treatment while tumor cells present in the lungs

TABLE 1. *Effect of LTI given at different dose rates on lung colony yield from 4×10^4 FSa tumor cells injected i.v. 24 hours after irradiation (16 mice/group)*

Dose Rate (rad/min)	Lung Colonies ± SEM	
	LTI 1000 Rad	LTI 1800 Rad
969	45.6 ± 6.9	58.4 ± 5.7
180	45.3 ± 7.0	—
15.1	44.9 ± 4.9	96.5 ± 11.5
Control	10.2 ± 2.2	

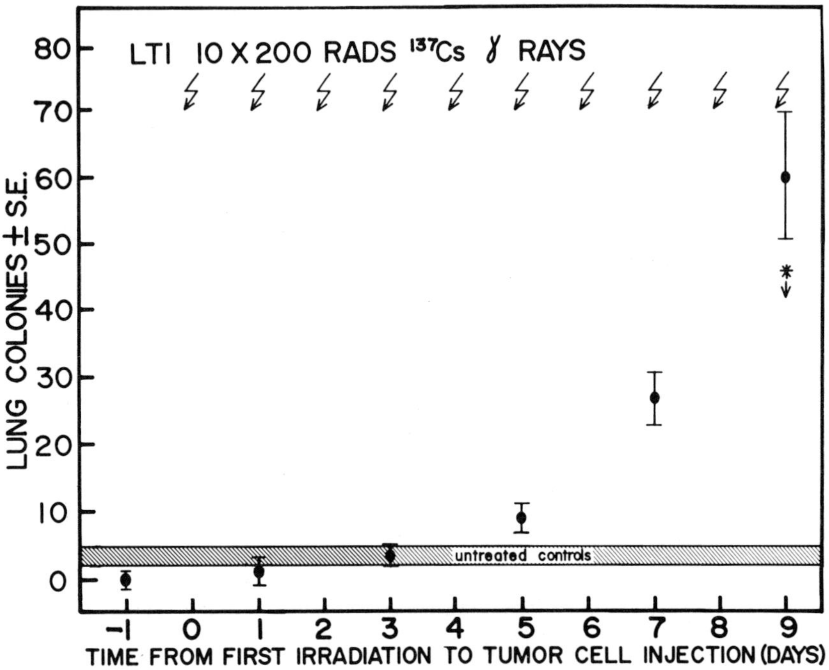

FIG. 6. Lung colony forming efficiency of tumor cells injected at various times during fractionated thoracic irradiation. Mice were exposed to 10 daily doses of 200 rad γ radiation. One day before irradiation was begun, at two daily intervals during irradiation, or immediately after its completion, 2 x 10^4 tumor cells were injected i.v. Only when tumor cells were injected in the latter half of the course of treatment was an increased yield of lung colonies observed; prior to this, cell killing by subsequent dose fractions more than offset the conditioning influence of the preceding doses. The effect of 10 $\times$ 200 rad delivered prior to tumor cell injection was greater than that predicted from a single dose of 2,000 rad (indicated by *).

prior to commencement of irradiation were effectively sterilized. Interestingly, the magnitude of enhancement produced by 10 $\times$ 200 rad given *prior* to tumor cell injection (~20-fold) is considerably greater than would be predicted from a single dose of 2,000 rads obtained by extrapolation of Figure 3, upper panel.

CLINICAL CORRELATIONS

Because of conflicting earlier reports concerning the incidence of ipsilateral metastases in patients receiving postoperative radiotherapy for breast cancer (Dao and Kovaric 1962, Chu et al. 1967), the records of 1,822 patients treated at M. D. Anderson Hospital from 1955-1978 were reviewed, using a coded computer file generated by Dr. Eleanor Montague. Patients with noninflammatory breast cancer who had had no evidence of distant metastases at the time of presentation were categorized into four prognostic groups as described by Fletcher (1973). The patients were then classified in terms of their treatment modality, and subgroups of those treated by surgery alone, surgery plus postoperative radiotherapy, or radiotherapy alone were selected for further study. In this series, essentially all radiotherapy was delivered by ^{60}Co teletherapy equipment. Patients in the postoperative radiotherapy group were restricted to those who received chest wall and lymphatic irradiation, since the volume of lung irradiated in patients receiving only "peripheral lymphatic" treatment is small (Fletcher 1973). With few exceptions, the "tumor dose" for postoperative treatment was 5,000 rad in 20 fractions over five weeks. Patients treated by radiotherapy alone for the most part received protracted treatment with an initial tumor dose of 6,000 rad in 40 fractions over eight weeks supplemented with a boost of 2,000–3,000 rad in two to three weeks to the primary lesion and involved lymph nodes. The maximum radiation dose to the lung in the apical segments (above the level of the second costal cartilage), to the lung underlying the internal mammary portal (usually extending 6 cm from midline), and to the slice of lung included in the tangential chest wall portals approaches the nominal tumor doses specified (excluding the boosts given with treatment by radiation alone). While the dose gradient from the tangential fields is steep, the direct anterior portals contribute a significant dose to the entire lung thickness. Overall, we estimate that perhaps one third of the lung volume could receive a dose in excess of 2,000 rad.

The distribution of disease categories in the three treatment groups is shown in Table 2. As would be expected, the patients treated by surgery alone had predominantly early stage disease, while those receiving radiotherapy alone had more advanced primary lesions. For each treatment group, the incidences of right, left, and bilateral lung metastases and pleural effusions was noted as a function of the side and disease category of the primary lesion (Table 3). We found no evidence of a predisposition to ipsilateral lung metastases in either of the irradiated groups, compared with patients treated by surgery alone. Interestingly, all groups showed a preponderance of right-sided lung metastases regardless of the side of the primary lesion. The overall ratio of right to left lung metastases was 1.29 : 1 (scoring bilateral metastases in both groups), while

TABLE 2. *Distribution of primary disease "categories" in MDAH breast cancer patients according to the type of treatment received*

	Disease Category*				
Treatment Modality	I	II	III	IV	Total
Surgery alone	363	69	90	44	566
Surgery + postoperative radiotherapy	293	38	221	320	872
Radiotherapy alone	5	7	124	248	384
					1822

* Definition of disease categories as in Fletcher 1973, p. 27.

TABLE 3. *Incidence of right- and left-side lung metastases and pleural effusions as a function of primary disease side and category, and of treatment modality*

Site of Metastatic Disease	Primary Side and Category									
	Left					Right				
	I	II	III	IV	Total	I	II	III	IV	Total
Grp I—Surgery alone										
Left lung	5	0	1	1	7	2	0	1	0	3
Right lung	6	2	0	1	9	5	0	2	1	8
Both lungs	5	1	2	1	9	8	1	0	2	11
"Lung"	1	0	0	0	1	1	0	0	0	1
Left P.E.	3	1	3	1	8	2	1	2	2	7
Right P.E.	4	2	1	1	8	6	0	0	1	7
Bilat. P.E.	0	0	0	0	0	0	0	0	2	2
Grp II—Postoperative radiation										
Left lung	0	0	3	0	3	0	1	3	3	7
Right lung	3	0	4	4	11	0	0	3	10	13
Both lungs	3	0	3	15	21	0	0	5	11	16
"Lung"	0	2	0	4	6	0	0	3	5	8
Left P.E.	1	1	4	14	20	0	0	3	4	7
Right P.E.	2	0	1	3	6	1	2	6	14	23
Bilat. P.E.	1	0	3	2	6	1	0	1	0	2
Grp III—Treated by radiation alone										
Left lung	0	0	1	2	3	0	0	3	4	7
Right lung	0	0	3	11	14	0	0	2	4	6
Both lungs	0	0	3	8	11	0	0	4	10	14
Left P.E.	0	1	7	14	22	0	0	0	2	2
Right P.E.	0	0	6	5	11	0	0	4	17	21
Bilat. P.E.	0	0	1	9	10	0	0	1	1	2

"Lung" indicates patients in which the side of lung metastases was not documented in the records.

the ratio of right to left primary lesions was 0.97 : 1. Although the incidence of ipsilateral lung metastases was not increased by irradiation, the possibility remained that the rate of development of metastases may have been accelerated. Figure 7 shows that this was not the case: those patients who developed lung metastases did so at the same rate, regardless of the treatment modality employed. In contrast to the findings for parenchymal lung metastases, a positive correlation was observed between the side of the primary lesion and the development of an ipsilateral pleural effusion, regardless of treatment modality. However, reference to Table 3 shows that pleural effusions occurred predominantly in patients with advanced local disease, suggesting that the pathogenesis of these effusions was related to direct spread of cancer through the chest wall lymphatics, as previously inferred by Weichselbaum et al. (1977).

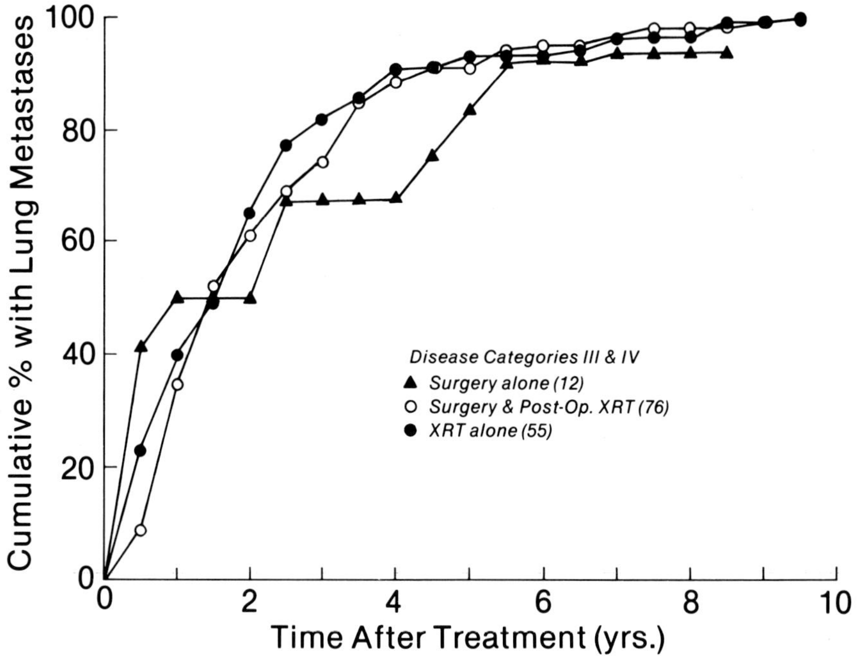

FIG. 7. Rate of development of lung metastases in all patients with disease categories III and IV who were at any time documented to have them, according to treatment modality. No acceleration in the rate of development of lung metastases attributable to irradiation can be detected.

DISCUSSION

Although the phenomenon of enhanced growth of tumor cells seeded to irradiated lungs has been repeatedly documented, its mechanism has not been established. Recently, we (Peters et al. 1978) reassessed the most frequently postulated theories about the effect and found none to be totally consistent with the available experimental data, as outlined below. Increased "trapping" of radioactively labeled tumor cells in irradiated lungs had been demonstrated by several authors, but the hypothesis that increased lung colony forming efficiency might be explained simply in terms of trapping was inconsistent with new data showing increased lung colony yields in irradiated lungs, even when the trapping differential was eliminated by concurrent injection of 10^7 unlabeled radiation-killed cells with small numbers of labeled viable cells. The theory that lung irradiation might act by inhibiting the cytotoxicity of lung macrophages was incompatible with the failure of reticuloendothelial blockade (Withers and Milas 1973) or of lysozymal enzyme inhibition to reproduce the effect of irradiation on lung. Further, resistance to lung colony growth induced by pretreatment of mice with *Corynebacterium parvum* (a biological stimulator of macrophage cytotoxicity) was not significantly impaired by lung irradiation. A possible inhibitory effect of thoracic irradiation on natural killer (NK) cell activity could not be reconciled with the Van den Brenk et al. (1973) description of enhanced lung colony growth in rats irradiated at the age of three weeks prior to maturation of the NK system. Evidence to support the notion that suppression of specific immune defense mechanisms might be responsible was even more meager, especially in view of the fact that the vast majority of cells destined not to form lung colonies lyse within 24 hours of injection, long before a specific immune response could possibly be mounted. For example, no correlation could be found between the immunogenicity of various tumors and the degree of enhancement of lung colony formation observed after lung irradiation; no mitigation of the effect was obtained by shielding the thymus of rats during lung irradiation (Van den Brenk et al. 1973), and treatment of mice with thymosin after thoracic irradiation completely failed to restore their resistance to lung tumor growth. On the basis of these studies, we concluded, as had Van den Brenk et al. (1973), that the most likely explanation for the effect was the production by irradiation of a local tropic influence supporting tumor cell survival and growth. Our present studies were directed toward obtaining more radiobiological data relevant to the effect in mice, not only for descriptive purposes, but also to gain new clues about the underlying mechanism.

Two unexpected findings that emerged from the experiments reported here were the relatively greater effect of split-dose or low dose-rate irradiation compared with single acute doses and the very low RBE values obtained for neutron irradiation. The only previous split-dose studies reported are those of Brown and Marsa (1978), who found no significant difference between the effects produced by two doses of 500 rads compared with a single dose of 1,000 rads. However, our data (Figure 3) show that it would be difficult to resolve a split-dose enhancement at this dose level, and the findings of Brown and Marsa are therefore not necessarily at variance with our results.

The lack of dependence on the interfraction interval for split-dose enhancement (Figure 5) makes it unlikely that induced parasynchronous progression into more sensitive cell cycle phases of cells surviving the first radiation dose could account for the effect. Moreover, all of the resident cells of the mouse lung have long turnover times (Gross 1977), and it is inconceivable that such cells would demonstrate a cell cycle–related increased radiosensitivity to closely spaced split doses.

Another possible explanation for an enhanced split-dose response is that restitution of target cells from outside the thorax occurred during the interfraction interval. This possibility, coupled with the similarity of $RBE_{n/\gamma}$ values for (single dose) lung colony promotion and for hemopoietic stem cell survival, raised the question of a circulating cell of hemopoietic origin as the target. However, such an explanation is inconsistent with other experiments in which the whole body of mice except the thorax was irradiated: this failed to produce any increase in lung colony forming efficiency (Peters et al. 1977) in spite of a profound suppression of circulating white blood cell counts. Moreover, the similarity of RBE values for single-dose lung colony promotion and for bone marrow stem cell survival could be fortuitous, in that the RBEs obtained for split-dose lung irradiation were somewhat lower. Nonetheless, the RBE for lung colony promotion very clearly differs from that reported by Field (1977) for respiratory death in mice, even when allowance is made for the different energy spectrum of the neutron beams employed, and we can conclude with some certainty that injury to the cells suspected of being the targets for respiratory failure (type II pneumocytes or vascular endothelial cells) is not responsible for lung colony promotion.

Whatever mechanism may underly the phenomenon, our experiments with fractionated lung irradiation show very nicely the critical influence of the timing of tumor cell seeding with respect to the amount of irradiation received. In Figure 6, it can be seen that unless tumor cells were seeded into the lungs after the sixth of 10 daily dose fractions of 200 rads, no increase in lung colony yield occurred. This is, of course, due to the fact that when cells are seeded early in a course of fractionated irradiation, the effect of cell killing by subsequent dose fractions exceeds the growth-promoting effect of the preceding fractions. The design of these experiments, in which equal numbers of cells were injected at various times during irradiation, is likely the most unfavorable for extrapolation to therapy for humans for the following reasons. When a primary tumor is being irradiated concurrently with normal lung tissue (as in the case of breast carcinoma), the number of viable cells in the primary tumor remaining available to seed the irradiated lung declines exponentially with increasing dose, and numerically, the great majority of tumor cells are sterilized by the first few dose fractions. Further, the mouse tumor we used grows much more rapidly than any human cancer, so that the dose required to offset tumor cell proliferation in the lungs in our experiments would substantially exceed that required in humans. It is apparent, therefore, that for an appreciable risk of enhanced metastasis to exist clinically, an uncontrolled source of viable tumor cells must be present outside the irradiated volume at the time of irradiation. Data obtained by Owen and Bostock (1973), who used dogs with a variety of spontaneous tumors, bear out this fact: when elective irradiation was given to one lung after successful ablation of the primary tumor, either an equal number or *fewer* metastases developed on the irradiated side. By contrast, when the primary tumor was uncontrolled, *more* metastases were found in the irradiated lungs.

On the basis of these studies, we did not expect to see an increase in ipsilateral lung metastases in patients undergoing radiotherapy, either as the definitive treatment or as a postoperative procedure, and the data presented in Table 3 confirm this expectation. A strong correlation did emerge, however, between advanced primary lesions and ipsilateral pleural effusions, suggesting that direct extension of the tumor through chest wall lymphatics was responsible.

Our failure to demonstrate an increased incidence of ipsilateral parenchymal lung metastases in irradiated breast cancer patients is in accord with the report of Chu et al. (1967), and offers no support for the original claim of Dao and Kovaric (1962) of a deleterious effect. In seeking possible reasons for this discrepancy, we have been struck by the extremely high incidence of radiation complications reported by these latter authors: in 50 consecutive cases they noted 18 patients with "permanent disability of the arm," 14 with bone necrosis, 25 with severe pneumonitis progressing to lung fibrosis, and 40 with skin changes ranging from desquamation to necrotic ulceration. These horrendous complications seem incongruous with the nominal dose delivered, i.e., 4,500 rad in three weeks using 200 to 400 kVp X rays, and provide clear evidence of biological overdosage. Since it has been well established that severe tissue trauma from any cause enhances the probability of metastatic tumor growth in the traumatized tissue (Fisher and Fisher 1959), it is possible that the results reported by Dao and Kovaric, as well as the historical case reports of Schürch (1935) and Schwarz (1935) are not analogous to the experimental phenomenon we have been investigating, in which enhanced tumor growth is seen in irradiated tissues prior to their manifesting any morphologic injury. Further, since in Dao and Kovaric's series, irradiation was administered to only 30 of 243 patients undergoing radical mastectomy, it may be assumed that irradiation was reserved for those patients with the most unfavorable disease who would most likely have foci of cancer outside the irradiated volume.

In conclusion, it seems clear that a significant risk of enhanced metastases in irradiated tissues exists only when relatively high doses of "prophylactic" irradiation are given without previous primary tumor ablation or its concomitant irradiation. Thus, protocols calling for elective irradiation of sites at risk of metastatic disease without local tumor control having been secured are rationally unsound and should not be employed. With an elective treatment, however, the therapist must be concerned primarily with the probability of *existing* micrometastases at the time of irradiation: if such metastases are present, the probability is that unless the tumor is highly radioresistant adequate treatment of these deposits would more than outweigh any conditioning of the irradiated tissue to subsequent reseeding by disseminated tumor cells or to regrowth of surviving cells within the treatment volume.

ACKNOWLEDGMENTS

We wish to thank Bailey Moore for construction of the neutron collimator used in these studies; Dr. Vic Otte for performing the neutron dosimetry; Jack Cundiff for making and calibrating the low dose-rate ^{60}Co irradiation apparatus; Dr. Howard Thames for statistical help; and Dr. Eleanor Montague for her kindness in allowing us access to her coded case history file of patients treated for breast cancer at this institution.

This investigation was supported in part by research grant CA-17769 and CA-06294, awarded by the National Cancer Institute, Department of Health, Education and Welfare.

REFERENCES

Brown, J. M. 1973. The effect of lung irradiation on the incidence of pulmonary metastases in mice. Br. J. Radiol. 46:613–618.

Brown, J. M., and G. W. Marsa. 1978. Effect of dose fractionation on the enhancement by radiation or cyclophosphamide of artificial pulmonary metastases. Br. J. Cancer 37:1020–1025.

Chu, F. C. H., J. C. Lucas, J. H. Farrow, and J. J. Nickson. 1967. Does prophylactic radiation therapy given for cancer of the breast predispose to metastasis? Am. J. Roentgenol. 99:987–994.

Dao, T. L., and J. Kovaric. 1962. Incidence of pulmonary and skin metastases in women with breast cancer who received postoperative irradiation. Surgery 52:203–212.

Dao, T. L., and H. Yogo. 1967. Enhancement of pulmonary metastases by x irradiation in rats bearing mammary cancer. Cancer 20:2020–2025.

Field, S. B. 1977. Early and late normal tissue damage after fast neutrons. Int. J. Radiat. Oncol. Biol. Phys. 3:203–210.

Fisher, B., and E. R. Fisher. 1959. Experimental studies of factors influencing hepatic metastases. III. Effect of surgical trauma with special reference to liver injury. Ann. Surg. 150:731–744.

Fisher, E. R., and B. Fisher. 1969. Effects of x irradiation on parameters of tumor growth, histology and ultrastructure. Cancer 24:39–55.

Fletcher, G. H. 1973. Management of localized breast cancer, *in* Textbook of Radiotherapy, G. H. Fletcher, ed. Lea and Febiger, Philadelphia, pp. 457–493.

Gross, N. J. 1977. Pulmonary effects of radiation therapy. Ann. Intern. Med. 86:81–92.

Hill, R. P., and R. S. Bush. 1969. A lung colony assay to determine the radiosensitivity of the cells of a solid tumor. Int. J. Radiat. Biol. 15:435–444.

Hranitzky, E. B., P. R. Almond, H. D. Suit, and E. B. Moore. 1973. A cesium-137 irradiator for small laboratory animals. Radiology 107:641–644.

Milas, L., N. Hunter, K. Mason, and H. R. Withers. 1974. Immunological resistance to pulmonary metastases in C_3Hf/Bu mice bearing syngeneic fibrosarcoma of different sizes. Cancer Res. 34:61–71.

Oliver, G., and B. Moore. 1970. The neutron shielding qualities of water-extended polyesters. Health Phys. 19:578–580.

Owen, L. N., and D. E. Bostock. 1973. Prophylactic X irradiation of the lung in canine tumours with particular reference to osteosarcoma. Eur. J. Cancer 9:747–752.

Peters, L. J. 1974. The potentiating effect of prior local irradiation of the lungs on the development of pulmonary metastases. (Letter) Br. J. Radiol. 47:827–829.

Peters, L. J., K. Mason, W. H. McBride, and Y. Z. Patt. 1978. Enhancement of lung colony-forming efficiency by local thoracic irradiation: Interpretation of labeled cell studies. Radiology 126:499–505.

Peters, L. J., W. H. McBride, and K. A. Mason. 1977. Pitfalls in the use of lung colony assay to assess T cell function in irradiated mice. Br. J. Cancer 36:386–390.

Schürch, O. 1935. Über hautmetastase im bestrahlungsfeld bei pyloruscarcinom. Z. Krebsforsch. 41:47–50.

Schwarz, G. 1935. Über die nachbestrahlung bei operiertem mammakarzinom. Strahlentherapie 53:674–681.

Suit, H. D., and A. Kastelan. 1974. Immunologic status of host and response of a methylcholanthrene-induced sarcoma to local x irradiation. Cancer 26:232–238.

Thompson, S. C. 1974. Tumour colony growth in the irradiated mouse lung. Br. J. Cancer 30:337–341.

Van den Brenk, H. A. S., W. Burch, C. Orton, and C. Sharpington. 1973. Stimulation of clonogenic growth of tumour cells and metastases in the lungs by local x irradiation. Br. J. Cancer 27:291–306.

Von Essen, C. F., and J. Stjernswärd. 1978. Radiotherapy and metastases, *in* Secondary Spread of Cancer, R. W. Baldwin, ed. Academic Press, London, pp. 73–99.

Weichselbaum, R., A. Marck, and S. Hellman. 1977. Pathogenesis of pleural effusion in carcinoma of the breast. Int. J. Radiat. Oncol. Biol. Phys. 2:963–965.

Withers, H. R., and L. Milas. 1973. Influence of preirradiation of lung on development of artificial pulmonary metastases of fibrosarcoma in mice. Cancer Res. 33:1931–1936.

Potential Therapeutic Applications

Radiation Biology in Cancer Research, edited by
Raymond E. Meyn and H. Rodney Withers.
Raven Press, New York © 1980.

Hypoxic Cell Radiosensitizers, Present Status and Future Promise

John F. Fowler

Gray Laboratory of the Cancer Research Campaign, Mount Vernon Hospital, Northwood, Middlesex HA6 2RN, England

There are two bases for the rationale of hypoxic-cell radiosensitizing drugs (Adams 1973, 1976). First, they radiosensitize hypoxic cells only and not oxic cells (Figure 1). This provides a differential effect not available with other types of radiosensitizers, nor indeed with other treatment agents of most kinds, because tumors contain hypoxic cells but few normal tissues do. Second, the radiosensitizing drugs are not used up by metabolism in the cells through which they diffuse, as oxygen is. Therefore, they penetrate to the hypoxic cells that lie beyond about 130 μm radially from blood capillaries (Thomlinson and Gray 1955, see Figure 2); oxygen has by then been used up. Figure 3 illustrates that even if the radiosensitizer is not as efficient as oxygen at the concentrations present in the blood, its concentration is certainly higher than that of oxygen at the position of the hypoxic cells.

CHEMICAL PROPERTIES OF THE RADIOSENSITIZERS

Another valuable property of the hypoxic cell radiosensitizers is that their sensitizing ability can be predicted accurately from their electron affinity. Figure 4 shows the radiosensitizing ability of several nitroimidazole compounds plotted against the one-electron reduction potential determined by pulse radiolysis. There is an excellent correlation (Adams et al. 1979). Few types of drugs have such a well-known structure-activity relationship, and this helps in the design of new radiosensitizers. A similar correlation has been found between the cytotoxicity to oxic cells and electron affinity, so it is expected that compounds much more electron affinic than misonidazole will be too toxic to be useful.

It is not this cytotoxicity that is the limiting factor in clinical use, however, but neurotoxicity. For this there is as yet no known correlation with molecular structure. Factors that affect neurotoxicity include the half-life of the drug in the animal and lipophilicity. In order to reduce neurotoxicity, a shorter serum half-life than that of misonidazole is required, and probably a less lipophilic compound.

The requirement for a shorter half-life is obvious for the radiosensitizing

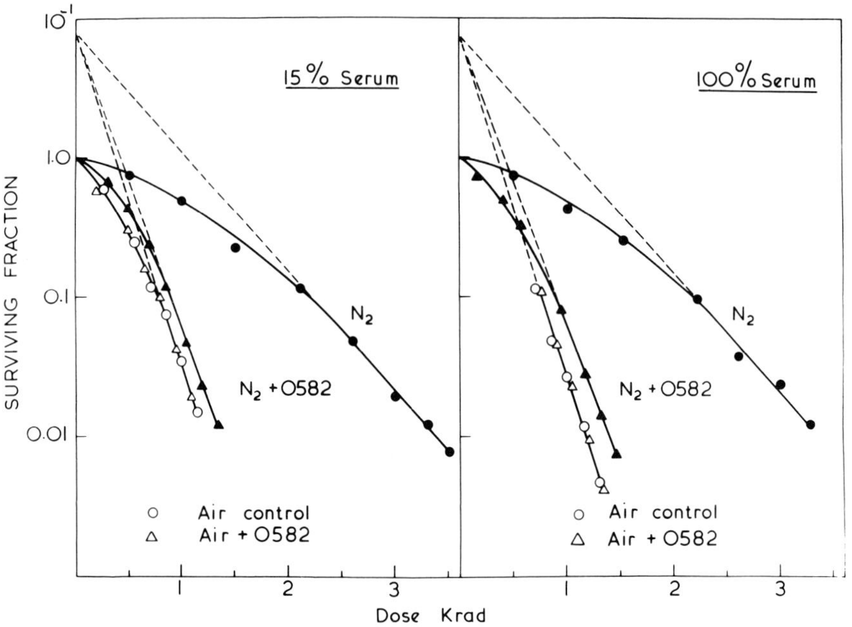

FIG. 1. Cell survival curves showing hypoxic cells (closed symbols) are radiosensitized but oxic cells (open symbols) are not. Reproduced from Asquith et al. (1974) with permission of Academic Press.

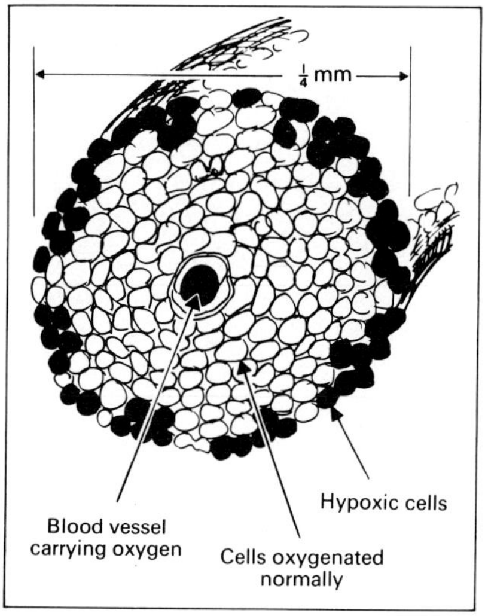

FIG. 2. The hypoxic cells are located at a distance from each capillary at which the oxygen, diffusing radially outwards, has been depleted by metabolism. Reproduced from Fowler et al. (1976a) with permission of Spectrum, Central Office of Information, London.

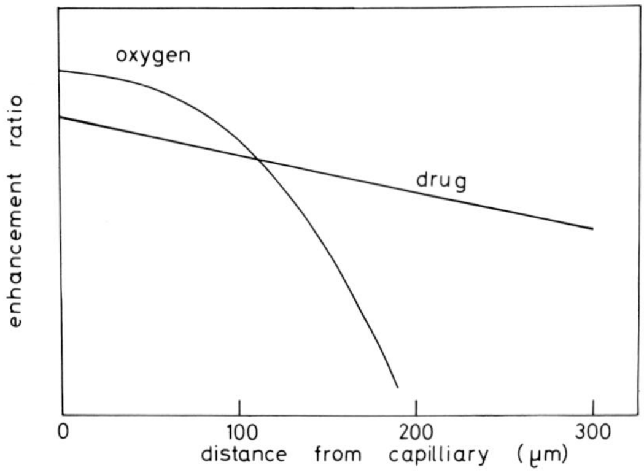

FIG. 3. The radiosensitizing drug will diffuse further than oxygen because it is not used up by metabolism, as oxygen is. Reproduced from Adams (1976) with permission of Plenum Publishing Corp.

and neurotoxic aspects of the drug. However, there is another way in which these drugs work that requires longer serum half-lives: a specific cytotoxicity to hypoxic cells (Sutherland 1974, Stratford and Adams 1977). The graph of cell kill against time of exposure to a constant drug concentration has an inefficient shoulder of many hours duration before the cytotoxicity becomes appreciable (Figure 5). Thus, a long serum half-life, or repeated topping up, is necessary if this method of eliminating hypoxic cells is going to be effective. The requirement would then be for multiple small drug doses instead of a few big ones—the opposite of the requirement if radiosensitization alone is considered to be the major effect.

In practice, useful compromise schedules are available because even as many as 25 or 30 small daily doses are likely to give sufficient concentrations of misonidazole in tumors to provide significant sensitizer enhancement ratios (SER), exceeding 1.2. The total amount of drug administered clinically is limited

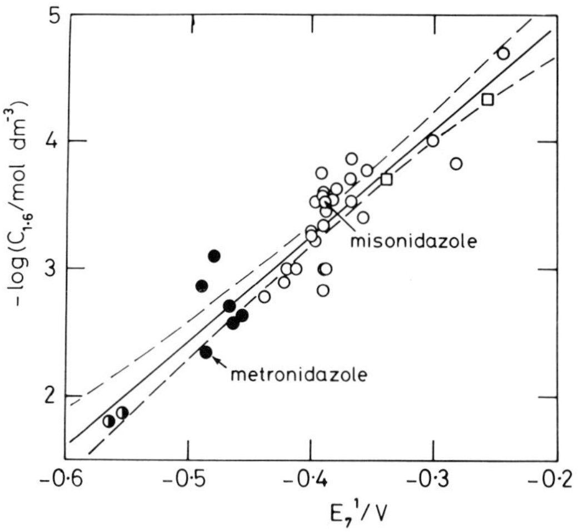

FIG. 4. The relationship between electron affinity (one-electron reduction potential, $E_7{}^1$) and sensitizing efficiency (reciprocal of the drug concentration required to achieve a sensitizer enhancement ratio of 1.6). ○, 2-nitroimidazoles; ●, 5-nitroimidazoles; ◑, 4-nitroimidazoles; □, 5-nitrofurans. Dotted lines are 95% confidence limits. Reproduced from Adams et al. (1979) with permission of *International Journal of Radiation Biology*.

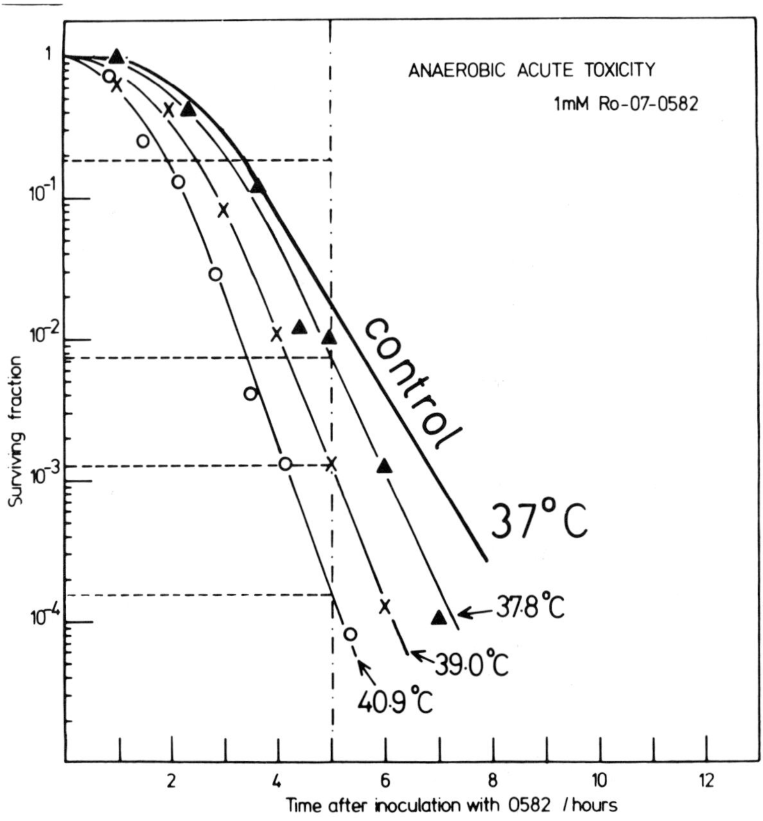

FIG. 5. Differential cytotoxicity of misonidazole to hypoxic cells, and its enhancement by moderate hyperthermia. Redrawn from Stratford and Adams (1977).

to 12 g/m^2 in three to four weeks or 15 g/m^2 in six to eight weeks (Dische et al. 1977).

MOUSE TUMOR EXPERIMENTS

Figure 6 shows SER plotted against the measured concentration of misonidazole. The curves are for cells in vitro and the points are for four types of solid tumor in mice (McNally et al. 1978). The concentrations relevant to clinical applications are 15–20 μg/ml for 25–30 small daily doses and 60–70 μg/ml for six doses in, for example, three weeks. These correspond to SER in mouse tumors of about 1.3 and about 1.7, respectively. The latter value is similar to the hypoxic gain factor obtained with fast neutrons. Figure 7 illustrates the

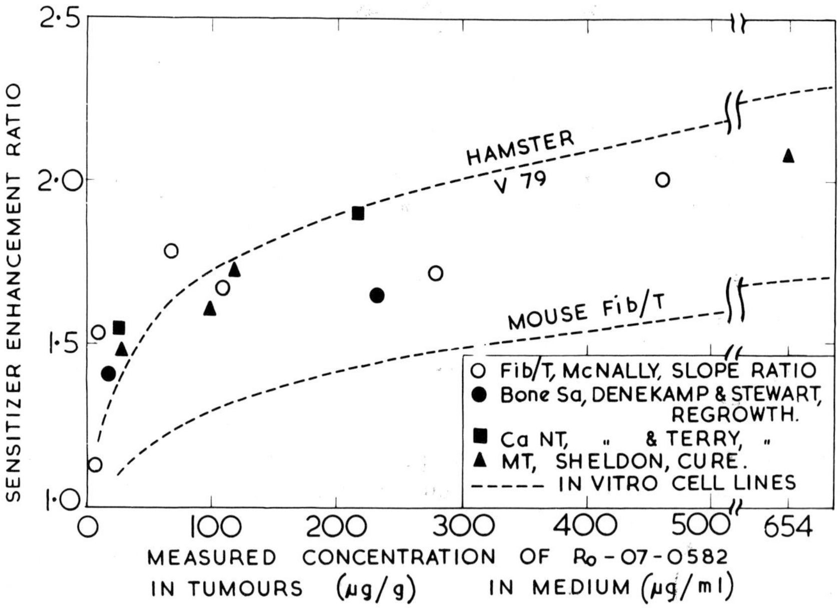

FIG. 6. SER versus measured concentration of misonidazole in solid murine tumors (points) and in cells in vitro (dotted curves). Redrawn from McNally et al. (1978).

dramatic improvements in local tumor control obtained with single doses of X rays at such values of SER.

CLINICAL RESULTS—HYPERBARIC OXYGEN

How important are hypoxic cells in human tumors? An indication is given by the recent results of the Medical Research Council clinical trials of hyperbaric oxygen. Table 1 gives the results for carcinoma of the uterine cervix (Watson et al. 1978, Dische 1978). The six-fraction schedules gave a poor result in air,

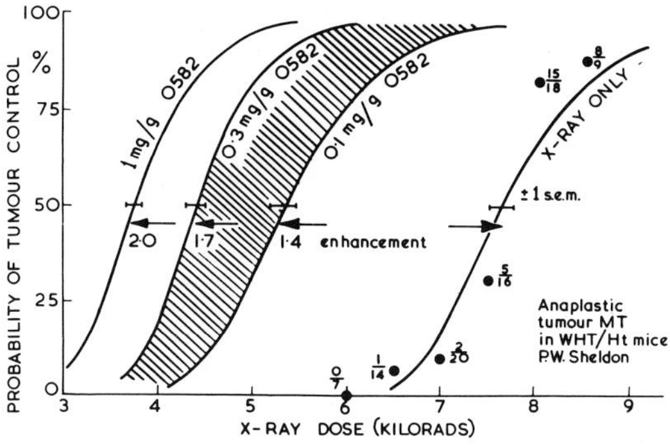

FIG. 7. Local control of MT tumors (cure of WHT mice) versus single X-ray dose. The shaded band represents clinically obtainable concentrations of misonidazole. Reproduced from Fowler and Denekamp (1979) with permission of Pergamon Press Ltd.

TABLE 1. *MRC Hyperbaric oxygen trial, carcinoma of cervix uteri, stage III*

		HBO	Air	Comments
5-Yr Survival				
6F) Portsmouth	(37 pts.)	42%	17%	In-air results
6F) Oxford	(23 pts.)	46%	8%	were poor
25F) Glasgow	(127 pts.)	50%	37%	Combined
30F) Mt. Vernon	(56 pts.)	39%	28%	P<.05
Local Control 5 Yrs				
Glasgow		87%	60%	Highly significant
Mt. Vernon		76%	50%	differences
Severe morbidity				
Bowel		12%	4%	

(Watson et al., 1978)

which was improved by hyperbaric oxygen to a value similar to the best obtained with conventional "daily" fractionation. The overall gain factor (SER for hyperbaric oxygen) was about 1.2. A significant improvement in local control was obtained, exceeding 20%; the increase in five-year survival was just barely significant. Table 2 gives the results for advanced carcinoma of the head and neck. Again, local control was increased by more than 20% by hyperbaric oxygen, and in the second trial there was also a significant improvement in five-year survival (Henk et al. 1977, Henk and Smith 1977).

Hyperbaric oxygen causes vasoconstriction and may therefore be self-limiting. It is expected that hypoxic-cell radiosensitizers would give somewhat larger gain factors than hyperbaric oxygen does, unless reoxygenation already copes with hypoxic cells in most human tumors, which is not considered likely.

CLINICAL RESULTS—HYPOXIC SENSITIZERS

Figure 8 shows the increase in the survival time of patients treated by Urtasun et al. (1976) for glioblastoma multiforme, using metronidazole with a low-dose

TABLE 2. *MRC hyperbaric oxygen trial, carcinoma of head and neck*

		HBO	Air	Comments
1st Trial (276 pts.)				
Local control	(5 yr)	53%	30%	P<.001
Survival	(5 yr)	40%	40%	Saved by surgery
2nd Trial (103 pts.)				
Local control	(2 yr)	65%	47%	P<.05
Survival	(2 yr)	71%	50%	P<.01

No difference in complications.
10% lower dose to larynx fields only.
(Henk et al. 1977, Henk and Smith 1977)

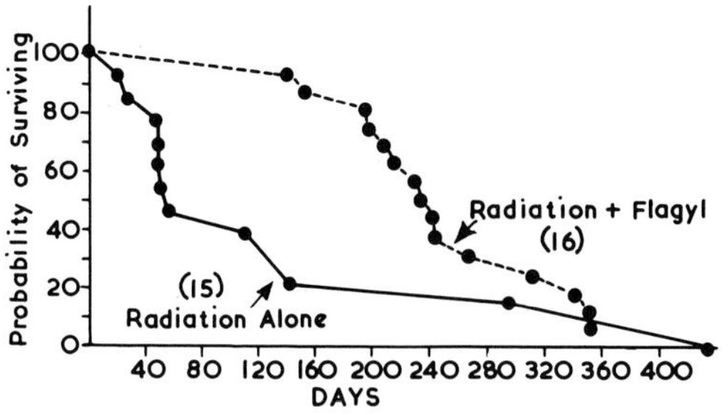

FIG. 8. Probability of survival of glioblastoma patients treated with minimal radiotherapy with and without metronidazole. Redrawn from Urtasun et al. (1976).

radiotherapy schedule of 9 × 330 rads in three weeks. The improvement was significant, and can only be due to the increased effect on hypoxic cells when metronidazole was used. However, it brought the survival time up to about the values reported from other good clinical trials using 30 fractions of radiation plus CCNU. This improvement of a poor X-ray-only schedule up to the level of the currently best available is similar to the effect of hyperbaric oxygen with six fractions shown in Table 1. Metronidazole is inferior to misonidazole as a radiosensitizer.

Figure 9 shows the regrowth delay curves obtained for skin nodules secondary to cervix carcinoma by Thomlinson et al. (1976). An SER of 1.2 was found. It was calculated, from the dose of X rays and of misonidazole used, that the nodules contained between 5 and 40% hypoxic cells, probably 12–20% (Denekamp et al. 1977a).

Recently Dr. Dan Ash, at the Royal Marsden Hospital, Sutton, has reported an SER of about 1.3 for regrowth delay of human lung nodules treated with 10-fraction radiotherapy (personal communication). Professor Sealy (1978), from Cape Town, reports promising results in advanced head and neck cancer using misonidazole combined with hyperbaric oxygen. In a historical control series he obtained complete tumor regression in none of the cases with X rays only, in 3/14 with hyperbaric oxygen (HBO) radiotherapy, and in 8/11 with both

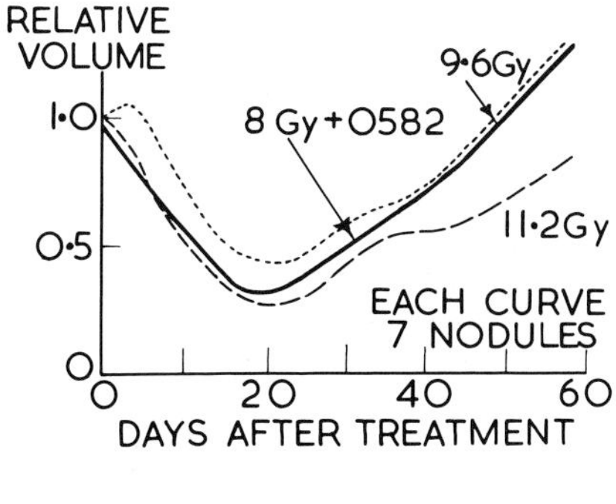

FIG. 9. Regrowth delay of three groups of skin nodules secondary to carcinoma of the cervix, two groups without and one with misonidazole. Redrawn from Thomlinson et al. (1976).

misonidazole and HBO. There were 9/19 cases of peripheral neuropathy but none was reported serious.

A number of clinical trials are being initiated in various countries and definitive results are likely to be available within a few years. Let us return to experimental mouse tumors to consider how much gain we might achieve.

MOUSE TUMOR RESULTS WITH FRACTIONATED SCHEDULES AND RADIOSENSITIZER

We have carried out comprehensive sets of fractionation experiments on two contrasting types of mouse tumor. The results illustrate the future promise of radiosensitizers in an interesting way.

First, a transplanted mammary tumor in C3H mice was used because it was known to reoxygenate rapidly and extensively and was therefore a challenging test for misonidazole. Figure 10 shows the results of several X-ray-only fractionation schedules, one for each point. Local tumor control is plotted for a constant mouse skin reaction. It can be seen that if the overall time is long (18 days is three or four volume doubling times in this C3H mouse mammary carcinoma), the cure rate is low. At intermediate times of 9–11 days all the fractionation schedules gave fairly good results. At short overall times the results were very variable, depending on how poorly the tumor had reoxygenated.

Figure 11 shows the corresponding results when misonidazole is given before irradiation with several of the same schedules. Single doses and the poor X-ray-only schedules were greatly improved. There was no improvement at the previous optimum of nine days overall time. Thus, all the schedules with misonidazole were brought up to about the same level as the previous best; the originally worst exhibited, of course, the biggest improvement. The short overall time schedules were particularly well enhanced (Fowler et al. 1976b).

This tumor was a rapid reoxygenator, and well before 9–11 days most of the hypoxic cells had been reoxygenated away. The use of misonidazole enabled shorter X-ray-alone schedules to be brought up to the same level as the better fractionated schedules. Just such an improvement was seen in Urtasun's clinical trial of misonidazole (Figure 8) as well as with the six-fraction hyperbaric oxygen

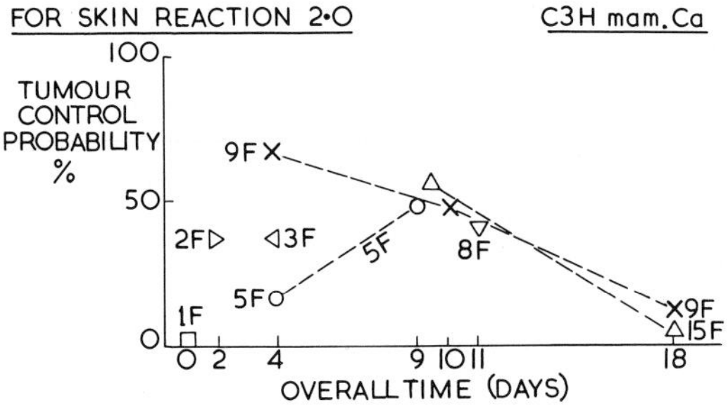

FIG. 10. Local control of mammary tumors (cure of C3H mice) as a function of overall time of fractionated X-ray-alone schedule, all of which cause the same level of acute skin reaction on mouse legs. Redrawn from Fowler et al. (1976b).

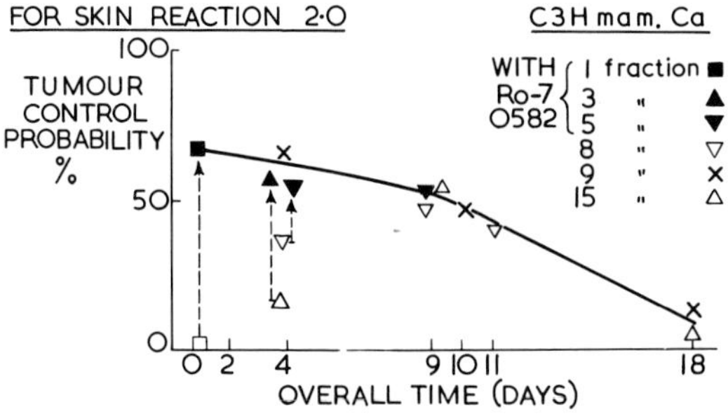

FIG. 11. As in Figure 10, but with misonidazole given before each dose in the single dose, 3 fractions/4 days, 5 fractions/4 days and 5 fractions/9 days schedules. Reproduced from Fowler and Denekamp (1979) with permission of Pergamon Press Ltd.

(Table 1). Eliminating hypoxic cells took the criticality out of the fractionation. Since we cannot be sure that we know the optimum fractionation schedule for any human tumor, misonidazole may improve any treatment. One practical advantage might be that short schedules, which at present cause too much normal tissue injury, would become safer and more effective.

The second mouse tumor we investigated was a slowly shrinking sarcoma (MT) that reoxygenated much less than the previously described C3H tumor. The results are shown in Figure 12, again all for estimated equal normal tissue damage. With X rays only (full lines), 20 fractions were better than 5 fractions at all overall times. The shorter times were better than the longer times because this tumor grew fast, with a volume doubling time of just under two days (Sheldon and Fowler 1978).

When misonidazole was given before each fraction, the 20-fraction and

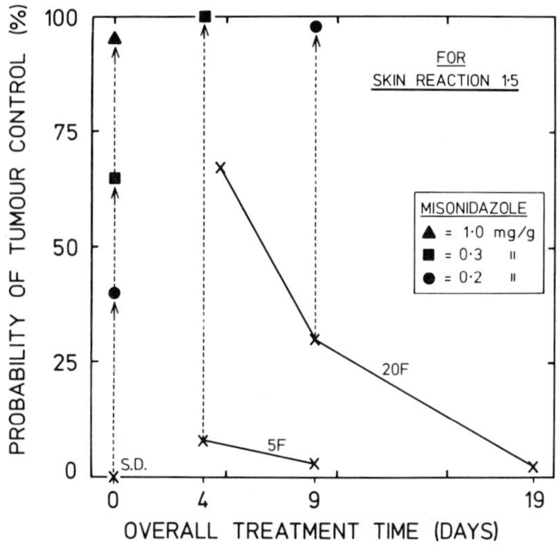

FIG. 12. As in Figures 10 and 11, but for the MT tumor (cure of WHT mice). Crosses and full lines: X rays only. Dotted lines and solid symbols: misonidazole before each X-ray dose. Reproduced from Sheldon and Fowler (1978) with permission of *British Journal of Cancer.*

5-fraction schedules were both improved (dotted lines). The poor 5-fraction schedule was improved to a higher level than the best 20-fraction X-ray-only schedule, as was the mediocre 20-fraction schedule. This is an illustration of something we have yet to see demonstrated clinically: a mediocre X-ray schedule being improved beyond the best that the optimum X-ray-alone schedule could achieve. If more human tumors are like the second than like the first type of mouse tumor, then significant clinical improvements will be observed readily by the use of hypoxic-cell radiosensitizers.

HYPOXIC-CELL SENSITIZERS WITH NEUTRONS

Neutrons plus misonidazole give a larger therapeutic gain than either neutrons alone or X rays with misonidazole, as expected because each gives an enhancement ratio or gain factor that is less than the full oxygen effect. This has been demonstrated in mouse tumors (Denekamp et al. 1977b).

ENHANCEMENT OF RADIATION INJURY TO NORMAL TISSUES

It is clear that misonidazole does not sensitize oxic cells. Adams (1976) has given reasons why no enhancement of effect would occur in tissues that are normally oxygenated, even though hyperbaric oxygen would enhance the injury by the equivalent of increasing the radiation dose by 3% to 10%. However, if any normal tissues have hypoxic regions, more radiation damage will occur than with X rays only. The only human tissue known to be sufficiently hypoxic for this to occur with multiple small fractions is the laryngeal cartilage (Henk et al. 1977). In animal experiments large single doses of X rays have shown significant enhancement of effect with misonidazole in skin, rodent tail, and spinal cord. These experiments are not necessarily relevant to radiotherapy because large single doses would show the problem whereas multiple small ones might not, and because anesthetics were used. Nevertheless, clinical caution is at present advisable when spinal cord, brain, and cartilage are irradiated in conjunction with misonidazole administration.

CYTOTOXICITY TO HYPOXIC CELLS

Figure 5 illustrated the specific toxicity of misonidazole to hypoxic cells. Oxic cells would not be depleted in number significantly over the whole time scale of that diagram. In principle, therefore, this is another way of eliminating hypoxic cells from tumors. It can, of course, only be used together with radiation, because the oxic cells have to be killed off, too. It is then difficult to justify not giving the radiosensitizer before irradiation to achieve both radiosensitization and direct cytotoxicity of hypoxic cells. The choice exists because if cytotoxicity is important, multiple small fractions would be better than a few large fractions of sensitizer. We have already seen that multiple small fractions can work well,

TABLE 3. *Cytotoxicity and radiosensitization: Summary of mouse tumors. Enhancement ratios if 1mg/g misonidazole is given before or after irradiation*

After	Before	Tumor	Reference	After	Before	Tumor	Reference
1.2	2.3	C3H Ca	Brown 1975	0.9	2.1	MT1/MT	Sheldon and Hill 1977
1.3	2.2	Sq Ca D	Hill and Fowler 1977	1.0	2.1	Fib/T	McNally (unpub.)
1.3	2.1	Ca NT	Denekamp and Harris 1975	1.0	1.8	B Sa 2	Denekamp (unpub.)
1.1	1.9	FibroSa	Stewart (unpub.)	1.0	1.8	C3H Ca	Sheldon et al. 1974
1.4	1.9	Rhod Ca	Denekamp (unpub.)	1.0	1.7	Disc Ca	Hirst (unpub.)
1.3	1.7	Sa F	Begg (unpub.)	1.0	1.7	FF Sa 1	Hirst (unpub.)
1.2	1.6	FF Sa 2	Hirst (unpub.)	1.0	1.5	Sa F	Begg 1977
			7, Some cytotoxicity				7, No cytotoxicity

(Denekamp, 1978)
Drug given before irradiation demonstrates both cytotoxicity and radiosensitization.
Drug given after irradiation demonstrates only cytotoxicity.
The references in this table are given in full in Denekamp (1978).

provided that each does not become *too* small. I would define "too small" as that amount of sensitizer that gives an SER of less than 1.2 per fraction, because that is about the smallest SER that can be detected with a clinical trial of reasonable size, for example 200 to 300 patients total (Tables 1 and 2).

How large the contribution of direct cytotoxicity is to the whole effect is a matter of argument. Figure 5 suggests that it is large, perhaps even larger than the radiosensitization. However, there would have to be no change in the hypoxic status of the cells. If hypoxic cells cycle through periods of being oxic, the effectiveness of the cytotoxicity would be enormously reduced because of the long initial shoulder. Similarly, if cells remain hypoxic for only a short time before becoming necrotic or being reoxygenated during treatment, the effect will be small.

In mice, little or no cytotoxic effect would be expected because the serum half-life is only about one hour—ten times shorter than in man. In spite of this, seven of 14 types of mouse tumors showed significant cytotoxicity, as judged by enhancement when misonidazole was given *after* X irradiation (Table 3) (Denekamp 1978).

Further evidence is required before we can assess the importance of this phenomenon. It is interesting, however, that modest hyperthermia can speed up the cytotoxic effect greatly (see Figure 5).

CONCLUSIONS

Hyperbaric oxygen results have demonstrated that hypoxic cells in some types of human tumors are a problem. Reasons are given for expecting radiosensitizing drugs to penetrate better than HBO to the hypoxic cells, which exist in solid tumors exceeding 1 or 2 mm in diameter. Several clinical studies have demonstrated significant effects on human tumors of misonidazole or metronidazole. Concentrations of misonidazole in human tumors that were 40–110% of those in plasma, usually 70–90%, have been measured. The studies also indicated that peripheral neurotoxicity sets a limit of 12–15 g/m^2 of misonidazole administered within a few weeks. The expected dose-modifying factors (SER values) are from 1.2 for multiple small fractions to 1.7 for six large doses, as compared with 1.7 for neutrons alone.

A number of clinical trials using misonidazole are being initiated in several countries. Definitive clinical results can be expected within the next five to seven years. These results will clarify the importance of hypoxic cells in causing resistance to radiotherapy.

ACKNOWLEDGMENTS

I take pleasure in thanking my colleagues, past and present, for stimulating discussions and help in compiling the data referred to, especially Prof. G. E. Adams, Drs. S. Dische, J. Denekamp, S. A. Hill, N. J. McNally, R. Sealy,

P. W. Sheldon, R. H. Thomlinson, R. C. Urtasun, P. Wardman, and C. L. Wingate. I should like to thank also Mrs. G. F. Mason for preparing the script. The work of the Gray Laboratory is supported by the Cancer Research Campaign.

REFERENCES

Adams, G. E. 1973. Chemical radiosensitization of hypoxic cells. Br. Med. Bull. 29:48–53.

Adams, G. E. 1976. Hypoxic cell radiosensitizers for radiotherapy, *in* Cancer: A Comprehensive Treatise, Vol. 6, F. F. Becker, ed. Plenum Press, New York, pp. 181–219.

Adams, G. E., E. D. Clarke, I. R. Flockhart, R. S. Jacobs, D. S. Sehmi, I. J. Stratford, P. Wardman, M. E. Watts, J. Parrick, R. G. Wallace, and C. E. Smithen. 1979. Structure-activity relationships in the development of hypoxic cell radiosensitizers: I. Sensitization efficiency. Int. J. Radiat. Biol. 35:133–150.

Asquith, J. C., M. E. Watts, K. Patel, C. E. Smithen, and G. E. Adams. 1974. Electron affinic radiosensitization: V. Radiosensitization of hypoxic bacteria and mammalian cells in vitro by some nitro imidazoles and nitro pyrazoles. Radiat. Res. 60:108–118.

Begg, A. C., P. W. Sheldon, and J. L. Foster. 1974. Demonstration of hypoxic cell radiosensitization in solid tumours by metronidazole. Br. J. Radiol. 47:399–404.

Brown, J. M. 1975. Selective radiosensitization of the hypoxic cells of mouse tumours with the nitroimidazoles metronidazole and Ro-07-0582. Radiat. Res. 64:633–674.

Denekamp, J. 1978. Cytotoxicity and radiosensitization in mouse and man. Br. J. Radiol. 51:636–637.

Denekamp, J., J. F. Fowler, and S. Dische. 1977a. The proportion of hypoxic cells in a human tumour. Int. J. Radiat. Oncol. Biol. Phys. 2:1227–1228.

Denekamp, J., and S. R. Harris. 1975. Tests of two electron affinic radiosensitizers in vivo using regrowth of an experimental carcinoma. Radiat. Res. 61:191–203.

Denekamp, J., C. Morris, and S. B. Field. 1977b. The response of a transplantable tumour to fractionated irradiation. III. Fast neutrons plus the radiosensitizer Ro-07-0582. Radiat. Res. 70:425–432.

Dische, S. 1978. Hyperbaric oxygen—the MRC trials and their clinical significance. Br. J. Radiol. 51:888–894.

Dische, S., M. I. Saunders, M. E. Lee, G. E. Adams, and I. R. Flockhart. 1977. Clinical testing of the radiosensitizer Ro-07-0582. Br. J. Cancer 35:567–579.

Fowler, J. F., G. E. Adams, and B. D. Michael. 1976a. Spectrum, Central Office of Information, London 137:8–10.

Fowler, J. F., and J. Denekamp. 1979. A review of hypoxic cell radiosensitization in experimental tumours. Pharmacology and Therapeutics (in press).

Fowler, J. F., P. W. Sheldon, J. Denekamp, and S. B. Field. 1976b. Optimum fractionation of the C3H mouse mammary carcinoma using X-rays, the hypoxic-cell radiosensitizer Ro-07-0582, or fast neutrons. Int. J. Radiat. Oncol. Biol. Phys. 1:579–592.

Henk, J. M., P. B. Kunkler, and C. W. Smith. 1977. Radiotherapy and hyperbaric oxygen in head and neck cancer. Final report of first controlled clinical trial. The Lancet 2:101–103.

Henk, J. M., and C. W. Smith. 1977. Radiotherapy and hyperbaric oxygen in head and neck cancer. Interim report in the second clinical trial. The Lancet 2:104–105.

Hill, S. A., and J. F. Fowler. 1977. Radiosensitizing and cytocidal effects on hypoxic cells of Ro-07-0582 and repair of X-ray injury, in an experimental mouse tumour. Br. J. Cancer 35:461–469.

McNally, N. J., J. Denekamp, P. W. Sheldon, and I. R. Flockhart. 1978. Hypoxic cell sensitization by misonidazole in vivo and in vitro. Br. J. Radiol. 51:317–318.

Sealy, R. 1978. Hyperbaric oxygen in radiotherapy, *in* Proceedings of the XI International Cancer Congress, Buenos Aires. Pergamon Press, Oxford, England.

Sheldon, P. W., J. L. Foster, and J. F. Fowler. 1974. Radiosensitization of C3H mouse mammary tumours by a 2-nitroimidazole drug. Br. J. Cancer 30:560–565.

Sheldon, P. W., and J. F. Fowler. 1978. Radiosensitization by misonidazole of fractionated X-rays in a murine tumour. Br. J. Cancer 37(Suppl. 3):242–245.

Sheldon, P. W., and S. A. Hill. 1977a. Hypoxic cell radiosensitizers and tumour control by X-rays of a transplanted tumour in mice. Br. J. Cancer 35:795–808.

Sheldon, P. W., and S. A. Hill. 1977b. Further investigations of the effects of the hypoxic-cell radiosensitizer, Ro-07-0582, on local control of a mouse tumour. Br. J. Cancer 36:198–205.

Stratford, I. J., and G. E. Adams. 1977. Effect of hyperthermia on differential cytotoxicity of a hypoxic cell radiosensitizer, Ro-07-0582, on mammalian cells in vitro. Br. J. Cancer 35:307–313.

Sutherland, R. M. 1974. Selective chemotherapy of non-cycling cells in an in vitro tumor model. Cancer Res. 34:3501–3503.

Thomlinson, R. H., S. Dische, A. J. Gray, and L. M. Errington. 1976. Clinical testing of the radiosensitizer Ro-07-0582. III. Response of tumours. Clin. Radiol. 27:167–174.

Thomlinson, R. H., and L. H. Gray. 1955. The histological structure of some human lung cancers and the possible implications for radiotherapy. Br. J. Cancer 9:539–549.

Urtasun, R. C., P. Band, J. D. Chapman, M. L. Feldstein, B. Mielke, and C. Fryer. 1976. Radiation and high dose metronidazole (Flagyl) in supratentorial glioblastomas. N. Engl. J. Med. 294:1364–1367.

Watson, E. R., K. E. Halnan, S. Dische, M. I. Saunders, I. S. Cade, J. B. McEwen, G. Wiernik, D. J. D. Perrins, and I. Sutherland. 1978. Hyperbaric oxygen and radiotherapy: a Medical Research Council trial in carcinoma of the cervix. Br. J. Radiol. 51:879–887.

Radiation Biology in Cancer Research, edited by
Raymond E. Meyn and H. Rodney Withers.
Raven Press, New York © 1980.

Rationale for Use of Charged-Particle and Fast-Neutron Beams in Radiation Therapy

Herman D. Suit and Michael Goitein

Department of Radiation Medicine, Massachusetts General Hospital, Department of Radiation Therapy, Harvard Medical School, Boston, Massachusetts 02114

There are two basic approaches to improving the results of radiation therapy: (1) modification of treatment strategy to increase the differential in the biological response of tumor and the critical surrounding normal tissue(s); and (2) employment of improved distribution of radiation dose so that a higher dose may be administered to the tumor or a lesser dose to the surrounding normal tissues (or both). If one or both of these goals are achieved with a new treatment modality, the results should be that a higher proportion of patients are cured of the primary tumor and are free of complications or treatment-related morbidity.

The present intense interest in the clinical study of fast-neutron and charged-particle beams is based on an expectation that one or both of these advantages will be realized. The term charged-particle beams includes beams of protons helium, carbon ions, neon ions, argon ions, and negative pions (π^-). A biological advantage is anticipated from high linear energy transfer (LET) radiations, such as fast neutrons (FN), heavy ions (C,Ne,Ar), and π^- (negative pions). Charged-particle beams have depth-dose patterns that are qualitatively different from those obtainable using photon or fast-neutron beams. For selected anatomic situations, the dose distributions that can be generated with the charged-particle beams are expected to represent a major advantage over those obtainable with photon or neutron beams. Charged-particle beams may be low LET, e.g., protons, or high LET, e.g., heavy ions or pions. Helium ion beams have an intermediate LET.

Radiation Biological Advantages

A biological advantage is expected from the use of high-LET beams, viz. FN, heavy ions, and negative pions. The radiobiological characterizations of these particles reveal: (1) relative biological effectiveness (RBE) is increased (Barendsen et al. 1966, Raju et al. 1978); (2) oxygen enhancement ratio (OER) is reduced (Barendsen et al. 1966, Raju et al. 1978, Hall 1978, Chapman et al. 1977, S. Curtis, personal communication; (3) repair of radiation damage produced by the high-LET radiation is less than that of damage produced by photons (sublethal—Chapman et al. 1977, Hall et al. 1975; potentially lethal—Shipley et al. 1975; and slow classes of repair—Field 1977); and (4) the age response function is suppressed markedly for most cell systems (Bird and Burki 1975, Raju et al. 1975) but may be affected only modestly in others (Withers et al. 1974).

Although the biological effectiveness of these particles is greater than that of low-LET radiations, the clinician is interested only in the prospect of a greater differential effect between tumor and normal tissue that favors normal tissues. In general terms, the OER is the parameter judged by us to be of greatest significance to the formulation of a rationale for the clinical study of high-LET radiation therapy. This is so because normal tissues tend to be well oxygenated while some of the cells of most epithelial and mesenchymal tumors are reckoned to by hypoxic and hence relatively radiation resistant. Accordingly, a therapeutic gain should be realized by the use of radiation characterized by a lower OER. The OER values for the high-LET beams designed for radiation therapy tend to be in the range of 1.5–2.4; this range is significantly below the average value of ≃3 for low-LET radiations. In Figure 1 are shown the OER values for many of the FN beams being used in clinical studies as determined by Hall (1978). These data indicate that the OER of FN beams is low (≃1.5–1.7) and essentially independent of FN energy over the energy range being studied clinically. The OER values for Chinese hamster V79 cells irradiated by C, N, Ar, and π^- beams with the cells positioned at the midportion of the spread-out Bragg peak were 1.5–2.2 as determined by Raju et al. (1978). These low OER values are corroborated by unpublished data of Curtis (personal communication, 1979) who has determined the OER values for R1 cells (a rat rhabdomyosarcoma-derived cell line) for C and Ne beams of 22–23 cm range in water and with a spread-out Bragg peak (SOBP) of 10 cm. For cells positioned at the midpoint of the SOBP, the OER values were 1.96 and 2.11 for the C and Ne ion beams, respectively, for doses that produce survival fractions of 0.5.

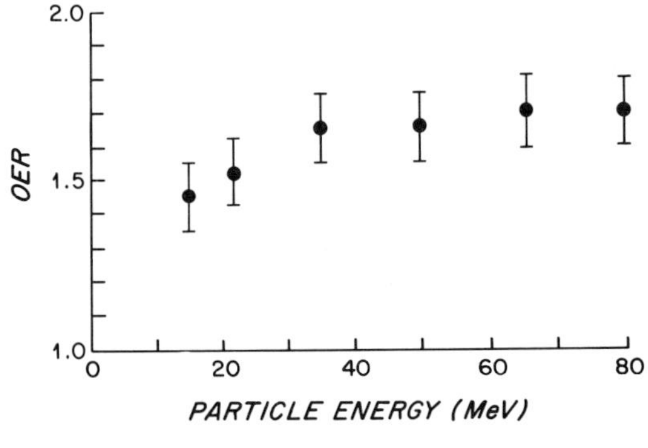

FIG. 1. Scattergram of measured OER for irradiation of Chinese hamster cells in vitro by fast neutrons produced by deuterons or protons accelerated to various energies. The OER showed very slight dependency on neutron energy over the range of 16–80 MV of the accelerating particle (Hall 1978).

There is no proof that an important cause of failure of radiation therapy to achieve local control of tumors in man is the hypoxic tumor cell. However, there are data from laboratory and clinical studies that demonstrate that hypoxic cells are critical determinants of the radiation response of at least some tumors. Investigations of laboratory animal tumor systems that have shown that respiration of oxygen at high pressure significantly increased the tumor response include the Ehrlich ascites tumor (Scott 1953, van den Brenk et al. 1962), a benzpyrine-induced fibrosarcoma (R1B5) (Thomlinson 1960), mammary carcinoma (DuSault 1963, Suit and Maeda 1967, Suit et al. 1977b), a fibrosarcoma, and squamous cell carcinoma of the C3H mouse (Suit and Suchato 1967).

Recently, convincing data from prospective clinical trials have been published that show for some human tumors that hypoxic cells are indeed important determinants of the radiation response. Bush and co-workers from Toronto in 1978 evaluated the role of transfusion in patients with carcinoma of the cervix, stages IIB and III. In their trial, the control patients were not transfused unless their hemoglobin level dropped below 10 gm %. For the test group, transfusions were administered as required in order to maintain a hemoglobin level of ≥12 gm %. The results (Table 1) show that pelvic recurrence rates were significantly higher in patients whose hemoglobin levels were allowed to remain in the 10–12 gm % range.

The Medical Research Council of Britain has sponsored several trials of hyperbaric oxygen and radiation therapy. Results demonstrate for carcinoma of the head and neck region (Henk et al. 1977, Henk and Smith 1977) and carcinoma of the cervix (Watson et al. 1978) but not for carcinoma of the urinary bladder (Cade et al. 1978) that higher local control rates were achieved in the hyperbaric oxygen groups. Results from two of the trials are presented in Tables 2 and 3. The benefit of hyperbaric oxygen was impressive for the head and neck group: local control and survival rates were higher and the level of damage to normal tissue was judged to be the same in air and hyperbaric groups. In the trial of carcinoma of the cervix, control results were superior in the hyperbaric oxygen groups but there was a slightly higher morbidity rate. Other trials have also been performed, and local control has tended to be higher in the hyperbaric

TABLE 1. *Pelvic recurrence in patients with carcinoma of cervix, stage IIB-III treated by radiation alone*

	Control		Transfused
Hgb Level (gm %)	<12	>12	>12
Pelvic recurrence	10/20 (50%)	11/48 (23%)	11/67 (16%)

Data from Bush et al. (1978).

TABLE 2. *Hyperbaric oxygen and radiation therapy in treatment of head and neck cancer*

Cardiff Second Trial Dose: (HPO) 400 rad × 10; 22 days (Air) 200 rad × 30; 6 wks	HPO	Air
No. patients	51	52
Local control of primary lesion (2 years)	65%	47% $0.1 > p > .05$
Survival (2 years)	71%	50% $p < .02$

Data from Henk et al. (1977).

oxygen groups, but not significantly so. This subject has been reviewed recently (Dische 1978, Suit and Scott 1979).

Two other radiobiological features of high-LET radiation may be clinically important for selected tumor-normal tissue situations. The reduced repair of radiation damage produced by high-LET radiation should provide an advantage when, following photon irradiation, the tumor cells are more effective at repair than are the cells of the critical normal tissue in the treatment volume. A representative sample of experiments that show the near absence of repair of sublethal damage is presented in Figure 2. This seems to us to be an unlikely basis for improving the results of treatment for the more common neoplasms. At a practical clinical level, most of the solid tumors for which local control is a problem are of mesenchymal or epithelial origin and the dose-limiting normal tissues are mesenchymal or epithelial. There are no data that demonstrate that for low-LET radiation a greater repair by the malignant than by normal cells derived from those tissues occurs. If the tumor cells were less efficient in repair than the adjacent normal cells, then use of high-LET radiation might yield a poorer result. Pertinent to this consideration is that one of the common explanations for the eradication of tumors by low-LET fractionated radiation therapy is that there is a differential repair capacity between the tumor and the normal

TABLE 3. *Hyperbaric oxygen and radiation therapy for carcinoma of uterine cervix, stage III, 3-year local control results*

	No. Patients		
Center	HPO	Air	p Value
Portsmouth	58	18	
Oxford	46	13	
Glasgow	87	60	.01
Mt. Vernon	76	46	.04

Data from Watson et al. (1978).

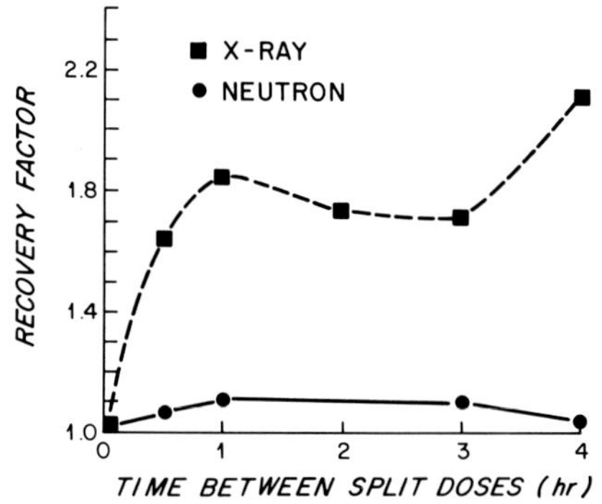

FIG. 2. The recovery factor for Chinese hamster cells irradiated with 210-KV X rays or 35-MeV neutrons. The recovery curves for the fast-neutron study show negligible recovery with intertreatment intervals up to four hours. This is in marked contrast to the very sharp recovery seen by one hour for 210 KV irradiation of these cells (Hall et al. 1975).

tissue that favors the normal tissue. If this latter were true then the reduced repair of damage by high-LET radiation would not be a justification for the clinical use of high-LET radiations.

The flatter age-response function (ARF) would be an advantage only if the degree of flattening were greater for the tumor cell population than for the cells of the critical normal tissues. This may not obtain in all situations as evident by the findings that for at least one normal tissue system (mouse jejunal crypt cells) the ARF for fast-neutron irradiation is highly structured and is only slightly suppressed in comparison with that for ^{60}Co photon irradiation (Withers et al. 1974) (see Figure 3). This may partially explain the relatively good tolerance of patients to fast-neutron irradiation of the abdominal and pelvic regions. An assessment of the impact of a flattening of the ARF on the therapeutic ratio must also consider explicitly the age density distributions of the tumor

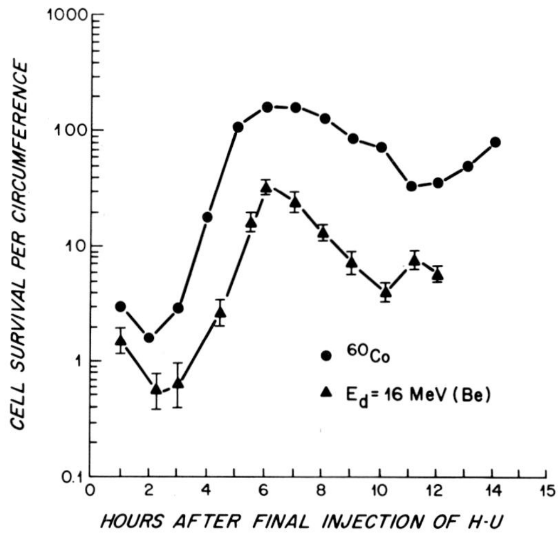

FIG. 3. Cobalt-60 and 16-MeV neutrons were employed to irradiate mouse jejunal crypt cells in vivo, and cell survival circumference of the jejunum was measured at various hours after final injection of hydroxyurea. In this particular system, the age response function was only slightly suppressed (Withers et al. 1974).

and normal cells during the course of fractionated irradiation. This subject has been developed by Withers and Peters (1979).

Fortunately, these various rationales are supported by results of laboratory and clinical radiation therapy using high-LET radiations in that improved "clinical" results are being obtained. This refers principally to studies using fast-neutron beams. In an extensive series of experiments comparing FN and X rays in the treatment of a C3H mouse mammary carcinoma, Field and Jones (1968) compared the average skin reaction at the TCD_{50} dose levels for several fractionation schedules. For only one fractionation schedule (five fractions/nine days) were X rays as effective as FN. The implication from their work was that a better "clinical" result would on average be obtained with FN.

Surely one of the strongest rationales for further clinical study of high-LET radiations is the encouraging results that have been achieved with FN radiation therapy. The initial clinical experience with fast neutrons was that of Stone (1948). A subsequent analysis by Sheline et al. (1971) of the records of those patients revealed that local control of tumor was achieved in nearly all patients, but that late tissue damage was quite severe. Dose levels employed by Stone for fractionated FN therapy are judged to have been excessive as the consequence of his use of an RBE value for the *acute* response of human skin to large *single* doses. Recent radiobiological research has shown that RBE is highly sensitive to dose per fraction, viz. RBE increases sharply with decreases in dose (Field 1976). Thus, the RBE derived from studies using large single doses would be lower than the RBE for small doses. Further, the RBE for late damage is slightly higher than that for acute damage (Withers et al. 1978). Accordingly, the RBE value employed by Stone was low.

Despite the serious problems encountered with the severity of late changes in normal tissue, Stone's results remain interesting because of the high frequency of control of the treated, often locally advanced, tumor. Catterall et al. (1977) have described results of a prospective clinical trial of fast-neutron therapy for head and neck cancer that demonstrates a higher local control frequency in the neutron-treated patients. Recent reviews of the current status of fast-neutron therapy are available in the Particle Proposal of the Committee for Radiation Oncology Studies (1978) and the Proceedings on Fundamental and Practical Aspects of the Application of Fast Neutrons and Other High-LET Particles in Clinical Radiotherapy (1978). In summary, there appear to be serious radiobiological bases for expecting that high-LET radiations would have some advantage over low-LET radiations (photons) and that the available results of experimental and clinical radiation therapy tend to support that expectation.

We have not been able to develop a radiobiological rationale for predicting qualitatively different cellular responses to high-LET radiations for the different particles (FN, heavy ions, or pions) provided their LET values are comparable. That is, at present, the radiobiological characteristics of a particle beam appear to be determined by the average LET and not by the fine structure of the

LET. Accordingly, we would predict no purely radiobiologic advantage of any of the high-LET charged particles over fast neutrons.

Improved Dose Distribution

General Considerations

The rationale for the clinical study of protons, heavy ions, or π^- is that the dose distributions achievable with these beams are qualitatively different than those achievable with photons or neutrons and that for certain clinical problems these appear advantageous. An improved dose distribution means that the treatment volume approximates more closely the target volume. Target volume is defined here as the volume of tissue known to be involved by tumor together with those tissues suspected as a clinically important probability of involvement by tumor. Treatment volume is the volume of tissue included within the high-dose contour used to specify the radiation dose administered to the tumor. The treatment volume will always be larger than the target volume because allowances must be made for set-up error, movement of patient during an individual treatment session, and changes in the target volume during the course of treatment.

In Figure 4 a treatment and a target volume are shown schematically. One sensitive structure is located within the treatment volume but outside the target volume, and a second sensitive structure is located partially within the target volume. If a therapy plan were devised that permits a reduction in the treatment volume in any or all dimensions, then that new dose distribution would be described as "improved." This is so, in a general sense, because the volume of normal tissue being irradiated to a high dose level would be reduced. Tolerance of radiation by the new smaller volume would be higher. Inherent in the presentation in Figure 4 is the suggestion that the relationship between reduction in

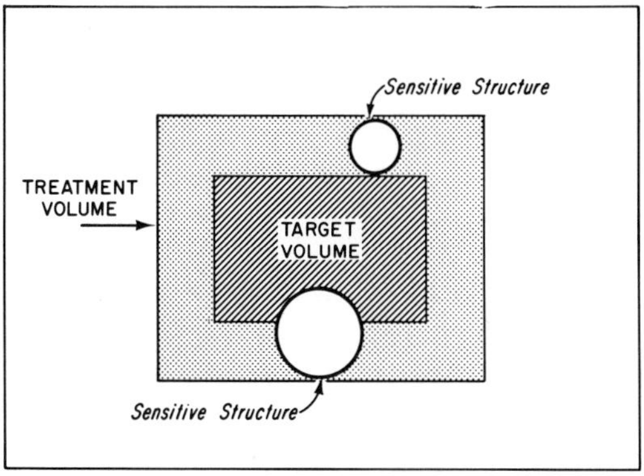

FIG. 4. Schematic representation of patient contour with outlined treatment volume (stipple area) and target volume (thatched area) with two sensitive structures included within the treatment volume.

treatment volume and increase in tolerance is not a simple volumetric consideration. Major changes in tolerance would be achieved with only slight reduction in the treatment volume, provided that this reduction occurs at the site of the sensitive structure(s). If the sensitive structures shown in Figure 4 were excluded from the target volume, then tolerance would be increased sharply. In fact if the tolerance dose is determined by the sensitivity of one or both of these sensitive structures, reduction of treatment volume without excluding all or part of such structures might not affect the tolerance dose. We emphasize that such a model suggests that the effort in improving dose distribution can be highly rewarding if there are reductions in *critical* dimensions of the treatment volume. This is so even though the change in total treatment volume is slight. In order to accomplish this goal there must be a clinical definition of the critical regions of the treatment volume and then a serious technical effort directed toward reducing treatment volume at those specific points.

The tolerance of the target volume itself will set a limit to which the dose to the target may be increased as a consequence of improving dose distribution pattern. We do not know if reduction of treatment volume will modify the tolerance of the target structure itself.

The biggest impact of improved dose distribution should be with those clinical problems for which dose levels with conventional photon techniques are low and are limited by sensitive structures outside the target volume. With reference to the target volume itself, it is likely that the tolerance dose of the target structure for most anatomic sites would not be higher than 7,000–8,500 rad when administered at 200 rad for 5 days per week. This limit will be a function of the specific site, target volume, and extent of destruction of target tissue by tumor. In some cases, we approach tolerance of the target very closely with current treatment techniques. This is shown by the fact that nonhealing or very slowly healing necrosis may appear in the target, with no symptomatic changes outside the target volume.

Consideration of the clinical importance of employing radiation therapy technologies that realize superior dose distribution is dependent upon three relationships for each anatomic site and tumor stage: (1) tolerance and treatment volume reduction; (2) radiation dose and complication frequency; and (3) radiation dose and tumor control probability. For human tissues, data on these relationships are scant except for skin and skin carcinoma.

Current Strategies for Improving Dose Distribution

There are a variety of approaches to developing improved dose distributions. These include: (1) combining external beam and interstitial techniques; (2) high-technology photon techniques using fixed fields (rigid immobilization, confirmation of patient position before each treatment, and use of special collimators for all fields, compensating filters, mixed beams, etc.); (3) dynamic photon treatment (motion of gantry, jaws, and table top during each treatment session);

(4) intraoperative beam therapy, particularly with electron beams; and (5) charged-particle therapy. Each of these is especially attractive for treatment for certain anatomic sites. The efficacy of these various approaches is being investigated by several groups. One effect of these investigations is likely to be stimulation of major up-grading of the accuracy and precision of radiation therapy in general. In particular, the intensity of the diagnostic evaluation and effort to define the target volume in precise anatomic terms will be greater. Overall this should result in a higher proportion of patients who are cured and a greater likelihood of cured patients being free of complications. We do not fear that reasonable efforts in this direction are likely to be complicated by higher frequencies of marginal misses. Results of radiation therapy have improved with the introduction of supervoltage external beam techniques, combined external beam and interstitial techniques, and elaborate secondary collimation (e.g., cerrobend cut-outs), etc. There does not appear to be a sound basis for thinking that we have already achieved the ultimate in terms of clinically practical dose distribution.

Dose Distribution Patterns with Charged Particle Beams

Charged particles have a finite range in matter. This physical property is the basis for the qualitatively different depth-dose curves and dose patterns that may be achieved with charged-particle beams as compared with photon or fast-neutron beams. This is illustrated in Figure 5, which presents a depth-dose curve for a 42-MeV X-ray beam and a modulated energy 135-MV proton beam (Suit et al. 1977a). For photons and fast-neutron beams, dose decreases approximately exponentially with depth in tissue, beyond the build-up region. In contrast, for a modulated energy particle beam the dose is essentially uniform across the SOBP and then falls precipitously at the end of the particle range. For the example shown, the dose decreased from 80% to 20% in 4 mm; there is a similarly rapid decrease in dose at the lateral edge of beam. Surface dose is usually in the range of 50–80% depending on the particle, the distance between the surface and the proximal end of the SOBP, and the particle range. For high-LET charged-particle beams the RBE is >1 and varies across the SOBP.

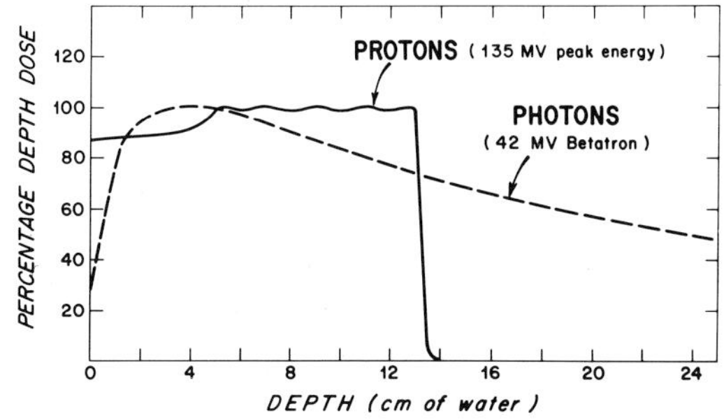

FIG. 5. Depth dose curves for a 42-MeV X-ray beam and a modulated energy 135-MeV proton beam.

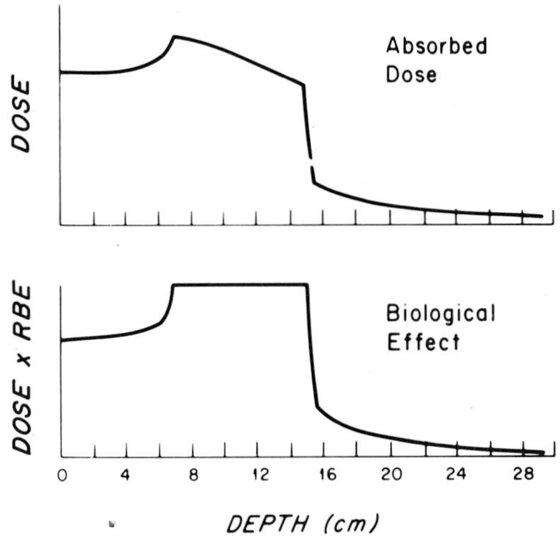

FIG. 6. Plot of the depth-dose curve for modulated energy neon beam in rad, upper panel. In the lower panel is the depth-dose effect curve, which is derived by multiplying the physically absorbed dose in rad times the RBE at that particular point in the depth-dose curve. Thus, for specially modified depth-dose curves, dose-effect curves can be obtained that are flat throughout the spread-out Bragg peak range.

Hence, the physical dose over the region of the SOBP is planned in order that the "dose effect" be uniform. This is illustrated schematically in Figure 6 for a modulated energy neon beam used at the Lawrence Berkeley Laboratory (J. Lyman, G. Chen, and J. Castro, personal communication).

Charged-particle beams like photon or neutron beams may be employed in multiple field techniques in order to achieve even better dose distribution patterns. Comparative dose or dose-effect distributions for two-field parallel-opposed technique using neon ion beams and a four-field box technique using 25-MV X-ray beams are presented in Figure 7. There are clear advantages with the particle beam technique. A detailed analysis of the comparative treatment planning with charged-particle beams and X-ray or fast-neutron beams is not included.

Modulated energy proton beams are low LET like photons and, hence, are comparable radiobiologically to photons. Therefore, the goal of the clinical study of protons is to evaluate the clinical improvement that can be achieved by use of dose distributions obtainable with such beams (i.e., no radiobiologic advan-

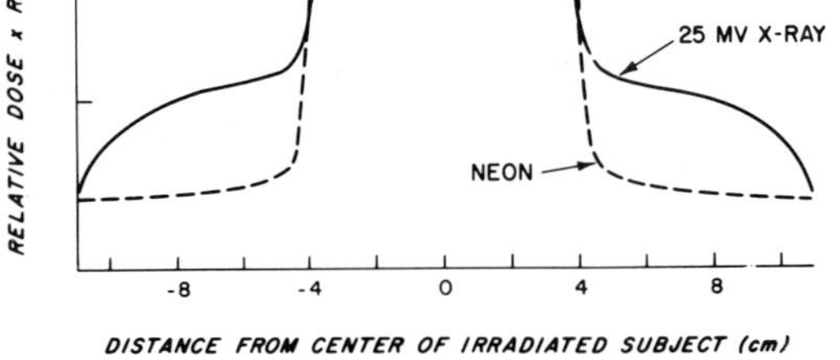

FIG. 7. Comparison of isodose effect curves for treatment of an 8-cm target located centrally in a 22-cm thick subject by parallel opposed (AP-PA) modulated energy neon fields or by a four-field box technique using 25-MV X rays.

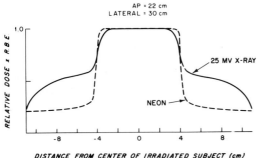

tage). Heavy ion and pion beams are high LET and have radiobiologic characteristics that are qualitatively similar to those of fast-neutron beams. Accordingly, the aim of a clinical investigation of high-LET charged-particle beams is the definition of the improvement of results of radiation therapy by using the dose distribution patterns of charged particles rather than those of fast neutrons. This means that no radiobiological advantage(s) for the high-LET charged-particle beams over fast-neutron beams is posited. Therefore, if protons have a therapeutic advantage over photons because of dose distributions then the high-LET charged particles should have a comparable advantage over fast neutrons. There is a further dose distributional advantage of high-LET charged-particle beams: there is less increase of dose in fat than obtains for fast neutrons.

Dose Response Curves for Tumor Control

Provided the superior dose distribution means that patient tolerance is improved, higher radiation doses can be employed. The strength of the rationale for the use of these improved dose distributions is a direct function of: (1) the extent that dose or dose × RBE may be increased to the target (our estimate is that this is likely to be in the range of 10–20% for most treatment situations) and (2) the impact of such a dose increment on tumor control probability (TCP). Specifically, our basic question is what will a 10–20% dose increment mean in terms of higher TCP? To provide a reliable answer, complete dose-response data for at least some human tumors are needed; such data are not now fully available. We consider here selected dose-response curves from studies on a model tumor system, spheroids of Chinese hamster cells in vitro, experimental animal tumors, a canine tumor, and limited clinical data.

For the tumor model, we employed this relationship: $TCD_p \simeq D_0[\ln M + \ln n - (\ln - \ln p)]$ where TCD_p is the radiation dose that on average yields tumor control probability p, D_0 and n are the mean lethal dose and the extrapolation number from the multi-target single-hit model, and M is the number of viable tumor cells (Suit et al. 1965). An increase in TCD from 0.2 to 0.8 requires a dose increment of from 24% to 11% for tumors comprising 10^3 to 10^7 cells (extrapolation number of 6). This increase corresponds to $1.97 \times D_0$. These response curves for tumor control represent the steepest feasible curves, as the response function for this model is entirely probabilistic. For these maximally steep dose-response curves there is no variation in cellular radiation sensitivity, number of viable cells per tumor, or host reaction against tumor among the tumors used in the dose-response assay. Experimentally determined dose-response curves for destruction of 21-day spheroids of Chinese hamster cells (Durand 1975, Durand, personal communication) and 8-mm diameter isotransplants of MDAH-MCaIV (an extremely weakly immunogenic mammary carcinoma of the C3H mouse) (Suit et al. 1978) treated by single radiation doses are presented in Figure 8. To raise TCP from 0.2 to 0.8 for these two systems, the dose must be increased by factors of 1.7 and 1.3, respectively. For an immuno-

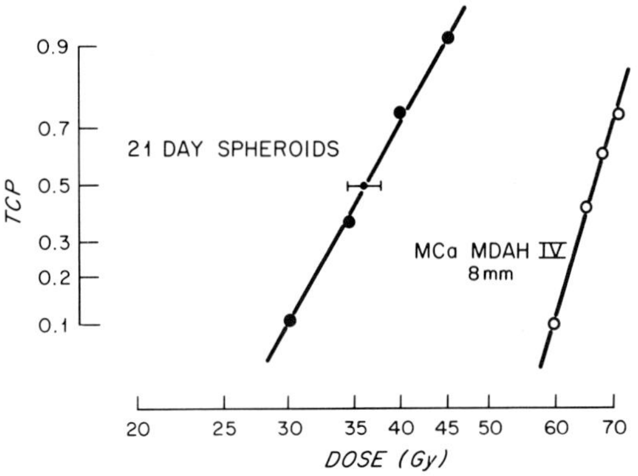

FIG. 8. A plot of tumor control probability versus dose on a probability log grid for 21-day spheroid of Chinese hamster cells and an 8-mm diameter mammary carcinoma (MCa-MDAH-IV). Tumor control probability for spheroids was determined by inactivation of the spheroid and tumor control probability for the mammary carcinoma was assessed at 240 days after irradiation (Durand 1977, Suit et al. 1978).

genic tumor, the status of the host is a highly significant determinant of the slope of the response curve. This is particularly so because there are important variations, even between members of a highly inbred strain of mice, in reaction to a constant tumor antigen challenge. For a striking illustration of this effect, examine Figure 9. Here, early generation isotransplants of a methylcholanthrene-induced fibrosarcoma (Suit and Suchato 1967) were treated by single radiation doses on the day that the tumor reached a mean diameter of 8 mm. The three curves are dose-response curves for recent experiments in our laboratory and were performed on tumor growing in normal, immune-suppressed (600 rad whole body irradiation 24 hours before transplantation), or immune-potentiated (350 μg of *Corynebacterium parvum* given i.v. on the day the tumor reached 5 mm in diameter) hosts. Thus, for this tumor the slope of the dose-response curve for local control was drastically affected by the immune competence of thc host.

Gillette (personal communication 1979) has determined the dose-response curve for spontaneous squamous cell carcinomas of the oral cavity of the dog. Twenty-five tumors were assigned dose levels of 3,000 to 5,000 rad at 500-rad increments. Treatment consisted of 10 equal doses given in three fractions per week over a total time of 21 days; ^{60}Co radiations were employed. The dose-response curves for tumor control and for tissue necrosis are shown in Figure 10. The slope of the curve for tumor control is clearly less steep than that for normal tissue damage. For the tumor control curve, the dose increment that raises TCP from 0.2 to 0.8 is 1,000 rad or a factor of 1.30 (4,300/3,300). If a dose increment of 10% were feasible by use of charged-particle beams instead of ^{60}Co technique, the TCP would be increased from 0.2 to 0.35. Because of the improved dose distribution, no increase or perhaps some decrease in the frequency of necrosis of normal tissue would be expected.

We have encountered only one clinical study in which patients were randomly assigned to different dose levels in order to examine explicitly the dose-response relationship. Morrison (1975) assigned patients with carcinoma of the urinary

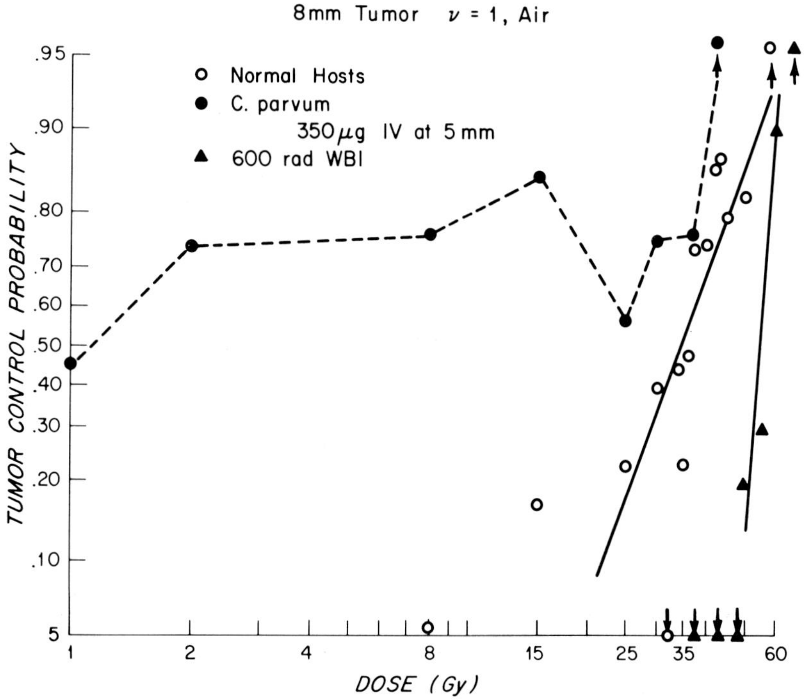

FIG. 9. Radiation dose–tumor control response curves for 8-mm isotransplants of FSaI growing in the right thigh when subjected to local irradiation under air-breathing conditions. Tumor control frequency is plotted against log dose for tumor growing in normal, *C. parvum*–treated and whole body–irradiated hosts.

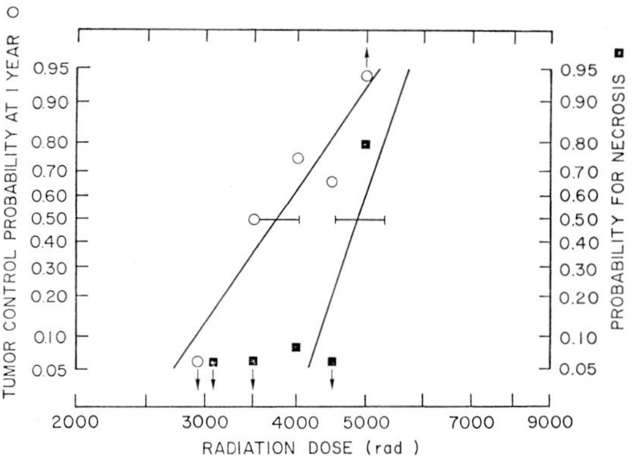

FIG. 10. Tumor control probability and probability for necrosis after treatment of oral squamous carcinoma in dogs (Gillette 1979). The open circles refer to tumor control frequency and the black squares are gross necrosis appearing in the oral cavity.

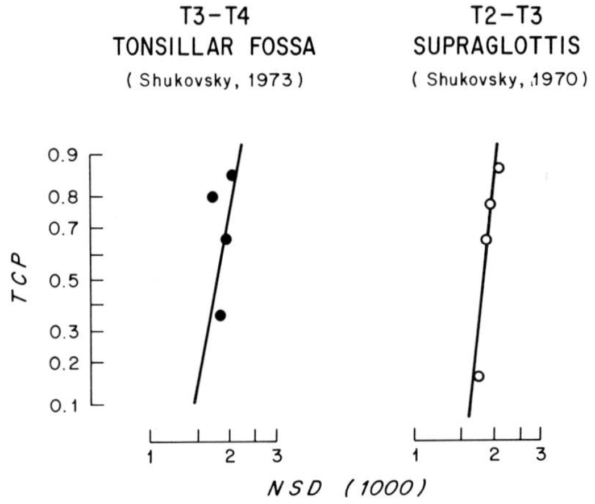

FIG. 11. Tumor control probability versus log NSD for T3–T4 squamous carcinoma of the tonsillar fossa and for T2 and T3 squamous carcinoma of the supraglottic region (Shukovsky and Fletcher 1973, Shukovsky 1970).

bladder stages T1–2 to receive 5,500 rad or 6,250 rad administered by a four-field technique using 20 fractions over four weeks. Similarly, patients with bladder carcinoma of stage T3 were assigned to receive 4,250 rad or 5,000 rad. Among the T1–2 patients the 13% dose increment from 5,500 to 6,250 rad yielded a control rate 19 percentage points higher; the dose increase for the second trial was 17.6% and the improvement in control was 17 percentage points.

Several reports have described retrospective analyses of tumor control and dose. In Figure 11 are shown local control curves for carcinomas of the tonsillar fossa (T3–4) and supraglottic region (T2–3) (Shukovsky and Fletcher 1973, Shukovsky 1970). For both of these tumors and clinical stages, the slope for the tumor control response is steep, viz. the $TCD_{.2} \rightarrow TCD_{.8}$ requires a dose increment of about 15%. The control of pelvic disease in patients with carcinoma of the uterine cervix treated by external beam therapy alone has been analyzed by Castro et al. (1970) and is given in Table 4. Although dose groups are broad there is an evidently steeply rising TCP with dose for carcinoma of the cervix stages I-IIIA. However, for the more advanced lesions, the effect of dose is not evident at levels greater than 6,000 rad. Figure 12 shows a dose-response curve for Hodgkin's disease (Kaplan 1966) that is rather similar in shape to that obtained for the immunogenic C3H fibrosarcoma growing in a

TABLE 4. *Local control, carcinoma of cervix*

Dose (rad)	Stage I-IIIA	Stage IIIB-IV
≤ 5000	0/2	0/30
6000	3/8 (37%)	9/20 (45%)
≥ 7000	12/14 (86%)	11/34 (32%)

Data from Castro et al. (1970).

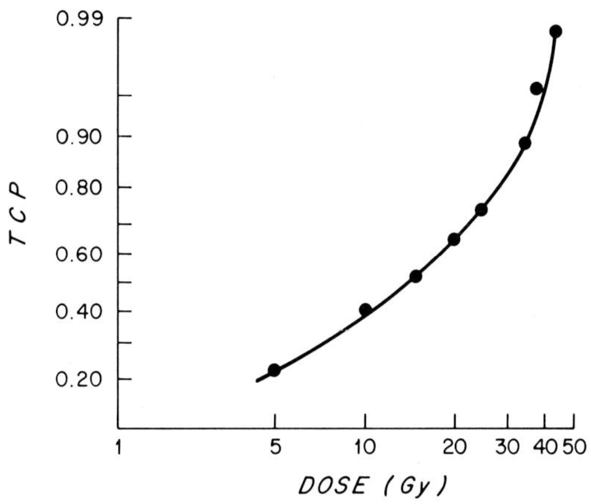

FIG. 12. Plot of tumor control probability versus log dose for local control of Hodgkin's disease (Kaplan 1966).

normal host as presented in Figure 9. The curve is complex with a broad range over which dose effect is modest; but, at high TCP values, small increments in dose have a profound effect on TCP.

There are data that are indicative of a much flatter dose-response relationship for control of human tumors, for example, human skin cancer (Hale and Holmes 1947). Conflicting data for local control of carcinoma of the glottis have been reported. Marks et al. (1973) found no convincing evidence of improvement in TCP with dose increase over a fairly wide dose range. In contrast, the data of Stewart and Jackson in 1975 indicated a steep dependence of control of glottic cancer upon dose. These various retrospective analyses cannot be accepted as necessarily true reflections of the slope for response curves of human tumors. This is so because the patients at the various dose levels were almost certainly not fully comparable with respect to tumor size, details of clinical presentation, and general condition (e.g., pulmonary function and hemoglobin level). We expect that the real response relationships are probably not greatly different from those found by Morrison for human bladder carcinoma or by Gillette for the spontaneous oral carcinomas of the dog; i.e., the curve is likely to be reasonably steep but less steep than found for model tumors or the C3H mouse mammary carcinoma. On balance, they do, however, support the proposal that dose increments of 10% to 20% should result in clinically important raises in TCP.

The second critical question is the extent to which tolerance will be improved by use of the superior dose distributions. This will determine whether we will be able to achieve dose increments of 10–20%. Data are less plentiful and refer mostly to skin. Figure 13 is a graph which relates "tolerance dose" of skin to area irradiated; these were derived from Paterson (1963). Tolerance is proportional to area $^{-.28}$. For example, a reduction in area by 33% is associated with an increase in tolerance dose by $\simeq$12%. Shukovsky (1970) plotted necrosis and severe edema incidences in patients treated for supraglottic carcinoma (Figure 14). Clearly, for those patients and their particular treatment schedules,

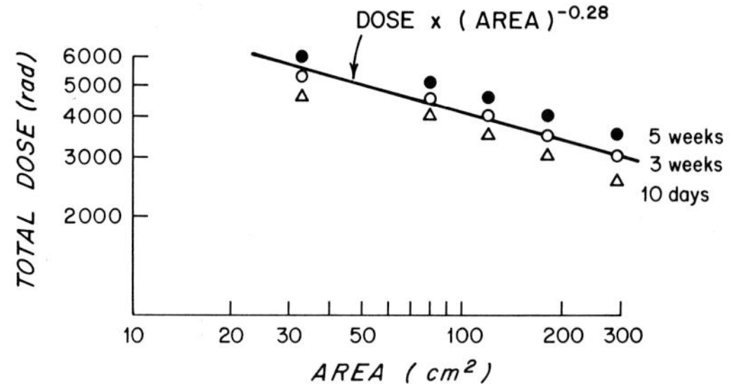

FIG. 13. Total radiation dose that produced tolerance reaction in the skin following orthovoltage radiation as a function of area of skin irradiated for treatment administered over 10 days, 3 weeks, or 5 weeks. This graph is derived from data published by Paterson (1963).

necrosis and severe edema were observed at progressively lower dose levels (NSD) as treatment area was increased. Actually, there have been few reported laboratory experiments relating tissue response to volume or area irradiated. One interesting experiment is that by Goffinet and co-workers in 1976, who reported that the myelitis 50 dose for mice was 7,400 rad and 5,900 rad for treatment of 6- and 12-mm lengths of the mouse spinal cord, respectively. Obviously, the length of the cord irradiated is a major determinant of response.

Although not proven by "proper" experiments, clinicians accept as fact that tolerance to radiation is increased appreciably as the treatment volume is reduced from covering all to only part of an organ or structure. This applies to the pharynx, rectum, extremity, etc. An example of this effect was shown in our use of the proton beam in the radiation therapy of carcinoma of the prostate as described by Shipley et al. (1979). By use of the proton beam for perineal boost technique the extent of rectum included in the high-dose region was sharply diminished. Radiation dose levels to the prostate and the sector of the anterior rectal wall in the treatment volume were increased by 500–600 rad. Even so, proctologic symptoms were drastically reduced in frequency and severity. With this new technique, tolerance is being determined at least in part by the urethra and not the rectum. Another pertinent experience with charged-particle beams

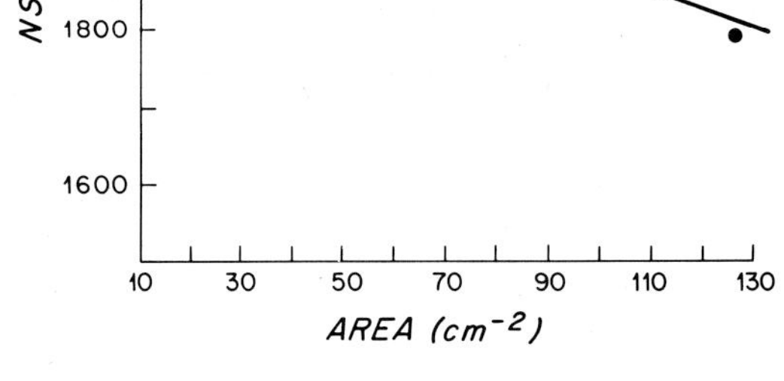

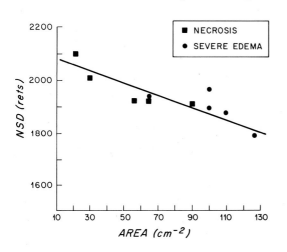

FIG. 14. Observed severe complications in patients treated by radiation therapy alone for supraglottic squamous cell carcinoma are plotted on an NSD versus field area (Shukovsky and Fletcher 1973).

is that of utilizing small field and high-precision proton beam techniques to treat malignant melanoma of the choroid. This experience has been successful in that in 21 of 21 patients, treated more than 9–40 months previously, partial regression of tumor has been achieved. Dose levels were 5,000–10,000 rad given in five fractions over eight to nine days. Reaction of the eye has been negligible, viz. partial cataracts in three patients. In contrast, 6,000–6,500 rad (30 fractions in six weeks) to the entire globe will result in severe changes and loss of function within one to two years in virtually every patient.

The effect on tolerance cannot be defined in advance for those treatment volume reductions likely to be achieved in clinical practice with charged-particle beams. Our own experience with proton beam therapy is that the reductions are clinically important. We have employed higher doses with proton beams than with photon beam techniques for several classes of patient problems and have observed no increase in normal tissue change. The potential for charged-particle radiation therapy must at this time be rated high.

SUMMARY

Fast-neutron and charged-particle beams are judged likely to provide important clinical advantages when employed in radiation therapy of human neoplasms. Radiobiological characterizations of high-LET radiations (fast neutrons, heavy ions, or negative pions) include: (1) a reduced OER; (2) reduced repair of radiation damage (sublethal, potentially lethal, and slow); and (3) suppression of the age-response function. In our opinion the most important of these is the lowered OER. Evidence is presented that for some human tumors hypoxic and viable cells are important in determining the results of low-LET radiation therapy. The pressing rationale for further study of high-LET radiation is the encouraging although not conclusively superior results being obtained with fast-neutron therapy in several centers. Charged-particle beams are attractive solely because of the superior dose distribution advantages they offer. There is an expected dose distributional advantage for the low-LET proton beam vis a vis photon; a comparable advantage should be realized in the use of heavy ions or pions vis a vis the fast-neutron beam. There is no known radiobiologic advantage of protons over photons or of the high-LET charged particle beam over fast neutrons, i.e., any clinical advantage should be explicable in terms of dose or dose × RBE distribution. The extent to which improved dose distribution will permit superior clinical results will depend upon the degree to which the treatment volume can be reduced to approach the size of the target volume and particularly the extent to which critical sensitive structures can be excluded from the treatment volume. The impact of such treatment volume reductions on tolerance is defined as the increment in radiation dose that can be administered. The ultimate limit in radiation dose is the tolerance of the target structure itself. The history of radiation therapy has already demonstrated that as techniques have permitted improved dose distributions so that higher doses could be administered, superior results have been achieved. We expect the same to obtain by application of even further refinements in dose-distribution patterns through the use of charged-particle beams.

REFERENCES

Barendsen, G. W., C. J. Koot, G. R. van Kersen, D. K. Bewley, S. B. Field, and C. J. Parnell. 1966. The effect of oxygen on impairment of the proliferative capacity of human cells in culture by ionizing radiations of different LET. Int. J. Radiat. Biol. 10:317–327.

Bird, R., and H. J. Burki. 1975. Survival of synchronized Chinese hamster cells exposed to radiation of different linear-energy transfer. Int. J. Radiat. Biol. 27:105–120.

Bush, R. S., R. D. T. Jenkin, W. E. C. Allt, F. A. Beale, H. Bean, A. J. Dembo, and J. F. Pringle. 1978. Definitive evidence for hypoxic cells influencing cure in cancer therapy. Br. J. Cancer 37(Suppl. III):302.

Cade, I. S., J. B. McEwen, S. Dische, M. I. Saunders, E. R. Watson, K. E. Halnan, G. Wiernik, and D. J. D. Perrins. 1978. Hyperbaric oxygen and radiotherapy: A Medical Research Council trial in carcinoma of the bladder. Br. J. Radiol. 51:876–878.

Castro, J. R., P. Issa, and G. H. Fletcher. 1970. Carcinoma of the cervix treated by external irradiation alone. Radiology 95:163–166.

Catterall, M., D. K. Bewley, and I. Sutherland. 1977. Second report on results of a randomized clinical trial of fast neutrons compared with x or gamma rays in treatment of advanced tumors of head and neck. Br. Med. J. 1(6077):1642.

Chapman, J. D., E. A. Blakeley, K. C. Smith, and R. C. Urtasun. 1977. Radiobiological characterization of the inactivating events produced in mammalian cells by helium and heavy ions. Int. J. Radiat. Oncol. Biol. Phys. 3:97–102.

Committee for Radiation Oncology Studies. 1978. Particle Proposal.

Dische, S. 1978. Hyperbaric oxygen: The Medical Research Council trials and their clinical significance. Br. J. Radiol. 51:888–894.

Durand, R. E. 1975. Cure, regression and cell survival: A comparison of common radiobiological end points using an in vitro model. Br. J. Radiol. 48:556–571.

DuSault, L. A. 1963. The effect of oxygen on the response of spontaneous tumours in mice to radiotherapy. Br. J. Radiol. 36:915–920.

Field, S. B. 1976. An historical survey of radiobiology and radiotherapy with fast neutrons. Curr. Top. Radiat. Res. 11:1–86.

Field, S. B. 1977. Early and late normal tissue damage after fast neutrons. Int. J. Radiat. Biol. 3:203–210.

Field, S. B., and T. Jones. 1968. The relative effects of fast neutrons and x-rays on tumour and normal tissue in the rat. Br. J. Radiol. 41:597–607.

Goffinet, D. R., G. W. Marsa, and J. M. Brown. 1976. The effects of single and multifraction radiation courses on the mouse spinal cord. Radiology 119:709–713.

Hale, C. H., and G. W. Holmes. 1947. Carcinoma of the skin: Influence of dosage on the success of treatment. Radiology 48:563–568.

Hall, E. J. 1978. Radiobiology for the Radiologist, Ed. 2. Harper and Row, Hagerstown.

Hall, E. J., L. Roizin-Towle, R. B. Theus, and L. S. August. 1975. Radiobiological properties of high energy cyclotron-produced neutrons used for radiotherapy. Radiology 117:173–178.

Henk, J. M., P. B. Kindler and C. W. Smith. 1977. Radiotherapy and hyperbaric oxygen in head and neck cancer. Lancet 2:101–103.

Henk, J. M., and C. W. Smith. 1977. Radiotherapy and hyperbaric oxygen in head and neck cancer. Lancet 2:104–105.

Kaplan, J. 1966. Evidence for a tumoricidal dose level in the radiotherapy of Hodgkin's disease. Cancer Res. 26:1221–1224.

Marks, J. E., L. D. Lowry, I. Lerch, and M. L. Griem. 1973. An analysis of recurrence as related to dose, time, and fractionation. Am. J. Roentgenol. 117:540–547.

Morrison, R. 1975. The results of treatment of cancer of the bladder. A clinical contribution to radiobiology. Radiology 26:67–75.

Paterson, R. 1963. The Treatment of Malignant Disease by Radiotherapy. Edward Arnold Publishers, Ltd., London.

Raju, M. R., H. I. Amols, E. Bain, S. G. Carpenter, R. A. Cox, and J. B. Robertson. 1978. A heavy particle comparative study. Br. J. Radiol. 51:712–719.
Raju, M. R., R. A. Tobey, J. H. Jett and R. A. Walters. 1975. Age response for line CHO Chinese hamster cells exposed to X-irradiation and alpha particles from plutonium. Radiat. Res. 63:422–433.
Scott, O. C. A. 1953. The response of tumours and normal tissues of the mouse to X-radiation delivered to animals breathing oxygen. Br. J. Radiol. 26:643–645.
Sheline, G. E., T. L. Phillips, S. B. Field, J. T. Brennan, and A. Raventos. 1971. Effects of fast neutrons on human skin. Am. J. Roentgenol. 111:31–41.
Shipley, W. U., J. A. Stanley, V. D. Courtenay, and S. B. Field. 1975. Repair of radiation damage in Lewis lung carcinoma cells following in situ treatment with fast neutrons and γ-rays. Cancer Res. 35:932–938.
Shipley, W. U., J. E. Tepper, G. R. Prout, L. J. Verhey, O. A. Mendiondo, M. Goitein, A. M. Koehler, and H. D. Suit. 1979. Proton radiation as boost therapy in patients irradiated for localized prostatic carcinoma. J.A.M.A. (in press).
Shukovsky, L. J. 1970. Dose, time, volume relationships on squamous cell carcinoma of the supraglottic larynx. Am. J. Roentgenol. 108:27–29.
Shukovsky, L. J., and G. H. Fletcher. 1973. Time-dose and tumor volume relationships in the irradiation of squamous cell carcinoma of the tonsillar fossa. Radiology 107:621–626.
Stewart, J. G., and A. W. Jackson. 1975. The steepness of the dose response curve both for tumour-cure and normal tissue injury. Laryngoscope 85:1107–1111.
Stone, R. S. 1948. Neutron therapy and specific ionization. Janeway Memorial Lecture, 1947. Am. J. Roentgenol. 59:771–785.
Suit, H. D., and O. C. A. Scott. 1979. Hyperbaric oxygen and irradiation. Review of laboratory experimental and clinical data. Proceedings of the International Meeting for Radio-Oncology. University of Vienna, Austria, 1978.
Suit, H. D., M. Goitein, J. Tepper, L. Verhey, A. M. Koehler, E. Gragoudas, and R. Schneider. 1977a. Clinical experience and expectation with protons and heavy ions. Int. J. Radiat. Oncol. Biol. Phys. Suppl. 2:115–125.
Suit, H. D., A. E. Howes, and N. Hunter. 1977b. Dependence of response of a C3H mammary carcinoma to fractionated irradiation on fractionation number and intertreatment interval. Radiat. Res. 72:440–454.
Suit, H. D., and M. Maeda. 1967. Hyperbaric oxygen and radiobiology of a C3H mouse mammary carcinoma. J. Natl. Cancer Inst. 39:639–652.
Suit, H. D., R. S. Sedlacek, L. Fagundes, M. Goitein, and K. Rothman. 1978. Time distribution of recurrences of an immunogenic and nonimmunogenic tumor following local irradiation. Radiat. Res. 73:251–266.
Suit, H. D., R. J. Shalek, and R. Wette. 1965. Radiation response of C3H mouse mammary carcinoma evaluated in terms of cellular radiation sensitivity, *in* Cellular Radiation Biology. Williams & Wilkins Co., Baltimore, pp. 514–530.
Suit, H. D., and C. Suchato. 1967. Hyperbaric oxygen and radiotherapy of a fibrosarcoma and of a squamous-cell carcinoma of C3H mice. Radiology 89:713–719.
Thomlinson, R. H. 1960. An experimental method for comparing treatments of intact malignant tumours in animals and its application to the use of oxygen in radiotherapy. Br. J. Cancer 14:555–576.
van den Brenk, H. A. S., K. Elliott, and H. Hutchings. 1962. Effect of single and fractionated doses of X-rays on radiocurability of solid Ehrlich tumour and tissue reactions in vivo for different oxygen tensions. Br. J. Cancer 14:518–534.
Watson, E. R., K. E. Halnan, S. Dische, M. I. Saunders, I. S. Cade, J. B. McEwen, G. Wiernik, and D. J. D. Perrin. 1978. Hyperbaric oxygen and radiotherapy: A Medical Research Council trial in carcinoma of the cervix. Br. J. Radiol. 51:879–887.
Withers, H. R., K. Mason, B. O. Reid, N. Dubravsky, H. T. Barkley, B. W. Brown and J. B. Smathers. 1974. Response of mouse intestine to neutrons and gamma rays in relation to dose fractionation and division cycle. Cancer 34:39–47.
Withers, H. R., and L. J. Peters. 1979. The application of RBE values to clinical trials of high LET radiations. Eur. J. Cancer (in press).
Withers, H. R., H. D. Thames, D. H. Hussey, B. L. Flow, and K. A. Mason. 1978. Relative biological effectiveness (RBE) of 50 MV (Be) neutrons for acute and late skin injury. Int. J. Radiat. Oncol. Biol. Phys. 4:603–608.

Radiation Biology in Cancer Research, edited by
Raymond E. Meyn and H. Rodney Withers.
Raven Press, New York © 1980.

Clinical and Experimental Alterations in the Radiation Therapeutic Ratio Caused by Cytotoxic Chemotherapy

Theodore L. Phillips

Department of Radiation Oncology, University of California, San Francisco, California 94143

Although radiotherapy was first attempted soon after the discovery of roentgen rays and radium, it did not come of age until the development of adequate radiation sources in the 1930s and 40s, and has been recently perfected with the development of a wide range of megavoltage units and improved brachytherapy techniques. Chemotherapy began in the 1940s with the discovery of nitrogen mustard, and has rapidly expanded over the ensuing 30 years through the introduction of a wider range of active compounds, with mechanisms of action ranging from alkylation through antimetabolic action to direct DNA binding of various types.

The improvement of the chemotherapeutic armamentarium and the training of significant numbers of medical oncologists coincided with the training of large numbers of full-time radiation oncologists. These two groups have finally come together in attempts to improve the local control and ultimate cure of the cancer patient. The first systematic attempts to improve cure were carried out in childhood tumors with agents shown to be active on their own against extensive metastatic disease. These were then applied in conjunction with radiation and surgery to treatment of patients with more limited disease. These applications have led to a large improvement in the survival of patients with most types of pediatric cancer, including acute lymphocytic leukemia.

The success of combined radiation, chemotherapy, and surgery in pediatric malignancies has led to the use of this approach in adult malignancies, with a certain amount of success, but with problems remaining, both in improving results and minimizing toxicity. A recent conference tried to bring together all of those working in the experimental and clinical field of chemotherapy/radiotherapy. This Conference on Combined Modalities, held at Hilton Head Island, South Carolina, November 15–18, 1978, involved presentations on the pharmacology, biochemistry, in vitro results, in vivo experimental tumor results, experimental normal tissue results, and clinical trials for a range of drugs including cyclophosphamide, Adriamycin, cis-platinum, nitrosoureas, bleomycin, and new and other interesting drugs. This paper is an attempt to present the current state of my work and review some of the recent literature and summarize certain presentations at the Hilton Head meeting. Particular emphasis will be placed on in vivo tumor and normal tissue experimental data and clinical data.

TERMINOLOGY

It becomes quite clear when one attends meetings on combined modalities or when one reads papers in this area that there is confusion concerning the terminology to be applied when two or more agents interact and result in cell kill in tumor and normal tissues. Steel in two sequential papers (Steel 1979, Steel and Peckham 1979) has proposed that a range of terms be utilized in describing the interactions that occur, primarily those in vivo. Dewey (1979) has proposed a somewhat different terminology for in vitro interactions, and I previously proposed (Phillips 1977) a somewhat simplified set of terminology based on Valeriote and Lin's system (1975). In Table 1 can be seen my approach to terminology at this time. It incorporates the scheme proposed by Steel, but attempts to simplify it. The major interactions between radiation effects and those of chemicals can be divided into independent action, enhancement, and protection. Independent action is either a spatial separation of the site at which radiation is effective and the site at which the drug is effective, such as in irradiation of the brain in patients with leukemia, where the drugs are not able to reach the brain because of the blood-brain barrier. Independent action may also occur when there is toxicity independence, i.e., there is no drug effect in the organs irradiated, even though the drug may be present and active against the tumor in the radiation target volume.

Enhancement occurs when the response is greater than that which would be observed with radiation alone. This enhancement may be subadditive, additive, supraadditive, or exhibit true sensitization. Steel has gone to great lengths to define how one might determine from the complete survival curves for radiation and for drug in a particular tissue whether subadditive, additive, or supraadditive

TABLE 1. *Terminology for cytotoxic drug-radiation interactions in vivo*

Independent Action
Spatial cooperation—independent sites of action
Toxicity independence—no drug effect on tissues irradiated
Enhancement
Subadditive—subadditive, interference, antagonism
Additive—within envelope of additivity
Supraadditive—potentiating or sensitizing—greater than sum of both agents—synergism
Sensitization—increased radiation effect caused by an inactive agent
Protection
One agent inactive (nontoxic), reduces effect of other, allows higher dose of active agent for same isoeffect

effects have occurred. In general, injury is subadditive when there is less than a numerical addition of the kill that would be expected from the survival curves at that level for each agent alone. Interference and antagonism occur when the cell kill is less than that of the more active agent or less than that of the least active agent, respectively. Supraadditivity involves potentiation or sensitization and includes kill greater than the sum of both agents. The term *synergism* has also been applied. Because these terms have become so confused, I agree with Steel that the term supraadditivity is preferable for this phenomenon.

The term *sensitization* should be reserved for increased radiation effect caused by an inactive agent, such as misonidazole, in concentrations not causing cell kill but enhancing radiation effects on hypoxic cells.

Protection occurs when an inactive agent reduces the effect of an active agent, i.e., chemical protectors against radiation, and allows a higher radiation dose with the same isoeffect. The term *protection* should be used only in this situation; situations in which a cytotoxic drug may interfere with the cell kill by radiation should be termed as *subadditive.*

Dewey (1979) and others working with in vitro systems have used the term independent to describe cell kill by a drug that does not change the shape of the radiation survival curve. He has used the term additive when the prior drug treatment removes the shoulder from the radiation survival curve and the term sensitization or potentiation when the slope of the radiation survival curve is changed. It can be seen that these terms are somewhat different from those proposed here and by Steel. One should clearly differentiate between those terms used exclusively in the in vitro situation in which the complete survival curve may be studied and those terms applicable in vivo.

REASONS FOR THE OBSERVATION OF ENHANCEMENT OF THE SUPRAADDITIVE TYPE

In Table 2 can be seen a summary of possible reasons why radiation and drug may interact in a given tissue or tumor, yielding supraadditive response. The cause may be physiologic with improved blood supply and better drug access or decreased cell-cell interaction, as one sees in single-cell suspensions. A drug may yield tumor shrinkage, which leads to reoxygenation and then greater sensitivity to radiation.

Cell kinetics may be markedly altered by prior irradiation or drug treatment before the second agent is given. The cell kill occurring may increase the number of cycling cells or the percentage of cells in sensitive phases of the cell cycle for a phase-specific chemotherapeutic agent. A drug may block progression in a phase of the cycle more sensitive to radiation and vice versa.

One treatment may interfere with the repair of injury caused by another treatment. A limited number of drugs appear to change the shoulder on the radiation survival curve and inhibit sublethal damage repair after irradiation. In rare situations the slope of the radiation-response curve may change, as is

TABLE 2. *Possible causes of supraadditive response or potentiation*

Physiologic
Improved blood supply = better drug access
Decreased cell-cell interaction
Reoxygenation
Cell Kinetic
Increased % cycling cells
Increased % cells in sensitive phases
Radiation Damage or Repair Modification
Change in shoulder ± inhibition of SLD & PLD* repair
Change in slope of radiation response curve, i.e., inherent sensitization of target
Drug Damage or Repair Modification
Radiation lesion sensitizes to drug lesion
Radiation and drug lesions compete for or block repair processes

* SLD, sublethal damage; PLD, potentially lethal damage.

seen with the incorporation of thymidine analogs containing halides into the DNA.

Radiation injury may interfere with drug damage accumulation or drug damage repair. The radiation lesion may make the target molecules more sensitive to the same level of drug if they are administered prior to repair of the radiation lesions. Radiation and drug lesions could compete for repair enzymes or processes. Examples of each of these interactions may be seen in some of the normal tissue and tumor experiments reported below. The number of alterations that can occur is obviously large and thus the interaction of radiation and chemotherapeutic agents is extremely complex. When one considers that one is dealing with a wide range of normal tissues with different cell kinetic attributes within a radiation target volume, as well as with a wide range of tumors, even within the same histologic type and site of origin, one realizes the complexity of the situation.

EXPERIMENTAL DATA FROM ANIMAL MODELS OF NORMAL TISSUE

A wide range of animal models is available for study by radiation and thus for the study of the interaction of cytotoxic chemotherapy with radiation. The available systems were recently summarized by Field and Michalowski (1979). Potential systems include those measuring CNS damage with both brain lethality and spinal cord transsection end points, oral death, pulmonary lethality, esophageal lethality, cardiac alterations and lethality in the rabbit, renal function and renal lethality end points, intestinal crypt cell survival, gastric crypt cell and rectal crypt cell survival, skin reaction grade, bone marrow colony-forming units and other assays, skin reaction scoring, testicular atrophy and clonogenic survival, skin contraction, and various $LD_{50/6}$ or $LD_{50/7}$ and $LD_{50/30}$ end points. An end point is lacking for hepatic injury, as are assays for late injury to the small and large bowel.

Only a limited amount of information is available on most of these experimental systems for the interaction of radiation and cytotoxic chemotherapy (Guigon et al. 1978, Phillips 1978, 1979a, Phillips et al. 1975). Unfortunately, only a small number of papers were presented on this subject at the Hilton Head meeting, and many dealt with the intestinal crypt cell system or with assays of intestinal crypt survival and DNA synthesis.

The information available on three systems from our laboratory is presented in Table 3. Information on nine drugs administered to mice was evaluated using the intestinal crypt cell technique. The maximum tolerated dose (MTD) was used, i.e., that dose yielding only occasional deaths ($LD_{.01}$). It can be seen for intestinal crypt cells that all drugs (tested at the MTD) except vincristine caused enhanced cell kill. In the pulmonary system, enhanced kill was seen with actinomycin D and cyclophosphamide; studies are incomplete for the other agents. The bleomycin dose used was quite small, 3 mg/kg, which may explain why no enhanced response was seen. This emphasizes the need to use high drug doses, equivalent to the clinical situation, and to use multiple drug injections in evaluating the dose effect factor (DEF) for experimental normal tissues. Esophageal lethality is unchanged by actinomycin D and Adriamycin, in spite of clinical reports of enhanced effects, even though maximum tolerated doses were utilized. It is probable that fractionated irradiation and multiple drug exposures are required to elicit an effect. Cis-platinum II shows a very marked enhancement of esophageal lethality.

The preceding examples were all obtained from experiments in which the

TABLE 3. *Dose effect factors* for three normal tissues*

		D.E.F. Values†		
Drug	Dose (MTD) mg/kg	Crypt Cells (10)	Lung ($LD_{50/160}$)	Esophagus ($LD_{50/28}$)
Actinomycin D	0.3	1.2	1.0‡–1.6§	1
Adriamycin	8	1.2	—	1
BCNU	25	1.3	—	1.2
Bleomycin	100	2.0	0.9‡	1.1‡
Cyclophosphamide	250	1.1	1.3‡	1.05
5-Fluorouracil	140	1.2	—	—
Methotrexate	700	1.5	—	—
Cis-platinum (II)	13	1.3	—	1.5
Vincristine	3	0.9	1.2‡	1.0

* Dose Effect Factor = D.E.F. = Radiation dose for effect without drug/Radiation dose for effect with drug.

† Drug given i.p. 2 hr before irradiation.

‡ At less than MTD.

§ At more than MTD.

TABLE 4. *Dose effect factors for experimental normal tissues—other reports*

Drug	Dose mg/kg	Time Relative to Radiation	Tissue (Mouse)	D.E.F.	Reference
Actinomycin D	.77	−4 hr	$LD_{50/6}$ intestine	1.4	Concannon et al. (1973)
Adriamycin	10	−1 day	$LD_{50/7}$ intestine	1.6	Schenken et al. (1979a)
Adriamycin	10	−14 days	"	1.1	Schenken et al. (1979a)
Adriamycin	10	−49 days	"	2	Schenken et al. (1979b)
Adriamycin	7	−2 hr	anagen hair	1.8	Griem et al. (1979)
Adriamycin	10	±24 hr	$LD_{50/180}$ lung	1.7	Redpath et al. (1978)
Adriamycin	10	+7 days	"	1.25	"
Adriamycin	10	−1 hr	skin reaction	1.05	"
Adriamycin	97–202 mg/m²	varied	heart rabbit	>1	Eltringham et al. (1979)
BCNU	30	−2 hr	skin reaction	1.3	Lelieveld et al. (1979a)
BCNU	30	−2 min	skin reaction	1.6	"
BCNU	10	−2 min, −24 hr	spinal cord paralysis	1.0	"
BCNU	4	−48 hr	dog TBI	>1	Paterson and English (1979)
BCNU	40	−24 hr	intestine $LD_{50/7}$	5	Schenken et al. (1976)
Bleomycin	5 × 5	±5 & ±1 days	skin reaction	1.5	Leith et al. (1975)
Cis-platinum (II)	8	−2 min	intestine $LD_{50/7}$	2.7	Schenken et al. (1979a)
Cis-platinum (II)	2.5	−2 hr	rat skin reaction	>1	Douple et al. (1979)
Cyclophosphamide	200	+2 min	intestine $LD_{50/7}$	2.9	Schenken et al. (1976)
Cyclophosphamide	300	+24 hr	skin reaction	1.1	Steel et al. (1978)
Methotrexate	30	−24 hr	intestine $LD_{50/7}$	1.3	Schenken et al. (1976)

drug was injected intraperitoneally and the irradiation carried out two hours after drug injection.

A number of other experiments have been reported including end points other than those presented in the prior table; among them are skin reactions, hair reaction, spinal cord paralysis, and rabbit heart injury. Selected other reports are summarized in Table 4. The doses were close to the MTD in most cases, although in some situations they were higher and in others lower. The majority of experiments were carried out in mice. Times of drug administration ranged from simultaneous administration to as long as 49 days separation between drug and radiation delivery.

Particularly interesting are the experiments reported by Schenken et al. (1979b) in which Adriamycin caused a profound suppression of intestinal proliferative capacity, as evaluated by the intestinal $LD_{50/7}$ with weekly radiation exposures or single-dose challenge. This finding was most marked with drug administration 49 days before split-dose irradiation. Adriamycin also caused a marked enhancement of effect on anagen hair, but not in resting telogen hair, as reported by Griem et al. (1979). In their experiments with pulmonary LD_{50} using Adriamycin, Redpath et al. (1978) did find an enhanced response in the lung, but not in the skin. BCNU has not caused a major effect in most of the tissues studied, with the exception of administration just prior to radiation in experiments of Lelieveld et al. (1979a), administration 12–2 hours before irradiation in our experiments (Goldstein et al. 1979), and in the intestinal $LD_{50/7}$ end point as reported by Schenken et al. (1976).

The prior two tables illustrate the broad range of activity of cytotoxic chemotherapeutic agents in normal tissues as evidenced by the enhanced response seen with radiation end points in these studies. It is of interest to explore the effect of time of drug administration on the degree of enhanced response observed. Table 5 summarizes the information available on the time of maximum enhanced response to drug plus radiation observed in intestinal crypt cells. There is a wide variation in the time interval between drug and radiation for maximum cell kill. For experiments performed in two different laboratories, there is significant deviation in the time of maximum effect, with the exception of cis-platinum for which both laboratories found maximum effect with simultaneous drug and radiation administration. The times for maximum effect for BCNU and cyclophosphamide were quite different. Schenken employed the intestinal crypt assay without a correction for cell number, and we employed Wither's formula for correcting for multiplicity (Withers and Elkind 1970), which yields results in terms of surviving crypt cells. Since the end points were taken at relatively modest survival levels, the lack of this correction in the Schenken material should not lead to these discrepancies. Further comparative studies are needed.

It is interesting to review the degree of enhancement that occurs, which can be exemplified by the ratio of the maximum number of crypts at any time of administration, from ±24 hours, to the minimum number of crypt cells surviving. Particularly large ratios are seen with BCNU, bleomycin, and cis-platinum, in

TABLE 5. *Time of maximum enhanced response to drug plus radiation relative to time of irradiation in mouse intestinal crypt cells**

Drug	Dose mg/kg	Time Interval for Minimum Survival	Max/Min†	Author
Actinomycin D	0.3	−6 hr	2.4	‡
Adriamycin	8	+12 hr	4	Ross et al. (1979)
BCNU	40	−24 hr	2.6	Schenken et al. (1976)
BCNU	25	−2 hr	25	Goldstein et al. (1979)
Bleomycin	100	−6 hr	50	Phillips et al. (1979b)
Cis-platinum (II)	8	0	5	Schenken et al. (1979a)
Cis-platinum (II)	13	0	20	Luk et al. (1979)
Cyclophosphamide	200	+2 min	1.5	Schenken et al. (1976)
Cyclophosphamide	250	−18 hr	5	Phillips et al. (1979a)
5-Fluorouracil	140	+24 hr	5	‡
Methotrexate	700	−12 hr	366	‡
Vincristine	3	+24 hr	10	‡

* For time range −24 to +24 hr.

† Maximum no. surviving crypts/crypt cells per circumference divided by minimum no. −24 to +24 hr.

‡ Unpublished personal data.

which a marked enhanced response is seen with administration of drug prior to irradiation for BCNU and bleomycin, and for simultaneous administration with cis-platinum. The changes are not large for the other drugs, with the exception of methotrexate and vincristine, which cause synchronization of surviving intestinal crypt cells at high doses, resulting in fluctuations in radiation survival as a function of time between drug and radiation administration.

Since most radiotherapy and chemotherapy treatments are given as multiple doses, it is important to investigate the effect of chemotherapy administration on the repair of radiation-induced sublethal damage. Although a number of in vitro experiments have been reported, little information is available from in vivo evaluations. Table 6 summarizes the information available from our laboratory on the D_q and on survival ratios for split-dose studies with three-hour intervals between two radiation doses. These split-dose experiments were conducted with two-thirds the maximum tolerated chemotherapeutic doses, given two hours before irradiation. Drug kill was estimated from drug survival curves with drug doses ranging up to the MTD delivered three hours after irradiation and two hours before irradiation. The D_q was estimated by extrapolating radiation survival curves back to zero dose, which were obtained two hours following drug administration at the MTD. The number of crypts that should be expected at zero radiation dose was extrapolated from curves available for high linear energy transfer (LET) radiation (argon and neon) and from the experiments reported by Masuda et al. (1977). It can be seen that the D_q values were similar to those for radiation only, although slightly suppressed in all cases with the exception of cyclophosphamide. This change may be due to an incomplete correc-

TABLE 6. D_q and survival ratios for intestinal crypt cells

Drug	Estimated drug kill S/S^0 at M.T.D.	Estimated D_q	3-hr Split Dose Survival Ratio
Radiation only	10^{-3}	400	3–6
Actinomycin D	0.4	300	16
Adriamycin	0.3	310	8.5
BCNU	0.35	270	11
Bleomycin	0.4	0, 400	4
Cyclophosphamide	0.6	400	15
5-Fluorouracil	0.1	300	3.8
Methotrexate	0.1	300	2.1
Cis-platinum (II)	0.2	280	3
Vincristine	0.7	320	6.5

tion for cell kill and does not prove any inhibition of sublethal radiation damage accumulation. If the cell kill information for bleomycin administered three hours after irradiation is used to make corrections, an estimated D_q of 0 is found; if the cell kill for administration two hours before irradiation is used where bleomycin markedly enhances the kill, then the D_q is unaltered at 400.

A limited number of two-dose experiments have been carried out in B6AF1 mice. (The survival curves from which the drug kill and estimated D_q were obtained were performed with LAF_1 mice.) These survival ratios, i.e., the ratio of the number of crypt cells surviving after two radiation doses to those surviving after one radiation dose with a three-hour interval, are shown in the third column of Table 6. Variation in survival ratios is seen, but in no case is an increase in the number of crypt cells with fractionation absent. Studies of split-dose survival are being performed with time intervals ranging from 0 to 24 hours and multiple replicate points; they indicate no evidence for impairment of sublethal damage with 5-fluorouracil and bleomycin.

The information on time of maximum and minimum response could be used to predict when administration of a chemotherapeutic drug relative to irradiation would be particularly dangerous and when it would lead to minimum intestinal crypt cell kill. If information were available for tumors that indicated a different time of maximum response, then treatments could be carried out that would lead to enhancement of the therapeutic ratio.

EXPERIMENTAL DATA FROM ANIMAL TUMOR MODELS

Although a number of experiments have been reported in the literature and were presented at the Hilton Head meeting concerning the chemoresponsiveness of mouse and rat tumors, few investigators have performed studies with more than one or two tumors and very few with more than one or two drugs in an individual tumor type. Thus it is very difficult to compare drug responsiveness across a range of mouse tumor models, particularly when performed by the

TABLE 7. *Chemoresponsiveness of selected mouse tumors** †

Tumor	Actino-mycin	Adria-mycin	BCNU	Bleomycin	Cyclophos-phamide	5-Fluoro-uracil	Metho-trexate	Cis-platinum (II)	Vincris-tine
EMT6/SF	2/.95	0/.8	5.5/.001	3/.4	10/.001	4/.4	2/.9	2.5/.3	1.5/.6
EMT6/St	0.7/	3.7/	5.9/	0/	4.5/			2.4/	
KHT	0.9/	0.3/	10.8/	0/	12/			2/	
RIF-1	1.5/	1.7/	2/	1.2/	11/			9.6/	

* Days growth delay (4X) / surviving fraction at M.T.D.
† (Begg et al. 1979a, 1979b, 1979c, Fu et al. 1979, Lelieveld et al. 1979b, Twentyman et al. 1979a, 1979b, 1979c, 1979d, 1979e.)

same investigator, and it is particularly difficult to evaluate experiments with one tumor across a wide range of drugs to evaluate that tumor for chemoresponsiveness. Although a considerable amount of information on the subject exists in the chemotherapeutic literature, most of the tumors used are not well characterized in terms of their radiobiologic response and many are poorly characterized in terms of cell kinetics. In addition, many of these tumors are highly immunogenic and many of the end points have been only semi-quantitative. This review will concentrate on a limited number of tumors for which the data are available for a number of chemotherapeutic agents. These tumors have been well characterized radiobiologically and are of limited immunogenicity, with exceptions noted. The chemoresponsiveness of the tumors is summarized in Table 7. The greatest amount of information is available for the EMT6/SF, EMT6/St, KHT, and RIF-1 tumors (Table 7).

There is a wide range of chemoresponsiveness among these tumors, with the EMT6 tumors being most responsive to BCNU and cyclophosphamide, as judged both by the in vivo/vitro assay and the regression-regrowth method. The EMT6 tumor is markedly immunogenic in the St subline and of limited immunogenicity in the SF subline. Reasonable agreement in results is noted with the exception of the response to Adriamycin, which was greater in the EMT6/St. There is also reasonable agreement between regrowth assays and in vivo/vitro cell survival assays, with the exception of BCNU in which cell kill is much greater by the in vivo/vitro assay than shown by delay in regrowth, and for bleomycin and 5-fluorouracil where regrowth delay is greater than one would expect from the cell survival data.

The KHT tumor seems most sensitive to BCNU and cyclophosphamide, whereas the RIF-1 tumor is most sensitive to cyclophosphamide and cis-platinum II. The Lewis lung tumor (Kovacs et al. 1979) was sensitive to cis-platinum II, as was the MTGB (Douple and Richmond 1979); the CBSAF and WHFIB tumors were sensitive to cyclophosphamide, but resistant to Adriamycin (Martin and McNally 1979).

It is clear that a greater range of mouse tumor studies is required and a wide range of tumors is needed. Tumors are required that are well characterized in terms of their doubling times, cell kinetics, TD_{50}, and immunogenicity, as well as chemoresponsiveness. As will be seen later, it is unlikely that one will see significant radiation-drug interactions in tumors that are resistant to a given cancer chemotherapeutic agent.

From the studies summarized in Table 7, one can determine the time of drug administration relative to radiotherapy for maximum tumor response. This information is summarized in Table 8. The responses were judged by both the in vivo/vitro assay (V) and the regrowth assay (R). Responses were judged as additive, subadditive, and supraadditive from the regrowth assays and survival curves. The judgements were made primarily by me based on the written papers from the various experiments reported at the Hilton Head meeting.

Actinomycin was not particularly effective in any of the tumor systems evalu-

TABLE 8. *Time of drug administration relative to radiotherapy for maximum tumor response in selected mouse tumors using effective drugs**

Tumor Type	Actinomycin	Adriamycin	BCNU	Bleomycin	Cyclophosphamide	Cis-platinum (II)
EMT6/SF	—	-1 hr^{V} A	$+1$ hr^{V} A $+1$ hr, $+8$ hr^{R} supra	-1 hr^{V} -8 hr^{R} A, sub	$+1$ hr $+8$ hr$^{V\&R}$ A, supra	—
EMT6/St	-6 hr^{R} supra	0 hr^{R} A	-24 hr^{R} $+6$ hr^{R} A	-6 hr^{R} A, sub, supra	None^{R} A	None^{R} sub
KHT	None^{R} A	None^{R} A	-2 hr^{R} supra	None^{R} A	None^{R} sub	None^{R} A
RIF-1	None^{R} A	None^{R} A	None^{R} sub	None^{R} A	$+3$ hr^{R} supra	-24 hr^{R} supra

V, in vivo/vitro assay; R, regrowth assay; A, additive; sub, subadditive; supra, supraadditive.
* (Begg et al. 1979a, 1979b, 1979c, Fu et al. 1979, Lelieveld et al. 1979b, Twentyman et al. 1979a, 1979b, 1979c, 1979d, 1979e.)

ated. There was no point of maximum response for any tumor except the EMT6/St, where maximum kill was seen with drug administration six hours before irradiation. Adriamycin was not effective in any of these tumors, but did show a slight additive increase in cell kill in the EMT6 tumor when administered one hour before or simultaneously with radiation.

BCNU was particularly effective in the EMT6 sublines and in the KHT tumor, with supraadditive effect seen in the EMT6/SF and KHT, with administration at −2 hours for the KHT and at +1 and +8 hours for the EMT6/SF. Bleomycin was rather ineffective in all of these tumor lines, although some enhanced response was seen at −8, +6, and +1 hours in the EMT6 sublines. In most cases this effect was subadditive, although there was one experiment reported in which the effect was supraadditive.

Cyclophosphamide was particularly active in the EMT6 and RIF-1 tumors. In the EMT6/SF tumor, there was a supraadditive effect as judged by regrowth assay at +1 and +8 hours, and there was a supraadditive effect in the RIF-1 tumor at +3 hours.

Cis-platinum II showed no significant response enhancement in the EMT6/St or the KHT tumors, but did show a supraadditive effect when administered at −24 hours in the RIF-1 tumor.

A limited amount of information was available for rat tumors, in particular the 9L gliosarcoma (Wheeler et al. 1979) and the R1 rhabdomyosarcoma (Kal et al. 1979). For the 9L gliosarcoma growing in the flank, as measured by Wheeler et al. (1979), drug administration at −16 hours (which is also +8 hours with these five daily fraction experiments), produced maximum effectiveness and a marked improvement in animal survival as compared to radiation or BCNU alone. A clearly supraadditive effect was noted. Kal et al. (1979) reported results with the R1 rhabdomyosarcoma in which vinblastine, which was rather ineffective when used alone, gave a supraadditive effect with radiation at −4 and +3 days.

There are a limited number of situations in which a favorable therapeutic ratio could exist, based on these studies. Those would be situations in which a time of administration could be chosen at which one would see no or little enhanced normal tissue response as evaluated in the intestinal crypt cell system and during which one would see, if possible, supraadditive effect in the tumor (Table 9). Obviously, the intestinal crypt cell system does not represent all tissues in the body and is probably valid for only its own cell type, thus predicting only for acute small bowel responses. Likewise, these particular tumors used are probably representative only of themselves and not even of a class of human tumors. Much further work is needed before one can make generalizations concerning the predictive value of specific experimental normal tissue and tumor systems.

It is possible to generalize, as shown in Table 9, on some factors. No situation with a favorable therapeutic ratio occurred, except in situations when the tumor was responsive to the drug. Indeed, a favorable situation could be found for

TABLE 9. *Situations in which a favorable therapeutic ratio would exist in mouse tumor treatment if intestinal crypt cells were limiting normal tissue*

Tumor Type	Drug	Favorable Times for Drug Administration Relative to Radiation (hr)
EMT6/SF	BCNU	+1, +8
	CTX	+8
EMT6/St	BCNU	+6
	CTX	+24
KHT	BCNU	+6
	Cis-PDD	+24
	CTX	Simultaneous
RIF-1	CTX	+3
	Cis-PDD	−24

CTX, cyclophosphamide.

each tumor using those drugs to which it was particularly responsive. Examples include enhanced response of EMT6/SF and St to BCNU when drug was given at intervals from one to eight hours following irradiation. In addition, cyclophosphamide was particularly effective in these two tumors when given 8–24 hours following irradiation.

In the KHT tumor, BCNU and cis-platinum II were effective when given 6–24 hours after irradiation, as was cyclophosphamide when given simultaneously with irradiation. In the RIF-1 tumor, cyclophosphamide showed an enhanced therapeutic ratio when given three hours after irradiation and cis-platinum II when given 24 hours before. These times were selected for additive or supraadditive cell kill during a time period when there was minimal effect on intestinal crypt cells.

Although these favorable treatment times for an improved therapeutic ratio may be valid only for mice bearing EMT6 tumors in the abdominal cavity, they do indicate that it is possible to predict optimum times for drug-radiation interaction from experimental data in tumor and normal tissue systems. Much more work is required before this type of study can be extrapolated to the clinical situation. Indeed, the experiments presented have only involved intestinal crypt cell kill and tumor regression and have not looked at tumor cure with multifraction radiation experiments involving tumors implanted on the abdominal wall, etc. These kinds of experiments are also required.

CLINICAL INFORMATION ON NORMAL TISSUE DAMAGE WITH COMBINED RADIATION AND CHEMOTHERAPY

A large number of reports have dealt with the enhanced response of radiotherapy seen with combined radiotherapy and chemotherapy. I have presented a number of reviews of this subject (Phillips 1977, 1978, Phillips and Fu 1977), and a conference held in San Francisco in March 1978 and edited by J. M. Vaeth addressed the combined effects of chemotherapy and radiotherapy in normal tissue tolerance. A small number of papers were also presented on this subject at the Hilton Head meeting. The clinical examples of increased normal tissue response with systemic single-agent chemotherapy and radiation are summarized in Table 10. Although a large number of examples of enhanced normal tissue response to combined radiotherapy and multidrug combination chemotherapy are available in the literature as well, it is not clear which of the drugs in these combinations is causing the enhanced damage. Thus, the presentation of the single-drug information is more valuable.

For the central nervous system, Byfield (1979) presented a review that clearly indicates the marked enhancement of CNS damage by combined radiation and methotrexate. Methotrexate alone causes CNS damage and leukoencephalopathy, but the combination is particularly more damaging and causes a number of syndromes, particularly those related to the brain and brainstem (Bleyer 1977). There is some evidence that vincristine and Adriamycin may enhance spinal cord radiation response and that a number of cord transections have been seen at doses below that expected, particularly with drug combinations containing Adriamycin. CNS damage has occurred with widely varying intervals between drug and radiation and often with multiple drug administrations before, during, and after radiotherapy, and thus no time for maximum responsiveness is evident.

Enhanced response in the lung occurs particularly with actinomycin D, cyclo-

TABLE 10. *Clinical examples of increased* normal tissue response with systemic single drugs and radiation†*

Tissue	Drugs	Timing for Maximum Responses
CNS	MTX, VCR?, Adria?	—
Lung	Act D, CTX, Adria, Bleo	C, B, A
Heart	Adria, Mito, Act D?	B, C, A
Intestine	Act D, Adria, 5-FU	C
Liver	Act D, VCR	C
Kidney	Act D	C
Skin, mucosa	Act D, Adria, MTX, Bleo, 5-FU, Hu, 6MP	C
Bone and soft tissue	5-FU, Bleo	C
Esophagus	Act D, Adria	C
Bladder	CTX	C
Eye	5-FU	C

C, concurrent; B, before; A, after; MTX, methotrexate; VCR, vincristine; Adria, Adriamycin; Act D, actinomycin D; CTX, cyclophosphamide; Bleo, bleomycin; mito, mitomycin-C; 5-FU, 5-fluorouracil; Hu, hydroxyurea; 6MP, 6-mercaptopurine.

* Severe or fatal.

† (Aristizabol et al. 1979, Bagshaw and Doggett 1969, Billingham 1979, Bleyer 1977, Byfield 1979, Cassady et al. 1979, Catane et al. 1979, Eltringham 1979, Fu 1979, Muggia et al. 1978, Phillips 1978, Phillips 1979b, Phillips and Fu 1976, 1977, Ransom et al. 1979, Ruckdeschel et al. 1979.)

phosphamide, Adriamycin, and bleomycin, and a major effect occurs if drug and radiation are administered concurrently. There is no clear-cut relationship between the total amount of drug delivered and the development of radiation pneumonitis, but it is clear from the review of Catane et al. (1979) that simultaneous administration is most dangerous.

In the heart, Adriamycin is clearly implicated in enhanced response. The histologic lesions caused by radiation and Adriamycin differ, with radiation causing mainly capillary damage and Adriamycin myocyte damage, but the result is increased incidence of cardiac failure when the two agents are used concurrently with, before, or after each other. There is a clear dose-response relationship, with greater incidence of heart injury occurring with increasing radiation and increasing drug dosage, as has been beautifully described by Billingham and her colleagues at Stanford (Billingham 1979). In a limited number of reports, mitomycin C and actinomycin D have also been noted to enhance cardiac damage (Eltringham 1979).

Small intestine injury has been noted with actinomycin D, Adriamycin, and 5-fluorouracil, particularly when they are administered simultaneously with radiation. These enhanced responses are usually nonfatal, and successful combinations of 5-fluorouracil and Adriamycin have been utilized in treating abdominal malignancies.

Information on the liver and kidney is available, particularly from children treated for Wilms' tumor, and it is evident that actinomycin D and vincristine enhance liver response, while actinomycin D enhances kidney response (Cassady et al. 1979).

Extensive information is available on the skin and mucous membranes, where a wide range of chemotherapeutic agents cause mucositis or skin reaction when used alone and where they increase the degree of radiation mucositis or dermatitis. Some of the rare quantitative information on clinical dose-effect factors is available from Bagshaw and Doggett (1969) for a number of the compounds listed in Table 10 for mucosal reactions. Reactions in the head and neck tissues have been summarized by Fu (1979) and indicate that actinomycin D, Adriamycin, methotrexate, bleomycin, 5-fluorouracil, hydroxyurea, and 6-mercaptopurine all enhance the degree and the time of onset of mucositis with concurrent administration. 5-FU and bleomycin also seem to increase bone and soft tissue injury in the head and neck region.

Injury to the esophagus has been seen, with some strictures following simultaneous radiation and Adriamycin administration. The known damage to the bladder mucosa caused by cyclophosphamide leads to enhanced response of combined radiation and cyclophosphamide on the bladder; this combination should be avoided. Some reports of increased injury to the eye by simultaneous intraarterial 5-FU chemotherapy and radiotherapy have been published.

All of the cases summarized in Table 10 are for marked, severe, or fatal injuries. Minimal and doubtful increases in reaction have been seen in almost every tissue with every drug and may not be of significance.

It can be concluded that clinical cytotoxic chemotherapy enhances radiation response, particularly in those organs in which the drug itself has demonstrated toxicity when administered alone. A second conclusion is that actinomycin D appears to enhance radiation response in almost every tissue, whereas the other drugs are selective and affect primarily those organs where they cause injury alone. Adriamycin may be an exception to this rule, but does not appear to enhance damage nearly as much as actinomycin D does. A third conclusion is that concurrent drug-radiation administration is in most cases the most dangerous in terms of enhanced normal tissue response.

CLINICAL STUDIES OF TUMOR RESPONSE AND CURE WITH COMBINED RADIATION AND CHEMOTHERAPY

Although a huge number of pilot studies and uncontrolled clinical trials have been carried out combining radiotherapy and chemotherapy, with or without preceding or subsequent surgery, only a limited number of trials have been carried out in a randomized, controlled fashion that have shown positive results. These trials are summarized in Table 11 from data presented by Muggia et al. (1978) and recent publications.

It is noteworthy that combined chemotherapy and radiotherapy have led to enhanced survival and local tumor control in a number of pediatric malignancies. Not listed in Table 11 but also included is acute lymphocytic leukemia in children, in which it is apparent that whole-brain radiation adds to the long-term survival by preventing central nervous system relapse in a more reliable fashion than intrathecal methotrexate. All of the pediatric trials have involved concurrent drug administration and multiple cycles after radiotherapy. All have led to a decrease in distant metastases, and in the case of Ewing's sarcoma and rhabdomyosarcoma an improvement in local control.

In adults, studies have clearly indicated that survival is improved in oat cell carcinoma when chemotherapy is added to radiotherapy, with administration sometimes before, often concurrently, and always in multiple courses after radiotherapy for limited disease. There has been improvement in local control and in particular lessening of the incidence of early failure due to distant metastases. Chemotherapy is becoming so effective that it is even argued by some that radiotherapy may be unnecessary in oat cell carcinoma, but this is doubted by many others who note the high incidence of local failure at the primary site if radiation is not given.

In the gastrointestinal region, studies have shown that 5-FU with and after radiation for colorectal cancer and for pancreas carcinoma significantly improves mean survival time. Both local control and the incidence of distant metastases have been influenced. It is not clear whether there has been a major influence on long-term survival, since most of these trials have been reported as early results based on mean survival times.

TABLE 11. *Improved results with combined radiation and chemotherapy in controlled clinical trials**

Tumor Site	Drug(s)†	Timing‡	No. of Courses	Site of Effect§
Wilms' tumor	Act D, VCR	C, A	6–12	M
Ewing's sarcoma	VAC	C, A	4–12	L, M
Rhabdomyosarcoma	VAC	C, A	4–12	L, M
Oat cell carcinoma	POCC, CMC, CC, CAM	B, C, A	4–8	L, M
Colorectal carcinoma	5-FU	C, A	2–12	L, M
Pancreas carcinoma	5-FU	C, A	4–12	L, M
Head and neck carcinoma	5-FU	C	1	L
Head and neck carcinoma	Bleo	C	1	L
Gliomas	BCNU	C, A	6–12	L
Gliomas	Hu + BCNU	C, A	6–12	L
Breast cancer, 1st relapse	CAF	C, A	12–24	M (L)

* (Levin et al. 1979, Lokich et al. 1979, Muggia et al. 1978, Nervi et al. 1979, Perez et al. 1979, Phillips 1978, Walker and Strike 1976.)

† Act D, actinomycin D; VCR, vincristine; VAC, vincristine, actinomycin D, and cyclophosphamide; POCC, procarbazine, vincristine, CCNU, and cyclophosphamide; CC, cyclophosphamide and CCNU; CAM, cyclophosphamide, Adriamycin, and methotrexate; 5-FU, 5-fluorouracil; Bleo, bleomycin; Hu, hydroxyurea; CAF, cyclophosphamide, Adriamycin, and 5-fluorouracil.

‡ B, before; C, concurrent; A, after.

§ L, local; M, metastasis.

In carcinoma of the head and neck, two studies of concomitant drug and radiation administration have shown improved local control and survival, one with 5-FU given intraarterially and one with bleomycin given systemically.

For malignant gliomas, BCNU was first shown by Walker and Strike (1976) to enhance the mean survival of patients using concomitant and multicourse chemotherapy after irradiation. Levin et al. (1979) have recently shown that the addition of hydroxyurea to BCNU further improves the mean survival in patients with glioblastoma multiforme. Preliminary studies indicate that combined multidrug chemotherapy and irradiation of all known sites of disease in breast cancer patients on the first sign of relapse may improve disease-free survival (Nervi et al. 1979, Perez et al. 1979).

A number of clinical trials are under way evaluating combined radiotherapy and chemotherapy, including studies of malignant gliomas, medulloblastomas, and ependymomas. Studies in the head and neck region with cis-platinum II and bleomycin prior to surgery; bleomycin combined with radiotherapy, followed by bleomycin and methotrexate; and bleomycin alone are under way in the United States and in India. A number of trials are ongoing to study the best combination of radiation and chemotherapy in limited oat cell carcinoma. Further studies of pancreatic carcinoma continue, involving 5-FU with conventional radiotherapy compared to 5-FU with high-dose radiotherapy using betatrons or charged particles. Treatment of liver metastases is under investigation with infusion vs. perfusion chemotherapy combined with radiation.

Carcinoma of the rectum is currently being studied in two large-scale trials nationwide. Postoperative radiotherapy with or without chemotherapy is given to operable patients and chemotherapy with radiotherapy for inoperable cases. Studies are also ongoing for ovary, cervix, bladder, and prostate, evaluating the effect of chemotherapy vs. radiotherapy or chemotherapy following radiotherapy and surgery on the control of micrometastases.

The majority of improved clinical results have occurred primarily because of additive cell kill by chemotherapy to that of radiotherapy. In none of the trials listed above has there been clear-cut evidence for a supraadditive effect on local radiation results. Improved survival has occurred primarily due to chemotherapy kill added to that of radiation, particularly with multiple courses of chemotherapy given over many months. Exceptions to this may be the concurrent administration of 5-FU or bleomycin during radiotherapy in head and neck cancer and the concurrent administration of BCNU and hydroxyurea with radiation in malignant gliomas.

The results to date have been essentially purely empirical, although the clinical experiments with hydroxyurea and to a certain extent with bleomycin have been based on some laboratory data. These empiric experiments in the clinical situation have led to a significant number of normal tissue injuries, as well as to improved survival. The future of combined radiotherapy and chemotherapy is bright, but could be brighter if better laboratory approaches to the prediction of enhanced response could be developed.

ACKNOWLEDGMENTS

The research was supported by the National Cancer Institute Grant CA 20529, and CA 17227 and in part by the UCSF Laboratory of Radiobiology, U.S. Department of Energy Contract No. EY-76-C-03-1012. Ms. Myra Goldring helped make this manuscript readable.

REFERENCES

Aristizabol, S. A., M. R. Manning, R. C. Miller, J. T. Leith, and E. G. Mayer. 1979. Combined radiation-Adriamycin effects on human skin, *in* Frontiers of Radiation Therapy and Oncology, J. M. Vaeth, ed. University Park Press, Baltimore, p. 103.

Bagshaw, M. A., and R. L. S. Doggett. 1969. A clinical study of chemical radiosensitization, *in* Frontiers of Radiation Therapy and Oncology, J. M. Vaeth, ed. University Park Press, Baltimore, p. 164.

Begg, A. C., K. K. Fu, L. J. Kane, and T. L. Phillips. 1980 Single agent chemotherapy of a solid murine tumor: Comparison of growth delay and cell survival assays. Cancer Res. (in press).

Begg, A. C., K. K. Fu, D. C. Shrieve, and T. L. Phillips. 1979a. Combination therapy of a solid murine tumor with cyclophosphamide and radiation: The effects of time, dose and assay method. Int. J. Radiat. Oncol. Biol. Phys. (in press).

Begg, A. C., K. K. Fu, J. Vennari, and T. L. Phillips. 1979b. Combination therapy of a solid murine tumor with 1, 3 bis (2-chloroethyl)-1-nitrosourea and radiation. Int. J. Radiat. Oncol. Biol. Phys. (in press).

Begg, A. C., L. J. Kane, K. K. Fu, and T. L. Phillips. 1979c. The effects of combined bleomycin and radiation on a solid murine tumor. Int. J. Radiat. Oncol. Biol. Phys. (in press).

Billingham, M. E. 1979. Endomyocardial changes in anthracycline treated patients with and without radiation, *in* Frontiers of Radiation Therapy and Oncology, J. M. Vaeth, ed. University Park Press, Baltimore, p. 67.

Bleyer, W. A. 1977. Methotrexate: Clinical pharmacology, current status and therapeutic guidelines. Cancer Treat. Rev. 4:87–101.

Byfield, J. E. 1979. Central nervous system toxicities from combined therapies, *in* Frontiers of Radiation Therapy and Oncology, J. M. Vaeth, ed. University Park Press, Baltimore, p. 228.

Cassady, J. R., S. C. Carabell, and N. Jaffe. 1979. Chemotherapy-irradiation related hepatic dysfunction in patients with Wilms' tumor, *in* Frontiers of Radiation Therapy and Oncology, J. M. Vaeth, ed. University Park Press, Baltimore, p. 147.

Catane, R., J. G. Schwade, A. T. Turrisi, B. L. Webber, and F. M. Muggia. 1979. Pulmonary toxicity after radiation and bleomycin—A review. Int. J. Radiat. Oncol. Biol. Phys. (in press).

Concannon, J. P., M. H. Dalbow, R. F. Hagemann, J. C. Frich, Jr., S. E. Hodgson, C. S. Weil, and R. Martinelli. 1973. Radiation and actinomycin D mortality studies in C57BL/6 mice. Cancer 32:553–561.

Dewey, W. C. 1979. In vitro systems: Standardization of endpoints. Int. J. Radiat. Oncol. Biol. Phys. (in press).

Douple, E. B., W. L. Eaton, Jr., and M. E. Tulloh. 1979. Skin radiosensitization studies using combined cis-dichlorodiammineplatinum II and radiation. Int. J. Radiat. Oncol. Biol. Phys. (in press).

Douple, E. B., and R. C. Richmond. 1979. Radiosensitization of hypoxic tumor cells by cis- and trans-dichlorodiammineplatinum (II). Int. J. Radiat. Oncol. Biol. Phys. (in press).

Eltringham, J. R. 1979. Cardiac responses to combined modality therapy, *in* Frontiers of Radiation Therapy and Oncology, J. M. Vaeth, ed. University Park Press, Baltimore, p. 161.

Eltringham, J. R., L. F. Fajardo, J. R. Stewart, and M. R. Klauber. 1979. Investigation of cardiotoxicity in rabbits from Adriamycin and fractionated cardiac irradiation: Preliminary results, *in* Frontiers of Radiation Therapy and Oncology, J. M. Vaeth, ed. University Park Press, Baltimore, p. 21.

Field, S. B., and A. Michalowski. 1979. Endpoints for damage to normal tissues. Int. J. Radiat. Oncol. Biol. Phys. (in press).

Fu, K. K. 1979. Normal tissue effects of combined radiotherapy and chemotherapy for head and

neck cancer, *in* Frontiers of Radiation Therapy and Oncology, J. M. Vaeth, ed. University Park Press, Baltimore, p. 113.
Fu, K. K., A. C. Begg, L. J. Kane, and T. L. Phillips. 1979. Interaction of radiation and Adriamycin on the EMT6/SF tumor as a function of tumor size and assay method. Int. J. Radiat. Oncol. Biol. Phys. (in press).
Goldstein, L. S., G. Y. Ross, and T. L. Phillips. 1979. The interaction of radiation and BCNU in intestinal crypt cells. Int. J. Radiat. Oncol. Biol. Phys. (in press).
Griem, M. L., G. S. Dimitrievich, and R. M. Lee. 1979. The effects of X-irradiation and Adriamycin on proliferating and non-proliferating hair coat of the mouse. Int. J. Radiat. Oncol. Biol. Phys. (in press).
Guigon, M., E. Frindel, and M. Tubiana. 1978. Effects of the association of chemotherapy and radiotherapy on normal mouse skin. Int. J. Radiat. Oncol. Biol. Phys. 4:233–238.
Kal, H. B., G. W. Barendsen, and H. C. Janse. 1979. Tumour growth delay and cell survival after treatment with radiation and vinblastine. Int. J. Radiat. Oncol. Biol. Phys. (in press).
Kovacs, C. J., L. L. Schenken, and D. R. Burholt. 1979. Therapeutic potentiation of combined cis-dichlorodiammineplatinum (II) and radiation by ICRF-159. Int. J. Radiat. Oncol. Biol. Phys. (in press).
Leith, J. T., B. S. Lewinsky, and W. A. Schilling. 1975. Modification of the response of mouse skin to X-irradiation by bleomycin treatment. Radiat. Res. 61:100–109.
Lelieveld, P., J. M. Brown, D. R. Goffinet, S. L. Schoeppel, and M. Scoles. 1979a. The effect of BCNU on mouse skin and spinal cord in single drug and radiation exposures. Int. J. Radiat. Oncol. Biol. Phys. (in press).
Lelieveld, P., P. R. Twentyman, R. F. Kallman, and J. M. Brown. 1979b. The effect of time between X-irradiation and chemotherapy on the growth of three solid mouse tumors: VI. BCNU. Int. J. Radiat. Oncol. Biol. Phys. (in press).
Levin, V. A., C. B. Wilson, R. Davis, W. Wara, T. L. Pischer, and L. Irwin. 1979. Preliminary results of a phase III comparison study of BCNU, hydroxyurea (HU) and irradiation (BHR) to BCNU and irradiation (B.R.). Int. J. Radiat. Oncol. Biol. Phys. (in press).
Lokich, J. J., C. G. Moertel, and P. T. Lavin. 1979. Comparative therapeutic trial of radiation with or without chemotherapy in pancreatic carcinoma. Int. J. Radiat. Oncol. Biol. Phys. (in press).
Luk, K. H., G. Y. Ross, T. L. Phillips, and L. S. Goldstein. 1979. The interaction of radiation and cis-diamminedichloroplatinum (II) in intestinal crypt cells. Int. J. Radiat. Oncol. Biol. Phys. (in press).
Martin, W. M. C., and N. J. McNally. 1979. The cytotoxic action of Adriamycin and cyclophosphamide on tumor cells in vitro and in vivo. Int. J. Radiat. Oncol. Biol. Phys. (in press).
Masuda, K., H. R. Withers, K. A. Mason, and K. Y. Chen. 1977. Single-dose-response curves of murine gastrointestinal crypt stem cells. Radiat. Res. 69:65–75.
Muggia, F. M., H. Cortes-Funes, and T. H. Wasserman. 1978. Radiotherapy and chemotherapy in combined clinical trials: Problems and promise. Int. J. Radiat. Oncol. Biol. Phys. 4:161–171.
Nervi, C., G. Arcangeli, F. Concolino, and M. Cortese. 1979. Improved survival with combined modality treatment for stage IV breast cancer. Int. J. Radiat. Oncol. Biol. Phys. (in press).
Paterson, A. H. G., and D. English. 1979. The effect of BCNU combined with total body irradiation or cyclophosphamide on survival of dogs after autologous marrow grafts. Int. J. Radiat. Oncol. Biol. Phys. (in press).
Perez, C. A., C. Presant, G. Philpott, and G. Ratkin. 1979. Phase I-II study of concurrent irradiation and multi-drug chemotherapy in advanced carcinoma of the breast. A pilot study by the Southeastern Cancer Study Group. Int. J. Radiat. Oncol. Biol. Phys. (in press).
Phillips, T. L.: 1977. Chemical modification of radiation effects. Cancer 39:987–999.
Phillips, T. L. 1978. Rationale for the selection of combined treatment schedules using fractionated radiation and chemotherapy, *in* Proceedings of the 11th International Cancer Congress, Buenos Aires, 1978.
Phillips, T. L. 1979a. Combined chemo/radiotherapy of cancer: Present state and prospects for use with high LET radiotherapy. Eur. J. Cancer (in press).
Phillips, T. L. 1979b. Effects on lung of combined chemotherapy and radiotherapy, *in* Frontiers of Radiation Therapy and Oncology, J. M. Vaeth, ed. University Park Press, Baltimore, p. 133.
Phillips, T. L., and K. K. Fu. 1976. Quantification of combined radiation therapy and chemotherapy effects on critical normal tissues. Cancer 37:1186–1200.

Phillips, T. L., and K. K. Fu. 1977. Acute and late effects of multimodal therapy on normal tissues. Cancer 40:489–494.

Phillips, T. L., G. Y. Ross, and L. S. Goldstein. 1979a. The interaction of radiation and cyclophosphamide in intestinal crypt cells. Int. J. Radiat. Oncol. Biol. Phys. (in press).

Phillips, T. L., G. Y. Ross, L. S. Goldstein, and A. C. Begg. 1979b. The interaction of radiation and bleomycin in intestinal crypt cells. Int. J. Radiat. Oncol. Biol. Phys. (in press).

Phillips, T. L., M. D. Wharam, and L. W. Margolis. 1975. Modification of radiation injury to normal tissues by chemotherapeutic agents. Cancer 35:1678–1684.

Ransom, J. L., R. W. Novak, A. P. M. Kumar, H. O. Hustu, and C. B. Pratt. 1979. Delayed gastrointestinal complications after combined modality therapy of childhood rhabdomyosarcoma. Int. J. Radiat. Oncol. Biol. Phys. (in press).

Redpath, J. L., R. M. David, and M. Colman. 1978. The effect of Adriamycin on radiation damage to mouse lung and skin. Int. J. Radiat. Oncol. Biol. Phys. 4:229–232.

Ross, G. Y., T. L. Phillips, and L. S. Goldstein. 1979. The interaction of radiation and Adriamycin in intestinal crypt cells. Int. J. Radiat. Oncol. Biol. Phys. (in press).

Ruckdeschel, J. C., D. H. Baxter, M. F. McKneally, D. A. Killam, S. L. Lunia, and J. Horton. 1979. Sequential radiotherapy and Adriamycin in the management of bronchogenic carcinoma: The question of additive toxicity. Int. J. Radiat. Oncol. Biol. Phys. (in press).

Schenken, L. L., D. R. Burholt, R. F. Hagemann, and S. Lesher. 1976. The modification of gastrointestinal tolerance and responses to abdominal irradiation by chemotherapeutic agents. Radiology 120:417–420.

Schenken, L. L., D. R. Burholt, R. F. Hagemann, and C. J. Kovacs. 1979a. Combined modality oncotherapies: Cell kinetic approaches for avoidance of gastrointestinal toxicity, *in* Frontiers of Radiation Therapy and Oncology, J. M. Vaeth, ed. University Park Press, Balitmore, p. 82.

Schenken, L. L., D. R. Burholt, and C. J. Kovacs. 1979b. Adriamycin-radiation combinations: Drug induced delayed gastrointestinal radiosensitivity. Int. J. Radiat. Oncol. Biol. Phys. (in press).

Steel, G. G.: 1979. Terminology in the description of drug-radiation interactions. Int. J. Radiat. Oncol. Biol. Phys. (in press).

Steel, G. G., and M. J. Peckham. 1979. Exploitable mechanisms in combined radiotherapy-chemotherapy: The concept of additivity. Int. J. Radiat. Oncol. Biol. Phys. 5:85–91.

Steel, G. G., R. P. Hill, and M. J. Peckham. 1978. Combined radiotherapy-chemotherapy of Lewis lung carcinoma. Int. J. Radiat. Oncol. Biol. Phys. 4:49–52.

Twentyman, P. R., R. F. Kallman, and J. M. Brown. 1979a. The effect of time between X-irradiation and chemotherapy on the growth of three solid mouse tumors. I. Adriamycin. Int. J. Radiat. Oncol. Biol. Phys. (in press).

Twentyman, P. R., R. F. Kallman, and J. M. Brown. 1979b. The effect of time between X-irradiation and chemotherapy on the growth of three solid mouse tumors. II. Cyclophosphamide. Int. J. Radiat. Oncol. Biol. Phys. (in press).

Twentyman, P. R., R. F. Kallman, and J. M. Brown. 1979c. The effect of time between X-irradiation and chemotherapy on the growth of three solid mouse tumors. III. Cis-diamminedichloroplatinum. Int. J. Radiat. Oncol. Biol. Phys. (in press).

Twentyman, P. R., R. F. Kallman, and J. M. Brown. 1979d. The effect of time between X-irradiation and chemotherapy on the growth of three solid mouse tumors. IV. Actinomycin-D. Int. J. Radiat. Oncol. Biol. Phys. (in press).

Twentyman, P. R., R. F. Kallman, and J. M. Brown. 1979e. The effect of time between X-irradiation and chemotherapy on the growth of three solid mouse tumors. V. Bleomycin. Int. J. Radiat. Oncol. Biol. Phys. (in press).

Valeriote, F., and H. Lin. 1975. Synergistic interaction of anticancer agents: A cellular perspective. Cancer Chemother. Rep. 59:895–900.

Walker, M. D., and T. A. Strike. 1976. An evaluation of methylCCNU, BCNU, and radiotherapy in the treatment of malignant glioma. (abstract) Proc. Am. Assoc. Cancer Res. 17:163.

Wheeler, K. T., K. Kaufman, and M. L. Feldstein. 1979. Response of a rat brain tumor to fractionated therapy with low doses of BCNU and radiation. Int. J. Radiat. Oncol. Biol. Phys. (in press).

Withers, H. R., and M. M. Elkind. 1970. Micro-colony survival assay for cells of mouse intestinal mucosa exposed to radiation. Int. J. Radiat. Oncol. Biol. Phys. 17:261–267.

Radiation Biology in Cancer Research, edited by
Raymond E. Meyn and H. Rodney Withers.
Raven Press, New York © 1980.

Cell Biology of Hyperthermia and Radiation

William C. Dewey, Michael L. Freeman, G. Peter Raaphorst, Edward P. Clark, Rosemary S. L. Wong, Donald P. Highfield, Ira J. Spiro, Stephen P. Tomasovic, David L. Denman, and Ronald A. Coss

Department of Radiology and Radiation Biology, Colorado State University, Fort Collins, Colorado 80523

Hyperthermia at temperatures above 41°C has been used sporadically as an agent for cancer therapy since the early 1900s. Although there were encouraging results, there was not enough consistency to encourage continued and sustained efforts in applying hyperthermia either by itself or combined with radiation or other agents. However, during the last five years, results from studies of cell cultures and animals, as well as from a few preliminary clinical attempts, have provided some enthusiasm for systematically investigating the potential of using hyperthermia in cancer therapy. The results of Stehlin and others, who use both hyperthermia and chemotherapy in regional perfusions for treating melanoma, are particularly encouraging. See the proceedings (Robinson and Wizenberg 1976, Streffer et al. 1978) of the last two international hyperthermia conferences for pertinent references.

Results in vitro have provided a rationale for considering the use of hyperthermia in cancer therapy. First, hyperthermia at temperatures above 41°C kills mammalian cells and sensitizes them to X irradiation, with the degree of killing and radiosensitization varying greatly with only a 0.5°C change in temperature (Sapareto et al. 1978a, b). Therefore, with inadequate temperature control and measurement in clinical studies in the past, consistent results could not have been expected (see Hahn and Pounds 1976, Robinson et al. 1978, Witcofski and Kremkau 1978, for examples in animals). Second, hyperthermia selectively kills and radiosensitizes cells that are relatively resistant to ionizing radiation; these populations are those in the process of synthesizing DNA and those existing in the hypoxic compartment of tumors (Dewey et al. 1977a, Westra and Dewey 1971). These hypoxic cells are probably also at a low pH, which sensitizes cells to hyperthermia (Freeman et al. 1977, Gerweck and Rottinger 1976, Overgaard 1978, von Ardenne and Reitnauer 1968). Therefore, the action of hyperthermia by itself or interacting with X irradiation should selectively kill those tumor cells surviving a dose of radiation. This same rationale has been offered for considering the use of high-LET (linear energy transfer) radiation and electronic affinic compounds in radiation therapy (Robinson et al. 1974). Third, hyperthermia has been shown to eliminate or reduce recovery from sublethal (Ben-Hur et al. 1974) and potentially lethal (Li et al. 1976) radiation damage and thus may have a selective effect on tumor cells existing in the G_0 compartment. Finally, the toxicity of electron affinic compounds for hypoxic cells (Stratford and Adams 1977) and the toxicity of several chemotherapeutic agents (Hahn et al. 1977) can be enhanced greatly by hyperthermia. Therefore, the basic studies of cell cultures indicate that there are reasons to believe that hyperthermia may enhance the therapeutic efficacy of radiation and chemotherapeutic agents used in the clinic.

Studies in animals and a few clinical investigations during the last two years have indicated that there may be some merit to the rationale proposed above. For example, hyperthermia has been shown to have a selective effect on the radiation response of chronically hypoxic tumor cells (Dewey et al. 1977b, Thrall et al. 1975) and, under certain conditions, can improve the therapeutic ratio (Stewart and Denekamp 1978), i.e., the radiation response of the tumor relative to that of normal tissue. However, as predicted from studies of cell cultures (Sapareto et al. 1978a), variations in the sequence between administration of hyperthermia and radiation may greatly alter the effectiveness of hyperthermia for improving the therapeutic ratio. Beneficial clinical results are still largely anecdotal but do suggest that careful clinical studies need to be carried out in which variations in sequencing are considered as the temperature profiles in tumors and normal tissues are monitored. Also, physiological changes, such as changes in oxygen tension and pH, occur from hyperthermic treatment (Reinhold et al. 1978, Song 1978) and should be considered in relation to the effectiveness of hyperthermic treatments. Therefore, preliminary data from animal and human studies indicate that hyperthermia may indeed enhance the effectiveness of radiation and chemotherapy, but these data also indicate that various parameters must be monitored and carefully evaluated in order to determine the reasons for both successes and failures.

In this presentation, we shall review the cellular and molecular effects of hyperthermia, either when delivered alone or combined with X irradiation. In certain cases, concepts will be considered in relation to effects observed in vivo that have implications for cancer therapy.

EFFECTS OF HEAT ALONE

Kinetics of Heat Inactivation and Effects of pH

Results are shown for Chinese hamster ovary (CHO) cells, although similar results have been reported for other cell lines (Robinson and Wizenberg 1976, Streffer et al. 1978). The logarithm of survival plotted versus duration of heating resembles X-ray survival curves, with the exception that thermal tolerance develops after three to four hours of heating (Figure 1). Note that the amount of

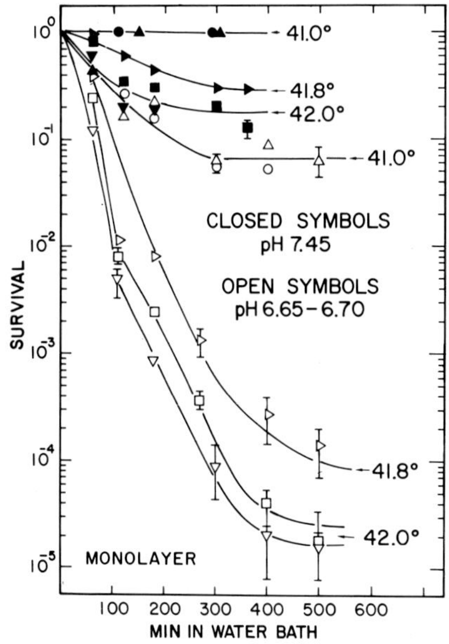

FIG. 1. Survival of CHO cells continuously heated for up to 500 minutes. Open symbols–acidic medium; closed symbols–alkaline medium. The cells were maintained with CO_2 at low pH for 30 minutes prior to heat treatment until 120 minutes after treatment. (Data taken from Freeman 1978.)

cell killing increases greatly as the pH is decreased from 7.4 to 6.7, and that when cells are heated at a given temperature, 42°C for example, the fraction of cells demonstrating tolerance is greatly reduced. This increase in cell killing at low pH (Bichel and Overgaard 1977) is probably the main factor responsible for reports that hypoxic cells are more heat sensitive than oxygenated cells (Bass et al. 1978, Robinson et al. 1974, and see Dewey et al. 1977a for other references). However, some thermal sensitization may be attributed to chronic deprivation of O_2 and/or nutrients (Gerweck et al. 1979). As shown in Figure 2, the amount of thermal sensitization induced by treatment at low pH may be greatly increased when the cells are maintained at a low pH after heating. These observations associated with low pH may have important implications in the clinic because many tumors appear to be at a lower pH than surrounding normal tissue (Ashby and Cantab 1966, Gullino et al. 1965, Kahler and Robertson 1943). In fact, in vivo studies (Crile 1963) have demonstrated that leaving the tumors in situ for a period of time prior to removing the cells for assay of their viability greatly enhances hyperthermic killing. This observation, which may relate to the increase in lethality as the low pH is maintained for a period after heating, may explain, at least in part, the observation (Marmor et al. 1977) that hyperthermic cell killing as determined by the in vivo to in vitro assay was insufficient to account for the cure of the tumors.

The effect of pH on the kinetics of heat inactivation is illustrated by the Arrhenius plots in Figure 3. Treatment at low pH shifts the curve upward, i.e., increases inactivation rates, but has little effect on the slope that represents the activation energy (about 140 kcal/mole). As indicated before, an increase

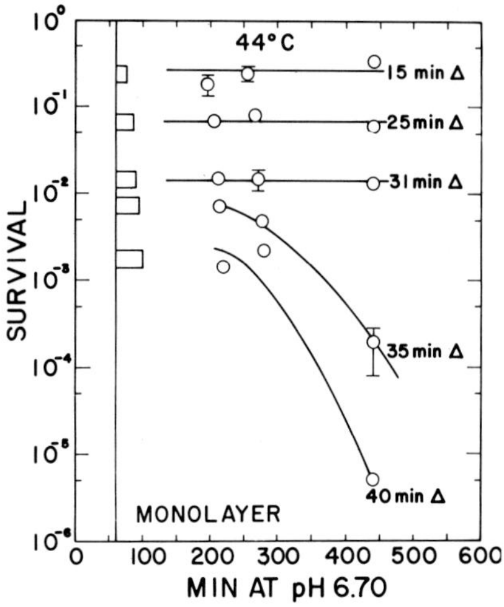

FIG. 2. Survival of CHO cells heated in monolayer at 44°C. At time zero, the cells were flushed with CO_2 to pH 6.70. The solid vertical line represents the initiation of heating. The boxes represent the duration of the heating interval (Δ) indicated for each curve. (Data taken from Freeman 1978.)

in the duration of incubation at low pH after thermal treatment causes a further increase in the inactivation rate for temperatures greater than 42.5°C. Thus, the break in the Arrhenius curve commonly reported to occur at about 43°C (Connor et al. 1977, Sapareto et al. 1978b, Schrek 1966) does not have a unique value at the low pH but instead depends on the length of time the cells are maintained at the low pH after heating. These curves are consistent with reports for several in vitro and in vivo systems (Dewey et al. 1977a, Overgaard 1978, Suit and Shwayder 1974): for a twofold increase in inactivation rate, the temperature must be increased by 1°C for temperatures above the break in the Arrhenius curve and by about 0.5°C for temperatures below the break.

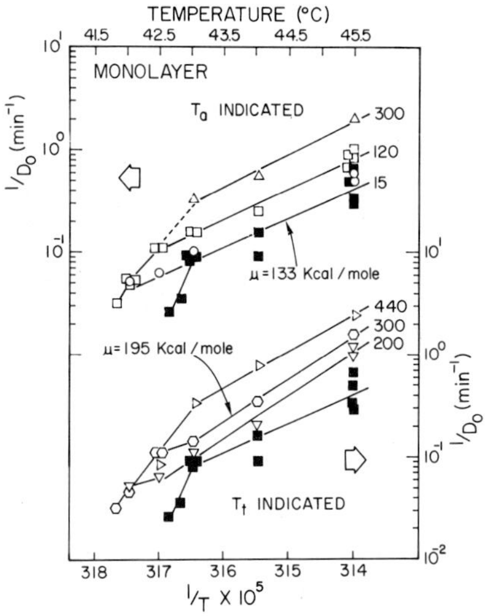

FIG. 3. An Arrhenius plot of heat inactivation for cells heated in acid (CO_2-adjusted, open symbols) or alkaline (closed symbols) medium. This figure was compiled from several experiments. In the upper curves, T_a, the period the cells were held at low pH after heating, was held constant. In the lower curves, T_t, the total period that the cells were held at low pH, was held constant. The numbers by each curve represent the appropriate T_a or T_t. The dashed line for T_a of 300 minutes was constructed because at 42.5°C and below, the points for T_t = 300 to 400 minutes apply. O, T_a = 15 minutes; □, ■, T_a = 120 minutes; △, T_a = 300 minutes; ▽, T_t = 200 minutes; ⬡, T_t = 300 minutes; ▷, T_t = 440 minutes. (Data taken from Freeman 1978.)

Thermal Sensitivity of Different Cell Lines

Various cell lines differ greatly in their thermal sensitivity (Dewey et al. 1977a, Raaphorst et al. 1979a, and Figure 4). In this experiment, all cell lines were cultured as monolayers in the same medium and were heated during exponential growth. Utilization of the same medium is important because studies have shown that thermal sensitivity can vary with nutrient levels and different types and amounts of sera (Hahn 1974). In general, the thermal sensitivity increases as the body temperature of the species from which the cell line was derived decreases (see legend of Figure 4). Reports indicate that transformed cell lines are more thermally sensitive than untransformed cell lines (see Dewey et al. 1977a, Suit and Shwayder 1974 for references), but the opposite also has been reported (Kachani and Sabin 1969, Ossovski and Sachs 1967). Therefore, malignant cells may not be intrinsically more heat sensitive than normal cells; instead, any increased sensitivity of malignant cells in vivo may be associated with low pH.

Our studies (unpublished data and Raaphorst et al. 1979b) have shown that in general cells respond independently to heat treatment without any interaction between the cells; a density effect was observed only after very long heating times as thermal tolerance was developing or when heating at certain densities under low pH conditions.

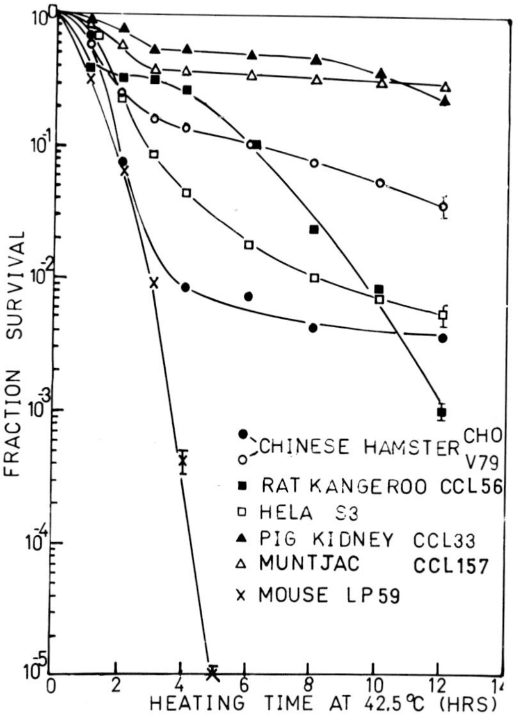

FIG. 4. Survival curves of seven cell lines exposed to 42.5°C. The body temperatures of the donor mammals were 39.4°C for pig kidney, 38.5°C for muntjac, 37°C for HeLa, 36.1 to 38.3°C for Chinese hamster, 36°C for rat kangaroo, and 35.7 to 37.7°C for mouse. (Data taken from Raaphorst et al. 1979a.)

Thermal Tolerance

Thermal tolerance is demonstrated in Figures 1 and 4 as cells were heated continuously for greater than three hours. In addition, tolerance is observed in terms of cell lethality when a second heat dose is delivered 4 to 24 hours after incubating the cells at 37°C, following an initial thermal treatment (Henle and Dethlefsen 1978, Sapareto et al. 1978b). In this type of fractionated dose experiment (Figure 5), two 10-minute treatments at 45.5°C were separated by incubation at 37°C, and tolerance was demonstrated by the observation that two treatments separated by 12 or 24 hours had less effect than only one treatment in terms of an increase in association of nonhistone protein with DNA. A similar type of tolerance is shown in Figure 6, in which a 15-minute treatment at 45.5°C was followed immediately by a 10-hour treatment at 42°C. Note, by the similarity between the A and B points and lines, that tolerance was observed in terms of cell lethality, delay in progression from G_1 into S, and inhibition of RNA and protein synthesis. Also note that tolerance was not observed (D points and lines) when the long treatment of 10 hours at 42.0°C immediately followed the 15-minute treatment at 45.5°C. Thus, tolerance is observed in terms of several parameters, including release of RNA precursors from the cell (Reeves 1972), when cells either are maintained for at least three to four hours at 42.5°C or are incubated at 37°C following an initial acute thermal treatment at 44 to 46°C (Henle and Dethlefsen 1978, Harisiadis et al. 1977, and Figures 1, 4, 5, and 6). Apparently, thermal tolerance can develop at 37 to 42.5°C after or during a thermal insult. An Arrhenius analysis of thermally tolerant cells (Bauer and Henle, in preparation) indicated that the curves were shifted down but essentially were parallel to the curve for the original population, i.e., the activation energy was not changed. However, the break in the Arrhenius curve for thermally tolerant cells occurred at 45°C, which was about 0.5°C higher than where the thermally tolerant cells manifested a biphasic response after three to four hours of continuous heating. There is also considerable evidence that tolerance occurs for several in vivo systems (Henle and Dethlefsen 1978). Furthermore, for both in vivo and in vitro systems, tolerance appears to be lost when the cells or tissue systems are maintained at 37°C for about 72 hours following the initial induction of tolerance.

The mechanism of thermal tolerance is not understood, although some degree of metabolism, possibly protein synthesis, must be involved because maintaining the cells at 0°C or inhibiting protein synthesis with cycloheximide inhibits or greatly diminishes the appearance of thermal tolerance (Henle and Dethlefsen 1978). The fact that incubation at 37°C following treatment with alcohols, which supposedly act on the plasma membrane, induces tolerance to heat and that heat induces tolerance to alcohol (Li et al. 1978) suggests that changes in the plasma membrane result in thermal tolerance. Thus, the observation of thermal tolerance for the several end points discussed above, as well as the loss of thermal

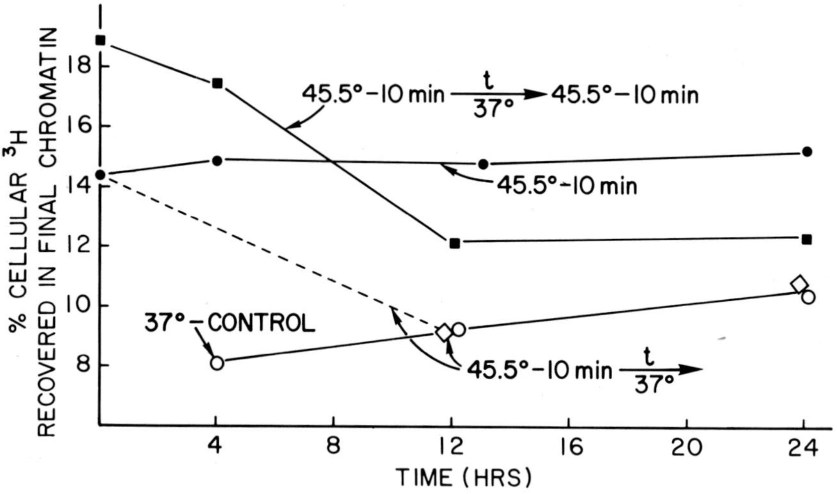

FIG. 5. At zero time, cells had been labeled for 24 hours with both ^{3}H-tryptophan and ^{14}C-TdR; label remained in the medium until the cells were processed for chromatin. Closed squares (■) represent the percent of cellular ^{3}H recovered in final chromatin (ordinate) after various times of incubation at 37°C between two single heat doses of 45.5°C for 10 minutes (abscissa). Open circles (O) represent controls (incubated at 37°C) that were processed at various times (abscissa). Closed circles (●) represent samples that received a single heat treatment (45.5°C for 10 minutes) at various times (abscissa) and processed immediately. Open diamonds (◇) represent samples heated (45.5°C for 10 minutes) at zero hour and processed after varying times at 37°C (abscissa). See Tomasovic et al. 1978 for methods of isolating chromatin.

tolerance in cells treated with polyamines (Figure 7), may relate to modifications in a primary target for heat lethality. This primary target could be the plasma membrane, although in our opinion, important thermal effects on the nucleus or other organelles have not been excluded.

Cell Cycle Delays

Hyperthermic treatments produce much longer delays in the cell cycle than an X-ray dose producing a comparable amount of cell killing (Kal et al. 1975, Sapareto et al. 1978b, Westra and Dewey 1971). These long delays can lead to a redistribution in the cycle with more cells in G_1 and $G_2 + M$ than in S phase. An interesting observation (Coss et al. 1979) was that after CHO cells recovered from a four- to five-hour delay in G_2 during a continuous treatment at 41.5°C, there was an uncoupling of mitotic events such that re-formation of the nuclear membrane occurred around metaphase chromosomes. Although metaphase was lengthened from 25 to 100 minutes, the re-formation of the nuclear membrane was not delayed until the cells entered anaphase or telophase.

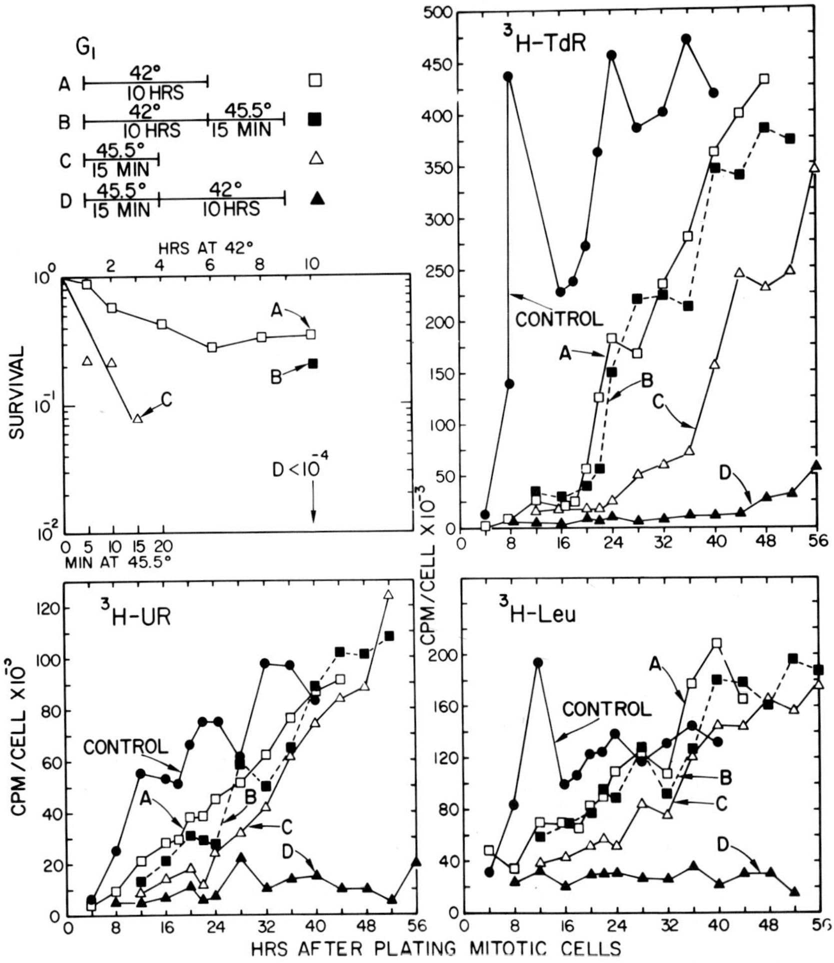

FIG. 6. The development of thermal tolerance as assayed by cell lethality and inhibition of DNA, RNA, and protein synthesis. Synchronous CHO cells were obtained by mitotic selection and were plated into medium containing 4 μCi/ml of either ^{3}H-thymidine, ^{3}H-uridine, or ^{3}H-leucine. Heating in waterbaths was begun two hours after plating, when the cells were in G_1; and after the desired heating period of incubation at 37°C, the labeled medium was discarded, the cells were rinsed and trypsinized from the surface, and the radioactivity per cell was determined by Coulter counting and liquid scintillation counting of cold trichloroacetic-insoluble material. Flasks used for survival determinations were heated concurrently with those labeled and were fixed for colony assay after incubation at 37°C for eight days. Groups A and B were initially heated for a continuous 10-hour period, at which time group A was placed at 37°C and group B was immediately heated for an additional 15 minutes at 45.5°C and then transferred to 37°C. Groups C and D were initially heated for 15 minutes at 45.5°C, at which time group C was placed at 37°C, and group D was immediately heated at 42°C for an additional 10 hours and then incubated at 37°C. Each survival point represents eight replicate flasks per group; note that survival for group D was less than 1×10^{-4}. In the top left panel, the abscissa at the bottom represents the duration of heating at 45.5°C for group C only. The abscissa at the top represents duration of heating at 42°C for groups A, B, and D. Correlations between flow cytometer studies and labeling with ^{3}HTdR (Sapareto et al. 1978b, and unpublished data) and study of acid-soluble pools after labeling with ^{3}HTdR, ^{3}HUR, and ^{3}H amino acids (Henle and Leeper, in preparation, and unpublished data) indicate that hyperthermia had no great effect on transport or pool sizes.

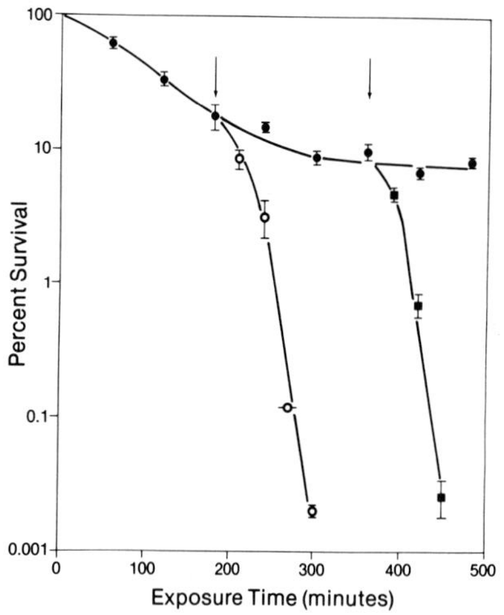

FIG. 7. Survival response of log-phase Chinese hamster ovary cells to 42°C exposure. Cultures were maintained and treated in McCoy's 5a medium supplemented with 20% fetal calf serum. Cells were either treated continuously at 42°C for up to 480 minutes (–●–) or for 180 (–○–) or 360 (–■–) minutes. In the last two cases, the medium was adjusted to 1×10^{-3}M spermine, after the initial heating intervals at 42°C. Cultures were then reincubated, now in the presence of spermine, at 42°C for up to 120 additional minutes. (Data provided by E. W. Gerner, D. G. Stickney and D. K. Holmes, in preparation.)

Target for Heat Killing

Evidence for the plasma membrane

As discussed under the section *Thermal Tolerance,* there is considerable evidence that damage to the plasma membrane is involved in heat killing. Structural changes in the plasma membrane as observed by light and electron microscopy have been reported (Bass et al. 1978, Lin et al. 1973) and are particularly evident in cells entering mitosis during a continuous seven-hour treatment at only 41.5°C (Coss et al. 1979). Membrane blebbing and a lack of cortical microfilaments next to the plasma membrane were evident. The positive correlation (Figure 8) between an increase in cholesterol content of several cell lines and a decrease in their thermal sensitivity (Cress et al. 1978) suggests that the plasma membrane may be involved in heat killing, because an increase in cholesterol has been shown to decrease the fluidity of the plasma membrane (Sinensky 1978). Other evidence (illustrated in Figure 9) is that treatment of either bacterial or mammalian cells with a membrane-active local anesthetic, procaine hydrocholoride, greatly enhances thermal sensitivity. R. Coss (unpublished data) has observed similar heat sensitization of CHO cells treated with 7 to 10 mM procaine HCl prior to heating at 42 to 43°C. In fact, treatment with 7 mM causes lethality at 43°C to resemble that at 45.5°C for control cells. Anesthetic potentiation of hyperthermic damage also has been observed in vivo in terms of prolonged survival of tumor-bearing mice when the local anesthetic lidocaine is injected into the tumor prior to heating (Table 1). Therefore, membrane-active agents may play an important role for utilizing hyperthermia in the clinic.

Other evidence summarized in Table 2 includes: (1) thermal sensitivity in-

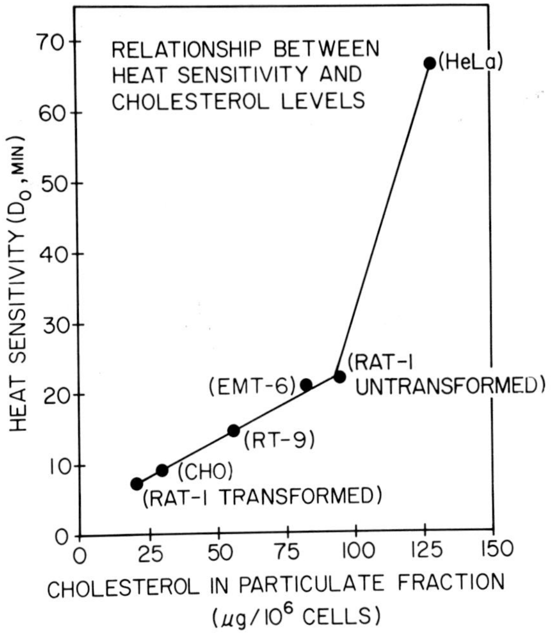

FIG. 8. Relation between heat sensitivity and cholesterol content in the particulate fraction of unheated cells. Results are for cultures growing under standard conditions for each cell line. These include growth at 37°C in medium 199 plus 10% fetal calf serum (HeLa cells), Waymouth's medium plus 15% fetal calf serum (EMT-6), minimum essential medium (MEM) plus 15% fetal calf serum (RT-9), and Hsu's modified McCoy's 5a plus 20% fetal calf serum (CHO). The Rat-1 transformed and untransformed cells were grown at 35°C and 39°C, respectively, in Dulbecco's modified Eagle's (DME) medium plus 10% fetal calf serum. Cells were treated at 43°C by immersion in a temperature-controlled (± 0.1°C) waterbath for times up to five hours. The reciprocal slope, or D_0, of the exponential portion of the resulting log-linear survival curves was then used as an index of heat sensitivity. Cholesterol content in unheated cells was determined, after sonication, in the particulate fraction after a 50,000 × g × 10 minute centrifugation of the cell homogenate. The pellet was then extracted with chloroform:methanol (2:1), and cholesterol was measured in the presence of $FeCl_3$ at 558 nm. (Data provided by A. E. Cress, P. Culver, and E. W. Gerner, in preparation.)

creases as the ratio of unsaturated to saturated fatty acids in the membrane increases (Yatvin 1977); (2) RNA precursors (Reeves 1972) and polyamines (Gerner and Russell 1977) leak from cells following hyperthermic treatment; (3) hyperthermia increases membrane permeability of Na^+-dependent amino acid transport (Kwock et al. 1978); (4) alcohols and polyene antibiotics thought to act on membranes sensitize cells to hyperthermia (Hahn et al. 1977, Li et al. 1978), and, as discussed before, alcohol can induce thermal tolerance; and (5) the pH thermal sensitization reported previously also may be related to membrane changes that occur at low pH (Goekoop et al. 1978) because Raman nuclear magnetic resonance studies indicate that structural changes in the membrane occur at a lower temperature when the extracellular pH is reduced (Wallach 1978).

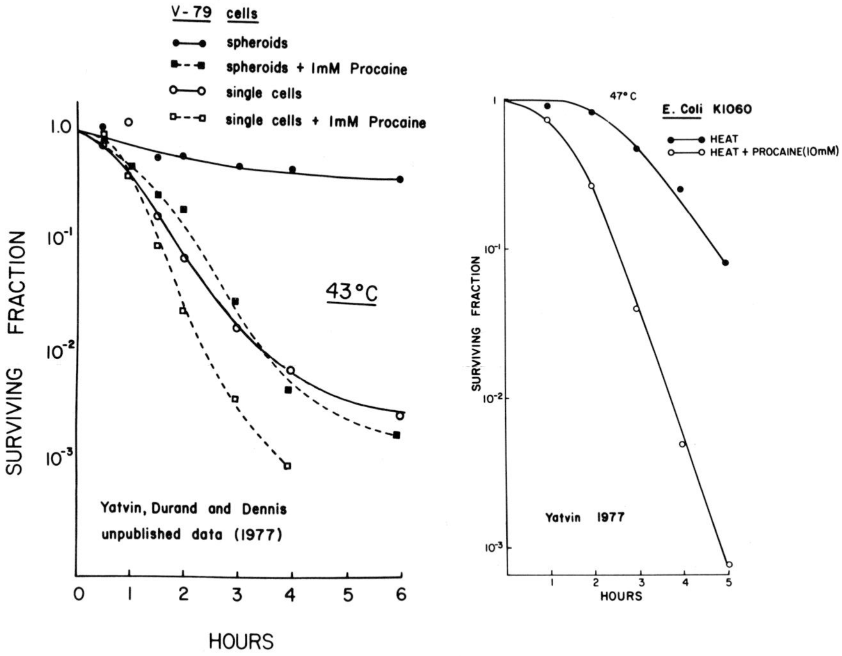

FIG. 9. Heat sensitization by adding procaine to the medium prior to heating *Escherichia coli* at 47°C or V-79 Chinese hamster single cells or one-day-old spheroids at 43°C. *E. coli* data were provided by Yatvin (1977), and the hamster data were provided by M. B. Yatvin, R. E. Durand, and D. W. Dennis (unpublished data).

TABLE 1. *Posttreatment mean survival time* of tumor-bearing mice*

	Injection		
Heating–one hour	None	Saline	Lidocaine
None	12.8 ± 0.90 (25)	12.7 ± 0.70 (17)	14.6 ± 1.21 (17)
42.0°C	12.6 ± 1.36 (10)	16.5 ± 1.61 (10)	13.6 ± 0.89 (10)
43.5°C	18.5 ± 0.88 (33)	19.4 ± 1.08 (30)	>37.3 ± 6.21 (31)

Some of the mammary adenocarcinomas in the legs of BDF_1 mice were injected with lidocaine or saline, and the tumor-bearing legs either were not heated or were heated in a waterbath at 42.0°C or 43.5° C for one hour. Statistical analysis indicates that the increase in survival when lidocaine is present during heating at 43.5°C for one hour is significant ($P < .05$) from all of the other eight treatments (four of 31 animals were still alive at 60 days). Heat alone at 43.5°C also significantly prolonged survival. (Data provided by Yatvin et al. 1979.)

* Days ± SE (n).

TABLE 2. *Evidence for plasma membrane as target for heat damage*

Structural Changes—Light and Electron Microscope:
Blebs
Cortical Microfilaments Diminished
Increase Membrane Fluidity and Increase Heat Sensitivity:
Increase Unsaturated Fatty Acids
Decrease Cholesterol
Heat Increases Membrane Permeability:
Amino Acids, RNA Precursors,
Polyamines, Antibiotics, and Chemotherapeutic Agents
Low pH Increases Heat Sensitivity and Decreases Raman Nuclear Magnetic Resonance and Electron Microscopic Membrane Structure
Sensitize to Heat with Membrane-Active Agents
Alcohols and Local Anesthetics
Alcohols Induce Thermal Tolerance and Heat Induces Alcohol Tolerance

See text for further explanation and references.

If the membrane is indeed the heat-sensitive target, the question still remains as to whether the lipids, proteins, or both are primarily involved. The activation energy of about 140 kcal/mole for both cell inactivation and denaturation of many proteins (Dewey et al. 1977a) would suggest that thermal effects on membrane proteins would be most important. In fact, both Bowler et al. (1973) and Kwock et al. (1978) believe that membrane structural integrity depends on the stability of lipid-protein complexes and that possibly an unfolding of sulfhydryl proteins rather than melting of lipids is responsible for thermal damage to membranes.

Evidence for targets other than the plasma membrane

Other possibilities for critical targets in heat inactivation, however, should be mentioned. First of all, hyperthermia induces alterations in the nucleolus by changing the granules from about 200 Å to 500 Å (Simard and Bernhard 1967, and unpublished data). This is associated with inhibition of synthesis and processing of ribosomal RNA (Warocquier and Scherrer 1969), and this alteration probably results in the high frequency of breaks at the secondary constriction of the X chromosome (nucleolar organizer) (Dewey et al. 1971). Also, for cells heated during the heat-sensitive S phase, there is a log-linear relationship between survival and chromosomal aberrations (Figure 10). In fact, the relationship between survival and chromosomal aberrations for heat treatment during S phase is almost identical to that observed for treatment with ionizing radiation, an agent that supposedly kills cells by damage to DNA in the chromosomes. This correlation, as well as the additive interaction between

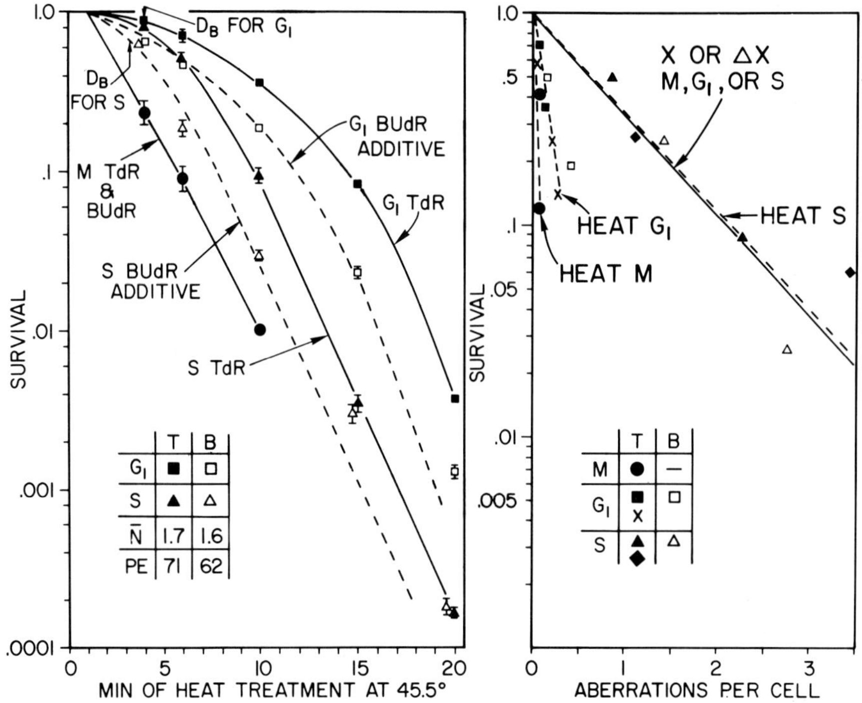

FIG. 10. Left panel: Survival versus length of heat treatment during mitosis (M), G_1, or S for TdR-control (T) and BUdR-treated (B) cells. The plating efficiency for BUdR-treated cells was 87% of that for TdR-control cells, which gives values for D_B as indicated (i.e., 3·2 and 3·9 minutes for S and G_1, respectively). As explained in Dewey et al. 1971, the values of D_B, which represent BUdR toxicity in terms of minutes of heat treatment, were used to construct theoretical curves (G_1-BUdR additive and S-BUdR additive) for the assumption that the effects of BUdR and heat were additive. Note that the observed points (□, △) for the BUdR-treated cells fell close to the theoretical curves. For mitotic cells, the data for TdR and BUdR treatments were combined, since there were no significant differences. (Data taken from Dewey et al. 1971.) Right panel: Survival versus aberrations per cell for TdR-control (T) cells and BUdR-treated (B) cells heated in mitosis (M), G_1, or S. The relationship plotted for X irradiation (X) or heat plus irradiation (Δ X) (data points not shown) was reported previously (Dewey et al. 1978) and derived from data illustrated in Figure 16. The relationships plotted for heat treatments (■, □, ▲, △) were obtained by relating survival data with aberration data.

heat and bromodeoxyuridine (BUdR), suggests that damage to chromatin is involved in heat killing of S phase cells. Heat killing of G_1 cells and mitotic cells, however, probably involves another process. In fact, heat effects on mitotic cells appear to involve the mitotic apparatus (Inoué 1952, Rieder and Bajer 1977, Westra and Dewey 1971), possibly including the centrosome (Barrau et al. 1978). An interesting observation was that heat caused the disappearance in the centrosomes of the virus-like particles which may be associated with transformation.

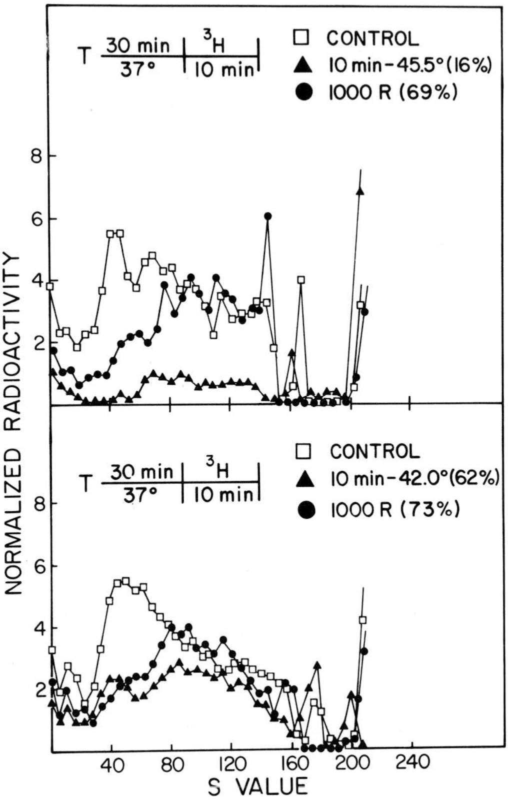

FIG. 11. Selective inhibition of DNA replicon initiation as indicated by alkaline sucrose gradients of DNA pulse-labeled following heat or X-radiation treatment (T). Top panel: Monolayer cultures of CHO cells were either heated for 10 minutes in a 45.5°C waterbath (▲), given 1,000 rads at room temperature (●), or received no treatment (□); then the cells were incubated at 37°C for 30 minutes and pulse-labeled with 10 μCi/ml ^{3}H-thymidine (20 Ci/mM) for 10 minutes prior to trypsinization. The procedure of Painter and Young (1976) was utilized, with a few modifications, to process the cells and to carry out the alkaline sucrose gradient sedimentation analyses. Data were normalized by multiplying the percent of the radioactivity in each fraction by the percent acid insoluble radioactivity in treated cells relative to controls (16% for heated and 69% for irradiated). To compare these data with those in Figure 12, 40S and 120S correspond to about 1.6×10^7 and 2.5×10^8 daltons, respectively. Bottom panel: CHO cells were either heated for 10 minutes in a 42.0°C waterbath (▲), given 1,000 rads (●), or received no treatment (□). All further processing of the cells for analyses and normalization of the data were identical to that described for the top panel. (Data taken from Wong and Dewey, in preparation).

The chromosomal aberrations induced during S phase do not appear to result from heat-induced DNA strand breaks since none were observed in alkaline sucrose gradients following heat treatment (unpublished data, Clark et al. 1976, Corry et al. 1977). These results probably rule out any significant effect from depurination. However, aberrations may result from heat inhibition of DNA synthesis (40%, 84%, and 90% inhibition of incorporation of ^{3}HTdR (tritiated thymidine) following 10 minutes at 42°C, 10 minutes at 45.5°C, and 15 minutes at 45.5°C, respectively), for which the effect at the replicon level is primarily on initiation, with some effect on elongation as well (Figure 11). This is illustrated by the shift of the alkaline sucrose gradient profiles toward higher molecular weights after heat treatment. Also, there is a 15-minute delay in forming double-stranded molecules from single-stranded molecules synthesized immediately

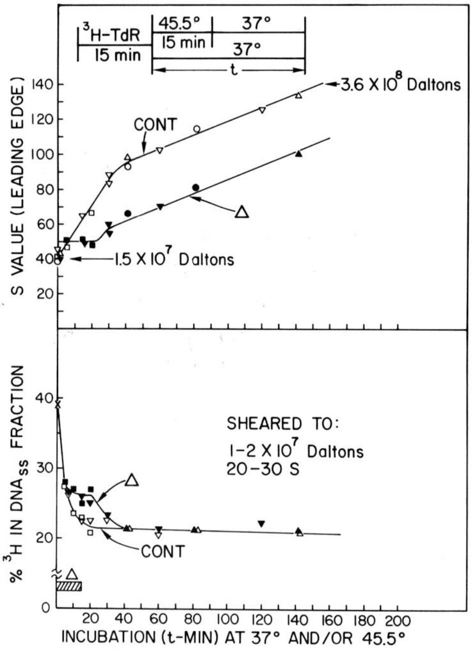

FIG. 12. Top graph: Effect of heat treatment (△) on processing of pulse-labeled DNA into high molecular weight molecules. Monolayer cultures of CHO cells were pulse-labeled with 2 μCi/ml ^{3}H-thymidine (20 Ci/mM) for 15 minutes, washed twice with medium containing 10 μg/ml stable thymidine, and then either incubated at 37°C (Cont.– O, △, □, ▽) or heated (△) in medium at 45.5°C for 15 minutes prior to incubation at 37°C for various periods (●, ▲, ■, ▼). Cells were trypsinized and prepared for alkaline sucrose gradient sedimentation analyses as described in Figure 11. The different symbols represent four different experiments.

To obtain a curve that depicts the rate of processing of small DNA molecules synthesized during the ^{3}H pulse into large molecules, the unnormalized alkaline sucrose gradient sedimentation profiles were used to obtain S values of the leading edges of the main peaks (ordinate) after various chase times (t-min, abscissa).

Bottom graph: Effect of heat treatment (△) on the decrease in the amount of pulse-labeled DNA found in the single-stranded fraction (DNA_{ss}) analyzed by benzoylated naphthoylated DEAE cellulose chromatography following chase at 37°C.

CHO cells were treated as described for the top graph. Control cells (△, □, ▽) were incubated at 37°C only following the ^{3}H-pulse, while heated cells (△) were treated for 15 minutes at 45.5°C (▲, ■, ▼) and then incubated at 37°C for various times. Cells were trypsinized and washed with Hanks' balanced salt solution, and DNA was isolated utilizing pronase and RNase treatments and extensive dialysis against Tris-EDTA buffer. The DNA samples were sheared to 1–2 × 10^7 daltons and less by passage five times through a 20-gauge needle. The molecular weight was determined by neutral sucrose gradient analysis. Benzoylated napthoylated DEAE cellulose chromatography was carried out according to Scudiero and Strauss (1974). The radioactivity eluting in the DNA_{ss} fraction contained both single-strand molecules and duplex molecules with single-strand regions. Since the percent of ^{3}H found in DNA_{ss} decreased as the DNA was sheared to smaller molecules, the percent ^{3}H in DNA_{ss} was plotted versus molecular weight, and the various samples were compared at the size (1–2 × 10^7 daltons) corresponding to 39% for the control sample analyzed immediately after the 15-minute ^{3}H pulse. (Data taken from Wong and Dewey, in preparation.)

prior to heat treatment, as well as some inhibition and a 20 to 30 minute delay in ligating molecules of about 1.5×10^7 daltons into large molecules of about 4×10^8 daltons (Figure 12). Possibly, the delay in rejoining discontinuities in DNA could enhance the frequency of chromosomal aberrations because movement of broken ends of DNA during the interval when rejoining is delayed might result in an increase in illegitimate reunion, i.e., exchanges between DNA molecules. These heat effects on DNA synthesis may result from either direct thermal effects on chromosomal proteins, including polymerases (see section *Target for Heat Radiosensitization*), or from indirect effects (Lin et al. 1978) such as alterations in intracellular ions, including pH. The dramatic effect of only 42°C for 10 minutes on initiation of DNA replicons (see Figure 11) is difficult to reconcile with direct thermal effects on chromosomal proteins. Furthermore, Dube et al. (1977) have shown that the α-polymerase involved in semiconservative DNA synthesis is relatively heat stable.

Other observations suggesting that heat damage may be in organelles other than the membrane, although the damage may result indirectly from membrane alterations, are tabulated (Tables 3 and 5): (1) The diffusion coefficient for ascites tumor cell membranes has an activation energy of 20 kcal/mole, which is greatly different from 140 kcal/mole for cell killing, and the Arrhenius plot does not break between 24°C and 46°C (Strom et al. 1973). (2) Anisotonicity, especially hypertonic treatment, increases thermal sensitivity (Raaphorst and Dewey 1978). (3) The thermally induced increase in nonhistone protein associated with DNA correlates, in general, with a decrease in survival as the hyperthermic treatment is increased (Tomasovic et al. 1978, Roti Roti and Winward 1978, Roti Roti et al. 1978). This increase in nonhistone proteins is nonspecific, with more of an increase in high molecular weight than low molecular weight molecules (Tomasovic et al. 1978). (4) Thermal tolerance observed in terms of the nonhistonc to DNA ratio (see Figure 5), as well as the other parameters involving

TABLE 3. *Evidence for various organelles as targets for heat damage*

Structural Changes in Nucleolus: Inhibits Processing of Ribosomal RNA
For S Phase—Correlation between Chromosomal Aberrations and Lethality
Inhibits DNA Synthesis: Initiation and Delays Elongation
Increases Nonhistone/DNA Ratio: Ratio and Lethality Increase with Amount of Heat
Thermal Tolerance Observed for Lethality; Increase in Nonhistone/DNA; and Inhibition of DNA, RNA, and Protein Synthesis
Mitotic Apparatus Is Heat Sensitive
Heat Causes Activation of Lysosomal Enzymes

See text for further explanation and references.

macromolecular synthesis (see Figure 6), may suggest targets other than the membrane. (5) Other observations that should be related to the mechanism of hyperthermic killing are the effects of electronic affinic compounds (Stratford and Adams 1977), 5-thio-D-glucose (Kim et al. 1978b), and glucose deprivation (Kim et al. 1978a) on increasing the thermal sensitivity of hypoxic cells. (6) Also, increased sensitization and loss of thermal tolerance by treatment with polyamines may or may not be related to membrane damage (see Figure 7). (7) Finally, heat-induced lysosomal enzyme activation has been reported (Hume and Field 1977) and appears to correlate with hyperthermic radiosensitization.

In summary, the weight of the evidence suggests that the plasma membrane is at least one of the primary targets in hyperthermic killing. However, the importance of other possible effects in the nucleus and cytoplasm should not be excluded at this time.

EFFECTS OF HEAT COMBINED WITH X IRRADIATION

Radiosensitization, pH, O_2, and LET Effects

The first quantitative in vitro studies (Belli and Bonte 1963, Westra 1971) showed that heat caused radiosensitization. These and other studies have shown that the slope of the X-ray survival curves can be increased and the shoulder reduced (see Dewey et al. 1977a for references). In particular, hyperthermia during low dose-rate irradiation has a dramatic effect on increasing the slopes of the curves (Ben-Hur et al. 1974) and is probably related to fixation of damage which results in inhibition of recovery from sublethal and potentially lethal X-ray damage (Li et al. 1976). Also, hyperthermic radiosensitization is much greater for cells irradiated with low-LET radiation than with high-LET radiation (Gerner and Leith 1977).

The radiosensitizing effect of hyperthermia on CHO cells heated and irradiated under either alkaline or acidic conditions is illustrated in Figure 13. Note that treatment at low pH greatly increased thermal cell killing and the amount of thermal radiosensitization (compare the heat curve for pH 6.7 with the dotted curve drawn parallel to the heat curve for pH 7.4); i.e., the slope of the radiation survival curve was increased more by heating at pH 6.7 than at pH 7.4, due at least in part, to the more rapid repair at pH 7.4 compared with pH 6.7 of heat damage interacting with the radiation dose delivered three hours later (see Figure 17). However, when radiation precedes heating, thermal radiosensitization may not be greater at pH 6.7 than at pH 7.4. Furthermore, the extent of increased radiosensitization at low pH cannot be quantified until synchronous cells are studied. In this study, increased cell lethality resulting from hyperthermia under low pH should tend to selectively kill S phase cells (see Figure 10), and thus the radiation response should be more characteristic of the normally radiosensitive G_1 cells, whereas the response for unheated cells or cells heated at pH 7.4 should be more characteristic of the normally radioresistant S phase

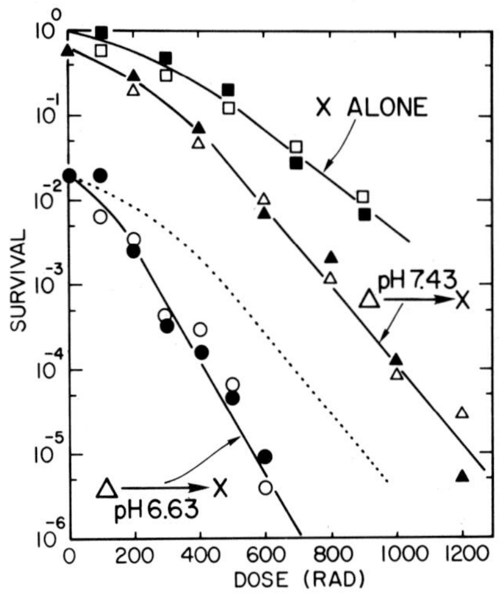

FIG. 13. The survival of CHO cells heated (for 60 minutes at 42.5°C) 180 minutes prior to irradiation. The treatments (heat [△] and irradiation [×]) were administered in medium at pH 7.43 (■) or pH 6.63 (□) (CO_2 adjusted); the cells were maintained at the respective pH values for eight hours prior to heating. All irradiations were at 37°C. In order to keep the total cell number relatively constant between $5(10)^4$ and $1(10)^5$, $5(10)^4$ lethally irradiated (1,800 rad) feeder cells were inoculated into flasks when the number of treated cells was less than $2(10)^4$. The number of treated cells added to the flasks were $1.5(10)^3$ to $1(10)^5$ for (○, △); $3(10)^2$ to $2.5(10)^4$ for (○); $3(10)^2$ to $5(10)^4$ for (▲); $1(10)^2$ to $3(10)^3$ for (□); and $1(10)^2$ to $7(10)^3$ for (■). Multiplicity was 1.37, and the plating efficiency was 0.68 for pH 7.43 and 0.77 for pH 6.63. (Data taken from Freeman, Raaphorst, and Dewey, in preparation.)

cells. Thus, redistribution of viable cells in the cycle may be partially responsible for the increased slope observed for the low pH. Furthermore, many of the studies in the literature report a reduction in OER when cells are heated and irradiated under hypoxia (see Dewey et al. 1977a for references); however, these findings may be associated more with the pH effect than with hypoxia per se (Bass et al. 1978, Bichel and Overgaard 1977, Gerweck et al. 1979). Nevertheless, this selective effect of hyperthermia on killing and radiosensitizing asynchronous cells at low pH (Figure 13) might be expected to extrapolate to a tumor, which in many cases would be expected to be at low pH and under nutritional deprivation.

Hyperthermic Radiosensitization of Seven Different Cell Lines

Differences in the hyperthermic radiosensitization of seven cell lines correlate in general with the differences in sensitivity to heat alone (Raaphorst et al. 1979a, Figure 14, and Table 4). The thermal enhancement ratios varied from 1.14 to 2.18 as determined by the dose of X irradiation divided by the dose of X irradiation plus heat when survival was reduced to 10^{-2}. These different thermal enhancement ratios correlate reasonably well with the different thermal sensitivities, which varied from a survival level of 0.9 to 0.3 following one hour at 42.5°C. However, there was no correlation between the thermal enhancement ratios and the different X-ray sensitivities of the seven cell lines.

Tolerance for Hyperthermic Radiosensitization

Another correlation between thermal killing and thermal radiosensitization is shown in Figure 15; as the CHO cells became tolerant to heat alone after three to four hours of heating, they also acquired a maximum degree of radiosen-

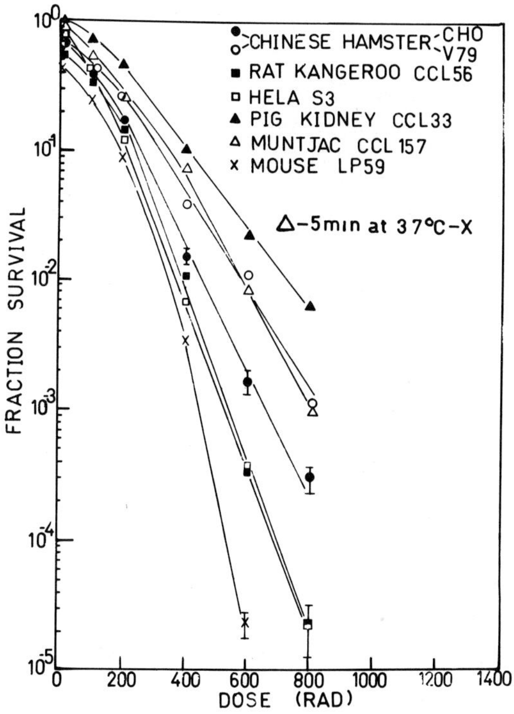

FIG. 14. Cells were first given a 1.0-hour heat exposure at 42.5°C, and 10 minutes after the termination of the heat treatment, cells were irradiated at 37°C. (Data taken from Raaphorst et al. 1979a).

sitization (Freeman et al. 1979). Since this study was carried out with asynchronous cells, the plateau in thermal radiosensitization may or may not be attributed to tolerance in terms of radiosensitization. Because of an expected redistribution of viable cells into the relatively radiosensitive G_1 phase following hyperthermic treatment, the X-ray survival curve for the cells heated for 6 to 12 hours may actually be the same as for the unheated cells, when redistribution is taken into account. Detailed studies with synchronous cells, including their cell progression after thermal treatment, are in progress to determine if thermal tolerance for cell killing corresponds to thermal tolerance for radiosensitization.

Time-Temperature Relationship for Thermal Radiosensitization

In general, the relationship between temperature and time observed for heat killing (Dewey et al. 1977a, and Figure 3) also is observed for heat radiosensitization (Sapareto et al. 1978a). However, studies indicate that the activation energy for heat radiosensitization (Brannen 1975, Law et al. 1978, Loshek et al. 1977) may be either somewhat greater or less than the activation energy for thermal killing (Table 5). When the duration of thermal treatment was varied precisely in order to obtain a constant amount of cell killing for temperatures of 42.0°C to 45.5°C, there was maximum radiosensitization at 42.5°C to 43°C (Sapareto

TABLE 4. *Comparing X-ray sensitivity with heat sensitivity and heat radiosensitization*

	Survival Curve Data						
	X Ray		Heat + X Ray				
	D_o (rads)	n	D_o (rads)	n	Survival–One Hour at 42.5°C	TER*–One Hour at 42.5°C	X Ray DMF†
Pig kidney CCL33	148	3	149	2	0.9	1.14	1.22
Muntjac CCL157	128	5	96	5	0.8	1.32	1.29
HeLa S3	117	3	67	3	0.7	1.70	1.49
CHO-10	109	15	91	2.5	0.7	1.62	1.24
Chinese hamster V79	156	7	106	3	0.6	1.62	1.0
Rat kangaroo CCL56	124	11	67	6	0.4	2.14	1.11
Mouse LP59	137	4	39	80	0.3	2.18	1.22

* Thermal enhancement ratios at $S = 10^{-2}$ are (dose for X)/(dose for $\Delta + X$), normalized for heat killing.
† The dose-modifying factors at $S = 10^2$ are (dose for V79)/(dose for cell-lines).
(Data taken from Raaphorst et al. 1979a.)

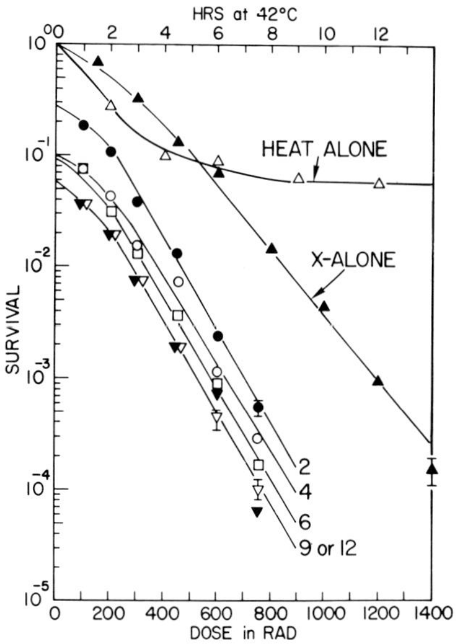

FIG. 15. The effect of heat and/or radiation on CHO cell survival: △, heat alone (42°C); ▲, radiation alone; ●, ○, □, ▽, ▼ heated for two, four, six, nine, or 12 hours at 42°C, and then 9.5 minutes after the end of the heating interval, the cells were irradiated at 37°C. The numbers by the curves represent the duration of heating. (Data taken from Freeman et al. 1979.)

TABLE 5. *Correlation between hyperthermic killing* (Δ) *and hyperthermic radiosensitization* (Δ*X*)

	Δ	ΔX
Sensitivity Increases as Body Temperature Lowers	+	+
Activation Energy (kCal/mole)	119–179	91–250
Change in Kinetics @ 42.5°–43°C	+	+
Thermal Tolerance	+	+?
Selective Sensitization of S Phase	+	+
Lethality Relates to Chromosomal Aberrations		
For S Phase	+	+
For G_1	−	+
For M	−	N.A.
Increase Sensitivity with:		
Low pH	+	+?
Hypoxia	−	?
Chronic Hypoxia	+	?
Low Nutrients	+	?
Glucose Deprivation or Glucose Analog and Electron Affinic Compounds for Hypoxic Cells	+	?
Anisotonicity (Hypertonic)	+	?
Polyamines	+	?
Membrane-Active Agents	+	?

See text for discussion and references. The symbol (+) indicates that the particular observation was made, and the symbol (−) indicates it was *not.* The symbol (?) indicates either that experimental data are unavailable or that the results are equivocal. The (+?) indicates a positive finding with some questions still remaining. N.A. means not applicable because no radiosensitization was observed.

et al., in press), which corresponds to the break in the Arrhenius curve plotted for heat only (see Figure 3).

Cell Cycle Effects—Survival and Chromosomal Aberrations

Another positive correlation between heat radiosensitization and heat killing is observed in the selective killing and radiosensitization of cells heated during S phase (Gerweck et al. 1975, Westra and Dewey 1971, and Figure 16). This selective radiosensitization of S phase effectively eliminates, or reduces at least, the variation in radiation response during the cell cycle. Also note that for cells irradiated in G_1 or S there was a positive correlation between radiosensitiza-

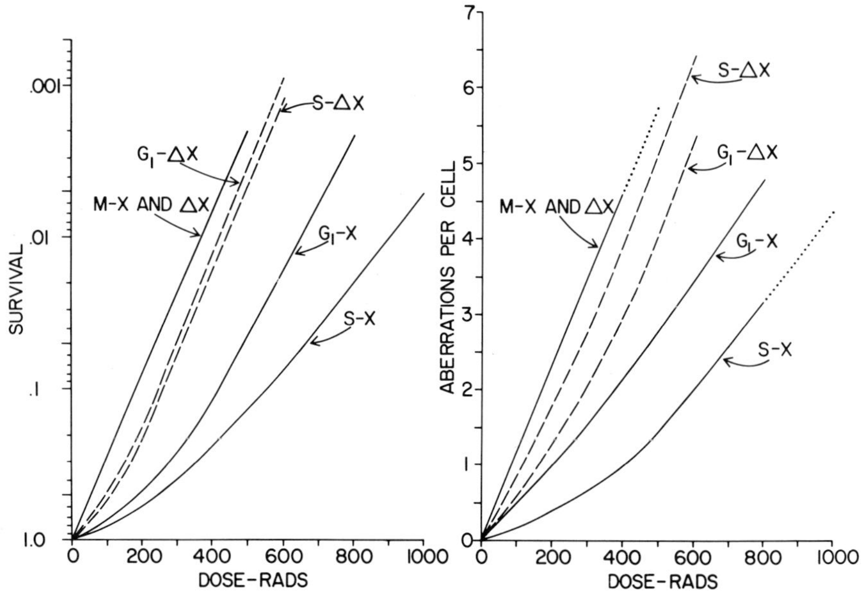

FIG. 16. Synchronous CHO cells were either X-irradiated (X) or heated and immediately X-irradiated (Δ X) when the cells were in mitosis, G_1 (two hours after plating), or S phase (10 hours after plating). The duration of heating at 45.5°C (four minutes in mitosis, 11 minutes in G_1, or seven minutes in S) was adjusted to give about the same survival (0.30 to 0.45) from heat alone. To illustrate the similarity between lethality and aberration frequency, the ordinate for survival starts at 1.0 on the bottom and decreases upward to a survival of 0.001. The values for survival and aberrations per cell for heat plus radiation have been normalized to 1.0 and 0.0, respectively, for zero radiation dose. This normalization of data was done by subtracting the aberration frequency for zero dose (heat only) from each aberration frequency and by dividing each survival value by the survival value for zero dose. Note that the radiosensitization indicated by comparing the dashed lines with the solid lines is similar for lethality and chromosomal aberrations and that S phase was selectively radiosensitized. (Data taken from Dewey et al. 1978.)

tion as observed by cell killing and that observed by radiation-induced chromosomal aberrations (Dewey et al. 1978). In fact, there is a log/linear relationship between survival and aberration frequency (see △ X line in Figure 10).

Effect of Varying the Sequence between Heat and Radiation

The sequence between heat and irradiation is quite important in determining the amount of lethality (Sapareto et al. 1978a). The greatest effect is observed when X irradiation is delivered during the heating interval. Changes in survival as a function of varying the sequence are also dependent on the pH and the amount of thermal killing (Figure 17). If there is a considerable amount of heat killing, the thermal effect will last for several hours (also see Gerweck et al. 1975).

The effects of different sequences also have been observed in vivo (Figure 18). These in vivo experiments conducted by Stewart and Denekamp (1978) using anesthesia and Overgaard (in press) without use of anesthesia compared the thermal enhancement ratios for tumor regrowth delay or tumor cure in 50% of the animals (TCD-50) with the thermal enhancement ratios for moist desquamation of the skin in 50% of the animals (DD-50). The thermal enhancement ratio is defined as the dose of irradiation without heat divided by the dose with heat required for a given biological effect. A comparison between the in vitro data in Figure 17 and the in vivo data in Figure 18 may be made by assuming that the response for cells at pH 7.4 represents the response for skin and that the response for cells at pH 6.7 represents the response for tumor. With this assumption, the in vitro and in vivo data agree quite well when thermal treatment follows irradiation by more than one hour, i.e., thermal enhancement is greater for tumor or cells at low pH than for skin or cells at normal pH. However, when heat and irradiation are administered simultaneously or when heat precedes irradiation, the enhanced thermal effect observed in vitro for cells at low pH is not observed in vivo for tumor. This failure to detect an enhanced thermal effect on tumor compared with skin when heating precedes the X-ray dose (includes the simultaneous treatments) may be related to heat-induced physiological changes such as alterations in blood flow, O_2 concentration, and pH. Such physiological changes in vivo resulting from thermal treatments should be carefully investigated, as well as the importance in vitro of O_2, pH, and nutrition on the interaction between heat and irradiation. Particularly important are in vivo studies involving fractionated radiation-heat treatments in which the end points are tumor cure and late effects in critical normal tissues. Nevertheless, both the in vitro and in vivo studies (Figures 17 and 18) indicate that repair kinetics may be different in tumor than in normal tissues and that administering the heat treatment after the radiation damage is repaired in both tumor and normal tissue may give a therapeutic gain because of the enhanced thermal killing observed at low pH.

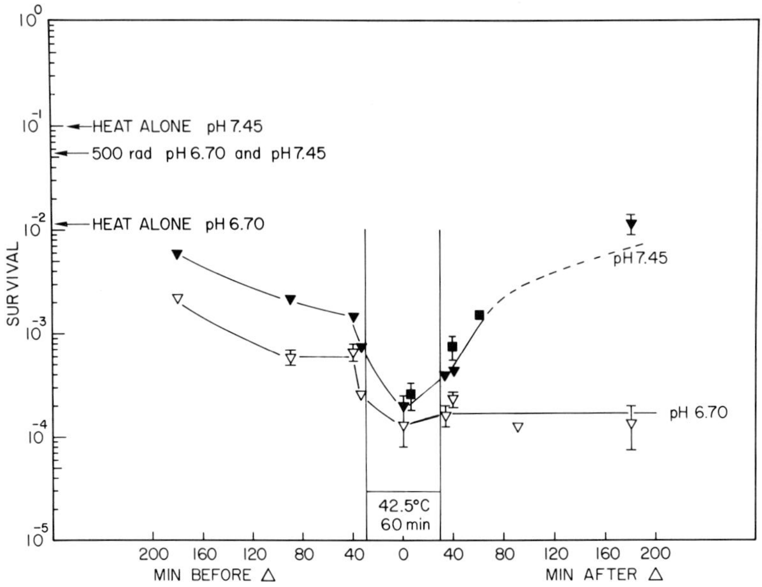

FIG. 17. The effect on CHO survival of varying the sequence between heat and radiation. Radiation (500 rad) was administered either during hyperthermia (points within the 60-minute period at 42.5°C), before heating (points to the left), or after heating (points to the right). The pH was maintained (with CO_2) either at 7.45 or 6.70 before and during the treatments. (Data taken from Freeman 1978.)

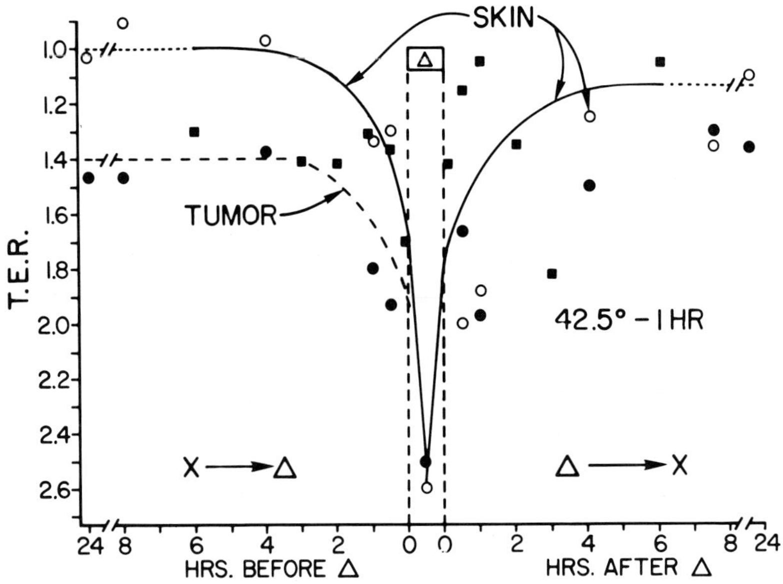

Target for Heat Radiosensitization

Heat radiosensitization and heat killing should be discussed separately because the two end points could involve different mechanisms (see Table 5). As illustrated previously (Figure 16), heat radiosensitization during both G_1 and S phase appears to be related to an increase in chromosomal aberrations. This increase in aberrations cannot be explained by heat increasing the number of radiation-induced DNA strand breaks (Figure 19, and from other studies: unpublished data and Corry et al. 1977, in which heat treatments of 45.5°C for 10 to 15 minutes reduced the molecular weight by 20% or less immediately after irradiation). However, the enhanced radiosensitization during S phase and the induction of chromosomal aberrations by heat alone during S phase may be related to thermal inhibition of DNA synthesis (see section *Target for Heat Killing,* and Figures 11 and 12). Furthermore, thermal radiosensitization during both G_1 and S may relate to the observed delay in repair of X-ray-induced strand breaks in DNA (Clark et al. 1976, Corry et al. 1977, and Figure 19). Note that there is a good correlation (Figure 20) between the delay in strand break rejoining and increase in lethality when radiation is delivered at various times after hyperthermia (Clark 1978). In other words, heat (45.5°C for 17 minutes) can interact with radiation to cause both an increase in lethality and a 20 to 30 minute delay in rejoining X-ray-induced strand breaks; for both end points, thermal damage is repaired over 12 to 18 hours with the same kinetics. Note that heat-induced delay in ligating DNA molecules synthesized during S phase (see Figure 12) is similar to the delay observed in rejoining X-ray-induced strand breaks (Figure 19). Thus, discontinuities or breaks in DNA would occur normally during S phase and would be induced by radiation during both G_1 and S, and the delay in rejoining these discontinuities or breaks might enhance the frequency of illegitimate reunions, i.e., exchanges between DNA molecules that would result in chromosomal aberrations.

Other evidence that hyperthermia interferes with the repair of X-ray-induced damage in DNA is seen in the studies of Warters and Roti Roti (1978) who found that hyperthermia (45°C for 15 minutes) greatly delays the excision of X-ray-induced base damage (Figure 21). These thermal effects on inhibiting or delaying repair of both strand breaks and base damage in DNA could be related to the thermal denaturation of nonhistone chromosomal proteins (Roti Roti and Winward 1978, Tomasovic et al. 1978), which could include repair

---

FIG. 18. The effect on the thermal enhancement ratios (TER) for mouse tumor and skin when the sequence between waterbath heat (△, 42.5°C for 60 minutes) and radiation is varied. See text for definition of TER. For comparison with in vitro data in Figure 17, TER increases from top to bottom. The solid line for skin reactions and (■) points for tumor regrowth delay were taken from Stewart and Denekamp (1978); these results for a mouse fibrosarcoma were essentially the same for a mammary carcinoma and a squamous cell carcinoma, and the TER for skin reactions was also observed for damage in the gut and cartilage. The points for skin reactions (O) and the points for 50% control of a mouse mammary carcinoma (●) were taken from Overgaard (in press).

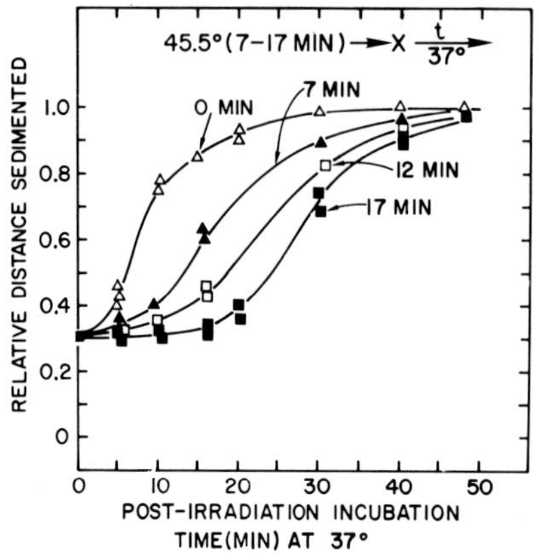

FIG. 19. Hyperthermic delay of the rejoining of DNA single breaks in asynchronous Chinese hamster ovary cells following irradiation. The cells in monolayers were heated at 45.5°C for 7 to 17 minutes immediately prior to receiving 1,000 rad at 0°C. Then, medium at 37°C was placed on the cells and incubation continued at 37°C until saline at 0°C was added to the cells, prior to lysing them on alkaline sucrose gradients. (Data taken from Clark 1978.) The relative sedimentation distance is inversely related to the number of unrejoined DNA strand breaks but is not a linear function of that parameter.

enzymes and a variety of other enzymes and structural proteins. Warters and Roti Roti (1979) have additional data from reconstruction experiments indicating that thermal inhibition of excision of X-ray-induced base damage is caused *not* by inactivation of the repair enzymes but by damage to the chromatin such that chromosomal proteins may interfere with access of the enzyme to the damaged base. From cell fusion experiments, similar conclusions were reached by Clark (1978) for the mechanism of heat-induced delay in rejoining X-ray-induced breaks in DNA. However, the preliminary data (Figure 22) for thermal inactivation of β-polymerase, a putative repair enzyme, suggest that a repair enzyme itself may be quite sensitive to hyperthermia. Although the

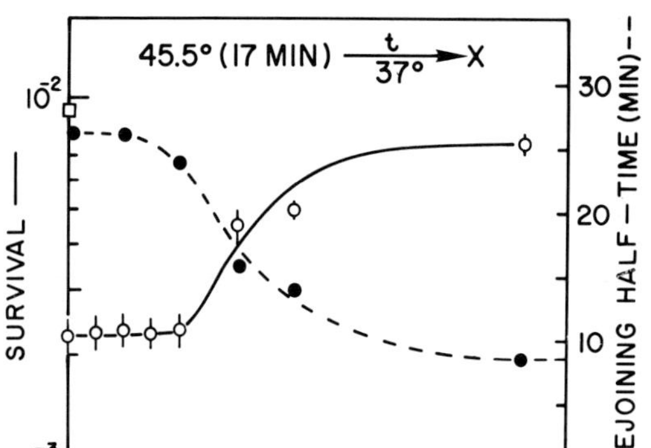

FIG. 20. Recovery of hyperthermic damage as cells were incubated at 37°C after heating (45.5°C for 17 minutes). Recovery is evident as an increase in survival when CHO S phase cells received 200 rad 18 to 24 hours after heating. Recovery is also evident as a decrease in rejoining half-times postirradiation, when asynchronous cells received 1,000 rad 18 to 24 hours after heating; note that by 48 hours, the rejoining half-time of eight minutes is the same as for unheated cells (see Figure 19). The (□) point represents survival from heat alone; survival from 200 rads alone was 0.78. Standard errors of means are indicated. Data taken from Clark (1978), who also observed (personal communication) similar recovery kinetics for survival of cells treated in G_1.

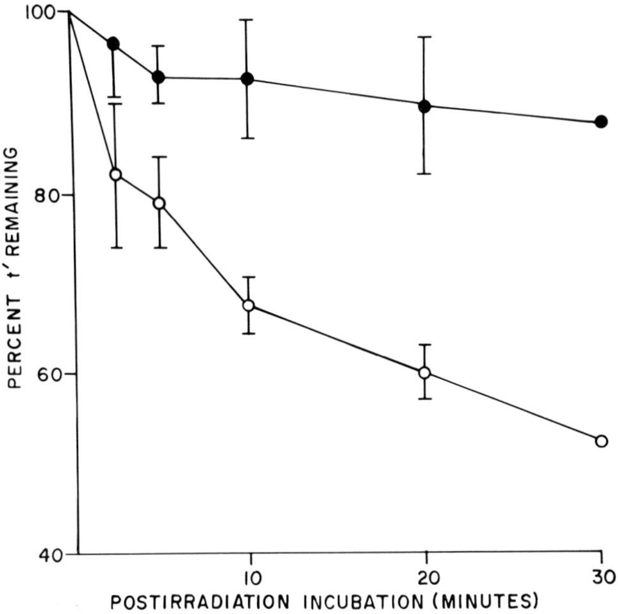

FIG. 21. Excision of thymine damage in heated and unheated cells. CHO cells were heated at 45°C for 15 minutes (closed circles) or left at 37°C (open circles). The cells were exposed to 100 krad of X irradiation at 4°C, after which they were incubated at 37°C for the time indicated, and the fraction of t′-type product remaining in their DNA determined. The average of at least five separate experiments is plotted, except for the 30-minute values which are the average results of two separate experiments. The error bars represent ±1 standard deviation. (Data taken from Warters and Roti Roti 1978.)

reduction in activity of β-polymerase, assayed with lysates of heated cells and calf thymus DNA as a substrate, may result indirectly from heat effects on other cellular components, an indirect effect is unlikely because the kinetics for heat inactivation of the partially purified enzyme (Dube et al. 1977) were very similar to those illustrated in Figure 22 for heated cells. Thus, evidence suggests that hyperthermia has a direct effect on repair enzymes and chromatin, thus causing an increase in the frequency of radiation-induced chromosomal aberrations which leads to an increase in radiosensitization as assayed by cell lethality.

Although thermal effects manifested within the nucleus have been demonstrated (see Figures 10, 11, 12, 16, 19, 20, 21, and 22), secondary effects resulting from heat damage to the plasma membrane still cannot be excluded. For example, membrane changes may cause changes in intracellular ionic concentrations, which in turn may modify protein structure or function (Lin et al. 1978) and thus affect the normal events in the nucleus or enhance hyperthermic effects occurring directly in the nucleus. Such a possibility was evident in the study of Williams et al. (1974) who observed that treatment with toxic agents, whose

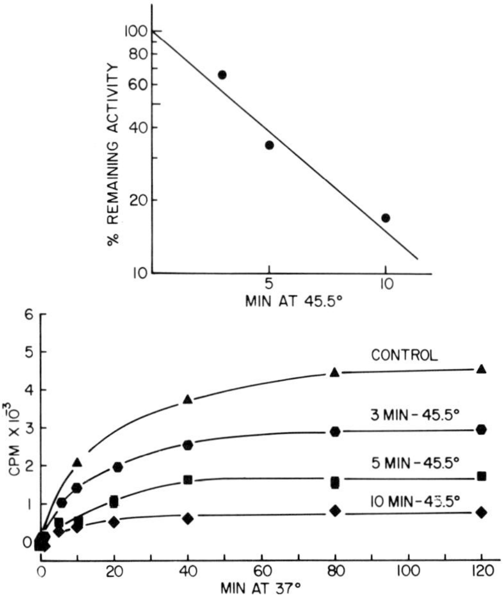

FIG. 22. Hyperthermic inactivation of DNA polymerase-β. Whole cells were heated for three, five, or 10 minutes at 45.5°C. Cells were then disrupted by repeated freezing and thawing and incubated for 60 minutes at 4°C with N-ethyl-maleimide to inactivate DNA polymerase-α and DNA polymerase-γ. The crude cell extracts were combined with a reaction mixture containing $MgCl_2$, dATP, dGTP, dCTP, TTP(^{3}H), and activated calf thymus DNA. Lower panel: Incorporation of TTP(^{3}H) into the exogenous DNA (ordinate) was measured as a function of incubation time at 37°C (abscissa). Upper panel: Percent of control levels of TTP(^{3}H) incorporation (ordinate) versus time of hyperthermic incubation (abscissa). Fractional activity was determined from the data in the lower panel at 10, 40, 80, and 120 minutes; the values that differed by less than 15 to 20 percent were averaged.

initial action is believed to affect the plasma membrane, produced DNA degradation in Chinese hamster cells. Finally, there may be two primary effects: one causing hyperthermic killing and the other causing hyperthermic radiosensitization. (See Table 5 for correlations between the two heat effects.) Further experiments are needed to study these effects.

SUMMARY—RELATIVE TO POSSIBLE CLINICAL APPLICATIONS

This review presents a rationale for considering the use of hyperthermia in cancer therapy. First, the S phase, which is relatively radioresistant, is most sensitive to heat either in terms of heat killing or heat radiosensitization. This could have a selective effect on cycling tumor cells. Second, the relatively radioresistant hypoxic cells, which are probably nutritionally deficient and at low pH, are selectively sensitized to hyperthermic killing and possibly to hyperthermic radiosensitization. Third, hyperthermia can inhibit recovery from sublethal and potentially lethal X-ray damage. This may or may not be selective for tumor in relation to normal tissue. Fourth, there is evidence that repair kinetics in tumor may be different from those in normal tissue. The present data suggest that radiation followed two or three hours later by hyperthermia may improve the therapeutic ratio. Finally, use of ultrasound, radio frequency, or microwaves may provide the possibility of obtaining higher temperatures in tumor than normal tissues.

Since the toxic effects of radiation, radiosensitizers, or chemotherapeutic agents are greatly enhanced by hyperthermia and since heat and radiation act in a complementary way (e.g., killing of hypoxic and S phase cells), the combined modality approach appears to offer a great potential for use of hyperthermia. Experimental combined modality studies involving hyperthermia with chemotherapy, radiosensitizers, radioprotectors, and membrane-active agents, in particular, should be considered.

The questions to be resolved or confirmed are the mechanisms of action for heat killing and heat radiosensitization, the importance of temperature and duration, the importance of sequence (heat before irradiation or radiation before heat), the best interval between combined heat-X-ray fractions, how to heat the tumor with precise temperature control, the recognition and possible utilization of thermal tolerance for heat alone and possibly for heat combined with radiation, heat effects on the immunological response and on destroying or spreading metastases (Schechter et al. 1978), and physiological changes that are especially important for heat before radiation and that may involve changes in pH, nutrients, oxygen consumption (Durand 1978), blood flow, and oxygen concentration. Thus, when heat is delivered before X-irradiation, physiological changes could increase or decrease the hypoxic fraction in the tumor. The most important observation to be made, however, is whether or not there will indeed be a differential effect between tumor and normal tissue when both are at the same temperature. If there is no differential effect, then the tumor will have to be heated selectively in order to achieve any therapeutic efficacy. These studies of therapeutic gain or efficacy are most important for fractionated doses and when tumor cure is compared with late effects in normal tissue.

The immediate challenge before us is to develop and make available heating and thermometry equipment that is essential for conducting definitive experiments in small and large animals and in clinical studies. Then, as basic mechanisms involved in heat inactivation and heat radiosensitization are understood, results obtained in vitro and in small and large animals, as well as in Phase I and II clinical trials, should lead to meaningful Phase III clinical trials. Hopefully, Phase III clinical trials will demonstrate that hyperthermia, when utilized properly, is a beneficial agent for cancer therapy, i.e., will provide an improvement in the therapeutic index defined as percentage of a group of treated cancer patients who remain both free of recurrence and free of severe complications over a given period of time.

ACKNOWLEDGMENTS

This investigation was supported in part by Grants CA-08618 and CA-18334, awarded by the National Cancer Institute, Department of Health, Education and Welfare. Appreciation is expressed to Ms. Clare Brien for her technical assistance. We wish to express our sincere gratitude to Ms. Carole Clark for her dedicated secretarial and administrative assistance over the past fourteen years. Her participation, sometimes in the laboratory, has been vital for our research program in hyperthermia.

REFERENCES

Ashby, B. S., and M. D. Cantab. 1966. pH studies in human malignant tumors. Lancet 2:312–315.

Barrau, M. D., G. R. Blackburn, and W. C. Dewey. 1978. Effects of heat on the centrosomes of Chinese hamster ovary cells. Cancer Res. 38:2290–2294.

Bass, H., J. L. Moore, and W. T. Coakley. 1978. Lethality in mammalian cells due to hyperthermia under oxic and hypoxic conditions. Int. J. Radiat. Biol. 33:57–67.

Belli, J. A., and F. J. Bonte. 1963. Influence of temperature on the radiation response of mammalian cells in tissue culture. Radiat. Res. 18:272–276.

Ben-Hur, E., M. M. Elkind, and B. V. Bronk. 1974. Thermally enhanced radio-response of cultured Chinese hamster cells: Inhibition of repair of sublethal damage and enhancement of lethal damage. Radiat. Res. 58:38–51.

Bichel, P., and J. Overgaard. 1977. Hyperthermic effect on exponential and plateau ascites tumor cells in vitro dependent on environmental pH. Radiat. Res. 70:449–454.

Bowler, K., C. J. Duncan, R. T. Gladwell, and T. F. Davison. 1973. Cellular heat injury. Comp. Biochem. Physiol. Part A 45:441–450.

Brannen, J. P. 1975. A temperature- and dose-rate-dependent model for the kinetics of cellular response to ionizing radiation. Radiat. Res. 62:379–387.

Clark, E. P. 1978. Hyperthermia and the rejoining of x-ray-induced DNA strand breaks. Ph.D. dissertation. Colorado State University.

Clark, E. P., S. Robinson, J. T. Lett, and P. Corry. 1976. Inhibition of DNA strand break rejoining by hyperthermia, *in* Proceedings of the International Symposium on Cancer Therapy by Hyperthermia and Radiation, J. E. Robinson and M. E. Wizenberg, eds. American College of Radiology, Chevy Chase, Maryland, pp. 27–33.

Connor, W. G., E. W. Gerner, R. C. Miller, and M. L. M. Boone. 1977. Prospects for hyperthermia in human cancer therapy. II. Implications of biological and physical data for applications to man. Radiology 123:497–503.

Corry, P. M., S. Robinson, and S. Getz. 1977. Hyperthermic effects on DNA repair mechanisms. Radiology 123:475–482.

Coss, R. A., W. C. Dewey, and J. R. Bamburg. 1979. Effects of hyperthermia (41.5°) on Chinese hamster ovary cells analyzed in mitosis. Cancer Res. 39:1911–1918.

Cress, A. E., T. S. Herman, and E. W. Gerner. 1978. Cholesterol content and thermal sensitivity of mammalian cells in vitro. (Abstract). Radiat. Res. 74:477.

Crile, G., Jr. 1963. The effects of heat and radiation on cancers implanted on the feet of mice. Cancer Res. 23:372–380.

Dewey, W. C., L. E. Hopwood, S. A. Sapareto, and L. E. Gerweck. 1977a. Cellular responses to combinations of hyperthermia and radiation. Radiology 123:463–474.

Dewey, W. C., S. A. Sapareto, and D. A. Betten. 1978. Hyperthermic radiosensitization of synchronous Chinese hamster cells: Relationship between lethality and chromosomal aberrations. Radiat. Res. 76:48–59.

Dewey, W. C., D. E. Thrall, and E. L. Gillette. 1977b. Hyperthermia and radiation—A selective thermal effect on chronically hypoxic tumor cells in vivo. Int. J. Radiat. Oncol. Biol. Phys. 2:99–103.

Dewey, W. C., A. Westra, and H. H. Miller. 1971. Heat-induced lethality and chromosomal damage in synchronized Chinese hamster cells treated with 5-bromodeoxyuridine. Int. J. Radiat. Biol. 20:505–520.

Dube, D. K., G. Seal, and L. A. Loeb. 1977. Differential heat sensitivity of mammalian DNA polymerases. Biochem. Biophys. Res. Commun. 76:483–487.

Durand, R. E. 1978. Effect of hyperthermia on the cycling, noncycling, and hypoxic cells of irradiated and unirradiated multicell spheroids. Radiat. Res. 75:373–384.

Freeman, M. L. 1978. The effect of environmental pH on hyperthermic and/or radiation-induced cell killing. Ph.D. dissertation. Colorado State University.

Freeman, M. L., W. C. Dewey, and L. E. Hopwood. 1977. Effect of pH on hyperthermic cell survival: Brief communication. J. Natl. Cancer Inst. 58:1837–1839.

Freeman, M. L., G. P. Raaphorst, and W. C. Dewey. 1979. The relationship of heat killing ana thermal radiosensitization to the duration of heating at 42°C. Radiat. Res. 78:172–175.
Gerner, E. W., and J. T. Leith. 1977. Interaction of hyperthermia with radiations of different linear energy transfer. Int. J. Radiat. Biol. 31:283–288.
Gerner, E. W., and D. H. Russell. 1977. The relationship between polyamine accumulation and DNA replication in synchronized Chinese hamster ovary cells after heat shock. Cancer Res. 37:482–489.
Gerweck, L. E., E. L. Gillette, and W. C. Dewey. 1975. Effect of heat and radiation on synchronous Chinese hamster cells; killing and repair. Radiat. Res. 64:611–623.
Gerweck, L. E., T. G. Nygaard, and M. Burlett. 1979. Response of cells to hyperthermia under acute and chronic hypoxic conditions. Cancer Res. 39:966–972.
Gerweck, L., and E. Rottinger. 1976. Enhancement of mammalian cell sensitivity to hyperthermia by pH alteration. Radiat. Res. 67:508–511.
Goekoop, J. G., F. Spies, C. Bierman-Van Steeg, R. Vrielink, and G. M. J. Van Kempen. 1978. pH-dependent behaviour of erythrocyte membrane elevations. Cell Biol. Int. Reports 2:139–145.
Gullino, P. M., F. H. Grantham, S. A. Smith, and A. C. Haggerty. 1965. Modifications of the acid-base status of the internal milieu of tumors. J. Natl. Cancer Inst. 34:857–869.
Hahn, G. M. 1974. Metabolic aspects of the role of hyperthermia in mammalian cell inactivation and their possible relevance to cancer treatment. Cancer Res. 34:3117–3123.
Hahn, G. M., G. C. Li, and E. Shiu. 1977. Interaction of amphotericin B and 43° hyperthermia. Cancer Res. 37:761–764.
Hahn, G. M., and D. Pounds. 1976. Heat treatment of solid tumors: Why and how. Appl. Radiol. 5:131–144.
Harisiadis, L., D. II Sung, and E. J. Hall. 1977. Thermal tolerance and repair of thermal damage by cultured cells. Radiology 123:505–509.
Henle, K. J., and L. A. Dethlefsen. 1978. Heat fractionation and thermo-tolerance. A review. Cancer Res. 38:1843–1851.
Hume, S. P., and S. B. Field. 1977. Acid phosphatase activity following hyperthermia of mouse spleen and its implication in heat potentiation of x-ray damage. Radiat. Res. 72:145–153.
Inoué, S. 1952. The effect of temperature on the birefringence of the mitotic spindle. Biol. Bull. 103:316.
Kachani, Z. F. C., and A. B. Sabin. 1969. Reproductive capacity and viability at higher temperatures of various transformed hamster cell lines. J. Natl. Cancer Inst. 43:469–480.
Kahler, H., and W. B. Robertson. 1943. Hydrogen-ion concentration of normal liver and hepatic tumors. J. Natl. Cancer Inst. 3:495–501.
Kal, H. B., M. Hatfield, and G. M. Hahn. 1975. Cell cycle progression of murine sarcoma cells after x-irradiation or heat shock. Radiology 117:215–217.
Kim, J. H., S. H. Kim, and E. W. Hahn. 1978a. Killing of glucose-deprived hypoxic cells with moderate hyperthermia. Radiat. Res. 75:448–451.
Kim, J. H., S. H. Kim, E. W. Hahn, and C. W. Song. 1978b. 5-thio-D-glucose selectively potentiates hyperthermic killing of hypoxic tumor cells. Science 200:206–207.
Kwock, L., P.-S. Lin, K. Hefter, and D. F. H. Wallach. 1978. Impairment of Na^+-dependent amino acid transport in a cultured human T-cell line by hyperthermia and irradiation. Cancer Res. 38:83–87.
Law, M. P., R. G. Ahier, and S. B. Field. 1978. The response of mouse ear to heat applied alone or combined with x-rays. Br. J. Radiol. 51:132–138.
Li, G. C., R. G. Evans, and G. M. Hahn. 1976. Modification and inhibition of repair of potentially lethal x-ray damage by hyperthermia. Radiat. Res. 67:491–501.
Li, G. C., E. C. Shiu, and G. M. Hahn. 1978. Thermal tolerance and tolerance to Adriamycin induced by ethanol. (Abstract). Radiat. Res. 74:479.
Lin, P.-S., L. Kwock, K. Hefter, and D. F. H. Wallach. 1978. Modification of rat thymocyte membrane properties by hyperthermia and ionizing radiation. Int. J. Radiat. Biol. 33:371–382.
Lin, P.-S., D. F. H. Wallach, and S. Tsai. 1973. Temperature-induced variations in the surface topology of cultured lymphocytes are revealed by scanning electron microscopy. Proc. Natl. Acad. Sci. USA 70:2492–2496.
Loshek, D. D., J. S. Orr, and E. Solomonidis. 1977. Interaction of hyperthermia and radiation: Temperature coefficient of interaction. Br. J. Radiol. 50:902–907.
Marmor, J. B., N. Hahn, and G. M. Hahn. 1977. Tumor cure and cell survival after localized radiofrequency heating. Cancer Res. 37:879–883.

Ossovski, L., and L. Sachs. 1967. Temperature sensitivity of polyoma virus, induction of cellular DNA synthesis, and multiplication of transformed cells at high temperature. Proc. Natl. Acad. Sci. USA 58:1938–1947.

Overgaard, J. 1978. The effect of local hyperthermia alone, and in combination with radiation, on solid tumors, *in* Cancer Therapy by Hyperthermia and Radiation, C. Streffer et al., eds. Urban & Schwarzenberg, Munich and Baltimore, pp. 49–61.

Overgaard, J. 1979. Combined hyperthermia and radiation in vivo. 12th International Cancer Congress, Buenos Aires, Argentina, Oct. 2, 1978 (in press).

Overgaard, J., and P. Bichel. 1977. The influence of hypoxia and acidity on the hyperthermic response of malignant cells in vitro. Radiology 123:511–514.

Painter, R. B., and B. R. Young. 1976. Formation of nascent DNA molecules during inhibition of replicon initiation in mammalian cells. Biochim. Biophys. Acta 418:146–153.

Raaphorst, G. P., and W. C. Dewey. 1978. Enhancement of hyperthermic killing of cultured mammalian cells by treatment with anisotonic NaCl or medium solutions. J. Thermal Biol. 3:177–182.

Raaphorst, G. P., S. L. Romano, J. B. Mitchell, J. S. Bedford, and W. C. Dewey. 1979a. Intrinsic differences in heat and/or x-ray sensitivity of seven mammalian cell lines cultured and treated under identical conditions. Cancer Res. 39:396–401.

Raaphorst, G. P., S. A. Sapareto, M. L. Freeman, and W. C. Dewey. 1979b. Changes in cellular heat and/or radiation sensitivity observed at various times after trypsinization and plating. Int. J. Radiat. Biol. 35:193–197.

Reeves, O. R. 1972. Mechanisms of acquired resistance to acute heat shock in cultured mammalian cells. J. Cell. Physiol. 79:157–170.

Reinhold, H. S., B. Blachiewicz, and A. Berg-Blok. 1978. Decrease in tumor microcirculation during hyperthermia, *in* Cancer Therapy by Hyperthermia and Radiation, C. Streffer et al., eds. Urban & Schwarzenberg, Munich and Baltimore, pp. 221–235.

Rieder, C., and A. S. Bajer. 1977. Heat-induced reversible hexagonal packing of spindle microtubules. J. Cell Biol. 74:717–725.

Robinson, J. E., G. H. Harrison, W. A. McCready, and G. M. Samaras. 1978. Good thermal dosimetry is essential to good hyperthermia research. Br. J. Radiol. 51:532–534.

Robinson, J. E., and M. J. Wizenberg, eds. 1976. Proceedings of the International Symposium on Cancer Therapy by Hyperthermia and Radiation. American College of Radiology, Chevy Chase, Maryland, 305 pp.

Robinson, J. E., M. J. Wizenberg, and W. A. McCready. 1974. Combined hyperthermia and radiation suggest an alternative to heavy particle therapy for reduced oxygen enhancement ratios. Nature 251:521–522.

Roti Roti, J. L., K. J. Henle, and R. T. Winward. 1978. Kinetics of the increase in chromatin protein content in heated cells and its relationship to thermal cell death. (Abstract) Radiat. Res. 74:509.

Roti Roti, J. L., and R. T. Winward. 1978. The effects of hyperthermia on the protein to DNA ratio of isolated HeLa cell chromatin. Radiat. Res. 74:159–169.

Sapareto, S. A., L. E. Hopwood, and W. C. Dewey. 1978a. Combined effects of x-irradiation and hyperthermia on CHO cells for various temperatures and orders of application. Radiat. Res. 73:221–233.

Sapareto, S. A., L. E. Hopwood, W. C. Dewey, M. R. Raju, and J. W. Gray. 1978b. Effects of hyperthermia on survival and progression of Chinese hamster ovary cells. Cancer Res. 38:393–400.

Sapareto, S. A., G. P. Raaphorst, and W. C. Dewey. 1979. Cell killing and the sequencing of hyperthermia and radiation. Int. J. Radiat. Oncol. Biol. Phys. 5:343–347.

Schechter, M., S. M. Stowe, and H. Moroson. 1978. Effects of hyperthermia on primary and metastatic tumor growth and host immune response in rats. Cancer Res. 38:498–502.

Schrek, R. 1966. Sensitivity of normal and leukemic lymphocytes and leukemic myeloblasts to heat. J. Natl. Cancer Inst. 37:649–654.

Scudiero, D., and B. Strauss. 1974. Accumulation of single-stranded regions in DNA and the block to replication in a human cell line alkylated with methyl methane sulfonate. J. Mol. Biol. 83:17–34.

Simard, R., and W. Bernhard. 1967. A heat-sensitive cellular function located in the nucleolus. J. Cell Biol. 34:61–76.

Sinensky, M. 1978. Defective regulation of cholesterol biosynthesis and plasma membrane fluidity in a Chinese hamster ovary cell mutant. Proc. Natl. Acad. Sci. USA 75:1247–1249.

Song, C. W. 1978. Effect of hyperthermia on vascular functions of normal tissues and experimental tumors: Brief communication. J. Natl. Cancer Inst. 60:711–713.

Stewart, F. A., and J. Denekamp. 1978. The therapeutic advantage of combined heat and x-rays on a mouse fibrosarcoma. Br. J. Radiol. 51:307–316.

Stratford, I. J., and G. E. Adams. 1977. Effect of hyperthermia on differential cytotoxicity of a hypoxic cell radiosensitizer, Ro–07–0582, on mammalian cells in vitro. Br. J. Cancer 35:307–313.

Streffer, C. et al., eds. 1978. Cancer Therapy by Hyperthermia and Radiation. Urban & Schwarzenberg, Munich and Baltimore, 341 pp.

Strom, R., A. S. Santoro, C. Crifó, A. Bozzi, B. Mondovi, and A. Rossi-Fanelli. 1973. The biochemical mechanism of selective heat sensitivity of cancer cells. IV. Inhibition of RNA synthesis. Eur. J. Cancer 9:103–112.

Suit, H. D., and M. Shwayder. 1974. Hyperthermia: Potential as an antitumor agent. Cancer 34:122–129.

Thrall, D. E., E. L. Gillette, and W. C. Dewey. 1975. Effect of heat and ionizing radiation on normal and neoplastic tissue of the C3H mouse. Radiat. Res. 63:363–377.

Tomasovic, S. P., G. N. Turner, and W. C. Dewey. 1978. Effect of hyperthermia on nonhistone proteins isolated with DNA. Radiat. Res. 73:535–552.

von Ardenne, M., and P. G. Reitnauer. 1968. Selektive krebszellenschädigung durch proteindenaturation. Dtsch. Gesundheitswesen 23:1681–1685, 1738–1744.

Wallach, D. F. H. 1978. Action of hyperthermia and ionizing radiation on plasma membranes, *in* Cancer Therapy by Hyperthermia and Radiation, C. Streffer et al., eds. Urban & Schwarzenberg, Munich and Baltimore, pp. 19–28.

Warocquier, R., and K. Scherrer. 1969. RNA metabolism in mammalian cells at elevated temperature. Eur. J. Biochem. 10:362–370.

Warters, R. L., and J. L. Roti Roti. 1978. Production and excision of 5′,6′-dihydroxydihydrothymine type products in the DNA of preheated cells. Int. J. Radiat. Biol. 34:381–384.

Warters, R. L., and J. L. Roti Roti. 1979. Excision of x-ray-induced thymine damage in chromatin from heated cells. Radiat. Res. 79:113–121.

Westra, A. 1971. The influence of radiation on the capacity of in vitro cultured mammalian cells to proliferate. Ph.D. dissertation. University of Amsterdam.

Westra, A., and W. C. Dewey. 1971. Variation in sensitivity to heat shock during the cell cycle of Chinese hamster cells in vitro. Int. J. Radiat. Biol. 19:467–477.

Williams, J. R., J. B. Little, and W. U. Shipley. 1974. Association of mammalian cell death with a specific endonucleolytic degradation of DNA. Nature 252:754–755.

Witcofski, R. L., and F. W. Kremkau. 1978. Ultrasonic enhancement of cancer radiotherapy. Radiology 127:793–797.

Yatvin, M. B. 1977. The influence of membrane lipid composition and procaine on hyperthermic death of cells. Int. J. Radiat. Biol. 32:513–521.

Yatvin, M. B., K. H. Clifton, and W. H. Dennis. 1979. Hyperthermia and local anaesthetics: Potentiation of survival of tumor-bearing mice. Science 205:195–196.

Radiation Biology in Cancer Research, edited by Raymond E. Meyn and H. Rodney Withers.
Raven Press, New York © 1980.

Thermal and Nonthermal Effects of Ultrasound

George M. Hahn, Gloria C. Li, Jane B. Marmor, and Douglas W. Pounds

Department of Radiology, Stanford University School of Medicine, Stanford, California 94305

The interaction of ultrasound with biological tissues has intrigued investigators for many years. For example, in 1933 Szent-Györgyi described attempts to treat tumors with sound waves. The results of his investigation were essentially negative as were those of an early American study (Beck and Krantz 1940). During World War II and shortly thereafter, several German publications noted beneficial effects from the irradiation of animal and human tumors by ultrasound. A representative work is that of Dittmar (1949) who utilized 800 KHz radiation at an intensity of 3–4 W/cm^2 at the skin. In order to attempt to separate thermal from possible nonthermal effects, Dittmar radiated ultrasound energy into water that was at a controlled and fixed temperature, with secondary coupling from the water to the tissue of interest. He found that the effects caused by ultrasound (as observed by pathological examinations) depended critically on the temperature of the water bath. As this temperature increased above 37°C, the sonically induced damage to the tumor became more and more pronounced. He treated two rat tumors, the Walker carcinoma and the Jensen sarcoma. Ultrasound proved much more effective against the Jensen sarcoma, whose growth could be inhibited. After long irradiation, and at elevated water bath temperatures, many of the Jensen sarcomas completely regressed. Dittmar suggested that, in addition to direct damage to tumor cells, the vasculature of the tumor was at risk and that in the Jensen sarcoma the capillary system was easily destroyed. Also discussed by him was the possible contribution of cavitation in inducing the observed effect, but the investigator concluded that if this phenomenon played a part in creating the observed effects, it was a minor one. His major, if tentatively stated, conclusion was that the primary effect of ultrasound was to induce elevated temperatures and hence that the ultrasound effects on the tumor were primarily thermal in nature.

We have described Dittmar's work in some detail, because many of the questions posed by him (e.g., Are there nonthermal effects? Do all tumors respond to heat in a similar fashion? What is the role of vascular damage in tumor eradication?) are still with us. Answers to these questions may tell us much about the possible role of hyperthermia in cancer treatment and about ultrasound as a treatment technique.

ARE THERE NONTHERMAL EFFECTS INVOLVED IN INACTIVATION OF TISSUE CULTURE CELLS BY ULTRASOUND?

In order to attempt to answer this question, we exposed cells to ultrasound under conditions that insured that the cells' macroscopic temperature was not inflated by the ultrasound. Chinese hamster cells (HA-1) were grown as monolayers in specially designed dishes, each consisting of a stainless steel ring and a Mylar bottom (thickness 1 μm) (Li et al. 1977). When the cell density was about 5×10^5 cells/cm^2, these dishes were immersed in a temperature-stabilized water bath. One side of the Mylar was in direct contact with the water, and the temperature in the medium in the dish was monitored by a thermistor. The water bath and dish temperatures never differed by more than 0.2°C, even during sonic exposures. Cells were irradiated by an ultrasound transducer (diameter 3.5 cm), which was placed in the water at a distance of 13 cm from the target. The signal characteristics were: frequency, 1 MHz; pulse modulation, 1,000 cps with a duty cycle of 60%. Pulse modulation was achieved by frequency-sweeping the signal through the transducer crystal's resonance at the appropriate rate. The field was 1.6 cm in diameter, and over that area the sonic intensity varied by ±21%. The quoted intensities are estimates of the average value. Survival of cells is maximum in areas seeing minimum ultrasound intensities; hence, the true survival values at specific intensities may be somewhat lower than those shown in Figures 1 and 2. However, these considerations do not change the conclusions derived from the results.

Two distinct effects of ultrasound were seen. First, there was an immediate

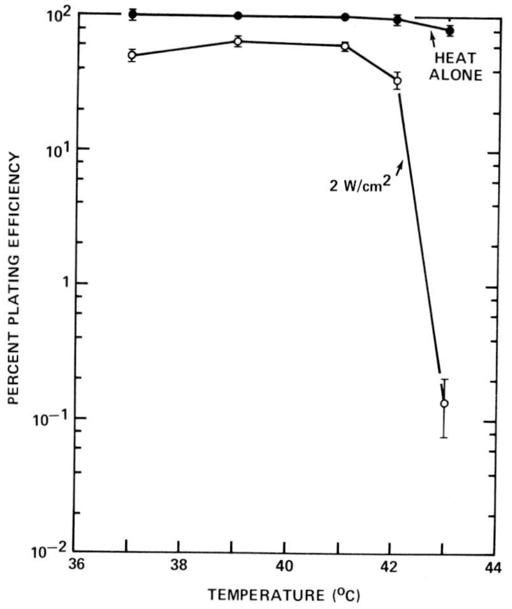

FIG. 1. Plating efficiencies of cells exposed to heat or to heat plus ultrasound. In these experiments, the cell temperature was determined by the hot water bath; the nonthermal effect of ultrasound (1 W/cm^2; 30 min) is therefore reflected by the difference between the two plating efficiencies at each temperature. Data from Li et al. 1977.

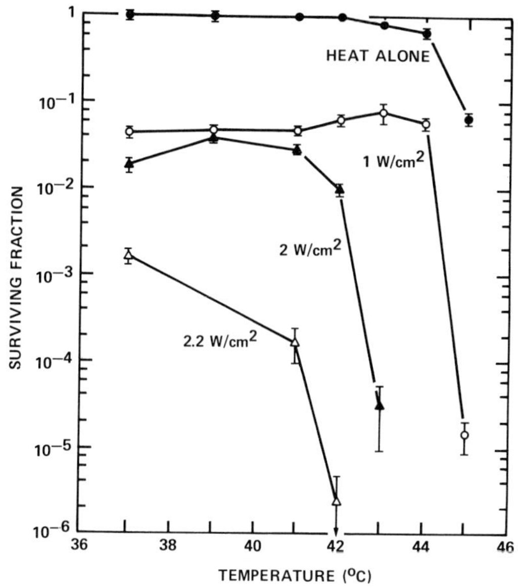

FIG. 2. Survival of cells exposed simultaneously to heat and to ultrasound. There are two components to this killing: an immediate lysis of cells, which dominates at the lower temperatures and which in magnitude is temperature independent; and change in plating efficiencies, which is temperature dependent, as was shown in Figure 1. Data from Li et al. 1977.

lysis of cells that occurred independently of the waterbath temperature. At low average intensity (<0.75 W/cm^2), this lysis affected few if any of the cells, but rose to 95% at 2 W/cm^2. Second, many of the cells that remained on the Mylar (and which retained their normal morphological appearance) lost their ability to form colonies. This is shown by the plating efficiency curves of Figure 2. These are characterized by threshold effects. Cells exposed to ultrasound at intensity or temperature levels beyond the threshold values were rapidly inactivated. The intensity threshold values themselves clearly varied with temperature. At temperatures above 37°C, cells became highly sensitive to ultrasound irradiation at intensities that did not affect survival at 37°C (Li et al. 1977).

While these results show clearly that, under the right conditions, cells are inactivated by ultrasound at a rate far exceeding thermal killing when such a comparison is based upon the measured, macroscopic temperature, the possibility remained that the observed increased inactivation resulted from microscopic, localized heating of specific heat-sensitive targets. To test this possibility, we measured the age-response function (i.e., the survival response of synchronized cells exposed to identical treatments at various positions in the cell proliferative cycle) of HA-1 cells exposed to either heat alone or to ultrasound at 37°C. Time-heat combinations as well as ultrasound intensity-time combinations were chosen so that the survivals of asynchronous cell populations after either the heat or ultrasound treatments were similar. Age-response functions show a characteristic pattern for most treatments and, therefore, constitute a "signature" of that treatment (Mauro and Madoc-Jones 1970). If the age-response functions after heat and after ultrasound were identical or even similar, this would constitute good circumstantial evidence that heat caused cell death in both cases. Cells were synchronized by mitotic shake-off (Terasima and Tolmach 1963, Tobey et al. 1967) and allowed to attach to Mylar at 37°C in the incubators. The cells were then exposed at various times after attachment to the indicated heat regimens (Figure 3a) or to ultrasound (Figure 3b) at 43°C. The heated cells show the age response for heat as had been demonstrated by others (e.g., Kim et al. 1975): resistance in early G_1 with maximum heat sensitivity late in the S phase. However, the cells exposed to ultrasound showed little, if any, variation in sensitivity throughout the cell cycle. This difference in the age responses strongly implies a different mechanism of cellular inactivation by heat from that of ultrasound, or at least the involvement of different "targets."

What was the mechanism involved in the nonthermal aspect of killing by ultrasound? An obvious candidate was cavitation, i.e., the formation of resonant air bubbles that act as efficient localized energy transducers. At an ultrasound frequency of 1 MHz, cavitation takes about 1 msec to establish itself (Gavrilov 1974). If cavitation played a dominant part in the observed effects, then two things should be true: the amount of cell killing should be a function of pulse

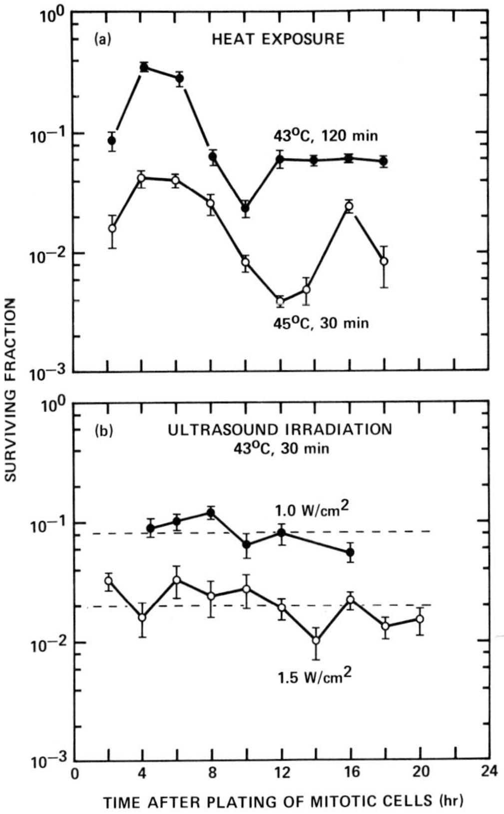

FIG. 3. Age responses of HA cells exposed to elevated temperatures or to ultrasound. In panel a, synchronized cells were treated by the indicated heat regimen but in the absence of ultrasound. In panel b, temperature was 43°C, but owing to the short (30 min) exposure time, the cell killing was effected by ultrasound (1.0 or 1.5 W/cm²). The fact that the shapes of the two curves are so different is another indication that high-intensity ultrasound killing is likely to be nonthermal.

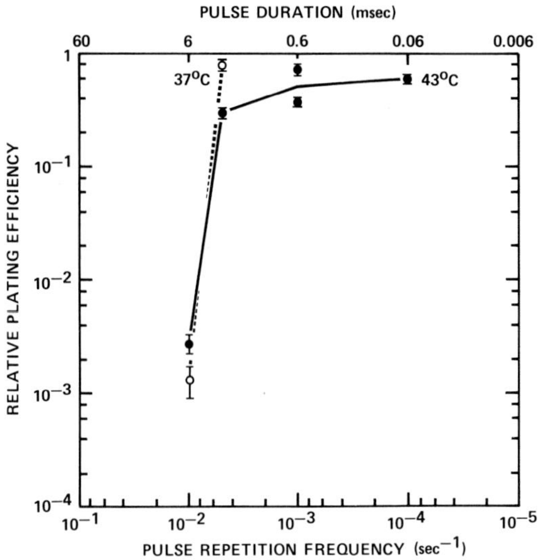

FIG. 4. Effect of ultrasound pulse duration on cell survival. Cells were exposed to ultrasound at fixed peak and average power. The pulse repetition frequency and the pulse duration were varied as indicated. The increased cell death seen as the pulse duration is increased beyond a few msec probably is the result of cavitation. There appears to be no temperature dependence of the cavitation effect.

duration, increasing rapidly as the pulse duration became longer; and the effects of cavitation should be temperature dependent. To test this, we irradiated cells as described, keeping average and peak energy constant, but varying both the pulse duration and the pulse repetition frequency. This was achieved by changing the rate of the sweep across the resonant frequency of the transducer. One series of experiments was carried out with the waterbath at 37°C, another with the bath at 43°C. Results are shown in Figure 4. The first of our expectations was clearly fulfilled; as the pulse duration was increased beyond a few msec, cell survival was reduced appreciably. However, there was little difference between the 37°C and 43°C curves, indicating that whatever role cavitation played, it was not the major determinant in the temperature-dependent aspects of cell killing.

At the present time we are left with the following picture: at intensities above about 1.5 W/cm², 1 MHz ultrasound kills cells at a rate over and above that resulting from heat inactivation. This cell killing has two components: a temperature-independent part that manifests itself as immediate cell lysis, and a temperature-dependent component that at elevated temperature efficiently impairs cell reproductive ability. The origin of the former is very likely cavitation; the origin of the latter aspect of cell inactivation by ultrasound is currently not understood.

INDUCTION OF HYPERTHERMIA IN TISSUES

The passage of a pressure wave at ultrasonic frequencies through tissue results in a transfer of energy from the wave to tissue. The rate of energy transfer is determined by the intensity of the wave and its frequency and by the mechanical characteristics of the tissue. Absorption coefficients increase linearly with fre-

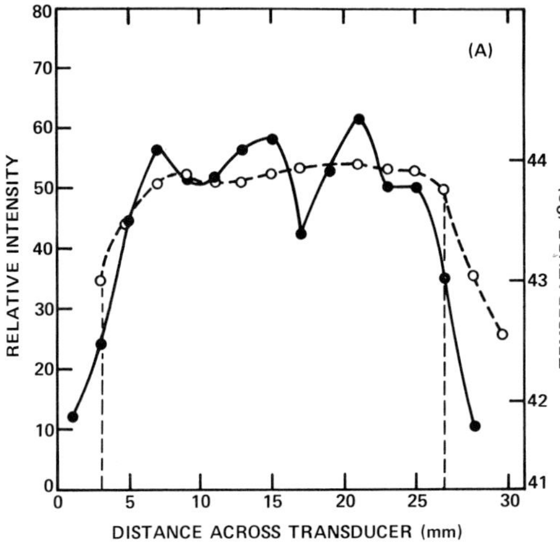

FIG. 5. Ultrasound intensity and temperature distribution produced by a 2.5 cm diameter transducer. The intensity distribution (closed circles) was measured in a water tank. The temperature distribution (open circles) was obtained in vivo by inserting a thermocouple into tissue and then slowly passing the transducer over the temperature measuring device. Although the intensity distribution was far from uniform, the temperature measurement in vivo was uniform. Very likely blood flow quickly equilibrates the temperature and tends to mask the irregularities of the transducer intensity pattern.

quency, so that for deep heating (>10 cm) frequencies below 1 MHz appear to be essential. The intensity distribution in tissue can be improved over that generated by a plane wave either by overlapping the output of several beams or by focusing the output of individual large transducers. The distribution can perhaps be optimized by scanning arrays of coherently excited transducer elements. As far as variations in absorption coefficients from tissue to tissue are concerned, ultrasound absorption is high in skin and bone, intermediate in muscle, and low in fatty tissue.

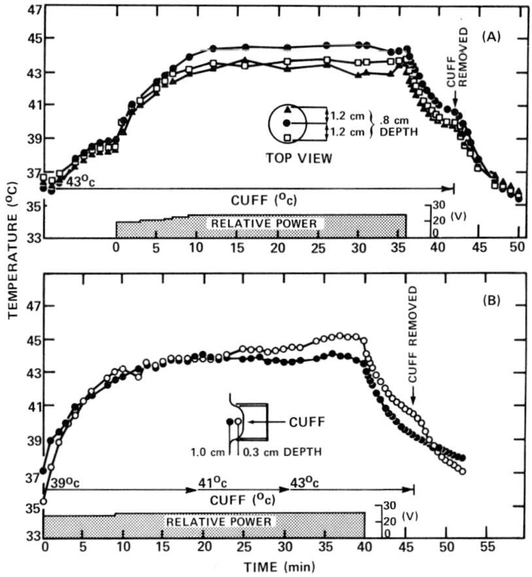

FIG. 6. Temperature distribution in a human tumor during hyperthermic treatment. In panel a three thermocouples were used to measure temperature in a plane parallel to the skin. The lower temperature found in the periphery is typical of temperature distributions in tumors heated by transducers having a relatively uniform intensity pattern. Panel b shows the vertical temperature distribution, which is a composite of the cooling effect of the water and of the heating of the tumor by deposition of energy from the sound wave.

The temperature distribution arising from this energy deposition is also affected by the rate at which heat is lost, be it via radiation from the skin, heat conduction, or heat convection (by blood flow or, in the lung, by breathing) or perspiring. Blood flow seems to play a central role; perhaps as much as 80% of heat deposited in tissue is carried off by blood. During heating by ultrasound, tissue having low blood flow should, therefore, rise to higher temperatures than adjacent tissue more effectively cooled by blood flow. Tumors are frequently reported as having lower rates of blood flow than normal tissue (see Hahn 1978, for review).

Ultrasound equipment developed by us specifically for the induction of hyperthermia utilizes conventional signal generators and amplifiers (Hahn and Pounds 1976). Electrical energy is converted to ultrasound by a piezo-electric crystal. The ultrasound so generated then passes through a water jacket that has two functions. Degassed water acts to keep the deformable membrane at a preset temperature so that the skin, which is in direct contact with the membrane, can be kept at a safe temperature. The other function of the water is to introduce some distance between the transducer and the tissue. Rapid oscillations of ultrasound intensity in the near field of the transducer could introduce dangerous hot spots in cutaneous or subcutaneous regions. These hot spots are avoided by introducing the water path between the transducer and the skin.

The actual ultrasound profile from such a transducer (as measured in a test tank) is compared in Figure 5 to the temperature rise produced by the same transducer in vivo (Marmor et al. 1979b). In spite of considerable point-to-point variation in the intensity of the irradiation, the temperature profile is uniform. Probably, redistribution of heat by blood flow and some scattering of the sound wave mask the intensity nonuniformity of the transducer. Figure 6 shows the typical heat distributions measured in many superficial tumors. The center of the tumor rises to a temperature somewhat higher than that of

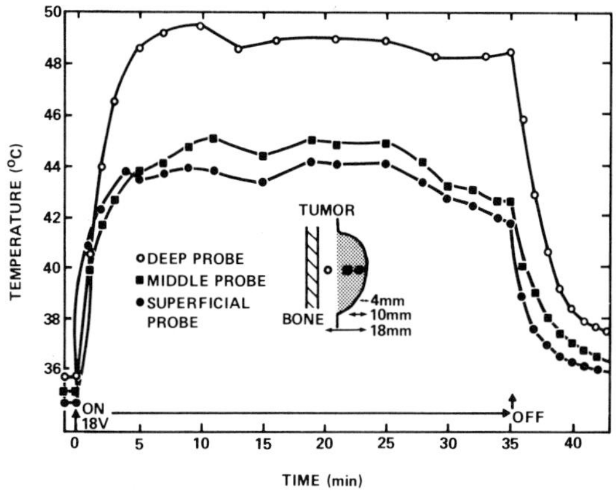

FIG. 7. Modification of temperature distribution by bone. Reflection of the ultrasound wave results in additional deposition of energy above the bone. As a result, temperature in this region rises above that encountered in the tumor. This temperature rise is frequently accompanied by sensation of pain during the treatment (see Table 5).

the periphery, probably because there blood flow is more efficient in providing cooling. The variation of tumor temperature with depth away from the skin is minimized by appropriate cooling of the tumor surface by the circulating water. In Figure 7, we show the effect on the depth profile of a bone underlying the tumor. In this case the temperature below the tumor can exceed that of the tumor, owing to the additional energy deposited by the reflected wave.

EFFECTS OF ULTRASOUND-INDUCED HYPERTHERMIA ON TRANSPLANTED RODENT TUMORS

Several lines of experimental evidence indicate that hyperthermia, independently of how it is induced, should have beneficial effects against tumors: (1) Malignant cells are sometimes (although certainly not uniformly) found to be more heat sensitive than their normal counterparts. (2) Tumors very likely contain cells that are at low pH (<7.0) and in regions containing inadequate nutrients; such cells have been shown to be very heat sensitive (Gerweck and Rottinger 1976, von Ardenne 1978, Hahn 1974). (3) As has been discussed earlier, low blood flow in tumors may cause them to rise to higher temperatures than surrounding normal tissue when both tumor and normal tissue are heated by ultrasound (or electromagnetic) waves.

These expectations are met in practice. Several reports exist on eradication of some rodent tumors by hyperthermia (Overgaard and Overgaard 1972, Mendecki et al. 1976, Dickson et al. 1977, Marmor et al. 1977, 1979a). Other tumors cannot be cured; however, regression or growth delay is observed almost universally. Thus, in our own investigations, we found that two tumors syngeneic in Balb/cKa mice (EMT-6 and KHJJ) were cured easily by a single heat exposure, while two other tumors (KHT and RIF-1), syngeneic in C3H mice, were cured only rarely even when exposed to multiple heat treatments.

Of major interest to the present discussion is the question of whether or not the mode of heat induction influences cure rates. In particular we ask: does ultrasound have demonstrable nonthermal antitumor effects? In order to attempt to obtain an answer, we heated EMT-6 tumors by two modalities: 13.65-MHz electromagnetic diathermy and 3-MHz ultrasound. In each case, the tumor temperature was monitored during heating; other parameters such as duration of heating, etc., were all matched so that meaningful comparisons could be made. The results are summarized in Table 1 and clearly show that ultrasound was *not* more effective than the other form of heat in curing this tumor. In fact, the cure rate utilizing RF heating is clearly superior to that seen with ultrasound, if the comparison is made at maximum tumor temperature, though most of the difference is removed if *minimum* tumor temperature is used. It should be pointed out that in these studies the ultrasound intensity never exceeded about 1 W/cm^2, i.e., it was well under the threshold for nonthermal effects seen in the tissue culture study. The intensity value is dictated by energy absorption requirements (typically 300 mW/cm^3) to maintain tempera-

TABLE 1. *Cure rate of EMT-6 tumors*

Mode of Heating	Tumor Temperature (°C) Maximum	Minimum	Duration (Min)	Cure Rate (%)
RF	44	44	30	100
US	44	43	30	30
RF	43.5	43.5	30	85
US	43.5	42.5	30	30
RF	43.0	43.0	30	50
US	43.0	42.0	30	0
RF	43.0	43.0	45	80
US	43.0	42.0	45	20
RF	43.5	43.5	45	85
US	43.5	42.5	45	60
RF	44.0	44.0	45	100
US	44.0	43.0	45	75
US	44.5	43.5	45	95

RF, 13.56 MHz capacitively coupled heating; US, 3.0 MHz ultrasound heating.

ture. Absorption coefficients are much lower at lower frequencies; hence, to obtain such a level of absorbed energy, much higher incident sound intensities will be required. We are currently developing an ultrasound unit to operate below 500 KHz. It will be interesting to see if we will be able to demonstrate nonthermal antitumor effects with this instrument.

The high rate at which EMT-6 tumors were cured by elevated temperatures was not entirely a consequence of direct killing of tumor cells (Marmor et al. 1977). This was shown by experiments in which cells from tumors, excised immediately after being given a curative heat exposure, were prepared as single-cell suspensions by appropriate enzymatic treatment. Cells were then plated, and their ability to form colonies assayed. It was demonstrated that at that time about 10% of these cells still retained their capacity for unlimited proliferation. Based on experiments with other modalities, such high cell survival should have resulted in only a transitory delay of tumor growth. For example, treatment of this tumor with X irradiation did not lead to cures until cell survival was reduced to below 10^{-3} (Rockwell and Kallman 1973). A careful study could not demonstrate any differences in the abilities of mice "cured" of their EMT-6 tumors by either X irradiation or heat to reject subsequent challenges with EMT-6 cells (Nager, Marmor, and Hahn, unpublished data). Thus, the attractive hypothesis that heat invoked a heightened cell-specific immune response was not substantiated. However, a morphological and ultrastructural study of heated tumors showed that, in addition to tumor cell damage (and particularly damage to plasma membranes), there was evidence of extensive destruction of blood vessels within the tumor. These results indicate that probably at least some cell killing after heating resulted from ischemia secondary to vascular damage (Fajardo et al. 1979).

EFFECTS OF ULTRASOUND-INDUCED HYPERTHERMIA ON SPONTANEOUS TUMORS IN PETS AND IN HUMANS

The eradication of transplantable mouse tumors is frequently only of marginal relevance to the treatment of cancer in man. This is particularly true for localized hyperthermia. Tumors can be implanted into sites where heating is convenient; small rodent tumors are easily raised to relatively uniform temperatures. Finally, some mouse tumors appear to be exquisitely sensitive to heat. For these reasons, we initiated a study of the responses of surface-accessible tumors in dogs and cats to hyperthermia induced by ultrasound (Marmor et al. 1979b). Tumor responses from that study are summarized in Table 2. In Table 3, we show the results of a modified phase I study in humans suffering from a variety of advanced malignancies (Marmor et al. 1979c). The two results are markedly similar, showing that the dog model has excellent predictive value. This is also shown by the comparison of adverse effects seen in the animal study when these are compared with those seen in the clinic (Tables 4 and 5).

CONCLUSION

Cell killing by ultrasound appears to have both a thermal and a nonthermal component. Only part of the latter can be attributed to cavitation phenomena.

In tumors the observed response after ultrasound heating is similar to that seen after heating by other modalities, provided the minimum tumor temperature is used for comparison. Direct cell killing by hyperthermia is followed by additional cell inactivation owing to ischemia, at least in the EMT-6 tumor system. Thus, the suggestion by Dittmar made in 1949 appears to be borne out by the current results.

TABLE 2. *Treatment of spontaneous tumors in dogs and cats with ultrasound*

Pathology	Courses	Animals	Complete Regression	Partial Regression	No Effect
Squamous cell carcinoma	8	7	3	4	1
Fibrosarcoma	5	5	0	1	4
Perianal/adenocarcinoma	3	3	0	1	2
Mastocytoma	2	1	0	2	0
Osteogenic sarcoma	1	1	0	0	1
Mammary carcinoma	1	1	0	0	1
Liposarcoma	1	1	0	0	1
Hemangiosarcoma	1	1	0	1	0
Total	22	20	3	9	10
Percent (all courses)			14	41	45

TABLE 3. *Objective tumor responses in human patients following localized ultrasound hyperthermia of superficial human neoplasms*

Histology	Primary Site	No. Patients	No. Courses of Treatment	Responses		
				CR (%)	PR (%)	NE (%)
Squamous cell carcinoma	Head and neck	9	12	1 (8)	7 (58)	4 (33)
	Lung	1	1	0	1	0
Adenocarcinoma	Ovary	2	2	0	1	1
	Breast	1	2	0	0	2
	Uterus	1	1	0	1	0
	Nose	1	1	0	0	1
Lymphomas	DHL	2	2	0	0	2
	Mycosis fungoides	1	2	2	0	0
Melanoma		1	1	0	0	1
Medullary carcinoma	Thyroid	1	1	0	1	0
Neurofibrosarcoma		1	1	0	0	1
Total		21	26	3 (12)	11 (42)	12 (46)

CR, complete response: clinical disappearance of tumor within the heated area; PR, partial response: >50% decrease in tumor volume within the heated area; NE, no effect: <50% decrease in tumor volume, stasis, or growth of tumor within the heated area; DHL, diffuse histocytic lymphoma.

TABLE 4. *Toxicity of ultrasound during treatment of dogs and cats*

Tumor	Site	Treatment Resulting in Burn	Peak Tumor Temperature (°C)	Frequency (MHz)	Toxicity	Duration (weeks)	Healing
Squamous cell	Nose	4th	45	2.970	Superficial burn-small	2	good
Hemangiosarcoma	Flank	1st	46	0.708	Superficial burn-small	<2	good
Fibrosarcoma	Neck	2nd	46	5.000	Superficial burn-small	2	good
Osteogenic sarcoma	Maxilla	2nd	46	5.000	Superficial burn-small	1	
Squamous cell	Lid*	2nd	>46	2.970	Superficial burn-small	3	good
Squamous cell	Maxilla	2nd	49	0.910	Superficial and deep burn	1½	good

* Heating done over styrofoam contact lens.

TABLE 5. *Toxicity of localized hyperthermia in humans*

Toxic Effect	Incidence	(%)
Superficial burn	5/26	(19)
Extensive central tumor necrosis ("shelling out" of tumor)	4/26	(15)
Significant pain: Site		
Skull	2	
Maxilla	1	
Mastoid	2	
Supraclavicular	1	
Arm	2	
Chest wall	2	
Total	10/26	(34)

The future for hyperthermia, and particularly for ultrasound-induced hyperthermia, appears promising. The clinical data obtained by us are especially encouraging because of the minimal side effects associated with the treatments. It must be remembered that clinical hyperthermia currently is at the stage radiation therapy occupied in Roentgen's days. We are only now beginning to develop appropriate equipment for the heating of nonsuperficial tumors. Furthermore, preferential heating of tumors by appropriate manipulation of blood flow is quite feasible. Another area totally unexplored is the utilization of "thermal tolerance" (i.e., the ability of cells surviving one heat exposure to become much more resistant to a second exposure) for the protection of critical normal tissue. Our preliminary experiments indicate that it is quite feasible to induce tolerance in such tissue by a heat dose that does not kill any cells (e.g., 41°C for 30 minutes), followed by tumor treatment four hours later at a time when normal tissue tolerance has reached its maximum expression. The development of new heating equipment and utilization of such favorable biological factors should greatly increase the effectiveness of hyperthermia to the point where it might become a tool in routine clinical practice. However, most likely it will find its niche in combination treatments, with either radiation therapy or chemotherapy.

ACKNOWLEDGMENTS

We are grateful for the help of Dr. Leonard J. Tolmach who participated in the tissue culture studies. We also acknowledge the excellent technical help of D. Kozak, F. Hilerio, N. Hahn, and E. Shiu. This work was supported by grants CA 04542 and CA 19386 from the National Cancer Institute/National Institutes of Health, Department of Health, Education and Welfare.

REFERENCES

Beck, F. F. and J. C. Krantz, Jr. 1940. The effect of ultrasonic vibrations on the growth and glycolysis of Walker sarcoma. Am. J. Cancer 39:245–250.

Dickson, J. A., S. A. Shah, D. Waggot, and W. B. Whalley. 1977. Tumor eradication in the rabbit by radiofrequency heating. Cancer Res. 37:2162–2169.
Dittmar, C. 1949. Uber die Wirkung von Ultraschallwellen auf tierische Tumoren. Strahlentherapie 78:217–230.
Fajardo, L., B. Egbert, J. B. Marmor, and G. M. Hahn. 1979. Effect of hyperthermia in a malignant tumor. Cancer (in press).
Gavrilov, L. R. 1974. Physical mechanisms of the lesion of biological tissues by focused ultrasound. Soviet Physical Acoustics 20:16–18.
Gerweck, L., and E. Röttinger. 1976. Enhancement of mammalian cell sensitivity to hyperthermia by pH alteration. Radiat. Res. 67:508–511.
Hahn, G. M. 1974. Metabolic aspects of the role of hyperthermia in mammalian cell inactivation and their possible relevance to cancer treatment. Cancer Res. 34:3117–3123.
Hahn, G. M. 1978. The use of microwaves for the hyperthermia treatment of cancer: Advantages and disadvantages, *in* Photochemical and Photobiological Reviews, Vol. 3. Plenum Press, New York, pp. 277–301.
Hahn, G. M., and D. Pounds. 1976. Heat treatment of solid tumors: Why and how. Applied Radiology/Ultrasound 5:131–134.
Kim, S. H., J. H. Kim, and E. W. Hahn. 1975. Enhanced killing of hypoxic tumour cells by hyperthermia. Br. J. Radiol. 48:872–874.
Li, G. C., G. M. Hahn, and L. J. Tolmach. 1977. Cellular inactivation by ultrasound. Nature 267:163–165.
Marmor, J. B., N. Hahn, and G. M. Hahn. 1977. Tumor cure and cell survival after localized radiofrequency heating. Cancer Res. 37:879–883.
Marmor, J. B., F. J. Hilerio, and G. M. Hahn. 1979a. Tumor cure and cell survival after localized hyperthermia induced by ultrasound. Cancer Res. (in press).
Marmor, J. B., D. Pounds, N. Hahn, and G. M. Hahn. 1979b. Treating spontaneous tumors in dogs and cats by ultrasound-induced hyperthermia. Int. J. Radiat. Oncol. Biol. Phys. 4:967–973.
Marmor, J. B., D. Pounds, T. B. Postic, and G. M. Hahn. 1979c. Treatment of superficial human neoplasms by local hyperthermia induced by ultrasound. Cancer 43:188–197.
Mauro, F., and H. Madoc-Jones. 1970. Age response of cultured mammalian cells to cytotoxic drugs. Cancer Res. 30:1397–1408.
Mendecki, J., E. Friedenthal, and C. Botstein. 1976. Effects of microwave-induced local hyperthermia on mammary adenocarcinoma in C3H mice. Cancer Res. 36:2113–2114.
Overgaard, K., and J. Overgaard. 1972. Investigations on the possibility of a thermic tumor therapy. Eur. J. Cancer 8:65–78.
Rockwell, S., and R. F. Kallman. 1973. Cellular radiosensitivity and tumor radiation response in the EMT-6 tumor cell system. Radiat. Res. 53:281–294.
Szent-Györgyi, A. 1933. Chemical and biological effects of ultrasonic radiation. Nature 131:278.
Terasima, T., and L. J. Tolmach. 1963. Variations in several responses of HeLa cells to X-irradiation during the division cycle. Biophys. J. 3:11–33.
Tobey, R. A., E. C. Anderson, and D. F. Petersen. 1967. Properties of mitotic cells prepared by mechanically shaking monolayer cultures of Chinese hamster cells. J. Cell. Physiol. 70:63–68.
von Ardenne, M. 1978. Prinzipen und Konzept 1977 der Krebs-Mehrschritttherapie: Pysiologische Grundlagen des neuen Timing. Selectotherm-Localhyperthermie. Arch. Geschwulstforsch. 183:211–215.

Radiation Biology in Cancer Research, edited by
Raymond E. Meyn and H. Rodney Withers.
Raven Press, New York © 1980.

Approaches to Clinical Application of Combinations of Nonionizing and Ionizing Radiations

Peter Corry, Barthel Barlogie*, William Spanos†, Elwood Armour, Howard Barkley†, and Mario Gonzales†

*Departments of Physics, † Radiotherapy, and *Developmental Therapeutics, The University of Texas System Cancer Center M. D. Anderson Hospital and Tumor Institute, Houston, Texas 77030*

The potential application of nonionizing radiation in combination with ionizing radiation in clinical cancer therapy has been a subject of intense interest for many years. Although some nonthermal mechanisms have been proposed for the interaction of both ultrasound (Li et al. 1977, Corry et al. 1976) and UHF radiations (Holt 1977) with biological tissues, there appears to be little concrete evidence that such radiations potentiate the effects of ionizing radiation by mechanisms other than those associated with elevated temperatures, or hyperthermia, in the 41°–50°C range. The basic radiobiology of the hyperthermia-radiation combination has been extensively reviewed by Dewey et al. (1980, see pages 589–621, this volume) and will not be further discussed here.

UHF (434 MHz) radiation has been used by some investigators in combination with ionizing radiation (Holt 1977, Hornback et al. 1977): However, in these studies thermal dosimetry was not attempted, and evaluation of this work is, at best, difficult. Other investigators (Kim et al. 1978) have used induction diathermy for the treatment of human melanoma with radiation and have achieved encouraging results in this traditionally radioresistant tumor. Ultrasound has for some time been under development in our laboratories and others' (Hahn et al. 1980, see pages 623–636, this volume; Marmor et al. 1979) for the treatment of human cancer. It has been demonstrated that this modality can be applied to large numbers of human tumors with few deleterious side effects to generate regional hyperthermia up to 48°C, provided adequate measures are taken for skin cooling. This modality, used by itself, has shown an overall objective response rate of approximately 50%; however, our experience has been that these responses have been of relatively short duration (four to eight weeks) (unpublished data). For these reasons we have initiated a study combining ultrasonic regional hyperthermia in combination with ionizing radiation for human tumor therapy.

There is considerable controversy concerning the heat-radiation sequence that will optimize the therapeutic ratio. Basic radiobiological data (Dewey et al. 1980, see pages 589–621, this volume), show that maximum potentiation of radiation by hyperthermia occurs when the two modalities are applied simultaneously. Stewart and Denekamp (1978) have shown that for mouse tumor–bearing limbs immersed in a water bath the optimum therapeutic ratio (skin effects vs. tumor regression) is obtained when the radiation is administered prior to heating, an effect that is attributed to decreased repair capabilities in the tumors. It should be noted, however, that in this latter case the normal tissue temperatures were greater than, or equal to, the tumor temperatures at all times, maximizing normal tissue reactions relative to antitumor effects. It is interesting to note that in one of the few studies in mouse tumors in which methods were employed to generate a differential local tumor hyperthermia (Overgaard 1977), thermal enhancement ratios of up to 3.5 were observed with little effect on normal tissues; those employing water bath immersion in many cases report values of less than 1.0 (Stewart and Denekamp 1978). Corry et al. (1977) have demonstrated that, for temperatures as low as 41°C, repair in cells heated prior to radiation under simulated tumor conditions is markedly reduced compared to repair in cells heated under normal tissue conditions, in which repair proceeds at the same rate as in 37°C controls. This observation, coupled with the fact that the methodology described generates a tumor temperature of 43°–44°C while normal tissue temperatures seldom exceed 40°C, led us to choose the heat-radiation sequence rather than the reverse. Simultaneous heat and radiation were precluded for practical reasons.

METHODS

Specially designed ultrasound transducers operating in the 1–3 MHz range with diameters of 2.5–7.5 cm (Figures 1 and 2) and with the appropriate electronic circuitry (Figure 3) were employed to generate the local tumor hyperthermia. This system does not differ significantly in principle from that described by Hahn et al. (1980, see pages 623–636, this volume). The transducers are strapped in position over the tumor-bearing area and held in place for one hour immediately prior to each fraction of radiation administered. The tumor

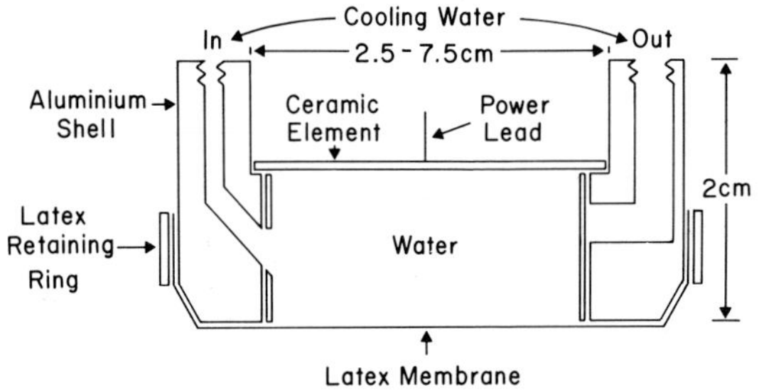

FIG. 1. Diagrammatic representation of ultrasound transducers used for local tumor hyperthermia. Transducer elements were PZT-4 ceramic discs with resonant frequencies between 1 and 3 MHz.

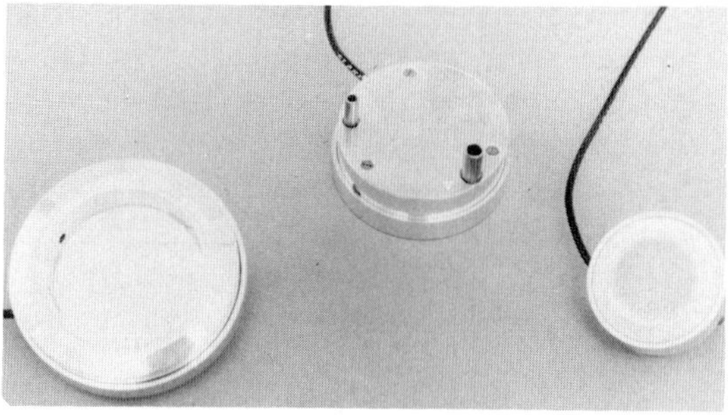

FIG. 2. Pictorial representation of the 2.5-, 3.75-, and 5.0-cm transducers.

temperature was maintained at 43.5°C ± 0.5°C as monitored by implanted thermocouples. Input power was adjusted to maintain this temperature for the one-hour period, usually resulting in an output power of 0.3–0.6 W/cm^2 depending upon tumor size and location. Skin cooling was accomplished by pumping water (30°C) over the surface of a latex membrane in contact with the skin, which also served to remove the skin from the near field of the transducer and cool the transducer itself. Transducer efficiency was approximately 50%.

Ionizing radiation from clinical ^{60}Co sources was administered at 400 to 500 rad per fraction on alternate days, three days per week, for a total dose of 2,400 or 2,500 rad unless otherwise indicated. Hyperthermia was administered immediately prior to each radiation fraction. The total dose was conservative because insufficient data are available in the literature to determine the possible toxic side effects of this combined modality. The heat-radiation sequence was

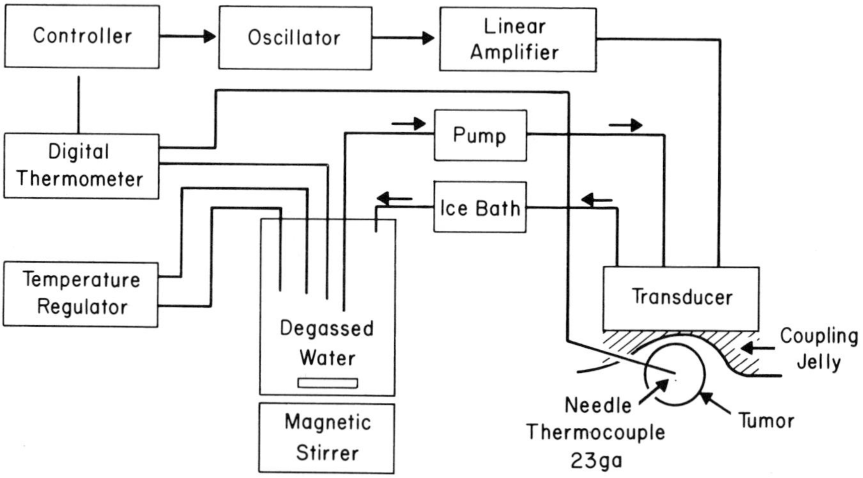

FIG. 3. Schematic diagram of the entire system used to drive the ultrasound transducers. Degassed water regulated to ± 0.2°C is pumped through the transducer to effect skin cooling and to cool the element itself.

chosen partly for practical reasons (cobalt machine availability) and partly from radiobiological factors discussed above. Tumor masses ranged from estimated sizes of 50 gr to 2 kg. It was therefore impossible in many cases to heat the entire tumor in one session. In these cases the transducer was moved to different positions prior to each radiation fraction, and the effective tumor volume heated prior to each radiation fraction varied from 20% to 100%.

RESULTS

To date, nine tumors in eight patients have been treated with the regimen described. All eight patients had failed to respond to conventional modes of cancer therapy. The types of lesions and a summary of the responses are given in Table 1. The responses of two large lesions, one leiomyosarcoma and one melanoma, are shown in Figures 4 and 5, respectively.

The patient with a leiomyosarcoma (Figure 4) was treated twice with ultrasound alone (six one-hour treatments at 44°C over two weeks). He responded objectively on each occasion but relapsed within six to eight weeks. Hyperthermia was administered with a 5-cm, 3-MHz transducer to approximately 40% of the tumor at each treatment session. Upon failure of regional hyperthermia alone, combined therapy of 2,500 rad in five fractions in a 12-day period was administered. The lesion showed a dramatic response, qualifying as a partial response (PR, 50% reduction in mass) within two weeks of beginning this regimen and a complete response (CR, nonpalpable disease) within six weeks. The treated area remained free of palpable disease for eight months; however, some disease has now occurred just superior to the margins of the treatment field. This patient showed immediate and marked relief of previously intractable pain.

The patient shown in Figure 5 suffered from disseminated, highly pigmented

TABLE 1. *Summary of patients treated with US and XRT*

Diagnosis	# Responses (CR)/Treated Lesions		
	US + XRT*	XRT*	US
Melanoma	5 (2)/5	2 (0)/5	3 (1)/4
Leiomyosarcoma	1 (1)/1	—	1 (0)/1
Adenocarcinoma, lung	1 (0)/1	1 (0)/1	0 (0)/2
Adenocarcinoma, cervix	1 (1)/1	—	—
Giant cell sarcoma	1 (TE)/1	—	0 (0)/1
Total responses	9 (4)/9	3 (0)/6	4 (1)/8
Percent responses	100	50	50

* XRT regimens equivalent in dose and schedule. XRT alone lesions in same radiation field as lesions treated with US + XRT. US, ultrasound; XRT, radiotherapy; TE, too early for evaluation for complete response.

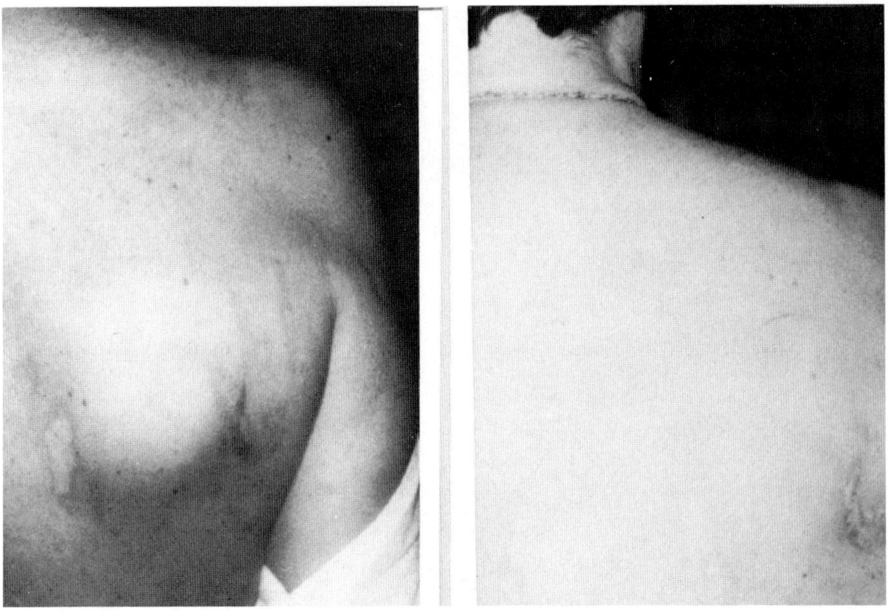

FIG. 4. A large, 10 × 8 × 6 cm leiomyosarcoma treated with combined therapy. A, left, immediately prior to treatment; B, right, six weeks subsequent to therapy.

malignant melanoma. One lesion (5 × 4 cm) was treated with ultrasound and showed a complete response. A large neck lesion was treated with combined ultrasound and radiotherapy of 2,400 rad in six fractions immediately preceded by a one-hour treatment with a 2.5-cm diameter transducer operating at 1.7 MHz. This lesion showed immediate subjective signs of response, local edema, and tenderness, and completely regressed over a 10-week period following cessation of therapy. Two other lesions treated by radiotherapy alone (6 × 400 rad) failed to qualify as PR. The two lesions demonstrating complete regression remained unpalpable until the patient's death six months later from pulmonary metastases. No postmortem examination was conducted. This response was particularly remarkable in view of the fact that this patient was treated at a time when our largest available transducers were 2.5 cm in diameter, effectively heating only 20% of the tumor prior to each radiation fraction.

The largest tumor treated, a giant cell sarcoma, weighed approximately 2 kg. This tumor was treated by two 5-cm ultrasound fields, effectively heating 20–25% of the mass, prior to each 400-rad radiotherapy fraction. The total dose in this case was 3,200 rad. Two months following therapy, this tumor measures 25% of its original mass and is continuing to regress.

Six of the treated lesions showed acute (in three to six days) local edema and tenderness in the treated area, which usually disappeared within one week of cessation of therapy. One patient showed significant skin blistering, which

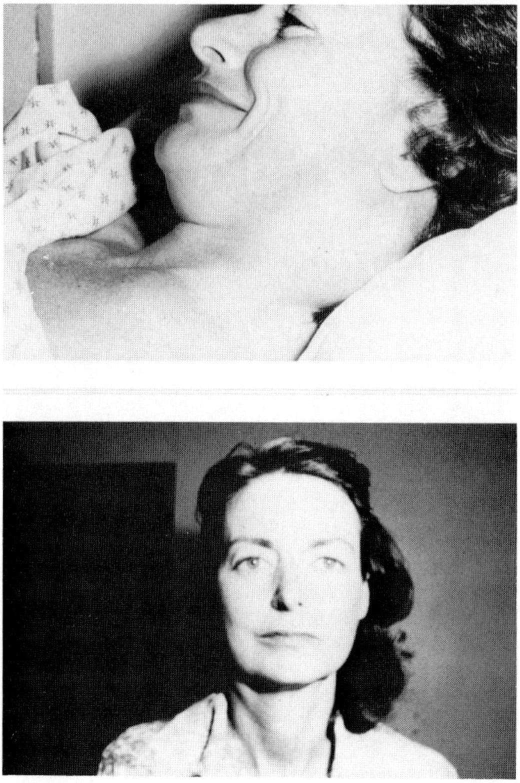

FIG. 5. Results of treatment of a highly pigmented subcutaneous melanoma. A, top, prior to therapy; B, bottom, six months subsequent to combined therapy. The $4 \times 4 \times 5$ cm supraclavicular mass seen in A was treated with ultrasound alone and completely regressed. The large neck mass was treated with combined therapy. At six months the treated areas were free of obvious palpable disease.

was the result of a technical problem with the skin cooling system. Three of the patients with lesions overlying bone experienced pain during ultrasound application that disappeared shortly after cessation of therapy. No significant acute or chronic complications have been observed to date.

Table 1 also shows the results of single-modality therapy with radiation and ultrasound for those patients receiving combined therapy. Both modalities showed an objective response rate of 50% by themselves. The response rate for ultrasound alone is comparable to that observed in larger series of patients by both ourselves and Marmor et al. (1979). Interestingly, some of the highly pigmented melanomas respond very rapidly to ultrasound therapy, whereas lightly pigmented melanomas respond similarly to other unpigmented lesions, with a lower incidence and speed of response. The duration of response for the ultrasound-treated lesions was short, four to six weeks, while the median duration of response for tumors treated with combined therapy has not yet been reached for the five evaluable patients (with responses beginning two to seven months ago). Three patients (all with melanoma) were lost to follow-up due to death within six weeks of therapy. We may, in fact, be underestimating the complete response rate at this time because of the long-term shrinkage of treated lesions. All patients treated have been included in these figures.

DISCUSSION

The responses seen with the combined treatment involving ultrasound heating and ionizing radiation have been quite remarkable considering the relatively low radiation doses employed. This is particularly true for the classically radioresistant melanoma and in the larger tumors where only a fraction of the tumor received adequate hyperthermia at any given session. While only estimates are possible, the tumor dose-modifying factors would appear to exceed 3 in some cases, a value that could not be predicted from existing published radiobiology results alone. Thermal enhancement ratios are not obtainable in our series since no quantifiable skin or other normal tissue toxicities have been observed to date.

One possible explanation for the apparently high dose-modifying factor involves considerations of tumor physiology. Bulky, solid tumors tend to have hypoxic centers with a lower pH as well as a lower nutrient supply than normal tissue. These conditions become more pronounced as tumors become larger. When hyperthermic exposures are carried out at a pH between 6.5 and 7.0 (Dewey et al. 1980, see pages 589–621, this volume) and under lowered nutrient supply (Corry, unpublished data), an increase in radiation sensitivity is noted for temperatures between 41° and 44° C. These data, however, do not seem to account for dose-modifying factors much in excess of 2.0. Enhanced local immune responses may account for some of the observed antitumor effect.

Ultrasonic hyperthermia may be safely applied in previously irradiated fields without acute or chronic normal tissue toxicity provided adequate measures are taken to cool the skin. This fact, coupled with the good responses we have observed at low doses of radiation for combined therapy, may afford further treatment possibilities for patients who have local recurrence following radiation therapy. We are now investigating the treatment of such lesions with ultrasound alone as well as with reirradiation plus ultrasound. Another promising approach is the use of regional hyperthermia in combination with drugs such as *cis*-platinum, which have been shown to be synergistic in vitro and in animal tumors (unpublished data).

ACKNOWLEDGMENTS

The authors wish to express their appreciation to Janet Cook and Alice Armour for their perseverance in the administration of the ultrasound therapy, and Bailey Moore and Robert Goga for the construction of ultrasound transducers. This research was supported by Grants CA 17891 and CA 11520 from the

U. S. National Cancer Institute. Barthel Barlogie is a junior faculty Fellow of the American Cancer Society.

REFERENCES

Corry, P. M., J. E. McGinness, and E. Armour. 1976. Semiconductor properties of the melanins related to preferential killing of melanoma cells. Pigment Cell 2:321–326.

Corry, P. M., S. Robinson, and S. Getz. 1977. Hyperthermic effects on DNA repair mechanisms. Radiology 123:475–482.

Dewey, W. C., M. L. Freeman, G. P. Raaphorst, E. P. Clark, R. S. L. Wong, D. P. Highfield, I. J. Spiro, S. P. Tomasovic, D. L. Denman, and R. A. Coss. 1980. Cell biology of hyperthermia and radiation, *in* Radiation Biology in Cancer Research (The University of Texas System Cancer Center 32nd Annual Symposium on Fundamental Cancer Research), R. E. Meyn and H. R. Withers, eds. Raven Press, New York, pp. 589–621.

Hahn, G. M., G. C. Li, J. B. Marmor, and D. W. Pounds. 1980. Thermal and nonthermal effects of ultrasound, *in* Radiation Biology in Cancer Research (The University of Texas System Cancer Center 32nd Annual Symposium on Fundamental Cancer Research), R. E. Meyn and H. R. Withers, eds. Raven Press, New York, pp. 623–636.

Holt, J. A. G. 1977. Fundamental differences between hyperthermia and 434 mHz effects, thermic and non-thermic effects. Proceedings of a Conference on Clinical Prospects for Hypoxic Cell Sensitizers and Hyperthermia. Madison, Wisconsin.

Hornback, N. B., R. E. Shupe, H. Shidnia, B. T. Joe, E. Sayoc, and C. Marshall. 1977. Preliminary clinical results of combined 433 mHz microwave therapy and radiation therapy on patients with advanced cancer. Cancer 40:2854–2863.

Kim, J. S., E. W. Hahn, and N. Tokita. 1978. Combination hyperthermia and radiation therapy for cutaneous malignant melanoma. Cancer 41:2143–2148.

Li, G. C., G. M. Hahn, and L. J. Tolmach. 1977. Cellular inactivation by ultrasound. Nature 267:163–165.

Marmor, J. B., D. Pounds, T. B. Postic and G. M. Hahn. 1979. Treatment of superficial human neoplasms by local hyperthermia induced by ultrasound. Cancer 43:188–197.

Overgaard, J. 1977. The effect of local hyperthermia alone and in combination with radiation on solid tumors. Second International Symposium on Cancer Therapy by Hyperthermia and Radiation, Essen, West Germany.

Stewart, F. A., and J. Denekamp. 1978. The therapeutic advantage of combined heat and x-rays on a mouse fibrosarcoma. Br. J. Radiol. 51:306–316.

Radiation Biology in Cancer Research, edited by
Raymond E. Meyn and H. Rodney Withers.
Raven Press, New York © 1980.

Symposium Summary

John F. Fowler

Gray Laboratory of the Cancer Research Campaign, Mount Vernon Hospital, Northwood, Middlesex HA6 2RN England

When the princes of ancient Europe (before it was united and became Common) sent envoys to other courts they instructed them carefully. They asked the envoys to find out two things: What people are arguing about and what they are agreed upon. It was easy to find out the first, but not the second. Nobody is interested in discussing what is agreed upon.

And so it is in radiobiology, usually. This meeting was a useful exception. As well as the arguments, there have been some masterly reviews of what we are supposed to be agreed on. I have only time to pick out certain points that strike me personally. I apologize to other authors in advance and hasten to refer the reader to the appropriate chapters. This summary can only present a few random comments condensed from the cloud of ideas that floats in our collective consciousness like DNA in intermitotic cells, awaiting synthesis and visualization.

SURVIVAL CURVE MODELS

Dr. Alper, in her keynote paper, states firmly that all survival curves are basically one-hit exponentials. Shoulders at low doses were due to repair of sublethal injury, the repair decreasing with dose as if some "Q-factor" is used up. Any shape survival curve could be explained by various rates and efficiencies of using up or saturating this repair. Drs. Tobias and Goodhead supported this idea strongly, Dr. Tobias with a detailed model. I feel that this type of "repair" model is preferable to the "physical" models such as multitarget or $\alpha D + \beta D^2$. The new, and definitive, evidence comes from Dr. Goodhead. Ultrasoft X rays from a carbon target, generating a 300-keV electron with a range of 7 nm, were *more* effective than gamma rays. If cooperation between two sublesions several hundred nanometers apart were necessary, as in the dual-action theory, the ultrasoft X rays would have been *less* effective. So the dual-action theory appears to be excluded for the killing and mutagenesis of Dr. Goodhead's HF 19 cells. However, Dr. Rossi defended the dual-action theory. Why is there so much data that fits the dose-squared relationship? Is it the trivial reason that a polynomial fits most dose-response curves and that two terms are enough? Or could it be that the decrease of repair with dose requires a second event in some of the processes? If the latter is true, our intellectual gap is closed, but that requires more detailed working out. Although the repair models seemed preferable, the dual-action theory will not die easily. The point was made on the first day that no theory can ever be finally proved; it is only that some can be disproved.

BIOPHYSICAL TARGETS

I was pleased to hear Dr. Chapman say that he thought the critical target volume for cell killing was about the size of a nucleosome (5 nm) and to hear Dr. Goodhead's evidence that the target diameter was much smaller than the "interaction distances" of 0.2 to 1 μm often mentioned by Drs. Rossi, Kellerer, Neary, and others. He calculated 3 nm from his own elegant results using ultrasoft X rays. I was pleased because Barendsen, Walter, Fowler, and Bewley (1963) found, from track segment experiments, that the thickness of targets for sterilizing human kidney cells in culture was about 5 nm, i.e., about nucleosome size. Dr. Rossi also appeared to give this small size his seal of approval, saying ". . . recent evidence points to a distribution of interaction distances which is highly skewed towards short ranges of interaction." That is a remarkable convergence, but it's not yet full agreement between the physical two-process and the repair models.

Small volumes are the important ones. Even for indirect damage from water radicals the range is only a few nanometers, as Dr. Chapman writes here and Dr. Hutchinson did some years ago. We are not observing cooperation between tracks at low linear energy transfer (LET), but damage caused by a few hundred electron volts at the ends of electron tracks, within a small volume of 3–5 nm, followed by repair. The repair processes are highly active for single-strand breaks (SSB) and rather less so for double-strand breaks (DSB): 0.75% and 10% remaining "unrepaired" after low-LET irradiation, 10% and 40% after alpha-particle irradiation (Dr. Cole).

MEMBRANE-DNA COMPLEXES

Nuclear-membrane-associated DNA complexes have been the popular choice for critical radiation damage for some years, following Dr. Elkind, Dr. Arthur Cole, and Dr. Alper. DSB in DNA are often said to be the main lesion, with misrepair being one apparent mechanism causing the injury. The "total number of unrejoinable breaks" is a more useful concept than the initial number of SSB or DSB. The papers by Drs. Alper, Elkind, Painter, Cole et al., and Tobias dealt with these matters. Dr. Tobias reminded us that some of the cells with misrepaired damage die, but others may survive and become mutants or malignant cells. A new meaning was given to "U" and "non-U," U meaning "uncommitted" here.

TIME SCALE OF CHROMOSOME DAMAGE REPAIR

Dr. Hittelman and co-workers describe applications of a beautiful technique in which interphase cells were fused with others that were just about to divide, so that premature condensation of the chromatin occurred. This has obviously opened the door to studies of the time course of chromosome damage. This approach has confirmed that only a small proportion of the breaks in DNA result in chromosome aberrations: the rest are repaired or in redundant DNA. He describes a fast component of SSB repair—six minutes—which would act during an irradiation of ordinary duration.

Dr. Tolmach describes how caffein seemed to shield radiation-damaged sites from being recognized by the cell, so that division delay was less with caffein. But the expression of the injury was only postponed, not prevented.

DNA BREAKS IN VIVO

Dr. Meyn and co-workers show that fibrosarcoma cells irradiated in vivo had fewer breaks than similar cells irradiated in vitro, in agreement with Durand and Sutherland's earlier (1972) observation of wider shoulders for cells irradiated as spheroids than as single-cell suspensions. It also agrees with Dr. Alper's and others' observations in 1963 that repair of sublethal X-ray injury in vivo was larger than in vitro (Fowler et al. 1963). Meyn and co-workers also referred to the inhibition of DNA breaks in hypoxic cells in vivo, which must be a more relevant result to mammalian radiobiology than the lack of oxygen effect for DNA irradiated "bare" in vitro. The technique of alkaline elution is an exciting one for investigating SSB. We look forward to neutral elution, which might enable DSB to be measured. As Dr. Cole writes, we can expect rapid advances in knowledge on these topics within a few years.

OXYGEN EFFECTS

Drs. Ewing and Powers dazzle us with the virtuosity of three oxygen effects, each being counteracted by a different protective compound. This work was done using both bacterial spores and mammalian cells and follows Dr. Powers' distinguished unraveling some years ago of several oxygen effects, separated by moisture content, LET, etc. The question now is: Does oxygen sensitize all cells by the same multiple pathways?

INSECT RADIOBIOLOGY

Why are cells from the cabbage looper insect 30 times more radioresistant than cells from hamsters? Dr. Koval invokes a greater degree of unscheduled DNA synthesis in insects, but he has to add redundant DNA, or *error-free* replication, to explain the whole thing.

LOW DOSES AND DOSE RATES

Low dose rates, between 55 and 154 rad/hour, yielded new results in the tour de force by Drs. Bedford and Mitchell. After a thorough investigation of HeLa and V79 cells in exponential growth and plateau phase, they also investigated six other cell lines in vitro. They found that the threshold dose rate required to stop cell division correlated only with the division delay caused by an acute dose of radiation, and not with cell cycle time, chromosome number, or acute radiosensitivity. They observed *more* cell killing at a *lower* dose rate, for a given dose, if cells continued to proliferate to any extent and so accumulated in a sensitive phase.

Dr. Brooks used hepatectomy to trigger G_0 liver cells into proliferation after known doses of various nuclides distributed in the tissues. He found a constant radiosensitivity of 3×10^{-4} chromosome aberrations per rad for low dose rates ranging from 0.5 to 250 rads per day. The situation was much more complex in normally proliferating tissues. The relative biological effectiveness (RBE) for alpha particles was 20. A cancer incidence of 18% followed 0.3–0.6 aberrations per cell.

Dr. Gaulden reviews the data on effects of low doses of X rays on embryos and the reasons for concern if a human embryo was irradiated during the first six weeks of development. The effects on the central nervous system, leading to mental retardation, were the most important because many chromosomes contributed. She puts forward a strong working rule: if there is less than 1 rad exposure to the mother's abdomen, forget it. If there is more than 10 rads, abort the fetus. Dr. Gaulden also gave, in response to a question, cogent reasons for continuing the 10-day rule for radiological examinations of women of child-bearing age.

In the chapters on mutagenesis (Drs. Russell and Kohn) and carcinogenesis (Drs. Little and Kennedy and Dr. Ullrich) we have agreement that there are no simple models relating the two to dose. Quantitative estimates of risk are premature. The reasons given are that a multiplicity of factors intervene between radiation damage and malignant transformation; that few mutation loci have yet been studied in mice; and that big variations have been found. Premature or not, someone must set down something about the risks to live by and so far this has been accomplished by International Commission on Radiological Protection and the National Council on Radiation Protection and Measurements in their pragmatic way.

TUMORS

Tumor radiobiology is notorious for its wide spectrum of results. Dr. Barendsen reports that his tumor lines, which grow both in vivo and in vitro, differ by a factor of three in acute radiosensitivity to high doses of radiation, but by a factor of five in their initial slopes, on the "shoulder" between 100 and 300 rad. The factor of five variation applies, therefore, to fractionated irradiation. He gives interesting results for the hypoxic fraction during fractionated irradiation: it was much the same at the end as at the start and often higher in the middle. He points out that "a large number of parameters have to be investigated to provide an adequate interpretation of tumor response," and he lists their relative size. However, this list contains no comparison or contrast factor with the same parameters in normal tissues, and if it had, hypoxic cells in tumors would have scored higher. As it was, "repair" scored highest.

Dr. Weichselbaum and co-workers present another tour de force with 12 different human tumor cell lines grown in vitro. Although the acute X-ray cell survival curves were all similar, the amount of repair of potentially lethal damage (PLD) was very different, and this correlated well with clinical radioresistance. In our laboratory, McNally and Sheldon (1977) showed that repair of PLD in the MT murine tumor occurred in chronically hypoxic cells only and was an enormous artefact in the interpretation of experimental results. This leads us to suggest that no animal tumor treatment results can be believed unless the effect of PLD has been measured and allowed for in that system.

Another technical feat is described by Dr. Grdina. Using cell elutriation on a density gradient, he found that the lightest cells were the most clonogenic and that heavier cells appeared to be the hypoxic ones. It would indeed enable many important questions to be answered if we could measure, reliably, how many cells in a tumor were hypoxic at various stages of fractionated treatments using, for example, neutrons or radiosensitizers. How fast does reoxygenation occur? How big a problem are hypoxic cells in tumors?

Drs. Stephens and Steel demonstrate that their experimental solid tumors repopulated more slowly after cyclophosphamide than after X rays or nitrosoureas that killed the same proportion of cells. The cells proliferated more slowly too. It was clear that after X irradiation, repopulation of tumor cells can be faster than volume growth. This has to be distinguished from the influx of cells into the tumor from outside. In their B16 melanoma, more than half the cells were macrophages that were "topped up" from outside very quickly after depletion. It was clear that no simple relationship exists between cell survival and tumor growth delay, although a given tumor system may be calibrated after enough experience and data.

Drs. Kallman and Dethlefson described the problems of assessing how many cells in a tumor were proliferating (P) or quiescent (Q). Dr. Kallman described some impressive automation for counting labeled colonies; we look forward to the outpouring of results. Dr. Dethlefson's "quest for the quaint quiescent cell" was quite queerly quantivalent in its quondam quality but his query included both a quatrain (verse of four lines) and a quaternion (set of four aspects). It was in fact nearest to a quintain: the mediaeval practice of tilting, with a lance, at a sandbag hanging from a post, in which there is a risk that the bag will swing round and knock the knight off his horse.

NORMAL TISSUE RADIOBIOLOGY

Vascular Injury

From tilting at sandbags we now move over to a bullfight. Dr. Hopewell demonstrates a good correlation between late damage to pig skin and the degree of *initial* vascular impairment. He suggests that focal occlusions in arterioles at three months might be "the primary vascular lesion responsible for the development of late normal tissue damage."

Dr. Withers and colleagues conclude, however, that "although vascular injury may play a role in the late manifestations of radiation injury, it is unlikely to be the sole, or even major factor." Work done recently in our laboratory by Drs. Hirst, Stewart, Travis, and Denekamp leads them to agree with Withers et al. Hirst, Denekamp, and Travis (in press) found that vascular injury occurred very late in the mouse mesentary, about 15 months after irradiation; but Stewart, Denekamp, and Hirst (in press) found it in mouse bladders at about 10 months, when proliferation in the bladder epithelium had become pathologically fast. Endothelial cells do not show damage at the appropriate late time but smooth muscle in arterioles does. Dr. Hirst suggests that whatever the critical tissue is it cannot be endothelial cells; perhaps it is smooth muscle. This fits in well with Withers and co-worker's conclusion that "most likely the late onset of injury in these tissues reflects the slow proliferation kinetics of the parenchymal or connective tissue cells."

Dr. van der Kogel sails in between these protagonists, as becomes a countryman of Admiral Tromp and Admiral Ruyter. He describes two types of lesion in the spinal cord of rats, depending on dose level. At doses in the range of 1,700–2,000 rad single dose, i.e., in the lower part of the "clinical tolerance" range of dose, where radiotherapists prefer to be, the late neurological symptoms seemed to show a vascular origin in capillaries or small venules. He suggests that the endothelial cell was the most probable target for these "late vascular lesions." After higher doses, above 2,000 rad single dose, the damage was indeed parenchymal, to the glial cells. He nearly bridges the gap.

Early versus Late Damage

Drs. Hornsey and Field describe the Hammersmith discovery of "slow repair" as being like Elkind repair but with a time constant of 100 hours instead of about 1 hour. It occurs in some slowly proliferating tissues (e.g., lungs) but not in others (e.g., bladder or spinal cord). It does not occur after cyclotron neutron irradiation. The effect is, of course, to increase the RBE for late damage and to cause more residual injury many months after neutrons than after X rays. Dr. Hornsey also reports dissociation between early and late damage if the overall time was very long. Acute damage was avoided but late damage was not. Re-treatment after neutron irradiation appeared to be more dangerous than after X rays; but from the papers by Drs. Hahn and Corry, hyperthermia could be used instead for re-treatment.

Dr. Hussey and colleagues describe the dissociation of late and early radiation injury. They report results from several end points in monkeys and in pigs, with consistent sets of valuable data. Dissociation happens after neutron irradiation when late RBEs are higher than early ones. However, it does not always occur. Dr. Hussey reports that a narrow band of neutron dose exists between too much late clinical injury and too much local recurrence. The dependence on dose of both these parameters is steeper for neutrons than for X rays.

Other Normal Tissue Studies

Dr. Clifton describes another promising innovation: the use of mammary tissues transplanted into other mice and assayed as "alveolar units" in a biopsy, and thyroid tissues transplanted to give "follicle units." There is clearly much scope for experimentation using these systems, and not only in carcinogenesis, which Dr. Clifton and his colleagues have already started on.

Dr. Peters and his colleagues have explored further the alteration of normal tissues by radiation or cytotoxic drugs so that they would trap and grow metastases better. The time course of the alteration was different after different agents were administered. Radiation gave peak effects at a few days and also at about four months after treatment; cyclophosphamide gave only the early peak and bleomycin only the late one. He gave reasons, along with both laboratory and clinical evidence, that this should not be a big problem in therapy, *provided that the primary tumor is effectively ablated first.* Dr. Van den Brenk and coworkers investigated lowering the resistance of lungs to metastases by measuring the increased number of colonies of tumor cells in the lungs after injection of adrenergic agonists, and inflammatory agents in 1976, including the mechanical stress of bouncing a mouse around inside a rotating box.

Dr. Kerr convinces us that apoptosis is a real and important phenomenon. It is the noninflammatory, natural process by which, for example, tadpoles resorb their tails or mammalian embryos their gills; or by which the prostate is resorbed after castration. Apoptosis occurs within a few hours of changing the hormone status and also soon after irradiation. It can easily be missed by light microscopy. It is a phenomenon that deserves more study.

POTENTIAL THERAPEUTIC APPLICATIONS

The final chapters concern the radiobiological topics closest to clinical application. I shall therefore deal with them more briefly. I review the prospects for hypoxic cell radiosensitizers, mentioning the recent results of the British MRC hyperbaric oxygen clinical trials, which suggest a gain factor of about 1.2 for cervix and ENT cancer. The gain factors expected for misonidazole, within the 12 to 15 g/m^2 limiting dose set by neurotoxicity, are in the range 1.3 to

1.7. This approaches the gain factor for fast neutrons of 1.7. We are agreed that the electron-affinic radiosensitizers are a most important clinical development from basic radiation chemistry and radiation radiobiology.

Drs. Suit and Goitein, in describing the rationale for charged-particle beams in radiotherapy, claim strongly that "a smaller treatment volume means that there is some increase in tissue tolerance to radiation." The worthwhileness of continuing improvements in dose distributions is the subject of much controversy between radiotherapists. Some of my radiotherapy colleagues are concerned that if treatment volumes are too small, malignant cells spreading out from the edge may be missed.

Dr. Phillips shows us how chemotherapy and radiotherapy might be used at critical intervals of a few hours, so as to cause more damage to tumors and less to normal tissues. Multiple fractions, of course, need further study. It should be noted that chemotherapy is always used, in the applications that give the best clinical results, concurrently with or after radiotherapy. This is consistent with Dr. Peters' argument that the primary tumor must be thoroughly treated first to prevent further seeding of metastases.

Dr. Dewey shows us, in a splendid review, that thermal tolerance was lost in vitro after 72 hours. In vivo, however, it takes longer to die away: very recent published work from Hammersmith, London, shows that in the skin of mouse ear over a week is required (Law et al. 1979). Dr. George Hahn and co-workers describe significant nonthermal effects of ultrasound and recount impressive clinical and veterinary experience of hyperthermia. Both he and Dr. Corry and his co-investigators explain that heat can be used to re-treat patients who had earlier been treated to full doses of X rays, with apparently no untoward effects and surprisingly good local tumor regression. Dr. Corry also presents preliminary results of some clinical treatments that agree with Dr. Hahn's. I believe that hyperthermia is like radiation in the 1920s: much remains to be done to achieve optimum temperature distributions and to measure them reliably. Results from different laboratories disagree. Dr. Peters pointed out, as chairman of the session, that techniques of delivering the heat dose need to be concentrated on. But it is clearly a very important modality, alone or in combination with radiotherapy.

CONCLUSION

I should like to present an assessment of my own of the amount of extra cell kill in radiotherapy that we might expect from various modalities (Figure 1). The large gain from eliminating hypoxic cells occurs if they are not removed by reoxygenation, and Dr. Barendsen's results suggested that they may not be. Circadian rhythms and chemotherapy could obviously be added to any of the other methods.

One recurring theme in this monograph, as in many other scientific monographs concerning cancer, is that the phenomena are too complex for practical

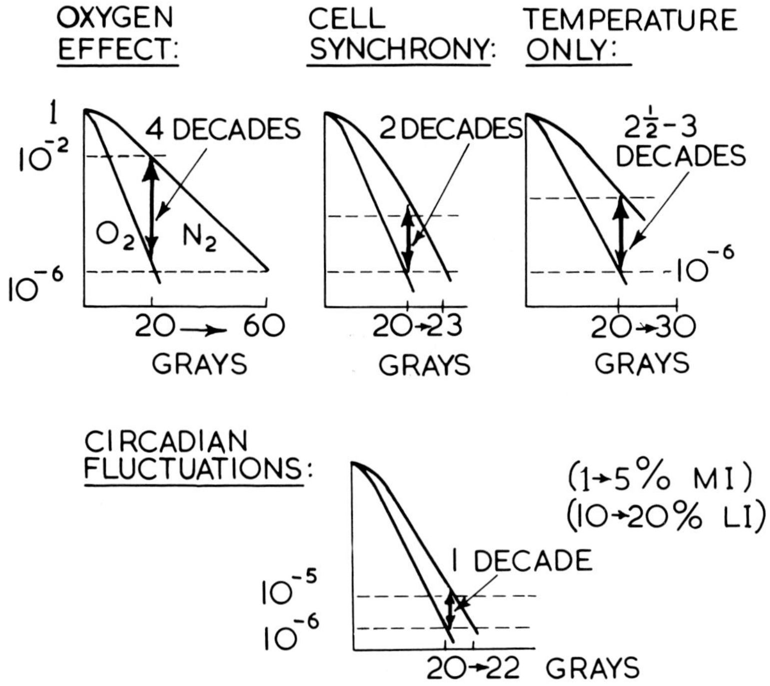

FIG. 1. Arbitrary estimates of the amount of extra cell kill using various "new" modalities for cancer treatment. Chemotherapy and diurnal variation could obviously be used together with any of the other methods. Dr. Phillips presents evidence at this symposium that chemotherapy and radiotherapy can be combined with optimum intervals of a few hours.

conclusions to be drawn until we have more complete data. Dr. Rossi reminds us that, even so, each factor is a result of simple, basic laws of nature and the more we know the simpler the explanations become.

To paraphrase Dr. Elkind's remark in the Bertner Memorial Lecture, mankind needs answers to questions of radiation biology, both for protection against hazards and for improving the treatment of cancer. We have made much progress since the 1964 symposium on radiobiology sponsored by M. D. Anderson Hospital, but there are still many questions awaiting answers. We shall leave this symposium feeling encouraged, by the recent advances and the new methods, to go out and get them.

ACKNOWLEDGMENTS

I should like to thank the authors (almost all of them) for sending their abstracts in time for the conference organizers to show them to me before the meeting, and then for being so much more interesting in their talks than in their abstracts that I had to rewrite this summary day by day throughout the conference. I thank the organizers for the privilege and pleasure of attempting to summarize a very full meeting.

REFERENCES

Barendsen, G. W., H. M. D. Walter, J. F. Fowler, and D. K. Bewley. 1963. Effects of different ionizing radiations on human cells in tissue culture. III. Experiments with cyclotron-accelerated alpha-particles and deuterons. Radiat. Res. 18:106–119.

Durand, R. E., and R. M. Sutherland. 1972. Effects of intercellular contacts on repair of radiation damage. Exp. Cell Res. 71:75–80.

Fowler, J. F., D. K. Bewley, R. L. Morgan, J. A. Silvester, T. Alper, and S. Hornsey. 1963. Dose-effect relationships for radiation damage to organized tissues. Nature (London) 199:253–255.

Hirst, D. G., J. Denekamp, and E. L. Travis. 1979. The response of mesenteric blood vessels to irradiation. Radiat. Res. 77:259–275.

Law, M. P., R. G. Ahier, and S. B. Field. 1979. The effect of prior heat treatment on the thermal enhancement of radiation damage in the mouse ear. Br. J. Radiol. 52:315–321.

McNally, N. J., and P. W. Sheldon. 1977. The effect of radiation on tumor growth delay, cell survival and cure of the animal using a single tumor system. Br. J. Radiol. 50:321–328.

Stewart, F. A., J. Denekamp, and D. G. Hirst. 1979. Proliferation kinetics of the mouse bladder after irradiation. Cell Tissue Kinet. (in press.)

Van den Brenk, H. A. S., M. G. Stone, H. Kelly, and C. Sharpington. 1976. Lowering of innate resistance of the lungs to the growth of blood-borne cancer cells in states of topical and systemic stress. Br. J. Cancer. 33:60–78.

Author Index

Subject Index

R